R. Hoe & Company, about 1900 / (*cover*) Press erection shop / (*i*) Hoe Scharf Gutenberg medal / (*iii*) Grand Street entrance with Gutenberg statue / (*vii*) Pattern shop / (795) Blueprint shop

Photographs courtesy Bronx County Historical Society

By the same author

AGE OF GREAT GUNS: Cannon Kings and Cannoneers who Forged the Firepower of Artillery

BOOKS FOR THE MILLIONS: A History of the Men whose Methods and Machines Packaged the Printed Word

Chronicles of

R. HOE & COMPANY AND THE

Genius and Folly

PRINTING PRESS AS A SERVICE TO DEMOCRACY

Frank E. Comparato

LABYRINTHOS / Culver City

Chronicles of Genius and Folly was set in Linotype Fairfield with Caledonia for display. The line casting machine will soon represent that short but hopeful era, between hand composition and computerized projection, when men first saw that, thanks to patient genius, many manual arts and skills could indeed be mechanized.

Typographer Paul A. Bennett, long associated with the Mergenthaler Linotype Company, wrote: "Fairfield, a slightly decorative original and contemporary old style, was designed for Linotype by Rudolph Ruzicka, the distinguished American wood engraver. Sharply cut, as though the letters came from the artist's graver rather than pen, Fairfield was designed for reading by one of the most knowledgeable men in the country about letter forms and their style." "Dignified but friendly," the Linotype company added.

Caledonia was designed for the Linotype by William A. Dwiggins as a modern face of Scottish origin.

The book was offset printed on S. D. Warren's "1854" text paper. It was completely manufactured by The Book Press, Brattleboro, Vermont. From within the Labyrinthos, December 1978.

Library of Congress Catalog Card No. 77–90647

Printed in the United States of America

LABYRINTHOS, 6355 Green Valley Circle, Culver City, Ca. 90230

For Charlie

MINOS

CONTENTS

Foreword

During my brief years at R. Hoe & Company—like many of the workforce of almost two thousand who were sons and daughters of past employees—I had numerous occasions to ponder the history of the company. Its greatest captain of industry (my great grandfather, Robert Hoe III) literally revolutionized the world of printing, but for all his genius he was not particularly democratic. When two workers had proudly painted his private apartment at the shop they approached him to ask how he liked the job; they were fired for daring to address him in the factory yard. Another employee was dismissed for sharing a ride with him in the company elevator.

Despite the persistence of his deep streak of elitism, Robert Hoe had the family's hearty sense of humor. Once when Joseph Pulitzer bought somewhat cheaper presses for his newspaper Hoe responded on black-bordered mourning stationery, lamenting Pulitzer's lack of foresight and subtly predicting dire maintenance problems. They probably materialized.

In 1964, at my father's suggestion, I applied for work at R. Hoe & Company, and almost to our surprise was hired. I was sequentially stereotype sales trainee, (acting) stereotype sales manager, assistant purchasing agent (and concurrent fleet leasing manager)—and manager (without title) of purchasing equipment for Hoe Leasing Corporation customers. At first I earned about $100 weekly and found myself, perhaps because of my name, a somewhat unpopular addition to the firm. Towards the end of my career my pay was $150, I got some overtime (always helpful to newlyweds), and found myself assigned to assist everybody. Nevertheless, unlike my great grandfather, from my experiences at the plant, ending as assistant to the controller, I managed to get an interesting business education—and a broad overview of capitalism itself.

The company's reputation was perhaps its greatest legacy from the old family. The men who made the presses were craftsmen who took great pride in their work. Sons followed their fathers into the shops. Highly competent engineer-salesmen (like Fred Adame, who managed the Metal Decorating Division, which had the lion's share of the world market, and Jim Crowe, a superior technician-salesman) believed in their products and represented the very best in a highly competitive sales arena. But things had gone wrong nonetheless.

The late 'fifties and early 'sixties framed a management struggle that ultimately destroyed the firm (although its real genius for leadership may have died in the nineteen twenties). Much of the peripheral press equipment manufactured by competitors was successfully marketed only because Hoe, in somewhat typical fashion (it had acquired a kind of corporate elitism), refused to renew its own patents. The company had survived recent decades by flukes and wars, the latter providing precious income when most needed, but without furnishing direction to the business of making printing equipment. In the absence of strong moral leadership various individuals soon stepped in to promote their own schemes.

There was a great power struggle at Hoe during those post-war years, fostered in part by the confusion following Mr. Auer's death, and the caretaker management of R. Palmer Hollister. Hollister had been the son of a Hoe machinist, proudly telling Ruth Hoe Sterling, during a tour of the shop, that he had Americanized his father's name. But he seemed no match for shrewder minds. Arthur Gordon, Treasurer, and William Henderson, Vice President—Sales, both competed for the coveted presidency. Of the two, Henderson appeared more the gentleman. Had he won the competition Hoe might still be in the press business today.

Henderson and Gordon were both promoted rapidly. In late 1965 Gordon became Vice President—Finance, and Henderson Vice President—Operations. Gordon was responsible for accounting, cost accounting, personnel, purchasing, accounts receivable and payable, and payroll; Henderson was responsible for operations (the press erection floor), including sales, engineering, production, and advertising. These two men essentially controlled the company after Auer's death.

Although apparently an Auer protegé, Henderson seemed regarded by Auer's old Board enemies as "too young" to assume the presidency. They certainly would not favor appointment of an Auer man under any circumstances. Gordon displayed a considerable engineering and accounting background, having taught for a time at a New Jersey college. Within six to ten years he rose from junior cost accountant to president, seizing opportunities and filling every management vacuum he was able to find. Gordon ultimately circumnavigated the Hoe sales department and was himself leasing and selling presses. Having worked for another press manufacturer he also had strong competitive marketing experience.

Ambitious and aggressive, Gordon was not without a streak of eccentricity too. He would not drive a car; hence an employee picked him up at his New Jersey home each morning. As he worked long hours the return trip could be as late as midnight, or he might stay overnight at the Bronx factory. Gordon berated all who would not devote as much time toward rebuilding the firm as he seemed to do—a strategy that was not lost on the easily impressed Board of Directors.

For a brief time things seemed to go well, perhaps helped along by the Gordon-Henderson rivalry. Henderson and his sales manager developed a thoroughly professional staff (most of the salesmen had strong engineering backgrounds). The Hoe letterpress ink pump (a patented automated ink flow computer which could make finite adjustments in the metering of ink) proved to be highly marketable; the Satellite offset press (developed by Aller of Denmark) was being manufactured under license with some success; the German MAN stereotype equipment was selling under a commission agreement; and the company was jointly developing (with the inventor) and marketing an offset platemaker with great potential. Under the able direction of Burt Polglase the staff seemed to be thriving.

But then executive competition escalated—in a particularly shocking way. Through the Hoe Leasing Corporation, a "catch-all" financial subsidiary, the company Treasurer found a way to rival the Sales Vice President. A Canadian promoter signed a lease with Gordon for a complete newspaper plant in Vancouver. I can attest to the fact that Hoe Leasing provided virtually every piece of equipment imaginable (with the possible exception of paper clips): we leased platemakers, offset cameras, electric motors, a million-dollar Satellite press, cold composition equipment, *and* we sent *our* subcontractor to install facilities as an *additional* monthly expense above the lease figure. All this took well over a year, during which time the Treasurer executed other leases and made some actual sales as well.

Naturally, these stunning arrangements were well publicized. The pleased Board of Directors overlooked the company books that did not often balance, the Treasurer's frequent lunches with auditors, and the fact that in the overburdened shop most orders for future delivery had long waits indeed. Hoe manufacturing facilities were backlogged for years, but to the public all seemed prosperous.

Hoe common stock rose from below a dollar to over $138 between 1964 and 1966, partly fueled by reported large sales orders (many of them based on old labor rates and low profit margins), released to the media by Hoe's advertising department. Brokerage houses would call for more information on such sales, and Hoe executives would say, coyly, that "management had no comment to make at this time"—inciting brokerages and their mutual fund clients to buy even more. Stock prices were thereby driven upward even by the firm's silence.

Most of Gordon's sales seemed to involve friends in outside press erection firms who furnished valuable help. But contracts were signed without benefit of accurate cost studies. Henderson became so frustrated he left the firm. Assigned in my job to analyze these transactions, I determined that one large sale (Newhouse Newspaper Group) had been made on a one percent margin in 1966, based on 1962-1964 labor rates even though the presses were to be manufactured in 1967 and on into the 'seventies. Strangely enough, the investigation was undertaken at Gordon's request: when I presented the results of my study he lambasted my youth and inexperience. Quite colorful language, as I recall!

With the Treasurer himself functioning as sales department the company's marketing program was largely unattended. Soon inventory problems arose, routine audits (which at Hoe always seemed to drag on) now took eight to twelve months, and considerable shuffling of figures appeared necessary to arrive at desired financial results. In the midst of reestablishing the amounts to be added to the out-dated Vancouver lease rates, I learned that the Canadian lessee was already defaulting on monthly payments. I knew then that tragic consequences were ahead.

Yet, to the end, the tremendous employee devotion to the firm never waivered. Frank Stellaccio, Hoe's purchasing agent, was one of the best and kindest men I have ever worked for—his uncle was one of the painters my great grandfather had fired. I will never forget Harry Parchman, who, in his late seventies, *still* worked for the company. He was a graduate of the apprentice school at age fourteen, giving him sixty-plus years of service. The stories he used to tell! He exemplified the esprit and pride in craftsmanship so common among Hoe employees. He was always a cheerful soul, and I loved him.

Arthur Gordon was a rough individual, but he had a style and charisma that perhaps blinded his followers to reality. King Thorpe, Controller, was a good man but badly misused by Gordon. Carl Ringe, a German scientist rescued by the American government during World War II, was an engineer-patent attorney-philosopher, but also miscast and underutilized at Hoe. The tragedy in the demise of this great firm was the loss of great numbers of men and women who were *really* the embodiment of the company. People with ability to make it all work were there, in the shop, but the fierce competition—directing their loyalties to individuals instead of the firm—consumed their energies.

Bob Corrao, a Hoe veteran and former accounting department manager, is now president of the "new" R. Hoe & Company, manufacturers of industrial wood saws and cutters. "America's Oldest Saw Manufacturer Since 1828," says their Annual Report. The firm is on its way to new success and prosperity.

Mr. Comparato's remarkable work tells all of us industrious Americans how we got to be who we are. You will not only meet the captains of industry, but the dynamic men who toiled ceaselessly to provide for their own self respect as well as their families. Since the words of the hundreds of individuals who breathe life into this narrative are their own, I think you will feel you are actually meeting them—the ideas they express are so vibrant and alive they seem current today.

Richard March Hoe

Tulsa, Oklahoma, 1979

PREFACE

Talent has been described as hitting a mark which ordinary men cannot reach; genius as sighting a point which others cannot see. But in business as in invention—both realms of the profitable and the practical—one must have both: the great enigmas must be solved and their usefulness made available to the public. Yet making money and dreaming dreams do not always work together.

Seldom in the archives of genius (or folly) have determined men meticulously charted their own successes, most of them too modest—or too busy—to try. But for the Hoes there was a way. Quiet and unassuming all their lives, shunning publicity and personal attention, the indefatigable brothers left behind a chronicle so rich and detailed we can know—almost to the hour—what transpired almost every day in more than a half century of early American business serving the nation's most vital sinews of communication (and democracy)—the printing industry.

It may be argued that the Hoes were not particularly ingenious nor inordinately talented; that in the pursuit of profit they merely functioned as would any other manufacturer or merchant. Indeed, their true strength was a loyalty to each other and to a father they had scarcely known except as children. But there is no disputing the firm's impact on a new branch of their industry that hardly existed before they gave it life—and speed. The Hoes literally created the newspaper press from a few borrowed mechanical concepts, financed it through lean times with unstinting credit (and mortgages), and shaped its compelling product over the century with progressively complex technology. They introduced the highest standards of quality and precision and instituted almost compulsive service and maintenance.

The company became an American institution. "The history of this house, originally established by his father," wrote publisher J. Henry Harper of his father-in-law, Richard M. Hoe, "and carried on from one success to another by his father's sons, is the history of the evolution of the art of printing not only in America but throughout the civilized world." Sixteen years later the claim was more succinct, but poignant: "The subsequent history of the development of the printing press is really the history of the Hoe firm," said the *New York Times* in 1928, unaware that the epoch was coming to a close.

But the newspaper was a greater institution still. Its effect in educating the peoples of the world, in furnishing them a platform for political discussion, and in encouraging free and meaningful suffrage did not at first necessarily follow. In solving the problems of fast and cheap papers the Hoes rendered useless all the conventional means of suppressing thought.

A contribution to the democratization of the free world can be added to their accomplishments and is not too bold a conceit.

In the United States, where pragmatic businessmen demonstrate a studied irreverence for history and a constant impatience with old paper trash, it is remarkable that the company's commercial and personal letters and records have survived at all. The stability of R. Hoe & Company, signaled by a century in one spacious location in downtown New York, was nevertheless instrumental in preserving materials in which later managements had no interest whatsoever. Considering the ravages of dust, water, mold and mildew, insect and chemical damage—all of whose marks are present, their physical survival is noteworthy. Although today incomplete and scattered among several libraries, the records are a priceless corpus of American achievement. With them (and the well known but—until 1972—unpublished Tucker manuscript) the history of the firm was already on paper, such as it was.

To retain the integrity of the correspondence and portray something of its visual impact—handwritten throughout, with all the illegibilities and ambiguities of uncertain spelling and unsteady hands, only minor editing was done. Most of the letters require little explanation, although it is perhaps inevitable that some personalities are beyond identification, some issues beyond settling, some comments beyond analysis. Fortunately, the obligation of salesmen in the field (Hoes among them) to make frequent reports, and their natural garrulity, furnishes many vital sequences. Where they occur (and they are very rare) a few irrelevancies—politics, weather, illness, family matters—suggest that for brief moments, at least, the Hoes could indeed take their minds off their chores. Surprisingly enough, there was never any profanity, obscenity, or libel.

A few stylistic comments are necessary. (1) Throughout the correspondence and other displayed matter outside quotation marks are omitted so as to preserve the original punctuation; aberrations are indicated by a superscript dagger (in place of "sic") to minimize the intrusion of that convention, and where an error is repeated in a letter, only the first instance is noted. (2) The brothers' (and others') disdain for the possessive apostrophe has been corrected (without note) only where its absence is confusing. (3) The Hoes' persistent indifference to the spelling of customers' surnames remains, perhaps, our only intimation of an inner haughtiness. Finally, (4) family inheritances and net worth are variously given as reported at the time in the press—not a reliable source in such matters. While considerable financial family data exist, many bequests were long in litigation and many intangibles subjected to questionable appraisals. It is sufficient here to concede that, by century's end, the Hoes were extremely wealthy. Similarly, corporate income and loss and other business statistics also come largely from the press and from unhappy stockholders. It is abundantly plain, however, that much was lost.

What does it all mean? How could three uneducated brothers and a conceited nephew run a large firm for a century, when large staffs of specialists—trained in every aspect of design, engineering, sales, finance—and almost unlimited reputation and credit could not do so for half that time? Ruthless mismanagement on the part of disinterested bankers had something to do with it; "money hunger" (Henry A. Wise Wood's words) on the part of corporate officers and mediocre directors, recruited from outside the graphic arts industry, had something to do with it, as did in-fighting and almost deliberate inattention to world conditions, technical developments, and trends. The resulting corporate dinosaur—a tremendous animal with tiny brain—emerged, incapable of rapid adaptation to new environments and unable even to hold its own terrain.

Throughout its later life one fatal misconception persisted: that the great old firm, with unfathomable industrial competence and unquestioned trade prestige—just as the brothers had fashioned it, could not really be hurt by petty machinations and intrigues. But the loss of one asset after another spelled slow atrophy. Weighted down at the end with only mythological attributes, the awkward giant, paradoxically, became unfit for survival.

Like Charles Darwin in his nature studies, we are here privileged to see a dynamic organism grow and function, dominating its world. But we see more in the very same chronicles; we detect ailments and aging as well, and sadly its final demise. Life and death are here together in the records.

FEC

ACKNOWLEDGMENTS

I am especially grateful to the staffs of the Library of Congress, Washington, D.C., Columbia University, New York, Yale University, New Haven, University of Michigan, Ann Arbor, New York Public Library, New York, and the Bronx County Historical Society, for their help. I am also indebted to Mr. Arthur J. Hoe for making available to me materials from the family's papers, and to Mr. Richard M. Hoe and Mr. Robert M. Corrao for their help and encouragement.

This work was made possible through the assistance of a research grant from the National Endowment for the Humanities, Washington, D.C. The findings and conclusions presented here do not necessarily represent the views of the Endowment.

The publisher wishes to express profound appreciation to the prepublication subscribers whose patience was heavily taxed but whose investment (he believes) will be amply returned.

R. HOE & CO.

Printers' and Binders' Warehouse,

29 & 31 GOLD STREET, NEW-YORK,

Manufacturers of

Type Revolving, Single and Double Cylinder

PRINTING MACHINES

ADAMS' PRESSES,

Bed and Platen Job and Card Presses,

HAND PRESSES AND INKING MACHINES; HYDROSTATIC PRESSES FOR PRINTERS AND BINDERS;

STEREOTYPE BLOCKS, ROLLER MOULDS,

IMPOSING BEDS of Marble and Iron, with Tables of Black Walnut or Cherry.
TYPE CASES of thoroughly seasoned stuff, and finished in the most durable and workmanlike manner.
CABINETS of all sizes and design in stock, or made to order, of Pine, Cherry, Black Walnut, or any Ornamental Wood.
CASE STANDS of Iron and Wood.
BRASS RULE and LABOR SAVING RULE of Beautiful Designs and finish.

CHASES, FURNITURE, QUOINS, COMPOSING STICKS,

BODKINS, MALLETS AND PLANERS, PROOF AND LYE BRUSHES, SETTING RULES; GALLEYS, DOUBLE AND SINGLE, PATENT LINED; EMBOSSING PRESSES, PAPER CUTTERS, RULING MACHINES, PIERCING MACHINES, LEAD AND RULE CUTTERS, AND PATENT MITERING MACHINES.

COMPLETE OUTFITS

Furnished for NEWSPAPER, JOB and LITHOGRAPHIC PRINTING ESTABLISHMENTS, STEREOTYPE OFFICES of PLASTER, CLAY, or PAPER; also, ELECTROTYPE OFFICES.

R. H. & Co. manufacture all the above-mentioned Machines and Wood Work on their premises and under their own personal attention. Estimates furnished in detail, and Type furnished from any Foundry in the Country.

ALL KINDS OF CAST STEEL SAWS OF THE BEST QUALITY.

Printed by Francis Hart & Co., on R. Hoe & Co's Four-Roller Book Printing Machine.

★ ★ ★ EXTRA! ★ ★ ★

The priceless "extra" of freedom is reaffirmed with each new edition that hits the streets. And National Newspaper Week reminds thoughtful Americans that their free press is no accident. It took a fight to make it free. It takes initiative, courage and responsibility to keep it that way. Newsboy, publisher, mechanical superintendent, pressman, reporter, editor—all share in the work—and together exhibit a virtuosity in the art of communications that is equalled no where else in the world. Hoe joins in the applause for a job well done! R. Hoe & Co., Inc., 910 E. 138th St., N.Y. 54.

HOE

Koenig's London *Times* machine / The Treadwell bed and platen /
The London Literary Gazette, 26 October 1822

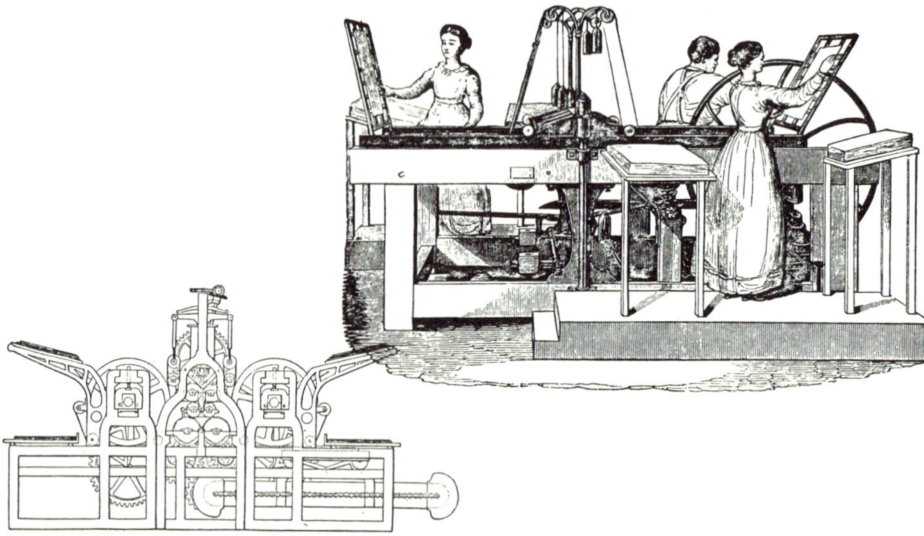

Part I

EVOLUTION IN PRESSWORK: THE FIRST MECHANICAL ART

Joseph Moxon, Hydrographer to Charles II of England, published his *Mechanick Exercises* in 1683-1684, the first treatise on the art of printing to appear in English. Understandably, the secretive craft began with careful construction of the press itself, a chore Moxon considered essential among the duties of a master printer who would want to contemplate its "harmonious design" as well as remedy any later defect or deficiency:

> An understanding *Press-man* therefore knows not only how to direct a Printers Joyner to Set up and Fasten a *Press* when it is made, but also how to give a strange Joyner and Smith instructions how to make a *Press,* and all its parts, in a Symetrical† proportion to any unwonted size, if in a strange place he shall have occasion to use it. . . .

†The dagger is used throughout in place of "sic"; see Preface.

THE LONDON LITERARY GAZETTE,

AND

Journal of Belles Lettres, Arts, Sciences, &c.

This Journal is supplied Weekly, or Monthly, by the principal Booksellers and Newsmen throughout the Kingdom; but to those who may desire its immediate transmission by post, we beg to recommend the LITERARY GAZETTE, printed on stamped paper, price One Shilling.

No. 301. SATURDAY, OCTOBER 26, 1822. PRICE 8*d*.

BENSLEY'S PRINTING MACHINE.

A Boy is represented as laying on
A The sheet of white Paper.
B The Cylinder which prints the first side of the Paper.
C Intermediate Cylinders over which the Paper travels to
D The Cylinder which gives the final impression.
E The Inking Rollers under which the Form (i.e. the types) is in the act of passing.
F The Reservoir of Ink, from which the Inking Rollers are supplied.
G The Form, receiving its last inking before it goes under the Printing Cylinder.
H A Sheet is seen just being delivered into the hands of another Boy, whose business it is to keep the sheets, as they come out, in a *heap*.
The *lines* at top of the Machine represent the Tapes. which run round the Cylinders and secure the sheet.

ARTS AND SCIENCES.

PRINTING BY STEAM.

WE may well range this article under the head of Arts and Sciences, for it relates to a prodigious improvement and application of ingenious and powerful agency to that which is the matrix and preserver of all arts and sciences,—PRINTING. It would be a waste of time and of these very means were we to expatiate on the subject. The process by which our Gazette and many other works are printed by Mr. B. BENSLEY, is one of the most remarkable sights that can be witnessed;* and we have endeavoured to convey to the public an idea of it by the following Description and Print of the Engine.

THE Cylindrical mode of printing, which, in contradistinction to the old prosess by the Press, is called *Machine Printing*, was invented by the late Mr. NICHOLSON, well known in the scientific and literary world, who took out a patent in the year 1790, though it does not appear that his plans and experiments ended in any actually practical result.—Whether M. KÖNIG, who at a later period more successfully attempted to print by Machinery, was indebted to Mr. NICHOLSON for his elementary principles, or whether almost the same ideas spontaneously occurred to each individual, is a question that can only be satisfactorily solved by the former. Thus much is certain, that M. KÖNIG's labours were the first which produced any fruit:—and surely more is due to him who, after years of persevering toil, succeeds in the application of hitherto unapplied principles, than to one of whom we can only say that he was simply the first to suggest ideas—since no evidence is offered of their ever having been acted upon.

M. KÖNIG, by birth a Saxon and by occupation a printer, many years ago conceived it possible to print by Steam, though he then expected no more than to be able to give accelerated speed to the common Press, to which end his first efforts were bent. As from the nature of such an undertaking, considering the state of scientific pursuits in his native land, he could calculate on little success unaided by others, and failing in his application for encouragement and support at the hands of the most eminent Printers in several of the continental Capitals, he turned his eyes towards England. Arriving in London about 1804, he submitted his scheme to several Printers of repute, who, not being disposed

* The Editor of the *Literary Gazette* accompanied the brothers of the Emperor Alexander in their examination of the original Machine, and explained its movements to them: he may assert that no spectacle in Britain surprised or gratified them more.

Under the direct supervision of the pressman the joiner (worker in wood adroit at making joints) assembled all components, the former insuring that, in the process, moving parts would be smeared with soap or grease. And the professional joiner had to be restrained from going too far:

> Joyners that Work to Printers have got a Custom to place a strong Piece of Timber between the middle of the Cap [top cross member] and the Ceiling or Roof of the Room, which can do no service there, unless they intend to support the Roof: For the weight of the *Press* alone will keep it close to the Floor, and the strength of Stuff [felt or cardboard cushions or stops] between the Mortesses in the *Cheeks* [channels in the upright sides] and the ends of them, are intended to be made strong enough to resist the Rising of the *Head* [in which the

spindle was embedded]: For should that strength of Stuff start [rise], neither their strong Piece of Timber, nor the strength of the Roof, would resist the Rising of the *Head*: but *Head* and *Cap*, and Timber and Roof too, would all start together, as by experience I have seen. . . .[1]

The work of establishing a level and sound foundation, raising the cheeks perpendicular and stable, and setting the cross members and "stuff" (packing) for a truly horizontal head from which the platen would bear parallel to the type bed, were some of the areas in which the press builder demonstrated his skill. When he did his work well the joiner (along with the "Inck-maker") was invited to the Waygoose, the yearly printers' feast at which such guests were liberally toasted —but nevertheless expected to "open their Purse-strings and add their Benevolence." The relationship of joiner to pressman (who, although not the shop's owner, engaged the man of his choice) was thus well cemented over centuries of common interest and regard.

Two presses Moxon described: the "old fashion" and the "new." The first was what Gutenberg himself had used: nothing more than a frame in which a heavy block was brought down by screw action to force paper against type—actually a cheese press, say some historians. The latter was the improvement of Willem Janszoon Blaew of Amsterdam, whose 1620 modification differed only in small details: vertical guides for the platen to prevent twisting (resulting in slurs) upon contact with the type, and springs to raise the platen when pressure against the arm was released. Both presses, essentially wooden, were in constant need of repair. They furnished carpenters and joiners throughout Europe and America a steady and growing business.

Little improvement was made to Blaew's press well into the eighteenth century when (in 1798) the Earl of Stanhope designed a hand press manufactured entirely of cast iron. More pressure was needed for woodcut printing and for larger job formats demanded by the growing world of commerce. A prolific and inventive politician, Charles Mahon, 3rd Earl of Stanhope (1753-1816), had regarded the printing press as one of the best instruments for insuring political liberty; he therefore entered notice of his invention at the British patent office but did not actually patent it. The Stanhope, first used at William Bulmer's Shakespeare Press, was immediately recognized as a significant advance in the art of printing.

In both Europe and America imitations followed, some better, some worse. Robert Hoe III noted:

The Earl of Stanhope reserved the screw, but caused to be added a combination of levers to assist the pressman in gaining greater power, when giving the impression, with less expenditure of energy. . . . The printers, seizing upon this new idea of a combination of levers to

> increase the power, were induced to place them upon their wooden presses, the improvement resulting generally in the destruction of the latter. . . .[2]

Between 1700 and 1800 the population of England almost doubled, and change from rural to urban (industrial) economy was already at work. It would be a period of rapid adjustment, and the simple press soon yielded rapidly to technological progress. When British patent agent William Nicholson invented his cylinder press and registered it in 1790, wrote Colin Clair, "The hand press was doomed."

Nicholson's brilliant ideas, however, never materialized. He died in debtor's prison, leaving his patent for the world to ponder. Practical critics promptly regarded Nicholson's work "insufficient and superficial," but, Clair continued,

> Nevertheless, his name deserves a mention in the history of printing, since his specifications embodied many of the essential features of the later printing machines, such as the reciprocating table [bed] with geared impression cylinder; the inking of the forme by rollers instead of ink balls; the distribution of ink upon the rollers by means of subsidiary rollers; an ink duct for supplying ink in pre-determined quantity at each revolution; and grippers for carrying round the sheet. . . .[3]

Nicholson also envisioned curved stereotypes, but his genius at foreseeing developments was not helpful without concrete details.

Like Nicholson, Sir Rowland Hill (1795-1879), originator of Britain's penny postal system, suggested both web feeding and wedge-shaped types for printing from a curved cylinder. But his 1835 patent was similarly more theory than practice. The New York Mechanics Institute (perhaps under the pen of young Richard M. Hoe, on the Board of Directors) in its 1837 *Proceedings* commiserated:

> The greatest difficulty to which Mr. Hill has had to surmount in the construction of his machine . . . [is] that of fastening the small pieces of type upon the surface of a cylinder, and with firmness to retain their places even when they are turned upside downwards by the revolution of the cylinder, at which time their gravity combines with their centrifugal force in tending to displace them, and to effect this without throwing new difficulties in the way of correction, revisal, etc. . . . Each piece of type is slightly wedge-like in its form.[4]

The "impossibility" of this mode of printing would soon challenge other inventors, and both Nicholson and Hill served only to whet their thoughts. But as Robert Hoe III explained, dreamers who did not produce were like thieves of grandeur: whether in the real world of business or the heady annals of invention success was clearly not a matter of good intentions. With the proliferation of schemes, however, many new ideas surfaced.

Bryan Donkin, an English engineer, had built papermaking equipment; with Norwich printer Richard Bacon he developed an unusual press in 1813 with "square" cylinder containing four flat forms of type, turned by square gears. These collaborators are credited with using the first glue and molasses (composition) rollers for inking, although they admitted the formula was old and well known. Donkin and Bacon reportedly got 800-1000 impressions per hour, but their press was not successful.

Frederick Koenig was born 17 April 1775 at Eisleben, Saxony, and apprenticed as compositor and printer in Leipzig, where he also attended university classes. In 1802, 28 years old, he first began improving printing equipment, making a press with movable carriage for type forms. Local printers were not interested. Koenig worked in Germany—including a stint at stereotyping—until 1806, when he was invited to reorganize the Russian State printing office at Petersburg, a chore that soon bogged in petty bureaucracy.

With the Napoleonic Wars in full fury, Koenig went to England, where, despite his poor command of English, he found an audience for his projects. His employer, printer Richard Taylor, introduced him to Thomas Bensley, another Fleet Street printer, who in March 1807 invited him to describe his idea for a mechanized press. The result was an agreement to finance the inventor, who, toiling three years in a little shop on Frith Street, finally developed a high-speed machine especially appropriate for newspapers of large circulation. Unfortunately, the imperious London *Times,* largest of them all, was not impressed.

John Walter II's "Old Thunderer" (so nicknamed because an editorial once began, "We thundered forth the other day . . .") nevertheless was probably the one paper that needed Koenig's machine. Walter's father had started a paper in 1785 to advertise his print shop but ran it quite indifferently. The son, however, believed in journalism as an institution for public good; he assumed the *Times* editorship in 1802 at the age of 26 and nursed the paper to a fine reputation. With a daily circulation of 5000 by 1813, Walter depended on two Stanhopes (with speeds not much better than wood hand presses) to get issues onto the streets at dawn. He had to set outside pages twice so that both presses could print last-minute news.

Koenig worked on. Assisted by Andrew F. Bauer of Stuttgart, a friend and excellent mechanic, Koenig finished a prototype and obtained a patent for it (29 March 1810). Although imperfect, the machine was put to work a year later; it was a screw-and-platen press with self-inking rollers, driven by steam power. The slight advance of 400 impressions hourly (over the Stanhope's 250) was not particularly encouraging, and the inventor threw the whole idea aside.

The Germans next approached the prospect of cylinder printing. Although Nicholson had proposed it for all kinds of cloth and paper presswork, he had never actually made a working machine. Adam Parkinson of Manchester subsequently developed a cylinder press for cloth (1805), but printing on paper had not really been explored. Koenig quickly patented a cylinder press 30 October 1811 and finished a working model in December of the next year (improved by further patents in 1813 and 1814). The radically different press worked fine, and Koenig pointedly distinguished his success from Nicholson's:

> The inventor of a machine is not he who has thought out a plan, more or less imperfectly, by which a thing can be accomplished; it is he who resolutely sets to work and conquers the difficulties and shows the machine in operation. . . .[5]

With several partners—Thomas Bensley, Richard Taylor, and George Woodfall—Koenig set up a factory on Whitecross Street to fabricate the machines. It was rumored that London pressmen, understandably hostile to mechanization, "threatened destruction to Koenig and all his traps," obliging the inventor to swear his workmen to utmost secrecy. James Perry, proprietor of the *Morning Chronicle,* and John Walter, finally persuaded to look, came down to see the single-cylinder book press at work: Perry called it "a mere gimcrack," but Walter liked it. Koenig promised that it would obviate the necessity of double forms. Walter promptly ordered two, at a cost of £1400 each, but he wanted them newspaper size—and to be perfectors, printing both sides of the sheet.

Like Koenig, Walter took no chance of sabotage. He remembered Thomas Martyn, one of his compositors, who in 1804 invented an automatic press in which his father had invested heavily—the machine so infuriated his pressmen the inventor had to flee in disguise. Walter consequently insisted on continued secrecy during two long years of development and construction—right next door to the *Times* office on Printing House Square. So revolutionary was the Koenig press John Walter's startling announcement in the *Times* of 29 November 1814 actually lacked a technical vocabulary to describe the wonder of the age:

> Our journal of this day presents to the public the practical result of the greatest improvement connected with printing, since the discovery of the art itself. The reader of this paragraph now holds in his hand, one of the many thousand impressions of *The Times* newspaper, which were taken off last night by a mechanical apparatus. A system of machinery almost organic has been devised and arranged, which, while it relieves the human frame of its most laborious efforts in printing, far exceeds all human powers in rapidity and dispatch. That the magnitude of the invention may be justly appreciated by its effects,

> we shall inform the public, that after the letters are placed by the compositors, and inclosed in what is called the form, little more remains for man to do, than to attend upon, and watch this unconscious agent in its operations. The machine is then merely supplied with paper; itself places the form, inks it, adjusts the paper to the form newly inked, stamps the sheet, and gives it forth to the hands of the attendant, at the same time withdrawing the form for a fresh coat of ink, which itself again distributes, to meet the ensuing sheet now advancing for impression; and the whole of these complicated acts is performed with such a velocity and simultaneousness of movement, that no less than eleven hundred sheets are impressed in one hour. . . .

The novelty of the inventing process itself could not go unobserved:

> That the completion of an invention of this kind, not the effect of chance, but the result of mechanical combinations methodically arranged in the mind of the artist, should be attended with many obstructions and much delay, may be readily admitted. Our share in this event has, indeed, only been the application of the discovery, under an agreement with the Patentees, to our own particular business; yet few can conceive,—even with this limited interest,—the various disappointments and deep anxiety to which we have for a long course of time been subjected.
>
> Of the person who made this discovery, we have but little to add. Sir Christopher Wren's noblest monument is to be found in the building which he erected [St. Paul's Cathedral, London]; so is the best tribute of praise, which we are capable of offering to the inventor of the Printing Machine, comprised in the preceding description, which we have feebly sketched, of the powers and utility of his invention. It must suffice to say farther, that he is a Saxon by birth; that his name is Koenig; and that the invention has been executed under the direction of his friend and countryman Bauer.[6]

To forestall trouble, Walter had slyly kept his regular pressmen waiting for "late" news, and presented them with Koenig's finished sheet.

With war raging in France the *Times* had been losing thousands of readers as its Stanhopes fell behind. Using Nicholson's cylinder concept, the Germans had put a form of type on an oscillating flat bed, reducing some nine hand-press operations to three: (1) placing the paper, (2) taking the impression, (3) pulling off the sheet. Robert Hoe captured the sophistication of their mechanism:

> These men also devised what has proved, even to this day, to be a most efficient reciprocating motion of the type bed. It consists of a pinion carried on the inner end of a long shaft which is turned by gearing from the outside of the press frame and has in its length a universal joint, allowing an up-and-down motion of the pinion as it revolves. To the outer end of the shaft the wheel connecting with the impression cylinder is attached. Underneath the bed and fastened to it is a "rack," or a row of teeth, with a crescent-shaped segment

> of hard metal at each end. In this rack, in addition to the teeth, are pins, or studs, at each end. The wheel before referred to, at the outer end of the shaft, being set in motion revolves the pinion and moves the bed by means of the teeth in this rack. At the proper moment, calling for the reversal of the bed, the pinion turns around over one of the pins or studs, against the segment on the rack, and immediately re-engages its teeth in the opposite side of the rack, so carrying the bed back again. . . .[7]

That this degree of engineering and machinework was advanced for its time cannot be exaggerated. Similar work was utterly beyond the capability of any shop in America and unknown to many in England.

Surprisingly enough, agent William Nicholson had duly registered Koenig's patent claims, and seemed to raise no objection that they trespassed on his own work seventeen years before. Yet, with this design and its skillful execution, said James Grant, Koenig—not Nicholson—performed "nothing less than a revolution in journalism—so far as related to its mechanical condition." [8] But the Germans won few contemporaries to their novel system.

In an England unsettled by war with France and related economic crises, the workingmen of London were understandably unhappy with encroachment everywhere of the machine. The *Times* had not easily placated its reassigned pressmen. In a December 1814 editorial Walter (addressing an upper-class public, to be sure) responded promptly to labor arguments:

> There is another class of men [in addition to scoffers] from whom we receive dark and anonymous threats of vengeance if we persevere in the use of this machine. These are the Pressmen. . . . What right have they to make us print it [the paper] slower and worse for their supposed benefit? [9]

Koenig remained in charge of the Whitecross factory, in 1814 building a perfecting book press for Bensley (later used for *Literary Gazette*) and one for Richard Taylor for printing *Philosophical Magazine*. But patent infringements were almost inevitable. Any clever mechanic studying the diagrams could understand in a day what had taken Koenig six years to perfect. In March 1817 the inventor issued a prospectus offering three types of machine, available for a high purchase price plus an annual premium "calculated," *Literary Gazette* surmised, "upon a principle most effectually to prevent any one from venturing to erect a machine. . . . It need scarcely be added that after the issuing of this Prospectus the original Patentees never sold a single machine!" [10]

Bensley, furthermore, so enjoyed his advantage—almost a monopoly—he refused to help Koenig protect his patents or let him make presses for others. Distraught, the German abandoned any hope of satisfaction in British courts and resolved to quit England in August

1817. Taylor (years later) wrote of these unpleasant affairs:

> Mr. Koenig left England suddenly, in disgust at the treacherous conduct of Bensley, always shabby and overreaching, and whom he found to be laying a scheme for defrauding his partners in the patents of all the advantages to arise from them. Bensley, however, while he destroyed the prospects of his partners, outwitted himself, and grasping at all, lost all, becoming bankrupt in fortune as well as in character.[11]

The English popular press with gradual admiration for the age of steam printing consequently failed entirely to associate Koenig's name with it; the public soon considered it a British triumph exclusively.

In Oberzell am Main, Germany, Koenig and Bauer hopefully set up a new factory, well aware that in the German Confederation of States there were few printers ready for fast presses. Among their first orders, nevertheless, was one from the great Leipzig book publisher Brockhaus, and one for an improved London *Times* machine. Koenig was obliged to reveal to Walter the discouraging state of the machinist's art in Germany:

Closter Oberzell bey Würzburg,
20 D'cebr 1818.

Dear Sir,

Your letter of Novbr 6 would have been answered sooner if I had not been unwell since 3 weeks.

In Spite of our exertions it has been impossible to complete your work in the time agreed upon with Bauer, and if you knew all the difficulties, we have to struggle with, you would not be severe upon us. The principal reason is, that this is not a manufacturing country, good workmen are extremely scarce, and even inferior ones are not to be had in sufficient number. Now, if you insist to have a thing well done by indifferent workmen, it goes very slow. . . .

At present all our hands are employed for you, though we have other work equally urgent. . . . After this explanation, though you may grumble still, you cannot be angry with us.

I thank you for your kind enquiry after my health and prosperity. My health is entirely restored, and as to my circumstances, my way goes still up hill, owing principally to Bensley, who has bilked me completely. Yet, I am gaining ground, and, perhaps, I may still have the sad satisfaction of dying rich, after having lived poor. I remain, Dear Sir,

Yours very truly
Fr. Koenig.

John Walter Esqr
Printing House Square, London.[12]

Koenig died in 1833 and Bauer carried on alone. By 1829 the firm had manufactured only 51 machines for German printers, although in 1865 it celebrated its fiftieth anniversary with its one-thousandth press. The company continues to this day to manufacture printing machines for a world market.

In 1814 Koenig had patented a continuously revolving cylinder press free of the three stop motions. Part of the cylinder not used for the impression was slightly reduced in diameter so as to allow the form to return under it. By this time, however, Bacon and Donkin were at work on a similar machine, as were others (Cowper, 1816, 1818; Applegath, 1818). The prestigious installation at the *Times* attracted visitors from around the world—and at the same time invited duplication and imitation.

When Koenig left England Walter perceived he would need a similarly accomplished mechanician to superintend the sophisticated steam pressroom; he also wished to add his own mechanical improvements to the equipment. Edward Cowper was engaged for the job. With his own brother Cowper had manufactured printing presses for several years and was a capable printer and engineer. He also made several technological contributions, working with his brother-in-law Augustus Applegath (1788-1871). Applegath and Cowper addressed themselves to the idea of a curved printing plate, and in 1816 patented a process for bending stereotypes; it was employed by the Bank of England for printing noncounterfeitable notes but it was not altogether successful.

With their improvements to steam printing Cowper and Applegath managed to increase the speed of a Koenig (at Bensley's office) from 1000 to 2000 impressions per hour. Another Cowper modification, reported in John Nicholson's *Operative Mechanic* (1825, England; 1826, United States), a machinist's handbook, "at one stroke removed forty wheels" in simplifying the ink distribution system. It appeared the technological revolution could be sustained well enough with British talent and techniques.

In 1821 Applegath set himself up as printing press maker, Duke Street, Blackfriars. He constructed a machine for Bensley (known famously as "Bensley's press") among others, and two for the London *Morning Herald*. By 1826 a dozen British newspapers were printed by machine, all of which took advantage of steam power's ability to handle large forms. The format of newspapers consequently grew dramatically.

One day John Walter paid Applegath a call. Years before Koenig had described to him a novel eight-cylinder press, promising eight times the output of an ordinary machine. Delighted with his Koenig two-cylinder perfector, however, Walter was not then interested, but now (about 1823) he remembered the concept and gave Applegath an assignment to make one. Some economy in space and personnel seemed feasible over simply employing a battery of eight single machines. Applegath accepted the task.

Apparently preferring to approach eight cylinders in smaller stages, Applegath drew up plans for a four-cylinder flat-bed machine, 14 feet long, capable of 4200 impressions. As an engineering feat it was impressive, the inventor himself commenting,

> When working at its quickest speed, in order to print 6,000 sheets per hour this heavy mass [the half-ton reciprocating bed] travels 76 inches backwards and forwards 3,000 times, . . . and the concussions attendant upon such rapid changes of motion are very considerable. . . .[13]

Vibration and buffeting would be a constant adjustment problem.

Applegath engaged millwright and engineer Thomas Middleton, Lowman Street, Southwark, to help on the project; with this introduction Middleton himself later manufactured four-feeders for other papers. But by 1826 Applegath, in financial difficulty, found it necessary to sell his Duke Street shop to William Clowes, a book printer and old customer. Clowes had been a pioneer user of steam—his 26 Cowper cylinder machines powered by "stinking engines" had so annoyed his neighbor, the Duke of Northumberland, he had been forced to find new quarters. At Duke Street Clowes printed *Penny Magazine,* the 1851 Exhibition catalog, and other important works.

In 1827 Applegath finished the revolutionary *Times* machine, although its completion had evidently been ruinous to him financially. He later complained that he and his partner had not adequately been reimbursed for the work, which was never even registered because of high patent fees. The press nevertheless enabled the *Times* to increase its circulation from 11,000 in 1830 to 29,000 in 1846. (Included were supplements to accommodate swelling advertising pages; Old Thunderer's rates were the same as the other London papers, but its circulation was greater than all competitors combined.) In journalistic and technological impact no other British paper approached the *Times,* but Applegath, like Koenig, did not prosper along with the Walters. He gave up full-time press building and turned to the printing of silk and other specialty work. Applegath did not return to newspaper problems for another twenty years, when the *Times* would call upon him again.

Far from these London developments another chain of events was similarly sparked by Koenig's genius: a yankee mechanical wizard would also make something of the new technology.

Daniel Treadwell (1791-1872) was orphaned as a boy and went to Boston to seek his fortune. Gifted mechanically, he worked as a silversmith and also developed several ingenious machines, among them screw- and nail-making equipment. As a result of studying anatomy with intentions (at the age of 27!) of becoming a physician, Treadwell learned that the muscles of the legs are far more powerful than those of the arms, a phenomenon he wished to test. "After some deliberation," he wrote, "I selected the printing-press, as connected with one of our most useful arts, and as well fitted to illustrate the principle assumed." In 1818 he fabricated a press with stationary form and rising platen, using a foot treadle and toggle joint. (Treadwell acknowledged that John Wells might have been first to use the toggle joint in a printing press, but believed the device originated around 1800 in nail-making machinery.)

Treadwell's "foot lever" press generated interest in Boston, and some were built for local printers. But to realize greater fame and fortune the ambitious inventor sailed November 1819 to England, where a more lucrative environment would mean commensurate rewards. After leaving specifications with David Napier, whose firm were accomplished machinists, Treadwell went on to visit France during the construction of his press. He soon returned to London to begin his hopeful sales campaign. He wrote a friend:

> *. . . My press will at most make its way but slowly, as it will have to conquer the prejudices of stubborn men, whose old habits of working are not to be altered in a day, if ever. . . .*

Treadwell resigned himself, consoled somewhat by another compensatory observation. His letter continued:

> *The English are wonderfully ignorant of all that relates to America (which they affect to dispise); this is decidedly to our advantage, for we learn from them, and have our own improvements in knowledge to boot. . . .*

Treadwell returned to Boston September 1820, having indeed learned something of value from the British. He began immediately to build a press on an entirely different principle: "After an examination of the steam cylinder press while in England, I concluded that a better steam or 'power press' for book-work might be constructed by using the platen instead of the cylinder for the impression." [14] Treadwell had surreptitiously brought true printing mechanization to America.

David Napier (1758-1873) came from an extraordinarily canny Scottish family, a branch of which almost singlehandedly developed the British steam shipbuilding industry. Trained as a general engineer, David went to London in 1808 where, upon the advice of Henry Maudslay, he hoped to set up a machine shop. Maudslay, pioneer industrialist and fabricator of the earliest machine tools in England, had similarly helped other great mechanics, James Nasmyth and Joseph Whitworth among them. Napier's stay with Maudslay was short: within a year the young engineer opened his own doors on Lloyds Court, St. Giles, prepared—as Maudslay must have warned —for any and all assignments. The tiny shop was in a poor but auspicious neighborhood:

> Maudslay himself had once been in business in Margaret Street, not a very great distance from St. Giles. In Soho Square, on the far side of Charing Cross Road from Napier, William Nicholson, who in 1790 had taken out the patent which fore-shadowed most of the essential ideas of cylinder printing-machines, had once kept a "school of science." [15]

Even Koenig's Frith Street works would be close by.

While Maudslay's generation had utilized laborious hand techniques to construct machinery with exquisite patience and precision, such craftsmen became increasingly rare. Napier was in a "second phase" of the industrial revolution, according to Wilson and Reader—"the machine tool phase, where the problem to which so many engineers addressed themselves was to make things more accurately, more plentifully and more cheaply than they could be made by hand." Struggling under strange new technical and business problems, Napier necessarily deemed no job too small, no application too remote. He had as yet no idea where to specialize.

Koenig's 1814 cylinder press triumph had come to Napier's attention, as must have Nicholson's earlier schemes. But the young Scotsman was quite unfamiliar with the printing process—until Daniel Treadwell wandered in. Treadwell had innocently brought Napier an assignment which incorporated some printing expertise (however slight) along with his diagrams. The "treadle" press was no challenge to Napier's craftsmanship, but it was inspirational: "Napier no doubt undertook the building of it in the ordinary way of business and took the opportunity to learn what he could from it," Wilson and Reader wrote. Ironically, as Treadwell profited from his trip to England, so also would Napier profit from the American's assignment.

With Treadwell's designs (and another 1819 prototypal machine of an inventor named Rutt) Napier soon boldly offered the public his own small manual press, using as an advertising slogan, "one man at this machine would be more than equal to two at the common press." The machine duly came to the attention of the Society of Arts, an organization usually laudatory on such new devices, but its judges found "its merits . . . by no means substantiated the assertions of the circular. . . ." [16] The Napier required much prior set up, and forms were hard to change.

Through his announcement, however, Napier came to the attention of Thomas C. Hansard, printer to the House of Commons (*Hansard's Parliamentary Debates* had begun 1806), who for some "ingenious and exact engineering" was seeking a capable machinist. Collecting material for his forthcoming *Typographia* (1825), Hansard had followed all the mechanical developments in press manufacture—in America as well as England—right up to Treadwell's visit. Of the "Treaddle,"† he wrote, "there is certainly great originality in the construction of this press,"[17] although he was quick to notice the large allocation of space it needed—very much a disqualification in crowded London.

Hansard commissioned Napier to build a large perfecting press especially suitable for the periodical work in his shop. His specifications, along the lines of a Koenig press, did not, however, preclude improvements that Napier himself perceived: (1) paper guiding and gripper apparatus, replacing the system of tapes on which sheets traveled with steel "fingers" actuated from within the two impression cylinders. Said Hansard:

> This beautiful mechanism is contained in the interior of the impression cylinders, which have openings along their circumference, through which the *gripers*[†] perform their operations, and upon their action depends that important desideratum of press-work, ACCURATE REGISTER, or the backing of the pages on the paper, and this purpose is so fully effected, that from the many thousands of sheets which have passed through my machine, without the smallest deviation after register was made, I venture to call them, infallible.

(2) Napier's "rocking" cylinders, rising and falling automatically when the type bed returned underneath them after an impression had been made. The machine was tested for six months in 1824 and found to produce an average of 2000 (perfect) impressions per hour. With great joy Hansard facetiously named it the "Nay-Peer" and described it at work:

> A little boy performs the business of the layer-on, but which requires very little considerable precision, as the rapidity of the machine will be interrupted if his attention is in the least degree diverted; if any corner should be doubled, or otherwise require particular time for placing, he will be obliged either to throw it to one side, or cry out "halt," and the machine must stop. . . . Should a sheet escape by being laid improperly for the gripers,[†] it is liable, instead of being carried round the cylinders, to be dropped upon the forme, from whence it will be conveyed to the inking rollers, and, from its adhesion to the [composition] substance and the ink, will take some time in being stripped off. . . .[18]

The imminent appearance of Hansard's typographic manual with glowing details of this machine soon made both Napier and steam presswork popular. For other kinds of printing—newspapers—Napier built two models (not perfectors): a single-cylinder machine (1500 per hour) called the "Imperial," and a double-cylinder called the "Double Imperial" (3000 per hour) using its two cylinders to make two impressions per cycle. They were offered for sale at £600-700. Hansard himself expected machine printing to appeal only for "works of extensive sale," that is, long runs, but this would soon cease to be a criterion.

Although the first Nay-Peer was not registered, Napier took patents in 1825 and 1830 for improvements, and did much to simplify the cylinder press mechanism. His 1830 double-feeder power-platen press for bookwork was exactly the motion Koenig first attempted before the *Times* steam press. Between 1836 and 1863 Napier constructed at least 87 perfectors (and another as late as 1873) worked by hand or steam; their prices came down to £450-600. Although one in 1844 was installed at the *Illustrated London News,* the perfectors went mostly into book and job shops.

For large newspapers Napier made the Double Imperial and the "Desideratum" (a simplified single-cylinder press), shipping over a few decades at least 140 of the latter and 23 of the former. In 1838 he built another machine, a "large quadruple" for the New York *Courier & Enquirer.* James Perry of the *Morning Chronicle* bought two Desideratums in 1834, following in 1835 and 1838 with a large quadruple; despite keen rivalry with Walter he had at last converted to modern presswork.

Through 1820-1850 printing presses were the mainstay of Napier's shop. He moved to larger quarters in 1836 at York Place, Lambeth. Had he chosen to manufacture printing equipment exclusively Napier might have secured for himself the enormous business of all England and much of America as well. But David Napier was thrifty and fussy, and proud of his other ideas: in the 1830's he turned to railroad work, ordnance machinery, large weighing scales, heavy-duty cranes, and other equipment. He brought his son James into the firm in 1847; both observed—not without pique—the continued expansion of the printing industry, especially newspapers. But somehow both father and son lost interest.

In the late 1830's Napier had begun experimenting with a rotary press he patented in 1837; progress was slow, as the engineering was complicated and capital expenditures at that time seemed ill advised. A decade later, when Applegath and Cowper—alone challenging the incursions of printing machinery from the United States—built their 1848 press, the Napiers scarcely followed its technical evolution. Walter of the *Times* next approached the firm to develop a fast rotary, but Napier would not accept the job unless the newspaper bore half the financial risk. Walter declined.

The Napier son, like his father, was brilliant, seldom compromised, and became increasingly choosy about new work; he further grew reclusive, shying from employees and peers alike. His thrift became self-defeating: he refused, he said, to wear out his machine tools making heavy equipment—the rotary presses—for which newspapers were clamoring. With Napier England effectively surrendered its claim to dominate the world's printing machinery business.

The proud father and son nevertheless had displayed three machines at the exposition of 1851—two steam-driven perfectors and a cylinder press for bookwork—and won a medal for one of them. The senior Napier retired in 1866 and the son carried on; the family firm continued until 1895 to make excellent specialized equipment—precision platen presses for printing paper money, postage stamps, and bank notes. The company has since developed high-performance engines for automobiles, boats, and aircraft, in which position today it enjoys foremost rank.

Robert Hoe in 1902 was justifiably magnanimous about Napier's position in the annals of printing:

> But the most ingenious and practical device in connection with the movements of a flat bed and a cylinder for printing machines was patented by Napier in 1828 and 1830. He was the first who introduced "grippers," or "fingers," for the conveyance of the sheets around the cylinder during the impression, and for delivering them after printing. Tapes or strings had previously been employed for this purpose. He was also the first to manufacture presses in which the impression cylinders are of small size and make two or more revolutions to each sheet printed, and he devised the toggles for bringing the cylinders down to print on the form and for raising them to let the form run back without touching.[19]

Before Hoe's grandfather could even find a Napier, however, Treadwell's novel press would quickly commandeer the American market.

Daniel Treadwell put aside his many theoretical activities and resolved to manufacture his new machine in quantity. Nathan Hale (nephew of the famous patriot), a Boston newspaperman, and Phineas Dow, an old friend and neighbor, gave him much advice and encouragement. Dow (1780-1869?), a machinist, handled the construction work; he had built fire engines, repaired printing presses, and fabricated several hand presses of local design. He later enjoyed a reputation in Philadelphia as "doctor" of "sick" printing machines. The results of the collusion, however, like most of Treadwell's ventures, were sadly premature:

> From the plans and descriptions furnished by Mr. Treadwell, Phineas Dow made first a model, and, subsequently, a series of working power-presses, which, so far as is known, were the first ever made or used in the United States; and, for this service alone, he deserves an honorable niche in the history of American printing. We can scarcely realize that Mr. Treadwell was so far in advance of his time, that, after his power-press was completed, he could find no purchasers—no newspaper proprietor ready to venture upon the doubtful experiment of printing more than a few hundred copies per hour—no book printer anxious to secure such facilities.[20]

Treadwell's bed and platen press, capable of 9 or 10 impressions per minute, was completed in 1821, its power (as steam engines were few) a horse. Early versions were somewhat rough on the animal. Said a witness:

> From some cause the inventor had failed to provide sufficient weight or momentum to the fly-wheel in order to straighten the toggle joint with ease by accumulated power; and whenever an impression was made the horse received a shock which nearly jerked him off his feet. . . .[21]

Until this was corrected the shrewd animal soon stopped just before the anticipated moment and had to be driven forward each time.

Treadwell's homely methodology was quite unlike Koenig's: he had found it impossible with American tools to execute the metal work required of a high-speed precision press. He recalled:

> There was not at that time [about 1821], I believe, a single steam-engine at work in any shop or manufactory in the old peninsula of Boston, and but a single one at the foundry at South Boston. There was not a lathe to be procured large enough to turn the face of an iron bed or of a platen, and I was obliged to construct these as in the Ramage press, the former of stone and the latter of wood.[22]

The wood-frame machine, although considered large and cumbersome compared to fashionably sculpted all-metal hand presses, nevertheless was commended—thanks to the toggle action—for the brute work it could do. It was also the first American press to use glue and treacle rollers. Treadwell's early progress had undoubtedly been quite unknown to Robert Hoe. Even 75 years later Robert Hoe III supposed only three or four Treadwells to have been constructed, but he was quite misinformed.

Treadwell knew he had a fairly profitable invention, but printers (in the United States as in England) were agonizingly slow to adopt new devices. He decided to build a second machine (for which he modified an old lathe so he could make an iron bed and platen) and conduct a printing business of his own as a practical demonstration. His stratagem was all too successful: "as soon as he was fairly at work issuing, with unprecedented rapidity, various popular books," said *Printers' Circular,* "his establishment was burned to the ground" [23]—probably by a hostile pressman. Treadwell promptly had more presses made, and Dow soon furnished them to others as well.

Although Dow's machine shop also burned (in 1830) and he removed to Philadelphia, he apparently accomplished some additional yankee inventing:

> *New York Febry 10, 1840*
>
> *Dear Sir,*
>
> *Mr. Conner has to day handed us your letter referring to your newly invented press & we received a communication from you on the sub-*

ject of it a short time since but in consequence of the absence of one of our partners who attends more particularly to the mechanical department of our establishment the letter was laid aside & was overlooked on his return. We desire to make & be interested in the best press and if yours is that one we would like to take hold of it with you. Before we can determine about its merits it will be necessary for us to see your drawings or model. If you think it best to send the model or come on with it we in that event could soon make up our minds as to its utility and decide about taking an interest in it.

Yours very respectfully
R. Hoe & Co.

To/
Phineas Dow Esq.
Worcester, Mass.

Nothing appears to have ensued, as by 1840 letterpress technology was much advanced—and already firmly in new hands. Treadwell, meanwhile, had seemed at last to fare better.

The ease with which early presses could be copied by intelligent workmen, and the infidelities of the new patent system (necessarily making public the very plans and specifications inventors preferred to conceal) apparently persuaded Treadwell not to advertise his presses for sale, but to grant exclusive rights to selected printers; thus his clients would be as anxious as he to prevent unauthorized manufacture or use. He did not patent the invention until 1826.

Treadwell's original advisors may have recommended this peculiar licensing arrangement, contemplating themselves among its earliest users. Timothy H. Carter, Boston printer and publisher (founder in 1828 of the famous "Old Corner" Bookstore), bought Treadwell's printing shop on Batterymarch Street, complete with presses, women and girls as operators, and rights for the entire state of Massachusetts. He recalled the enterprise:

> [T]he Treadwell Machine Press came into the market and Mr. Hale of the Boston Daily Advertiser, and myself, purchased exclusive right of use for this part of the country, and in a short time Mr. Hale had four of them running in his office by horse power and then by steam power, and we also had four jointly owned, running on the Mill dam by water power. The Treadwell Press did fine work but required much room and much power.[24]

The *Daily Advertiser,* founded 1813, had been the first successful daily in New England, lasting well over a hundred years; from 1829 it was printed on Treadwells—the first paper to do so. Some New York newspapers in 1829 also were licensed by Treadwell, but its employment there for bookwork became more noteworthy.

The famous Harper establishment in New York dated to 1818 when brothers James and John opened shop as printers. Although it had 37 hand presses as late as 1835, the Cliff Street plant in 1828 installed a Treadwell whose motive power was often promoted as more a curiosity than the press itself:

> The Treadwell press was connected by a system of gears to a vertical shaft which extended to the basement. A horizontal beam was then connected to the bottom of this shaft and at the far end of the beam a white draft horse was harnessed. As the horse walked steadily in a circular path, the shaft turned the gears upstairs and the presses flapped away as fast as the sheets could be fed to them.[25]

The brothers had perhaps agreed by this ploy—they never described the press—to safeguard their investment in what was becoming a very competitive printing town.

New Yorker Daniel Fanshaw (1789-1860) printed for the American Bible Society and bought two Treadwells in 1825—possibly the first such machines in New York for bookwork—to handle American Tract Society work as well. The Tract Society would publish over six million pamphlets in 1829 and eventually acquired its own presses. Fanshaw ultimately had no less than 20 Treadwells (with donkeys on the top floor providing power), and his large shop trained many well known American printers. According to R. P. Yorkston, a frequent contributor to *Inland Printer,* when a steam engine was first installed at Fanshaw's (1827) a yankee mechanic, Isaac Adams, ran it and also made press repairs. Samuel Bingham, a pressman there, was assigned full-time to make composition rollers. Yorkston himself, then an apprentice, soon assisted "in preparing for cremation the wooden parts of two Treadwell presses"[26] when they became obsolete.

Philadelphian Isaac Ashmead (1790-1870) had eight Treadwells, printing *American Sunday School* work on them. Others were located in Baltimore and Washington; all together, about 50 Treadwells were built until 1830, some of them a "double-acting" model, two presses back to back operated by a single drive shaft. By that time, however, larger competing machines were on the market, and oversize printing papers were available for them; as mechanical power became common, platen size no longer remained confined to what the pressman's arms could pull. Treadwell finally relinquished his interest in the machine in 1829, having realized about $70,000 on it. (Jonas Booth of New York had an 1822 power press like Treadwell's, but did not sell many.) Although the Treadwell soon thereafter declined in use, R. Hoe & Company in 1830 advertised as its "manufacturers," but this may only have been an invitation for repair or custom work.

Only 38 in 1829, Treadwell blossomed in inventiveness and scholarship. He developed rope-making machinery which completely displaced the hand-braided rope industry in England and America; he received an honorary master's degree at Harvard for his attention to hydraulics, mechanics, and railroad engineering. He edited and lectured widely, managing the affairs of the American Academy of Arts and Sciences, Boston Mechanics' Institution, and other august bodies. He invented type-setting machinery and became a disillusioned champion of built-up steel cannon rejected by American ordnance officers but (at the end of the century) possessed by every large army and navy in the world.* Occasionally Treadwell & Company executed minor machinework for Hoe.

Characterized in Longfellow's poem, "Wayside Inn," as a theologian, Treadwell, a deist, preferred instead to support Darwin, who was "setting *afloat* the old 'Plan of Creation' in the organic world, if he does not destroy it. . . . " As Darwin showed it possible to conceive the world as having no original design, Treadwell concluded *"that man, or the human mind, is the highest self-conscious intelligence (or intellectual power) in the universe. . . . "*[27] Treadwell left to others the fate of his printing press mechanism; two lads in particular, far more gifted than Jonas Booth, picked it up.

Employed as journeymen cabinetmakers in Phineas Dow's shop were brothers Seth and Isaac Adams. Engaged to make casting patterns, they soon proved versatile and hard-working; Isaac was highly inventive and Seth became a shrewd businessman. Together they could obviously succeed at any undertaking, even in the increasingly complex and confused state of the printing art. The brothers remained three years in Dow's employ, working on Treadwell's machines, and soon saw their opportunity.

The Adamses came from a poor Rochester, New Hampshire, family. Isaac (1802-1883) had worked in a cotton factory, and after cabinetry training spent time in several machine shops. By 1827 he developed a bed and platen press promising enough to exploit, and with his brother went into business. The "Adams Power Press" was almost immediately instrumental in turning the craft of printing into a skilled art. The first machines had a wood frame, later changed to iron. They printed a 30 x 40-inch sheet (a fairly large bookwork form) beautifully and rapidly, ending for all practical purposes the usefulness, commercially, of the hand press—even the best of them.

By 1830 the Adamses had improved their Treadwell-inspired machine, and did so again in 1836, at which time they organized a company to market the product. (Otis Tufts, a Boston machinist similarly once in Dow's employ, also began, in 1834, to offer a version of the Treadwell.) Principally by quality workmanship and a forthright policy of selling (not leasing) the machines, the Adamses became famous throughout the printing industry in America. Soon cast and machined entirely in metal, their press thereby escaped what Treadwell and others had always feared: counterfeiting by any gifted car-

* See *Age of Great Guns.*

penter, although Adams critics (like those who would soon carp at Hoe) even decades later would reduce the brothers' own contribution to little more than deliberate copying. Yorkston was such a critic:

> In this connection, let me add, it seems strange, in all that has been written about presses and their inventors, not one word has ever been said as to how and where Adams got hold of his ideas in building the admirable machine with which his name is so intimately connected; yet, that his service in the Fanshaw house, and close study of the Treadwell press, were the direct cause of his invention, there cannot be a doubt. . . .[28]

That is, nevertheless, the nature of invention; to improve, largely, rather than to conceive anew. The Adams in any case was a great step forward, if only for the skillful application of precision machine-work to effect excellent mechanical presswork. Robert Hoe III described it with admiration:

> In these machines, the type is placed upon an iron bed, after the usual manner of the hand press, and this bed is raised and lowered by straightening and bending a toggle joint by means of a cam, thus giving the impression upon the iron platen fixed above it, and firmly held in position by upright iron rods secured to the bottom bar, a strong crosspiece, at the base of the machine. The ink fountain is at one end of the press; the inking rollers travel twice over the form, in a movable frisket frame, while the bed is down; the paper is taken in by grippers on the frisket and carried over the form, when the bed rises and the impression is given; and finally the sheets pass forward from the frisket by tapes to a sheet flier, which delivers them on the flyboard.[29]

With a speed four times that of the hand press, the Adams appealed especially to book printers who, before the phrenetic rise of newspapers, alone bore the burden of staggering mountains of presswork. They rallied to the mechanized Adams, and for a long while would not believe anything else suitable for fine work.

Long in the Boston area were Welsh, Bigelow & Company, successors to the Cambridge, Massachusetts, University Press founded by William Hilliard. As printers to Harvard the University Press acquired a fine reputation for turning out books in foreign and scholarly languages. It was one of the first to install an Adams. The firm reorganized in 1859 with Albion K. P. Welsh and Marshall T. Bigelow as principal partners. Within a few years the shop had 15 Adamses and 3 Hoes, and Welsh, with a mechanical flair, knew them inside out; he soon began adding modifications of his own to both makes.

Scotsman John Wilson was another early printer in New England. He and his young son were running a thriving Glasgow shop when they decided to remove to the United States. In 1847, a year after their arrival, they commenced work with an Adams purchased from money jointly raised by Boston businessmen. Within twenty years, acquiring (in 1865) Welsh, Bigelow's old plant in Cambridge, John Wilson & Son's Cambridge Press possessed 11 Adamses and a Cottrell & Babcock stop cylinder. John Wilson, Jr., reminisced:

> It was on a Stanhope press, in 1840, that I first tried my hand as an apprentice, 2,000 copies being considered a good day's work. On that same press, about 1843, I assisted in printing a sixteen-page octavo tract of 100,000 copies on the "Corn Law or Free Trade Question," by Richard Cobden, whose memoir we lately printed on a hand press, for it took fifty days to print the edition. To-day, on a two-revolution press, we could print the edition in a little over five days, and on a Hoe perfecting machine in five hours.[30]

The Wilson shop played an important part in the revival of good printing.

Henry O. Houghton (1823-1895) started as a printer's devil; then, after a college education, he set type during the day and functioned at night as a reporter for the Boston *Evening Traveler*. He had just started business for himself in 1848 when he saw an Adams demonstrated, and knew it was what he wanted. Houghton revealed immediately both his hearty capacity (long before William Morris in England) for issuing superior presswork, and his uncanny recognition of sound labor-saving equipment (he later introduced both wire-stitching and machine thread-sewing for books). A Boston newspaper described his shop after a year's operation:

> Messrs. B. & H. [Bolles & Houghton] run four power presses, three of them Adams' patent with his recent improvements attached, and these with good ink and good paper, together with the care and attention which they personally bestow upon this part of their business, enable them, we think, to come as near perfection as possible in this difficult art. . . . The presses are driven by a steam engine of six horse power, built by Otis Tufts Esq., after a new model, which for compactness, beauty of finish, and economy of steam well deserves the attention of connoisseurs and economists of fuel. . . .[31]

In 1852 Houghton established the Riverside Press, doing excellent composition and presswork. Increasingly he became a prime prospect for R. Hoe & Company, but he would not for a long time abandon his Adamses.

John F. Trow (1809-1886) had begun with a New England apprenticeship to Flagg & Gould, specialists in printing oriental languages. Coming to New York in 1833 he tried various businesses until 1840, when he formed a publishing partnership with Jonathan Leavitt which included (beginning 1852) the New York city directories. Trow was among the earliest to use power presses—after a mule-powered Tufts, he was first in New York with nine Adamses.

He also pioneered with stereotyping (1840) and type-setting machines (Mitchell's, 1855). Trow was long a Hoe friend.

About 1840 Harper & Brothers acquired a two-roller, self-flying Adams which made printing history in the United States. Joseph A. Adams (1803-1870—no relation to the Boston brothers), a wood engraver of extraordinary talent and patience, approached the Harpers about a lavishly illustrated large folio Bible using his exquisite woodcuts—with the stipulation that he supervise the presswork. The brothers agreed, unaware how meticulous Adams could be. The artist thereupon "took off his coat, rolled up his sleeves," and began to put in two weeks of delicate makeready (adjustment of the impression under the blocks, usually with tissue silhouettes behind solid areas) before he would let the first form roll. The publishers were sufficiently dismayed, according to J. H. Harper, to abandon their conventional responsibilities:

> James supervised the mechanical operations, while John was the financier and made the purchases; Wesley read proof and conducted the correspondence of the House; Fletcher was in charge of the composing-room and gradually assumed control of the literary department. . . .[32]

Although makeready (before cylinder presswork) was almost unknown, Adams, who had been a printer and knew well what he was doing, would not yield to pleas to go promptly to press:

> The humorous brother joked, the serious brother expostulated, the amiable brother suggested the plan of starting the press on some given day, and then touching up the work a little, from time to time, as it progressed (but Mr. A. couldn't possibly see it in that light), and the judicious brother rolled up his clear and eloquent eyes in the agony of despair. . . . But they let him continue. . . .[33]

Presently, as "the most beautiful printing ever done in America" emerged, Harpers in 1844 announced it would serialize the magnificent work, which ultimately contained 1600 Adams illustrations. To handle the anticipated demand the brothers bought six new machines for its Cliff Street shop, especially designed by Isaac Adams with 6 inking rollers for superior ink coverage. Adams the engraver had started a reformation in bookmaking in America, and the Adams press—called by Ringwalt a "perfect machine"—had been the only one capable of executing it.

In 1853 tragedy struck the Harper establishment, a penchant for modernization in printing technology its downfall. With 34 Adamses (valued at $75,000) the Harpers had built a special fireproof zinc-lined room where rollers could safely be cleaned with highly inflammable camphene. On 11 December a plumber working near the roller washroom tossed a match into a pail of camphene, thinking it was water. The resulting fire destroyed the entire plant, including presses and $30,000 worth of sheets of Adams' Bible. (Fire and water also damaged the extensive Franklin Square chair manufactory of William Tweed—an ominous happenstance.)

The brothers lost no time; they telegraphed Adams in Boston an order for 20 replacement presses—by their clever use of the "wires" giving themselves a three-month lead over mail orders. Their loyalty to Adams, like Houghton's, transcended what must have been a growing professional relationship—and family involvement—with the Hoes, whose facilities and expertise were so close at hand. Two years later the new Harper plant was fully functioning, designedly housed in a novel cast-iron fire-proof edifice, the first of its kind in New York. The Harpers ironically had owned a building (A. B. Taylor's) in which a boiler explosion proved costly in lives and business; they were determined that no such accident would wreck their future home. It was now equipped with no less than 28 steam-driven Adams bed and platen machines.

Among those championing fine presswork, the Adams despite its limitations was revered through the century. Wrote De Vinne:

> Considering the larger surface printed, as well as its greater speed, the improved Adams press did in one day the work of ten hand-presses quite as well as it had been done before, for its provision for inking and exact register were of the best. For more than fifty years it was the machine preferred for book printing. Nor is it yet [1902] out of fashion. The Riverside Press has a large number in daily use. . . .[34]

Stanhope, Nicholson, Koenig, Applegath, Hill, Napier, Treadwell, Adams—what a flurry of activity in just a few decades! They and dozens of others entered the strangely uncertain and professionally debilitating world of printing press mechanization. Observers—even those with a mechanical bent—all acknowledged this peculiar obsession to have no widespread application, especially in rustic America:

> Since these experiments [in presses] were commenced, the splendid *foreign* invention of the steam press has been introduced into this country; but it is at once so complicated, cumbersome and costly, that it can be purchased only by large capitalists, and employed in extensive offices. . . .[35]

Yet they persevered, perhaps only because so much was not yet accomplished, not yet established, not yet dreamt!

Robert Hoe, a young New York carpenter and joiner, knew none of these plucky inventors before him and had none of the answers they sought. But all these things—and more—he quickly learned and showed his sons, for he had little time. All was accomplished within a decade.

ROBERT HOE

1784–1833

Courtesy New York Public Library

1.

Robert Hoe & Company, New York

CRISIS, CAREER, AND FAMILY / SMITH AND WASHINGTON HAND PRESSES / FIRST CYLINDERS IN AMERICA / DEATH AND REORGANIZATION / NEWTON'S TRAVELS / THE GREAT CONFLAGRATION

" 'O Mr. T., Mr. T., I shall die, I shall die—I never can stand this'; and he threw his brawny arms across the bed, as if going to grapple with death."

Grant Thorburn, a Quaker seed merchant, tried compassionately to comfort nineteen-year-old Robert Hoe, racked by the debilitating last stages of yellow fever.

" 'Die,' says I, 'Robert, to be sure, we must all die, but you are not going to die this week.' . . . I left him for fifteen minutes. On my return, I felt his pulse; said I, 'Robert, you are fifty percent better already; I hope to see you walk from the bed, and sit by the window tomorrow. . . . Death is nigh at hand at all times and in all places; but my impression is, that you will not die with this attack. I hope to see you a thriving master builder, married to one of the bonny Yankee lassies, *and to hold your grandchild in my arms*. . . .' "[1]

Fever left the young man within the hour, and, almost as quickly, the dreaded epidemic left New York City. There would soon be life again everywhere in the bustling town.

Robert Hoe was born 29 October 1784 in Hose, Leicestershire, a tiny hamlet in stern English farm country. His parents Ann March and Richard Hoe (whose surname was Anglo-Saxon for "high" or "hill"), although successful farmers, had a large family for which to provide, and Robert consequently got meager formal education in the village school. In his teens Robert was apprenticed to a nearby carpenter—the first of a series of fortuitous accidents which rescued, from the obscurity of British country life, a clan of gifted men better suited to serve mankind through the industrialization of printing. Robert took to his appointed calling fast and well.

Despite his ancestors' long association with the rural locality dating to 1581 (was not his name unmistakably related to that of the village?), Robert Hoe heard accounts of fame and fortune in yet another world. America, although incredibly far away, seemed to offer working men social and political independence, and—perhaps more enticing for a young bachelor—the promise of unknown challenges and munificent rewards. Before his apprenticeship expired Robert made his decision; he bought the remainder of his indenture, bade goodbye to his family and its long heritage, and booked passage to the United States. The land of golden opportunity itself was only 27 years old, scarcely as civilized as his native England, hardly as friendly as home and hearth. Once again, however, fate gave him courage and pluck to compensate for vague plans and dreams.

The *Draper* reached New York in September, 1803. Boldly—he was only a lad of 19—Robert Hoe disembarked, calling himself a professional carpenter and expecting ready employment in what was to him a gigantic metropolis (population 25,000) in obvious need of willing hands. Although yellow fever raged through the city and most businessmen had shut their shops to flee with their families (and tiny Greenwich Village to the northwest far enough away for most), the bewildered emigrant, penniless, strode hopefully through the crooked little streets. One contrary merchant, open for business, watched him with surprise and interest.

Grant Thorburn (1773-1863) had himself come to America after arrest in England for agitating parliamentary reform. A maker of iron nails displaced by new nail-making machinery, Thorburn turned eventually to seed farming, admitting in his memoirs he didn't know "a geranium from a cabbage" when he began. His unique family business, the first of its kind in the United States, lasted more than a century in New York, although he later lost a fortune in the chimerical scheme of cultivating mulberry to make domestic silk. With cheerfulness and Quaker philosophy Thorburn overcame his own cruel handicap (he was a short-legged dwarf—victim of a nurse's carelessness in infancy) and gained rare insight into the indispositions of others. Loquacious and odd in his Quaker clothes, he was nevertheless a controversial correspondent to New York newspapers and a prolific author. Thorburn recorded his strange encounter with Robert Hoe:

> In September, [1803], the yellow fever prevailed to a fearful extent in this city. As I never left town while it was raging, I was "sitting in my tent door in the cool of the day, and lifting up my eyes I beheld a stranger," a rare sight in fever times. He was moving from Cedar street

along Nassau, having his face set towards Maiden Lane. He walked in the middle of the street, and was reading the sign boards on the right and left. He paused in front of my door, and mine was the only store open in the block. As he stepped in, he said, "Mr. Thorburn."

"Where did you learn my name?" I enquired.

"I saw it over the door," said he. "I have just come on shore from the ship *Draper*, from Liverpool. I am a carpenter by trade, my name is Robert Hoe; I am now in my eighteenth year."

Says I, "Robert, were your indentures fulfilled before you left England?"

Says he, "I never was bound, I learned my trade with my father; I can't find work, I have no money; can you recommend to me a house in a healthy part of the city, where I may board till I get employment, when I will pay them honestly?" [2]

Young Hoe may be forgiven for shaving his age (if indeed Thorburn —who put this event in 1805—remembered correctly) and simplifying the details of his work history; it was no time to suggest that promises to pay board and keep might be broken as flagrantly as quitting an apprenticeship—a serious matter. Robert's own sister, however, intimates a conventional live-in apprenticeship.

In 1859 Ann Hoe Goodall, 70 years old, described her playful brother for Robert Hoe II, then visiting England ("[†]" omitted):

. . . I ham your Father['s] *one Sister. i dearley loved your Dear Father, we used to play tog*[et]*her till your dear Father went* [ap]*Prentis. . . . Bruther Robert hee was livley and full of is triks. . . .*

It is doubtful that if apprenticed to his father he would have ended his family pranks so abruptly.

Tiny lies were no matter; Hoe's beguiling manner did not betray them. Thorburn, himself only 30 and engaged in his seed nursery only about a year, nevertheless considered himself a swift and sure judge of character. He continued:

I knew the heart of a stranger, having been a stranger myself, and there was so much of honest simplicity in his speech and deportment, my heart warmed towards him; I gave him a chair, and ran upstairs; says I, "Gude wife, a stranger standeth at our door; shall we take him in?"

"If thee pleases," she replied.

"If he takes the fever, will thee help me to nurse him?"

"I will," she answered.

"Thank you, dear, for this; God will bless you." Now, says I, "Come and look on his honest English face."

The impression was favorable. Says I, "Robert, this neighborhood is accounted the most healthful in the city; you will lodge here; if you take the fever, my wife and I will nurse you, you shan't go to the stranger's hospital."

> His eyes spoke thanks more eloquent than words. As he had no business abroad, I advised him to stay at home. The fever seized him, however, in less than a week. I procured an eminent physician; my wife and I nursed him. In seventeen summers that I've nursed among the sick, I do not think that I ever saw a case so violent but it terminated in death, his only excepted. On the fourth day, generally the crisis, the burning fever was coursing through his veins, and drinking up his English blood. His skin was burning, dry, and yellow, heart sick, home sick, all round sick; and his spirits sunk down to his heels. . . .[3]

Considering its usual ten-day incubation period, Robert had contracted the disease almost as he stepped off the *Draper* gangplank. With darting, restless eyes on Thorburn sitting at his bedside, weak with pain, fever, and vomiting, Robert shrieked his mournful expectation of imminent death.

Thorburn later confessed that his impassive response was prompted not so much by any real observation of improvement as by recollection of Pope Pius' Bull on the end sanctifying the means. *"Fancy kills, and fancy cures,"* he thereupon thought to himself, and explained the unorthodox tactic: "He was under the influence of powerful medicine at this moment; I knew there would be a lull, as the sailors say, soon; and I meant to take advantage of the circumstance to persuade him to live, if possible." Thus the wily Quaker returned after a quarter hour, fabricated a steady pulse, and chatted glibly of master carpentry, bonny wife, and bouncing grandchildren. Thanks perhaps entirely to the psychology of this subterfuge, Hoe's crisis passed.

Although Thorburn published this account in 1852 and confused some of Robert's accomplishments with his son's, his modest appraisal of his charity and mercy suffered not:

> [N]othing in my past life affords such pleasing reflections as this act of duty and humanity to a stranger. When his aching head lay on my breast, as I held the cooling draught to his parched lips, I little thought that in his head lay the germ of a machine destined to revolutionize the world of literature, and shed light on the dark places of the earth, whose habitations are full of horrid cruelty.[4]

Comically underpinned with "short duck legs," Thorburn became a giant in Robert Hoe's estimation and was not forgotten. But he performed one further service for the recuperating boy.

Some historians have concluded that Hoe worked for Thorburn, but the seedsman's account contradicts this. Others have suggested that Thorburn's "bookish" avocations brought the printing press, made largely of wood, to Hoe's attention. More convincing, however, is the possibility that his benefactor helped in a more practical way: furnishing Hoe an introduction to a fellow Quaker who could indeed use the services of a young master carpenter. They quickly met.

Friend Underhill had a contract to build a bridge in Westchester County, north of New York city, and duly asked Robert if he was familiar with that kind of work. Hoe was no less aplomb than in his interview with Thorburn; of course he was—although he did manage to show some structural drawings of his own which were sufficient to convince the Quaker to place the entire project in the lad's hands. Hoe erected the bridge so successfully that Underhill admitted, "Thee can do better for thyself by taking work on thine own account."[5] He magnanimously relinquished to Hoe some other commissions for building homes in that undeveloped—but already desirable—suburb above the city.

As late as 1850 the town of New York extended only as far north as 34th Street; on Manhattan Bloomingdale, Yorkville, Manhattanville, and Harlem were remote, isolated Westchester County villages. Hoe may actually have journeyed only to the upper end of the island to perform the bridge and house work which occupied him for about two years until 1805. During this time he made the acquaintance of two men who were also carpenters. Matthew Smith and his son Matthew (Jr.) lived in North Salem, New York, and were engaged as printers' joiners and equipment makers. But something more than presses caught Robert's roving eye: Rachel Mead Smith, the pretty daughter, was that special bonny yankee lass Thorburn knew would come. They married 10 January 1805—and Robert once again fell into happy circumstances.

The Smiths quickly recognized Robert's ability and offered him a place in their business. His participation made it possible for them to set up separate shops for a more deliberate attack on New York's printing industry, already important. Matthew and Robert as partners began by soliciting woodworking business in general—even to making hobby horses—and printers' requirements in particular. Peter Smith, 15 years younger than his brother Matthew, educated at Yale (class of 1816), preferred to dabble alone in cabinetry rather than join the others.

In 1805 the partners were casually established in New York as Smith, Hoe & Company. For the firm's 1955 sesquicentennial George Hickman, of Hoe's engineering staff and a company historian, hunted methodically for traces of those early years:

> I searched through many records before I located the exact spot [89 Fair Street] where the first carpenter shop stood, which is the present No. 63 Fulton Street. This is on the north side of Fulton about 100 feet east of Ryders Alley. . . . From 89 Fair Street Robert Hoe moved to 41 Barclay, then to 76 Nassau, then 80 Nassau Street. Then as Smith & Hoe in 1809-10 they were at 11 Cedar Street. They again separated. . . .[6]

Longworth's New York City directories also kept vigilant track of them: Matthew, with home addresses at 80 Nassau, 82 Anthony, 8 Dutch, 7 Ann, 241 Pearl, and finally 61 Maiden Lane; Robert, at home at 76 Nassau, 80 Nassau, 19 Thames, William and Beekman, 19 Warren (business at 90 William Street), 90 William, rear (business at 84 Maiden Lane). From 1820 R. Hoe & Company is shown at 64 Pine Street until 1833, when 29 and 31 Gold Street are listed. The year 1805 was proudly if loosely acknowledged by the firm to be its founding date.

These moves, each never more than a few blocks, kept Robert very near to Grant Thorburn's own shop, making it difficult for the lad to forget the Quaker's generosity. "We met many times without number," Thorburn recalled, "his never-failing salutation was, 'Grant, as the instrument under God, I have to thank you for my recovery from that fever. . . .' "[7] At 122 William Street, a house built of small bricks brought from Holland, Robert and Rachel about 1828 established a riotous household by collecting their nine children—Mary (born 1805), Elizabeth (1807), Emeline (1809), Richard (1812), Robert (1815), Rachel (1817), Harriet (1819), Peter (1821), and Theodosia (1823)—in what could at last be called a proper home.

For the hopeful businessmen the beginning years were not easy. In 1809 Robert's family was living with Matthew's, and the wives, under Rachel's maiden name, took in washing. Peter ran a succession of cabinetmaker's shops until about 1822, when the success of his newly developed printing press encouraged him to open at 81 Maiden Lane as joiner and pressmaker. Robert and Matthew conducted separate operations until 1820, although considerable cooperation existed between them. It is likely that Hoe supervised press repairs at his address while Matthew ran a shop:

> Printer's Warehouse
> M. Smith,
> 241 Pearl Street, New-York
>
> Has constantly on hand, a variety of Printers' materials, to wit: Chases, Composing Sticks, Brass Rules, Standing Gallies,† Standing Racks, Composing Stands, Slice Gallies, and most articles requisite in the printing business. All orders attended to with promptitude and punctuality.
>
> New-York.
> November 26, 1818 [8]

Adam Ramage (1772-1850) in Philadelphia had come from Scotland in 1795. About 23 at the time, he made mechanical repairs and began building cheap, small presses. He improved the hand press (1796) with a triple-thread rapid-motion screw and all-metal construction: pioneers going west appreciated its light weight, but its manual operation was cruel drudgery. "Only those who commenced their typographical career with treading pelts [inking] and pulling on a Ramage press," said Joel Munsell, could fully appreciate Hoe's later machines.[9] (Hand presswork had lasting effects on lifetime printers: the right shoulder and foot became enlarged, the left shoulder sunken; the body moved sidewise, crablike, when walking.) Ramage successfully manufactured over 1200 presses between 1800 and 1820 (including the "Ruthven"), but by 1849 was content to let the Hoes introduce a new model:

> [*From Ramage, 31 October 1849*] . . . *I have concluded to send you one in a few days if you think you might have any call for a* [*Fools-*]*Cap or Proof Press and it does not interfer with any of yours. . . . I should like one of them to be in the hands of a good Printer in New York and being the first put it at a little less* [*in price*].
>
> *Your Mr Hoe can inform you of what little consequence a Press Maker is at nearly 78 years of age with his little shop kept for his recreation*[?] *more than any thing else. . . .*

F. Bronstrup, Philadelphia, acquired Ramage's business and in turn sold out (1875) to Robert Menamin.

After several careers, including cabinetry, George Clymer (1754-1834), another Philadelphian, had also developed an all-metal press (1788). He dispensed entirely with the screw in 1816 in favor of a powerful lever (toggle) system. Pressroom visitors ogled its cast caduceus, dolphins, and other fauna, all surmounted by an American eagle as counterweight—"undoubtedly the most spectacular machine ever devised; its ornamentation is riotous, whimsical, and classical," [10] said Lewis Allen in 1969, finding specimens still available for avid hobbyists. Clymer, puttering in his father-in-law's foundry, had purposely used the excessive embellishment just to attract attention. Brought to England where it sold well, the "Columbian" was at first manufactured by Richard W. Cope (who then countered with the "Albion"), and later by Clymer himself with a partner. The Columbian gradually replaced the Stanhope and its imitators, doing fair service even for newspaper work.

The *Connecticut Courant* (founded 1764) in Hartford had built its own paper mill in 1775 and three years later—newsworthy times, to be sure—reached a circulation of 8000, phenomenal for America or England. The paper bought a Columbian in 1815, and explained to its readers:

> The different parts of the press are of iron, not excepting the platen, which ensures its strength and permanency. . . . Being entirely composed of iron, it is far less liable to get out of repair and is obviously more durable than the common presses, and the impressions it gives are more uniform and distinct.[11]

Such all-metal construction was a powerful threat to carpenters and joiners—and it was just beginning. In 1828 the *Courant* bought John Wells' hand press, and in 1832 an Adams power press—the first in Hartford. After that the paper enjoyed a succession of Hoes (the company's 1892 web-perfector being called the "Courant" as a generic type), and in 1956, with 100,000 circulation, it had Hoes and Wood presses side by side. Like the *Courant,* however, many shops used the iron hand press so faithfully it seemed a permanent fixture even as steam and cylinder machines appeared. That noble instrument was indeed not so quickly doomed, said *Scientific American*:

> The mighty printing machine counts its hourly productions by thousands, the humble hand-press produces at the utmost but a few hundred impressions in the same time; such being its vast inferiority, its total disuse would appear inevitable, and yet, contrary to this inference, the hand press maintains its position, chiefly owing to its simplicity and consequent cheapness. . . .[12]

Not all shops, of course, enjoyed the volume of work required of a modern machine.

John I. Wells in Hartford had taken the iron press a step further. An ink maker, Wells in his travels witnessed the excruciating labors of pressmen, and recognized the "knee-joint" or toggle "levers" as an excellent substitute for brute force. In 1816 he began building an iron toggle-action press; he patented it (1819) and advertised it for sale at $350. But Connecticut proved too remote, and he did not prosper. Some of his presses, furthermore, were seen by persons not inclined to buy but to copy. Wells reported, sadly:

> In 1820 several of my presses were in the Bible office in N. York; and after Peter Smith had commenced the manufacture of presses, I was informed that he had examined them many times. His Patent is dated "29th Dec. 1821," of which I have a copy. He claims that his "improvement consists chiefly in applying the Wedge Power," whereas all his power is derived from his Levers. And since then, nearly all the inventors of printing presses have adopted the [toggle] principle, with some variation in the manner.[13]

Peter Smith's "Acorn" press—"an acorn shape, homely but strong," Tucker said of it—was announced to the trade in 1822, and even Robert Hoe III found it no great conceptual improvement: "The frame was of cast-iron, and in place of the screw with levers, [Smith] substituted a toggle joint, at once simple and effective." Tucker concurred: "The works gave a powerful impression and for years it was the best press of its kind."[14] R. Hoe & Company continued to market the machine through the century, the successful element obviously sales, sales, sales. Ralph Green recapitulated this process:

> Peter Smith has been given more credit than he deserves. On the strength of a patent of doubtful value, a press was produced which contained no new features worthy of comment. The business ability of the Hoe firm, together with good workmanship, and a heavy, strong design were probably accountable for the success of the press and its long life on the market.[15]

Upon the death of Matthew in 1820 Peter had succeeded to his brother's place in the partnership, and with the widow's approval Robert Hoe quietly became senior partner. The Acorn gave Smith and Hoe a genuinely useful article around which to organize sales campaigns, their first "name" product. In perhaps a flagrant over-simplification to dominate printing press history, the company later described inventor Smith as a "member" of R. Hoe & Company, an organization that did not yet exist. When Peter Smith apparently retired in 1822 Hoe was at last in absolute command—although (according to her daughter-in-law) upon becoming sole owner of the Smith interests Rhoda was for a time Hoe's principal partner. Rhoda Smith was unlikely to conduct the thriving business, of course, and as Smith-Hoe relations were always cordial, would not object to Robert's reorganization in his own name.

Now in possession of the firm, such as it was, Robert Hoe quickly removed all references to "Smith" and in 1822 it became "R. Hoe & Co." It was a time for celebration, but—not given to inane festivities—Robert awaited a proper occasion. On Friday, 4 November 1825, the opening of the Erie Canal gave him the promotional opportunity he wanted. On this day business ceased, public buildings and harbor ships raised flags and pennants, and a grand merchants' and craftsmen's parade snaked through lower Manhattan. On a cart drawn by four horses the Typographical Society of New York displayed two presses in operation, both making copies of a poem honoring the occasion. The *Evening Post* reported:

> Between the presses stood Dr. Franklin's arm chair, occupied by Mr. James Oram, the oldest printer in the city, surrounded by two Heralds and two Mercuries, who assisted him in folding the sheets and throwing them from the car to the public. The presses were handsomely gilt, and are beautiful specimens of American workmanship and taste. One of them is the invention of Mr. Rust, and the other of Mr. Smith.[16]

At first the enterprise was still very modest: Hoe employed only a "handful" of men in an old building located in the middle of a large block bounded by Maiden Lane, Pine, William, and Pearl Streets, served by the north-south Ryders Alley. The works were driven by four powerful English draft horses until steam was introduced in 1828. An extension of Cedar Street soon confiscated the property and Hoe moved to nearby 29 and 31 Gold Street, a five-story brick structure.

The new company included at least one Smith son, just as Hoe was bringing his own boys into the business. Matthew Smith III (1812-1841) was young (and frail) but impetuous. New local competition worried him immediately.

Samuel Rust, a New Yorker, in 1821 had developed an iron press similar to Smith's, but with a "figure-four" toggle joint. Called the "Washington," it was manufactured for about a dozen years by a rival firm, Rust & Turney. Rust improved it in 1829 by making the frame stronger, lighter, and easy to dismantle. Robert Hoe III commented:

> The frame, instead of being all of cast-iron, had the uprights at the sides hollowed for the admission of wrought-iron bars, which were securely riveted at the top and bottom of the casting. This gave not only additional strength, but greatly diminished the amount of metal used in construction.[17]

Dr. Thomas P. Jones, reporting in *Journal of the Franklin Institute,* was not impressed by the modifications: "several improvements are claimed," he wrote, "but they are generally for matters of detail, the principle of the press remaining unchanged." [18]

Young Stephen Tucker, familiarizing himself with this formidable Hoe competitor, judged the Rust mechanism even more powerful than the Smith, a fact he may have communicated to his new employers. And it was less troublesome: the Smith toggle tended to jump out of its socket if released too suddenly. The Washington was a most attractive property, and the wonder is that the Gold Street firm did not simply appropriate its features. But R. Hoe & Company were always patient.

Matthew Smith particularly liked the Washington press, and after a few years, without much consultation with his associates (partner Sereno Newton was traveling), employed a most unusual ploy to get it:

> [*Smith to Newton, Florida, 7 March 1835*] . . . *About a week ago we sent* [*John*] *Colby to Rust to learn on what terms his patent could be bought. Knowing that R. would be disposed to take all the advantages of* <u>*us*</u> *he could if he thought* <u>*we*</u> *wanted to purchase so we plan'd it that Colby should go to him on his own a/c & pretend that he was going to leave us & start opposition: this pleased R. very much & he*

was quite anxious to sell; he said his price for the patent Right, lease of shop, patterns, all his stock, tools, platen lathe † lathes, horsemill & horse, in fact his whole establishment—together with the good will of his business, &c, &c, was 3,000 dolls. This Colby objected to as being entirely too hight † and offered him $2000 and was to see him again. He went two or three times without being able to see him.

Yesterday Harpers came to us and wished to speak to us in confidence on a subject that might be advantageous to both, and made us promise to take no advantage of what they should say—it seems that they were negotiating with Rust (I suppose R. went to them after Colby's offer); they wanted to know what we would contract to build 50 Rust presses per year, saying that they were going to purchase his establishment for 3,000 dollars cash (we of course expressed no surprise). We told them we should be happy to make them and had no doubt but a mutually advantageous arrangement might be made, that we w[d] think of it and let them know. They said however they had not yet closed the bargain with Rust.

Immediately after they left we sent Colby to Rust—but—he was unable to find him. We thought that now at all hazards we must have the concern. Today Mr. Colby has seen Mr. R. and concluded a bargain with him for the whole concern for 3000 dollars cash—could not get it for any less. Rust is to pledge himself not to be engaged or interested directly or indirectly as agent or any other way in the making, contriving, selling or introducing any press of any kind for the term of his patent (8 years). They are going to Holden's [Hoe's lawyer] monday † morning to bind the bargain and if R. does not fly from it (which we think he will not do as he has pledged his word of honor to Colby) we shall have it snugly fixed. He makes about 40 presses a year besides you know he was always cutting under & ruining the business. . . .

John Colby, then a Hoe foreman, later (1846) worked as pressman on the New York *Commercial Register*, at about which time he was reengaged full-time to erect Hoe presses around the country. Sereno Newton was not greatly impressed, thinking $2000 would have done as well.

Alongside the Smith, nevertheless, the Hoe Washington quickly became a strong seller, some 6000 of them fabricated by the end of the century. Its silhouette became the insignia of the United States Government Printing Office, as well as that of the International Printing Pressmen and Assistants' Union.

Frequently modified and improved by Hoe designers, the Washington became the definitive hand press in America. "They were manufactured in great numbers," said Robert Hoe III, "and continue to be manufactured and sold at the present time for taking fine proofs. . . . They have been sent all over the world." [19] Indeed, as late as 1890 the Hoe catalog offered both Smith and Washington machines in no less than 10 sizes each, with little modesty:

> The celebrity which our Patent Washington and Smith Hand Presses have obtained, and their exclusive and constant use in almost every Printing Office in the United States and other countries, during the last forty years, renders any remarks upon their superiority unnecessary. They are elegant in appearance, simple, quick and powerful in operation, and combine every facility for the production of superior printing. Each press is tried at the manufactory, and warranted for one year.[20]

The toggle joint had become a powerful mechanism in many industrial applications, and few inventors could overlook it. John Wells was only the first to apply it to printing presses, but even those who copied from him were imitated: Smith's Acorn inspired three others —Phineas Dow, Otis Tufts, and Seth Adams, each of whom developed similar hand presses (1827, 1830, 1831). Abraham O. Stansbury, a New York bookseller, designed a toggle-action press (1821) manufactured by the Cincinnati Type Foundry; when the patent expired Hoe made a small Stansbury from 1867-1885. "This was yet another example," wrote James Moran, "of Hoe's taking over an existing press and making 'improvements.' "[21] But despite the excellence of its workmanship and the reliability of its printing production, the hand press was now at the end of its development. Its lack of speed was its fatal flaw.

Having observed, nevertheless, the quiet but pervasive intrusion of metal into the art of press making, Robert Hoe quickly realized that knowledge of—and capacity for—metal-working would be essential for press repair and construction. With little technical help available, he began assembling the necessary machine tools. In 1824 he devised a machine for planing cast-iron platens, where the casting was made to revolve while the cutting tool traveled from center to periphery. Forerunner of the modern turret lathe, Hoe's machine was judged very accurate in making a flat surface. He also developed a machine for planing metal of large dimensions, and he bought the first gear cutter made in New York. These were expensive and ambitious investments.

Robert Hoe's experimentation—dating to 1819—with steam-powered printing presses had indifferent success. It must be presumed that he, like others, was attempting simply to mechanize the rise and fall of a horizontal platen as sheets were fed and removed by attendants. The application of steam power from engines his firm would design and make became another necessary adjunct of the business. Only 36 in 1820, however, Robert was not the accomplished theoretician and innovator these advanced techniques demanded—that would be the domain of the next generation.

But America in the 1820's was not to be the scene of the next creative act in printing press development. Quite unknown to the little group off Ryders Alley Koenig and Napier were hard at work in England; yet word of their progress somehow crossed the Atlantic. Thanks to an insatiable desire to read, to learn the news, to keep up with the (old) world—in effect, to encourage newspapers, such new developments could not long remain hidden. And, of all things, there were indeed "large capitalists" in New York after all.

Two progressive newspapers in need of new equipment, the *New-York Daily Advertiser* (first American paper to begin—1 March 1785—as a daily) and the *New-York American*, had found nothing suitable in the United States to compare to the legendary capabilities of a Koenig or Napier; newspaper proprietors, it seemed, always wanted revolutionary equipment to symbolize their ultramodern editorial posture. John W. Walker, printer and owner of the morning *Advertiser*, evidently was determined to explore rumors that a new English press, commercially available, could produce 2000 impressions per hour. He persuaded his partner and the evening *American* owners to share the expense of such a machine.

In April 1825 Walker sailed for London to examine the situation: spending three months at the Napier shop watching its construction and learning its operation, Walker was satisfied to ship home one of the controversial Nay-peers. An *Advertiser* announcement was carried by a contemporary paper:

> The New York Daily Advertiser gives notice, that the publishers, in connection with the proprietors of the New York American, have sent to England for a press which will strike off two thousand sheets in one hour. This, they say, will enable them to keep the paper open several hours later, and yet have the press-work performed in so short a time as to have it distributed throughout the city as early as can be desired.

Robert Hoe, just opening shop under his own name, watched the news carefully; 2000 impressions per hour was a rate 8 or 10 times faster than the feeble presses in common use. In the fall the *Advertiser* itself described its acquisition:

> The press we have now imported is one of "Napier's Imperial Printing Machines," which have lately acquired great reputation, both in England and the neighboring countries, for the expedition and perfection with which they perform their work. It is to be moved by the strength of a man, although capable of having any other power applied to it, as the motion is first communicated to a large fly-wheel. . . . The form containing the type is moved from end to end of the machine at the rate of from 1600 to 2400 times in an hour, passing under the inking roller in the middle, and one of the pressing cylinders each time, the roller and cylinders being regularly raised to permit the paper to be carried in and out by the tapes and cords on which it is

> laid. The machine is of cast iron, and weighs about two tons. The Courier, British Traveller, Morning Post and Globe and Traveller of London are printed in Napier's presses; and during the time that our press was building, others were making to answer orders from Scotland, Ireland and France. . . .[22]

Walker began shopping for new joint headquarters; he anticipated doing the *Advertiser's* 2760 circulation (formerly almost 9 hours' work) in less than 2; the *American's* press time would similarly be under 1 hour. Bedding of the press, casting of rollers, and trial runs (limited by poor paper in New York to speeds around 1400) continued over several months. Walker had actually bought the Double Imperial, rated by Napier at 3000 impressions per hour, but he was not complaining.

In an *Advertiser* article of 30 December 1825 Walker described the press so minutely one could almost visualize all its throbbing motions. In its 10-foot length and 5-foot height and width, the machine contained a myriad of precision parts unknown to hand presses: a half dozen gears, 22 brass tape and cord reels, a universal joint, a rack and pinion, two hollow impression cylinders, and rollers, rachets, levers, and slides too numerous to mention. These betrayed the engineering sophistication (and incidental high cost) that made the installation of such a press news in itself. Walker admitted:

> The machine, it is true, is expensive compared with other Printing Presses; but the early dissemination of intelligence in the country, and the timely delivery of papers both morning and evening in the city, was of too much importance to the public and the proprietors, to let the price [$4120 including shipping and duty] stand in the way of the great advantages which the machine promised, and which it performs to our entire satisfaction.

It seemed quite beyond the ken—no less engineering facilities—of anyone in the United States. And for this reason it was an extraordinary affair from a nationalistic point of view.

To the country as a whole the significance of introducing an imported machine to the printing industry of America seemed a deed of noble proportions. In a petition to the United States government requesting relief from payment of import duty, the proprietors pointed out:

> That the introduction of it into the United States must be considered as an important event to the public at large, as well as to the publishers of newspapers; and the memorialists would respectfully suggest, that if they should not be considered as coming within the precise terms and provisions of the law which authorises the importation of *models of machinery and other inventions* free of duties, yet they view themselves as having rendered a more important public

> service by importing one of the most useful machines that has ever been invented, in a completely finished state, and free from the errors and imperfections which might attend a mere specification or model —one which exhibits to the observation, and by actual experiment, its great intrinsic and practical importance; and this also, without subjecting Individuals, or the public, to the inconveniences of procuring the right to use an exclusive privilege.[23]

By Treadwell's exclusive licensing two of his new machines had just been sequestered at Fanshaw's, perhaps never to be inspected by others—exactly the inventor's desire. Walker and his party, decrying the use of such "exclusive privelege" (had Treadwell earlier turned him down?), appeared to condemn that system and capitalize on their public obligation. It is likely they anticipated little inventors' traffic through their shop.

The House of Representatives focused on the delicate point of public service; it reacted favorably to the *Advertiser-American* memorial but took no chance that such advantage might become another secret monopoly. To accomplish this a House committee required only that one phrase be added to its bill: "Provided said Press be Subject to the public inspection upon application to the Proprietors." On 15 May 1826 the final bill was read in the House—but it failed to pass.

A successful appeal to Congress would have returned to Walker and his associates only about $760, but the snub was more irritating. Printing the two newspapers with dispatch, Walker had sufficient press time to do job work as well, and must shortly have recouped his entire investment. But it seems beyond question that he abandoned his public spiritedness; his coveted machine became as unobtrusive as any of Treadwell's. There is no indication that R. Hoe & Company had any immediate opportunity to see it—until repairs were needed.

In 1829 another Napier was imported, this time for the Washington, D.C. *National Intelligencer* (founded 1800). The modest four-page paper was respected for over 50 years, the first half of which it was methodically raked by other journals for its government news. In 1810 it had been acquired by Joseph Gales, Jr., and W. W. Seaton (Gales' brother-in-law), who joined him two years later. Although the partners' coverage of congressional proceedings was superb, their combined strength turned to weakness:

> Both editors . . . were highly regarded personally, and the handsome and aristocratic Seaton was long mayor of the city of Washington. Though their paper was very profitable in its best days, they were not good enough financiers to build it up on the business side; they never made an accounting to each other, and always helped themselves from the till when either side was in need of money.[24]

According to printing historian Henry Bullen, whose resources were copious if not authoritative, the Nay-peer ordered by the *National Intelligencer* was actually imported by Robert Hoe himself. Upon its arrival in New York it was reportedly withheld in customs because Gales & Seaton could not raise the duty for it. The firm had lost (to Duff Green) its government printing contracts and had to sell its headquarters to raise money, but to no avail. Gales & Seaton were bankrupt. The port surveyor, Major Noah, by coincidence was more than routinely interested in establishing the value of the foreign machine.

Mordecai M. Noah (1785-1851) had come to New York in circuitous careers: editor of a South Carolina newspaper, then American Consul at Riga, Latvia, and Tunis. Recalled in 1816, he began editorial management of the New York *National Advocate,* a paper founded 1813 as the Tamany Hall democratic organ. For his patronage Noah was appointed sheriff of the city and county of New York, but lost the office upon protests that it would be improper for a Jew to hang Christians. Noah left the *Advocate* in 1825—replaced by young James Gordon Bennett—to execute a grandiose plan to unite Jews and American Indians (considered by many early "anthropologists" to be descendants of Israel's lost tribes—else how could they and Blacks trace their roots to Adam and Eve?) in a utopian community at Niagara Falls.

Returning to New York city, Noah soon started the New York *Enquirer,* which he managed until 1832. In a grumbling 1828 letter to Martin van Buren (then a senator) he complained about publishing finances:

> . . . It is extremely difficult to carry on a political paper in a commercial community at an expense of $25,000 per annum without aid from the party & this can come in a more agreeable shape than subscriptions and contributions. . . .[25]

Noah went to Washington to get something "agreeable" for himself. To make newspaper careers seem as miserable as possible, he cajoled Washingtonians Francis P. Blair, Amos Kendall, and Isaac Hill (journalism's "worst looking—the lean, the halt, the blind" [26]) to plead for him. Although Congress at first objected, it finally approved President Jackson's appointment of Noah to a most lucrative post: for his party loyalty he was (again) rewarded, this time as surveyor (customs appraiser) of the port of New York. Again he held the position a short time, but was fortuitously on hand when the Gales & Seaton Napier lay stranded in his customs house. Major Noah (a title admittedly received at an 1813 election attended only by himself and two others) recognized a unique item and promptly called press joiner Hoe to assemble it for appraisal.

Robert Hoe was happy to oblige. Never before had a press manufacturer so complete an opportunity to examine a competitor's product. While inspecting the single-cylinder machine Hoe realized it was superior to anything he had ever seen in America; he made copious drawings and models of its parts. The Napier was duly returned to England, but its essential characteristics and specifications remained in New York. This outstanding machine brought home to Hoe the fact that in a highly specialized business like press-building, foreign—European—progress could not be ignored. Thanks to Noah great ideas began to circle in Hoe's mind. But wait.

Still another Napier reached these shores at this time. The New York *Commercial Advertiser,* a daily dating to 1793 as Noah Webster's *American Minerva,* was not important politically, but its advertisements (as in all the mercantile papers) were always compelling reading. Colonel William L. Stone, a capable writer, became the prosperous paper's editor in 1821 and brought it to a circulation worthy of an imported machine. Of this press Yorkston remembered rare details:

> The first cylinder press [in the United States] was the Napier. It was purchased by Col. Stone of the New York *Commercial Advertiser,* in 1828. Robert Hoe, printer's machinist, father of the present heads of that establishment, was the person engaged to erect it, and as he wisely intimated to Col. Stone that in the event of a break in the machinery, it would take at least four months to get the duplicate parts from Napier's works, and further suggested putting the main castings in the sand, and thus Hoe & Co. were enabled soon to construct printing-presses themselves; and from that day to the present a record of the productions of that house would form a complete epitome of everything pertaining to the printing-press in America.[27]

Hoe began to incorporate Napier's mechanisms into domestic presses almost immediately.

In the absence of international patent protection, there was no obligation on Robert's part to be more scrupulous than simply acknowledging a vague intellectual debt to the British manufacturer. Apart from financial considerations, however, the challenge of invention was much like survival of the fittest in nature. Indeed, there was so much satisfaction in being able to improve an earlier machine (every patented device by its very existence seemed to shout, "top me if you can!"), an innovator felt he was only building anew from a dead or dying carcass. Beating the smug British at their own game was a particular enjoyment for many yankee tinkerers.

Richard M. Hoe was just a lad when his father began to manufacture Napiers, but he was surprisingly frank a decade later about still another purloined design:

New York August 28, 1837

D[r] Sir,

Agreeably to your request we would state, that in the year 1829 we commenced manufacturing the Napier Printing Machines, having taken the model drawing from the only Single Napier then in this country, which was imported for your paper [*New York Courier & Enquirer*] *and in operation in your office at the time & that since then we have kept continuously employed in manufacturing Napier Machines from twenty five to thirty men & are enabled to sell them at a discount of fifty per cent, from the London price. Should you import any of Napier's improved machines, & furnish the model to our house (or any other machinists in the country) it will enable us to supply the different newspaper establishments with them at one half what it would cost to import them, and at the same time give constant employment to a great additional number of laborers and mechanics. In consequence of our commencing the manufacturing of the single Napier printing machines from the model procured at your office the value of your press was reduced one half.*

Resp[y] y[r] ob[t] Sr

To *R. Hoe & Co.*

James Watson Webb Esq.

So a fourth Nay-peer had also been subject to Hoe's methodical scrutiny.

While it is not clear why, verging on bankruptcy, Gales & Seaton would order an expensive press—indeed, the whole episode is considered apocryphal by some—nor what model went into Stone's *Advertiser* office, four importations of Napiers into the United States in 1825-1829 were more than enough for Hoe. Pasko claimed that the Walker machine had actually been ordered by Noah when he started the *Enquirer;* Timothy Dwight, Walker's partner, then took it out of bond and commenced printing the two New York papers. "It is said," Pasko added, "that it was by his repairs to this machine that Richard M. Hoe, then a boy, first acquired his knowledge of power-presses."

All four machines—or more—may collectively have been the necessary inspiration for what followed. Interestingly, the *idea* of a radically new press was not enough: all Napiers had to be slavishly copied "in the sand" for reproduction. Nevertheless, each account details a different aspect of the counterfeiting: Walker's, its national value transcending the stigma of theft; Gales & Seaton's, the irony of chance and circuitous involvement; Stone's, the rationale of making castings; Webb's, the economics of it all. Only as these four lessons were driven home would Hoe act so decisively; anything less would not likely have launched an American technological revolution—as no widespread demand for the sophisticated machines had yet been established.

Wesley Pasko had similarly favored "multiple" inspiration, citing Treadwell's work, that of Jonas Booth (a printer, he "remembered" seeing a cylinder press in England and had one built by founders and machinists Worrall & Company, New York, for the *Courier & Enquirer*), and Walker's importation as precursors: "These facts," said Pasko, "together with the favorable reports he had from London as to the working of machine-presses, induced Hoe to study their principles carefully, and he began to repair them when called upon. . . ."[28]

While Napier might have thought his creations too complex for easy counterfeiting, he did not bargain for a shrewd carpenter whose modest joinery had grown into a complete machine shop, nor could he expect a quiet American conspiracy determined to put the best machines, cheaply and locally made, immediately into printers' hands. On Gold Street specifications went right to paper and sand molds rolled into the casting room. The imitations were meticulous. Stephen McNamara, Chicago manufacturer of printers' rollers and prolific historian, in 1885 looked closely. "Upon examination the cut [a view of Napier's 1825 perfector] would seem to be the double cylinder Hoe, from the feed-board standards to the rock-shafts, yoke and lifting toggles, so perfectly has it been copied. . . . England had able mechanics," he concluded wryly, "America fortunately had Hoe."[29] The plucky American was quietly in the "Napier" press business.

Early claims that Stone's *Commercial Advertiser* machine was not imported, but actually the first domestic one, were recounted in 1886:

> Old Major Noah, editor of *Noah's Sunday Times and Messenger,* was Collector of the Port in those days, and being desirous of seeing how the Napier press would work, sent for Mr. Hoe to put it up. He and Richard succeeded in setting up the press and worked it to such advantage that two duplicates were ordered at once—one for the *Commercial Advertiser,* of this city, and the other for the Philadelphia *Chronicle.*[30]

Despite its historic press the *Chronicle* did not long survive. R. Hoe & Company's own first machine (perhaps not simply a "chinese" copy) was probably one sent to Albany for Charles Van Benthuysen's *Temperance Recorder,* 1830; it was designed to be powered by a man, but steam (reportedly the first in America to run a press) was substituted three years later. By the early 1830's ambitious printers apparently no longer had any difficulty obtaining steam engines.

Van Benthuysen (whose father Obidiah had started a literary paper in 1807) became Albany agent for R. Hoe & Company, as well as state printer. Considering the high cost of new machinery, even an agent pondered his purchases:

[Charles Van Benthuysen to Hoe, 12 October 1843] . . . I am hesitating between two opinions in relation to the Press, but if the Colonel will stick to his pledge to "make a Machine that will do as good work as the Tufts or Adams Press" I will go it. . . .

The Albany enterprise, making its own paper, ink, and type, had stereotype and electrotype facilities, and a large pressroom (1850): 3 Napiers, 4 Adamses, 2 Tufts, 1 Ruggles, and 6 hand presses. Van Benthuysen and the Hoes were long friends.

The surprising confusion of claims for importation versus local manufacture was—as aftermath of the bold appropriation itself—not entirely accidental. Scarcely 20 years after these events occurred the one man alive who knew the facts clearly was invited—urged—to record them carefully. William Van Norden, a New York printer and historian, had done all he could:

New York, May 1, 1849.

Dear Colonel:

I have the great pleasure to receive your letter of April 6, on the 28th. The gratification with which your friends have heard that your Great [Paris] Machine is at length in successful operation is very much increased by learning also of the steady, patent firmness and resolute perseverance with which you have met and overcome the immense difficulties by which you have been opposed. . . .

I have been for some time past seeking for particulars relative to the introduction of cylinder presses into this country and have been pretty successful. But there are many points upon which any information is defective, and which I think you will agree with me in considering of sufficient importance to justify no little labor and exertion to procure the most full and exact accounts respecting them. I allude to an accurate and workmanlike description of the different presses brought from England. Without this we shall be entirely unable to meet the charges that have been or may be made against us and those made here. . . .

I find that the first machine was brought over by Mr. Walker of the Daily Advertiser, for that paper jointly with the American: that it arrived in September 1825, and after much delay and difficulty, was got to work in December. It was a double cylinder and is described at some length in both papers of the 30th. December, but the description being drawn up by a person not accustomed to machines, it is very indefinite and unsatisfactory.

Major Noah, the proprietor of the New York Enquirer, had one brought over in 1828. This was a single cylinder but no description is given of it. Both these were driven by hand. . . .

My search for an account of the first cylinder machine made by your concern was for some time unsuccessful. The books of that date appear to have been burnt up in some of the many fires to which you have been subjected. The patents (if any) cannot be found, no one could recollect the time: but I have at length ascertained that it was put up and at work on the 8th. March 1830 printing the Commercial Advertiser. A particular description of this also is very desirable and you seem to be the only person who can furnish it. . . .

Be especially careful to preserve all that is published about the "Fast [rotary] Press" as it is understood that Napier lays claim to the original ideas in this also and this above all should not be left in doubt. . . .

Respectfully,
W. Van Norden

To: Col. R. M. Hoe,
Paris

While the company—in huge scrapbooks—did indeed keep all manner of clippings and memorabilia about its printing press progress, the materials were all usually popular, exaggerated, apocryphal. The Colonel apparently had no wish to correct them, nor in any other way preserve a more factual account, especially of the firm's acquisition of patents and machinery. Reconstructed entirely from Rollo Silver's scrupulous research, only John Walker's *Advertiser-American* story is fully documented.

Although the Napier lacked three desideratums of newspaper presswork—rapidity, perfecting impression, automatic feeding, it was a tremendous advance over the hand press. For Hoe it became a new and sensational line of merchandise well calculated to catch the public's fancy as would computers in another generation. Although printers with a heavy volume of work were quick to capitalize on these ultramodern machines, many were apprehensive. For one thing, at a time in the American economy when money was "tight," the cylinder press was expensive, also requiring (beyond its high initial price) an auxiliary source of power.

Nevertheless the company's "Patent Single 'Large' Cylinder Printing Machine," available by midcentury in three sizes for $960 to $1490, succeeded—as the Treadwell and Napier did not—in pushing the technology, especially for newspapers, beyond the woeful hand press. *The Printer* expounded:

> We naturally think more of a practical pressman's opinion in regard to the actual working of a press than of any other person's. . . . But, when our token-working friend undertakes to define the stopping place of improvement. . . . we are comically reminded of the progressive differences between an old creaking two-pull Ramadge† press, and a light but firm single-pull Stanhope—or between a rumble-come-tumble old heavy-rounced Columbian, and a nice easy-going Smith or Washington. The comparison certainly is comical; and, therefore, let us charitably hope that "ye antiente pressmen" may be forgiven for the gallons of beer they used to drink.

The paper in 1859 promised the printing industry an even more rapid evolution of the cylinder machine in its new environment:

> Seriously, however, if such beneficial changes as these could be brought forth in ordinary hand or bar presses, how can we set a limit to the extension of an entirely new and scientific adaptation of an undying principle (first broached to the attention of printers by the gifted genius of a Napier), especially when so ably and clearly enunciated for practical utility as in the admirably improved cylinder presses now manufactured by the Messrs. Hoe? [31]

To insure that his grasp on this new technology was a firm one, Robert Hoe sensed he must learn more, explore deeper, dictate—not await—improvements. The emerging American industrial community was full of conflicting claims and confusing results. Few mechanical wizards really understood precision machinery; fewer still could design superior equipment for routine assembly by others. Printers seldom knew exactly what they wanted or what was best for their work; they would at least expect trouble-free service for such expensive novelties. The new age of machine printing would ultimately depend upon sound engineering and the best designs; good workmanship and fair prices would follow.

To study the latest developments in the printing industry at its center of activity, Robert Hoe—reportedly at the suggestion of young Richard—decided to send a representative to England. Not the engineer to make the trip himself, he also knew a faithful emissary could be just as effective. He called a friend and fellow worker to his aid.

Sereno Newton (1801-1836), a foreman at R. Hoe & Company, a fine machinist and inventor, a scholar and mathematician, was an excellent choice. Newton had written a treatise on gear teeth-cutting reportedly the text in America on that subject for over forty years. His trained eye at a glance could fathom the inner workings of printing machinery; his trained mind could store such data until he was back home, if necessary. It is possible he was related to Messrs. William E. Newton & Son (also William), London patent agents who eventually handled Hoe patent affairs in England. They could certainly have assisted him in gaining entrance to machine shops and pressrooms.

The importance of the mission was obvious to Newton. Rare as it was in the early nineteenth century to ascertain foreign conditions before committing financial resources to a local product, Hoe must have found England's prestige to be enormous. He had already seen American printers, scarcely giving him a chance to demonstrate, ordering British equipment and gladly paying dearly for it. There was strong feeling in America that, although yankee genius was apparent everywhere, fashion and innovation were acceptable only from the Old World.

Sereno Newton, 31, went off to Europe in 1832 and performed his task admirably. He returned full of ideas on the art preservative of the arts, and together with his employer quickly transmitted thought to paper. On 26 February 1833 Newton received two patents for making the reciprocating cylinder-press mechanism capable of higher speeds without excessive vibration and shock. This work was quickly—and perhaps too bluntly—noted in the May issue of (New York) *Mechanic's Magazine* as "Rutt's Printing Machine, made by Napier, (Hoe's improvement.)." The magazine was promptly invited to Gold Street for a better look, resulting in a technical description of the works (June 1833) which was the first journalistic visit to the company's modest premises. The reporter's facts were straightened:

> The second story is occupied by artizans engaged in the construction of single Napier presses, as improved by Mr. Newton, one of the proprietors. . . .

Despite the death of the firm's founder just months before, no mention was made of that fact nor of any reorganization. The interviewer expounded:

> The punctuality, liberality, and well known business manner in which this establishment is conducted, has insured the proprietors a very large share of patronage, and, we must add, it is not misplaced. Such of our ingenious mechanics, and especially those who reside at a distance, will pass a pleasant hour in going over this establishment; and that pleasure will be considerably enhanced by the truly affable manners of the principals, who will accompany them and point out every thing deserving notice.[32]

It is likely the article was written before Hoe's death and printed without alteration.

The firm had indeed rewarded Newton promptly with a partnership in the business for his brilliant operation. Perhaps because of this new fame the New York Mechanics Institute got him to help organize its first annual fair (1835), but he did not stay long in town. Despite a troublesome cough Newton agreed to an ambitious sales trip through New York, the West, and the South to canvass hardware dealers as well as printers for their needs. From Albany and Schenectady he moved west, writing every few days to Matthew Smith and still explaining his mission:

> [*From Jamestown,* 7 *August 1834*] . . . *But my greatest object is to introduce our make of saws wherever I go & I find it can be done effectually by appointing in some of the largest towns some responsible enterprising merchant our agent for that section. . . . I should have stayed longer in Buffalo but they have the cholera there. . . .*

The trip south was uneventful, but lonely:

> [*From New Orleans,* 20 *January 1835*] . . . *Arrival extraordinary, one sealed package containing twenty four words in the english*† *language*

(mostly of one Sylable†) purporting to have come from your house on the 20th of Decm last & diverted to your humble servt. . . . My friends, if I do not receive letters from you before I leave this place I shall set you down for —— & give you such a blowing up as you never saw in black and white. . . .

Newton gleaned saw orders from Natchez and New Orleans, as well as inquiries about Adams versus Hoe presses which he was happy to settle. New Orleans, however, was a surprise:

[From New Orleans, 9 February] . . . This is a wicked city. Killing a man is no crime here, for instance in the house where I am, which is the best in the place (the City Hotel) there has been several of the inmates robed† of their money while asleep in their rooms; several of them been dirked in the house, one of them was shot in the street within a few steps of the house last week, . . . & another had his pistol burst in his hand (at the door of his room) which, instead of killing a man a few feet from him, as he intended, blew his own hand to pieces. . . .

Then disaster at home! Smith caught Newton at St. Augustine, Florida with bad news:

New York Feb 6, 1835

Dr Sir,

Your letter from Natchez was duly received. We presume you have heard of our misfortune by fire on the 24 Jany. We should have written you immediately but did not know where to address you. We are sorry to hear that your health is not improved and think it advisable for you to give it a fair chance in the southern climate and remain at one place long enough, exercising with moderation and paying particular attention to your diet &c, &c.

As to the fire, it is so much better than it might have been that we feel quite easy on the subject, though it will inconvenience us considerably—our loss will probably be about [$]2000--. The steel you took so much pains to select at Phil amounting to about 600 dollars was got to the shop the day before the fire & was burnt with the rest; we have put the saw makers to work in the ware room and removed our office upstairs right over the cartway. You have no idea what a tremendous hot fire we had. The head of our largest fly press was melted off and the bed of a sherman† press was melted so that it was all curled up & past gone and sheets of steel were so burnt that you could crumble it between your thumb & finger. Our main buildings were on fire several times and only [saved by] the exertion of our own men who poured water from the windows; the [fire] engines had no communication [with the flames] owing to the intense heat in the street. Minors building was not injured and prevented the fire from spreading towards John St.

Great exertion is now making† by Platt & others to widen Gold St. and we think they will succeed. The plan is to cut 17 ft off of our building and 10 ft from the other side. We shall of course oppose it strongly. There is no doubt but that we shall be driven from Gold Street and we are now looking round for a place. Dundee [St.?] will never answer, and we are inclined to think that our best course will be to buy 4 lots up town, somewhere near the river and build.

[Added in another hand] Feby 7th. Smith has written so far & left it for me to finish. We have calculated the advantages & disadvantages of buying in the country & concluded it was not best but think of putting up on some ground to be purchased a twenty or twenty five horse engine & moving our principal business there, keeping in Gold St. a ware room & office with one forge, lathe & 2 or three finishers. We have now a great quantity of work orders from good houses & are driving as sharp as possible. We have turned the ware room into a saw shop & the saw makers are again at work. Bells press will be put in operation about the 15th of this month so you will see we have not been idle.

We are all in good health & spirits & shall write you again soon.

Yours R. Hoe & Co. [RMH]

Hoe manufactured the 1831 hand press of Anson Sherman but it was not popular. Newton replied with lengthy and meticulous instructions on handling insurance claims, inventorying the shop's lost tools and raw materials, reordering stocks of lumber and metals, and other details which demonstrate that as the "mature" partners he and Smith were very much in command of the company.

On 25 February Sereno Newton sailed for Cuba. Though he wanted to race to New York to help reorganize, his nagging cough reminded him he had to cater first to his frail condition:

[From Matanzas, 23 March 1835] . . . My health at present is not as good as when I wrote you last, owing to my own imprudence in visiting the internal part of a large cavern near the city which has seriously affected me. . . . Previous to this I thought myself almost well, in fact I have gained 7 or 8 pounds in weight since I left New York. . . .

On his way home the following month he innocently encouraged a man who (40 years later) would give the firm a dreadful time:

[From Savannah, 23 April] . . . Enclosed I send you an order from Wm H. Bullock, the proprietor of the Savannah Georgian, a daily paper printed in this place. The house is perfectly good for the amount of the order; you will therefore ship it direct. The press must be the latest improved, with collums;† which I find are liked very much wherever they have been seen. You will think perhaps that I have been pretty liberal with Mr. Bullock in giving him such credit & taking his old press at $100--; but the press is a No. 1 Smith press in good order on which he now works his paper, & would continue to do so, were he not to inlarge† his paper, which he did not intend to do til† fall & therefore in order to make sure of the order I have

done what you se;† & shall do the like as often as I can on my way home. . . .

But Newton continued to cough up blood, and he suspected he would not recover. By the Fall of 1835 he was back in Ward, Massachusetts, where he hoped a bucolic routine would buy him time:

[From Ward, 23 August 1835] . . . *My cough, & the cause of it appear to be gradually leaving me, sometimes I hope forever; though I am affraid† that cold weather will bring it on again. I am very particular as to my diet, which is principally milk; & to avoid being exposed in bad weather. I exercise freely; every pleasant day you might see me cutting about with my horse & waggon. . . .*

Newton furnished directions for visiting and begged for more newspapers. He still had much shop and sales lore to pass on to his younger associates:

If you can possibly find time you would find it greatly to your interest to take our list of prices & go to all the principle† Hdware merchants in the city, & talk to them a little & leave our prices; & also to have some of our advertisements in a frame put up at every public house in the city where they are not; & now is the time to do it, for the merchants are now all on from the South, & are begining† to purchase their fall stock of goods. . . . Another thing, I do think it important that we have some of the various kinds of saw refinished up for the fare,† & exhibited at one or both of the fares in the city. . . .

Three months later, after his usual cheerless health report, Newton asked after young Richard:

[From Ward, 4 November 1835] . . . *I find I am failing every week; my cough increases & I raise more from my stomach. My appetite is not as good & I am more restless nights. . . . Does that young man attend attend† to business as he ought*[?] *Let him answer for himself (softly). He has my best wishes; therefore I cannot but hope he does; or will. . . .*

A Smith reply set him at ease. "I am heartily glad," Newton remarked, "to hear that Richard 'feels the importance of his own personal application' to Business; it will make a very material difference in our concern." Progress in the plans to move uptown now so excited Newton that—in the absence of promised letters and papers—fantasies devoured him:

[From Ward, 18 November 1835] . . . *My mind is so permanently & immoveably† fixed on the buisness† in which we are engaged, & in which your future prospects very much depends,† that I cannot keep it out of my mind, even, while I am asleep. Though I know at the same time that it must be very soon (by me) forever forgoten.†*

Last night I fancied that I was all over the new shop, which had just gone into opperation.† *I thought I examined minutely the different parts that had been put up since I left New York; & felt highly gratifyed*† *to se*† *how still & beautifully every thing moved. But Lo! when I awoke, I was in the same cold, dark, hapless room. . . .*

Newton apparently made occasional trips to New York to exercise his duties as a partner, but there was no mistaking his fatal condition. He died a year later, in 1836, age 35. Stephen McNamara fittingly regarded him, on behalf of the entire printing community, "one of the best mechanics of his day."

The foreign-inspired presses (with many Hoe improvements over the following decades, to be sure) became so much superior to the British designs they were to constitute Hoe staples in the trade for the remainder of the century. Admitting at first they were "primarily Napier inventions," the company officially designated them modestly:

(1) The "Single Small Cylinder (Napier's system)"

(2) The "Double Small Cylinder (Napier's system)"

(3) The "Large Cylinder Perfecting Press (Applegath and Cowper's system)" [33]

The company now became very possessive indeed of these properties, especially as improvements were added to them. The Hoes' lifelong trust in the patent system began all the more poignantly in the 1830's, when, under a devastating depression, they willfully paid to have some lost data reconstructed.

Upon the occasion of a December 1836 fire at the United States Patent Office which destroyed both models and records, R. Hoe & Company reminded the Commissioner of Patents of its early place in the annals of American printing invention:

New York September 20, 1837

Dear Sir

We have this day forwarded herewith to your address the patent rights and assignments as per list annexed; you will please return them to us after having them properly recorded, and any expense attending the recording of the assignments will be paid by us any way you may direct. At the same time you will oblige us by saying whether we shall furnish the Department with the models as per statement, left with you by our Mr. Hoe a few days ago, while in your city.

Respectfully
Your M^{o} *Ob*^{t} *S*^{t}
R. Hoe & Co.

To
Henry L. Ellsworth Esq.
Commissioner of Patents
Washington City, D.C.

[*List*]

Samuel Rust patent of Washington Press, assigned by him to Wm *Rowe, and by Rust & Rowe to John Colby & by John Colby to Matthew Smith.*

Sereno Newton's Patent on Double Cylinder Napier, assigned by him to Matthew Smith & Richard M. Hoe.

Robert Hoe's patent on standing press assigned by him to Matthew Smith & Richard M. Hoe.

Samuel Fairlamb & John S. Gilpin patent for Inking machine. Fairlamb's assignment to John S. Gilpin. Gilpin's assignment to Harper & Brothers & Harper & Brothers assignment of one half to Matthew Smith & Richard Hoe.

Sereno Newton's Patent for Register Apparatus applied to Napier press assignment to Matthew Smith & R. M. Hoe.

Peter Smith	*Patent Printing Press*	*expired*
Matthew Smith	*" Stereotype Block*	*do*
B. F. E. Brown	*" Standing Press*	*do*

Robert Hoe's contribution had been a 22 April 1828 improvement of Brown's hand press; Fairlamb's hand inker (about 1830) was followed by several other models, but they declined with the hand press itself. A year later Henry Shute, possibly a Hoe employee, petitioned the Secretary of State to grant a patent (necessary under special circumstances) for a one-piece finger for Napier and other presses. Thus the possession of a handful of pioneering patents up to 1840 placed an entire industry in the little firm's control. But other factors, of course, were also at work.

Machines—some of which remained in service 75 years—principally for the best book, woodcut, and job printing, were mute testimonials of the quality that R. Hoe & Company conscientiously put into their presses from the start. The brothers' solidarity perhaps obviated the debilitating suspicions and expensive safeguards that lesser partners might expect and demand of each other. Perhaps too each brother may unwittingly have been intimidated by the others to do his best work, or by their close and pervasive family recollections. Indeed, they divided the chores among themselves and pursued the business as if each were still personally responsible to his father. And so they were: the simple name "R. Hoe & Co.," never altered to admit the founder was no longer in control, became a symbol of family pride and progress the sons never cheapened with personal considerations.

On several occasions Robert Hoe II told customers that these concerns were not accidental:

> [*To R. Raphael, 13 March 1845*] . . . *As to our being contented with moderate profits we are sure that, for the quality of our articles, our prices are reasonable and will compare favorably with that*[†] *of any respectable machine works in this country. The fact is that work is generally done apparently with no attempt at accuracy, and prices are made to correspond. It is our intention to give only the best and to charge nothing more than a fair remunerating price therefor. In*

> ***the long run any experienced person will acknowledge that the best is the cheapest. . . .***

But even more than quality would be necessary.

These new machines were radically different for their time, both for the state of the machining industry and for the pressmen expected to use them. The challenge was fully as staggering as father and sons may have imagined. New tools were necessary, some to be built on the spot; new gauges and calipers were required, many to be imported from the best tool makers in England; new sales contacts were developed—often with payment schedules, penalties, and mortgages; and new sales techniques, with promises of prompt and continuing service and maintenance often involving the boarding of erector crews for days or weeks.

At first, of course, these problems manifested themselves only slowly. Without planing machines, slotting machines, shaping machines, or boring machines, the cylinder press nevertheless required tolerances and accuracy seldom possible in commercial shops of the day. Tucker, as an apprentice learning the techniques then in practice, described the problems:

> The ribs of cylinder presses were first chipped and then planed out with iron hand planes. The plane was about three feet long and fitted with face and side cutting tools. It was drawn forward by a man turning a windlass with the workman standing on the plane to give the required cutting pressure. The plane was carried back, the tool re-adjusted and another cut made. A great deal of care and skill was required at this work and it was surprising how accurately the ribs could be planed. A week was considered a fair time to plane a rib, which was not then faced with steel. . . .

The cylinders of the larger presses were built up from wide rings covered with sheet iron 3/8 inch thick, and the impression was adjusted by slipping pieces of tin, paper, and cardboard between bearings and side frames. Tucker added:

> The universal joint was a true ball and socket joint, the central ball being 3½ inches or 4 inches diameter and the jaws turned out inside to fit over it. This form admitted of all the work being done in the lathe.[34]

The universal joint had certain mechanical limitations, however, and the company strove often to find substitute mechanisms.

Only 48 in 1832, Robert Hoe was now fully immersed in work ideally suited to his disposition and talent. He was the prototypal technically oriented American businessman with indefatigable drive and ingenuity. His small dedicated company of workers was probably also above average, some of them staying on for decades, many distinguishing themselves in the progress of machinework or printing technology. A New York state history, describing native enterprises before the coming of the factory system, pictured the Hoe shop exactly:

> The artisan, whether master or journeyman, enjoyed almost complete economic independence. He was his own capitalist, the cost of his tools and buildings was small, he manufactured for a limited or local market, and in consequence his supply of raw materials and of finished products was measured by immediate needs. Moreover, no wide social gap existed between master owner and journeyman worker. Often working side by side, they knew and had mutual respect for each other; likewise their families usually mingled. . . .[35]

The prospects were excellent, but they encouraged too much. Robert Hoe's enthusiasm and drive were more than one man could contain.

Hoe frequently went without breakfast or lunch, so determined was he to finish one project or start another. He spent overly long hours at worktable and counting bench; his health began to weaken. In 1832 he reduced his activities and retired to the Salem, Westchester County, homestead. But the strength with which he had fought the fever many years before had deserted him. A year later (4 January 1833) he was dead.

Like soldiers in battle, despite a fallen leader the firm carried on. Within a few months cylinder presses of Newton's pattern and others as well were coming off the assembly floor—some destined for prestigious installations of which the founder would have been proud. Two large 1833 (Applegath and Cowper) cylinder perfectors were erected for Thomas H. Jackson & Company, at the nearby corner of Gold and John Streets, to print American editions of Knight's *Penny Cyclopedia* and *Penny Magazine*. In a preface the magazine acknowledged the timely mechanical asset (perhaps punning over its foreign origin):

> Without the aid of Machinery,—which has arrived in this country to great perfection, especially in the art of printing—the "Penny Magazine" could not be published at its present price. . . .[36]

Sadly, just as the hard-working father was laid to rest, crucial cylinder-press inquiries and—thanks to low domestic prices—orders began coming in with encouraging regularity.

In 1833 J. & J. Harper discarded its Treadwell to install a steam-powered Hoe, possibly the earliest book printer to do so. John Farrington, Harper's chief foreman, in the privacy of his home wept at the news.

> "They are going to do something dreadful at Harpers," he told his wife. "It will take the bread out of the mouths of so many of the boys. It is terrible to think of how much misery it will cause."

"Why, what can it be?" she replied. "Have they failed or lost money, and must they cut down the force?"

"No, it is not that," he answered, "but they have decided to put in a steam engine and get presses that run by steam. There's no telling how many of us will be thrown out. . . ."[37]

The offending press was operated by William K. Colyer, according to Harper historians, but others say it was run by Samuel Bingham (1779-1876), who in 1848 would buy a small roller-making shop and start his own business; the composition roller industry alone would soon engage thousands.

The Harpers were thereafter sold on Hoes, one of which a decade later actually saved the house from ruinous mammoth newspaper competition:

New York Jany 5, 1843

Dear Sir,

We annex invoice of the Stereotype Block,† *ordered in yr esteemed favor of 23*[d] *ult. We have also rec*[d] *yours of 29*[th] *ult. and on submitting it to Messrs H.B.* [*Harper & Bros.*] *they wrote the answers to yr queries which are on the opposite page* [*not present*].

In conversing with us about their machine they freely say that they have been able with it to retain that part of their business which some of the newspaper weeklies were fast taking from them. It is not for their interests (it is easy to discover) that similar machines should be in use in your city, as they have kept[?] *the field pretty much to themselves. They run their press from 4 o'clock on Monday morning to 11 Saturday night, without stopping for other purpose than change of form or blankets. We are certain that the machine we are now making for them will be perfect in every particular & so well satisfied are Messrs H. Bros. about it, that they are not troubled with any doubts relative to it.*

The great strength of the machine & the accuracy of all its parts render it not at all liable to the objections which may heretofore have been urged against cylinder presses. We shall have two that are ordered finished this month & if you will come on we will satisfy you of the correctness of what we say about our machine.

For the proprietor
Robert Hoe

To L. Johnson Esq.
Phila.

[*Invoice*]

New York Jany 5, 1843

To The "HOE" Printing Manuf[y] *Presses &c*
Mr. L. Johnson

for 32 Octavo Blocks	$60 --
" Boxing & porterage	63
	$60.63
Less 20% commo & disco	12 --
	$48.63

In 1834 Major Noah with a printer named Thomas Gill, formerly business manager of the New York *Evening Post,* established another newspaper, the *New York Evening Star,* dedicated (again) to Whig politics. Noah, whom the New York *Sun* fondly called "Grand Mogul of the Editorial Tribe," had not ceased surprising his friends and readers with his strange opinions:

> [From the *Sun*] . . . Major Noah's singularity—The Evening Star of yesterday comes out in favor of the French, lottery, gambling, and phrenology for ladies. Is the man crazy? [38]

In April of their first year Messrs. Noah & Gill Co. ordered from Hoe a double cylinder Napier, paying $3000 (plus $75.83 in interest on installments)—which seems no concession to the port surveyor who opened Robert Hoe's eyes to an industrial plum. When Gill died in 1841 the paper merged with *Commercial Advertiser,* as Noah was not disposed to run it alone.

Within a year, however, Noah was appointed a judge (to the intense dismay of his arch rival, James Gordon Bennett), in which capacity he soon had the supreme pleasure of sitting on a *Herald* libel trial. (Noah was also plaintiff at another Bennett libel suit where the latter was fined $300.) Major Noah, whose sons Manuel and Jacob were also newspapermen, had served brilliantly on several New York papers and also wrote notable plays of patriotic "military melodrama"; a close friend of the Hoes, his many faceted literary career was long and colorful.

A steady stream of correspondence began to answer queries and furnish quotations:

> *New York September 22nd 1834*
>
> *Dear Sir,*
>
> *A single cylinder Napier Press large enough to print a sheet say 30 by 20 or 21 inches would cost $1500—terms ½ cash x ½ credit of 6 months, the expense of boxing & packing to be an extra charge. It requires one man to turn at the crank (a common laborer), one pressman to put on the sheets & also to take charge of the press, & a small boy who sits behind the press to smooth the sheets as they are discharged from the press. The press will print faster than the larger ones & will throw off with ease 1200 per hour. The pressman should be one acquainted with a machine who would also be able to put it up, such an one receives here 10 to 12 dollars per week & boards himself. A set of moulds is included in the price of the press in which the press man will cast his rollers.*
>
> *Yr Obdt*
>
> *To Wm H. Wright Esq.* *Robert Hoe & Co.*
> *at J. P. Garcia & Co., 107 Front St., New York*

As the younger Hoes began to work in the office they assumed their father's name should be spelled out, and in some places "Robert

Hoe & Co." also appeared in print. But despite such slips (probably by Robert Hoe II as a clerk, who also for a time put the family name in quotation marks), Richard, whose identity could also be inferred by "R," apparently demanded the initial. Eventually the short "R. Hoe & Co." became a recognized mark everywhere.

Gradually, of course, the Napiers simply became Hoe presses, none happier than the Hoes themselves that the transition called little further attention to itself. As one model followed another only one early and faithful customer seems to have been trapped by the company's progress. When Samuel Bowles launched his daily Springfield *Republican* in 1844 the newly introduced Napier was the undisputed symbol of modernism in printing. Frederic Hudson 30 years later chided the paper:

> It has for its device, with the head-line, a cut of a Napier cylinder press capable of printing three thousand sheets per hour. Where the old style of papers inserted their mottoes, the *Republican* has "R. Hoe & Co." If that firm be its motto, the device should be changed to a ten-cylinder lightning press, or newspaper advertisers will think that the concern has not kept up in progress with its great and enterprising motto-firm and press-builders.[39]

In the Massachusetts paper's own history (it ceased publication 1946) it proudly noted that during its long life the typewriter, telegraph, telephone, teletype, linotype, stereotype, photoengraving, power press, wood-pulp newsprint, and the camera were introduced, all combining to turn an itinerant printer's gamble into a mass-production industry. Under Bowles the *Republican* was one of the great papers of America.

Although they seemed made for each other, the cylinder press was not always linked to steam power; the hand crank was very common, especially where no mechanical power was available, as in rural areas of the United States and in many foreign countries.

> *New York Jany 12th 1844*
>
> *Dear Sir,*
>
> *It is with much pleasure that we give you the annexed information* [*not present*] *relative to our single and double cylinder presses and other articles and we trust that you may be able to procure us orders for the same.*
>
> *Our single cylinder press is calculated to do not only newspaper work but also first rate book work and will print of ordinary size paper (say 40 inches by 28 inches) from 1200 to 1500 impressions pr hour. These presses are far superior in construction and workmanship to any English press.*
>
> *Our double cylinder press as represented in the cut will print from 3000 to 3500 impressions pr hour. This press takes two persons to drive it, two boys or girls to put on the sheets and two to take them*

off or fly them, as we term it. These machines are in universal use in this country for newspaper printing and are strongly built and not liable to get out of order. We should have remarked that our single press requires only one person to drive or propel it and one girl or boy to put on the sheets and one to take off the sheets. The prices of both double & single cylinder presses with the sizes you will observe are noted in our list of prices several copies of which we send with this.

We manufacture hand printing presses also with self inking machines attatched† *which are more desirable for strength, durability & cheapness than any English article of that description. Of these presses & self inkers we can only say that they are about the only kind that are in extensive use in this country.*

Copying presses similar to the one we send you we sell for $30.

German printers will be able by our circular of prices & drawings of the machines to form some idea of what we manufacture.

Should you procure us orders we will allow you 5% commission on the amount, which of course would have no reference to any commission that we might afterward think proper to allow the merchant through whom the order should be transmitted. Hoping that we shall have the pleasure to hear from you with some considerable amount of orders we remain

Your most obt svts

To Louis Mark, Esq. *R. Hoe & Co.*

U.S. Consul at Berlin, Germany

Years later another mode of power was still common, even in a large printing community ("†" omitted):

Boston February 4th 1850

Mr. Hoe Dear Sir,

I have sent you a paper some two or three times a week that you mite see how well your Press worked. The [Boston] Traveler folkes seame mutch pleased so far with it and in fact every one that has seene the work that it has don for the last three or four days have ben highly pleased with it.

I donot think mutch of the Water Power, for one day it will give me as high as 28 and the next probably I cannot get more than seventeene Revolutions per minute. I donot care to run mor than thirty but should like to have that mutch at least, and Mr. Huse has been promising to give me that many turns for some time but I have not seene it yet. As far as I am concerned I have got a firstrate set of feeders and they behave first rate and every thing workes well as I could wish. I have not had the least bit of troubel with the press.

Remember me to Mr. Blair and all other dear friends.

Respectfully H. Ditmars

To R. H. Hoe, New York

P.S. I wish you would send me what is due me for I am somewhat short.

[Added in another hand] Peter will please send a check of the amount due Mr. Ditmars; enclose it in the letter to H. Flanders & Co. RMH

Ditmars was long one of Hoe's press erectors.

The gradual increase in size of cylinder presses and other printing machines as steam power became generally available soon made the union of the two devices an invincible combination. This in turn transformed the printing craft into an enormous industry, capable of a wide variety of specialties and products. As a participant in the new technology, however, the purchaser of printing equipment had frequent questions and sought endless advice.

> [*To N. Lyman, Buffalo, 21 December 1844*] . . . *the very best way that we think of to ascertain which frame belongs to the No. 2 & 3 press is to open one of the boxes and observe the marks on the legs and the corresponding marks on the frame will set the matter right we have no doubt.* . . .
>
> [*To Buffalo Courier, 7 May 1850*] . . . *In cases where you print very fast it is better to have 2 blankets, one thick & one thin. But when you print slow or moderate & always of a regular impression it does not make any material difference. An India rubber blanket will bring up the impression better, does not require to be taken off for washing, the type leaves no impression in the blanket,* [*and*] *will last about five times as long. Price $1.00 per yard.* . . .
>
> [*To "Wright Freres," 16 May 1850*] . . . *In putting it on the cylinder it will be well to cut the blanket its entire length half its thickness where it laps over the 2 corners or edges. It should be cut on the under side. It should have woolen blanket or canvass sewed to it at each end to fasten to the hooks & pins. The holes in the rubber should be made with a shoe makers punch.* . . .

Wright had bought 50 inches of 36-inch rubber blanket, $8.38.

The firm had secured for itself two of the best hand presses in America, it had complete patterns for a promising line of Napier machines, and it had functioning foundry, factory, and salesrooms. Other printing accoutrements too had not been neglected: wood furniture (blocks and strips to fill blank spaces in the chase), type cases, compositor's banks, and makeup tables continued to be profitable long after wood presses disappeared. A contract for pairs ("upper" for capital letters, "lower" for small) of the compartmented trays was renewed periodically:

> *Memorandum of an Agreement between Robert Hoe* [*& Co.*] *& Francis Mean, New York, Nov. 24, 1834.*
>
> *Said Mean hereby agrees to make or cause to be made for said Hoe viz.: three thousand pairs printers type cases such as are ordinarily used by printers in a first rate workmanlike manner upon the following terms, viz. for the sum of sixty two & a half cents per pair payable in weekly instalments as the cases progress, & it is understood that Mean is to have free use, as heretofore, of the Steam power & machinery for making cases & in addition Hoe & Co. are to construct*

for his use a circular saw for cross cutting—he is also to be allowed bench room for himself & two men or boys in the same room where the said machinery runs, & where the cases are to be made. The materials are to be furnished by Hoe & Co. as heretofore & delivered in the yard & when a lot is to be got up in the shop Hoe & Co. are to furnish one man to assist in getting it up. Mean is to deliver the Cases into Hoe & Co. wareroom when finished. Mean is not to be taken off of cases against his consent.

Francis Mean
Witness *R. Hoe & Co.*
Alfred S. Bowen

The father's knowledge of carpentry and the skills of his many talented associates would make the company formidable enough for its architectural needs. Yet, from about 1820 on, another branch of Hoes, brothers of the first Robert, appeared in New York: Richard Hoe, mason, and William Hoe, carpenter (whose son incorporated as James C. Hoe & Sons, builders). These relatives may have participated in the earliest construction work; they were in fact employed by Robert Hoe II later on.

The February 1835 fire, however, had given little time for planning. Tucker failed to mention it—actual losses appear to have been minor—and gave it no weight in forcing the firm's first expansion:

> The business of Messrs. R. Hoe & Co. was now fast outgrowing their premises, and in consequence a new factory was built in 1835 on a plot of ground 100 feet square on the corner of Broome and Sheriff Streets, then owned by Mr. Benjamin M. Brown, President of the Butchers' and Drovers' Bank. The factory was built 'L' shaped on the two streets and was four stories and basement in height.[40]

Brown soon became a trusted friend and advisor. Richard, despite his precocious talents, was not, at the age of 23, solely responsible for this ambitious real estate venture and factory reorganization. His cousin Matthew Smith and Sereno Newton (despite failing health) appeared to monopolize the plans. Even as they finished their blueprints another threat menaced the struggling business.

During the 1835-1836 winter R. Hoe & Company began to transfer manufacturing operations from Gold Street to the new plant, a distance of about a mile. Suddenly to the south a few blocks the shrill horns and whistles of fire companies again awoke the town to the great dreaded fear. Just a year before, Philip Hone, former mayor of New York, had deplored the inadequacy of municipal fire equipment. Ladders couldn't reach the tops of lofty (five-story) buildings, nor could water be pumped that high. Summertime blazes in 1834 and a bad fire on Ann Street had been sufficiently threatening to make uncontrollable conflagration a possibility. Now a First Ward alarm, 17 December 1835, had struck under brutal conditions: with intensely cold weather water froze in the hydrants and firemen faced a stubborn all-night holocaust.

The flames started on Merchant Street—Pearl down to Coenties Slip caught fire—Wall Street to the river—all of South and Water Streets, and westward to Exchange and Post—Lords, William, Beaver, Stone Streets—all were an inferno. The worst in U.S. history (until the great Chicago fire), the New York blaze also became a strange boon to American journalism.

James Gordon Bennett had started his New York *Herald* in May 1835. He rushed to the scene of the conflagration, for a full week spending half the day poking around the smoldering ashes and the other half at his writing desk. Taking particular notice of losses to his competitors (of a large press visible in the rubble near Hanover Square he wrote, slyly, "We believe on this site there was a printing office . . ."), he recapitulated:

> The Newspaper Offices have suffered in the general calamity. Yesterday [18 December], all the three morning papers [*New York Gazette, Daily Advertiser, Journal of Commerce*] appeared, with the exception of the [New York] *Times,* which we have not seen. We know well how to sympathise with our brethren of the press. We have gone over that ground before them in the Ann Street fire. . . .[41]

More importantly, with the enormous destruction Bennett had sensed in his fellow citizens an intense curiosity and fascination equal to his own, and—with text, linecut of the burning Merchants' Exchange, and map of the destruction—he satisfied them as no other paper:

> He spread before the public reports so detailed, unconventional, and graphic, that a reader sitting at his ease in his own room became, as it were, an eyewitness of these appalling scenes. . . . American journalism was born amid the roaring flames of the great fire of 1835.[42]

A disaster for many businesses—almost 700 shops and stores valued at $20 to $40 million were lost, as well as most merchandise: horses and carts hastily engaged at astronomical prices to carry goods off did not unload far enough away. The flames soared for two nights and could be seen in Philadelphia and New Haven; they were finally extinguished to the north only a block from the Hoe office. Like Robert Hoe's provident recovery from yellow fever, the young firm—with two close calls—had narrowly escaped annihilation by fire. It would face one further worry—industrial sabotage—strangely uttered by the dying partner:

Ward, Dec^br^ 24th 1835

My Dear Matthew,

I have been expecting a letter from you for some time, particularly since the terrible conflagration in your city. I hope you were not so offended at my last as to let it interupt[†] *our correspondence, as it is now explained by Mr. Hoe, to my satisfaction; & I hope to yours; though from the time I recd your last, until I saw Mr. Hoe, I was a very unhappy man, on that account [possibly partnership reorganization]. My health for the last ten or twelve days has been better. I have slept better & my appetite is better; Still, I have no more grounds for hope, it is the nature of the disease. To day I am not as well as usual; I raised some blood last night by coughing. I do not cough as much as formerly. Hope to be better tomorrow. . . .*

I have not seen a New York paper since the fire; though the Spectator should come regular, I have gathered what information I could respecting it from Boston & Worcester papers. I think it an extraordinary, & to us, or you, a very fortunate circumstance that the Bowery Insurance Company were so little affected by the fire. I do not se[†] Rawdon Wright Hatch[†] & Cos. names among the suferers;[†] had they moved[?] How is the fire going to affect us taking it all round[?] I know there will be some new presses to make & other extra work to do. But will it not curtail the circulation of money among the manufacturers[?] Will it not increase our insurance[?] How much had we insured by offices that have become bankrupt[?] I wish you would write me a long list of particulars; how it will affect R. Hoe & Co., &c, &c.

You should keep our factory insured to the full amount, for you know we have made enemies in this country & in Sheffield by going into the saw buisness[†]; & I suppose you know there are men in the country that have been sent out by the manufacturers for the express purpose of Burning down Manufactories; & very probable[†] we may suffer by them. You may laugh at me; But look out.

I wish you would get up a Mutual Insurance Company among the manufacturers; it would be a mutual benefit. We have to pay double that of a merchant, for the same risk & I suppose they will charge us from ½ to 1 per cent more since the late fire. I would reduce the insurance by keeping a watch, a few buckets of water in each loft & keep the stock of finished work as light as possible at the factory. . . .

You will send on those papers relating to the dissolution soon, I suppose. Do write me soon, & dont forget the Inteligencer.[†]

I remain Your Friend & Servt

Sereno Newton

To Matthew Smith, Esq.
New York

Fire would ever after be a special concern of the company, its disruption almost more a calamity than the loss of a partner.

The unexpected death of the firm's founder might have stopped any ambitious venture, but the thrust of the business and the vitality of the partners and their families quickly carried them across the chasm. Here, the loss proved a catalyst for the youthful drive and pluckiness of the sons. Although Robert Hoe had scarcely begun to work closely with his offspring, *Scientific American* in 1858 recognized at R. Hoe & Company an uncommon asset in American business:

> One advantage possessed by European manufacturers over those of our country is the experience which has been accumulated during several generations, and which has been transmitted to the direct descendants of those who introduced various manufacturing arts. Each race has grown up with the business and acquired knowledge to conduct it in the most skilful† and economical manner; whereas, joint stock companies, so common with us, are necessarily subject to frequent changes, whereby the wisdom, skill and experience acquired by one set of directors are often lost to those who succeed them in business. The advantages, however, of accumulated experience, combined with skill, ingenuity and enterprise, belong to those who conduct the [Hoe] establishment. . . .

The newspaper's further observations were nevertheless cogent:

> The manufacture of presses and other printers' furnishings was founded by Robert Hoe, father of the present proprietors, and their uncle, Peter Smith, both very skillful and ingenious mechanics. At the father's decease in 1834,† his sons continued the business, and soon extended its proportions. At the period of time referred to, their annual sales amounted to about fifty thousand dollars, now they reach the sum of six hundred thousand; then a comparatively small shop in Maiden-lane sufficed for their operations, now two large establishments scarcely afford room for their purposes.[43]

"Hoe died," said Roger Burlingame (as recently as 1938), "leaving children who would create the business of making presses in America—a great family to carry on printing improvement for four generations and we may not know how many more. . . ."[44]

HOE BROOME STREET WORKS

From Graham's Magazine, June 1852

2.

Building the Fast Printing Machines

DEPRESSION AND ITS AFTERMATH / SHOP FIRES AND EXPANSION / SHEET FLYER AND PATENT PROBLEMS / THE START OF THE PENNY PRESS / MAMMOTH MACHINES, BLANKET SHEETS

Richard March Hoe was Robert's oldest son—welcome indeed after three daughters—born 12 September 1812 in New York. At about the age of 12 he left public school to work in his father's pattern shop. The year was 1823, just months after Peter Smith had died, and the young lad no doubt begged to assist a harried father reorganize the newly acquired business. Richard was indulged this request: the profession of pattern-making, although light work physically, would drive home a lesson his father had slyly contrived. Pattern making demanded a sound knowledge of mathematics and geometry; it required a respect for precision and a command of accuracy all the more delicately—manually—constrained by nineteenth-century tools of home manufacture. Richard applied himself with the rudiments of his elementary schooling and, to be sure, a rare gift for tackling just such problems.

After only 18 months in the shop, however, the boy returned abruptly to highschool, where he remained until he was 18. Richard thus acknowledged his need for whatever remaining education he could get. Yet, free to visit the nearby plant, the youth spiced his droll homework with practical benchwork and mechanical amusements only a bustling factory could offer. At 15 he watched his father's experiments with cylinder presses, and with paternal encouragement began to help repair machines around town. He learned well:

> The education of Richard M. Hoe was purely practical and much of it was obtained through the eye. He was ardently devoted to the busi-

ness to which he was born and was determined to deserve the bright predictions made of his future by his proud and doting father. When he was twenty years old he was thoroughly familiar with the details of manufacture, could set up and take down a press, understood the why and the wherefore of each part and was virtually the head of his father's establishment. . . .[1]

In 1830, after graduating, Richard returned to the plant better equipped all around, and took his place at the worktable at a kind of voluntary servitude that lasted over 50 years. The challenges were exciting:

Later on, in 1832, the discussion of the possibility of printing by steam was attracting the thought of inventors and scientists. In Mr. Hoe's shops arguments pro and con were heatedly urged, and in them the "young master" learnedly participated. His father was sanguine in the belief that steam would sooner or later be thus applied, and Richard gave the old gentleman aid and comfort by sharing that conviction.[2]

Accompanying him at his chores about this time was Richard's cousin, Matthew Smith (III), having inherited the partnership of his own father. Together—both 21 in 1833—they contemplated eventual responsibility of the establishment, although, perhaps like their fathers, Richard occupied himself with the manufacturing aspects, and by default if not by choice, Matthew found himself in the accounting end. "A man of uncommon business talent," [3] said the *New York Tribune* of young Smith, acknowledging him the equal in commerce to Richard's genius in invention. Matthew Smith, however, died prematurely in 1841, and although he had two sons (Sereno N., born 1836, and Eugene), the close family reorganization of R. Hoe & Company or their own interests elsewhere precluded the young Smiths' entry to the works. The firm thus became completely a Hoe enterprise, administered ultimately by three remarkable brothers.

Robert Hoe II entered the firm about 1836, age 21, with Sereno Newton's share of the business. Their younger brother Peter at age 20 joined in 1841, taking Matthew Smith's partnership. With Richard now clearly in control, the brothers dutifully took places relegated to them by their abilities: Robert handling business and finance, Peter supervising saw work, procurement of materials, and other routine projects. The family kinship that made the house of Harper strong and monolithic was equally at work at R. Hoe & Company:

Their business was conducted on the basis of absolute trust and confidence in each other. There was no system of checks between them, and no necessity for one. So close, indeed was the intimacy, and so unbounded the common confidence, that for many years no accounts were kept between the brothers. Each one took what he needed for himself, and the others neither knew nor cared to know how much each one drew out for his own use. . . .[4]

"The three would dispute with vigor about matters of management," James Collins wrote of the Hoes, "and then always act together. They did everything together, even to buying and using one carriage." [5]

Stephen Tucker enjoyed talking about machines more than the men responsible for them. He described the 1834 company succinctly:

> The works were in Gold Street, and the firm was then making Single Large Cylinder Presses, Two-Revolution Presses, Double Cylinder Presses, Large Cylinder Perfecting Presses (Applegath & Cowper's system), Small Cylinder Perfecting Presses (Napier's system), Smith Hand Presses, Inking Machines, Copper-plate Presses, Copying Presses, Type Cases and Stands, Stereotype Blocks, Paper Ruling Machines, Standing Presses, Hand Lithographic Presses; in fact, all the plant then used in a printing office, bindery, and stereotype foundry, even to making inking balls and composition rollers. On the opposite side of the street, in Ryder's Alley, was the Saw factory, in which were made the various kinds of Long and Circular Saws, Cane Knives, Trowels, etc., etc. . . .

Six foremen ran the departments under Richard's supervision:

> When I entered the establishment Mr. John Colby was foreman of the Hand Press and Jobbing rooms and forge; Mr. Alexander Drysdale was foreman of the Single Cylinder Press rooms; Mr. A. B. Taylor was foreman of the Double Cylinder Press rooms; and Mr. A. S. Bowen had charge of the Pattern and Wood-working Shops. Mr. William L. Colby was a leading man in the printing press rooms, and Mr. James Blair was then running a large slide lathe.[6]

These men were long faithful to the company and instrumental in the development of presses in America. Such loyalty may be all the more surprising when daily work conditions are considered:

> [*Matthew Smith to Sereno Newton, Oxford, N.Y., 16 September 1834*] . . . *Business is rather dull and orders very scarce. The Cholera has doubtless effected*[†] *business very much—the number of deaths for two days past has been 10 each day and very little is thought of it. Several of our men has*[†] *had it and old Murray, assistant engineer, has died of it. Murray* [*his son?*] *has had it and has not been here for several days. Mike has taken his place and* [I] *think he makes a better engineer than Murray. Shall discharge Murray & put Mike in his place. Curtis has been sick for two or three weeks, and it is doubtful whether he will be able to work again for some time yet. . . .*

> After the move to Broome Street there were new assignments:
>
> The new factory was occupied during the winter of 1835-6 and it was during the removal that the great fire of December 1835 occurred.

> Mr. A. S. Bowen was made general superintendent of the works; Mr. A. B. Taylor was foreman of the cylinder press department; Mr. Snaith was foreman of the hand press and miscellaneous work; Mr. D. Writner of the blacksmith shop and Mr. J. Wheatman of the saw shop. James Yetman had charge of the stable, etc., which was located on the opposite side of Broome Street. The general business office, however, was retained in Gold Street, and also a repair shop, the latter in charge of Mr. John Colby, which charge he retained until his death in 1858. . . .

The attractive facilities nevertheless had their problems:

> The new [Novelty Works] boiler was of the water-tube class and proved a source of great trouble and expense; the lime from the well water (the Croton was not yet introduced) filled the tubes and they then soon burned out. It was soon replaced by a set of five plain cylinder boilers, which did duty for nearly thirty years. The blacksmith shop was equipped with a large and a small trip-hammer which did excellent service for many years, until superseded by the direct-acting steam hammer. The blast for the forges was given by a blowing cylinder (for fan blowers were not then in use) and the pressure was partially equalized by a tub bellows. . . . The large Frost lathe was soon replaced by a much larger one, and went to the junk shop. The new lathe was designed by Mr. James Blair, and, somewhat modified since, is doing good service in the works to this day. . . .

Croton Reservoir was planned partly as a result of water main problems emerging in the December fire. Tucker chronicled the critical conversion from carpentry to machine shop:

> The [Gold Street] works were driven by a steam engine of 12 horse-power, built by the Sebastian Iron Works of this city; and the machine tools consisted of a few hand and slide lathes, a lathe for turning type beds, platens and similar work; drilling machines, a bolt cutting machine, a wheel cutting machine, and a few wood-working tools. Almost all of these tools were made in the works. . . .

Of all the machinery, however, Tucker was most impressed by a great gear cutter, crucial to the company's precision products:

> The Wheel cutting machine was built by Thomas Freeborn, an excellent, and then well known, mechanic of this city, and was really an extraordinary machine for its day. It would cut spur, bevel and spiral wheels up to six feet in diameter. The divisions of the index plate were originally laid out on a metal circle 20 feet in diameter, with the index plate in the exact center, and from these divisions circles of holes were drilled direct in the face of the index plate with numbers such that the machine would cut teeth of every description from 1 to 100; every even number from 100 to 200, and every fourth number from 200 to 400. There was also a circle of 1000 smaller holes drilled in the peripheral edge of the index plate, making a total

of about 18,000 holes, and the accuracy with which these holes were drilled was surprising. . . .

This elaborate machine had several cutting heads and positions so as to do a variety of work, but there was still room for human decision:

The shapes for teeth were not drawn out for the workman in charge of the machine, so he used his own judgment in a measure, but the general rule was, the smaller the wheel the more its teeth should be pointed, and vice versa. This machine was known far and wide, and quite a business was done in cutting teeth in wheels for the trade. . . . Mr. Sereno Newton compiled for this machine a 'Table of the Proportional Radii of Wheels,' from 10 to 400 teeth, and from ¼" to 3" pitch. . . .

Metal planing machines were unknown, as were slotting machines, shapers, and borers. It was a situation the sons began to remedy.

In 1839 Richard bought a second-hand planing machine from Messrs. James Nasmyth & Company of England. It planed an area 4 feet wide by 8 feet long. Tucker had seen nothing like it before, but he quickly found a way to enlarge its capacity to a width of 5½ feet. After that it was no longer a curiosity:

The table was drawn back and forth by chains which wound around a drum that was driven in opposite directions by a circular double rack and oscillating pinion. This put an end of planing press ribs by hand planes. This was our planing machine for several years, and the next one was a hand machine made by the Novelty Iron Works of this city. After that, planers of improved construction came fast into use, and in the course of time the old English machine was condemned to the scrap heap.[7]

Such planework was not always flawless. When Buffalo agent N. Lyman returned a foolscap Washington in 1844 Peter Hoe attributed to "witchcraft" any problems encountered. But Lyman was not buffaloed:

[*To R. Hoe & Company, 3 April 1845*] . . . *I have had to make so many complaints about the Presses last year that I presume you are quite wearied with it.*[†] *But be assured I have had a good deal of trouble with them. W. Thomas' Cap Press works well after I had it balanced at the Foundry* [*here*] *with their planes. . . .*

The catastrophic economic depression of 1837 proved a crisis worse than any fires. Many insurance companies, usually good sources of loans, had been ruined in the 1835 conflagration, and businesses could not find money to borrow. Philip Hone had been pleased to note the city's vitality after the great fire: "It is gratifying to witness the spirit and firmness with which the merchants meet this calamity," he wrote on 19 December 1835. "There is no despondency; every man is determined to go to work to redeem his loss, and all are ready to assist their more unfortunate neighbours. . . ." But he was soon obliged to record its real repercussions as banks were mobbed, stock market prostrated, and businesses idled:

> [10 May 1837] . . . I . . . witnessed the madness of the people,—women nearly pressed to death, and the stoutest men could hardly sustain themselves; but they held on as with a death's grasp upon the evidences of their claims, and, exhausted as they were with the pressure, they had strength to cry, "Pay! Pay!" . . .
>
> [11 May] . . . A dead calm has succeeded the stormy weather of Wall street and the other places of active business. All is still as death; no business is transacted, no bargains made, no negotiations entered into. . . .
>
> [26 May] . . . No goods are selling, no business stirring, no boxes encumber the sidewalks of Pearl street; stocks have fallen again, but not back to the prices at which they were before the suspension of specie payments. No remittances come from other states, and even where debtors are able and willing to pay, there is no means of getting the funds to N.Y. . . .[8]

Hone himself was hurt badly and obliged to return to business from retirement. Between speculative public-land sales and new government regulations requiring land payments in cash, the banks had collapsed early in the year, and most commercial firms suffered for a long time thereafter.

These acute economic difficulties made men do strange things. They were behind James Watson Webb's plea for relief from the Napier import duty, they were behind Benjamin Day's foolish sale of the *Sun* to his brother-in-law, they were behind Horace Greeley's earliest—panicky—advice to go west—all to escape financial ruin. For R. Hoe & Company they prompted the wildest schemes to make money (and were perhaps—for prestige—behind Richard's National Guard commission) bringing on an indebtedness the new Colonel would labor twenty years to eradicate. This depression—the nation's worst until 1929—long shaped the Hoes' every thought and plan.

The firm's own normally remunerative real estate investments were of no avail; the partners repeatedly begged unashamedly for more time:

> *New York November 3, 1837*
>
> *Dear Sir,*
>
> *Owing to the impossibility of collecting our rents we trust that you will favor us a little by drawing at 60 d*[*ay*]*s instead of at sight for the last quarter. It is nigh to impossible to get any thing from our tenants, indeed we have not yet received any thing on the last payment we made* [*to you*]*; you will therefore not be surprised at our request. Please write us on the subject.*
>
> *Very Respectfully Yours*
>
> *R. Hoe & Co.*
>
> *To/G. P. Darley Esq.*
> *Pokeepsie,† N.Y.*
>
> *P.S. We should like to have you draw at 60 ds from 10th inst. adding*

interest. Previous to that time our payments are as much as we can meet. Yrs, R. Hoe & Co.

The following year saw no improvement. Young Richard, with some bright ideas on saw manufacture, was hopefully dispatched to England to exploit them for whatever ready cash they would bring. While there his brother and partner both nagged unmercifully:

> [*Robert Hoe to Richard, London, 1 June 1838*] . . . *Dear Richard bring home money, remember the old woman's parting advice to her son.* . . .
>
> [*8 June*] . . . [*S*]*till we manage* [*somehow*] *to pay wages per week $400* [*for over 100 men and boys*]. . . .
>
> [*Matthew Smith, 13 July*] . . . *We are endeavoring to curtail our expenses as much as possible and are about selling off our horses and wagons.* . . .

Money, unfortunately, was not the only problem, although it was at the bottom of a malaise throughout the struggling shop:

> [*Matthew Smith, 16 August 1838*] . . . *You know that I am a person not easily discouraged or likely to "give up the ship" easily. And I do not write you this to make you feel uncomfortable, for I presume it will be no new thing for you to learn that Hoe & Co. are in trouble. You can have no idea how important it is that you should be here to attend to matters uptown. They are in a very unsatisfactory condition. Things do not go on as they ought to.*
>
> *The fact is we have not the right kind of men to oversee matters. Bowen is a good faithful man, and will and does look after our interests to the best of his abilities, but you know he lacks energy, and is not the person calculated to stand in the breech to meet any contingency. Taylor does not seem to care any thing for the concern. I cannot make any definite charge against him, but he & the men under him seem to move without spirit or interest in what they are engaged, and in fact everything there seems to go on slowly, and with a perfect indifference & hastelessness. As to the result, whether we make money or not on a job does not seem to concern any one there. I ought in justice to except Bowen from this remark.*
>
> *I have said enough that you may know what I mean in regard to uptown works, the whole uptown squad seems to have lost all vital parts; it requires some one to inspire them with new life & energy. This we know you can do on your return, by sticking to the shop and making that your home. We therefore say come home as soon as you can.* . . .

Vast numbers of Hoe employees would prove immensely faithful, of course, but the depression was a poor time to test loyalty.

Some men were gifted enough to strike out for themselves; all, fortunately, would not, like Taylor, go into direct competition with the company. Both Isaac M. Singer and Bernard Sheridan worked there.

Singer, by training a machinist, had instead pursued a theatrical career; but about 1835 his first job in New York was with Hoe, and he developed his sewing machine shortly thereafter.

Bernard Sheridan, a New Jersey cabinetmaker, entered R. Hoe & Company in 1830 as a patternmaker. He remained five years when he decided to establish his own business; locating himself just across Gold Street from Hoe, he began manufacturing embossing and letter copying presses and printers' rollers. But he aspired to more:

> [*Robert Hoe to Richard, London, 3 October 1838*] . . . *Sheridan has been bribing our men to get at our patterns, drawings &c &c & says to Kingsland, who is honest enough to tell us about it that there is one man in our employ who will give him at any time all the information he wants, & further that to said man we owe considerable more than we can ever pay & that we cannot last 3 months &c &c. Who he is we know not. Now he* [*Sheridan*] *also is making Wash[n] presses & ink[g] machines. We only want to find a responsible man using one of his make & then we will "kill two birds with one stone."* . . .

Somewhat thwarted, perhaps, Sheridan turned to compete with Frederick J. Austin, then the only New York maker of paper cutters, whom he eventually bought out. Sheridan's sons entered the business and carried on for many years. Today the firm is a component of the Harris-Intertype Corporation, Cleveland, Ohio.*

"Things don't go right up town or any where else where we have any interest at stake," Matthew wailed a year later (1 December 1839), with financial conditions still abysmal. "It may truly be said of us," he philosophized, "that we live upon hope and are in truth 'young men of great expectations' and I do hope & believe some of them will be realized." But, like Sereno Newton, Matthew Smith's strength was weakening, and he would soon attend to his poor health.

But Smith was nevertheless still in charge of the New York office, and something had to be done to alleviate the oppressive conditions. Without waiting or consultation, Matthew again (as with the Rust negotiations) acted impulsively to clear the company's indebtedness:

> [*Smith to Richard Hoe, London, 8 September 1838*] . . . *Since we last addressed you we have had again to resort to our truly valuable & kind friend Mr. E. F. Sanderson. We had as once before got very far behind hand in money matters and Mr. S. has generously loaned us about $7,000, which has made us comparatively easy. This we are to repay him by handing over to him such business paper as we get in. We shall have about $1,500 to hand him in a few days and shall pass all to him that we can possibly spare from time to time & we hope before many weeks in this way to refund him the whole am[t] he has so kindly loaned us in our time of need; we have never known an*

* See *Books for the Millions.*

instance in which the old proverb was so fully illustrated—"a friend in need &c, &c." We were just at the point of despair when we very reluctantly had resort to him, and he received & treated us in the kindest manner. . . .

Mr. B[rown] is very kind and respectfull,† his resignation from the presidency of the bank has released him from many cares & perplexities, and his bearing toward me is quite different from what it was—indeed I may say he now treats me kindly. But as to pecuniary assistance we have long since ceased to look to him for it. He seems to be willing to continue just as he has, without in any way increasing the am^t of his responsibility. This he told us you knew long ago, and he now lives up to it religiously. . . . The [Butchers and Drovers] Bank treats us very well. . . .

It was rumored Brown was about to be prosecuted for a $100,000 bank loss; hence his resignation. But wild accusations were also a sign of the times; Brown was apparently innocent. As for R. Hoe & Company, the retiring banker got along much better with young Richard, and after Smith's death their relationship improved considerably. The Sanderson indebtedness, of course, did not soon disappear; it grew, in fact, and would in 1842 engulf the firm.

The 1835 acquisition of the Broome Street property through Benjamin Brown (who remained for a time as landlord) was the start of a cordial friendship. Brown, who later received some political appointments, became a principal financier to the business, for which the Hoes reciprocated with frequent services:

[To Blair & Rives, Washington, 29 January 1840] . . . Please send your "Semi Weekly" Paper to Benj. M. Brown Esq., Mamaroneck, N.Y. during this session of Congress and charge the same to us.

To day we learn by the paper that the clerk has authority to employ printers to the House. We hope that the action just taken will be a sufficient inducement for you to direct us to send on the Napier. . . .

The brothers also performed little chores in New York for Brown, who, with increasing infirmities (and outbreaks of cholera) preferred to remain at home. R. Hoe & Company nevertheless relied heavily on Brown financially:

[To Brown, 11 November 1840] . . . We are so very short of money that we hardly know how to pay any thing upon the note due 14th inst. of $357.73. We have had lately several disappointments in money matters which have made us a great deal of trouble. We enclose our check 30th inst. for fifty dollars, and a new note for the balance with interest added which we hope you will make answer for this time. We will try & do better. . . .

Brown soon took such promises in stride, his involvement and friendship for the company making him a kind of silent partner:

[*Brown to Robert Hoe, 15 June 1841*] . . . *I do aprise*† *you, it affords me but little encouragement to endorse for you, if it takes you more than four years to return borrowed money, which was lent to be returned in three days.* . . .

Brown's indulgence and the enormous Sanderson loan had saved the day—but only momentarily. To remain solvent the brothers apparently continued to borrow heavily from commercial banks which would not be so tolerant. The firm's credit at last gave way, and in September 1842 the Hoes were obliged to announce a state of bankruptcy.

William Swain, upon hearing the news, knew exactly where it hurt the most:

Philadelphia, Sept. 5, 1842

My Dr Friend Richard—

I learned with equal surprise and momentary regret this evening . . . that your house had been compelled to yield to the burthen [*debt*] *which, like true phylosophers*† *you had been bearing up under for months past. I said "momentary regret," for while I can very well appreciate the unpleasant feelings experienced by being thus forced to yield I am not quite certain but it may be better for you in the end than to continue to bear the burthen for the benefit of those public cormorants the b—ks* [*banks*] *while you might be sinking in age under it.*

You will escape unscathed (except it may be perhaps in feelings, for who is not sensitive in any case of his own, however much phylosophy he may have for others) and whoever the cormorants be or how you may have been seduced into assuring the burthen that they might prosper, shall have had their day, and had passed to sleep among the "things that were," you may have occasion to rejoice that you dissolved the connection "whilst there was yet life and hope." . . .

But I am in haste, tired not well, and the pilot line has arrived. So good night. From your old friend

W[m] *M. Swain*

P.S. . . . If I can be of service to you command me. Swain

Swain's florid language, unfortunately, adds no details of the failure.

At this crucial point the Sandersons again came to Hoe's rescue. They quickly furnished the capital necessary to revive the unfortunate company. Now consisting of only the two young brothers in partnership, the firm was required (at least until 1843) to reveal its true nature atop every letter:

THE "HOE" PRINTING PRESS, MACHINE AND SAW MANUFACTORY,
29 & 31 Gold Street, and Corner of Broome & Sheriff Streets,
New York.
[Cut of Washington Press]

Presses of the most approved construction, and every article necessary for a Printing Office and Bindery, including Types, Ink, &c. always on hand or furnished at short notice. Cast Steel Saws warranted of superior quality. Machine work in general.

Richard M. Hoe and Robert Hoe beg to notify their friends, that although the above Establishment is now owned by Sanderson, Brothers & Co., yet they are engaged as the sole Machinists there; and they request their old friends to patronise the same. If such friends prefer to correspond with them in regard to the mechanical department of the business, and also in regard to orders, they can do so, and their communications will be promptly attended to—or they may address Mr. Joseph Steele, the receiving and paying clerk, and general agent to the proprietors. RICHARD M. HOE, ROBERT HOE.[9]

Hence the name "Hoe" in quotation marks: it was not really theirs. At any rate the Sheffielders were pleasant owners, and Joseph Steele too became a trusted friend and advisor. The Hoes by their inexhaustible energy and agility would eventually repay their British creditors—in their own country, in their own coin—for their life-saving generosity and confidence.

In 1843 the Gold Street headquarters was once more in flames, although Tucker does not refer to this fire. Possibly it was not nearly so inconvenient physically as exhausting in insurance transactions. In connection with their plans for future expansion the brothers resolved not to replace the shop as it was, but to make improvements. This the insurance people were reluctant to do.

New York January 20th 1844

Dear Sir

We had the pleasure of receiving your esteemed favor of 18th inst. yesterday. The "Bowery Ins. Co." paid us on Saturday last; the "North Amer^n Compy" have delayed on account of the absence of their President. They paid theirs this morning.

The only difficulty we now have is to settle with the "Mutual Safety." They are willing to pay us $2800-- which is Mr. Hurry's estimate of the extent of the sum necessary to put up the buildings as they were. On the other hand Mr. Price says that it cannot be done for any such money. Mr. Pell proposes that Mr. Hurry shall put up the buildings, we arranging with him, i.e. Hurry, for any alteration in the plan &c. This is all that we can get out of them. In thinking over the matter it seems to me that where a building is totally destroyed they cannot assume the right to rebuild it but must pay the loss the amount to be ascertained as may be agreed upon. I hope that your health will enable you to come in on Monday.

Yours very truly
Robert Hoe

To B. M. Brown
Mamaroneck, NY

In March there was colossal news to temper the insurance negotiations—the *New York Tribune's* momentous order for the first Hoe type-revolving press:

New York March 30th 1844

Dear Sir

I hope the sovereign people in Mamaroneck on next Tuesday week will find it necessary to order a second edition of those [*poll?*] *tickets. I am aware your town will not compare with London in population but still most any place ought to number over 250 Democratic voters.*

As to business matters, we have Greely[†] *and McElrath's order for one of those new presses, price $6000.--. We have rec*^{d} *several orders also for cylinder presses last week and have enquiries for several more, the orders for which we shall certainly receive in a few days. Had it not been for the rain to day we should have seen the first tiers of beams on Gold Street. The builders are not through with estimating up-town. There is nothing further of interest.*

Very Respectfully yours
Robt *Hoe*

To/Benjn *Brown Esq.*
Mamaroneck, NY

The loss of facilities on Gold Street nevertheless made it necessary to reexamine the desirability of moving additional departments uptown. The owner of one piece of adjacent property had already been solicited:

[*Richard Hoe to Mrs. F.C.G. McWhorter, New Haven, 1 September 1843*] . . . *Mr. Fowler informs use*[†] *that you own the lots of ground next adjoining our Factory & on the corner of Broome & Columbia. My object in addressing you is to enquire whether you will lease them for a long term of years & if not if you would feel disposed to put on such a building as I should desire & then lease the premises. The rent would be so secured that there could be no possibility of loss, the cost of the Building would not be over $2000. & on that out-lay we would be willing to pay you, say, 7 to 8 per cent, we paying taxes, insurance &c. I should be glad to hear from you on the subject.* . . .

The demand for a large foundry had grown steadily. The company had farmed this work to neighboring shops—and others hundreds of miles away—but with increasing dissatisfaction. Robert Hoe told a customer (24 February 1843), "Our plans [to put a flyer on your press] were deranged by several accidents that occurred to our works, and work which had consumed much time had to be thrown away in consequence of bad castings." A foundry was begun in 1842 in the old Gold Street smithy, with John McLaren as foreman. He turned out good work promptly, but the problem had not really been solved.

About 1842 A. B. Taylor, cylinder press foreman, left the company; William L. Colby (probably John Colby's son) replaced him and remained in charge until his death in 1881. Colby later developed a means of driving cylinder machines (1853) using bevel gears (which proved noisy) and (1858) with a universal joint, employed on several presses; he became a member of the firm in 1860. For the 1842 alterations the company closed for about a week, but business quickly resumed—and fell behind. Tucker explained:

> The foundry in Gold Street was now found too small for the work required of it and it became necessary to have a larger one. Accordingly in 1844, a plot of ground 100 feet square, on the corner of Columbia and Broome Streets, adjoining the factory, was bought and the present foundry building 40 feet x 100 feet was erected on it. This was a great advantage in every way.

At the same time the brothers finally got a satisfactory insurance settlement:

> *[Robert Hoe to Brown, 8 April 1844] . . . Having spent all the morning in Wall Street I have only time to say that we this day received $2975.-- in full from [Mutual] Safety Ins. Co. and that the assignment also is completed and the amount is paid by New York Ins. Co. . . . Our buildings are progressing finely in Gold St. although the weather has not been as good as we could wish. . . .*

A few years later the Hoes' hard work at maintaining viable shop facilities was again wiped away in smoke and cinders. A fire 6 February 1849 was more disastrous than the one before it:

> As the firm did not now occupy the whole of the Gold Street works, the remainder was let out to tenants. Early in 1849 a fire broke out in the premises of one of these tenants, which totally destroyed the building on the street, and partially destroyed the one connecting with the rear building. They were, however, immediately built in a more convenient form than before.[10]

Destroyed and damaged stock alone amounted to $14,000 for the company, which insurance companies promptly paid.

Real estate coverage of $8000 had been divided equally between two new firms, North Western Insurance Company and East River Insurance Company, the latter balking at payment. East River officers observed that in 1848 the property had been deeded to Robert Hoe, constituting a sale or transfer without their consent and therefore voiding their policy.

Not so, said Horace Holden, the Hoe lawyer, at a court hearing. The problem had started in 1829 when Robert and Rachel Hoe and Rhoda Smith mortgaged the 29 and 31 Gold Street parcel for $13,000, and again in 1833 Rhoda (now as sole owner) similarly obtained another $4000. As the aged Rhoda retired to upstate New York, and as Benjamin Brown also had a mortgage on the property, the 1848 transfer to Robert Hoe II—for $10—had only "a view to the greater facility in the management of the property (which was under bond and mortgage to a greater amount than its value)." [11] It was not a sale but simply a way to enable the company to act "more fully" with the property, Holden argued. The court ruled in Hoe's favor.

Brown was aging too, and now wanted to divest himself of the troublesome corner. The Colonel furnished a plan:

> [*Richard Hoe, Paris, to R. Hoe & Co., 2 March 1849*] . . . *I think with Messrs Sanderson that it is Mr. Browns part to rebuild, but that is not for us to say & now as regards purchasing the down town place, as you say Mr. Brown wishes it. I think it would be a good plan for Peter to buy it, but in order to clear the Title from all encumberances*† *Mr. Brown must foreclose and sell. This of course he cannot do except with Sandersons consent by their agreement. Messrs. S. must also give to Mr. B. the claim they pretend to have on the purchase for the assesment.*† *If all this could be done & you could get the title into our hands, & it would I think be best for Peter to buy it & keep our names out on a/c of Bankruptcy for the present. . . .*

The elderly Rhoda Smith, however, was not cooperative in this complex real estate matter:

> [*Brown to Robert Hoe, 22 April 1850*] . . . *I regret to learn that M*[rs] *Smith hesitates to sign the deed. She must of course be well aware that her interest & that of her late Husband was completely destroyed in the failure in 1842 & it was only to save you the expense of a foreclosure & to expedite the business that M*[r] *Schell suggested the deed from her. I am exceedingly anxious to close this matter, but I can not give a Lease for property on which I may have yet to foreclose. . . .*
>
> *Please say to M*[r] *Peter S. Hoe I thank him for his kind attention to my request for Crackers &c. May he live a Thousand Years, & then for ever. . . .*

Eventually, of course, the firm owned the Gold Street property freely and clearly, as it did the uptown works—and all the brothers their personal residences as well; the Colonel grew adamant about owning land before commencing to build. As they returned to the business of press-building, however, they were never far from mishaps:

> [*To J. Walker, Charleston, 24 November 1851*] . . . *A wall fell upon our blacksmith shop which killed 2 of our men and wounded 2 others. It was pushed out by the side pressure of grain in the [adjacent] building occupied as a brewery. In its fall it smacked one part of the shop which will put us to some temporary inconvenience. . . .*

But acquisitions went on:

> [*From F. Sheahy, 16 June 1853*] . . . *Mr. Chardon the owner of the premises No's 9 & 11 Columbia St. now informs us that you can*

have the property at the Same price he offered it to you ($10,000) if you wish it. . . .

"We are out of kilter by the miseries of moving," the Colonel wrote to Robert (17 May 1859), a groan that might have punctuated correspondence throughout these decades. A four-story machine shop had already been added on Columbia Street in 1850, followed by a new 100 horsepower Novelty Works engine to power the entire shop, equipped with oversize boilers and artesian wells for unlimited water. About the same time the foundry's output was doubled by new cupola and mechanical blower arrangements.

The prestigious *Graham's Magazine* visited in 1852, careful to assure gentle readers that the gigantic Hoe machineworks—with average annual midcentury sales never less than $400,000 and often over a half million dollars—did not in any way forebode mass unemployment:

> It is a pregnant fact, and one singularly corroborative of the soundness of the writer's view . . . relative to the effect of machinery in increasing rather than diminishing the number of hands employed, or likely to be employed, in the business of printing, in consequence of the daily augmenting demand for printed matter arising from its cheapness and perfection—that, since the introduction of the fast-printing machines the call for hand presses has greatly increased. . . .[12]

The Philadelphia journal regretted that the day's newspapers did not all reflect a tone as elevated as Hoe's quality workmanship.

Graham's learned the company possessed in New York, Boston, and London a total of 35 city lots—about 2 acres—and employed almost 600 hands. How best to keep so great an enterprise in constant work would be a gigantic problem for the Colonel.

Upon the death of Sereno Newton in 1836 Richard had become the mechanical expert of the firm. But tinkering was no longer a boy's pastime; it had to be dependable and consistent. Few others could be consulted in a professional way, but surprisingly enough Richard managed to find around him a number of mature, experienced businessmen who felt privileged to aid the young man. Having early lost his father, Hoe perhaps unconsciously sought the advice of Brown, Swain, Steele, the Sandersons—all of whom served him well, especially at first, when his doubts were paramount. They extracted no greater fee than his friendship, and all enjoyed the symbiotic bargain.

Richard's (1837) improvements to saw-grinding equipment, like his early press modifications, were ingenious and welcome, but they presented no clear direction for a full-time inventor. He revealed his apprehensions to Charles Sanderson in Sheffield, whose counsel—

despite evidences elsewhere of keen international industrial rivalry —was warm and honest:

> [*To Richard Hoe, 21 January 1841*] . . . *I notice with peculiar pleasure your remark respecting the "sudden success," what we call the "Royal Road" to fortune. Be assurd[t] that the encouragement of such ideas is a mere infatuation. There are many things and many inventions by which sums of money may be very easily realized, but there requires time to persuade a prejudiced world, there require industrious steadiness to carry out the mere invention, and there require plans laying down for your mode of action as well as a systematic arrangement of the whole, both on the producing side of the question, as well as the selling. All these do demand a mind of steadiness of purpose and a sort of singleness of reflection which few are master of & I do really attribute the nonsuccess of many valuable inventions to want of order & system on the part of the inventor. I sincerely hope you may succeed in bringing out at least some of your inventions and you may be assurd that it will give me pleasure at all times to render you any assistance in my power. I notice with great satisfaction the opinion given upon your new printing press* [*the "Pony" double cylinder*] *and I trust you wish to reap that pecuniary advantage from its sale which you so richly deserve. . . .*

A year later Richard's old friend William Swain gave him similar encouragement:

> [*5 September 1842*] . . . *Your knowledge of mechanics now that mechanical knowledge and mechanical industry are about coming to be made available, and to be turned to the account of their proprietor is an inexhaustable mine of wealth upon which you can fall back and bid defiance to foes* [*creditors?*], *whether they assail you under the specious garb of friendship, or with a more open & manly attitude. . . .*

Richard's formal efforts at this time were directed to an application (1843) of air springs to cushion the reciprocating bed of a cylinder press; this was less successful than A. B. Taylor's work, patented in 1846. Next he developed an 1844 adjustable type-high bearer on each side of the press bed, making the machine capable of better impression control. His methods in 1844 and 1846 for warming ink fountains to improve viscosity were of doubtful value, as was an 1847 double race and sliding pinion drive mechanism to replace the universal joint (although the company returned to this mechanism at the end of the century, at which time the Miehle Press Company also used it successfully). More conspicuous was the need for automatic collecting of sheets after printing, and the young inventor proceeded to tackle the problem doggedly.

Despite their amazing speed and efficiency, the new cylinder machines still required a boy to feed and another to remove sheets.

Several developments had already suggested that stacking could be done mechanically: (1) Napier's grippers showed a way to carry sheets with positive control; (2) improved drying of inks made piling possible without fear of offsetting (impression of one sheet soiling its neighbor); (3) gravity should be made to work for nothing, creating a stack as printed sheets fell from the press.

Richard had set to work about 1842 on a mechanical sheet delivery system or "flyer" for fast cylinder presses. A year later it was almost ready:

> [*To James Burgess, Charleston, 24 February 1843*] . . . *We this morning received your esteemed favor of 19th inst. and regret very much that you are so much dissatisfied about receiving the press without the apparatus to fly the sheets. The press was put on board the "Sutton" yesterday, after waiting several days, that it might not be at the bottom of the vessel. We regret that it was thought best to send the press without the flying appendage but we supposed the delay would be a very serious matter to you. . . . There is no doubt in our minds that we shall succeed with the flying apparatus. In a few days we shall put it to the test when you shall hear further from us in relation to it. . . .*

The device assured positive sheet control at high speeds: grippers on a small drum above the impression cylinder grabbed the sheet and transferred it to endless tapes, from which it fell onto "flies" (large wood-lathe combs), the flies swinging each sheet almost a half circle to drop it onto the pile. Despite continuing difficulties, brother Robert was broadcasting the advantages of the $250 flyer:

> [*To L. Johnson, Philadelphia, 27 January 1843*] . . . *We are making at this time 3 single cylinder book presses all of the same size and description, like the one now in use by Harper & Bro[s] with the addition of a flying apparatus. . . . We are in treaty with two other printers of this city and do not doubt that we shall shortly receive their orders. The price of the machine is $1800# without flying apparatus & $2000# with it. . . .*

Almost two years later the flyer's development finally came to a conclusion:

> [*To E. R. Jewett & Company, Buffalo Commercial Advertiser, 17 December 1844*] . . . [*A rebuilt press*] *has the addition of a very simple apparatus to fly the sheets which of course dispenses with one boy; consequently only one boy is required who feeds on the sheets. The "fly apparatus" takes up no additional room and is not half so likely to get out of order as one made of flesh and blood. We have now orders to put the apparatus to some 25 of our presses in this and the adjoining cities. . . .*

Hoe was understandably proud of a solution with almost universal application. But by this time he grew apprehensive that others, possibly Wilkinson and Beach of the *Sun,* were doing similar work. He had need of wise counsel—and fast patent service—if his invention were to survive. He wrote an old friend, Dr. Thomas Jones.

Thomas P. Jones (1774-1848) had come to the United States from England as a young man. In New York in 1835 he became editor and soon owner of *American Mechanic's Magazine,* and at the end of the same year, when offered a teaching post at the Franklin Institute, Philadelphia, he took his journal with him. He merged his paper with that of the Institute to become *Journal of the Franklin Institute,* which continues today. Jones, like many ambitious newcomers to America, enjoyed a lifetime of service to America in the pursuit of mechanical engineering.

In April 1828 Jones was appointed Superintendent of the United States Patent Office, and although he removed to Washington he still retained control of the Institute journal. His work was tireless:

> Under his editorship of twenty-two years, the *Journal of the Franklin Institute* became a valuable repository of information concerning scientific and engineering subjects. An important feature for many years (1826-59) was the list of patents, with descriptions, which had been filed in the government Patent Office. . . . In his editorial capacity he was [quoting an Institute memorial to him] "always ready to recognize and warmly encourage genuine invention; but he was equally watchful and uncompromising in showing the defects he believed to exist in any patented invention, especially in those cases in which the invention embodied no possible advantage or gain to the public. . . ." [13]

When the Patent Office was reorganized in 1836 Jones was appointed an examiner, from which position he resigned two years later to engage in private patent work. He was a most useful man in Washington for a prolific inventor who, like all creators, worried incessantly that his brainchild would be claimed by others:

> *New York December 31st 1844*
>
> *Dear Friend,*
>
> *I send you a model and a drawing of a Cylinder Printing Machine.*
>
> *The machine is the ordinary cylinder press. The improvements for which I wish a patent and for which you will find a caveat in the office, are first the manner of flying or delivering the sheets, the guides for the tapes and the rollers moveable on the shaft over which the tapes run so as to allow the points or ends of the frisket to pass under the sheet as it is turned out, or the sheet to pass over the ends of the frisket. 2nd, the simple lever & cams to move the grippers or fingers. 3rd, the manner of raising the cylinder when from any cause the impression is not desired. 4th, the arrangements of the springs to stop the momentum of the bed, on rods*

with toothed wheels turned by a wheel of the same or larger size. I prefer the wheel to be 4 to 6 times larger than the pinions to which the springs are attached, on the other end the springs are fastened to a rachet wheel which will allow the pressman to increase or diminish the resistance at his pleasure.

You will see this combination of springs is very important as the first contact may be very light and the resistances increase with the greatest rapidity. I should like the strings claimed in that combination for presses and also for any other use.

You will oblige me very much if you will make out a description in your usual and happy perfect manner using the drawing, & send and forward the same to me with the drawing as I have a lithograph made and wait to letter the stone again after I receive your description.

I have heard nothing of your bill. Please draw for the amount and it shall be honoured. Mr. Young is anxiously awaiting his patent.

Yours very truly
Rich[d] M. Hoe

To Dr. Tho[s] P. Jones
Washington City, D.C.

[P.S.] *Please give this business your immediate attention. RMH*

Young is not identified. The Colonel wrote again a week later, revealing his fears:

[6 *January* 1845] . . . *Since sending off the model & note enclosed for which the above* [*note*] *is a receipt, I have heard there are some of our persons in this city getting up a similar device. If it is possible I wish to secure the flying frame or frisket made to stand upright or on an incline in combination with a cylinder machine,* [*to*] *make this combination & make this so as if possible to exclude any & all persons from attaching a flying frame to the* <u>*cylinder*</u> *press. As there are others endeavoring to get up some machine, why you will give me as much speed in your operations as possible. . . .*

Dr. Jones was 71 in 1845, and perhaps not the sharp editor of briefs he might once have been. His submission of Hoe's claim was met with resistance at the patent office. Hoe was annoyed, but calm:

[25 *January* 1845] . . . *For my part I cannot see by what reasoning the examiner can escape passing such a claim, at any rate I know that should such a claim come before the judge of our district it would be sustained.*

It is certainly a useful improvement & ought to be secured to the inventor but there is now a person talking about putting the frisket on but arranging the movement by a different means. The claim would be the same as that in Isaac Adams of Boston [*for bed and platen presses*]*: claim which is in fact all a claim of combinations now. It is important for me to have this so inserted that if possible*

> *I may be secure in its use. So pray have a particular care with it. You will see that I not only wish the manner of combination but the combination itself to be claimed. Perhaps the claim might be made in this way: "I claim the combination of the flying sheet frisket in a position nearly vertical or above an angle of 45 degrees with a cylinder printing machine the cylinder of which has one or more revolutions for each impression."* . . .

Make a strong case, Hoe admonished, offering to come to Washington if necessary. "I would be perfectly willing to risk a suit here on it. Do my dear Doctor make a desperate effort. . . ." Whether Dr. Jones was simply dilatory, ill, or disposed cavalierly to find Hoe's claim less than novel (as would Isaac Adams) or of little gain to public interest, he was suddenly incommunicado. The Colonel's anxieties grew, goaded by his race with a phantom competitor:

> [24 *February* 1845] . . . *Since then* [*when specifications were first sent to you*] *there has been put in operation a machine very much resembling mine*. . . .

On 20 March Hoe begged a friend in Washington to call upon "patent agent [Jones] & ask him if he has rec[d] a letter from me. I fear he must be unwell. . . ." At last the Doctor's reply arrived, but it was bad news. The patent examiner found the flying frisket neither new nor useful. Hoe was now furious (24 March 1845) and showed it: "This is not the application of an old machine to a new purpose (which by the by I think should be patentable)," he insisted. He promised to appeal, but curiously must have surmised that Dr. Jones, possibly infirm, could little help in a strenuous fight. No further correspondence resulted. Yet Jones had perhaps done his work well:

> Mr. R. M. Hoe was granted a patent, May 1, 1845, for adapting the automatic sheet-flier to fast running cylinder presses. The arrangement operated so well in practice that the sheet-flier was immediately in great demand. We attached it to all new presses at once, as also to a great many old ones. In fact, it came into universal use, both at home and abroad and remains so at this time.[14]

To one early inquirer interested in a small $800 press and a medium one for $1100, Hoe (27 December 1844) replied: "Either of the above we should make to *fly their own sheets according to our new patent make* which is all the rage here. . . ." In February 1845 Richard gave final approval to Messrs. Newton & Son, London, to draw the flier specifications for British enrollment. Curiously, just the year before, he had instructed Newton similarly to patent for him a press with automatic feed mechanism (self-rising pile and spring clips on the cylinder to seize sheets); not in fact a good design, it got none of Robert's enthusiastic promotion.

One of Hoe's oldest business friends was perhaps understandably leery at first of rank novelties. Samuel N. Dickinson, Boston printer and type-founder, was blunt about it:

> [*To R. Hoe & Company, 28 June 1841*] . . . *I have said that I would never again buy the first machine made upon a new plan, because many improvements are most always made in the second or third that follows, and the purchasers of the last made always have the best machines. I presume there will be alterations in this much for the better, and I should not like to take it unless you would agree to add or alter it in any way so as to make it equally good with those that may be constructed hereafter.* . . .

But the company was persistent, and Dickinson yielded good-naturedly (15 July 1841): "The job cylinder is in good health, and will no doubt be glad to see its projector, when he comes to Boston. . . ." A few days later the Bostonian revealed further ramifications:

> [*24 July*] . . . *We cannot run the little press to advantage without steam, and for that trial I am waiting—reserving a very good job for the purpose of trial. In the course of 2 or 3 weeks I am in hopes to move several of my machines by steam. Besides this, Mr. Ruggles wants me to wait till I see a new press of his in operation—of a larger size, but of the same (or nearly the same) plan as his present press. I am favorably inclined toward your press.* . . .

Two years later Dickinson was so delighted with Hoe equipment for his "Fancy Book and Job Printing" he talked of showpieces:

> [*To R. Hoe & Company, 4 October 1843*] . . . *It is my wish to put your press in the room where my Adams presses now are, and to remove them into the next room. The room now occupied by my presses you will recollect is*[†] *a high ceiling, and lighted from the top as well as the sides: for this reason I am ambitious to have a splendid machine occupy the room. We have numerous visitors, and an elegant machine as well as an elegant person always commands more attention than those of an inferior aspect.* . . .

Financial troubles unfortunately later forced Dickinson to sell his shop, but he was by this time (1849) a U.S. Senator to whom the Hoes looked for help in securing certain steel tariff concessions.

One by one, nevertheless, the largest book and job printers all fell under the spell of cylinders clattering their unique "music of the press." Many of them little realized they were at the same moment willy-nilly equipping their shops and tooling up for a new genre of publishing—indeed, the most pervasive and lucrative of them all, newspaperdom! In 1871 an anonymous journalist agreeably credited the company for this revolution in graphic arts:

> [C]ivilization owes R. Hoe & Co. a large debt for what it has already given—the inestimable boon of a cheap and rapid dissemination

> of news and of knowledge. They have made their name immortal, and in securing this reward let us also rejoice that the material emoluments of labor are also theirs. . . .[15]

With Charles Knight's *Penny Magazine* for a physical model and witnessing other local but momentary attempts at cheap journalism, New Yorkers expectantly awaited a good penny newspaper. Benjamin Day (1810-1889), a yankee reporter, had served an apprenticeship on Bowles' Springfield *Republican*; he came to New York at 20, and impetuously resolved to start a newspaper only to exploit his sagging printery:

> In the summer of 1833 this young printer found himself in a rented room, twelve feet long and sixteen feet wide, with a printing press, paper and plenty of type, but no orders for handbills and letter heads. So, to push his printing business, Ben Day set that firm chin and mouth of his more firmly and got out the first issue of the *Sun*. Publisher, editor, reporter, compositor, pressman and mailing clerk all in one, he stayed up the night which ushered in Tuesday, September 3, 1833, setting type and preparing for the press work.[16]

Mott described the product of such self-conscious industry:

> The *Sun* appeared [by typographical error dated 1832] . . . with four pages, each about the size of a sheet of business letter-paper and divided into three columns. The paper was neat, the type small, the news concisely stated. The fresh, even flippant, style of the *Sun's* news items, and its emphasis on local, human-interest, and often sensational events caught the town's fancy; and that news treatment and the one-cent price brought a circulation of 2,000 in two months and 5,000 in four months.[17]

George W. Wisner, another young printer, offered to report daily police-court activities for the *Sun,* and his facetious coverage of drunkenness, petty theft, street walking, and sundry assaults created immense interest among common readers and established a new genre—called by Mott "domestic tragicomedy"—still the main stock of many papers. Wisner soon owned a share in the new paper, although political differences with Day made his stay short. Nevertheless, for Messrs. Day & Wisner Tucker, in his first year at Hoe (1834), recollected building a single large cylinder machine for printing the *Sun* and trailing along with the press gang to help erect it in Day's William Street basement.

After about a year the *Sun*'s circulation was 10,000, aided substantially by cunning street newsboys whom Day employed much as the London papers had done: at first four or five lads hired at $2 per week would each sell 100 papers a day—they paid 67 cents per hundred or 75 cents on credit. Advertising followed. In 1835 the paper with increased format was already one of the world's more imposing dailies; after Wisner's departure about that time Richard Adams Locke, its new editor, was writing crack copy, including the stunning hoax in August of that year, porportedly describing Sir John Herschel's discovery, by telescope, of life on the moon. Other papers, "scooped" by an event of which they could not obtain the slightest intelligence, were obliged unwittingly to reprint the *Sun's* copy—while the latter's circulation soared to 19,000. Although the public (in Mott's words) took this "greatest 'fake' of our journalistic history" [18] with considerable good humor, the lesson and the ominous foreboding of newspaper credibility was not missed.

With the economic panic of 1837, presswork problems (keeping up with still rising circulation), and a nagging libel suit against him, Day impulsively sold the *Sun* to his brother-in-law for $40,000; he himself later called the sale the silliest thing he ever did.

In a speech at an 1851 dinner honoring Richard M. Hoe, the founder of a new industry recounted those early days:

> It is true I originated the *Sun,* the first penny newspaper in America, and, as far as I have known, the first in the world. But I have always considered the circumstances as more the result of accident than any superior sagacity of mine. It was in 1832 that I projected the enterprise, during the first cholera, when my business as a job printer scarcely afforded a living. I must say I had very little faith in its success at that time, and from various causes it was put off. In August, 1833, I finally made up my mind to venture the experiment, and I issued the first number of the *Sun* September 3. It is not necessary to speak of the wonderful success of the paper. At the end of three years the difficulty of striking off the large edition on a double-cylinder press in the time usually allowed to daily newspapers was very great. In 1835, I introduced steam power, now so necessary an appendage to almost every newspaper office. At that time, all the Napier presses in the city were turned by crank-men, and as the *Sun* was the only daily newspaper of large circulation, so it seemed to be the only establishment where steam was really indispensable. But even this great aid to the speed of the Napier machines did not keep up with the increasing circulation of the *Sun*. . . .[19]

Regretting his impulsive sale, Day began searching for other printing schemes. Day's son Benjamin (1836-1916) in 1878 would invent the benday tone engraving process that bears that name, and their progeny were further immortalized by grandson Clarence Day's humorous episodes of American family life.

Moses Yale Beach (1800-1868) as a young Connecticut farmhand and apprentice cabinetmaker demonstrated an incredible capacity for work and invention (gunpowder engine, steamboat propulsion, rag-cutting machine for papermaking). At the age of 20 he met and married Day's sister Nancy, and eventually joined his brother-in-law a year after the *Sun* began. Despite little apparent experience in a printing shop, Beach became manager of the *Sun's* mechanical department, and he soon bought out Day's partner Wisner—and in 1838 Day himself. Beach became sole owner of a popular (30,000 circulation) but expensive novelty; profits were slim and unpromising.

Although first of the viable penny papers in New York, the *Sun* shortly faced competition from Bennett's *Herald,* Greeley's *Tribune,* and dozens of others, less ably run but loudly distracting. Beach knew what to do, however, to face these challenges: he established a ship news service (meeting incoming ocean steamers), horse express lines (for news from Albany), special trains for war and political coverage, and homing pigeons for other distant news. These activities naturally found him at the center of cooperative efforts to pool New York resources; both the Harbor Association (a fleet of ships) and the New York Associated Press resulted. Beach is also credited with inventing the syndicated article when in 1841 he distributed President Tyler's Congressional message to 20 other papers. His was also the first European edition of an American paper (1848).

On its tenth anniversary the *Sun* employed 8 editors and reporters, 20 compositors, 16 pressmen, 12 folders and counters, and 100 carrier boys; its circulation was 38,000. Installed not long before was the Colonel's "audacious" high-speed "Pony" Double Cylinder capable of 6000 hourly:

> [A]fter a delightful trip [to England, 1838, Richard Hoe] returned to New York with his head full of plans for a new double cylinder press. In a short time he announced that Hoe's double cylinder press was built and would on exhibition print 6,000 impressions an hour. The audacity of this assertion attracted general attention and disbelief, but criticism ceased and praise began after the promise was kept. The first press of that kind was bought by Mr. Beach, of the New York *Sun,* and when it was set up in his vault it became an object of curiosity and admiration to all the printers and scientific men of the city. . . .[20]

In 1848 circulation was at 50,000, whereupon Beach turned control over to his two sons, Moses Sperry and Alfred Ely Beach. Moses (1822-1892) had begun his career in his father's composing and press rooms. In 1845 he helped George Roberts manage the Boston *Daily Times*, but returned to New York after a few months to head the *Sun*. Alfred withdrew in 1852 to pursue engineering plans (particularly a pneumatic subway), and Moses, only 30, became sole proprietor. Except for a brief period (1860-1862) when it was a semi-religious paper, the *Sun* remained a working man's daily with a variety of news, features, and fiction. Beach was, like his father, according to a later owner, Charles A. Dana, "a business man and a newspaper manager rather than what we now understand as a journalist." [21] But—like his father, too—he had for his lack of news sense some interesting compensations.

Despite his apparent satisfaction with Hoe equipment, father Moses Y. Beach had met another very promising press inventor. Jeptha A. Wilkinson (1791-1873), a Brooklyn mechanic, began work in 1818 (in England) on a rotary printing machine but did not have a functioning model until 1839 when, in Providence, Rhode Island, he is credited by Rollo G. Silver with designing the first American web printing press. He produced a newspaper issue on it in May 1840 but it needed more work and the inventor needed more money. Moses brought Wilkinson to New York the following year. He then developed his machine as a rotary on which wedge-shaped types were to be secured to the periphery of a large cylinder. Outsiders soon heard about it:

> The machine was constructed in strict privacy in the *Sun* building. It was to print from a roll of paper and cut and fold the sheets, and its success was believed by its builders to be a certainty. Great expectations were based upon it, and Mr. Moses S. Beach patented it in Great Britain in 1842. Mr. Wilkinson openly boasted that R. Hoe & Co.'s career as press builders would soon be at an end. . . .[22]

Young Moses Beach himself apparently contributed its apparatus for roll feeding and also a mechanism for folding, cutting, and delivery of 300 copies per minute. But although well supported with *Sun* funds the Wilkinson machine refused to hold the types around its cylinder, and the inventor was stymied. The elder Beach stopped payments and demanded both American and British patents be in his (or his son's) name. Except for a trial issue in 1841, after two more years the expensive machine had failed to print a single salable copy of the *Sun*.

Seven years later, nevertheless, Wilkinson was still at work, promising a speed of 20,000 hourly. He received a United States patent in 1853, built another prototype using types especially cast with taper and indentations for sure lockup—but still could not get a well printed product from it. Three years later he petitioned for an extension on his patent, showing Congress the 1841 specimen. Wilkinson went to Europe (1862) on a selling trip, but gave up entirely when, back again, fire destroyed his shop. Young Beach carried on alone.

Dissatisfied with his own feed system, Moses S. Beach in 1853 announced a reward for the design of feed machinery for the problematical rotary; it went uncollected. He next devised a way in 1857 to turn a sheet on a press to print the other side: two sets of fingers grabbed the tail and rerouted it for a second impression. He attached his sheet feed system to a Hoe Lightning at the *Sun* (making it a perfecting press) and demonstrated it widely, patenting the modification in America and Europe. A year later the Beach-Hoe hybrid was declared "perfected," but it was no technical threat to the Hoes who (without Beach) were solving the same problems. Despite Tucker's reference to it, Silver raised questions about the Colonel's timely knowledge of Wilkinson's work:

> Ingenious as Wilkinson was, he could not compete with the apparatus and skilled help of the established press manufacturers. His type-revolving press appeared at least seven years before the type-revolving press of Richard M. Hoe. Whether or not Hoe knew about the Wilkinson press cannot be definitely determined even though it had been claimed that Hoe proposed to erect a plant for its manufacture. . . .[23]

The accumulating Wilkinson-Beach developments, however, may possibly have been a source of infringement for which the Colonel was later gathering evidence: "Your remembrance of Mr. Beachs visits to see the experiments are according to the facts," he wrote mysteriously to William Conquest in England (7 February 1859). "I am not sure that we shall need your testimony but if so will inform you. . . ." Termination of their work short of success may have made prosecution unnecessary. Wilkinson's anticipation of a Hoe proposal to manufacture, like his prediction of ruining Hoe, seems only a desperate inventor's conceit.

Benjamin Day's happy experience with penny journalism quickly made all his friends and associates aware of what they also could do with a case of type and a printing machine. Even his strongest critics were converted.

William M. Swain—although harboring strong doubts about penny journalism (he "nearly discouraged Mr. Day from issuing the first number," [24] said Frederic Hudson)—had become printing foreman at the New York *Sun.* Arunah S. Abell, originally a yankee printer, also regarded Day's unusual paper as a bad risk. When Swain and Abell saw the project prosper, however, they were quick to follow; with a third partner, Azariah H. Simmons, and strong encouragement from Gold Street, they raced to Philadelphia in 1836 to start a similar daily.

In its first year the *Public Ledger* explained penny press objectives with logic calculated, perhaps, more to convince its own formerly incredulous partners:

> In the cities of New York and Brooklyn, containing a population of 300,000, the daily circulation of the penny papers is not less than seventy thousand. This is nearly sufficient to place a newspaper in the hands of every man in the two cities, and even of every boy old

enough to read. These papers are to be found in every street, lane, and alley; in every hotel, tavern, counting-house, shop, etc. Almost every porter and drayman, while not engaged in his occupation, may be seen with a penny paper in his hands. . . .[25]

The *Ledger* was never so offensive as the New York press, but it had its share of crime news, scandal, libel, and mob threats. As editor for twenty years Swain reportedly read every line; as business manage he nursed circulation from 20,000 in 1838 to 44,000 twelve years later:

A chief reason for this success was Swain's unusual enterprise in taking advantage of the new means of speed in communication as they developed, and in installing the best new presses in order to improve and enlarge the paper.[26]

Swain's early interest in the electric telegraph he would soon transmit to the Hoes as an enterprise almost as rewarding as printing itself.

William Swain had left the $12 weekly foreman's job for a speculation in Philadelphia of unknown risk; when he retired in 1864 he was worth $3 million. During the paper's first years, however, the larger and more heavily committed R. Hoe & Company, still hurting from the depression, was obliged to badger Swain and his partners:

New York January 7, 1840

Gent.
We are going on with your order for Press &c and shall have it completed by the time specified. In the mean time it requires considerable outlay of capital, the principal part of the expense being in labor, which must be paid in cash. We find the times exceedingly dull and money difficult to collect. Under these circumstances we have thought you would be willing to advance us your note for $1000.-- at four months on account of the payment that will become due at the completion of the press. This would be an accommodation to us, as we could make use of it, and would put you to no inconvenience. You will please make the note payable in N. York, say at Mechanics Bank, and we will obligate ourselves to pay the exchange whatever it may be at maturity of the note; we annex our receipt for the note, and would esteem it a particular favor, if you will be kind enough to send us your note by return mail.

Your M^o Obt Sv^t
R. Hoe & Co.

Messrs. Swain Abell & Simmons

over

[*reverse*]

$1000.--
received of Messrs. Swain Abell & Simmons their note at four months from this date for One thousand dollars payable at the Mechanics Bank N.Y. which is on a/c of the cash payment which will be due upon completion of the Napier Press & Engine which we are now

making for them as per Contract, as the above named note is made payable in N.Y. City for our convenience, we agree to pay the difference of exchange upon the am of said note that may be, at its maturity.

R. Hoe & Co.

New York January 8, 1840

During this year Hoe shipped to Philadelphia seven vault lamps, "Rockwells Patent," at $7 each, and advised against laying the engine's foundation until it was on hand. Purchases continued over the next few years, and a truly satisfying friendship between the Colonel and Swain matured. In so cultivating his professional friends Richard had effortlessly reaped business ties—and business—that would remain for decades. Another such encounter, however, would put this kind of trust to a severe test.

In Boston George W. Roberts and William H. Garfield began the *Daily Times* in 1836, along the lines of the New York penny dailies. By this magical formula it was successful from the start:

> Like the New York *Herald,* the [Boston] *Times* was inclined to argue in grandiloquent phrases about the nobility of its mission in educating the common people, but its exploitation of the morbid details of [crime and scandal] made many of its contemporaries doubt its sincerity.[27]

With carriers sending the paper to all the large towns of eastern Massachusetts, the *Daily Times* ran up to 12,000 in its first four months. Although its circulation never went much above 20,000 (1851), a new Hoe double cylinder seemed indicated—and was bought on credit. Roberts had many schemes for the big press:

> *[To "Friend Hoe," 9 July 1842] . . . I assure you it has given me a great deal of pain to allow your recent notes and drafts to be returned to you dishonored, but I assure you I have been unable to pay them at any sacrifice. If you can have patience a little while longer I hope to be able soon to take up several of my protested notes which you now hold—for from my novel publishing and from my Mammoth [paper] which appears in August and from the government pap I mean to make a little harvest and bring up averages. [Jonas] Winchester has been here a few days past and from him I learn confidently that he has made nearly four thousand dollars nett† profit from the recent novels he has published. Of course I must have a finger in that pie and have embarked in the same experiment myself—Having no new books at hand I have taken Bulwer's "Night & Morning" and if I only sell half as many as Winchester . . . I shall make a pretty sum of it. . . .*
>
> *You have probably learnt that the "Times" has at length been selected as the advertising organ for the government in all the departments—Navy, Custom House, Post Office, &c, &c—the same having*

been taken from the Boston Atlas! I told you I should have this, and now I mean to have more. . . . We thought times bad a year ago—I only wish they were half as good now. . . .

The Colonel must have hoped secretly that Roberts would prove as good a customer as Abell and Swain in setting up prosperous dailies in continual need of printing machinery and supplies. But Roberts' troubles, thanks apparently to a gentle and trusting nature, started with that Hoe double cylinder and plagued Hoe as well:

[*To Roberts, 31 October 1843*] . . . *Mr. James A. Coburn has just called upon us in relation to his bill of board for the men we sent on to put up your presses*† *some two years ago. . . . Mr. Coburn says that he does not pretend to assay claim on us but he is determined to get it from the men if you dont pay. Now as those men are still in our employ and we feel a disposition to protect them when they have been on our business and followed our instructions we feel very solicitous that you should attend to the matter at once. . . .*

This oversight, however, was not sufficient to curtail business with Roberts, and an order for type became embarrassing:

[*To Roberts, 25 November 1843*] . . . *We stop the press of business to inform you that we are not a little put out about not being able to ship your type this week. The founder who is casting the agate* [*5 point*] *give*† *it as his excuse that his "rubbers" have been taken sick and that "it aint every body that can rub his small type." All very true no doubt but very inconvenient to you we fear and makes us feel very uncomfortable in not making good our promise also. . . .*

In 1845 Moses S. Beach became Roberts' partner in a subsidiary venture. This was the operation of a printing plant (no doubt reorganized to issue the *Times*), for which they would require much advice and material support. But the partners got off to a slow start: the Roberts & Beach orders were misdirected to the factory and, said the company later (1 August 1846), "for want of further information were neglected" for weeks.

But with the apparent return of prosperity Roberts, like Swain and Abell, came through splendidly, in 1847 taking type-revolving press No. 5. With some circulation improvement and this prestigious printing machine the partners dreamed of a modern headquarters. The Colonel advised:

New York, May 20/50

Friend Roberts

Your favor was duly rec[d]*. . . .*

Now if your friends have an idea that a Granite front is to be compared either in beauty or strength to Iron they are much mistaken & for so narrow a front the iron is in every respect superior.

Swain Abell & Simmons intend to build in iron in Baltimore a building 55 x 75 feet & intend to expend from 40 to 50,000 dollars.

Now if it was not good they would not go into it so strongly.

I had a conversation with Mr. Miller, Druggist in Broadway of our city, who had an iron front put in by Mr. Bogardus & he says he likes it so well that he intends to have the rear of his store iron also immediately. His front is 5 stories. . . .

There is also another advantage in an iron front. It will never be destroyed by lightning for being so good a conductor the fluid is dissipated as rapidly over it that no damage can be done by any possible storm.

I could give you a great number of arguments in favor of iron & also of iron beams so that your house would be entirely fireproof.

I expect to have the Baltimore Sun have iron beams. If you have iron beams with brick arches between each it would cost for each floor say about 300$† deducted from this. [Make] the beams of wood & you have the excess for a real fireproof building in which you need have no insurance &c &c.

I know I think more on this subject than some others, as it is one I have spent some time in examining.

I should like you to have a good building. While you are about it you would do well, I think, to pay [your] Architect's expenses here to examine what there is to be seen. I will show him up. I am going to Phil on Wednesday morning & shall be home on Friday.

Yours very truly,
Rich[d] M. Hoe

A few months later there was real trouble. The paper was not doing well, but even the proprietors did not recognize the situation. Roberts sent a plaintive call which brought the Colonel to Boston to "have a jaw," but more than an innocent chat took place. Hoe reported to another Roberts backer:

New York, Aug 19th 1850

P. T. Barnum, Esq.
Iranistan Bridgeport Co.
My Dear Sir,

Your favor of yesterday is at hand. I went to Boston last week at the request of Mr. R & on examining his affairs advised him as a creditor & friend not to attempt to go on. I then consulted with the lawyer: he has done business for our firm for several years past in whom we have great confidence, & who I found had been so friendly as to loan to R & Co. some loose cash. I also met Mr. H. Winser, paper dealer to whom the firm are indebted $5300--, & we all came to one conclusion that they had better stop and wind up. I believe that Mr. Roberts was in the firm conviction that his concern was under only a temporary embarrassment. At the same time I blame him for not knowing his real situation which I think has been in a measure brought about from too great trust in others. I think that if he can find some friend who will purchase the establishment from the assigners appointed by the court it would be made to yeild† from 8 to 9000 Dollars profit per *annum. I think that the assets as they stand will yeild 50% for the creditors. The amount owing to our firm is about twelve hundred dollars. We have concluded to send to Mr. Manlius S. Clark, attorney, State Street, Boston, a power of attorney to act for us to prove our amount & take our chance. The law of Mass. does not give a release where the creditor has not paid 50%. Mr. Roberts has been in N.Y. for two days past endeavoring to make some arrangements. He told me he intended to call on you on his way home.*

Yours very truly,
Rich[d] M. Hoe

Among his many minor commercial ventures P. T. Barnum in his memoirs did not record any involvement with Roberts. But his own talents and interests had been aided immeasurably by his association with such newspaper friends, not the least of whom were Webb, Noah, and (at first) Bennett, who all gave him unstinting publicity. Promoter Barnum recalled visiting their shops:

> Night after night, at the midnight hour and later, I crawled up these several newspaper staircases to put in these journals some fresh and startling announcements about my business. I even did this after the editor had gone home, but the foreman in the composing room had some authority then, and could often put the matter I offered in type or make an announcement for me. . . .[28]

(Barnum once outwitted Bennett in a real estate transaction, ending their relationship.) The master publicist also knew and respected tinkerers and mechanical geniuses like the Hoes, although he admitted, "I never had the slightest love for mechanism; on the contrary, I have a sort of abhorrence for complicated machinery." [29] Barnum was of course more like his explosive editor friends. He had started his career by publishing a small newspaper (1831-1834) which he left with a brother-in-law. A biographer saw in this early enterprise all the requisites to newspaper success:

> For three years P. T. Barnum was a journalist of mark and there is every reason for believing that had he chosen to make that his life work, "Barnum" would have been one of the great names in American journalism, ranking with Dana, Bowles, and his friend Greeley.[30]

In 1852 Beach and Barnum jointly financed a New York illustrated newspaper, with Briton Frank Leslie (whom Barnum had brought to America to produce a pictorial catalog of his famous museum) as managing foreman. Roberts may have aspired to print their paper, but the *Illustrated News* lasted only one year—and Leslie stayed on to start his own.

Although George Roberts' debt of $1200 was for those days a considerable sum, the Hoes were seldom disposed to foreclose on a favorite customer. Surprisingly enough, however, the Colonel was soon talking of new funds, new equipment, a new start:

> [*To Roberts, 16 September 1851*] . . . *Whatever I can do to advance your interests I will do but at present I cannot leave here "nohow."*
> *Should you conclude to start anew we have a double cylinder 50 x 31 in. which was used at the Boston Traveller office, I think not over 1½ years, and which in perfect order we will sell for 2500$--.*† *We have a column* [*vertical*] *engine, 2nd hand, of our make, 5 to 6 horses power we will sell in complete order for 600$--. It has been in use at Balt*ᵒ *Sun office. Perhaps one of our portable engines with boiler attached as represented on page 30 of our catalogue, a copy of which we send by mail, will suit you better for the present. The price you will observe complete is $600. This includes heater & pipes. This latter would be the most economical & does the work well & takes very little room. It will drive that press & another without any trouble. . . .*

On 12 November 1851 the company endeavored to clean up the delinquent Roberts et al. accounts, especially to get the $1000 cash for the press. Hoe reminded him, "you have the use of those flyers [evidently left on the machine] which will daily save you much money. All that we want is that you should pay us at least our outlay. . . ." The delinquent friend soon met still another businessman with whom he formed a partnership; perhaps here would be a more responsible manager of funds. The *Times* redesigned its format in the summer of 1853 and announced in December:

> With a view to enlarge the facilities and to enhance the public usefulness of the "Times' NEWSPAPER and JOB PRINTING ESTABLISMENT,"† we have entered into a co-partnership with that enterprising firm, (with whom we have been partially associated for the past three years,) the Messrs. J. H. & F. F. FARWELL,—and, on the First Day of the New Year, the combined establishments will be conducted under the name and firm of ROBERTS & FARWELLS.†
> The general policy hitherto pursued by the TIMES, will be maintained under the new arrangement.[31]

The former "general policy" was not what the innocent Roberts should have brought along with him. The *Times* faced increasingly stiff competition in Boston's dignified newspaper arena, and in 1857 was absorbed into the Boston *Herald*. But George Roberts had many other ideas, and would keep at them.

Horace Greeley (1811-1872) had begun a yankee printer's apprenticeship in 1826 at the age of 15—and soon yearned for the hustle of New York. After wandering aimlessly for a few years he resolved to have a newspaper of his own. With an associate, Jonas Winchester, he inaugurated *The New Yorker* in 1834, a journal of news, comment, and "general literature," most of it lifted from British papers. Greeley struggled with it for seven years, coping desperately with balky subscribers ("They took my journal, and probably read it. They promised to pay for it, and defaulted, leaving me to pay my paper-maker, type-founder, journeyman, etc., as I could.") as well as competition from the mammoth newspapers ("The great beasts murder me in the way of circulation . . .").[32] Finally, in 1841 he gave up the paper as a loss. Not until 1843 was there relief from the big "extras," when the post office finally began to charge them a higher postage rate.

Greeley described himself the year he failed as "tow-headed, and half-bald at that, . . . long-legged, gaunt and most cadaverous of visage, . . . slouching in dress; [he] goes bent like a hoop, and so rocking in his gait that he walks down both sides of the street at once. . . ."[33] With such a sense of humor Greeley was not licked. He next started the *New York Tribune* in April 1841 with a borrowed $1000, a staff of 10, and a list of 500 potential subscribers. He soon had a resounding success on his hands which he pushed to its limits. He added a weekly edition directed at suburban readers (reaching as far west as the frontier itself), and issued gigantic daily supplements which swelled the paper's encyclopedic coverage.

With his experience and sagacity Greeley matched the cut-throat competition with flair: in 1843 he secretly shipped compositors and type to Boston along with his reporter, Henry J. Raymond, to cover a speech of Daniel Webster; on the return trip they set Raymond's copy and had it ready to print as the boat docked. Astonished fellow reporters had been scooped. Greeley, like Day, now had a winning formula—he made his paper the medium of the laboring man, the business man, the farmer, the dreamer and inventor (he praised innovation as big news)—but he knew he had to guide the sheet personally, carefully, through very hectic years.

The *Tribune's* first office was as decrepit as the balky machine which groaned at the paper's 3500 copies. A fire in 1845 destroyed the premises, but the intrepid decision of the proprietors—Thomas McElrath, an attorney with good printing and business sense, was his partner—to publish "as usual" on the next day was a symbolic act that gave the public confidence in them. The *Tribune's* new headquarters, however, was not much better: in a five-story "dry-goods box" with presses "in an unwholesome dingy cellar," Greeley could be found in his second-floor study "driving a furious pen in weird hieroglyphics."[34] (His handwriting was acknowledged the most illegible in the industry.) So instrumental in effecting emancipation in the Civil War, Greeley became the "slave of a newspaper," he once said of himself. But it was voluntary servitude, interrupted only by passionate forays into politics itself. But he never forgot he was a newspaperman.

Consummate publicist Greeley said everything in 6 point type on the *Tribune's* stationery:

> Daily Tribune, $5 a year, Semi-weekly Tribune, $3. Weekly Tribune, $2. The Daily Tribune is issued every morning except Sunday, on a large sheet of fine white paper, forming eight pages of six columns each, or about the size of two average Country Newspapers. About six of these pages are new every morning; the aggregate of Editorials, News, and Miscellanies, being greater than that of any other paper in America, and greater than any but one or two in Europe, though its price is but *half* that of other first-class Dailies in this Country, and less than *one-fourth* of the cost of similar journals in England.—Two Evening Editions are issued, to supply each subscriber with the latest news up to the hour when his copy must be mailed. Terms.—$5 a year; $1.50 for 3 months.
>
> The Semi-weekly Tribune contains nearly all the non-advertising matter of the Daily, except such as is of local interest, or has been superseded by fuller & more exact advices before the Semi-weekly is issued. It is of the same size with the Daily, and has few advertisements. Terms.—$3 a year; 2 copies for $5; 10 for $20.
>
> The Weekly Tribune is issued every Thursday morning, & contains most of the matter of the daily, with more summary accounts of such Events & Proceedings as cannot be published in full.—We mean that no weekly shall surpass this in giving a full, graphic & faithful account of what the World is Doing, whereof it is Thinking, and how it is Progressing.
>
> The TRIBUNE has obtained an aggregate circulation of upwards of ninety-four thousand copies. Every subscription is paid in advance, & the paper stops when the advance pay runs out, so that no man need hesitate to subscribe from an apprehension of being dunned for arrears. Terms.—$2 a year; 3 copies for $5; 8 for $10; 20 copies (to one address) $20.
>
> [Pointing fist] Postmasters or others taking charge of and remitting us the money for a Club of Twenty, will be entitled to a Copy of the Weekly, gratis.
>
> Postage on the Tribune. Under the new law, postage to regular subscribers to the Weekly Tribune, one year, is .26; the Daily Tribune, one year, $1.56; the Semi-Weekly Tribune, one year, .52.[35]

In just 20 years the mechanics of issuing a daily paper had indeed become far more sophisticated than Benjamin Day's little gossip sheet. But the efficiency of the *Tribune's* encyclopedic prospectus was shattered in at least one instance (18 August 1853) on a clerk's note to R. Hoe & Company: "You had better make out another bill against Mr. Greeley. The one you left can't be found." The *New York Tribune* by 1865 had nevertheless solved a lot of little problems and was the third largest newspaper in America.

James Gordon Bennett (1795-1872) had come to the United States from Scotland at the age of 24; he tried teaching, but found, as a reporter for Charleston, Washington, and New York papers, that journalism was the challenge to his talent for understanding politics, finance, and human nature. Several newspaper projects failed, however, as Bennett's flair for argument repeatedly lost him the backing he needed. He was nevertheless still scrappy after a long and abrasive career:

> Thus, at forty years of age, Bennett was out of a job, with only $500 to his name. He had some reputation as a brilliant, though rather flowery, writer, with a dangerous talent for personal controversy. He was a wiry, energetic man, with squinting eyes which had been injured by too close application to books and papers. Nobody under estimated his Scotch shrewdness, and many feared him. Day and Wisner refused to employ him on the *Sun;* and when he suggested a partnership to Greeley, who had recently lost money on a newspaper venture, that young printer refused. Thereupon Bennett decided to start a penny paper of his own.[36]

Frederic Hudson (1819-1875), long associated with Bennett, credited the unpleasant man with originating the "Independent Press" under most extraordinary conditions. With a population of 270,000 in 1835, New York already had 15 dailies, 11 semi-weeklies, and 31 weeklies, whose combined circulation was over 13 million. It seemed no time to start another, but Bennett was not daunted.

The Scotsman approached a new Ann Street printery, Messrs. Anderson & Smith, who were already doing the mechanical work for the *Sun* and the New York *Transcript*—one of the *Sun's* most pernicious rivals. With his last $500 Bennett contracted to have his paper manufactured there too. The first issue of the *Morning Herald* appeared 6 May 1835, and with early journalistic novelties—after his map of the December 1835 fire his coverage of the 1837 Canadian revolt included the first "war" map—and Bennett's astute, expostulary "money" articles, it soon became required reading in New York.

Despite his lack of formal literary training, Bennett stunned his readers with his innovations. He had, said John Bigelow, "the sagacity to find in news and gossip a cheap substitute for brilliant leaders. . . ." [37] Fellow editor Henry Watterson added, "The modern newspaper, as we know it, may be fairly said to have been the invention of James Gordon Bennett, the elder. Before him there were journals, not newspapers. . . ." [38] Elmer Davis, too: "Bennett was the inventor of almost everything, good and bad, in modern journalism." [39]

Understandably unhappy sharing printing facilities with arch competitors, Bennett himself was apprehensive about the situation. Indeed, *The Printer* a few decades later found this moment in history worthy of some reflection:

> While speaking of newspaper work, we may be allowed to observe that Mr. John Thomas (of the firm of Wynkoop, Hallenbeck & Thomas) formerly had charge of the single cylinder presses which

printed the *Sun*, the *Transcript*, and subsequently the *Herald*, side by side, all in one pressroom, for Messrs. Anderson & Smith during 1834-5, in Ann Street, this city. There was risk enough in starting these papers, but the idea could never have been entertained at all, were it not for the facilities afforded by the single cylinder presses supplied by the Messrs. Hoe. We have no doubt that Brother Thomas remembers the circumstances clearly, and the large share the cylindrical principle and the Hoe presses had in originating and facilitating the success of the "penny press" generally, among those other newspaper enterprises which have now become permanent institutions in popular favor—an absolute necessity of the public mind—our daily pabulum from the press.[40]

Nervously watching Anderson & Smith's Hoes bring his dreams to reality, Bennett would of course soon want machines of his own. His little sheet, a four-pager one-quarter the size of the financial "blanket sheets" he would eschew, was nevertheless provocative and successful enough to be irritating to rivals; the *Sun* and *Transcript* expressed their annoyance by leaving Anderson & Smith, who were furious and tried fruitlessly to shake Bennett's contract. After five months, however, calamity settled the problem. A fire in August 1835, just months before the great conflagration, wiped out the Ann Street printing plant, and Bennett was crushed. Tragically, Anderson & Smith, young men who had just started business in 1834, were ruined, and both died shortly after.

In the nineteen-day enforced suspension of his paper Bennett revamped the format drastically (dropping "Morning") and at last proudly announced:

> We are again in the field, larger, livelier, better, prettier, saucier, and more independent than ever. The Ann Street conflagration consumed types, presses, manuscripts, paper, some bad poetry, subscription books—all the outward material appearance of the *Herald*, but its soul was saved—its spirit is exuberant as ever. . . .[41]

Bennett had fumed over the slow erection of a Hoe Napier at his own newly developing Broadway shop, just as the rival *Sun's* moon hoax captured the public's attention. Nevertheless in its second year, when Bennett raised the price from 1 to 2 cents its circulation was unaffected; with the additional revenue he enlarged the format once more, bought new equipment (especially telegraph connections), established nationwide and European correspondents, set up a Washington bureau, and launched a fleet of his own boats. From October 1835 on he operated from 148 Nassau Street.

Bennett had labored indefatigably, but he was still disappointed that his impact on the public grew too slowly to suit his expectations. When Benjamin Day derisively referred to the *Herald* as an unknown paper Bennett realized diabolically that the way to gain notoriety was

—of all things—to make the other journals notice him in their columns. Unwittingly they would furnish him exactly the publicity he wanted. He commenced ruthless and promiscuous attacks on them all:

> Day, Webb, Noah, Brooks, Weed, Greeley, Raymond, and every other editor would henceforth be forced to be forever on the lookout . . . for the bitter invective and telling sarcasm of Bennett's pen. . . .[42]

Thurlow Weed was once a pressman for William Cobbett and James Harper. He started several upstate newspapers, mixed into politics, and in 1824 was elected to the New York Legislature. From 1830-1862 he was proprietor of the Albany *Evening Journal,* but for a time was also editor of the New York *Commercial Advertiser.* Bennett's plan worked all too well; some of his antagonists responded with their fists as well as their newspapers.

Philip Hone almost immediately voiced society's growing objections to Bennett's vexing journalism in his diary:

> There is an ill-looking, squinting man called Bennett, formerly connected with Webb in the publication of his paper, who is now editor of the "Herald," one of the penny papers which are hawked about the streets by a gang of troublesome, ragged boys, and in which scandal is retailed to all who delight in it, at that moderate price. This man and Webb are now bitter enemies. . . .[43]

Hone the day before had witnessed a duel between Bennett and his former employer on the *Courier & Enquirer:* Webb had attacked him on Wall Street, "beat him, and knocked him down." While Bennett professed to enjoy such attacks as good publicity, his conduct was slowly antagonizing the whole city's reading public.

Shortly after perpetuating his moon hoax Richard A. Locke had left the *Sun* to found the *New Era,* a short-lived paper. He wasted no time lambasting Bennett, whose *Herald* had begun just a year before. Locke's remarks, however, had repercussions beyond a feud that he and Bennett might have enjoyed. Matthew Smith (III) cried out to defend a fragile printing press manufactory as well as his own personal interests:

> *New York, July 21, 1837*
>
> *Dear Locke*
>
> *It is with regret and no little surprise that I perceive in your paper of this morning an article attacking Mr. Bennet;† which is certain to bring down upon my head a shower of abuse from him through the columns of his detestable paper.*
>
> *I have as you are well aware been much annoyed by the personal attacks of Bennet which have in all cases appeared in reply to articles in the New Era. As I have repeatedly been assured that no allusion either directly or indirectly should again be made to B— I cannot for a moment suppose that the article of this morning appeared with your*

knowledge. But I must entreat you to take effectual measures to prevent a recurrence of this matter: besides there are reasons [for] which I independent of my own personal feelings . . . consider these attacks extremely annoying to me, as they are more than likely to affect my pecuniary interests—and in addition to the above allow me to say that in my opinion it is highly derogatory to the interest & respectability of your otherwise valuable & interesting paper to descend so far as to notice the fellow or his paper.

Your Ob[t] svt & friend
Matthew Smith

It is not clear why Bennett would harass Smith at all. The latter's "pecuniary interests" apart from the firm's business seemed to be the financing of printing equipment, illustrated by correspondence with another struggling paper.

The New York *Express*, founded 1836 by James and Erastus Brooks, was long a successful paper although, representing the city's commercial interests, it was highly critical of Lincoln and the Civil War. Its early years were strangely dependent on R. Hoe & Company for credit:

New York January 9, 1841

My Dear Sir:

You will doubtless remember that when I loaned you the name of our firm for the purpose of raising the money of U.S. Bank, for† to pay up the notes of Messrs. Brooks & Hudson held by Mr. [Benjamin?] Brown, it was with the express understanding that we should be secured by a mortgage upon the Napier Press. Owing to my long illness this has not been done.

I had supposed that by this time the note, if not paid, would have been materially reduced in amount, which I find is not the case.

As I am about to leave for Europe to spend some months abroad, you will oblige me by having the mortgage executed without delay. If you will send me word by the bearer, in whom the title of the press is invested, or in other words who is to execute the mortgage, . . . I will draw up the mortgage myself and send it to you for execution. Please send an answer by the Runner & oblige.

Your obt ss
M. Smith

Jas. Brooks Esq.
Daily Express

Smith's travels, alas, brought no recuperation. He died in July 1841 —the company ironically soon reselling the *Express* machine as second hand. The Brooks letter nevertheless suggests a similar loan arrangement with Locke.

Toward 1840 Bennett's many competitors and enemies almost spontaneously conspired to smear his name and wreck his paper. Reader boycotts were urged, and pressures applied to sabotage *Herald* advertising revenues. Noah, with little apparent reason to hate

Bennett, nevertheless in 1842 filed two libel suits against him, although he knew one of them to be quite insupportable. But in this "moral" war of only several months' duration, all of Bennett's foes had circulations aggregating 36,000; his alone was already 50,000. While he ultimately lost about a third of his readers at one point, Bennett was little perturbed. His post-fire reincarnation had resulted in a program of aggressive and efficient newsgathering; he parried enthusiastically with his attackers (especially Webb), and to publicize circulation increases—the one argument rivals could not deprecate—he also announced he "would make a contract with Hoe & Co. for a new cylinder press." [44]

Bennett had been first in New York to publish a Sunday edition (1835), but desisted after much local opposition on religious grounds. Again in 1838 he tried it, but not until 1841 did he persist, as his Sunday sales finally outdistanced "blue law" penalties. For the Sunday paper Bennett solicited formal press quotations:

R. HOE & CO.
Printers and Binders Warehouse
Saw Makers & Machinists
Works Corner Broome & Sheriff Sts, New York

Matthew Smith
Rich[d] M. Hoe
Robert Hoe

New York, June 28th, 1841

Dear Sir
We will make you two large double cylinder presses for your [*Sunday*] *weekly paper for three thousand and five hundred dollars, each. The small double cylinder presses for your daily, for three thousand dollars each. These latter presses to be of sufficient size to print your present daily paper with an additional column on each paper. We will also make a Steam Engine of our most improved kind for one thousand dollars. The above named price includes expenses of putting up in running order. We will receive in part payment your present double cylinder for thirty five hundred dollars.*

Yours Respectfully,
R. Hoe & Co.

To J. G. Bennet,[†] *Esq.*
New York

(In July 1841 this stationery continued in use with Smith's name crossed off.) Bennett did not furnish deposit for these presses until November, however, when Hoe quickly consummated the deal:

New York, November 29th, 1841
This agreement witnesseth that R. Hoe & Co. agree to make for me, James Gordon Bennett, Two Double Cylinder presses of improved construction, small size, for the sum of three thousand dollars each and Two Large Double Cylinder presses of improved kind for the sum of

thirty five hundred dollars each, also one steam engine and shafting for the sum of one thousand dollars all to be put up in the Basement of Building corner of Nassau and Fulton Streets—one small double cylinder to be put up as soon as the room can be made ready for it, the other three and engine to be put up and ready for operation by the first day of May next.—Boiler now used in Herald office to be set &c. The whole is to be done in the best manner and the alterations of building to be superintended by our R. M. Hoe.—Payment for the above to be made in the following manner. Ten thousand five hundred dollars in cash on the first day of May next and one Double Cylinder Machine press now used in printing the N.Y. Herald. R. Hoe & Co. also agree to keep said presses in order for five years at the rate of one hundred and seventy five dollars per year.

R. Hoe & Co.

Frederic Hudson, witness *James Gordon Bennett*

Enjoying a lively business among penny newspapers whose existence had hardly been contemplated a few years before, Hoe turned out small cylinder machines at a happy pace. But the older, conservative business press—despite its large and (to most) dull formats—were accommodated as well. The coupling of steam power with the large-bed press, with the company eager and fully equipped to furnish both, had indeed made giant newspaper formats possible, in fact, economical. No one knew how large a newspaper page could be and still remain functional. The New York commercial community would now find out.

James Watson Webb (1802-1884), trained as a soldier, resigned his commission in 1827 and sought a civilian occupation with similarly demanding qualifications. His father-in-law, wealthy department store owner Alexander L. Stewart, bought for him a $6000 share in the New York *Morning Courier,* founded that same year. Webb assumed his new calling with gusto and candor:

> We left the army, a mere boy, to take charge of a political press at the commencement of the [1828 Jackson] political campaign, . . . and, in a party point of view, we possessed not a solitary qualification for the position. We brought into political life the one leading characteristic of the army, a determination on all occasions to speak not only the truth, but the whole truth; . . . and to this we may justly attribute the somewhat well-known fact that we have been considerably the best-abused personage connected with the American press.[45]

Handsome, tall, well-built, Colonel Webb ran his paper with flair and enterprise; an attempt to mob and ransack his office was once met with a professional military defense. His pugnacious attraction for vituperation, street whippings, and duals would become scandalous.

In 1829 the New York *Enquirer* merged with the *Courier,* bringing Noah and Bennett (who continued to assist Noah) into association with Henry J. Raymond, Thurlow Weed, and Webb. Personalities of such magnitude, however, could not long work together, but Webb, in command, followed his principles and made no concessions. In February 1830 he bought a new press, and another in January 1832, the latter a two-cylinder Napier capable of 2000 impressions hourly. Webb wanted speed, but he also wanted magnitude: he had set his mind on stiff competition for his rival, the *Journal of Commerce.*

Like Gerard Hallock and David Hale's New York *Journal of Commerce,* the *Courier & Enquirer* was one of the first oversize "blanket sheets"—Bennett's lasting term for the phenomenon. When he started his first paper, the New York *Globe,* in 1832, Bennett in his prospectus pointedly slammed these papers already enjoying some journalistic notoriety:

> For years past the public has been cloyed with immense sheets—bunglingly made up—without concert of action or individuality of character—the reservoirs of crude thoughts from different persons who were continually knocking their heads against each other, without knocking any thing remarkably good out of them. . . . My sheet is moderate in size [12 x 17 inches], but neat and manageable, printed on fine paper and with beautiful type. . . . I shall avoid, as I would a pestilence, those enormous sheets—the pine barrens of intelligence and taste, which have been undoubtedly sent into the world as a punishment for its growing wickedness. . . .[46]

Perhaps for his own pompous wickedness, however, this early Bennett paper collapsed after a month. The large papers would go on without impediment among readers who apparently judged their money's worth by volume rather than quality.

Outside New York a few newspaper proprietors used the mammoth format as a lure for space-conscious advertisers, despite persistent jokes. "The weather is too cold to read it in the apple orchard," someone quipped about the Toledo *Blade,* "and the paper is too large to unfold in any ordinary house in Ohio." [47] The long lived Connecticut *Courant* in 1867 discarded its Adams for a new Hoe double cylinder with smaller capacity; its Hartford advertisers however were so unhappy competing for small inside pages they forced the paper to return to its 30 x 46 inch page. (When the paper built another headquarters in 1880 the 1867 Hoe was brought along and used until 1892; the *Courant* remained mammoth-size until 1887.)

Traditionally dedicated to New York's political and mercantile life, with fast schooners to intercept seaborne news and sleek pony express from Washington, the six-penny newspapers nevertheless claimed circulations about 4000—fairly high for the time. Reaching 4500 in 1833, the *Courier & Enquirer* became the largest paper in the country. Webb wanted it to be more.

Despite rough times following the 1837 crash—during which the Hoes sorely wished to sell presses, to be sure—Webb again contemplated a new machine. His requirement for faster presswork, not unlike John Walter's of the London *Times,* forced him to the same desperate solutions. Anxious to buy an improved Napier but perhaps lacking the wherewithall, Webb solicited Hoe's letter (28 August 1837) to substantiate some government concession. Richard complied generously, expecting at least some eventual business. The result was a surprise to all.

In the spring of 1838 Colonel Webb imported a four-feeder flatbed printing machine from London, manufactured by Napier in Applegath and Cowper's system for the London *Times* and other journals. Named by Webb the "Great Webster," the press was to run at 6000 an hour. It had an upper and a lower set of feed boards and fly boards, and it used tapes to convey sheets through, there being no grippers on the cylinders. But the feed system was designed for English white paper stock, however, and the *Courier* was soon in trouble using flimsy American cotton "rag."

> But it was found in running the machine that the thin, poor quality of paper used here would so wrinkle and tear and clog the press, and cause such a waste of paper and loss of time, that the use of the machine was impracticable, and after many months of trials and changes by the mechanic who was sent out with it to erect and run it, the machine was finally abandoned. . . .

The domineering newspaperman reluctantly admitted his error:

> A year or two later the *Morning Courier and New York Enquirer* gave an order to R. Hoe & Co. for one of their 'improved' double cylinder presses, which when erected, proved capable of doing as much work, and of a better quality than the English press and required only one half the number of hands.[48]

Tucker recollected helping remove the British press to the Hoe factory where, after some time, it was broken up for scrap.

The *Courier & Enquirer* had demonstrated the ideal case for Hoe superiority: the European model—so blindly insisted upon, found wanting; the Hoe (albeit European inspired) performing admirably. While the partners were not so crass as to retail such episodes to their customers' embarrassment, they were not at first beyond some personal conceit over such early victories. The company's pre-bankruptcy stationery had included an American eagle holding the banner, *"Detur Pulchriori"*—"Let it be given to the more beautiful." Years later, as the machines began to speak eloquently for themselves, the brothers dropped such bombast.

Henry J. Raymond became associate editor of the *Courier & Enquirer* in 1843, raising the paper's literary style somewhat, but

Webb was still interested more in producing a larger sheet than his competitors. By midcentury (having merged in 1845 with Charles King's *American*) the paper bragged of 68 square inches more than the erudite London *Times* itself. For these shinnanigans "serious" journalists had little patience. The New York *Sun* exploded over one such issue.

> Outrageous—the Courier and Enquirer of Saturday morning is just twice as large as its usual size. The sheet is now large enough for a blanket and two pairs of pillowcases, and it contains, in printer's language, 698,300 ems—equal to eight volumes of the ordinary-sized novels of the present day. If the reading matter were printed in pica [12-point] type and put in one unbroken line, it would reach from Nova Zembla [Arctic Islands] to Terra del Fuego [southernmost Argentina]. Such a paper is an insult to a civilized community.[49]

The *Journal of Commerce* at about the same time jumped to 1820 square inches—amounting to 61 less than the *Courier & Enquirer*, but soon made up for its near miss in 1853 with 76 square inches more than Webb's paper, proclaiming itself the "largest daily paper in the world." Webb was furious, but he had had enough.

In 1861 James Watson Webb sold the *Courier & Enquirer* to the New York *World* when, now a general, he accepted President Lincoln's appointment as Minister to Brazil. But there were other papers still in the race.

Robert Bonner was almost more an advertising genius than publisher. His modest printing office had contracted to print a few small newspapers, the commercially oriented New York *Ledger* among them. In 1851 he recognized a rare opportunity:

> It appears that the first proprietor of the *Ledger* had invented a printing-press which he thought would supersede Hoe's lightning machines, and wished to devote his whole time and attention to its development, for which success there was a fortune. With an eye to the main chance, Bonner did not like to lose the weekly job of setting the type of the *Ledger*, and this, as much as any other reason, induced him to purchase the paper.[50]

(Although Hudson disclosed that the mysterious inventor-owner was a former merchant who had founded the paper for country shopkeepers, he gave no name.)

Almost unsure how to proceed, Bonner inserted some family fare—fiction and poetry—much of it by the great literary personalities of the day. He began to advertise relentlessly in other papers, in 1858 buying seven solid pages in the New York *Herald* (largest paid advertisement for a long while) in which a tiny sales message was repeated thousands of times. "Bonner has made a decided hit in getting [statesman and author] Mr. Edwd Everett to write an article weekly," the Colonel noted that year (6 December 1858). ". . . He says he is receiving lots of yearly subscriptions even beyond his expectations. . . ." The *Ledger* soon specialized in gothic romances in a steady torrent of installments. Bonner also printed segments of his "cliff-hangers" in other journals to draw hooked readers to the main stream. By 1860, with an astronomical circulation of 400,000, the *Ledger* was the largest weekly miscellany in America.

Bonner became a millionaire lover of fast horses and had Hoe to thank in large measure for his good fortune. *The Printer* said it:

> As for long numbers, requiring continued day and night work, perhaps, we know of no equal to these single "large" cylinder presses, on account of their even running, uniform impression, and ready adjustment in case any changes or form liftings are required. . . . Hoe's single "large" cylinder presses, now in use for the *New York Ledger,* have thus been made expressly to order, with the view of using extra rollers all the time, day or night. . . .[51]

Bonner's headquarters became a marble palace. It had one of the finest pressrooms in the country, with four Hoe cylinders and four Taylors. And the phenomenon of weekly fiction attracted many imitators to an already busy field.

Two publications in particular had started years before issuing royalty-free British novels—in some cases entire "triple-deckers"—complete in one issue; they were *The New World* and *Brother Jonathan.* These mammoth newspapers, at first allowed cheap postage rates for their literary content, bombarded the public and provoked countless series of cheap books (paperbacks) by way of retaliation from desperate book publishers. Selling as low as 12½ cents, these weekly literary papers threatened the whole world of literature—even Harper & Brothers reduced its popular Family Library to 25 cents a volume—and bought Hoe presses—in order to compete with them.

New World appeared in 1840 under editors Park Benjamin and Jonas Winchester; pages were 4 feet long and up to 11 columns wide. But this was essentially only a copy of their earlier *Brother Jonathan* (named after Jonathan Trumbull, advisor to George Washington, who personified the United States before "Uncle Sam" became popular). Formed in New York in 1839 by Benjamin and Winchester expressly to pirate fiction from England, *Brother Jonathan* uniquely satisfied a growing public which, thanks to universal education, was just learning to read but could not afford the prices of bound books. Benjamin Day and James G. Wilson, a printing shop foreman, were engaged to print *Brother Jonathan,* and perceiving its popularity, quickly became its owners. The paper continued variously as a monthly and weekly under a succession of editors until newsprint became scarce (1862), when Day retired at last.

George Roberts in Boston, unable to make his conventional *Times* a great success, had, like his friends in New York, jumped into job printing, government work, mammoth papers, and whatever else he could find. But perhaps conservative Boston itself was at fault, and he could do little about that. If nothing else, he still had the warm friendship of R. Hoe & Company, and a colossal practical sense of humor. If people wanted blanket sheets quite beyond the reader's armspan he would give them something they would never forget. There was not yet an end to the novelties of the machine age, and he would capitalize on Hoe's well established reputation for gigantic presses—the "Last Fast" machines of who knew how many cylinders or what speed.

On the 4th of July, 1859, Roberts issued a commemorative publication, "The Illustrated Quadruple Constellation: The Mastodon of Newspapers—Eight Mammoth Pages, Thirteen Columns to each Page and Each Column Forty-Eight Inches Long." The editor-publisher promised to get out the *Constellation* once every hundred years, confident he would retain the record for size. Conceit that his firm would last 100 years, considering his recent financial failings, was perhaps a private joke only his creditors could appreciate. Roberts hoped at least to continue well into the future: "[The editor] thanks God that he was bred in a printing-office from the age of nine years [and had] no doubt that he will die with the printer's harness on his back." [52]

Of the *Constellation* 28,000 copies were printed, sold at 50 cents throughout the United States to a public genuinely baffled at the thought of a press that could print so immense a sheet. This was Roberts' sublime secret, to which the Hoes necessarily were a party:

> For some years past there had been published a popular paper called the *Brother Jonathan,* owned or edited by Mr. George Roberts. Messrs. R. Hoe & Co. had built a large double cylinder press for this paper a short time previous, which was of course duly announced to the public in its columns. On the 4th of July, 1859, the paper issued a holiday edition called *The Constellation* which was a great curiosity, as well as a great wonder, to the general public. This sheet measured 6 ft. by 8 ft. 4 inches, each of its eight pages being 36 x 50 inches, and much speculation and discussion were indulged in about their new monster press that could print such a sheet. As a matter of fact, it was folded to page size and printed on their ordinary press by refolding and running it through eight times.[53]

With the ominous rift of North and South over matters of national importance there would never again be occasion for such journalistic comedy. After the Civil War America—and its press corps—was too grown up for innocent games.

SERENO NEWTON[?]
1801–1836
Courtesy Arthur J. Hoe

3.

Early Machine Shop and Contract Work

SAW MANUFACTURER AND ENGLISH STEEL / NOTABLE OUTSIDE CONTRACT WORK / MIDCENTURY ARMS DISPUTE / TELEGRAPHIC ENTANGLEMENTS / MACHINE TOOLS AND TOLERANCES / TYPOGRAPHIC INTERESTS / GRAND STREET FACTORY

When American pioneers forged westward to face 900 million acres of virgin forest—with 525 different species of native tree—the circular saw quickly joined the traditional lumberman's long saw and the carpenter's hand saw to help clear the rolling land for farms, roads, and towns. Such woodcutting tools in many varieties were crucial in the early building of the nation, and even a citified *New York Times* could wax eloquent over them:

> Saws . . . buzz to-day among the forests of Maine, ripping up the great pines, and forming the lumberman's only music. In the wilds of Texas; along the banks of the Mississippi; on the borders of the [Great] Lakes; up among the mountains of the Delaware; by the banks of streams upon which float the logs of the raftsman; wherever timbered districts repay hard labor with rich returns, where trees grow and are felled,—in among the shipyards of the city—in the joiners' shops,—by the stand of the cabinet-maker,—in a hundred places besides, where work is done and saws are needed,—there is Hoe's saw. . . .[1]

The woodlands of the New World were vast indeed, and furnished ready materials for shelter, furniture, and vehicles.

Benjamin Cummings of Bentonville, New York, is credited with making the first American circular saw for his own mill in Schenectady, 1814. (England had had one dating to 1777.) About 1820 it came into widespread use, coupled to the versatile steam engine, although the cutting edge was not easily mass produced. To achieve speed and efficiency the delicate saw blade had to be strong, flexible, homogeneously made without flaws, cut absolutely round with center hole bored true, teeth angled perfectly and sharpened uniformly—all exacting hand operations. Considering these demands, even with a large amount of saw work in the shop Robert Hoe might have been content to continue purchasing saw blades (as well as files and rasps) from England. Made of fine Sheffield steel tempered in small batches, a circular or carpenter's saw was as critically made as a cavalryman's sword.

In 1828, however, Hoe determined he could fabricate his own circular saws from cast steel. He soon set up to do so, and began selling them as well. The company's early handling of steel for saws consequently acquainted the partners with the metallurgic arts they would sorely need for press building. The process became a fortuitous—if thorny—adjunct to the business:

> [*Matthew Smith to Sereno Newton, Jamestown ("care Gen[l] Harvey"), 29 August 1834*] . . . *We have rec your letter of the 21st containing orders for gang saws &c. The steel I shall order today of Messrs Jackson R*[*iddle*] *& Co.*
>
> *There is trouble in the saw shop; the men are dissatisfied with Nicholls*† *& are impatient for your return as I have told them that nothing could be arranged until you came back. I hope you will be as expeditious as your business will allow, & after remaining here two or three days you can leave the city again. The fact is things dont go right in the saw shop. Nichols says that two of the men do not earn 5$*† *per week (which may be the fact); they say that he throws obstacles in their way & that they have not had an opportunity to do what they are able. Under these circumstances it seems necessary that you should be here for a few days to put things to rights a little. . . .*
>
> *Dunham & Sherriden*† [*Bernard Sheridan?*] *are to work at the mill getting the stones* [*dressed?*] *&c. We expect to have them start in the course of next week. I have taken on a grinder by the name of Ward who does very well. It seems as if the very devil was in our saw makers and it is absolutely necessary that N should be removed. . . .*
>
> *The cholera is about the same and very little alarm is excited. . . .*

Newton obligingly returned to the shop before continuing on the southern portion of his trip. Aaron and Joseph Nichols, however, had developed for Hoe the processes involved in manufacturing the saws. It is not clear which one remained to supervise the work, but he was obviously no ordinary foreman. Yet he was let go.

Jackson & Riddle's domestic (Philadelphia) steel, furthermore, may have been as much at fault as a tough supervisor, and the company's problem was quickly described to Sanderson Brothers & Company of Sheffield, whose agent, Edward Sanderson, had a New York office. Strangely generous in many ways to these struggling yankees who would soon show John Bull how to run a machine shop, Sanderson offered the Hoes the services of John Wheatman, one of their most accomplished saw makers. Wheatman replaced Nichols.

When he returned to England for a brief vacation Wheatman was commissioned to recruit new men:

May 30, 1835

We hereby authorize Mr. John Wheatman to engage for us three saw makers & one handle maker to come with him, upon the following terms, viz.:*

The saw makers we will furnish with fall employment by the piece at the rate of five dollars to the S[aw] which is an advance upon the Sheffield prices for a year or longer. The handle maker will be paid at the same rate as the saw makers, and have full employment for one year or longer.

R. Hoe & Co.

* *Also two saw grinders upon the same terms as above.* R. Hoe & Co.

The rate seems very high, but may indicate the preciosity of this skill. Wheatman brought his men with him in the fall, and although his position was considered temporary, he actually remained many years in Hoe's employ.

Coincidental with Wheatman's authorization was a large order for saw blanks—the first of many—rushed to the Sandersons, who had previously supplied finished saws and files to Hoe.

New York May 30, 1835

Gent.: Please furnish us as soon as possible the following order of sheet steel of best quality. Viz—

12 cross cut saw plates ea 4 + 4½ ft long 15 guage†
24 " " " " " 5 + 5½ " " " "
. . .
20 plates ea 6½ + 7 ft long 15 in wide 14 "
25 " " " " " " " " " 13 "
. . .
10 " " 24 " 72 inches 12 "
10 " " 24 " 72 " 13 "
10 " " 24 " 72 " 14 "
10 " ea 15th 16th & 17th guage 18 by 72 inches
500 lbs 10½ x 26 inches 19 guage
. . .
100 " ea 20, 21, 22, 23 & 24 guage 12 in wide 32 to 40 in long

This last item of thin steel to be of extra good quality. We shall want at least ⅔rd of the above immediately after its arrival & probably the whole of it: and you would greatly oblige us by expediting this order as much as possible.

R. Hoe & Co.

To Messrs. Sanderson & Bro.
New York

Through Sanderson's local office and also directly at Sheffield Hoe was soon buying large quantities of files and rasps as well—flat bastard, hand, hand second cut, square bastard, round bastard. The firm would for many years boast it was using the very finest English steel, obtaining it by frank complaints and unsolicited advice despite the humbling financial arrangement then in existence:

New York Aug 29/[18]43

Dear Sir

I am sorry to be compelled to say that the last lot of [our!] Co's Saw Steel was of a quality so poor, that it would destroy our sales entirely if continued. If you remember I stated to you that the price had been increased, from the time we first used your steel without a comparative increase in quality. Now Lessup & Naylor, Wilson Hawkhurst & Moss are furnishing our competition with an article very good for from 60/ to 70/-. The last lot was not anything to be compared with what we formerly had of you. . . .

If we attempt to heat it (the steel) hot enough to harden sufficiently high it will fly & twist into all maner† of shapes rendering it impossible to hammer a plate straight. It would seem a very great sacrafise† after the amount of money that has been spent to obtain a high reputation & good business for saws that when we see success before us we should find ourselves obliged to relinquish it in consequence of not being able to procure steel of the best quality as low as any other in the trade. I beg you will remember that for the past (at any rate it will be so for a few years) our efforts are made with an eye to your benefit. . . .

In order to increase the sales we have made great discounts to the trade & I think a very large business can be done, for consumption is increasing every day. . . . I think I have made such arrangements at Boston with Saws as to enable us to work up the large lot of long mill saws . . . for the Maine Market. . . . I hope you will not consider me to† particular in urging upon you to furnish good steel at low prices for we must in order to continue making saws make none with our name on of an inferior quality.

In the other department of our business every thing is in a flourishing condition, in fact we have as many orders as we can execute. . . . Please give my warmest and best regards to Mr. James & E.F. S.[anderson] as† also accept for yourself my sincere esteem,

RMH

To John Sanderson
Sheffield, Eng.

[From Richard Hoe, 6 April 1844] . . . We were straightening a bar of your machine steel yesterday when it broke & shewed a triangular hole proving the steel not well made. Could you not use salts in the fire when heating the ingot, the same as we do in welding iron & steel[?] I have sometimes found a bar of steel full of seams, have put it into the fire, & as it was heating threw into the fire occasionally pulverised Borax & common salt mixed with beach sand or lime. . . .

The Colonel included what a local merchant had confided to him:

"We do not pretend to make better steel than Messrs. Sanderson's, but we are sure our steel has a handsomer appearance, & we will warrant it equal to that of any make in quality. We are able to sell it for a trifle less price, because my brothers are practical workmen & are superintendents of our works themselves. . . . Now Messrs S. are rich & you might give us a small order to try our article. . . . You will at the same time help along a young house." . . .

Examining the mechanical problems of sawmaking as he would soon scrutinize those of press-building, Richard had quickly developed a device for toothing saws, almost neglecting to have it patented:

New York August 24th 1837

Dr Sir,

You will oblige me by giving your opinion respecting my taking out a patent for a machine for toothing [cotton] Gin & Circular Saws; the said machine was invented by me two years since & has been in operation for toothing gin saws but not until lately have I perfected it for toothing Cir Saws. But I understand that a machine has been made similar to mine by a person at the eastward who got the idea from a man who left us & went in his employ & I believe has patented it as original. . . .

I expect to be in Washington City the first of next month & shall be pleased to hear from you before I leave.

Yours Respectfully
R. M. Hoe

To/Dr. Thos P. Jones
Washington City, D.C.

The device consisted of a bed on which the saw blank was mounted and which could be shifted (or rotated, for circular blades) the exact length of a tooth; for each tooth position a steam-powered punch descended to cut the tooth shape. "The power required for punching out a solid piece of thick steel for each tooth," said *Scientific American,* "is very great, but this machine—to use a term employed by Dr. Livingstone [of Africa] in reference to the lion's jaws—*crunches* them out as easily as if it were chopping windle straws." [2]

This was shortly followed (1840) by a saw grinder, almost a model of early automation for the interrelationship of moving parts:

It may be described as a vertical iron plate, with raised centre, upon which the saw revolves. A lap [disk], revolving with great velocity against this plate, in a direction perpendicular to its line of motion, distributes a compound of polishing material—composed of a mixture of flint and emery,—upon the surface of the saw, and grinds and polishes it with perfect precision of movement. This is one operation. The second, which is carried on at the same time, is a horizontal motion of the lap against the saw, which motion is so graduated at pleasure that the saw is rendered thicker or thinner at any point. . . . The third process drives the lap laterally along the surface of the saw;

> an operation which insures an equality of finish and an exact gradation of thickness. This "lap," therefore, performs not less than three distinct motions at one and the same moment. . . .[3]

Hoe had solved this problem only for long saws at first (caveat 1839); the circular version was patented 1842.

In the Broome Street saw shop, despite unprecedented mechanization, there were no less than 18 operations in making a saw blade, most of them still hand controlled: paring, drilling, toothing, flattening, filing, hardening, tempering, smithing, grinding, glazing, oiling, marking, blocking, trying, second glazing, second trying, setting, and finally, sharpening. For the long saws, grindstones about 6 feet in diameter weighed two tons apiece; they were imported from England, replaced every 8 to 10 weeks, and trimmed almost constantly:

> Several times in a day, each stone is chipped upon the grinding surface with an adze, in order to bring out the harder portions of the stone, so as to render the process more effective. Great care is exercised in running them, so that they rarely fly [to pieces], and we believe that but one or two instances have occurred in which workmen have been injured.[3]

In 1837-1838 Richard had gone to England to patent the saw-grinding apparatus. He was gratified to notice that all saw-making operations there were still manual, and without benefit of steam power, assuring him that for the vast American lumbering market he would have a very competitive position—both in location and in technology. The trip took place during the fearful 1837 depression and was emotionally charged with get-rich schemes as well as the naive hopes of youth. And the expedition was not without its intrigues:

> [*Matthew Smith to Hoe, London, 15 May 1838*] . . . *Matters move on in the usual way—no improvement, business does not come in, and with the exception of the affairs of the patent rights I have nothing particular to communicate. . . .*
>
> *FOJ Smith sails in the Europa and is determined from what I can learn to set aside all that you have done at the Patent Office. Their plan I suppose (for I know nothing about what course they intend to pursue) is to enter their caveat and obtain a patent for the stone dressing machine and cover in the same the right of grinding and polishing metals. . . . There has been a great stir at the Astor House to day in making preparations for Smith's departure, in preparing documents &c &c and I suppose he is well charged with affidavits stating that you have fraudulently taken that which belongs to others &c &c. . . .*
>
> *May 16th . . . Smith goes out in the vessell*[†] *with this* [*letter*], *and it causes us some uneasiness, still we cannot but think he will be unable to trouble* [*you*]; *you know he is an intreaging,*[†] *calculating yankee, and he will doubtless endeavor to accomplish by trickery and*

chicanery what he could not by open and fair dealings. He probably has letters of the highest political character. . . .

Maine congressman Francis O. J. (nicknamed "Fog") Smith was a force to reckon with. Although he may not even have intended any such conflicting business in England, he was a sharp entrepreneur.

[Smith to Hoe, 31 May 1838] . . . By the Bye should you compromise with FOJ Smith we have no authority from Goodell & Tiffany to resign their contract of the foreign patent and I think it can be done (if done at all) without affecting their contract in any way. . . .

After three weeks in London arranging documents with patent agents William Newton and Miles Berry, 66 Chancery Lane, Richard journeyed to Sheffield where Charles Sanderson gave him a warm reception. Hoe engaged Messrs. Booth's works nearby to make a demonstration saw grinder for the British trade; its fabrication (by an obviously disinterested management) was excruciatingly slow, giving the lad time to ruminate in a diary he had just started:

Friday June 8th [1838]—The time flies & I have not accomplished anything today as Mr. Booths man has not yet returned. As I did not find him I went with Mr. Chas S[anderson] to see Wheatman whom we found & Mr. S left me with him & we went to see some dray horses & then to the Barracks where I saw some good manouvering† of field Art[iller]y thinks I to myself Col of the N.Y. States Arty how strange a figure you would cut to be called on to drill those boys eh! We then went and took dinner. . . .

Monday evening, August 20th— . . . Mr. Chas S has just left me; we have had a discussion on magnetism, electricity & galvanism which we both concluded was one & the same thing. [Dr.] Sherwoods discoveries formed some part of our discussion & I wish I was more fully acquainted with them. Should they prove correct it will create a new era in science—but it is not best to be too sanguine therefore I will say with old Ritchie [Sanderson], "nous verrons."

When long conversations did not fill the time, there were daydreams:

. . . How anxiously my mind turns to the "land of the free" the land of my birth—prejudices & friends—but I shall soon have plenty to engage my whole time & energies in the disposal of my [saw] patent. Bright visions streaked with gold flit across my imagination. . . .[4]

But long after the machine had had ample inspection by local saw makers, it was plain there was no British gold to be got from it:

[Charles Sanderson, Sheffield, to Richard Hoe, New York, 16 September 1839] . . . None of your machines here are raising any interest & I will sell them (and/or patents) if you give me limits. Re "your millsaws—in our order book they are called slitting knives according to your request." . . .

The partners' last hope of striking it rich also evaporated. Matthew Smith, more than the others, had banked heavily on the brainchild of their friend, Dr. H. H. Sherwood, the "Dipping Needle & Naval Geometer," a navigational aid whose world-wide application might have been immensely profitable—even for a fractional share. "Tomorrow," Matthew had written (31 May 1838), "Sherwood goes to Wash with friends to ask Congress for a donation (Congress has just given Morse $30,000 for telegraph '—a mere play thing—') and will hopefully do as much for the needle, which has national importance." On behalf of the firm Matthew had bought a ⅛th interest, thinking it would "realize a large fortune to us. . . ."

Sherwood had gotten a caveat in Washington and Hoe secured rights in England, France, and Scotland. A letter of credit from Sanderson was only for saw expenses, but Matthew urged Richard to call the needle patent fees a "personal favor for a NY friend." Nothing came of the invention.

But the saw-grinding machines were something else. The great American lumber industry was not a chimera for those who would work at it. The company's investment in fees (British and Scottish saw patents alone had cost about $4000—advanced by the Sandersons) and code-named saw supplies were prudent after all. With the circular saw itself developed only a few years before, R. Hoe & Company was proud to reveal its own timely involvement:

> [*To J. Eliot, North Carolina, 5 December 1840*] . . . *We have considerable experience in the use of cir saw*[s] *having an extensive machine shop & Foundry as well as saw manufactory; we have used circulars for 15 years past.*
>
> *The largest circular saw ever made either here or abroad in one plate is 48 in. diameter; of these we have made several; with inserted teeth this plate would make a saw 51 inches. The flanches*[†] *would require to be nearly 12 in. in diameter or a little less than one quarter the diameter of the plate.*
>
> *We prefer inserted teeth only that they are a a*[†] *saving of expense; unless the teeth are well inserted, however, they are dangerous. They should not be wedging where they are inserted in the plate but straight & made to drive in tight without straining the plate; the rivet should be of the best steel & a small hole should be made below the tooth to enable you to take out the tooth easily. We have made saws of this size & inserted from 4 to 16 teeth but from the experiments we have seen & tried we think 8 teeth to be better than more or less. The widest plank that such a saw would be capable of cutting to advantage would be 15 inches. . . .*

The advantage of separately manufactured saw teeth was covered by the *New York Tribune* in 1869, when improved versions had justified themselves by long use:

> A new circular saw lately patented by [Hoe] is so admirable in principle, and so perfect in construction, that its manufacturer might justly claim it to be the best in the world. This saw is made with strong, circular, movable teeth, easily removed or inserted. These are so formed by an ingenious arrangement of groove and shoulder, that when they are turned into the sockets they become as firmly fixed as if they were a part of the plate itself, and this, too, without the use of any rivets or other objectionable appliances. The advantages of movable teeth in case of breakage are obvious, and this saw is so simple, strong, and perfect, that it seems destined to eventually supersede all others.[5]

It had earlier been recognized that the right kind of steel for circular saw blades, functionally, did not hold the sharpest cutting edge. Too, when the expensive saws struck foreign objects, although only one or two teeth were injured, all of them had to be refiled so that the blade would remain balanced. And repairs or replacement in remote lumbering areas were always costly. The solution seemed to be the use of detachable teeth of high quality steel, but various means of fastening them to the blade had not proved workable.

About 1840 a Sacramento sawmill superintendent devised a slotted groove socket which permitted quick and easy changing of uniformly machined teeth. In 1866 (and 1868) patents were granted to Warren P. Miller, the Californian, for improved tooth anchoring methods, and R. Hoe & Company acquired manufacturing rights to them in 1869. These gave the company its important lead in the market.

Miller in 1874 (and 1875) improved his system still further by making it necessary to replace only the tooth's cutting edge, called a "chisel-bit." "These chisel-bits," Tucker added, "are very inexpensive, and when worn out can be removed and replaced by a new set in a few minutes' time."[6] The company in 1875 sent complimentary blades to several mills for their comments, and shortly told all inquirers, "These saws are no longer an experiment, but are used in all sections of the country and give perfect satisfaction."[7] Miller's detachable tooth was assigned to Hoe in 1876 and was among the firm's most important legal properties. By the end of the century Hoe had bought the complete patent from Miller's widow.

In its encyclopedic way *Scientific American* added considerable detail to describe the specialty of saw manufacturing and marketing:

> Gang saws for large mills, single saws, and muleys [a vertically reciprocating circular saw mechanism] for small mills; long whip saws, for pits in shipyards, and cross-cut saws; in fact, every variety of large saws and minor sized saws are manufactured by Messrs. Hoe. The circular saws range from a few inches in diameter up to eighty inches—nearly seven feet.
>
> Circular saws are now very extensively employed for sawing every

description of timber, both in the log and in ripping stuff, such as board and planking. They are more compact and require less space than the long reciprocating kind, and their cutting action is uniform and constant from the beginning to the end of the operation. When first introduced into our country, for sawing logs, much prejudice had to be removed and considerable practice acquired to manage them. They are now coming into more general favor, and possess advantages, in certain circumstances, over all others. . . .

[A] correspondent who had erected several saw-mills in various parts of our country, says of Messrs. Hoe's saws:—"A great amount of time and money has been expended in bringing these saws to perfection, and those who have seen them in operation assert they are not exceeded by any in use. A saw of 54 inches will run 600 revolutions per minute, cutting three-quarters of an inch at each; and when rightly put up, they never heat nor require water to keep them cool like some other saws." The velocity, thus stated, at which these saws can be run without heating, amounts to more than a mile and a half at the edge per minute. This affords proof of their accuracy of construction, and the quality of their material. . . .[8]

Thanks to the company's steady solicitation of outside work, a number of saw inventions and improvements would come to its attention. There were, for example, automatic barrel-stave saws and joiners, various self-contained circular sawmills, and automatic shingle-cutting equipment. In one instance an outside inventor's old patent was deemed of merit: the 1860 circular saw grinder of C. W. Hubbard was better than the firm's own, and a shop right was bought. Hubbard machines built in 1866 remained serviceable for almost thirty years (incorporating 1868 and 1869 improvements by Tucker). In 1867 a customer's cutlery grinder was similarly assimilated: the machine was adapted to grind pairs of circular saws, and became another shop tool useful for years. About 1839 a veneer saw mill and two reciprocating saw devices had been made, the latter of which were for a steady customer and friend, Pearson Crosby.

A portable saw mill was built this year for Mr. Pearson Crosby. It was driven by a portable engine and was intended to be set up anywhere in the forests and moved from place to place as required, the trees to be cut down and sawed into lumber on the spot. Mr. Crosby said there woud be money in this, for it would save the expense of hauling the logs to a saw mill and carting the lumber back to where needed.[9]

Crosby—and Hoe—learned something else:

[*From Crosby, Bangor, 20 June 1862*] . . . [*F*]*or some 3 weeks after I arrived here the water rose so high the mills could not run, but as soon as the water had run down sufficiently I had my sons put in the best gang mill I have ever seen. They run tolably*[†] *well with light feed but on putting on the feed they usually do . . . they would shake, as they call it—make the lumber crooked.* [C]*onsequently I came very near making a failure.*

Swartz the saw maker here came [to] the mill and looked at the saws and said they were hammered to† light in the midle† [and] consequently would not strain sufficiently tight at the root of the teeth. . . . [He] said he could hamer† them so they would work much better so I thought best to let him make a trial. . . . I found them to work so much better I had him hamer them all and since that† I have been perfectly successful—have sawed something over three hundred thousand feet of lumber in a short time. . . .

Now . . . if it is a secret known only to [Swartz] I had better get one of his saws [to] send you so you can see how it is done. It certainly appears to be a very important matter. . . .

In 1858, when a new saw department head (salary, $20 weekly) was needed, Wheatman—himself with new employers—was commissioned once more to send help:

RUSSELL WORKS, KETHAM ISLAND, SHEFFIELD
Manufacturers of Saws, Files, Steel, etc. etc.†

Feby 5th, 1858

Messrs. Hoe & Co.
Gentlemen,
I write this to say that Mr. [John] Pinder, the Gentleman I have engaged for you, to take to the management of your saw business, will sail from Lpool on Wednesday the 10th Inst. for NY by the steamer, City [of] Washington, where I hope he will arrive safe. I think I may say, their† is little doubt, but that he will suit you in every respect. He has been manager at Mess. Horn & Co. for five years, and wich† he [h]as left, to enter upon your service. He is a good workman, & steady, wich, I know, is of much importance, in your large establishment. His experience as manager will (I hope) leave you nothing to complain off.† I feel assured that had I not been so fortunate as not† to have secured his services, I should likely have had considerable difficulty in finding a suitable person. He brings with him his two sons, both steady young men. I hope you will be able to find them work. Mr. P. having the feelings, naturall† to a father, [h]as a strong desire to have them with him. He spoke to me about them, and I told him I thought he might calculate upon you being able to find them employment. I am going with them to Lpool, to see them away, and shall find them money for their passage, some £30-0-0 on your a/c, wich I hope will be right. The Col^{n} did not say, what amt I was to find. I have been guided solely by a desire to serve you, and I hope what I have done will meet with your approval. Mr. P. thinks you should pay his passage out. Well, I do think you should, but I have not told him so. You will see after he [h]as been with you a while, and if he does as well for you, as I hope he will, I think you will pay his passage out. We shall send your Steel out next, or the following week. Give us fair play & no favour, and then we are certain it will not be the last steel order you will give us. I remain

Yours truly
Jno Wheatman

The company had second thoughts, however, telling Wheatman to wait. But with Pinder and family already at sea Wheatman begged the Colonel to honor his word.

The choice was a good one. The new saw manager was not only welcomed at R. Hoe & Company, but a year later he returned to England at the firm's expense; in typical Hoe fashion, however, while visiting relatives he was also burdened with a host of personal matters for his employers:

New York July 25th 1859

Dear Sir

Mr. Sanderson has given his house directions to hand you the amount you require, and has directed them to pay for and ship the Grindstones you have an order for. We find on refering† to our order book that Mr. Wheatman has already an order for back saws; if so you will please not order more.

We are very busy in the saw business and are fearful that we shall lose some of our best trade. The competition is increasing and if it continues we must either loose† business, or make a serious reduction in prices. When you return we can see wherein a reduction in cost of manufacture can be made.

By the last Steamer I wrote my brother in London that you would take charge of certain articles for him and to write you or see you.

I wrote to you about a black & tan Terrier dog, if you find one that good judges pronounce very superior you may bring him. You spoke of Fowls when I saw you last. The Colonel would like to have a trio (one cock, two hens) of each White Pile, Lefton & Earl Derby Game Fowls. All must be warranted to stand Steel; get some one to buy them who you can depend upon and who knows all about them. Send young birds. We want some thing better in every way than we can get here, the cost is no object.

In addition to the stones for which you have an order we want 50 Wickersley[?] Stones, 36 in. diameter and 7 inch full; these are for a manufacturer of shears.

Drop me a line and say when you expect to return. Hope you find your family well and are having a pleasant trip.

Truly Yours
Peter S. Hoe

Mr. John Pinder
29 Brightmore St.
Sheffield, Eng.

At midcentury the company was realizing the great sales potential it had labored for; annual sales in circular sawblades, mill saws, pit saws, and cross-cut saws amounted to $140,000, and still grew. It continued to build special machinery for saw fabrication, including a mammoth 1857 hydraulic press to flatten blades as they came from the tempering furnace (it was not successful). A steady stream of testimonials was comforting:

[From Little Falls, New York, July 1853] . . . Please send me by Express one 40 in. Cir. Slitting saw 1-5/8 in. hole with 26 teeth. . . . I am in great want of this saw and trust you will oblige me by forwarding as soon as possible. The streams are getting low and the season for sawing is short. . . .

[From Delaware, Ohio, July 1853] . . . [Send] two of your best cast steel Mulay saws 7½ feet long; let them be tolerably thick and such as will do credit to your establishment as I shall run them in a section of country where there are a vast number used but where another NY saw is in great demand and generally (I am told) preferred to yours. . . .

[From Archbald, Pennsylvania, March 1855] . . . I want more speed, & am informed that with the Mulay arrangement I can have what I wish. . . .

[From Brookfield, New Hampshire, February 1858] . . . I have run saws for about 15 years and find from practice and experience that your method of shaping the tooth and keeping it in order is the only proper method. I am in want of a saw to saw shingle, and wish to have one of your 32 in. diameter with 24 teeth. A great portion of the timber which I have to saw is spruce and hemlock and some knotty at that. . . .

For these rugged and independent lumberjacks R. Hoe & Company furnished rare patience, quality, and service. It routinely sent saw catalogs to rural correspondents, sometimes puzzling country printers who wanted press data; it reclaimed used blades by "gumming" (deepening the spaces between teeth) and selling gumming machines; it discarded iron gummings as scrap—11,000 pounds to Peter Cooper in 1844 brought $106.37; it offered 62½ cents a pound for unusable old saw blades. The company had its "Eagle" brand as "second quality" if its first line was deemed too expensive; it also employed Dun, Barlow & Company to check credit references and repossess equipment.

In 1858 the Hoes developed a scheme to obtain, from county postmasters, the names of local sawmill proprietors, and got good response. Unsolicited advertising, variously negotiated, often helped:

[From the Saginaw City Times, Michigan, 17 March 1855] . . . In this county are between 40 and 50 steam saw mills, using in the aggregate not less than 10 saws each, of various kinds. . . . Now, then, might not a little advertising here insure to your benefit? I am in want of one of your improved Lead & Brass Rule cutters—price $12. I shall be glad to pay half cash and half in advertising. . . .

The company had for many years been offering a similar inducement in the pages of *Scientific American* and other newspapers:

Hoe's Cast Steel Circular and Long Saws. The subscribers manufacture from the best Cast Steel, CIRCULAR SAWS, from two inches to five feet in diameter. These Saws are carefully hardened and tempered, and are ground and finished by machinery designed expressly for the purpose, and are therefore much superior in truth and uniformity of surface to those ground in the usual manner. They require less set, less power to drive them, and are not so liable to become heated, and produce a savings in timber.

They also manufacture Cast Steel MILL, PITT† and CROSS CUT SAWS and BILLET WEBS, of superior quality, all of which they have for sale at their Ware Rooms, Nos. 29 and 31 Gold Street, or they may be obtained of the principal Hardware merchants in the U. States.

R. Hoe & Co.

Printing Press, Machine and Saw Makers, 29 and 31 Gold Street

The following extract is from a report made by a committee of scientific and practical gentlemen, appointed by the American Institute:

"Your committee are of unanimous opinion, that in the apparatus invented by Mr. R. M. Hoe, for grinding Saws, he has displayed great ingenuity and tact in the adaptation of machinery to the production of results in the manufacture of Saws, which may with propriety be denominated the *ne plus ultra* of the art."

Publishers of Newspapers who will insert this advertisement three times with this note, and forward us a paper containing the same, will be paid in printing materials by purchasing four times the amount of their bill, for the advertisement.

New-York, July 19 [1853] [10]

With the discovery of gold in California in 1849 the American lumber industry had galloped along with prospectors to exploit the richness of the Pacific. The Colonel visited California that year and saw for himself the desirability of opening an agency there. But it took time for the gold fever to subside.

Feby 17th 1859

Dear Sir

We were duly favored with yours of 4th ulto. with order for Saws which is being attended to.

We are getting some Catalogues printed in the Style of the one shown to you . . . & will also send you some printed Show Cards & framed views of our Works which we doubt not you will be able to dispose of to advantage. . . .

We have [on hand] 6 Mandrels, suitable for large Saw Mills, with Cone pulleys &c. They were sent to a party in this city in payment of a debt, and we bought them at a considerable reduction—could you do anything with them[?] . . .

Mr. Kumbel declined to entertain your proposal of an Agency for Belting, but Messrs. Hoyt Bros. seemed favorable to it and promised to correspond with you on the subject. You might write to them explaining your views.

Do you think a gilt circular Saw, hung out as a sign over the door of your Store, would be of any service[?] If you wish it we will send you one.

Yours very truly
R. Hoe & Co.
per A. Prusman

Mr. Geo. Stead
San Francisco, Cal.

Setting up Stead on the Pacific coast was simultaneous with the purchase of the Adams shop in Boston; money would be extremely tight, but Stead was not informed about such considerations:

[To Stead, 24 March 1859] . . . The opposition [to Hoe prices] you allude to is not new. We have had to contend against it for many years and yet have never found it to affect our trade materially, and as the demand for our Saws is greater this Spring than it has been at any time previous, we do not feel the necessity of reducing our Price List at present.

If you think it necessary, we might pay the freight, but if it can be at all avoided, we would rather not even do that.

Your attention to our interests we highly appreciate and thank you for the trouble you have taken to lay this matter clearly before us, and trusting you will still be able to secure all of the San Francisco trade that is worth having, we are. . . .

Within a few years another enterprising Californian was seeking the patronage of San Francisco:

Aug 31st 1863

Gentlemen

We are duly in receipt of your esteemed favors of 6th and 7th inst. We wish to put our goods at a price that will enable you to dispose of them to the trade as you desire; at the same time we feel that you should look after our interest. Our previous connections in your city [Stead] were such as almost to decide us to abandon any similar arrangement. With your house we felt sure that we would be safely advised and that the trade would be more profitable. . . .

We are just preparing a shipment of Saws, which go forward in a day or two.

Yours truly,
R. Hoe & Co.

Messrs. J. Underhill & Co.
San Francisco, Cal.

This alliance too soon ran into difficulties, and a third agent—less knowledgeable about saws, perhaps, but plucky enough to make suggestions—was duly initiated:

[To H. L. Tatum & Co., San Francisco, 15 January 1877] . . . We are fully as anxious as you can be to secure a large share of the saw trade of the Pacific Coast, and shall do all in our power to reach this end.

Under no circumstances, however, can we consent to sending blank saws to have the hole drilled to suit, when ordered. You do not seem *to understand saw manufacturing, otherwise would not suggest this, or to send machinery to make over solid plates into chisel tooth saws. in both cases large and long experience is essential, and in the former the slightest variation in the holes, or springing of the blade, would throw the saw out of truth and render it unfit for use. In the latter, machinery costing at least $4000, a saw shop and men of experience would be necessary, so you see that these propositions are impracticable. What we propose is, to send you a small stock of saws, and will do so, if you state size and give full information regarding those that would be most saleable. In addition to this, we shall shorten the time of delivering on the saws here—in no case to exceed ten days after receipt of order and we are preparing a code where by you can telegraph orders at our expense, giving full particulars. This code will be ready in a few days.*

All our facilities for getting out these chisel tooth saws are now prefected and we can promise promptness in filling orders. We shall also, as soon as back orders are executed, keep a stock of regular sizes on hand. There is an immense amount of work on these saws, which you do not seem to appreciate. . . .

By midcentury over 300 patents had been issued for improvements in saws and saw mills, especially for sharpening and resetting teeth. There was much potential for infringement, and Hoe was particularly watchful. On one occasion Hoe's lawyer made an unusual recommendation:

[*From Low, Smith & Watkins, New York, 24 February 1870*] . . . *It seems to me that the most effectual, if not the only, weapons by which you can defend yourself are the ones by which you are attacked. You can meet advertisement by advertisement, and caution by counter-caution and offers of indemnity to purchasers. I must say candidly that in my opinion your chances of fronting the enemy* [*a sawmaker named Spaulding*] *in this way are better than they would be if you should assault him by an action at law. . . .*

This advice came from Eugene Smith, one of Matthew Smith III's sons, whom the firm retained for a time.

Gradually the competition in saws, as in printing presses, would reach a sensible balance of advertising claims and actual performance. Battles in court became too expensive for the risks involved. Henry Disston, a mechanic, had acquired the small Philadelphia saw shop in which he was employed. His saws were good, and, unlike Hoe, in 1855 he began making his own crucible steel. Hoe nevertheless stood firm (16 May 1878) when Disston seemed to intrude: "Our Mr. Mead saw you about your infringing on our chisel-tooth saw," said the company. "Despite your patents, we still think there is infringement. . . ." The two firms kept their distance.

In 1902 Robert Hoe admitted the saw works was still a highly lucrative branch of the business, that year alone calling for over 6 million of the perfected detachable teeth. He referred mysteriously to the development of the system as a "casual invention of one of the former members of the firm," [11] showing a reporter a piece of tree trunk with an embedded horseshoe sliced clean through by a Hoe saw—at a cost of just a few cents for ruined teeth. At century's end he seemed to consider the technological problems solved.

When first organized in the 1830's, however, the uncertain saw department obliged the Colonel to fight on still another front—Washington. Except for a later request that Congress extend his patent for the Lightning press, Richard had generally not found it necessary to lobby for special legislation. Yet the vagaries of steel's chemistry, the intricacies of foreign manufacture and import tariffs, and the unbridled enthusiasm of a struggling American steel industry showed the Colonel he needed more help:

> [*To John C. Rives, Washington, 28 December 1844*] . . . *When I had the pleasure of an interview with you in Washington, you were so kind as to say that you would present to the Hon^l Mr.* [*James T.*] *McKay a statement of my wishes respecting the article of sheet cast steel of which as you are aware we use in large quantities in the manufacture of cast steel saws. In the Hon^l Mr. McKay's report No. 306 (made 11th March 1844) page 54, cast steel and other steel in bars has a specified duty affixed. Now what I wish is to have sheet steel inserted in the bill with bar steel it being in fact the same article and it was doubtless the intention that it should pay the same duty. In consequence of sheet cast steel not being particularly named we have been subjected at times to much trouble and inconvenience by the different construction put upon the law by the different collectors, at times obliging us to raise the prices of our saws.*
>
> *We do not desire any increased duty upon imported saws but only that the raw materials, cast sheet steel which is not manufactured in this country should not have a duty so high put on it as to be prohibitory in its action and thereby break up this branch of our business in which you are aware we have spent much time and money in perfecting. We have no doubt that Mr. McKay as well as yourself will see the reasonableness of this position and trust that the proposed alteration will be made in the bill should any thing be submitted this session. . . .*

For Hoe (and others) it seemed essential to buy imported steel, as no American shop could reliably make it—a situation that prevailed almost until the 1870's. In 1846 the Colonel asked his friend Samuel Dickinson, then a U.S. Senator, to look into the new tariff bill which still did not distinguish among cast-steel in bars, plates, or sheets. Years later there were similar problems:

> [*To William E. Dodge, Washington, 5 January 1867*] . . . *A petition to Congress signed by the principal Saw Manufacturers in this Country was recently sent to Senator* [*Edwin D.*] *Morgan for presenta-*

tion. It asks that the duty on imported steel for Saws may not be increased, or, if increased for revenue purposes, that the duty on imported Saws may be advanced in the same ratio.

The American Steel manufacturers are making strong efforts to have the duty on imported steel largely increased. Being a comparatively new branch of manufacture, and in view of the present state of the currency, it may need some protection, but unless the duty on articles made from the steel is raised to correspond the result will be that articles now made here to some extent from American Steel will be imported at lower prices. . . .

The Colonel found he was caught between the desires of the developing American steel industry and his own viability:

[*To Dodge, 17 January 1867*] . . . *Indeed it would seem to be the design of these Amer^c^ Steel makers to drive out the foreign article by prohibitory duties which . . . would be injurious to the public interest. The duties on Steel are so expressed as to mystify one not initiated, intended undoubtedly by some persons to cover up the enormous rates proposed. . . .*

The Colonel himself had gone to Washington, troubled by the petition he had signed and yet unhappy (in Robert Hoe's words, 23 January 1867) that "the steel men of the west have made it dangerous ground for any one to invade their projects."

Correspondence with London agent William Conquest would soon demonstrate the circumlocutions of overseas orders for steel:

[*To Conquest, 16 March 1870*] . . . *"The Bed-shoe and Rib" may be a good Machinery-Steel, of medium temper so that it can be filed & planed.*

"The rounds" are ordinary machinery steel.

"The hexagonal" is intended for set screws and should be the same quality as bed-shoe steel.

"The homogeneous" is intended for Column-rules, and should be mild and tough so that it will bend without breaking.

Please ascertain at what price you could procure Cylinders for Hydraulic Presses, 12 in. bore, bored and tinned, and to insure our pressure should be forged from best Charcoal Iron. It is necessary that the interior of the bore should be perfectly sound and free from flaws. You of course will know all about this. . . .

Conquest explored the hydraulic cylinder problem but raised more questions than he solved:

[*To Conquest, 3 June 1870*] . . . *We have had hydraulic Cylinders made of Cast-Steel which do not answer. If forged of steel it is not necessary to make them so heavy as if made of iron.*

An objection to the steel is the extra cost & extra duty, as they pay 45% & the iron but 35%.

In shipping any machinery to us please bear this in mind, for if there is any steel in it it pays duty as though it were all steel. Parts of tools or other machines that are of steel had better be sent under separate invoice. . . .

Gradually, of course, American steel became fully reliable and sufficiently plentiful to obviate further importations, although the Colonel opined more than once that, if R. Hoe & Company were to be truly independent and progressive, it must make its own steel. This the company never attempted.

Throughout the nineteenth century, consequently, the brothers bought iron and steel from a number of firms: Sanderson Brothers & Company, Sheffield, J. G. Pierson & Brothers, Ramapo, New York, Albany Iron & Steel Works, Albany, New York, Messrs. Hotchkiss & Son, Thames Valley, Connecticut, Pompton Steel Works, New Jersey, Crane Iron Company, Philadelphia, Washburn & Moen Manufacturing Company, Worcester, Massachusetts, American Iron Works (Jones & Laughlin), Pittsburgh, Pennsylvania, Benedict & Burnham Manufacturing Company, Waterbury, Connecticut, Chrome Steel Company, Brooklyn, New York.

Quantities of coal to keep Hoe furnaces glowing were enormous—1200 tons yearly, usually delivered by barge at nearby East River wharves. (The company's endless consumption of water-borne raw materials may explain a tugboat named "Robert Hoe," seen in 1885 pulling two barges of Bleecker Street residents on a Sunday outing.) At midcentury annual deliveries were 600 tons of pig iron, 150 tons of bar iron, 150 tons of cast steel. Large quantities of lubricating oils were furnished by Samuel Leonard & Company, New Bedford, Massachusetts, some of which was whale oil. The firm also consumed annually 20,000 pounds of nails, 20,000 pounds of emery, and 134,000 feet of lumber, much of it choicest hardwoods.

The company had always paid special attention in its plant layouts to the constant traffic of wagons laden with materials and the accessibility of chutes and traps to basement storage bins and stalls. But what the firm could do daily with quantities of metal and wood was a greater demonstration of planning and magic. The brothers for a long time—out of necessity, to be sure—did not confine themselves blindly to presses and saws. No machine shop of the day could afford to be that particular.

With three times its former horsepower in the 1830's, and considerably more space in which to work, R. Hoe & Company found it desirable to solicit outside work—admittedly an intrusion in the quaint "apartments" into which the factory was divided. "Still," Tucker acknowledged, "as the new shop was large anything was taken in that was offered and seemed reasonably profitable."[12] Even during the 1837 depression there were 100 hands to keep busy. The decision nevertheless brought a succession of strange and fascinating mechanical experiences into the plant, their variety and complexity a challenge and a lesson to partners and employees alike. For Stephen Tucker especially—charged with making them work, if their original designers could not—they stimulated a fair amount of creative thinking and rethinking.

Among early contract work were 1835-1836 pin-making machines for the Howe Pin Machine Company; a friction calender for smoothing and glossing playing cards; Messrs. Rawdon, Wright & Hatch's banknote press for making duplicate plates from an original steel engraving; a roller embossing machine; printer William Applegate's mammoth two-revolution press (about 1837) for one-sheet posters (when Tucker saw that "dashing actresses" he so admired on huge posters "were nothing but a pine board carved out with a carpenter's gouge," he was "disgusted with the coarseness of the process. . . .").[13]

For Dr. A. G. Hull in 1837 the firm also built a machine for planing wood boards, and for him a year later four machines for manufacturing gimlet-pointed wood screws—an advance over expensive imported screws, but not entirely successful. But as the machine-made screw began to attract considerable interest among inventors, one of them knew exactly how to tackle the problem.

Thomas William Harvey (1795-1854) learned blacksmithing in Vermont, where he was born, and became a skilled mechanic at 20. During years of smithy work in Jamestown, New York, where he built factory machinery, Harvey chanced upon a mechanical device having countless uses:

> I am inclined to believe that I have discovered a new principle in mechanics, or rather a new application of an old principle, to wit: the knee-power press, by a rotary movement produced by the revolving of wheels, and by which a powerful press may be applied centrally upon any object desired in a more rapid manner. It is an immense power. . . .[14]

Harvey calculated he could derive up to 50 tons of pressure with his device, and do it as many as 2000 times a minute. He was apparently unconcerned that it was already at work, although manually applied, in printing presses. He foresaw its use in making nails, spikes, rivets, screws, bolts, pins, buttons, coins, and even bricks and printing types. He came to New York to exploit it.

First Harvey obtained proof of his convictions: about 1830 he brought the concept to Robert Hoe, who made a test model of the toggle action. It worked fine. In Ramapo, New York, he next arranged with Messrs. J. S. Pierson & Sons' Ramapo Manufacturing Company to develop the various applications he visualized. In Ramapo too he met Frederick Goodell, a cloth manufacturer, who similarly appreciated the novelty. Harvey and Goodell returned to Gold Street:

One of the first of the outside machines built was a coining press for Goodell & Harvey. It was the ordinary 'fly press' with screw, lever and balls, and was arranged to punch out the blank pieces as well as to coin them. The frame was of wrought iron so as to be strong and light, for it was said to be destined for the interior of South America, and would have to be transported on the backs of mules. This machine was somehow associated in my mind with the business of counterfeiting.[15]

For the Piersons Harvey developed a brick-making machine; this too Tucker built, finding it good but slow. But these peripheral applications were just a start. Upon obtaining a patent (27 April 1832) Harvey had told a friend, "The knowing ones at Washington have complimented the discovery as being more important than any discovery that has been produced in the last one hundred years. What I shall be able to make out of it I cannot tell. . . ." [16] But suddenly he knew.

About 1834 the Piersons, Goodell, and Harvey combined forces to use the knee-joint to manufacture gimlet-pointed screws. The wood screw (theretofore lacking a self-starting point) with this improvement would be an important hardware item in the building of America. Work led to new patents in 1837, and the emerging Poughkeepsie Screw Company soon became immensely successful. So competitive had the screw industry become, however, that a decade later, when Hoe was again called upon, there was a special element of secrecy:

During 1844-5 some very expensive wood-screw machinery was built for Messrs. Pierson & Co., under the direction of General Harvey. We only made certain difficult parts, which were taken away and fitted into the machines elsewhere.[17]

Active in the New York State Militia, Harvey had been commissioned a major general in 1833, but his discovery won him a more unusual title. His son, Hayward A. Harvey (inventor of "Harveyized" steel battleship armor), recalled:

The invention created a great deal of inquiry and interest all over the country; so much was this toggle-joint used, so identified was the inventor with it that he was known for years as General T. W. Toggle-Joint Harvey. . . .[18]

General Harvey performed pioneer experiments in electricity, drawing both Henry and Morse to his laboratory. He also developed iron- and steel-making operations in Putnam County, New York, but severely injured 1853 in a railroad accident, he died a year later. He had long been a friend of Newton and the Hoes.

Despite the slowness of business during the depression, another get-rich scheme materialized—and vanished. Dr. Alexander Jones, a versatile physician and tinkerer, had set up shop in New Orleans and generated a slew of ideas. One in 1834 was a way of transferring a freshly printed page to a soft iron plate which could be etched with acid, and used for printing. This he called "Stereotype metalagraphic† printing." Within a few years he turned to something a little more substantial:

> [*Alexander Jones, New Orleans, to R. Hoe & Company, 12 April 1837*]
> *... My Lithographic establishment is getting along pretty well. It has been a little put back, like every thing else, these d—nd crooked times, but it yet yields a pretty fair business, & would do much better, provided I could get honorable journeymen to work for it. That fellow Mooney, whose passage I paid out from N.Y. and advanced him money when he got here, feigned himself sick soon after landing, and after writing two circulars, cut [away], and returned to N. York in the Kentucky, the same ship that brought him out!! ...*

Jones also invented a cotton gin which used up to 80 circular blades, making its production in quantity an attractive assignment; it was so promising, in fact, the partners bought a half interest. For some reason, however, they apparently could not tool up for it:

> *New York May 20th 1837*
>
> *Dear Sir,*
>
> *Yours of 7th inst. is at hand and has recd from us that attention which its importance demanded. We have tried to devise ways & means to go on with the gin but so long as these times continue we cannot. You must be aware of the impossibility of getting any facilities for carrying on business at the present time. We have been compelled to discharge all but a few of our men and even now we find the greatest difficulty in collecting sufficient to pay them off, while we continue to meet our payments as they fall due....*
>
> *Any loss from the delay which you may experience we will bear with you as our interests are identified in the Gin. This we can arrange at any future time. ...*
>
> *You may depend upon every exertion being made on our part to advance the interests of the Gin. ... Nothing, we must again assure you, but absolute necessity, could have prevented our executing your orders, more we need not add & remain*
>
> *Very truly*
> *Your Obt Svt,*
> *R. Hoe & Co.*
>
> *To/Alexr Jones Esq.*
> *New Orleans*

Dr. Jones' new interest in communications and journalism quickly propelled him into activities he would soon perform with delight.

Among continuing routine commissions were a stone-dressing machine; a press for compressing medicine pills; a device for drilling holes in stone for blasting purposes. Then Tucker developed a hydraulically operated bookbinder's standing press. Early models had leaked at pressures above 5000 pounds per square inch, but then the wrought-iron cylinder castings were lined with copper. The results were gratifying:

> They would withstand a pressure of 10,000 pounds per square inch with safety, and that then became the standard pressure. The firm has continued to build these presses for many purposes, with rams from 4 inches to 20 inches diameter. . . .[19]

The standing press, for smoothing pages printed damp, was obsoleted in the 1880's with the advent of dry presswork.

Next a wood-floor planing machine for Hervey Law, later a prolific inventor of bookbinding equipment;* an automatic lathe designed by employee James Blair to make patent medicine boxes; a music-printing machine. Then followed a rare manufacturing opportunity the company let slip away.

A young Virginian, Cyrus H. McCormick, had in 1840 started making agricultural reapers of his own design in his father's smithy. He was soon ready to subcontract:

> *New York Jany 6, 1843*
>
> *Dear Sir*
> *Your esteemed favor of 25th Decr is at hand in which you wish to know at what price we will make the blades for your reaping machine. We think that if we make them in the best manner & from the best material (as we should do if you direct us to make them) they will be worth $4.38 each and we do not think we could make them for less and earn a living profit from the business. Should you wish any made at that price we should be glad to receive the order through Mr. Thomas R. Blair or any other respectable house in Richmond, V^{a}.*
>
> *Your mo obt svt*
> *for the proprietors*
> *Robt Hoe*
>
> *To Cyrus H. McCormick Esq.*
> *Steeles Tavern, V^{a}*

McCormick countered with $4.00, but the Hoes declined.

An interesting assignment from England similarly involved much correspondence and very little machine work: a riveting machine, for which Hoe had a customer. Beginning with shops in 1817 at the foot of Beach Street, New York, brothers William and Governeur Kemble consolidated their enterprise in 1838 at Cold Spring, on the Hudson River just opposite West Point. Their "West Point" Foundry thrived with ordnance work, for which they needed the latest heavy-duty tools. As American "agent" Richard Hoe built a toggle-jointed riveting machine from specifications furnished by William Fairbairn, Manchester, England:

> [*To Fairbairn, 16 October 1843*] . . . *After a long time & very great trouble & expense, I have succeeded in getting the Rivetting*[†] *machine*

*See *Books for the Millions.*

into operation; but although our boiler makers acknowledge the advantages they are unwilling to pay for the right so large a sum. In fact, before I could succeed in getting the one I purchased of you, into operation, I was obliged to sell it, for the small sum of $600—Six hundred dollars being less than it cost to build. . . . If you would consent to have me remit you say £50. fifty or not to exceed 60£ sixty pounds sterling for the right to use one machine I really believe I should be able to remit you more. . . .

Fairbairn, an influential member of Britain's engineering community, engaged in iron ship building. He approved Hoe's approach:

[*To William Kemble, 29 November 1843*] *. . . I have received an answer to my letter from Mr. Fairbairn & am prepared to sell you the right to make & use the patent Rivetting*† *Machine for Five Hundred dollars. . . . Mr. F. states that they are in general use throughout Europe & Stillman* [*Novelty Iron Works*] *told me a few days since that they should soon wish another. . . .*

Through 1844-1848 the firm made other novelties: a large paper calendering machine for E. J. Bartow & Company, New York; a street-sweeping machine which ran well in the shop but failed on New York's badly paved streets; a cylindrical ink mill; a locomotive with extra rack and cogwheel for mountain climbing; small ink fountains for New Yorker George F. Nesbitt to print several colors at one impression (and for him in 1852 three special presses to print U.S. postage stamps on envelopes); a fan blower for Peter Cooper's iron works; a press for punching holes in rails for the Harlem River and New Jersey railroads; a device for twisting telegraph wires into cable; sets of cooking kettles for public institutions; ruling machines for George F. Cooledge (press supervisor of D. Appleton & Company). During these years Henry Thompson became a frequent supplier and customer.

Henry G. Thompson (1818-1903) as a young man ran an iron foundry for Peter Cooper in Baltimore. Recommended to Hoe by Cooper, Thompson in 1844 was commissioned to forge six iron slabs which were found to be defective—and hopelessly late for work in hand. The lad apologized (20 January 1845): "Our Foreman came near losing an eye, our Coal gave out & we could not get a supply, & finally the breaking down of the Engine completed our mishaps. . . ." Thompson nevertheless furnished large quantities of "Catocton" iron with which the Colonel was fully satisfied.

In 1847 Tucker built for Thompson a lattice-work iron bridge 20 feet long which proved stiffer and stronger than any solid beam bridge of the same weight. Thompson next approached the company with plans for a revolutionary rotary steam engine he wished built. The Colonel advised caution:

[*Richard Hoe, Paris, to New York, 24 May 1849*] . . . *I hope you will be extremely careful in making a bargain with Mr. H. G. Thompson, not to take the job too low; it is much better not to do it than to do it without making as much as we could make by other work in the same place. . . . It is possible that the Rotary Engine may be made to work, but I think it will not be[†] until he obtains a Rotary Motion for his "Sliders" or valves. However he has exhausted all his wit to ascertain if I would endorse his Engine or not, & I have invariably evaded giving an opinion. . . .*

The Hoe prototype actually performed so well, said Tucker, there were expectations it would "entirely supersede the ordinary reciprocating engine." [20] Thompson then contracted with the West Point Foundry to build one for a Hudson River steamboat, but his finances failed and the project ended.

He was not discouraged. Although a prolific inventor, Thompson was also a manufacturing genius who could turn almost any idea into a salable product. After the Civil War he returned to Connecticut, where he collected patents for book-sewing and wire-stitching equipment, on which he created a large business. He acquired David M. Smyth's first book sewer as well as early wire staplers developed by Louis Goddu and Henry R. Heyl.

Years later R. Hoe & Company, newly interested in on-press periodical binding as much as newspaper folding, made inquiries. It wrote Thompson's Automatic Book-Sewing Machine Company, Milford, Connecticut, but the entrepreneur made no concessions:

[*13 March 1877*] . . . *We have no Book Sewing Machines on hand to dispose of. We have Book Stitching Mach[s] which use wire in place of thread as pr Circular enclosed* [not present]. *We do not make any discount from our regular prices to any one. You can see the Wire Stitching Machines at Harpers & the BS mach[s] at Appletons. . . .*

That same year the Singer Manufacturing Company, New York, had also been investigated but found wanting (8 January 1877): "[Our customer, who ordered one of] your Book Binding Machines some time since, is fearful that it will not answer his purpose and prefers to wait until they have been more thoroughly tested. . . ." Both Thompson's and Singer's book-sewing equipment were largely superseded by Smyth's vastly superior 1879 machine.* Thompson later specialized in metal-cutting saws and stapling machines of all description.

In the competition for oceanic steam transportation to match the speed of the clipper ships, one enterprising businessman turned to Hoe for help. The 1847 assignment was Sunday afternoon diversion:

* See *Books for the Millions.*

> During this year the firm made a model of a steamship for Capt. E. K. Collins, to determine the relative efficiency of paddle wheels and screw for propelling a vessel. The boat was about six feet long and the machinery was driven by strong flat coiled springs. We had many trials in a little lake in the upper part of the city. . . . But the paddle wheels invariably beat the screw, and the result of these trials was that when the Collins Line of Steamships were built they were propelled by paddle wheels.[21]

Collins' wooden sidewheelers were costly to build and operate, but they were fast and popular. The first, the *Atlantic,* sailed in 1850. R. Hoe & Company reciprocated by giving Collins—actually the New-York and Liverpool U.S.M.S.S. Co.—some business, notably an 1856 group consisting of William Conquest, James Blair, and six feed boys for Edward Lloyd's London printing establishment. Curiously, the Colonel did not care for Collins' sidewheelers, and the Hoes in their many Atlantic crossings avoided them.

Apart from Hoe's own 12 horsepower steam engine built by Sebastian Iron Works, New York, the company became sufficiently interested in manufacturing its own engines for sale, particularly to drive its new cylinder presses. It was tricky business: so many boilers had exploded mysteriously—often with appalling loss of life—this new source of energy seemed beyond even engineers' comprehension to control it. A congressional commission appointed to study the matter in 1839 bought of R. Hoe & Company one "water guage[†] to order" (price $176.35) and began making pressure tests.

About 1843 the company adopted a new pattern steam engine, perhaps reflecting the committee's findings. With a pump and a governor it was, said Tucker, "a very complete, compact and strong engine." [22] The first, 10 horsepower, went to S. W. Benedict & Company, New York printers, and many others, from 5 to 25 horsepower were built in the two or three next decades. "Steam Engine building is a very small branch of our business—in fact only an appendage to our printing press departments," an inquirer was told (3 May 1850). Yet at the same time an estimate was furnished to another client for a $1200 engine and flywheel with the comment (25 July 1850), "The N.Y. Herald, Sun, Tribune, Bible Society, Tract Society & others have these engines." The firm nevertheless eventually abandoned this business.

For the California gold rush Hoe made an ore-washing machine and a gold-separator. It built a hydraulic press for construction work; a rolling mill and punch press for coining money (strong and light for mule transport into Mexico); envelope gumming and folding machines; a screw press for making spaghetti; a hydraulic testing press for Peter Cooper capable of 350 tons pressure; a printing and gilding press for stamping hatbands; sugar packing machines; a hydraulic press for extruding lead pipe; a mechanical candle maker; machines for cutting and curving pen nibs; a rotary press for repeating trademarks on rubber sheeting; a thread-cutter for machine screws; a saw for cutting roofing slate.

And more: cardboard and paper processing machines for framing pictures and making tags and labels; a vibrating paper cutter for trimming playing cards; a press for printing cigar box covers; a rig for drilling oil wells; a pantograph for photoengraving work; a hydraulic press for the American Phototype Relief Printing Company, Philadelphia, capable of 1270 tons—the most powerful the company had made up to that time (1871). But the outside work was coming to an end at last.

The Civil War nevertheless demanded full attention:

> The war of succession had now commenced and business of all kinds was completely prostrated except for war supplies and materials, and the firm took hold of anything that seemed profitable.[23]

In the United States the production of armaments was generally situated in army and navy arsenals which seemed fully adequate to develop and produce the tools of war. Springfield Arsenal, Springfield, Massachusetts, the country's foremost manufactory, was designed by Eli Whitney to mass produce rifles with machinery—and an assembly system—superior to anything in Europe. There was little inducement on the part of ordnance officers to ask civilian firms to assist—an attitude few businessmen fought successfully.

R. Hoe & Company, despite its extraordinary manufacturing capabilities, did not solicit government work, and only routinely accepted civilian ordnance assignments as paying jobwork. A brass 3-inch howitzer had been cast in 1838 "for the United States Government," said Tucker, although its designer was probably not officially endorsed. A "centrifugal" gun made about a year later featured two long hollow shafts containing balls revolving at high speed; upon triggering they flew out the barrel. "It was said," wrote Tucker, "the balls were discharged with so little force in comparison with a musket ball that the device was useless."[24] Again, in 1857 Hoe built a small steel breech-loading rifle, 1.25 inch bore, but not officially sponsored. These were years of phrenetic gun development by the great European powers, but the Hoes were not really interested.

R. Raphael (Rafael Rafael y Vila, 1817-1882) was born in Barcelona and apprenticed to a printer and engraver. He came to the United States about 1836, running a small newspaper, *El Eco de Ambos Mundos* ("Echo of Both Worlds"), in which capacity R. Hoe & Company must have met him:

New York Dec 19, 1842

Dear Sir
Our price for a 12 horsepower engine with boiler & fixtures complete is two thousand dollars. It will take from 450# to 800# of coal per day of 10 hours, depending upon the work it has to perform. We will furnish necessary directions for putting up with drawings without any extra charge and also for running the engine so that one man of little experience would be able to attend it. On an engine of this size it is customary for the engineer to make and attend his own fires without assistance. The engine & all the fixtures shall be made in the best and handsomest manner. It will take us about 2 months to complete it after receipt of order.

Yours truly for the proprietor
Robert Hoe

To R. Raphael Esq.
New York city

In 1844 Raphael removed to Mexico City where he found (disagreeable) work with an old printer named Cumplido. He quickly hoped Hoe would let him organize a dealership, but Robert (18 January 1845) was justifiably skeptical: "[T]he accounts from Mexico at this time represent Mexico in such an unsettled & revolutionary state that it would not be prudent for us to propose such an enterprise. . . ." Perhaps to ameliorate this uncompromising reply he sent Raphael a Hoe price list and a small editorial assignment:

> [*19 March 1845*] . . . *We shall be glad if you will translate our list of prices into the Spanish language for us, as corrected herewith. We wish to print them in the Spanish language and we of course would like to have the technical terms well defined and the whole done in the cleanest & Most correct manner. Whatever you may charge us therefore we will be happy to pay.* . . .

Raphael had retorted (14 March 1845), "This country is always agitated, but what has that to do with printing materials? . . . If you repulse my offer . . . I will associate myself with some one else. . . ."

By this time, however, Raphael had started his own printery; his woodcutting skill and his fine presswork won him good literary and religious jobs in the capital. He also produced *El Universal,* a conservative daily. Robert Hoe was later (wrongly) more cheerful:

> [*13 March 1845*] . . . [*A*]*lthough we hear that affairs in Mexico have assumed something of a settled state yet the passage of the act to annex Texas in to the United States confederacy has alarmed many of our more peaceably disposed citizens and as the present Mexican minister has demanded his passports & rec^d them it is thought that war may ensue: although for my own part I don't feel particularly alarmed for the safety of any of your warriors or ours. It will probably be a paper war which is often times very fierce and terrible but generally unattended with any blood shed.* . . .

The war itself called for little from Raphael or Hoe (although much promotional copy in World War II suggested otherwise); the New York shop of course prospered with the nation's increased printing requirements. Raphael in 1851 commissioned Hoe to make a unique press, possibly of his own design; a solicited statement must have functioned in lieu of a Mexican patent:

> *New York October 28th 1851*
>
> *We certify that we are now making a machine for Mr. R. Raphael of the City of Mexico to print different colors of ink at one impression which we believe to be original with him as we know of nothing like it in this country.*
>
> *R. Hoe & Co.*

Repeated political demonstrations on Raphael's part, however, resulted that same year in his exile from Mexico. He left his business behind and relocated in Cuba, where he published *La Constancia* and *La Voz de Cuba.* He would return, nevertheless, to his fiery Mexico of guns and presses. Hoe meanwhile found other ordnance jobs.

Gilbert Smith, of Buttermilk Falls, New York, was one of those civilian gunsmiths who persisted beyond a single invention to hope for generous acceptance by the federal government. In 1847 Hoe manufactured for him 4 cast-steel rifles of large (1 inch) caliber; these were apparently conventional shoulder weapons mounted on tripods for sharpshooting. In practice, however, they were judged too cumbersome, and forgotten.

For Saltus & Company, Staten Island, Hoe about 1847 cast a crucible steel 3-inch cannon to be test-fired against conventional brass ordnance. Twelve years later Hoe cast two more steel howitzers, one 2.5 inches, rifled, the other 3 inches, smooth; Saltus' tests went on. Francis H. Saltus was awarded the French Legion of Honor and other decorations for his pioneering rifled steel ordnance.

In 1855 Hoe's old customer Raphael, now functioning as Mexican Consul, was directed by President Santa Anna to purchase armaments. Threatened by insurrectionists in the southern hills, Santa Anna needed cannon. Raphael turned to R. Hoe & Company as perhaps his only industrial connection in the United States. Peter Hoe communicated with Gilbert Smith, who proffered his patterns for gun and carriage, perhaps earning a small royalty for their use. Hoe produced 25 steel breech-loading rifled Smith cannons, bore about 1 inch in diameter, mounted on regular field carriages. They were apparently delivered. Old and weary, however, Santa Anna was reluctant to face his foes at Acapulco; after desultory frontal attacks and severe adverse weather he fled to Havana, abandoning the presidency—and the Smith guns. Raphael fared better. Slated for a place in Maximilian's government, he was back in New York in 1861 attempting to influence newspaper opinion in the emperor's favor.

Beyond the experimentation stage and yet not suitable for widespread adoption, the Smith guns had apparently been manufactured by Hoe only as a concession to Raphael. Such modest orders nevertheless encouraged Smith to continue his work:

> [*Gilbert Smith, Buttermilk Falls, to Peter Hoe, 9 February 1858*] . . . *I feel well pleased with my prospects at Washington. I am sure of orders from the Army and Navy departments as soon as Congress appropriates the necessary funds. I have commenced making experiments with a four-ball cartridge to be used in the regular service rifle. When I got in Washington I learned that Secretary* [*John B.*] *Floyd had given orders at the arsenal yard to have experiments made with this kind of ammunition, and they have been at it for some time, but with no success whatever. I think I have got hold of just the thing he wants.* . . .

Smith, alas, like so many disgruntled civilians during the Civil War, was given no encouragement, however promising his ordnance.

Springfield Arsenal, committed to a superb rifled shoulder arm, had already developed automatic machinery to make it in quantity. General James W. Ripley, U.S. Army Chief of Ordnance, quickly put Major Alexander B. Dyer in charge of the Massachusetts factory despite its earlier traditional direction by civilian heads. Dyer, a "pugnacious, red-headed Virginian," tore into his job.*

Although the Springfield shop was turning out 10,000 rifles monthly by January 1862, that figure was still woefully short of requirements. Ripley had instructed Dyer in 1861 to have more rifling machinery made, and R. Hoe & Company was apparently recommended for that equipment by an old friend, James T. Ames.

Ames (1810-1883) succeeded his brother Nathan as head of the Ames Manufacturing Company in 1847 and actively directed the firm until 1874. Established by his father as a smithy, the shop had relocated at Chicopee Falls, Massachusetts, in 1829, and fabricated fine cutlery and other edged weapons. Making swords for the U.S. Army and Navy and (from 1836) bronze cannon as well, the Ames organization prospered. Under James it also built turbines, toolmaking and cotton-processing machinery, sewing machines, bicycles, silverware, and statuary castings. The Ames brothers became the best known master mechanics in the United States and their work received the highest awards for design and workmanship.

The large Ames family was long intimately connected with Springfield Armory, which had been founded by David Ames (its first superintendent) under President Washington. Hoe had long dealt with several branches of the family, as early as 1844 buying slab iron from Oliver Ames (nephew of David), Falls Village, Con-

*See *Age of Great Guns*.

necticut. The Colonel bought tool-making machines from James Ames who, during hard times, gave very generous terms:

New York September 5th 1837

Gent.

We have again to ask you to renew one of our notes in part, the one which comes due on the 19th inst. for $1167 76/100. Every thing continues so very dull here that it is almost impossible to make any collections and we are reluctantly compelled to ask this favor. Please find enclosed our note for $850 the proceeds of which we trust you will forward in time to meet our note. Let it be in season—a delay of a day or two might seriously incommode† us.

As soon as we can get time to look into your account, say in a day or two, we will forward the notes as you request.

Very Respectfully Yours,

Messrs. D. & J. Ames *R. Hoe & Co.*
Sringfield, Mass.

[To D. & J. Ames, 21 November 1840] . . . It is our earnest desire to reduce our obligations to you as fast as we possibly can. We have already reduced it to a much less formidable amount than it once was. It is our purpose this writer, at all† wants to pay at least 10% and in the spring we have every reason to believe we can do much better. We had hoped that you would be able to give us orders for saws and other things of our manufacture and reduce the amount some considerable† in that way. We proposed this month to pay the note of 139 66/100 due 28th which would be more than 10% on the note of 1073 45/100 due 25th inst. and to renew the latter note which we hope you will accede† to. . . .

These nagging debts elicited the most apologetic correspondence of all Hoe obligations:

[To D. & J. Ames, 1 December 1840] . . . In urging the payment of our notes you must bear in mind that we cannot do more than we can. At present it is entirely out of our power to pay more than what amounts to 10% upon these notes. The parties who equally bound with us on these notes have failed and the whole burden & loss has fallen upon us. It is a hard case all around and we should be very glad to pay them at once and have them out of the way; a good six months business will enable us to accomplish it and we think before next summer they will be out of the way. At this season of the year business is slack and you must take that into consideration. . . .

Upon the national urgency of the Civil War Ames, who owned the patterns for the Springfield rifle barrel boring machine, induced the Hoes to make the additional machines and parts. Other firms in the North had already expressed their desire to produce rifles for the United States government: among them were James Ames himself, Hoard & Son, Watertown, New York, William Mason, Boston, Spencer Rifle Company, Connecticut. They needed the Hoe boring equipment fast.

The company, however, was plagued by tooling-up problems:

New York February 28th 1862

Mr. E. S. Allen
Master Armorer
U. S. Armory
Springfield Mass.
Dear Sir

We send to day per express the 6 composition nuts (9 of copper to 1 of tin) for your Barrel Tapping Machine. Although the composition nuts furnished with the machine may not have been of the above proportions, still they were of good metal & should have lasted much longer than they have. Is it not possible & probable that there may be a very slight difference in the pitch of our lathe screws & yours, so that the thread on your taps does not quite agree with the thread on the tap spindles[?] We mentioned this to Mr. Porter when he was here last week & he said at once that [it is] quite correct per inch, but he was uncertain whether it is too long or too short. The lead screws of our lathes are cut from a screw made expressly for this purpose by J. Whitworth & Co. of Manchester Eng. (the best toolmakers in the world) and is said to be correct. We think it would be advisable to put the tap spindles in the lathe you cut your taps in & see if you can detect any difference between the thread on them & on your taps, & if so the spindle can easily be cut to your thread & smaller nuts put on them.

Will you have the kindness to send us the drawings of the Tapping Machine as you proposed some weeks since. We will return them immediately. Mr. Porter when here spoke about making the slide or plunger of the small Punching & Shearing Machine of wrought iron instead of cast iron & also said something about a new drawing to be sent. Please advise us at once if any change is to be made.

We are making among other things for Mr. Hoard, 4 Clamp Milling machines for gun work. Will you please inform us if you think it desirable that any of them should have two driving pulleys so as to run backwards & forward, & if a friction clutch is preferable for this purpose to an ordinary clutch.

We hope to secure the patterns a memorandum of which was left with you by our Mr. P. S. Hoe. We have also an order for Iron Profiling Machine. Locks, Cone Seats & Triggers, which was† omitted on the memorandum, we would like patterns for these also. We are anxious to get these machines all under way as soon as possible and shall be pleased to hear from you at your earliest convenience.

Very truly yours,
R. Hoe & Co.

[P.S.] *You spoke of extra heads for Rifling machines; please send your order.*

Tucker recorded this assignment all too laconically:

Thus, during 1861-2 the firm built about sixty machines of various kinds, for working in both metal and wood, such as turning, boring, milling, stocking, porfiling, punching, squaring, tapping, threading, rifling and other machines, the drawings and patterns for which were furnished by the U.S. Armory.[25]

Much of this work was billed through Hoe's Boston plant, where some casting and final assembly were performed. In addition to the $1500 boring machines, Hoe made a large number of rifling rods, the borer's actual cutting tool. In March 1862 the company sent its final invoice to Springfield Armory for shipments "at sundry times" of rifling machines amounting to $33,306.13. Dyer exploded. He replied in great agitation, obliging Hoe to explain.

March 15th 1862

Dear Sir

We are duly in receipt of your favor of 13th inst. and regret that you deem our charge high for the Rifling machines. As you are aware we declined to fix any price, but offered to make them at [cost?], a copy of which was furnished Mr. Ames, to which 20% was to be added, and these terms were accepted, with the approbation of those in charge at the Armory. We stated that we labored under some disadvantage in not having had previous experience in these particular machines, that the quality of the work would be superior and that our price would no doubt exceed what you had been paying.

We should also here state that the cost has been much increased by working over time, as you directed to expedite the work in every possible way. The men were paid, as customary, more than for the same number of hours, beside continued night work is done at some disadvantage† disadvantage. Every economy has been used and no waste has been made. The account rendered to you was made up at precisely the rates & per cartage agreed upon. The quality of the work on examination surely will speak for itself. We will furnish bill of particulars to yourself or to Mr. Ames if desired. One of us will come on early in the week and will be happy to furnish any further information desired. We remain

Very respectfully Yours

R. Hoe & Co.

To Major A. B. Dyer
U.S. Armory, Springfield, Mass.

The Colonel would not budge.

To James Ames the company quickly described the delicate situation, necessitating the admission that Hoe was in an uncomfortable position—a position that, ironically, it had shortly before similarly found intolerable in a subcontracted order for British presses:

[*To James Ames, 11 April 1862*] . . . *The quality of workmanship is very superior and will undoubtedly give great satisfaction. Every economy has been used though of course we could not make this machinery as cheaply as if we had been fully prepared for this particular work. The cost has certainly somewhat exceeded our own expectations but the amount is just. . . .*

Dyer was not placated by Ames' intervention, and arbitration was recommended; Hoe, Ames, and Dyer each to appoint a referee. The Colonel quickly named Tucker and proposed a meeting at Massasoit House, Springfield, 7 May 1862 at 2 o'clock. The referees could not settle the matter, however, one of them (Ames') still cogitating two months later:

Springfield June 27, 1862

Messrs. R. Hoe & Co.
New York
Gentlemen
I regret exceedingly that I am still unable to make a decision in respect to the machinery matter, and as the subject stands now, since talking with your Mr. Robert Hoe, I am of the opinion if he and Mr. Ames should meet, that they would agree upon some basis of settlement or proceedings, and to this end I have informed Mr. Ames that I should address you, he holding himself ready to meet you here, or will arrange to meet you in New York the early part of next week. I shall not change my position as one of the referees until I know what is the result of this suggestion, without again meeting Mr. Tucker and Mr. Kendall. Your early reply at Boston will much oblige

Yours mo truly
John B. Taft

It is not clear how the dispute was settled—as in January 1863 Dyer was still promising to call and talk about it. In May 1862 Hoe had already returned the drawings and patterns to Springfield, but patriotism notwithstanding, the firm stood fast.

Following the rifling equipment were presses for making commercial ammunition. As these were Hoe inventions, Tucker had something more to say of them:

Early in 1862 the firm built two powder compressing machines, one for the Hazard Powder Co., and the other for the du Pont Powder Co., for pressing cannon powder into cakes to facilitate loading the guns, which at that time were all muzzle loading. The machines were built from our own designs and . . . would [each] turn out from 30 to 40 cakes per minute, from 3 to 6 inches in diameter. . . .

Another machine fashioned by Tucker did not work as well:

A small cartridge machine was built for the du Pont Powder Co., for compressing a charge of powder onto the base of a conical rifle bullet. . . . The base of the bullets was coated with a glutinous substance, and as the disk of the machine revolved they were dropped, point foremost, into conical holes. . . . The fatal defect of this style of cartridge was that the powder could not be made to adhere. . . .

For another group of contractors Hoe furnished a variety of hydraulic presses. For processing army rations the company made two large machines for compressing dessicated meat and vegetables to a fraction of their bulk. A. W. Denmead & Son, Baltimore, ordered two 12-inch cylindrical presses for forming sheet iron into navy Monitor hulls and turrets. Messrs. Cornell, New York, and Niles Tool Works, Cincinnati, also took two each of these. Along with them another hydraulic press was useful:

> Mr. R. P. Parrott, Cold Spring Foundry, ordered two 6-inch hydraulic presses and pumps for forcing the conical shot and shells for Parrott rifled guns through cutting or shaving dies to make them suit the bore of the guns. This was a very expeditious and successful operation, for the projections that enter the rifling grooves of the guns could be left on the projectiles.[26]

This Parrott ordnance was very effective in the Civil War—prompting wags to suggest the parrot replace the eagle as national symbol. The Cold Spring or West Point Foundry occasionally furnished castings to R. Hoe & Company.

In July 1862, with the Springfield matter still fuming, Richard had been approached to lend his presence to a grand war rally in New York's Union Square. On a sunny Saturday 100,000 spectators came to hear speakers at five massive stands swaddled in tricolor bunting and banners. Stand No. 1, Mayor George Opdyke officiating (and William F. Smith a "secretary"), was sponsored by a host of city luminaries, several old Hoe friends among them—Horace Greeley, George Putnam, Cornelius Vanderbilt, J. Pierpont Morgan, James Bennett, and Richard himself. Few of them were present; the Colonel certainly was not. But then, so much of the most routine company work—in peace or in war—was in the national interest: across the nation Hoe's presses spoke eloquently even when their designers were silent. The genius of another old friend had helped.

Samuel F. B. Morse developed his telegraph apparatus on New York's Nassau Street, just a few blocks from the Hoe Gold Street offices; some of his electrical experiments had also been conducted, coincidentally, on the site of Treadwell's Boston printing shop. Residing in Manhattan since 1815, the artist was probably long acquainted with the Hoe family, who supported his efforts through the National Academy of Design (of which Morse was first president, 1826-1842) to elevate the works of American artists. About 1832 Morse had begun electromagnetic telegraph trials culminating, in May 1844, with a test line calculated to clinch government support for his communication system. The New York *Sun* noted:

> The new invention is completed from Baltimore to Washington. The wire, perfectly secured against the weather by a covering of rope yarn and tar, is conducted on the top of posts about twenty feet high and one hundred yards apart. . . .[27]

Although Morse failed to sell the telegraph to a skeptical postmaster general, others immediately perceived its usefulness. Within a year the short demonstration line was open for message traffic, and entrepreneurs everywhere were scrambling to participate. Surprisingly, with a few exceptions, the newspapers failed to recognize the potential of these activities and gave the invention only passing notice.

The Magnetic Telegraph Company, in Maryland, was the first company to form for commercial purposes, announcing in 1845 its plan to connect Fort Lee, New Jersey (directly opposite New York City) with Philadelphia. Among the company's officers was William M. Swain of the Philadelphia *Public Ledger*. Like his partner Arunah Abell, Swain was also a close friend of Morse; both helped his cause immensely. Although himself not a charter subscriber, Swain undoubtedly saw the telegraph's value to newspapers far removed from New York and Washington. As agents proceeded to negotiate rights of way, linemen began insulating the long copper wires on simple glass bureau-drawer knobs.

William Swain was elected president of the enterprise; although "taciturn and cautious to excess," according to James D. Reid, the industry's historian who started with the Magnetic as telegraph operator, Swain attended personally to every detail (as he did on his newspaper) and even invented a proper glass insulator. To Swain the news-carrying telegraph became of crucial journalistic importance. "At the time of the Mexican War," said Reid, "we often slept on the same table together, as we waited in the office night after night the arrival of army news at Baltimore." [28] The wires of the Magnetic Telegraph Company were fully open to New York by January 1846.

Among unsure New Yorkers James Gordon Bennett ironically resisted Morse's lightning form of communication; the *Herald's* own collection devices were all too successful. Nevertheless, as papers gradually admitted its utility Morse himself corresponded with them (via their columns) to keep the public appraised. Dramatic episodes accumulated. "For example," wrote Morse, "the President's Mesage, entire, on the subject of the war with Mexico, was transmitted with perfect accuracy (exclusively for and at the expense of the Baltimore *Sun*) at the rate of 99 letters per minute. . . ." [29]

In a very short time Swain and Abell had infected Richard M. Hoe with the excitement and technological challenges of the electric novelty. While the Colonel was in fact lining up Swain's paper for his own brand of lightning—a radical new press he was putting together—the newspapermen corralled him for their project. Despite his energies so heavily—compulsively—engaged in press building, it is significant that the telegraph was the only outside business activity that seriously interested him. And with reason:

[*William Swain to Richard Hoe, Paris, 6 February 1849*] . . . "The

Magnetic Telegraph Co." at the last meeting of the Board of Directors declared a dividend of three percent on the capital stock of the co. for the last six months which gives to each share holder twelve per cent per annum upon the cost to him. One percent a month is a fair enough dividend upon investment, is it not, particularly upon a stock continually growing more valuable so far as the increasing business of the company is concerned[?]

The two wires now work from Washington directly into Hanover st. New York and the business is now so heavy upon the line that it is difficult to get along with it. It is important therefore to have another wire on as soon in the Spring as may be. But the patentees have offered to yield all claims for any more issue of stock to them on any after wires being put on provided two more wires are put up the ensuing Summer. The stockholders generally are I believe in favor of this as the interests of patentees and stockholders will then be mutual entirely. . . .

Our new Quincy granite building on Third st. above Chestnut, (reverse this) is now nearly completed. It looms large and magnificently —is the observed of all observers, [*with*] *the second story being fitted up for the different Telegraph lines, Eastern, Western, Southern and Northern. . . .*

In any event, subscriber Richard Hoe (36 shares; Peter had 20) was presently a director (and possibly treasurer) of the Magnetic, and, Reid added, not unkindly, "of a nature too practical to fail to see its possible advantages to himself." Indeed, the Colonel had a unique plan for his own telegraph system.

With the Broome Street factory over 40 blocks uptown from Gold Street, and both locations some distance from the 11th Street family residence, the Hoes had frequently deplored the wasted time of messengers in daily routes. In 1849 the Colonel decided upon a private company telegraph line. As the physical requirements became apparent to him he appealed to Royal E. House, telegraphy inventor and also a director, for help:

[*18 May 1850*] . . . *We contemplate connecting our Broome Street and Gold Street Factories by a Telegraphic wire and wish to ascertain whether we may use the poles of your company* [*probably the New York, Albany and Buffalo E.M. Telegraph Company*] *which run through William Street to terminate at Essex market from which place we would put up independent poles to our Broome St. factory and what you would charge us. . . .*

There was no difficulty, evidently, and the short stretch to Broome Street—mostly over housetops—was quickly completed. The line was equipped with conventional instruments at its terminals, making R. Hoe & Company the first to maintain a fully professional private wire. (Even the New York Police Department had no telegraph until 1853.) The Hoe network clicked: "Its value was at once evident," Reid reported. "The master was felt to be among his men although two miles intervened." [30]

Many years later the telegraph's staccato at Grand Street still fascinated visitors:

> This machine, which is of the Morse pattern, and stamps its messages with dots and dashes on a continuous strip of paper, was introduced about 20 years ago by Col. R. M. Hoe, himself one of the pioneers of American telegraphy. It is the first private telegraph ever used, and has proved a practical advantage which the proprietors would not give up for thrice the small sum which it costs them to maintain it. Another noticeable convenience is the set of ten speaking tubes communicating with the upper stories of the building, and with the various departments of the adjoining factories. . . .[31]

The Colonel, at the same time, had not relaxed his interest in commercial telegraph operations, although, as dozens of lines failed or merged, the industry quickly became confused with new developments and intrigues. Charles F. Wood, superintendent of the (now) New-York and Washington Magnetic Telegraph Company, was obliged to bring urgent matters personally to Gold Street or to the Colonel's home for his attention. Telegraph business acquainted Hoe with Henry M. Schieffelin, president of the New York & New England Union Telegraph Company; the Hoes and Schieffelins were long close friends and neighbors.

Richard soon became party to larger schemes to consolidate and strengthen the telegraph network throughout the country—and abroad (his friends Peter Cooper, Cyrus W. Field, and their associates began planning a transatlantic cable in 1854). Along with Cooper, Field, Hiram Sibley, Ezra Cornell, and others, Hoe worked patiently against contrary individuals like "Fog" Smith, who had more personal aggrandizement in mind. Smith, still a congressman in 1839 when he saw Morse's demonstrations, was (in Reid's words) "shrewd, vigorous, aggressive, and of a nature cold, sharp and imperious." He quit the legislature and bought a one-fourth interest in Morse's patent; "Never," said Reid, "was a partnership in all respects more unfortunate." [32]

Francis O. J. Smith became an officer in several independent telegraph companies, and having exasperated all who dealt with him, could best be approached by a cool pragmatist with unimpeachable integrity:

> *New York Septr 8th 1858*
>
> *Hon^{l} F.O.J. Smith*
> *My Dear Sir*
> *I have had several interviews with Mr. Cooper & Mr. Field & they are prepared to open negotiations on the plan agreed to by them & us objected to by you. If you remember it was a new company to be*

formed of the American, Union, & Magnetic stock to be apportioned by three parties to be selected one by each co., the individual to be acceptable to all. Will you now entertain any thing of the sort[?] *If so I will endeavor to get the Magnetic to take it again into consideration. Please let me hear from you soon.*

Yours very truly,
Rich[d] *M. Hoe*

This matter dragged through another year. In 1860 the emerging American Telegraph Company bought Smith's holdings in the New York & New England Union, and in turn became part of the resulting Western Union.

The Colonel's interest in the telegraph continued to be both corporate and personal. In 1877 he liked the idea of having customers telegraph press orders at the firm's expense and would print his own "telegram" order forms; for this he felt an "abatement" from Western Union appropriate. A year later, ignoring the concessions and considerations once granted him, he refused to let a similar favor become a habit:

[*To Atlantic & Pacific Telegraph Company, 10 April 1878*] . . . *We gave permission to attach the wire to our building for a few days while the pole was being placed, not longer, nor would we consent to have it there permanently. . . .*

Aside from personal experience with the lucrative communication system—increasingly falling everywhere into greedy hands—the Colonel could not fail to appreciate how the telegraph had affected the craft of journalism—and the prosperity of certain large newspapers. Horace Greeley in 1868 agreed:

A copy of a great morning journal now contains more matter than an average twelvemo volume, and its production costs far more, while it is sold for a fortieth or fiftieth part of the volume's price. There is no other miracle of cheapness which at all approaches it. The Electric Telegraph has precluded the multiplication of journals in the great cities, by enormously increasing the cost of publishing each of them. The *Tribune,* for example, now pays more than one hundred thousand dollars per annum for intellectual labor (reporting included) in and about its office, and one hundred thousand dollars more for correspondence and telegraphing—in other words for collecting and transmitting news. . . .[33]

In New York the reliance on telegraphic transmission of news became a microcosm of personal ingenuity—a sacred mission.

Dr. Alexander Jones of New Orleans had quickly been swept into the telegraph fever, even as he practiced and tinkered, and accumulated considerable experience in a very short time. He gathered news in Europe and America as a "free-lance" reporter (selling "stories" where he could) and devised the first telegraph cipher code to reduce message lengths. He sent one of the earliest news telegrams on record—an 1846 account, for a Washington paper, of the launching of a U.S. warship in Brooklyn.

When six New York papers (*Journal of Commerce, Herald, Tribune, Times, Sun,* and *Courier & Enquirer*) joined forces to form the Associated Press in 1848 to pool telegraphed news, they selected Jones as its first general agent. Seven days a week—including holidays—Jones, for $20 weekly, labored hard, recalling:

> Often on stormy nights in winter, when our errand boys were ill or absent in Jersey City [the New York terminal] have we gone around at twelve and one o'clock and delivered messages with a snow or sleet storm beating in our face; and having, at many of the offices, to climb three or four pairs of stairs to find the composing room. For months at a time, we seldom retired before one o'clock and then had to be on duty through the next day.[34]

But the strain finally reached the intrepid doctor, and after three years he retired.

Daniel H. Craig replaced Jones. A New Hampshire printer's apprentice, Craig had become fascinated with carrier pigeons as news transmitters; he soon inaugurated a service wherein he met ships off Boston, condensed the news, and sent it via pigeon to his subscribers. Moses Beach, James Bennett, and Arunah Abell, among others, signed up with him.

In 1851 Craig, now an Associated Press agent at Halifax (replacing Boston as first stop for transatlantic steamers), faced stiff competition. He reduced his pigeon post as telegraph lines penetrated almost everywhere, although of course the birds remained for ship-to-shore traffic. Craig nevertheless could beat telegraphers at their own game: he was probably the first to send dummy messages (often entire chapters of the Bible) to monopolize the line. As he took control at Associated Press its successes—and problems—were growing beyond a simple gentlemen's fraternity. Yet Craig's sense of news and his flair for organization led in 1856 to better efficiency among many rival and squabbling factions:

> Assisting in the arrangement of these matters, the general agent [Craig] did not increase the number of his personal friends; but amid all the trials he fought the battles mildly, pleasantly, and gentlemanly in conversation, but savagely, bitterly, and ruthlessly on paper. . . .[35]

Reid knew Craig as "cool, shrewd, indefatigable." After fifteen hectic years he was discharged from the explosive office in 1866—forcing him prematurely to reveal secret plans (backed by Western Union) to form a rival organization, the United States and Europe Telegraph News Association. Craig won support from a group of western newspaper owners (for whom he became general agent) also engaged in fighting the imperious New York Associated Press, as well as old friends who enjoyed his plucky news gathering methods.

Upon the removal of Craig James W. Simonton took the post. Simonton had started at 21 as law-court reporter on the *American Republican,* Leavitt & Trow's New York paper, then moved, after a stay on Webb's *Courier & Enquirer,* to the *New York Times.* He had been proprietor and editor of the San Francisco *Bulletin* and was also associated with other western papers, connections he did not sever. Simonton too was a man of rare perseverence.

As a reporter for the *Times* in 1857 Simonton had exposed a railroad "land grab" which would have given away all of Minnesota; although four congressmen implicated were obliged to resign, that august body censured Simonton for refusing to reveal the source of his information. Simonton was expected to dispell association confusion and suspicion with similar frankness, especially to counter increased public misunderstanding about the nature of collective newsgathering by telegraph. He established an agent in London to furnish European news, giving him better foreign coverage than Craig (whose domestic news, however, was unsurpassed). Although Simonton had never been robust, he would head the New York Associated Press fourteen years, retiring 1881 because of poor health.

Craig, however, suffered hard set-backs. The New York Associated Press and its western critics resolved their differences within a year and together created a new organization in January 1867. Western Union, for its part, interested only in playing one wire service against the other, withdrew its support. Once again Craig found himself alone and unwanted; he was disillusioned and bitter —but full of new schemes. Slowly he regained some resources with which to confront his arch enemies, the admittedly monopolistic associated press organization, and an interrelated network fast becoming a grander monopoly—Western Union. Part of Craig's assets was a collection of new inventions calculated to replace those of Morse and his workers. One such device was George Little's automatic telegraph recorder, reportedly capable of registering 60,000 words hourly. Applying such equipment to that branch of telegraphy most lucrative—transatlantic cabling—Craig would strike his foes where they would most feel it. But he needed backing.

Craig remembered Richard M. Hoe as a steadfast promoter of scientific achievement and a discrete gentleman in business. He wrote to the Colonel (11 January 1877) in confidence, describing the state of the art and especially his new transmission system. Transoceanic messages were then traveling 8 to 10 words per minute (for up to $10 per word), using a flashing light for detection.

The speed of electricity, however, theoretically permitted rates up to 300 words per minute. Craig's "metrical" device, an electro-mechanical improvement, could do 250. With financial support a system using it could be practical and profitable.

Craig wrote again at the end of the month to reveal that General William F. Smith had joined his enterprise. Smith (like F.O.J. Smith no relation to Hoe's partners) had taught mathematics at West Point and left the Army after a controversial Civil War career. Police Commissioner of New York and president (1866) of the International Ocean Telegraph Company, founded to lay a cable to Cuba, Smith was willing to travel to Europe at his own expense to conduct negotiations for Craig's patents. Surprisingly, Craig asked for a niggardly $2500 for Hoe's participation. The Colonel turned down this prospect but Craig tried once more:

> [*16 February 1877*] . . . *I dare not go to anyone in the telegraph business, or even to acquaintances outside of the telegraph circles, except in very rare cases, until my Patents should have been fully secured, and except yourself and Mr.* [Charles?] *McKesson, I do not know two other gentlemen who have ample means & with whom I should feel that my secret was safe.*
>
> *Aside from any profit which may accrue to you by co-operating with me in my cable working plan, you, certainly, ought to feel some interest in any arrangement which promises to smash those twin Monopolists, the Western Union Co. & the N.Y. Ass'd Press. . . . Against such a monopoly no new journal can possibly survive. Let me get my new system into use (reserving, as I shall do, its use for News Reporting) and I will smash both the Ass'd Press and the W.U. Co. in less than two years, and as far as the Ass'd Press is concerned, in less than 90 days. That act will create more new journals in one year, than we have seen in the last fifteen years. . . .*

In a postscript Craig added that royalties on traffic to and from London alone would run to £50,000 or more annually.

But the disgruntled telegrapher apparently failed to realize that Hoe's interest in the mechanics of telegraphic communication had declined as the press business prospered. Never social or political crusaders, the brothers were scarcely disposed to crush monopolies; they were in fact forging one of their own. Participation in a scheme to increase the number of new—small—journals could only work now against Hoe interests in serving the great papers already in existence. Expensive research in rotary, cylinder, and web equipment would only make the huge metropolitan sheets bigger and faster, and their mechanical needs more and more remunerative. The Colonel could not forget how the company for decades had nursed new and impecunious shops, an obligation he no longer welcomed. He gave Craig no encouragement.

The R. Hoe & Company private telegraph, first of its kind and probably one of the last to be dismantled despite the introduction of the telephone, in 1877 was augmented by a private automatic signal telegraph for fire alarm purposes. But its magic had long since disappeared.

For the great "Cable Jubilee" of 1 September 1858, celebrating the successful installation of the transatlantic cable, R. Hoe & Company had cooperated, as for the opening of the Erie Canal, by participating as one of the procession floats. A wagon (supplied by the company) with a new Hoe cylinder and an old hand press side by side (both run by men from Wynkoop, Hallenbeck, & Thomas) rumbled down the New York parade route printing a brief history of the telegraph as it went.

At a grand dinner at Delmonico's (30 December 1868) to honor Samuel F. B. Morse, 200 scientists, statesmen, and captains of industry gathered, but Richard declined:

Decr 29th† 1868

Jas. E. Reid, Esq., Secr^{y}
Dear Sir

I do not feel well enough to be present this evening to join in honor's† to my friend Proffessor† Morse, as I fully intended to have done & which I regret very much.

It would have given me much pleasure to join with his many friends on that occasion in expression of sincere respect & affection.

Few persons of genius, have had more difficulties & discouragements to encounter in prosecution of scientific inventions, than our friend, and none have lived to triumph over them all, more signally than he, & I am sure no one personally can be more worthy to bear & enjoy such world wide honors as he has won. Permit me to offer a toast "The Electric Telegraph" the friend and promoter of all other inventions & designs for the diffusion of knowledge.

Yours very truly
Richd M. Hoe

After a lifetime in telegraphy Reid had seen political and industrial empires grow and fall and watched the sciences tenuously assert themselves. His comment on Darwin was charming: "His researches are not pleasant. Few delight in such ancestry as he accepts as his own. . . ."[36]

The Colonel similarly passed up the 1871 inauguration ceremony for Morse's statue in Central Park. The Craig machinations which followed these proud moments must have brought his long association with telegraphy to a sad and painful end. No wonder he was ever anxious to return to the security and comfort of his own machine shops. The craft of press building had flourished almost in spite of itself—and without the schemes and intrigues the Colonel found at work elsewhere.

In the 1840's the problem of acquiring accurate machine tools was of concern throughout the United States. Few American firms were capable of recognizing their requirements clearly; fewer still could make tools whether for sale or their own use. For a time Hoe struggled alone:

> As the works had grown larger new tools had been added, but for the most part they were made in the factory and copied from older ones, although somewhat improved. We had no standard of measure but our ordinary two feet pocket rules; gauges were comparatively unknown, and the various pieces of work were fitted one to the other. The screw threads were as they had been originally picked up, and the size of our taps and reamers were[†] taken from the old ones or from our pocket rules.[37]

It was natural at first to seek close at hand the engineers and tool makers who could help put a machine shop together. But in early New York the industrial revolution was only fitfully appreciated, its few practitioners often celebrated more for their spectacular accomplishments than for genuine contributions.

Dr. Eliphalet Nott, president for 62 years of Union College, Schenectady, conducted numerous experiments on the application of heat as energy. He designed a steamboat about 1832 which he named the *Novelty,* because of its peculiar anthracite-burning steam engine. When the ship anchored for repairs and alterations at the foot of 12th Street, on the East River, the shore facilities took on its name. The Novelty Iron Works serviced the vessel and sought other jobs; managed by Thomas and James Stillman and then (1842) by Horatio Allen, a locomotive expert, the shop prospered. It soon became famous for marine and stationary engines and boilers, sugar-cane and cotton processing equipment, and machine tools. For Hoe the firm made a number of slide lathes and other tools, but the Colonel —contrary to his own principles—was unhappy buying "unknown," domestic machinery. Like many other unsure Americans he looked to England—"home" of the industrial revolution—for modern tool efficiency and precision. It may have been one of his few prejudices, but then, one Manchester firm did enjoy a world reputation.

Joseph Whitworth (1803-1887) was educated by his father until the age of 12, when he was enrolled for 18 months at a private academy. Immediately recognized as an extraordinary lad, Joseph was given the opportunity to develop his mechanical talents, spending six enterprising years in an uncle's cotton factory. Thus early in his life Whitworth became decisively committed to a machinist's career.

At the age of 18 Whitworth entered a Manchester machine shop. Three years later he was working for Maudslay in London, and shortly thereafter for Charles Holtzappfel, whose shop in the 1830's manufactured, among other things, small job printing presses. But the end product was not nearly so interesting as making it accurately: Whitworth had become a fussy craftsman completely preoccupied with machine tooling, its techniques, and its precision.

By 1830 Whitworth had so perfected his knowledge of tools and their handling he had managed to finish metal surfaces so flat and so smooth that blocks mysteriously adhered to each other when brought together. He soon found he could not master any engineering situation unless he had an array of precision equipment to insure minimal tolerances and deviations. Constructing many tools to his own specifications, Whitworth returned to Manchester and in 1833 went into business.

For many years Joseph Whitworth made accuracy and precision his watchwords. As late as 1856 he was still communicating his concern to others:

> I can not impress too strongly on this institution [Institution of Mechanical Engineers], and upon all in any way connected with mechanism, the vast importance of possessing a true plane as a standard of reference. All excellence of workmanship depends upon it. Next in importance to the true plane is the power of measurement. . . .[38]

Whitworth produced gauges which became the standard for British industry. He found that one 40,000th of an inch could easily be distinguished by touch, and developed measuring equipment that could detect a millionth of an inch—minute dimensions which, of course, were even affected by temperature variations. Where there had been a dozen incompatible machine-screw thread systems, he introduced workable industry-wide standardization. Between 1834 and 1849 he took only 15 patents (of which three were street-cleaning machines—the last reportedly making Manchester streets sparkle); as British patent applications were discouragingly expensive, he declined to register many ideas he nevertheless incorporated into his machine-tool business.

For the Crimean War Whitworth was approached by the British government to manufacture small arms, for which the need was critical. With his customary deliberation he pointed out that tests had never been conducted to ascertain optimum caliber, rifling system, projectile shape, and other factors; he would not commence production until such questions were settled. "In the matter of gunnery," said a biographer, "(like Darwin in some of his special investigations), he may be said to have proved all things in order to hold fast that which was good." His usual remark was "Let us try." [39]

Although the war had already begun—Russell in the Crimea was telegraphing its fury to the London *Times*—Whitworth carefully made his tests, using a long shooting gallery and tissue paper screens. The British Army carped for many years at his resulting hexagonal-bore rifle and eventually adopted another weapon altogether. Whitworth nevertheless entered the armament business and soon applied himself to ballistic research for large field artillery pieces and seacoast cannon. By 1863 he was selling ordnance to governments everywhere, including the American Confederacy.*

Whitworth subsequently approached the manufacture of steel with his usual thoroughness, developing (and in 1865 patenting) his own product, "compressed steel." As his earliest reputation reached them R. Hoe & Company were pleased—honored—to deal with him:

> About this time [1844] the firm bought from Messrs. Joseph Whitworth & Co., England, three hand lathes with slide rests, a set of standard measure ring and plug gauges from ¼ inch to 2 inches diameter, a complete set of taps, hobs and screw cutting tools for hand use, and a bolt-cutting machine. This introduced into the works the Whitworth or English standard screw threads, which we are still using. There was also bought a fine bronze standard yard measure very accurately divided, made by Potter of London. These tools were a revelation to us, and made a vast change for the better in our manner of working. We began making gauges to standard measure for the various parts of machines, so that when brought together there would be the least possible amount of fitting, thus doing more accurate work at less cost.

Having worked well, these Whitworth tools recommended others:

> The firm had, from time to time, bought more tools from Messrs. Whitworth & Co., hand and slide lathes, surface plates, a slotting machine, a shaping machine, a drilling machine, and latterly, a wheel-cutting machine with self-feeding and tripping motions; this last machine more particularly for cutting the large wheels of the type-revolving presses.

Fortunately, the state of the machinist's art in the New World was advancing rapidly, and foreign purchases declined:

> About this time Mr. A. M. Freeland, who had previously been a foreman in the Novelty Iron Works and who had lately visited England, to see the latest improvements in machinist's tools, started the making of machine tools in this city. Three slide lathes were bought from him in 1850 and they were found so much more convenient and better than any in the works from Whitworth & Co., or any other maker, that the firm then bought their lathes, planers, slotters, wheel-cutters, etc., almost exclusively from him until his death in 1870.[40]

Strangely enough, mounting difficulty and dissatisfaction with

*See *Age of Great Guns.*

some of Whitworth's equipment contributed to the acceptance of American tools. In 1846 Hoe had bought a Whitworth lead-screw to be placed on a lathe for cutting screw threads. The British firm claimed it to be absolutely exact in pitch, and Hoe used it carefully only as a master checking tool. Tucker compared it one day with the new Potter standard, "when to my surprise we found the screw to be just 1/20 of an inch too short in 6 feet of length." To compensate for this, Tucker developed a plan to stretch the stock as it was being threaded, calling for considerable modification of the machinery, but it worked:

> Thus for each revolution of the lead-screw the drawing-screw wheel made one revolution, plus one tooth, and when the lead-screw had made 144 revolutions the drawing screw wheel had gained just one revolution and drawn the lead-screw and its tool carriage endways 1/20 of an inch. The master-screw cut in this manner was as near an exact pitch as we could measure. . . .[41]

Despite the gradual improvement of commercial tool-making, the firm had also grown accustomed to making many major alterations to suit its special needs. Peter Hoe once complained:

> *[To A. S. Bowen, Boston, 25 August 1859] . . . The machine from Gray & Wood is not a success. The feed motion is not right. It does not take the strips through, the round band not being sufficient to drive it. We shall have to alter it as we have all the other machines in the place. We are obliged to put in new bolts &c. . . .*

Allied with this problem was the company policy to keep its equipment topmost:

> It had always been the practice of the firm to dispose of its older machines and tools and replace them with the best obtainable, and also to introduce all new styles of tools, from home or abroad, that seemed advantageous for this class of work. Many special tools were made to finish the small parts of the various machines manufactured, to expedite and cheapen their production.[42]

Such routine customization, however, was not nearly so demanding as recurring problems with the measurement and master checking equipment, on which all shop activities relied for their exacting execution.

With Whitworth as a Hoe subcontractor in 1856, revelations of sloppiness would be a matter of stupifying amazement to the Colonel and his brothers. By 1874 the company—abandoning both "Potter of London" and Joseph Whitworth—sought help from the United States government for its mensuration requirements:

> *July 23d 1874*
>
> *Dear Sir*
>
> *We are very anxious to secure a Rule between twelve and twenty*

feet long, or it might be a little longer, laid out in one inch divisions in accordance with the United States standard. Of course we desire it to be of the greatest possible exactness.

Mr. Wurdermann told our Col. Hoe that the Government had an accurate method of making such [a] rule or measure of, we think, 6 French meters in length. We wish one not less than twelve feet long, and can subdivide it our selves, but would prefer to have it accurately done in Washington.

Will you kindly favor us by calling at the U.S. Coast Survey Office, when convenient, and ascertain whether they can furnish such a Rule made either of steel, brass, iron or wood, what their charge will be and whether we shall send on the material or rod.

The Col. sends his kind regards and desires to say that he is still endeavoring to produce a perfect screw—and does not despair.

Yours very truly
R. Hoe & Co.

Edward H. Knight, Esq.
U.S. Patent Office
Washington D.C.

Knight, a prominent patent attorney, was also compiler of several mechanics' dictionaries and other reference works. The purpose of this rule was to make a rack-gear cutting machine large enough to handle the big cylinder presses of the 1870's.

Hoe consequently made two steel bars, 1 x ⅝ inches x 15 feet, duly sending one to Washington, and the other to machinists Darling, Brown & Sharpe, Pawtucket, Rhode Island, to be inscribed both in inches and centimeters. Tucker reported the results:

> Upon examination of the bars after division, the one divided by Darling, Brown & Sharpe was found to be far the better, in fact its few errors were scarcely discoverable. . . . The cutter-head was then so set that the division lines on the rule and on the bar, seen through a microscope, exactly coincided, and a cut was made across the rack at the proper angle. . . . This was an exceedingly delicate and tedious operation, occupying about three months, for not only were the racks allowed to cool after each cut, but the temperature of the room was also taken into account. . . .

Whitworth, however, was engaged in one last tool-making assignment for Hoe. Respect for the venerable British firm had deteriorated rapidly when another particularly irritating incident came to light:

> It was observed that the wheels on the large impression cylinders of the web presses ran as if they were in some way out of truth, and an examination showed that the teeth were not equally divided, and the fault was traced to the Whitworth Wheel-Cutter by which they were cut. Messrs. Whitworth & Co. were written to in the matter and they replied in effect that they were aware their dividing wheel was not perfectly accurate, but thought it sufficiently so for all practical purposes; however, they intended to make a new wheel in the near future and would then send a duplicate cut from it. When this dupli-

cate arrived a year or two later, it was found to be quite as defective as the wheel we were using and it was at once sent back. . . .

In an irate note (19 April 1878) Hoe demanded some explanation: "We presume you have received the dividing wheel by now, & we are expecting a report from you. . . ." Without awaiting either reply or report, however, Tucker had again taken action:

[I]n 1877 we decided to make a new wheel ourselves. . . . This was an operation requiring great skill and patience, but was very successfully accomplished. . . . This wheel is used only for dividing other wheels and similar work, and when so used it is put on the arbor of the Whitworth Wheel-Cutter in place of the ordinary dividing wheel. . . .

Tucker described the effects of temperature:

Now the operation of cutting will heat, expand and distort the wheel, so the next cut should be made at a distant point, say directly opposite, and then at the quarters and so on, but if the greatest accuracy is required, the wheel should be allowed to cool between each cut. All the dividing wheels on the wheel cutters in use in the factory up to a late date were divided by this wheel. It was undoubtedly the most accurate dividing wheel in the country and in all probability has no superior to this day.[43]

The company thereafter had little difficulty with precision work, and in fact—for itself and for government armament work in the world wars—executed some of the most demanding machine work known to the industry.

The Colonel was frequently called upon—a scientific committee of one—to make judgments in other areas of endeavor. This he did with some reluctance, but he could seldom entirely escape the chore. "[O]f course I do not wish you to express any opinion which you do not consciously believe," said an 1842 inventor, "for I know your candor in the expression of opinions in such matters." This frankness and impartiality brought him high regard among friends and colleagues, especially in related fields.

In the area of type founding Richard at first participated willy nilly. Holding at least one patent for a stereotype block, the company was sorely tempted to engage in the specialty. With friends already in the business, the Colonel became involved almost effortlessly.

Arthur Pell, one of the company's early Gold Street tenants, had tackled an inventor's obligations in admirable if blustery style:

New York May 22 1833

Sir

I claim to have invented the use of a single cylinder for the purpose of drying Paper in preference to the use of two or more, and that the use of this Cylinder is a great and manifest improvement over any other method.

I claim also the invention of a cutting Machine having great and decided advantages over any other, heretofore used.

These claims are, through the medium of the Patent Laws, in such a shape as to have their validity established, or made of no effect.

I therefore respectfully notify you that if you sell, or Cause to be used, either of these machines built by any other person than myself, I shall immediately proceed to have an injunction laid upon your using them until the question be legally decided. Nor shall I relinquish my claims, until I have tried every means the law will give to make them good.

Respectfully yrs
Ar Pell & Brother
To Mr. Paddington *W. S. Graves* [*clerk?*]

This letter was also sent to "Hector Greg," both offenders unidentified. As many nineteenth-century devices were indeed easy to copy, an inventor who believed in patents had always to be on guard against infringement, theft, and unlicensed use. Arthur Pell unquestionably furnished young Richard Hoe with an understanding of these problems he would soon need for his own work.

No doubt R. Hoe & Company made Pell's machines (before Tucker's arrival—he does not refer to them), and from the timing of their appearance they may have been byproducts of Hoe's own cylinder printing mechanisms. Arthur's brother was probably Alfred Pell, an investor who earlier formed a typefoundry with George Lothian; William Hager later bought Lothian's interest. A loose alliance between the Pells and the Hoes developed.

Pell soon passed along stereotype orders to Hoe, perhaps to make use of Hoe's (Smith) patent base:

[*Arthur Pell to T. M. Shine, Auburn,* 17 *June* 1833] . . . *When we received your order for the block,*† *we handed the letter to Messr. Hoe & Co. to have them made. They now say they made them no larger than is usual in all cases, when the size of the plate is given. I think they never make a Block exactly the size of the plate, & just so that the plate will go on it, unless special directions are given.* . . .

An unidentified client received an unusual shipment:

[*Pell,* 10 *September* 1833] . . . *By tomorrows Steam Boat we will send you one of the machines for rubbing type and a young woman with it who understands its workings. We are so certain that you will approve it that we accept the instruction you make in one condition, namely, that if you do not approve of the machine you may return it, and draw on us for the amount.*

We will therefore draw on you at 3 *mos for* $1000 *dollars—the conditions of payment as we before stated, namely* $250. *to be paid quarterly.* . . .

A few years later Pell and Hoe engaged another interrelated firm, suggesting the existence of a partnership not heretofore recognized:

New York July 5th 1837

Gent/

Mr. Pell just handed us your letter of the 23d since relating to the set of plates stereotyped by Smith & Valentine, and forwarded by them to you about 2 weeks ago. And now before we go any further, we will explain to you why we address you on this subject in order that you may not think strange of it. Mr. Smith (of the firm S&V) is a brother of one of our house [Matthew III?] & we have assisted them by advances to transact their business, and in this case have advanced to them the amount of the last set of plates sent to you, and your note when received by them is to be handed over to us. Neather† ourselves nor S&V know any thing of the contract Mr Pell effected with you in this matter and so far as S&V are concerned they do not wish to continue the work and are perfectly willing to relinquish their contract to Mr Pell.

When Mr. Smith was in Philad^a^ at your store conversing with a member of your firm on the subject of the last note that was sent on for them, some conversation took place about discontinuing the volumn† [a periodical in stereotype?] (the last one shipped) nearly completed and was assurder† that it would of course be received by you when finished, as you did not wish to have a volumn stop't in an unfinished state.

We think their† is some misunderstanding in the matter & we trust upon inquiring into it and calling to mind the conversation above alluded to you will forward your note for the amount of the last bill without delay, and thereby greatly oblige Mess. Smith & V as well as

Your M^o^ Obt Svt
R. Hoe & Co.

To Messrs. Desilver† Thomas & Co.
*Philad^a^*

Richard C. Valentine was for many years a practicing New York city stereotyper. But stereotyping was patently not a lucrative business for the Hoes, and they needed little persuasion to drop it.

James Conner (1798-1861), a contemporary of the first Robert Hoe, had established a New York typefoundry which enjoyed considerable success. Popular and influential, Conner was an officer in Tammany society, an elected clerk of New York County, and a major in the Washington Greys, a prestigious National Guard unit. His closeness to the Hoes probably enlisted young Richard in this military organization, and possibly—more subtly—convinced the brothers that printing press manufacture was separate and distinct from type founding and stereotype work. This the partners conceded, although some involvement in type-casting gadgetry was unavoidable.

As early as 1843 the typesetting machine of a Swede, Frederick Rosenberg, was on display nearby—at 49 Gold Street—and the Colonel must have studied it thoroughly, as he would others in a long line of developments. Such early efforts, however, considering the state of the machinist's art, were premature, but in a few decades the subject would be as fervid and secretive as wood-screw machinery had been; many were at work on the problem.

One inventor, a New York printer, commendably—but perhaps too persistently—came to Hoe for his endorsement:

May 23rd 1874

My Dear Sir,

You ask my opinion as to the probable success of your Type Casting & finishing machine.

I am very much pleased with the improvement you had made in simplifying the machine since I first saw it & it seems to me worthy of further expenditure & use to prove its value which I think must be very great.

Yours very truly
Richd M. Hoe

To Mr. C. H. Westcott
New York

A year later Westcott needed something still more definitive:

88 John St.
New York July 6, 1875

My Dear Sir

I dislike to annoy you with requests to examine my type machine, but I am endeavoring to get it into such position that I can manufacture the machines for use; and I am constantly asked "Has Col. Hoe seen it—What does he think of it?"

I would not ask you again if I supposed you had any objections to examine it, but I presume your time is occupied to such an extent that it passes out of your mind when you are in this part of the city.

A superintendent of a large house in Boston was here Saturday, and others from the same house are coming on here this week for further examination and they have already asked the usual question and I hope to be able to tell them that you have examined the present machine.

Under these circumstances may I not ask you to look at it in the course of Wednesday or Thursday of this week[?]

You will do me a great favor if you can do so.

Yours truly
C. Westcott

To Col. Richard M. Hoe

Yes, yes, of course, the harried Colonel said to himself, and may actually have scribbled something complementary. The Westcott Type-Casting and Setting Machine was displayed at the Philadelphia Centennial: its reservoir of molten metal (with pump), water-cooled molds, and copper matrices made its casting operation much like

Tolbert Lanston's 1889 Monotype, although justification was done by hand. Hoe seemed free at least to accommodate another, a paying client who entered the shop with fresh drawings.

Joseph Thorne, from a Quaker family, had been second engineer on the *Arctic;* his last voyage had just preceeded the ship's fatal 1854 accident. He invented a typewriter, a sewing machine (working with Elias Howe), and in 1878 commissioned R. Hoe & Company to make typesetting and distributing equipment. Tucker commented:

> The machine seemed quite simple in principle, and I have heard that quite a number have been made since, but they require to be kept very accurately adjusted, and I do not think that they have ever become very popular.[44]

Many were nevertheless employed in the United States and England. But here was Westcott again.

> [*To Richard Hoe, 25 August 1878*] . . . *I have a note from Mr. Evarts saying "I am glad that Col. Hoe has been to see your machine in operation, and I hope we may secure his cooperation in our plans."*
>
> *Will you do me the favor to give me a letter stating your opinion of its utility*[?] *I am constantly met with objections from Type founders, and those having adverse interests, and your endorsement of the utility of the machine, will do away with a vast amount of prejudice, with which I have to contend. I am wasteing*[†] *time and earning nothing, with expressions which (under the circumstances in which I am placed) are very heavy, and I am extremely anxious to get to work.* . . .

Others were succeeding too, as Westcott might have noted.

Ottmar Mergenthaler (1854-1899) had come to the United States in 1872, and in Baltimore was becoming a naturalized citizen even as Westcott wrote. The German studied typewriting and typecasting devices in existence—and improved some of them; by 1883 he struck upon the line-casting feature that, unlike all others (which usually attempted to pluck precast type from trays or slots), made a workable system possible. After much refinement Mergenthaler's first machine appeared in 1884. Two years later Whitelaw Reid turned the switch on one in the *New York Tribune* office—and gave it its descriptive name, "Linotype." The future for commercial typefounding was thereafter uncertain indeed.

James Conner's firm passed to three sons, William C., James M., and Charles S. Conner. They expanded the works considerably, and with style. In 1865 they engaged John Henry (1818-1889), an Irishman who in 1858 had founded *The Printer* (first American trade journal not associated with an equipment maker) and who for the Conners edited *Typographic Messenger*. The latter periodical, an unvarnished advertisement for the typefoundry, by some undisclosed arrangement was very generous with space to R. Hoe & Company. After three years Henry left Conner to manufacture his own printing machines:

Peter Hoe, Esq.

Dear Sir—I have perfected two new series of Drum Cylinder Presses —one for three form rollers for the general run of Job Work, Book Work, and ordinary Wood Cut Work; the other for six form rollers, and applicable to the highest order of Press-work of all Kinds—embracing landscape Painting in blended Colors. . . . I shall also issue a semi-annual Magazine same size as the "Aldine," with a broad Illuminated border to each page. . . .

This magazine will illustrate the truth of the register in the colored border & tinted corner, and the complete & satisfactory Distribution of the plain Black printing of letterpress & cuts within the colored border—fine colors blending. . . .

Now, dear Sir, I have only regretted it once (and that is all the time) that I did not act upon your twice repeated hint six years ago on two chance meetings in the City Hall Park; but at this time my negotions† *with Robertson were far advanced—so far that I could not honorably back out, and I was obliged to appear not to understand your drift. This man and myself closed, [and] soon after formed a Company; but within the first three months there-after I found he had expended every dollar of the Working Capital for his own purposes, except some $1,500 legitimately expended on the Drawings, Patterns, and building of the first Country Press. &*† *It is but just to say that he afterward refunded the money; but it came back in "bits & scraps," generally weeks or months behind time to meet expenditures which sadly embarrassed business. We were thus unable to build any but the Country Press, and only accomplished the partial introduction of it (leaving my then regular series of Presses untouched) during the whole five years of the Co.'s sickly existence. He also assumed the reins of business, of which he had not the slightest conception—no two weeks adhering to any adopted policy. Under such circumstances, of course, nothing was accomplished; but as things now stand it turns out better for me that the higher classes of Presses were left untouched by the Company. My present complete and perfect machinery will come out under much better auspices than if the defunct Company had ever to do with other than the Country Press.*

I am desirous of an interview with you on the subject of their manufacture and the Magazine above referred to. As you could build them, these presses will produce a revolution in Book Printing and afford important additional facilities and advantages to printers in all the ordinary classes of commercial printing. I would be much pleased, therefore, if you'd drop me a line appointing an interview at your office for a free conversation on the subject.

Very truly yours
John Henry

No. 93 Grand Street
Jersey City, Dec. 7/[18]74

The Henry country press reportedly had a good ink distribution system but its designer could not interest A. B. Taylor or Hoe. It was produced in New Jersey until the end of the century, at which time a manufacturer named Whitlock acquired it. Cottrell bought some of Henry's patents.

The remaining years of a vast American typefounding industry were closing fast. William Conner had been the first founder to set up an agency in California (the United States Type Foundry, San Francisco), managed by Jerome B. Painter. Hoe soon designated Painter its agent as well, an arrangement of some duration. With the immense success of the Linotype and the Monotype, however, the uncertain future of the professional typefounder would become obvious to all. The Hoes returned to their presses.

The Hoe factory itself had never been opulent:

> The sharp-eyed visitor who approaches the establishment from Broadway will first espy, at the corner of Grand and Columbia-sts., a four-story office building, and then, adjoining this, a long five-story factory extending down Sheriff-st, toward Broome-st. These constitute the new portion of the works, and present a lofty and imposing Sheriff-st. frontage of 200 feet. The new factory adjoins the old one, which is four stories high and runs along Broome-st. to the "Iron Foundery,"† that stands at the Coulumbia-st.† corner, and is one story in hight,† with sloping roof and four tall ventilating chimneys. . . . All the present buildings are of brick, and have just been painted a light stone color; but there is nothing very striking in their appearance, no special attempt having been made at architectural elegance. . . .

The factory's great merits, of course, were hidden inside:

> Upon entering, the visitor finds himself in the airiest and freshest of offices. . . . The press makers have been busy for the past week or more in removing from their old quarters to the new factory, which already begins to present that formidable and bewildering array of huge belts and revolving shafts, complicated machines and clattering tools, dust-begrimed workmen and mysterious work. . . . The two upper stories are appropriated to the manufacture of single and double cylinder printing machines and hydraulic presses; the third story to the development of improvements, to the making of certain

> newly-patented steam lithographic presses, and to the construction of the lately invented 5-roller and 8-roller book and wood-cut printing machines. . . . By continuous invention, and the purchase of patent rights, Hoe & Co. have extended their stock until it now includes, with scarcely an important exception, every desirable press in the market. . . .[45]

This new machine shop virtually doubled the capacity of the firm, and was the scene of considerable modernization. During these resettlements the Colonel strove to reorganize the flow of work among the largest lathes, planers, and drills, whose output—efficiently executed—were the real strength of the company.

> The principal business office of R. Hoe & Co. had been in Gold Street for fifty years past, but business offices were now gradually going further up town, and printing offices were also beginning to move up, so it was decided to put a large building on the corner of Grand and Sheriff Streets, for offices, warerooms, drawing office and light manufacturing, and also to establish there the principal business office and eventually to move up the entire Gold Street works. Accordingly, during 1873 a building fifty feet by eighty-eight feet was erected, seven stories high with a wing connecting it to the factory, and the Gold Street office gradually settled in it.[46]

Benson J. Lossing, long an excellent wood engraver and also a chronicler of American history, cut blocks for the Hoes and noted their success with interest. He admired the big shop:

> The growth of Hoe's establishment is a conspicuous example of the mighty expansion of business in the city of New York during the last fifty years. . . .
>
> The total surface of floor-room amounts to over 200,000 square feet, or over four acres; in all the work-shops the floor room is equal to five acres in extent.
>
> The total number of persons employed in the several establishments at the beginning of the year 1883 was over 1000; the yearly amount of wages paid was $750,000, and the number of apprentices was about 250. . . . At a recent visit of the writer to their establishment there were over 200 machine printing presses in course of construction, most of them already ordered. The tools used in their business are valued at $1,000,000.[47]

Grand Street, New York, had become a commercial wonderland. It was now noted for fine apparel and dry goods shops and department stores, made all the more accessible by an East Broadway horsecar line ending at the Williamsburg, Brooklyn, ferry dock. It was an exciting thoroughfare, as was East Broadway itself, perhaps the most popular mercantile avenue on the East Side. But the community had changed dramatically. It was now teeming with immigrants, newly arrived, struggling for a place in the uncompromising metropolis. Hobbled by accents, customs, and fears, they found the lower East Side a trial, a trap. Only the local waifs (the Colonel must daily have watched them play) enjoyed the din:

> Grand Street . . . was as good as a show. It was the great shopping center of New York's lower East Side forty years ago. But not all those who thronged the sidewalks came out to make purchases in Ridley's, the biggest department-store in that part of the city, nor to buy of the peddlers whose little carts lined the gutters, block after block, from the Bowery to Essex Street and beyond, spilling over into the side-streets. . . . But for the boys who were out for a lark it was a riot of fun. . . . Not infrequently the feuds of the gangs were fought out on Grand Street, sometimes to the great consternation of the merchants of the carts, the contents of which were tumbled into the street. . . .
>
> Tammany Hall's influence on the East Side was, and is, largely due to the very human qualities shown by its representatives. They not only know every one who lives in the block, but they know about his domestic and economic and social needs. . . .[48]

While the company's earlier constructions had all been utterly practical and efficient foundries, shops, and warerooms, there was now some justification for flair and beauty in the design of his new building. The Colonel had little difficulty deciding upon a key feature for it: with a deferential nod to the company's master collector of sundials (and ever himself the promulgator of punctuality), he proposed to give the East Side a great clock tower.

> *June 25th 1874*
>
> *Dear Sir,*
>
> *We accept your proposition to put up the work for Clock Tower, as per Mr.* [O. P.] *Hatfield's drawings, for the sum of Eight hundred dollars, $800. You will please push the work forward with all possible dispatch, and oblige*
>
> *Yours very respy*
> *R. Hoe & Co.*
> *Stephen S. Hoe*
>
> *Mr. Jas. O. Morse*
> *c/o G. John Sr., N.Y.*

O. P. Hatfield may have been related to the brothers' old friend, R. G. Hatfield. Other details contributed to the security:

> [*To Mess. Herrings & Fanel, 8 July 1874*] . . . *We will accept your proposition to put doors on our vault with combination locks, and furnish the second hand Banker's closet examined by our Mr. P. S. Hoe, for our three old safes and $325. . . .*
>
> [*To N.Y. Fire Department, 9 July*] . . . *We are desirous of having an alarm station at our building, cor. Grand & Sheriff Sts. This is a*

> *prominent position, and while the alarm would be a safe guard to us, it would add to the efficiency of your Department. . . .*

But two more years went by:

> [*To Mess. Howard & Co., 24 February 1876*] . . . *Please make the dials for our office clock of black, instead of white marble, with gilt figures. . . .*

And another:

> [*From P. J. Kruger, 28 February 1877*] . . . *Concerning the big clock dial which I have here yet, your Mr. Depew was here some time ago taking measure for a Case for that clock, and he can tell you to what size the dial should be cut down; this depends on the case—in proportion to it.*
>
> *There are here yet the Dial, the heavy pendulum tube filled with quicksilver, and the wooden planks & iron framework for the clock. Please send your wagon for these as soon as possible, as I am in expectation of removing in a few days.*
>
> *I will come to set up your clock if you will send me notice to that effect. . . .*

Eventually, after its face was redesigned smaller in a different color scheme, the clock began to mark the passage of time for East Side dwellers. Not everyone liked it—but lesser members of the firm could not outvote the Colonel on his pet project. It nevertheless became a neighborhood landmark.

Among the many tinkerers who courted his professional counsel, one fertile mind—he had already (1868) built a ten-passenger steam-driven automobile—furnished the Colonel a surprise without the slightest intimation (for posterity) of the nature of his brainchild. Hoe, of course, must have known its species and chuckled at a good joke:

> RICHARD DUDGEON
> Patentee and Manufacturer of Improved Hydraulic Jacks,
> Punches, Steam Hammers, and Patent Roller Tube Expanders,
> 24 Columbia St., N.Y. and 114 Fenchurch St., London
>
> *Peacock Point, L.I. June 24th 1875*
>
> *Dear Col^{l}*
>
> *Maby† this will be long—But dont be scaired.† Consider where it comes from. Or do with it what you once advised me to do with ajients,† Put it in your pocket. And further console yourself You haint† had manny† such letters.*
>
> *I did not know that I was within a year of writing to you, an hour ago. But I am glad I dont pretend to know much for I am maid† to feal† daily that I dont know what I will do for sure two hours ahead.*

I wonder if there is any much diference[†] *between women finding themselves in the family way and scheamers*[†] *finding themselves pregnant with some scheam. Conseption*[†] *with both is the best part I guess, Judging from my limited experience.*

All this preface to tell you that I am afrait[†] *I am in the family way with a Dummy and the Devil take You I believe You are the father of it.*

Did not cost you much did it! Lik[†] *half the fathers in the world.*

Well not as if I ment[†] *to swear the child on You if I am so, or as if I wanted any help before I know for sure whether I am so or not. But mind this. If I am safely delivered of something that goes well and does well no matter if it is not covered with splatches of Brass and shiny iron. You own it And I'll call it The Col*^{l}*.*

Truly
R. Dudgeon

GRAND STREET AND EAST BROADWAY

c. 1885

From Outlook, 24 March 1926

4.

A Personal Brand of Salesmanship

PRINTING PRESS COMPETITORS / MARKETING IN THE PRINTING TOWNS / WESTERN AGENTS AND ITINERARIES / CHECKING AN AMBITIOUS ENTERPRISE

Like other metropolitan shops, R. Hoe & Company relied heavily on friends, relatives, and fellow merchants to reciprocate whatever business and favors would bring them into social and commercial intercourse. In this way considerable good will and broadening experiences—as well as serendipitous sales—would result.

As Arthur Pell introduced Richard Hoe to the ways of patent protection, so also his relations must have furnished early guidance in merchandising. The Meads (sister Elizabeth in 1834 married Enoch M. Mead—from his mother's family) recounted their dry goods sales and problems when the young businessman visited Messrs. Leander, Mead & Company's shop—"batts" of cloth lost in shipment, delinquent accounts, warehouse damage. These reminded the Hoe partners (if nothing else did) they must look constantly to petty sales and customer satisfaction as well as the back-room agonies of disgruntled employees, payroll preparation—and the technical riddles so much more attractive to tinkerers.

With business fairly active in the early 1830's there was no need to be particularly resourceful about formal advertising. In 1836, eighteen years after Matthew Smith had placed an advertisement in Van Winkle's first edition, the firm took two pages in the second, with essentially the same copy:

> [Left page]
> Washington Press. [cut]
> R. Hoe & Co. New-York,
>
> Manufacture, and have generally on hand, every article necessary for a Printing Office and Bindery, and a variety of Machinery for other purposes.
>
> [Right page]
> Smith Press. [cut]
> R. Hoe & Co.
>
> Manufacture, and have on Sale, the following articles. Machine presses, double and single cylinder, for newspapers or book work, [cotton and tobacco presses, bookbinders' tools, rollers, type, etc.].
>
> [Pointing fist] The above establishment has been in operation many years, and possesses every facility requisite for manufacturing purposes. The proprietors are therefore prepared to execute any order in their line with promptness and despatch; and flatter themselves, that by personal and unremitting attention to business, they will merit a continuance of the liberal patronage which has heretofore attended their exertions. . . .

Competition in the 1830's, of course, was correspondingly lively. In 1818 I. Hoit, 32 Burling Slip, had filled three Van Winkle pages with small type to advertise his "Printers' warehouse," but he was gone a generation later, replaced by White & Hagar's Printers' warehouse (Elihu White being Hoit's successor). White and Hagar handled Hagar's foundry type and all the popular presses, including Hoe's:

The Smith Press.		The Rust press.	
Medium	$230	Medium	$230
Super Royal	240	Super Royal	240
Imperial No. 1	250	Imperial No. 1	250
" No. 2	260		
" No. 3	275	Nos. 3 and 4	275
" No. 5	300		

William Hagar took orders for the Napier and also designed his own hand press (the "Hagar"), manufactured in Germany.

Modest advertising, patient word of mouth, meticulous workmanship and service—and a superior product—were slowly paying off their investment. Printing historian Cornelius Van Winkle himself described the situation in 1836:

> [W]ithin a few years, however, the cast-iron one-pull press, called the Smith press, manufactured by Robert Hoe & Co., and that made by Mr. Rust, called the Washington Press, have almost entirely superseded the use of the screw press of Mr. Ramage, as well as the more complicated press of Mr. Clymer.

> The Napier double-cylinder power press is getting into general use in the United States, for newspapers that have extensive circulation.[1]

As the Hoes were soon making all three popular types of press as well as the usual accoutrements, printers everywhere would make their acquaintance.

The company was not above soliciting business by mail:

New York Jany 13, 1843

To Messrs. P. R. Freas & Co.
Proprietors of the "Germantown Telegraph"
Germantown, Philadelphia Co., Pa.
Gents
We have rec^d^ a number of your paper in which you advertise for a second hand Napier press, bed not less than 44 x 32 inches. At present we have no press on hand of that description. It would afford us pleasure to furnish you with a new one, our price for the size, No. 2, Bed 46 x 31 inches you will perceive by the annexed circular of prices is $2100. On one of these presses you would be able to do first rate book work as it would have a registering apparatus which could be thrown off or on at pleasure. The cylinder presses we now make are far superior in every respect to those formerly made, being much stronger and so arranged as to prevent any jar to the frame, consequently they will not be subject to getting out of order and will with careful usage last many years without any expense except for Blankets. If you will come on here we can readily satisfy you that it will be far more economical in the long run to buy a new press than an old one. Our terms are half cash & half 6 months for approved paper, any deviation from these terms interest to be added. The present made presses have a strong iron foundation upon which the frame is fastened. We shall be happy to see or hear from you on the subject and remain

Your mo Obt Svt
for the proprietors
Robt Hoe

It was perhaps fortunate that the young brothers, only modestly trained academically, had a flair for letter writing. Dozens of little notes had to be dashed off daily, as we today make countless telephone calls. For sales messages and elaborate quotations nothing was as cheap, durable, and emphatic as the U.S. mails, reaching remotest communities as well as big towns. Through the 1840's and 1850's a constant stream of estimates for newspaper and job shops went to the hinterlands—Texas, Virginia, Ohio, Louisiana, the Carolinas.

Through these years the company had comported itself publicly as though competition scarcely existed. Ostensibly, even the formidable Adams brothers were no great threat:

> *[To Charles Palgrave, Montreal, 30 May 1850] . . . "Down East" people are more famous for making things cheap than good & substantial. If Mr. Adams can afford to sell as good a press as we make for less money our customers have not found it out. We have sold hand presses for years in Boston but we hear of none being introduced into New York [from there]. . . .*

The Colonel and his associates of course had not closed their ears to noises from New England—and elsewhere, which indeed reached New York. Early in its business one development came very close to Gold Street and might have been a very serious worry.

Treadwell's novel idea of a foot treadle to motivate presses came to the attention of another Bostonian, Stephen P. Ruggles (- 1880). He immediately built a large-cylinder power machine on which one person could easily get 750 impressions per hour; Putnam & Hunt, Boston, printed *Ladies' Magazine* on it with good results. He next designed a powerful platen press, but like Treadwell, could not find adequate tool machinery in Boston to make one.

Taking his plans to New York, Ruggles managed to have a prototype built. It was a compound toggle-jointed platen whose double-frisket carriage efficiently put one sheet on the bed as another was fed. But Ruggles was badly hurt by what might have been vicious rivalry: "Poverty," said *The Printer,* "and a groundless rumor that something vastly superior was about to appear in New York, alone prevented the successful introduction of this improvement as early as 1827." [2] Both Williams' type-revolver and Hoe's Napiers appeared in New York at this time—both indeed commanding attention, but for vastly different work; other inventions of course may also have been touted. But rumors are always hard to fight.

Again in 1830 Ruggles had a machine constructed, in D. K. Minor's New York machine shop (where his 1827 press must have been built); it was his first card press, but it too was not manufactured. Strangely, something was not yet ready for this Boston genius.

Familiar with Ruggles' requirements and objectives, machinist D. Kimball Minor might have profitably undertaken the production of his machines and become an important Hoe competitor. His nearby shop had stopped their 1835 Gold Street fire. Like them, Minor had early made machinework his career, but unlike them he had a literary bent: a part owner of the *New-York American,* he also founded and published, among several trade papers, the *Mechanics' Magazine and Register.* At the *American* office Minor must have perceived the engineering importance of Walker's imported Napier which—had he been so inclined--he could have manufactured even before the Hoes saw it. Minor also owned a famous hotel in Philadelphia.

Thus divided among many interests, Minor did not give his shop the main thrust of his energies. Richard Hoe recounted a delightful story at his expense:

[To Sereno Newton, 13 January 1834] . . . S. E. Morse & Co.—the [New York] Observer—went to Miner† & had his presses moved & put up; the Double press he got together wrong. Screwed the cylinder down so hard that he could not get them to move up and down. Set the pinion on intermediate roller wrong & drove the thread off the box & screw. Drysdale [a Hoe employee] was at work at Anderson & Smith & Morses man stopped & wanted him to go & see it. He went not knowing that Miner was there & that he had been moving the machine; when he found Miner & his man Miller there he would not say a word but walked round the press & laughed. Miner asked him if we had the tap that cut the thread in the box for the intermediate roller; D asked him how he thought the box would have got there if we had no tap. Morse met him [Drysdale] on the steps going out & asked him if he could set the press running; he told him yes, without jamming it all to pieces, but "three doctors will kill a patient." But says Morse that will not get our paper off. No, says D, nor will it get worked off then to day & left them. We have not yet heard from them. . . .

Newspaperman Sidney E. Morse was Samuel F. B. Morse's brother.

In 1833 Ruggles was commissioned by the Perkins School for the Blind, Watertown, Massachusetts, to develop equipment for the education of the blind. This involved experimentation on form and size of raised letters, types and paper suited to emboss them, and an efficient machine to print quantities of much needed literature. Although he remained on this assignment until 1838, he quickly designed a press that was turning out the Bible and other works for the blind within a year. Ruggles also constructed a large globe of the world for the blind, still on display at the school.

The Ruggles embosser mechanism was soon available commercially as his "Engine" press (1839); it was self-inking, treadle driven, and capable of 1200 per hour. The first one was installed in the shop of Samuel N. Dickinson, Boston, and although continually improved, it was not popular. The inventor tried to interest the Hoes:

New York Decr 22nd 1842

Dear Sir

Your letter of 9th inst. was duly received and we have taken some pains to ascertain the minds of printers in our city as respects some points in your press. The only objection urged is the type having to be with faces down which as far as we can ascertain is a difficulty in their minds which we fear it would be impossible to overcome.

It is so uncertain whether RM Hoe will soon be in your city that we would not have you defer any negotiations on our account.

Yours Very Truly
Robert Hoe

To/
Mr. S. P. Ruggles
Boston, Mass.

The chase unfortunately was unalterably positioned in the press face down, forever raising fears that a loose form would destroy the machine with pied types. As was so common in nineteenth-century industry, disaster struck Ruggles:

> *[R. Hoe & Company to R. Raphael, Mexico City, 13 March 1845] . . . We have no knowledge of who now manufacture the Ruggles job press. His workshop in Boston was burned out some months ago. As he works the type face downward we should not wish to take any responsibility in regard to them,† for every printer that we have conversed with here objects to them on that ground. . . .*

The intrepid inventor nevertheless returned to presses once more. He improved his printing "engine" and developed rotary card and jobbing machines, and other devices. R. Hoe & Company were quick to respond to an infringement problem:

> *[To Ruggles, Boston, 1 June 1850] . . . We are in receipt of your favor of 29th ult. We have three or four of the paper & card cutters on hand but will make no more before seeing you. We were not aware that you was† the inventor. . . .*

Ruggles' crowning achievement, according to *The Printer,* especially safe for "careless boys, or other persons of little skill or judgment," was the Combination Job Press:

> This press is worked by a treadle, with the foot or steam power, the same as his other presses are, and throws off with astonishing rapidity the best quality of work. An important and valuable feature in this press is a simple arrangement by which the impression may be increased or diminished with great ease and expedition, and another which insures the perfect working of the form, however small in size it may be, whether it is placed in the center or in one corner of the bed.[3]

The Ruggles Power Press Manufacturing Company, Boston, was established about 1865 (capitalization $500,000), fittingly very near the birthplace of Benjamin Franklin. "It appears that someone has bought the Ruggles patents & patterns & is going on with the manufacture of small presses," wrote nephew Theodore Mead (27 September 1865) to the Hoes. Produced were job and card presses in several sizes and combinations, many of them well received in the trade.

Despite a high price, requiring Ruggles (like Treadwell) to include territorial rights with each machine, sales continued. He was forced to buy back exclusive rights as metropolitan demand increased, and, in Boston at least, made a spectacular showing—it was claimed that George Gordon, even offering free installation, later could only introduce one of his Franklin presses there. Ruggles retired with well over a half million dollars. The Hoes had not only heard the "music" of his presses—with nothing comparable they very much wanted a machine like the Ruggles, acknowledged the prototype of all modern job platen presses. But then came Gordon.

George P. Gordon (1810-1878) was a yankee printer and inventor with strong spiritual leanings. He settled in New York and from about 1835 long tinkered with a card press, seeking no patent on it for almost 15 years. His 1851 "Alligator" clamshell machine proved to be most dangerous to operators' fingers, but improvements followed, and the resulting "Yankee" job press was more popular. After several other designs (including the "Firefly"), about 1858—in a dream—Benjamin Franklin described a better mechanism to him, and Gordon put it to work. He named the new device the "Franklin," demonstrated it at the 1862 London Exposition, and even saw it manufactured in England as the "Minerva." In the United States it became immensely successful as, simply, the "Gordon"; it featured slowly rotating ink disks which assured excellent ink coverage.

In addition to his mystical aid from Franklin, Gordon was not ashamed to credit other help: in a letter from Gordon to Ruggles, according to Henry Bullen, "he acknowledges his obligation to Ruggles as the inventor of the basic principles utilized in the Gordon presses." [4] Gordon ultimately built more than 100 kinds of presses, some manufactured in Rhode Island, but most of them made later in his 1872 factory at Rahway, New Jersey. Gordon also became quite wealthy and gave Rahway a magnificent opera house; his plant was bought in 1901 by Chandler & Price. The Hoes forgot Ruggles and reckoned they had to compete with Gordon for the lucrative job shop market.

About 1857 the Colonel recognized Franklin L. Bailey's job press, much like Gordon's Firefly, as a strong contender. Tucker spoke highly of it:

> The firm brought out at this time a Card press under a patent granted to Mr. Franklin L. Bailey June 23rd, 1857, and assigned to Messrs. R. Hoe & Co. This press was so convenient and light running that it at once became popular and soon superseded our 'Machine Card Press' that had been in use for the previous fifteen years. . . .

In 1860 Bailey patented another machine which Hoe believed useful in the job field. Tucker found it a serious setback:

> A new bed and platen job press was built at this time and called the 'Caxton Press.' It was made under a patent issued to Mr. Franklin L. Bailey, February 21st, 1860, and assigned to Messrs. R. Hoe & Co. It was driven by a treadle and would print a sheet 9 x 12 inches. The form was vertical and the platen, which was pivoted to the frame, fell back to a state of rest at such an angle as to be very convenient

to lay a sheet on. The impression was given through a crank and connecting rod, the platen was arranged to stop and start without jar and the impression could be thrown off as well. But the press never was a favorite. Printers objected to the pivoted platen of the press meeting the form at an angle. The Gordon Press was already in the field and liked, and its platen approached the form squarely. . . .[5]

Tucker suggests the company was ready to quit this costly race with Gordon—especially as the Colonel was making phenomenal news elsewhere, but Bailey was taking the challenge personally. When Robert Hoe queried some point in their relationship Bailey vented his anger and frustration:

New York Jan. 17th 1863

Mr. Hoe,
Dear Sir—

Enclosed you will find, on two half-sheets, a description of some proposed changes, and perhaps, improvements in the half Medium Press; and some suggestions which you may make the most of. I have just made these drawings or sketches, and a neighbor at my elbow placed his signature on them as witness of their date &c. These suggestions are of no use to me, but may be to you. . . .

I wish to make a proposition which I hope you will think has some sense in it, and will, in view of all that has been said and done, consider it in good faith, [at] once send me an answer. I have determined upon it only after, at last, being convinced that we are not congenial spirits. . . .

From the fact that so small a sum was paid me, and that a portion of that was to be paid after one hundred presses had been made and sold, you must conclude, as I believe, that there should be no delay in the[ir] production and perfection. But with the most important press of the whole, there has been a delay of about two years. And from the same small price paid me Mr.† Col. Hoe could not have expected me to spend six months or thereabouts in perfecting a job press which was better than he bargained for. . . . And from the fact that you do [not], or have not seemed inclined to look over the whole of the patents and from them make as much as possible; and that you do not wish or seem [dis]inclined to encourage improvements, and consequently [neglect] the keeping [of] the present plans intact against the encroachments of others; . . . I feel inclined to cut our connections (such as they have been) off, by most respectfully requesting that you consider [paying] the remaining Five Hundred dollars as in justice due me at the present time. . . .

What did I mean in making such a singular bargain? What did your brother Col. Hoe Mean? Why was he so indifferent? As compared with your other business all my aggitation† of this matter is much ado about nothing. . . . As it is and has been, I sold my experience for the small sums stated; virtually it is so. What did Col. Hoe mean by leaving the whole in so indeffinite† a manner? What did I mean? You must judge from my past actions. But view [it] in dollars and cents, and see if you cannot come to the conclusion that I have, within a few hours, but not hastily—that the whole transaction ought to be ended. I will promise you that when I make any more sales to you that† there will be no misunderstanding; they will be [in] "black and white."

F. L. Bailey
No. 3, West Washington Place

[*sketches follow—not reproduced*]

Like other inventors, Bailey had been much flattered by the Colonel's attention and was eager to see R. Hoe & Company push his equipment in the marketplace. When his press did not live up to expectations he was doubly disappointed—cheated by the company of both fame and rewards he might have obtained had he retained possession and sold his own presses. Few inventors, of course, were good salesmen; this dilemma usually had only academic importance. But Bailey did not give up.

At work on still another machine, Bailey corresponded with the Boston office. Mead sent a memo to New York:

[2 *January* 1864] . . . *Please find herweith a communication* [*not present*] *from F. L. Bailey respecting his patents. You will I hope give them a careful examination as they will come into competition with our presses immediately if not purchased*. . . .

But all knew by this time that Gordon had run away with the market, and although he labored well into the 1870's Bailey's fury was spent. The Gordon enterprise had become unconquerable.

Algernon Sidney Doane, a Gordon nephew, had studied law and become a patent agent; he functioned as general agent and on Gordon's death became manager and trustee of the property. His knowledge of the press business was appreciable. Gilbert K. Harroun, another prodigious inventor, developed a numbering machine for Gordon's railroad ticket presses, increasing their usefulness considerably. Also at the Gordon plant was Frederick O. Degener, an expert machinist who worked for years as draftsman and general designer. About 1859 (still in Gordon's employ) he invented his own machine, the "Degener" (or "Liberty"). It was patented 1861 and very successful. Thus Gordon and his associates had a firm grasp on America's small commercial job shops, and "even the formidable Hoe & Co. gave way to him," [6] said James Moran. And Gordon won over Ruggles' Boston as well:

[*Mead to R. Hoe & Company, 17 February 1866*] . . . *Gordon's agent here told me that he sold, last year in New England, Gordon presses to value of $25,000*. . . .

Little perturbed by Hoe's competition in the small press field, George Gordon proved himself genuinely gracious:

96 Fulton Street,

New York, July 3, 1868.

Col. R. M. Hoe:
My Dear Sir:—Understanding that you have made application to Congress for an extension of the patent granted you on your Rotary Printing Press, and regarding, as I do, your invention of the same as the leading and most prominent improvement of the century in printing machinery, I cheerfully and willingly add my testimony to its value, and trust that your efforts to procure an extension may be crowned with success.

Surely a just recognition of the merits of your great improvement demands it.

Very truly yours,
Geo. P. Gordon [7]

Gordon had likewise made himself a popular and valued friend to the whole industry.

Upon his death in 1878 the printing trade of New York city resolved to give his widow an engraved tablet, recognizing that "his family has lost a devoted husband and father, and the 'art preservative of all arts' a craftsman and inventor of rare genius and skill." [8] Andrew Little, of the typefoundry Farmer, Little & Company, and Stephen S. Hoe, of R. Hoe & Company, made the presentation. Imitation again was flattery: in 1884 Schneidewend & Lee, Chicago, began to make "Challenge" presses and in Cleveland Harrison T. Chandler and William H. Price in 1886 also produced Gordon-style machines.

Merritt Gally (1838-1910) was apprenticed to a printer at 11, but went to a theological seminary and was ordained a minister. In 1869 he nevertheless invented and patented a heavy duty "Universal" press which Colt's Patent Fire Arms Manufacturing Company, Hartford, Connecticut, began producing in 1872. Gally was also inventor, in 1876, of the self-playing pianola and aeolian organ, both motivated by perforated paper rolls; he also improved commercial telegraph apparatus and had 500 other patents to his credit. With improvements by John Thompson, manager of Colt's press division, over 2000 of Gally's machines were turned out by 1886, soon advertised (from 1885 on) as "Colt's Armory." Thompson (with more than 40 of his own patents) bought the division from Colt in 1902 and set up to make jobbers at a new plant in Long Island City, New York.

Nathan Babcock (1824-) grew up on a large Connecticut farm. At the age of 15 he attended school in Westerly, Rhode Island, and then apprenticed to a nearby machinist, with whom he remained for 13 years. During that time he mastered the mechanic's trade and was ready for a career of his own. With Calvert B. Cottrell (1821-

1893), whose education and professional background were quite similar, Babcock in 1855 leased a Westerly factory, for Charles Potter, Jr., to make an oscillating printing press designed by New Yorker Merwin Davis.

Owner of a small printing shop, Davis had developed his machine in 1854; he offered it to A. B. Taylor, who saw no merit in it. He then took the plans to Potter (1824-), an enterprising New York salesman who agreed to manufacture it. The Davis won a gold medal in 1855 upon its introduction to the trade, and sales were promising until the Civil War intervened. Potter's assignment persuaded Cottrell and Babcock to manufacture their own presses; they soon turned out stop-cylinder, two-revolution, lithographic, two-color, country, job, and newspaper machines which were well received. Cottrell invented an air spring to cushion flat-bed reciprocation, and with the universal availability of steam power developed larger presses —not without almost inevitable scrapes with R. Hoe & Company.

The Cottrell & Babcock firm failed at one point (1876) whereupon Potter blamed their undoubtedly ultraconservative "mode of doing business," and Robert Hoe (29 August 1876) commiserated: "I don't see how they can continue, especially since we ourselves must lower prices & have instructed Munsen[†] & Philipp to commence suit against [them] for infringing on Marinoni [lithographic press] patent. . . ." But the implacable yankee partners were refinanced. Finally, in 1880, Babcock sold his interest to undertake another venture: with George P. Fenner and Charles B. Maxson he formed the Babcock Printing Press Company, New London, Connecticut, to make large cylinder machines in a half dozen models.

Cottrell (who himself accumulated 125 U.S. and foreign patents) took his three sons, Edgar, Charles, and Calvert, Jr. into the business, and later a fourth, Arthur. As C. B. Cottrell & Sons the firm would soon specialize in a newly emerging field—four-color (process) printing presses, in which it engaged successfully for many years. But Potter's genius for salesmanship continued to complicate the situation for the yankee press builders.

Charles Potter functioned as New York agent for the partners, although without conspicuous results until 1864, when their improved drum-cylinder began to attract country newspapermen and fine book printers. Potter claimed he could sell it even better if it had his name on it, and so the "Potter" was manufactured until 1867 (at which time Cottrell & Babcock opened their own New York office). Potter in 1865 formed a partnership with J. F. Hubbard (later the Potter Press Company) and with works at Plainfield, New Jersey, did very well. "The Potter single cylinder press is a great competitor for us at present with the two-roller presses," Robert Hoe wrote his brother (10 November 1871). "Is there anything worth adopting that you see abroad to improve them[?]" Potter subsequently made and sold Scott presses.

Among Cottrell & Babcock's many early machines was a "chromatic" power clamshell invented by 17-year-old boilermaker George Babcock (distantly related to Nathan). Babcock (1832-1893), however, no simple boilermaker, was born in upstate New York into a family of professional inventors. In 1844 the Babcocks had moved to Westerly where George, feeble with consumption, took up the novelty of Daguerre photography. Surprisingly, his health recovered fully —thanks, he believed, to the continual inhalation of iodine fumes. In 1851 he started a printing office to issue *The Literary Echo,* a weekly, but three years later sold the shop and newspaper to engage in photography full time. With his father he invented a system of multicolor printing which failed commercially only because of its advanced technology. Babcock next developed a job printing press which Potter and others undertook to manufacture; it reportedly won a prize at the London Exhibition of 1851, but George Babcock had not yet reached his stride.

Babcock and his father resumed control of their paper, renaming it *Narragansett Weekly,* and at about the same time (1858) the young inventor offered his press to R. Hoe & Company. Babcock insisted the machine was already bringing him profits of $1000 monthly, and he would not let the patent go for less than $7000. The Colonel's response was not enthusiastic. Cottrell & Babcock eventually acquired it.

In 1860 George Babcock moved to Brooklyn, New York, where he worked daytime for a patent attorney and at night as drafting instructor at Cooper Union. The Civil War sent him back to New England making warships and shrapnel shells, but toward the end of the conflict he met an old Westerly friend, Stephen Wilcox, with whom he formed a long and successful partnership in 1867 to manufacture steam boilers and engines. Babcock & Wilcox continues to this day.

With uncannily gifted competitors like these one after the other putting together (in Leigh Hunt's words) "a business of screws and iron wheels," the Hoes might understandably have dreaded each new issue of the *Patent Office Gazette.* Some of the newer or impecunious tinkerers offered their wares to the Colonel, perhaps satisfied enough with an audience if a contract did not ensue. Richard very often had little time to investigate the merits of unsolicited ideas, and reacted negatively as a matter of habit. He was seldom wrong.

Joel G. Northrup of Cortlandville, New York, got a patent in 1842 for a machine meant to serve between the hand press and larger cylinders. He recollected:

I made a small model with wooden frames and cylinder, and exhibited it at the fair of the American Institute at Niblo's in October, 1842. It attracted a great deal of attention from the crowd, as I printed a diminutive sheet upon the little press. An old press builder by the name of Fairman† [Fairlamb] became interested in it and induced the Hoes to inspect it. Mr. Hoe having remarked that he could see nothing in it worthy of note, Mr. Fairman said to him:

"Sir, notwithstanding the crudeness of that model it contains a most valuable principle we as press-builders have overlooked, and which will some day be used to print from a continuous web of paper. . . ."[9]

The following winter Northrup designed a medium cylinder press which was built and sold (under royalty) by Wheeler, Young & Company, Bridgeport, Connecticut. Several were installed. In response to typefounder George Bruce's 1851 contest for a new press ($1000 premium), Northrup improved his novel machine, on which entire flat beds rotated under an impression cylinder. The Phoenix Foundry, Pennsylvania, built the improved version but could not finish it before the deadline. Bruce's judges recommended it anyway as the winner, and about 200 were built over the following three years.

Northrup continued inventing, his 1886 project a combination job press and folding machine, but history passed him by. The Colonel need never have regretted his decision.

One rival press maker, however, promised to give the Hoes a brisk race, trained, as he was, in the Gold Street shop. Knowles & Anthony, Providence, Rhode Island, dared to ask the Colonel his opinion of this competition:

[*To Hoe, 7 May 1856*] . . . *We contemplate enlarging the* [*Providence*] *Journal on the first of July next. We have been using hitherto the Adams (improved) press. Our increased circulation renders more rapid execution desirable. . . .*

We notice in Bruce's specimen book an advertisement of a press with the name "A. B. Taylor & Co." upon it. [*How does it compare with yours and what are your terms?*] . . .

But we suppose you keep the lead in the manufacture of this kind of machinery and as the latest circular we have of yours is 5 or 6 years old, we conclude you will now be able to give us all the "modern improvements." At least all that are really such. . . .

This was only a ploy to get a Hoe catalog promptly; an order resulted. While it is possible that Knowles & Anthony had not heard of A. B. Taylor's misfortune, the Colonel was at least relieved of rendering an account of it.

Alva Burr Taylor (1803-1889) was born in Westport, Connecticut, and worked in his father's smithy making horseshoe nails. By trade a blacksmith, Taylor moved to New York 1825. Four years later he entered Robert Hoe's employ and in 1832 became foreman of the double-cylinder rooms and then general superintendent of the whole cylinder press department. He was long a capable employee and faithful friend.

About 1840 Taylor left to take charge of the printing plant in a Connecticut state prison, although after three years he returned to New York city to engage in the manufacture of his own printing presses. His experience in this new industry was ample, of course, and his intimate knowledge of the Hoe establishment and its techniques made him singularly competent.

The A. B. Taylor Printing-Press & Machine Company, Hague Street, New York, was founded 1842; it was soon, like Hoe, making drum-cylinder presses based on Napier's (and Hoe's) designs in nine sizes, for which the demand was lively. Taylor did well for several years, developing an 1846 air plunger and cylinder which nicely cushioned the shock of reciprocating bed motion in his presses. Then tragedy struck.

In February 1850 a 200-horsepower steam boiler in the shop's basement exploded and demolished the entire six-story building. For the newspapers it was news of the most blatant kind. A *New York Tribune* reporter rushed to the scene:

> We never saw a wreck so complete: the walls, floors, large timbers and roof lie in one compact pile, all on fire, and seething, smoking and steaming beneath the flood of water poured upon them. . . .[10]

Windows, doors, walls, and foundations all along Hague Street had been jarred and smashed.

The blast occurred 7 a.m. on a Monday morning, largely as a result of the new but second-hand boiler being over-fired after a cold weekend. More than 50 of Taylor's 55 workmen and boys had just arrived, as had girls who made hats for a haberdasher on the first floor. (Taylor occupied floors 3 through 5.) The call went out for police and firemen. William M. Tweed, assistant foreman of Americus Engine Company No. 6, first fireman to arrive, promptly saved five persons singlehandedly. After 12 hours workers had retrieved 32, of whom 10 were dead. New Yorkers began to realize it was the city's worst disaster. Neighbors and friends rallied to help. Taylor's old business associates did not forget him and his employees:

> *New-York, Feb. 4, 1850*
>
> *Dear Sir:*
>
> *We have learned, with the most painful feelings, of the sad catastrophe at your establishment this morning, and cannot retrain*[†] *from tendering you our most sincere sympathy.*
>
> *If we can be of any assistance to you in the emergency we shall be happy to render any aid in our power.*

We shall be glad also to have you disburse the enclosed $100 to the sufferers or their families in such a way as you think proper.

Yours very truly

To Messrs. A. B. Taylor & Co., Present[11] *R. Hoe & Co.*

After the second day of digging and searching 22 victims were found alive in the rubble, and 31 more bodies, some of which were mutilated beyond recognition. Again the community was aghast, following the newspaper accounts. At the Gold Street shop John Colby took up another collection and turned $60 over to the victims' fund; at Broome Street A. J. Bowen collected $78. Throughout the city thousands of dollars were similarly contributed. The grisly work went on.

After five days 7 more bodies were recovered, totalling 62 dead. A Grand Inquest, formed to determine the cause of the explosion, heard master printer Daniel Fanshaw:

> I have employed steam for printing purposes for the last 20 years; I have used two boilers for the last 17 years which resemble in part, the boiler in question; I think the arch at the bottom of said boiler tends to weaken it; I have had a number of partial explosions. . . .[12]

Authorities from the Novelty Iron Works also testified. Fanshaw though the pipes must have frozen, and one witness blamed the engineer (whose body was last to be recovered—a week later) for throwing several buckets of horse dung into the boiler the previous Saturday to stop leaks.

Of Taylor's workers 30 were dead and 15 injured; among the other tenants 32 were dead and 15 injured. Five bodies were never identified. Taylor lost $50,000 in tools and equipment and could scarcely be expected to continue with any vitality. Remarkably, he managed nevertheless to do so. In addition to a small output of stock presses he produced several important machines: one for *Harper's Weekly* was a fine testimonial of the publisher's confidence; another, in 1858 for *Frank Leslie's Illustrated News,* was an automatic perfector which "fed itself with paper, printed both sides at once, and stacked the finished sheets at the rate of 1,200 per hour, all without the assistance of human hands." [13]

A. B. Taylor, Son & Company continued in business until the 1870's, having at one point (1865) also constructed three fire engine pumpers. The Hoes slyly seemed to prefer him as a rival:

> [*Richard Hoe to Robert, London, 10 May 1859*] . . . *I saw A. B. Taylor a few days since, & proposed to him to arbitrate the matter of infringement of the Adams* [*flyer*] *patent according to our understanding with Mr. Adams; he assented to the arbitration, but proposed to consolidate or to sell out to us. I told him we did not wish to increase our business or obligations; he wished us to consider the*

> *proposition. I told him we would do so but I did not think we should think favorably of it. I think & so does brother Peter that we had much better have his competition than that of some better man. . . .*

The following year Hoe invited Taylor to settle by paying $10,000 for all past infringements, and $250 for each fly attached in future work. These were costly terms for a small firm fending off other yapping press builders.

Taylor's older son George W. (who invented an automatic press feeder with vacuum fingers) in 1867 formed a partnership with S. P. Rounds to make a "Chicago Taylor Press," but he died two years later and the family business fell to grandson A. B. Taylor, Jr.

Present at Taylor for some time was Andrew Campbell (1821-1890), a "mechanical genius of great ability" who had joined the firm in 1853. Raised in a poor New Jersey family, Campbell was apprenticed to an Illinois coach builder and employed at odd mechanical jobs in St. Louis. When he returned East to see the New York Crystal Palace Exhibition in 1853 his admiration for the modern state of printing machinery compelled him to remain to participate in its exciting progress. While at Taylor's he constructed the Frank Leslie and *Harper's Weekly* perfectors, and was long the shop's foreman. When Campbell left in 1858 the struggling proprietor offered a partnership to Stephen Tucker—but Hoe would make a hasty counteroffer the young man could not refuse.

Campbell in 1858 set up his own shop to make "country" cylinder presses, two-revolution machines, and oscillating and lithographic presses. His man-powered country press appeared 1861 and was very popular; in 1876 he found the first one he made still running, bought it, and displayed it at the Philadelphia Centennial—where he also had a modern machine printing newspapers at 10,000 per hour. Campbell was a good friend of William Bullock, to whom he gave some design help. In 1866 he built an $80,000 Brooklyn factory which, among others, furnished a special web-perfector to the Cleveland *Leader* the following year.

But for Campbell too competition was brutal. At one point (1870) he had only a handful of men at work and was struggling to pay them; five years later Campbell and Hoe both raced to court in England with similar folding mechanisms, but, providently discouraging a long and costly fight between two foreigners, the judge uniquely recognized both claims and issued "two good [provisional] patents."

Despite encouraging sales, however, disagreements with his stockholders and nagging patent suits forced Campbell to retire early, and the company continued prosperously without him. In the 1880's John T. Hawkins was president and chief designer; a new factory at Taunton, Massachusetts, employed between 400 and 500 hands. The firm would present a strong challenge to Hoe salesmen wherever they met.

Before such competition in the field became a vexing—and sometimes comical—display of agent tomfoolery, however, R. Hoe & Company with deadly seriousness strove to stop rival inventors in the courts. After the Civil War the state of the machinist's art was such that almost any clever gadgetry could quickly be put on the market; with few satisfying pre-war precedents in patent infringement, many impetuous inventors threw caution to the winds. It was good yankee sense to create; to refrain because someone else may already have done so seemed unconscionable. So many elements, motions, and mechanisms in an endless variety of presses—all of which had to accomplish pretty much the same work—made patent infringement almost constantly inevitable.

No organization was better situated than Hoes' in wording its own patents with consummate skill, watching competitors and their products, following the domestic and foreign engineering trends, and pursuing infractions promptly, thoroughly. As disputes became increasingly complex—and critical in time and net value—the company engaged Munson & Philipp, Broadway, foremost patent and copyright attorneys (both having resigned in 1875 as Patent Office examiners). With many industrial giants among their clientele partners H. T. Munson and M. Bernard Philipp had excellent reputations. The Colonel put them right to work and tolerated no procrastination. Regarding the British provisional folder patent (versus Campbell): "Our agent informs us that whoever completes his patent first, that is, has it *sealed* first, has the priority. . . ." Then on to another:

> *[R. Hoe & Company to Munson & Philipp, 5 January 1877] . . . as to Messrs Cottrell & Babcock there should be no delay. They have copied us in every possible way & now we insist upon the suit being prosecuted with all vigor to the close. . . .*

Although Campbell throughout his career was reportedly scrupulous in not imitating Hoe work—and Cottrell and Babcock no less honorable men—occasional scrapes were unavoidable even if culpably unintentional.

Toward the end of the century younger press designers nevertheless managed to make room for themselves. Robert Miehle developed the first two-revolution machine (1887) and manufactured presses in Chicago. His success led to a vertical model which became the mainstay of modern letterpress job shops. William M. Kelly in 1911 was 42 when he developed his flatbed cylinder machine; very popular, it appeared 1914. Scott, the Goss brothers, and others were similarly determined to earn a portion of the trade.

Notwithstanding the large and lucrative business of job, book, and periodical press sales, the Hoes, having almost singlehandedly nurtured the penny newspaper into a mammoth industry, had struck a vein of gold they considered rightfully theirs. Robert (26 November 1875) reiterated the company's long general desire "to clear off all encumbrances and owe nothing," but added, significantly, "the great thing for us is to retain all the newspaper press business. . . ." Gradually Hoe left the small and medium job shops considerable variety in their selection of platens and flatbeds —by 1894 over fifteen different press manufacturers had offices in New York clamoring for attention. By the turn of the century R. Hoe & Company—with Bullock and Walter out of the way and not yet troubled by Goss and Scott—would commit itself strongly to the demands of the largest newspapers, and by one incredible advance after another, meet them handsomely.

But almost from the beginning Richard and Robert Hoe had demonstrated uncanny sales ability. Having established sound footing in the New York area the brothers lost little time seeking markets in neighboring communities of whatever population. This invariably involved long and taxing journeys with well-thumbed timetables and atlases in hand. Matthew Smith had described such a trip:

> [*To Sereno Newton, 29 August 1834*] . . . *I returned from Phil yesterday—whither I had been on some business.* [*Charles*] *Alexander is entirely broken up, he expects however to make some arrangement to buy the press before many days. Business in Phil*[a] *is looking up a little. I sold a No. 3 press there to go south. Mr.* [*John*] *Rives has called to day & ordered a medium press for* [*Francis*] *Blair &c &c. . . . The cholera is about the same and very little alarm is excited. . . .*

Where towns of some size possessed fellow tradesmen of established reputation the formation of an "agency" was far better than personal trips. Newton had perhaps inaugurated this procedure among hardware dealers, and with especial good luck and sense the company cemented an early friendship with a most industrious printing ally, L. Johnson & Company of Philadelphia.

Lawrence Johnson (1801-1860) had apprenticed to a British printer before emigrating with his parents to the United States in 1819. He worked for a year at Bunce & Gray's New York shop, often setting type 16 to 18 hours a day. Johnson became interested in the new art of stereotyping, and to learn its intricacies worked for some months for B. & J. Collins, New York, then one of its few practitioners. Still only 19 in 1820, he removed to Philadelphia to engage in his own stereotyping business.

With limited capital Johnson, over a decade, built a thriving stereotype foundry. In 1833, with George F. Smith, he bought the typefoundry of Archibald Binny and James Ronaldson. Despite—or because of—his indefatigable work habits Johnson's health began to fail, and in 1845 he brought together as associates Thomas Mackellar, John F. and Richard Smith (all Smiths unrelated to Hoe's partners), and Peter A. Jordan. The firm of Mackellar, Smiths and Jordan soon turned the foundry into a prospering business, issuing a myriad of new typefaces and a journal, *Typographic Advertiser,* to help publicize them. Mackellar himself, however, was the firm's best salesman.

Thomas Mackellar had apprenticed at Harpers; he became foreman of the Johnson stereotype plant, editor of *Typographic Advertiser,* and compiler of the company's grand specimen book. R. Hoe & Company also respected Mackellar's scholarship: the long footnote history in his typographic manual was commended by Peter Hoe as perhaps the best "account of our concern, and the origin of my father."

Johnson naturally functioned as jobber and distributor of Hoe equipment. An exchange in the 1830's demonstrates he had much to teach the young brothers:

> [*Date lost*] . . . [*A*] *French lady, who called at our office . . . has already purchased a similar press* [*Washington Foolscap*] *from you, for which you charged only $130. . . .*
>
> *In the spirit of our agreement (I have not referred to the letter of it) it is stipulated that should you at any time reduce the prices of your articles, or any of them, that they shall be reduced in like manner to me; and you very well understand, that unless this is done, on the same terms as yourselves, that I can obtain no customers, for people will always go to you if you sell lower than I. . . .*
>
> *Now this is of more importance than it may seem to be, for you are well aware that the tendency is to a permanent reduction, although possibly you thought of it only as applying to the particular sale. Such things cannot long be concealed, and when known will have their effects, the tendency of which is necessarily injurious to you and all who deal in your manufactures. . . . Stability, and confidence, in these matters are indispensable. . . .*

A few years later, thanks again to Johnson, another lesson in discounting was brought home:

> [*Robert Hoe, to Richard, 14 April 1837*] . . . *I don't know whether you are aware of the arrangement which has been talked about with the type founders and ourselves. Matthew* [*Smith*] *told Johnson when he was on here that in case they would buy our articles exclusively & not get the small articles of any one (such as Cases & Compo*[*sing*] *sticks) who might undersell us, we would make to the trade a disco of 15% instead of 10% as heretofore but on no other consideration. In his letter this morning he* [*Johnson*] *says "The Type Founders all agreed that they would accept your terms so far as it might apply to every article they purchased from you, but they would not bind them-*

selves not to purchase cases &c elsewhere although they felt assured they would not so long as you do not raise the price & keep a good supply & a good article." Now you see this is not the thing at all as it leaves us just where we now are—minus 5% . . . .

I met with quite a disaster; the Whippletree broke and my horse tumbled down. No bones broken, however. . . .

Such discount machinations, however, seemed only to materialize during depression times, when manufacturer, salesman, and customer all felt the miseries of debt and paucity. At most other times sales activity was at a euphoric high:

[*Matthew Smith to Newton, 16 February 1835*] . . . *Wm Jackson we think will pay—we have made an arrangement by which we feel secure. We put Bells press up on Saturday last; it answers well. We have plenty of work, more than we can do at present. . . . The screw & lever copying press*† *sell like wild fire and our only trouble is to get them finished fast enough. . . .*

Only a week later the roster of work in progress was all new:

[*Richard Hoe to Newton, 20 February 1835*] . . . *We are driving ahead with as much work as we can stagger under. We are finishing for* [*Van Orden?*] *the Standing Press; 5 stave* [*making*] *machines are under way all to be* [*paid*] *cash. . . . We are making 3 Embossing presses with a frame like a fly press* [*tiny diagram*] *& the smith*† *power bar & joints like sherman*† *press—one for Ricker, one for Griffen & one for Colton & Jenkins at $250 each. We are also making 2 Sherman presses for Scatchard & Adams*[?] *who were burnt out over Dearborn. We are all well. . . .*

and a month later—

[*10 March*] . . . *Business is too pressing; we have to turn away work. . . . I am so tired & sleepy that it is impossible for me to write a straight sentence much less a straight line, so good night. . . .*

Within a few years both Sereno Newton and Matthew Smith had left the scene; Richard was busy with a lifetime's activity of invention and shop improvement, and Robert and Peter were desperately covering sales matters from the field and the office, respectively. In addition to typefoundries and hardware stores the services of agents in Buffalo, Boston, Baltimore, Washington, Richmond, Atlanta, Charleston, New Orleans, and San Francisco were not simply adjuncts but a practical necessity. By midcentury arrangements with them were fairly clear:

New York November 16th 1849

Dear Sir:

Your favor of the 11th instant is received. We regret that our Mr. Hoe was not more explicit in his arrangement with you in regard to the terms upon which you were to become our agent for the District of Columbia.

We are writing to allow you the same comm[s] as we allow to the Type Founders in the city of Phil[a] & Balt[o], which is fifteen per cent upon all articles except type, charging of course for boxing, shipping expenses including Insurance; these charges are invariably made & they are paid by the purchaser (our prices, being for the articles in our wareroom; we do not even deliver our presses in this city without an extra charge). During the arrangement in Exchanges, we have always charged the current rate of exchange in N York at maturity, of payments, in all cases making our draft in accordance with such arrangement, i.e. "pay with current exchange at its maturity.". . .

To our Agents generally (excepting the above named) we allow 10 per cent commission, subject to all the conditions above mentioned. It is true that we allowed Mr. [Garrett] Anderson [Washington typefounder] a Comm[n] of 15% but this was upon a credit of four months, and we sent him nothing upon Comm[n], all that we sent him was upon actual orders; for what we drew upon him at time of shipment "with current rate of exchange."

It is not our intention to consign articles on commission (except in a very few instances) and then a small assortment of such as are constantly in demand by printers, and of which they would not be likely to order a large quantity, such as Brass Rule, Furniture, &c &c.; consequently the business growing out of the agency would be merely the reception & transmition† of orders; the above comm[n] in all cases includes the guaranty, and upon shipment of an order to any of our agents we are at liberty to draw for the am[t] at six months from date of shipment. As it regards consignments made to you to be accounted for when sold, we would not make any charge for boxing & Insurance but all other charges we would expect you to pay.

Should you conclude to accept the Agency upon the Terms now stated, you will please inform us at your earliest convenience. In the mean time

We remain your Obt &c.
R. Hoe & Co.

To R. Farnham Esq.
Washington City, D.C.

P.S. As you decline to guarantee P. Force's order, we have concluded to ship direct to him. R. Hoe & Co.

Peter Force published a paper and had been mayor of the capital, 1836-1840. Refusing him credit was a bad start for Farnham, who apparently did not long survive as agent.

Agents in distant cities were useful, of course, for many tasks and chores beyond selling. But favors were invariably reciprocated.

[To Johnson, 2 November 1844] . . . You will much oblige us by purchasing for us if it is to be had in your city, six copies of Nicholsons operative mechanic,† the latest edition. It was published by Mess. Carey & Lea in two volumes. Please send it [express] by Adams & Co. . . .

A few days later, with a customer requesting a "flyer," Johnson asked what was meant. Richard in 1844 was indeed working on his mechanical "fly" to remove printed sheets from a cylinder press, but strangely this was not the wanted article. "We think by flyer your correspondent means a fly-wheel," said the company (19 November 1844). "Originally we made smaller flywheels than now. A new fly-wheel costs $100." In a subsequent letter Robert Hoe offered the services of a mechanic on his way to Washington who could stop two or three days at Philadelphia to exchange flywheels (and allow Johnson 12½ per cent commission).

A month later Johnson himself was in New York, shopping:

> [*Robert Hoe to Richard, Washington, 6 December 1844*] . . . *Mr. Johnson is here this a.m. and as its late and I but a moments time to write therefore only say a few words. Oliver Ames has offered the shafts or hydraulic rods at 8 cts pr th* [*each*] *& I have written him to make them. Says he can have them done in a few days. The pattern for Kemble* [*West Point Foundry*] *has been sent out with the order.*
>
> *I have sold the Tuft*[†] *press to Mr. L. Johnson for 500$ 6 mos. net.*
>
> *Ludwig's boiler was put in yesterday; the engine goes over to day.*
>
> *Have that coal Peter wrote about sent on immediately or let us know . . . whether to buy it here of Del*[*aware*] *& Hudson. Nothing new by mail. . . .*

Johnson's clientele extended throughout the South and even to "Campeache," where interminable postal delays called for impulsive sales techniques: when Hoe asked whether the Mexican printer wanted an 1848 inking machine or was ordering from the 1848 catalog, Johnson replied (7 April 1858), "Our friends through whom the order comes . . . think we had better send the [most likely] machine, as they say it will take three months to get the information. . . ." On his frequent trips to New York Johnson must occasionally have left his wife and pretty daughters at one of the Hoe households while he talked shop with the brothers. The two families became very close.

Although still a small town, Washington, D.C. became tremendously active commercially as telegraph and newspapers (reporting government affairs) began to reach beyond the district's boundaries. But from the beginning business there was a frustrating experience demanding the Colonel's own exertions regularly. No agent could really handle that territory.

Duff Green, a St. Louis lawyer and friend of Andrew Jackson, became U.S. government printer (1825, 1827, 1829, 1831). He was also editor, from 1823 on, of the Washington, D.C. *United States Telegraph,* which for a time was "Old Ironsides' " official spokesman. By 1837, however, Green had lost favor and suffered heavily in the depression:

New York October 2nd 1837

Dear Sir

Enclosed please find your note for $1615.63/00 due & paid by us this day for which this morning we rec^d^ your two notes. It really seems to us that you might pay something upon them from time to time. The press would avail us nothing at present and very likely should we take it as you propose we would be obliged to advance the amounts of the mortgage. We want the money or such security as will enable us to realize the amount at an early date. We look for the assignment of your mining stock to us as agreed as early as may be. We have every disposition to accommodate but really this acco has been so long standing & so frequently have the notes been renewed that we again have to express our belief that through extra exertion if necessary at least 10% might or at least should be raised as they fall due. Do try to effect this if nothing further, in future.

We remain with sentiments of respect

Yr Obt Svt
R. Hoe & Co.

To Duff Green Esq.
Washington City, D.C.

Green attempted some wily money-making schemes, among them a boys' printing school (to obtain cheap labor) and unsurpassed production, hopefully, from Charles Williams' unique rotary press. Shortly R. Hoe & Company would recognize in all transactions with Washington printers the elements of speculation, favoritism, and vacillation, until a proper government printing office ended the intruding vicissitudes of politics.

Thomas Allen, government printer 1837-1840, 1841-1842, was editor of the *Madisonian*, which soon became an influential daily. For the paper Allen ordered much equipment and, like Green, paid little. Hoe was no longer content to rail by correspondence, however, and sent A. S. Bowen, one of his senior employees, to camp on Allen's doorstep:

New York April 25, 1840

Dear Sir

We were quite disappointed in not hearing from you today, as we were quite certain we shld^t^ receive some money; however we know that some good reasons exist for yr not writing. We have this day forwarded to Mr. T. Allen the balance of the order (excepting the type) which will go early in the coming week. The amo of the bill now sent is $149 ^50^ as you will find by invoice sent Mr. Allen. We are expecting to hear from you to morrow certain. Write us every day so long as you are detained. Nothing of interest has occurred & all as usual.

Yours very truly
R. Hoe

To A. S. Bowen Esq.
Washington City (Care of Mr. T. Allen)

[To Bowen, 28 April] . . . We have this a.m. rec^d^ yours of 26th

> *(Sunday) and am sorry to hear that you have so much difficulty in getting the money, for we never wanted it more as you know very well. We hope you will stick to it until a final arrangement is made. Mr. Hogan [Hogan & Thompson, Philadelphia] has his men working nights in order to complete the balance of the type as soon as possible. We wrote you yesterday that it would be sent on by Friday of this week. It will be a long time for you to stay from us if you wait till the balance of the type comes on but we think it advisable for you to settle the whole before you take final leave of Washington City. We need you home very much but still we know the importance of yr being in the spot to secure the whole amo without delay. We should be willing to take Mr. Anderson's dft on Duff Green for the amo of our claim against Mr. A. . . .*

Incredibly, days turned into weeks:

> *[To Bowen, 5 May 1840] . . . We are waiting most impatiently for money from you, we have heard nothing since Thursday last and are very anxious to know how you get along. We have to day forwarded as per recpt annexed the balance of the type, a bill of which you have above am° $561.66. We have charged the 100 lbs leads in bill but have not sent them as they were ordered document length; please send a pattern to cast them by. We hope you will be able to close the whole business immediately on recpt of the type so that you can come home. If possible get them to close it without keeping you waiting for its arrival. Col. will leave to night for Albany & we want you at the works very much, notwithstanding this we are anxious you should make a final settlement with the Committee & Mr. Allen even if it should detain you several days longer. Write every day. In much suspense. . . .*

> *[To Bowen, 13 May] . . . We have forwarded the Leads charged in our former bill as per annexed line recpt. We have also put in the box the sorts and the cut of eagle, a bill of which is above to Mr. Allen's debt $4.52. This morning we were in confident expectation that we should receive a considerable remittance from you. If you can send no money we wish you would write by every mail as we are watching the Post Office daily in eager expectation of hearing from you at any date. It is too bad for them to keep you waiting so long. We should think your patience would be sorely tried with such treatment. We are out of all patience with them. We have made our calculations upon having the greater part of it by this time & can assure you we have found it most inconvenient to find that our calculations and plans amount to nothing. You must make them toe the mark in some way. Don't fail to write by every mail. . . .*

The cost to Hoe of this deliberate harassment—for a rather small sum—must illustrate the serious financial difficulties of the struggling firm. It also became obvious that Bowen was not the driving bill collector the brothers hoped for. Some discrepancy in the type order added complications:

[*To Bowen, 16 May*] . . . *The type we will make all correct with Mr. Allen. Mr. Hogan says in hurrying it off, his hands deceived him but in future he will see to the founts himself. If Mr. A. will make up an account of the deficiencies & return the surplus quantities we will have them all made correct. The size of the old Press bed of Bennets*[t] *is 50 x 31 inches. It is a first rate machine & would be put in complete order. B.'s new one is to be 57 x 34 in. Perhaps you might sell this by going to Richmond.* . . .

Your folks are all well; your wife says it is a great shame for you to be kept so long from home. However I suppose she will be more happy to see you when you come home which we do not wish you to do until you have closed all your business. . . .

The side trip to Richmond was fruitful:

[*To Bowen, 23 May*] . . . *We have yours of May 17*[th] *& suppose you are now in Richmond. I have written to Mr. Allen a letter which you will read & if you think it advisable hand to him & if not so keep it to yourself. Mr. Longtree has been on here; you had better call and see him. He saw the Fairlamb press & wants one; he sent us the parts of two hand presses & copper plate press also parts of a standing press & ordered them & some chases to be put in order which we are doing, the cost of which repairs I think will not exceed $300 to $350. He said some parties were to hand him some money & as soon as he rec*[d] *it he would give us an order for his entire office and order a Fairlamb—*<u>*that*</u> *you can talk to him about.*

If you should get to the orders from a committee for him do not leave it as I left Allens affair but have a responsible name for the [*signatory*]. *Your folks send a letter.* . . .

Lamentably for Hoe, Washington business involved interminable delays not so much because the editor-printer would not pay, but because political investors—the "Committee"—were seldom quick to make decisions, no less advance campaign funds. But Allen's supporters were especially contrary:

New York May 23rd 1840

Dear Sir—

We have just received a letter from Mr. Bowen stating that he has been unable to collect any more money from the committee on a/c of the Madisonian Office. If you recollect the committee rejected my proposition to give them credit & I made a large disc[t] *for cash & in expectation that after such men as the Gents. composing that committee had given their promises; for in the note which they passed in my presence they pledged themselves to make up the deficiency of the collection, & to do so immediately & there would be no delay. I wrote home & my folks hurried the articles out at an extra expense & a great outlay of cash expecting a speedy return.*

Now the manner in which the committee treat this affair does not appear fair or just.

If they do not get the money from collections they should do as they agreed by some other means.

If Mr. Grinell[†] [*Henry Grinnell, New York shipper and merchant*] *would give his draft at 30 or 60 days on his house here or his own note for the balance we would take it as cash.*

Something must be done & done quickly as we are in real want of the money. Will you be so kind as to see the members of the committee & show them this letter if you can have them all see it. . . .

Very respectfully your friends
R. Hoe & Co.
Rich. M. Hoe

To/Mr. Thos. Allen
Washington City D.C.

Considering the very hectic political scene replete with intrigues and conspiracies between Jackson supporters and detractors, it is hard to estimate whether the company was ever suitably reimbursed for its generous terms and its exasperating collection endeavors. A year later the Colonel personally handled the Washington accounts, sufficiently exercised over them to chafe at the loss of a most extraordinary business day:

Washington D.C. Dec. 25th 1841

Dear Bob,
To day is nothing but another Sunday here & I can do nothing & see nobody & therefore have nothing to write on Monday. I shall endeavor to accomplish all I have to do here & start toward home in the evening & if possible get through all I have to do in Bal[*timore*] *& on Teusday*[†] *move toward Phil & come home Wednesday morning. . . .*

(The 25th that year was actually a Saturday.) A few years later his Washington trip was again full of the usual matters to be settled:

[*To Richard, 18 December 1843*] . . . *I have this morning yr letter from Washington D.C. with Greer's note by same mail; also rec*^{d} *bill of Lading from Mr. Greer of the Adams press. Nothing new. . . . Dont fail* [*in Philadelphia*] *to get that note from C. Alexander; also have the* [*roller*] *stocks & moulds sent on for the old dbl cylinder we rec*^{d} *from Swain as Mr. Trow wants them. . . . We are all very well barring a cold. . . .*

Attention to petty details punctuates all the correspondence:

[*To Richard, 15 December*] . . . *I expected certainly to have a line from you this morning but no sign of it yet. Since I wrote on Wednesday I sold to Mr. Trow the single cylinder for $2300. terms 300$ in cash & the remainder in notes 100$ pr month, interest after 6 months.*

We have this morning an application from Mr. Darlington of Lancaster Penn for a 2nd hand cylinder press to do fair book work, in payment for which he proposes to give us $500 & that Stewart *press we sold him. (He says it has performed very well.) We have no single cylinder 2nd hand unless you have made an exchange in some way with L. Johnson for his. Write at once about this as Mr. D. wishes an early answer to his letter. Bowen is to work to day getting in Harpers boiler & we are also putting up Osborn's Hydraulic. The men were at work yesterday at the factory & have made considerable progress in putting up those machines.*

We had some 20 persons to dine with us yesterday and after dark sent all the children and womankind home to Brooklyn [the Ely residence] *in the lumber waggon, down East fashion (a round dozen in the load). I hope you will succeede*† *in selling a Napier or two. Just collect the annexed bill from J. B. Jones; & do write even if you have done nothing in the way of selling. Have that Wm Greer matter settled up strong & what can be the reason that we dont get a bill of Lading of the Adams press? I would see it shipped myself if it is not already on board. The river to Po'keepsie will soon be closed* [by ice] *& then perhaps it may not be wanted by Platte & Co. Write to me about it. . . .*

And over the early years (as always) a myriad of other problems came in the mail:

[July 1853] . . . *My business has suffered much for the last nine weeks in consequence of sickness in my family. I am obliged this week to send my wife East for her health, which uses all the funds I have. . . .*

[March 1855] . . . [Two Cincinnati partners] *have done their best to "rope" me in by my assuming the* [the financial] *responsibility,—they are slippery fellows* [but their newspaper shop is being sold by the sheriff]. . . .

[April 1858] . . . *I am in trouble again—I want a fly-wheel and stand, so to working*† *my press by man power . . . and I want it immediately. . . .*

[June 1858] . . . [I can't pay the draft] *as my clerk, Mr. Bigelow (who selected the last press) . . . was scalded on the steamer Eclipse . . . and his recovery is very doubtful. . . .*

There was no end to them:

[February 1858] . . . [T]*he propeller Atalanta with the copying press on board . . . is ashore north of Hatteras and likely to go to pieces. . . .*

[April 1858] . . . [W]*e have no currency to speak of but State Script. . . .* [Minnesota bonds] *cannot be issued, until we are admitted as a State. . . .*

[June 1858] . . . *Send me yours* [letter] *on who keeps the Best Paint in New York. I want enough to Paint mothers house* [in Mississippi]. . . .

[June 1858] . . . *The* [Iowa] *streams are all very high and the roads are in an awful condition. After 4 days very hard work I succeeded in making near 70 miles by swimming several streams against a strong current endangering my own life and that of the horse . . .*

> *and shall not venture to try it again for two or three weeks.* [*Since I can't collect*] *can see no way of paying. . . .*

Somehow the firm bore all these trials and invariably made friends—and a sale—in the process. While the job shops around the country grew or failed with the times, and their proprietors quit or moved on, it was soon evident that the newspaper establishments, with a more compelling investment in their communities, tended to settle down and prosper almost in spite of themselves. The penny press rushed relentlessly into the smaller towns of America and its formula still seemed to work.

George W. Kendall hailed from Vermont and in New York reportedly worked for Horace Greeley. But when he won $700 at cards he was on his way again, heading South to start a paper. In 1837 he launched the New Orleans *Daily Picayune,* boasting with his partner, "already are our thoughts turned toward a double cylinder Napier press!"[14] The depression and yellow fever made the prospect unlikely.

Kendall soon returned to old friends in New York; Greeley introduced him to George Bruce, who gave him credit on fonts of type, and possibly to Richard Hoe who (in 1839) furnished the coveted press. In 1845 his paper introduced steam power, and a year later acquired another Hoe. Adventures in Texas and Mexico made Kendall the first "war correspondent" and his paper the foremost authority on those frontier areas.

In 1850 when fire destroyed the *Picayune* office, the old Hoe double cylinder, protected by a wall, was dug up and put to work in four days. Finally, however, in 1856 the paper was at last ready for a new four-cylinder Lightning, and for other Hoes in 1885 and 1892. Thanks perhaps to men like Kendall, New Orleans—with its lively port activity—promised to be an important printing town. Hoe entertained thoughts of an agent there and asked for an opinion of a local printer, Thomas Rea:

> [*To Arthur Robb, New Orleans, 20 November 1851*] . . . *He* [*Rea*] *proposes to become our agent in New Orleans, for the sale of printing materials, on commission. The property we should send him would probably not amount at any time to more than 2000 to 3000 dollars. He would make returns every 3 months & give his notes at 6 months. We should be much obliged if you will ascertain whether he has the character & business talents for such a trust and how much capital he is supposed to have already accumulated of his own. . . .*

Rea was found acceptable and given the agency (24 November 1851); he got 15 percent commission on hand presses, inking machines, and small articles, but large presses and steam engines "must be the subject of a special bargain at the time of sale." The large cities of the South all enjoyed increasing business, but a decade later Hoe's patient endeavors below the Mason-Dixon Line were lost:

June 12, 1862

Messr. R. Hoe & Co.
Dear Sirs:

Enclosed please find inventory for your goods surrendered to the Confederate States Receivers for the City of New Orleans, on the 1st of October 1861, under a compulsory law, the penalty of noncompliance being $5000 fine and one years imprisonment.

Knowing the object of the law being to squeeze money out of northern men, I surrendered the goods, rather than claim them, and pay the C.S. Receiver for them; believing they would not be able to make much out of them.

Your goods were sold on the 11th of March, 1862, the sequestration law having been changed into a confiscation law. At that sale I bought in all the goods at the following prices.

1	Imperial	No.	5	Hoe Press		$ 31.00
2	"	"	3	" "	@ $50	100.00
1	"	"	3	[ink] distributor		5.00
1	"	"	2	"		5.00
1	"	"	1	New Chase		1.00
4	"	"	2	" "	@ $1	4.00
2	"	"	3	" "	"	2.00
4	"	"	4	" "	"	4.00
1	"	"	5	" "	"	1.00
1	"	"	6	" "	"	1.00
						$154.00

By some unaccountable oversight, the auctioneer did not have the other items of your inventory on his memorandum from which he made the sale, and they were not sold. Of course, I did not trouble myself to point them out, and so they escaped. . . .

George H. Clinton

During the war R. Hoe & Company lamented its loss of the promising Southern market. But historians later have determined that of 800 Confederate printers in 1863, almost 500 had volunteered or been drafted into the Army within a year; only 20 daily newspapers were still operating when Lee surrendered. Nothing really was lost, and there would be many other avenues to explore.

Upon the termination of the Civil War, nevertheless, the vast but impoverished Southland was a market the Hoes could not ignore. Thomas Cochrane was the Hoe "clerk" (agent) thrown—much to his dismay—into the forlorn territory. As a concession, perhaps, he retained Pennsylvania and environs. With competitors' agents similarly at work among dozens of printing and publishing firms reorganizing and refitting, some fancy salesmanship would be required.

Cochrane, despite a rustic hand at letters, was the right man for the job. In Pittsburgh on one occasion he clearly turned the tide:

OFFICE OF PITTSBURGH EVENING TELEGRAPH,
No. 116 Smithfield Street.

H. B. Swoope,† President — John H. McMasters, Treasurer
J. H. Miller, Secretary — Thos. MacConnell, Jr., Business Manager

Pittsburgh Feb. 3, 1874

Messrs. R. Hoe & Co.
Gentlemen

Owing to the fact that Mr. S[tephen] Hoe failed to respond to our proposition on his return to New York some time since, we opened negotiations with the Bullock Press Co. but after considerable persuasion on the part of your Mr. Cochrane who called on us yesterday, we have concluded to give him the order, regardless of other arrangements with them. The price for press is to be $15,500-- & as was estimated $300-- for boxing and putting up. In consideration you are to accept double cylinder press now in use by us at $4640-- one third of balance cash, & remainder in two properly & satisfactorily secured notes drawn at not longer than four and eight months after date of payment of cash, with 6 percent interest.

I have said to Mr. C to have such contract drawn as may be satisfactory to you regarding purchase and putting up of press, and have it sent us for consideration & signing.

Respy
Thos. MacConnell Jr.

At the Washington office of the *New National Era,*

[From Cochrane, 13 February] . . . I have had an interview with Judd & Detweiler and enjoyed a hearty laugh over the correspondence. The result is that the[y] expect to give an order for a larger press to R. Hoe & Co. . . .

Back in Philadelphia:

[From Cochrane, 21 February] . . . I have done the best that I could with Mr. Jones [National Publishing Company]. . . . He has been run down by all most every press house in the Streets.

As I was closing the mater† this morning in walks Tom Senior as the[y] call him in the interest of Cambel.† And he was notifyed† that the mater was just beeing† closed with Mr. Hoes agent. It did not take him long to put his hat on. . . .

James Ross 228 Chesnut† Street—no such man there. Pile pronounces him a consumate† swindler. Turn him over to some one else; you have lost your cattaloge† on him. . . .

Then Knoxville, Tennessee:

[From Cochrane, 25 September 1874] . . . Rule & Ricks proprietors of the Chronicle last June put in a Cotral† & Babcock $1600 dollar

press & the[y] are now sorry the[y] done so. If the Communications were attended to at once promtly† I think the above would not have been the Case.

The[y] will write you next week in reference to the chromatic half medium press which I offered to them at $350. $50-- cash, the ballance† in monthly payments of $100. I think the[y] will take it on those terms.

Mr. Ricks introduced me to all the hardware houses in this city. But distant† [Disston?] has the foothold and [I] could not obtain an order for cash or short time. . . .

I leave tonight for Atlanta and Charleston. I am sick of the South and hope you will not send me here again to the next time—I shall do my work well and leave nothing undone.

I shall send you a half rate message every night and a dispach† will reach me in care of the western union.† I enquire at the office on my arivel.† If there is any more matters† to attend to in the South let me do it now that I may not have to come back to this Godforsaken Country. . . .

Later, once more in Philadelphia:

[From Cochrane, 6 May 1875] . . . I learned in new york† that Allan Lane & Scott had given an order to Campbel† for two of his Doubel† enders—size of bed 27 x 36 Catalage† price $3,200-- 20% of[f]. I lost no time in getting here and found that he is putting them on trial—So the[y] Say.

I have left nothing undone to burst the arangement.† I have told them that the[y] will not get them in time, that the[y] can not rely on his word and that the[y] had better give us the order.

I would suggest Mr. Robert Hoe come over tomorrow. . . .

ST. CLOUD HOTEL
G. W. Mullin & Bro., Proprietors
Arch Street, above Seventh

Philadelphia May 7, 1875

R. Hoe & Co.
Gentlemen,

Your dispatch received and hearst† [hasten] to reply.

Mr. Scott [Allen, Lane & Scott] will be over tomorrow morning. If not tomorrow he will be over early in the begining† of next week—to arrange for Presses. I met Mr. McGlaughlin† [Frank McLaughlin, Philadelphia Times] to day according to promise. He told me he spent tow† nights at the [N.Y.] tribune† office. I asked him how he liked the press. He said he was not going to give any opinion of it.

And then quietly remarked that he could not say that he disliked it. He said he was coming over next week. That is all I got out of him. It strikes me that he is trying to build up the reputation of his paper by saying he is going to get a Hoe press. . . .

[Thomas Cochrane]

[From Cochrane, 11 May 1875] . . . Mckensie† I have seen and he is working like a beaver for R. Hoe & Co. Scott was over and Campbel† agreeed† to have one press in on Saterday.† Brown told me he could not have it in under three weeks.

Through Mc[kensie?] I am in hopes to burst the trade. He is to be with me at the Hotel this evening. I have seen McGlaughlin† on third† st. this afternoon. The delay is simply in getting a sutabel† location; once the[y] are suited in that matter then the[y] will talk press. . . .

[From Cochrane, 21 May] . . . The Luisvill† matter is beyond redemtion.† The order is in Philad in writing.

Kilburg† [John Kelberg of Bullock] is gowing† to San Francisco on Monday. I am trying to find out what he is after there. Shall know tomorrow all about it. . . .

There is† two Potter presses gowing into Hunters on 6th Street to day. All the type founders here complain of business. McKillar Smith & Jordon† have discharged Six of there† foremen in the different departments. . . .

I am suffering from diptheria† and must return home tomorrow if not better. . . .

[From Cochrane, 24 May] . . . The Allen Lane & Scott matter is beyond redemption:† The cammels† [Campbell] have come. . . .

Mr. Sloane [apparently a bankrupt] I have seen and he is unable to do anything. You will have to decide what cource† is best. . . .

[From Cochrane, 24† May] . . . I have seen Mr. McGlaughlin† on third† st. He says on the 1st of July the[y] pay $50,000 dollars on there† new property, 8th and chesnut† st. at which time the[y] take the title—to the same. No press will be ordered before that time. Old Williams [Bullock] pays a visit every other day to the times† office.

I understand the times† of chicago† have ordered two Bullocks. I have a hope of straitning† Sloans† matter up tomorrow.

[From Cochrane, 25 May] . . . I have succeeded this afternoon in making arangements† with Church & Taylor, 127 South 3rd st. Philad to take Sloans† Hoe press, on the following conditions: $300 [on] 10th of August, [$350 in] Ninty† days from 10th of August with interest aded.† . . . Church & Taylor are good and responsabel.† The[y] steated† to me that the[y] are out of debt and there is no bill of Sale on there† office. . . . The[y] paid $2000 for there office and are in business Several years. The[y] have several gordens† but no cylinders. . . .

[From Cochrane, 29 May] . . . There is no alternative but to take Sloans press out and ship it over to New York. The parties that I sold [it] to have backed out. Notwithstanding the[y] signed a contract to take it.

Sloan has given permision† to take it out at once. Important that it be done at once as he owes considerabel† money. . . .

And so it went daily. Thomas Cochrane's phonetic spelling and his cryptic identifications nevertheless got the job done. Always at the right place when an order was pending, he seemed equally persuasive even after an adverse decision was made. As other territories also became immensely promising, the company would wish it had more men like him.

Chicago in only a few decades grew noisily into a thriving Illinois metropolis. It started (in Mott's words) as a "little boom-town in the mud" [15] plagued by shifting sand dunes, sprawling shanties, and unfettered livestock wandering over the wooden sidewalks. Horace Greeley may never actually have said "Go west, young man," but prophetically he did tell Joseph Medill (1823-1899) to start a penny newspaper there.

Born in Canada, young Medill had practiced law in Ohio and soon enjoyed politics and journalism together. With several partners he bought a place in 1855 on the Chicago *Daily Tribune* (founded 1847) and gradually became its driving force. Fond of Abraham Lincoln, Medill championed that potential president and helped fashion a vigorous Republican party for his election. The year Medill started, the *Tribune* bought the first power press in Illinois—a steam-driven Hoe; the paper's circulation was 3000. During the Civil War, with sensational battle news and inflammatory editorials, the Medill paper was issuing 45,000 copies daily. Its new four-cylinder Hoe was replaced in 1864 by an eight-cylinder machine. The *Tribune* soon became a powerful national paper.

Another personality in the city's growing printing community was Sterling P. Rounds, agent for Johnson of Philadelphia and—presently—Hoe of New York. Born in Vermont, Rounds (1828-1887) came west with his father's family and settled in Chicago 1856. That year he started a printing journal, *Printers' Cabinet,* as an organ for his supply house and electrotype foundry. Hoe in 1862 persuaded the Chicagoan to advertise and sell Hoe equipment to meet expenses, and a year later Rounds put his campaign together.

May 14th 1863

Mr. S. P. Rounds
Chicago, Ill.
Dear Sir,
Your favor of the 11th inst. is at hand. We prefer that in our advertisement you should say "Sold by S. P. Rounds" instead of naming yourself as our agent. The Chicago Type Foundry people are our authorized agents and would be sure to feel aggrieved.

The Caxton press sent you is one of the first made and is not finished in all respects as those which we are now turning out. The frame is to be made somewhat heavier.

The No. 3 Single Large Cylinder Press will be sent to Mess.

Dunlop Sewall & Spaulding about the first of July.

The price is	$1490
2 *Extra fountains* [@] 45.	90
8 " *roller stocks* [@] 5.	40
3 " *cloth rollers* [@] 6.50	19.50
	30
	$1669.50

We will send the guides wanted, instead of our registering apparatus. Interest will be charged over our regular terms of ½ cash & ½ @ 6 mos.

We will in this case allow you the 5% commission, though we had supposed your efforts would be directed to new trade instead of our old customers.

Do you want a Caxton Jobber or a Caxton Card Press @ $80 for Dunlop Sewall & Spaulding?

Your obt svt
R. Hoe & Co.
per T. H. Mead

Later, when he established a printing press manufactory with Taylor, Rounds ended his relationship with Hoe, but he was completely burned out in the Chicago fire. About to ship 7 presses to western customers when the blaze occurred, Rounds reportedly ordered them uncrated for desperate Chicago houses. He later became United States Public Printer (1881-1886); more than any other westerner, said James Eckman, Rounds outfitted scores of pioneer printers from Chicago to the Pacific Ocean.[16]

When the Colonel was obliged to make alternate plans for the coverage of the Chicago market, there was no hesitation: Kerkhoff was the man.

William H. Kerkhoff was a tireless traveler and worker—the perfect Hoe salesman to match the indefatigable New York brothers. About 1868 he was assigned the whole Midwest as his territory, and lest he ignore some portion of that great wilderness, his employers consulted their atlases for some exhausting itineraries:

Aug 1st 1870

Dear Sir,

We wrote you at P.O. Detroit, as in your former letters you gave no directions where to address.

Yours from Fond du Lac is received this morning. We like the trip you have made thus far, & desire you to go to Chicago. Thence West & Sou. West, through Ill. touching at the principal places, to St. Louis then to Memphis & E. through Indiana, stopping at Louisville, Columbus, Cincinnati. Canvass Ohio, striking northward so as to return to Detroit, & go again over the same ground you have visited. Stop on your route at every place where business may be had, & let people know you will return soon, especially the Hardware Trade. Canvass Chicago thoroughly & see all the Printers there & everywhere you go. We are printing a new Catalogue & hope to have it done in time so that [you] can circulate it. . . .

Give us your address from time to time that we may write. Enclosed please find check to your order for $100.

Yours very truly
R. Hoe & Co.

Mr. W. H. Kerkhoff
Madison P.O., Michigan

Throughout his peregrinations, however, no territory would match Chicago itself as the hub of midwestern printing activity—or the *Tribune* as Hoe's most promising customer.

The busy town, however, kept a rustic appearance. As late as September 1871 with Medill himself running for mayor, the *Tribune* berated Chicago for its precarious condition:

> Chicago is a city of everlasting pine, shingles, sham veneers, stucco and putty. It has miles of fire-traps, pleasing to the eye, looking substantial, but all sham and shingles. Walls have been run up 100 feet high and only a single brick in thickness. . . .[17]

Within a month the devastating Chicago fire of 8-9 October 1871 proved the point.

William Bross, a *Tribune* partner (and one-time lieutenant governor of Illinois), rushed to the paper's office when he learned of the fire's seriousness:

> He found the paper humming, even though some of its plate-glass windows had cracked from the heat when the blocks fronting it on Dearborn and Madison streets burned. . . . The fire seeped across the street, under the wooden pavement and sidewalk, and began to burn in the woodwork of a basement barbershop in the Tribune Building. . . .

Men rushed to the flames with water and extinguishers and went back to work. The roof too was repeatedly afire from flying brands, and Medill himself stamped out sparks for hours.

> Forms for the edition went to the composing room. Bross ordered the men to let the press roll as soon as the city room sent down a paragraph bringing the spread of the fire up to date. But even the fireproof Tribune Building, which Bross had thought would "stand a thousand years," was not proof against the intensity of the Great Fire.[18]

A *Tribune* reporter picked up the story:

> In the pressroom . . . a sweating crew was racing against time. The composing room had delivered type forms for the inside pages—2 and 3—and these were being run off. Meanwhile, the compositors were

> laboring over the forms of the outside pages—1 and 4—which would contain the latest news of the fire.
>
> Never, probably, in newspaper history has a press crew spent such a night. Above the whir and thumping of the Hoe cylinder type presses rose the roar of flames and the unnerving cries from the streets of the stricken city.
>
> As the last sheet—printed on one side—came off the press, buildings directly opposite the Tribune burst into flames. . . . The pressmen vowed they would get the paper out even if the roof over their heads should take fire.
>
> At that instant, however, their battle was ended by the simultaneous bursting of gas and water mains, depriving the plant of light and water for the boiler of the steam engine that drove the presses. A few minutes more and the Tribune's building was ablaze. . . .[19]

By 7 a.m. the composition rollers had melted on the presses and the edition of 9 October, insides printed, could not be finished to document the conflagration. Workers, some stumbling with books and records, fled unceremoniously. The uninsured building and its contents—valued at $325,000—was totally demolished.

With incredible strength and enthusiasm Joseph Medill rallied the people of Chicago with stirring editorials that won him the approaching election for mayor: "All is not lost. Though four hundred million dollars worth of property has been destroyed, Chicago still exists. . . ."[20] For the destitute he grandly directed gigantic relief and reclamation projects. For the *Tribune* he acted just as deliberately.

Within days Medill had leased an old job shop and bought a nearby press just moments before it was spotted by an agent for another paper. He recalled his staff and slowly began to produce his newspaper, working in partnership with another sheet and carting forms several blocks to its press. By telegraph Medill then bought a four-cylinder Lightning he located in Baltimore and had it shipped to Chicago within hours. Bross raced to New York to refit.

> [*Robert Hoe to Richard, Paris, 17 October 1871*] . . . *Our thoughts here have all been engrossed by the Great Chicago Fire. The calamity so immense and dreadful to contemplate shocks everyone. The Newspapermen from there are now dropping in to see us and to provide for resuming business. Fire proof buildings shared the fate of wooden houses, all being destroyed or burned up. Mr. Bross of the Chicago Tribune was here yesterday. He seemed to feel his losses very much and was too worn out to say much. . . .*

Robert bemoaned the fact that the company had almost no stock on hand with which to fill orders quickly. A new Hoe machine for the *Tribune* did not arrive until December.

In addition to his own prowess Medill gladly acknowledged some (but not all) offers of help received. In a "Brethren of the Press" editorial he commented:

> No time will be lost in getting to work again at Madison and Dearborn. The Cincinnati Commercial sent us a complete font of type, distributed in cases. The St. Louis Democrat sent us 65 bundles of paper. The Cincinnati Gazette and Enquirer offered us anything we need. The Aurora Beacon offered to print our paper. Among our brethren of the press a spirit of liberality is equalled only by the magnificent charity which has flowed. . . .[21]

The munificence of Horace Greeley (whose unfortunate presidential campaign was strangely supported by the *Tribune*) apparently went unnoticed. Medill's managing editor had worked for Greeley, and on his behalf the latter, with some disdain for "fireproof" buildings, nevertheless offered to help:

> Ever since *The* [*New York*] *Tribune* office was burnt out, in 1845, we have kept always on hand duplicates of every thing needed for the publication of the paper, forming in fact a complete duplicate office. Mr. Thomas N. Rooker, our foreman, yesterday telegraphed, on behalf of the [Tribune] Association, to Mr. Sidney Howard Gay, Managing Editor of *The Chicago Tribune,* placing this office at his instant disposal for the use of *The Chicago Tribune,* or any other Chicago newspaper. . . .[22]

By November 1872, nevertheless, the paper was fully installed in its new building, within a year arising phoenix-like on the site of the old. "Time works wonders," said the *Tribune* on the fire's anniversary. "The same press in the same room which issued *The Tribune* of October 8, 1871, and would have printed the issue of the 9th had the gas burned a few minutes longer, has issued *The Chicago Tribune* of October 9, 1872. Editors, reporters, printers and pressmen are in their old places. . . ."[23]

A year later a lonely eight-cylinder Hoe commanded both basement and first floor of the new $30,000 plant, but the paper apologized for constant delays because "our printing machinery has been wholly inadequate." The *Tribune* had left space for three more presses and just ordered "one of the latest improved Bullock printing machines with double delivery and other improvements, a capacity of 15,000 impressions an hour. . . ."[24] Its circulation was 140,000.

Conrad Kahler had been *Tribune* pressroom foreman, a post he held more than 20 years. In 1878 he developed a folding and pasting machine (capable of 12,000 papers hourly) which helped the *Tribune*'s Hoe press, two Bullocks, a paper wetter, two engineers, four stereotypers, nine pressmen, and eight mailers to dispose of its "immense" editions. Stronger than ever, the *Tribune* boldly consoled subscribers there was no conflict between Christianity and Darwinism: "There is no science of the soul, any more than there is a prayer in mathematics." [25] Medill's daughter Katherine married Robert S. McCormick (nephew of Cyrus), bringing him into the firm and insuring another generation of dynamic family newspaper management.

Like the Chicago *Tribune* partners, William H. Rand and Andrew McNally in 1871 also watched the approaching flames with dread. Before their 1864 merger, however, Rand with bravado had faced such calamities alone. His first trial—the ruinous financial crash—found him establishing a line of credit for a new business:

148 Lake Street
Chicago, April 7, 1858

Messrs R. Hoe & Co.
Gentlemen,

I want a Hoe cylinder press, super royal size—that for which you charge $1490—but cannot accede to your published terms of sale. Will you send me one for

¼ cash
¼ note, 3 months
¼ " 6 "
¼ " 9 "

Notes secured & mortgage on the press.

For my responsibility—I beg to refer you to D. L. Dodge, Chicago Type Foundry, S. P. Rounds, Printers Furnishing Warehouse, Chicago, & to Messrs C. H. & L. Laflin, Paper Dealers, Chicago. Also to Messrs. Phelps & Dalton, Type Founders, Boston. I have no business acquaintance in New York.

If you can accept of this proposition how soon can you deliver the press?

Your early attention to this will confer a favor.

Respectfully yours,
Wm H. Rand

A week later, fearing he had come on a bit strong, Rand wrote again (13 April): "I did not wish to be understood as *ordering* a press, but merely as making inquiry about terms. . . ." Hoe, meanwhile, had inaugurated one of its most comprehensive checks:

[From Phelps & Dalton, 12 April 1858] . . . We had formed a favorable opinion of the man from personal observation, and had learned from inquiry that his business prospects were considered very good indeed. . . .

[From C. H. & L. Laflin, 14 April] . . . We consider Mr. Rand a responsible man. . . .

[*From Rounds & Langdon, 14 April*] . . . *Should not hesitate, myself, to sell him the Press.* . . .

Two weeks later (30 April), however, Rand terminated the exchange: "I shall be unable to order the press we have been corresponding about, at present." Rand had apparently given up altogether the idea of going into his own business.

Rand instead joined the pressroom of the Chicago *Tribune,* and in 1861 (when the paper was incorporated) he was listed as an owner and head of the mechanical department. Without personal financial worries, Rand fearlessly bought the paper's 1861 and 1864 Hoe presses as well as the eight-cylinder, but with command of such power his thoughts returned once again to his own shop. Although from 1865 he still drew income—$22,491 that year alone—from the *Tribune,* he had already met McNally.

Andrew McNally came to the United States from Scotland, learned hand composition, and gravitated to Chicago in 1858, where he tried to start a paper. He too abandoned his plans and became foreman of the *Tribune* job office. When the paper in 1864 resolved to discard its jobbing department, Rand and McNally saw their chance and took it.

The partners lost heavily in the 1871 fire, prompting McNally to buy a working shop immediately while an assistant was sent to New York and Philadelphia to procure new presses. By their swift concerted action once again they held on to their thriving business—the printing of maps, atlasses, railway guides, and textbooks. A year after the fire, however, Rand McNally was still desperately cleaning up its old accounts: To R. Hoe & Company (on exotic purple stationery advertising "This letter Head is Printed in Copying Ink") the Chicagoans offered a $500 parcel of county bonds. Hoe, accustomed to strange mediums of payment, probably took them.

Reuben R. Donnelley began his Lakeside Press in 1864, naming it out of admiration for Houghton's Riverside Press. After the fire the going was rough:

THE LAKESIDE PUBLISHING AND PRINTING COMPANY
Clark and Adams Sts.
R. R. Donnelley, Manager

Chicago, February 17, 1874

Messrs. R. Hoe & Co.
29 & 31 Gold Street, New York
Gentlemen,

Our company is in trouble,—and we have delayed writing you on account of the very uncertain shape matters might take. We thought it possible our Mr. Donnelley might have to go East & explain matters personally—But as our creditors there were few in number, and most of them have accepted the best and only term we could offer, that *step has been thought uncalled for. . . . After the great fire we were very undecided what to do, as you may remember our Mr. D. stated when in New York. But when the smoke had cleared away our stockholders gathered together and consulted with the architect inspecting the costs of re-erecting the building; he took time to make what we supposed were careful estimates, and opined that from $105,000 to $115,000 at the outside would give it to us again. . . .*

The actual construction bill was a staggering $222,000, and Donnelley was understandably struggling to stay alive. Robert Hoe III had personally visited Donnelley and apparently liked what he saw:

June 10, 1874

Dear Sir,

Since our return from Chicago, we have concluded to accept in settlement of our claim against the Lakeside Ptg & Pub Co.† *the offer they make of 30% cash and 50% in stock of said company.*

You will please receive and keep for us the stock, as we do not wish it at present in our hands. In view of any changes you may make in your connection with the company, we beg to ask that the Hand Stop Cylinder Press now in the office may be made our property beyond doubt.

Your stop Cylinder we hope to finish by July 1st and will forward as instructed.

We remain

Very truly yours
R. Hoe & Co.

R. R. Donnelly† *Esq.*
Chicago, Ill.

Admitted a partner on 2 November 1861 (with a 1/14 ownership), Robert Hoe (who used "Jr." until his father died in 1884; he was the third of that name) found himself touring the Midwest to acquaint himself—and his firm—with its rediscovered market potential.

The Hoes had so exhausted themselves in the British arena they had quite forgotten America's expansion westward. The Chicago fire dramatically awakened their attention, the senior Robert applauding the city's courage in rebuilding and returning to business (with fat fire insurance checks). But he was further startled to find numerous Bullock presses being installed because Hoe deliveries were slow. Tucker, too, was suggesting the firm start a Chicago factory rather than one in London.

Between their European commitments and increasingly cautious business projections, Robert did not explore the scene himself until 1873. He was again surprised:

[*To Robert Hoe III, 8 May 1873*] *. . . Nothing of the sort could be more wonderful than the rapidity and magnificence with which the best business portion of this City* [*Chicago*] *is being rebuilt. How they* *have got together the material and labor, to say nothing about the money to do so much, in so short a time is truly astonishing. There is a vital force existing in these Western men of business and a push that beats the world over. We need to be in intimate connection, personally, if possible, with all these newspaper people and the job and litho printers. . . .*

Robert had to return to his post in England, but his son would do perfectly in exploiting the Western business.

Robert Hoe III in May 1874 bundled his wife Olivia and the baby (Thyrza, their sixth daughter; the five older girls were left behind) and went off to establish a Chicago office. Within a few days he leased a store ($1200 per year) and put Kerkhoff in charge. "My opinion was—is—that with a place there properly managed," he wrote his father (2 June), "we can almost controll[†] the Milwaukee, St. Louis & Cinna trade. . . ." He foresaw no imminent competition. Within a month he had relocated:

UNION TELEGRAPH COMPANY

Dated Chicago, Ill. May 25, 1874
Received at 610½ Grand St.

To R. Hoe & Co.
504 Grand St.
Am entirely satisfied of importance of starting office here and propose renting store centrally located twenty by sixty five [feet] with Enterance[†] for carts on rear rent thirteen hundred dollars possible something less answer.

Robert Hoe Jr

36 paid half rate Pacific Hotel
Saty 26

By June the centrally located 84 Dearborn Street was functioning. Kerkhoff was nominally in command, but instructions from New York were voluminous:

[*To Kerkoff, 6 June 1874*] . . . *Until your books arrive you can keep what accounts you may have so as to transfer them regularly at the proper time.*

We are getting out your samples and hope to make a shipment very shortly. In the mean time you canvas amongst the saw men and then let us know what stock you are likely to want. The painting we presume is being done and we hope you have secured a proper young man as an assistant.

Mess. Brady & Mellette, Muncie, Ind. enquire for a Large Cyl. Press 2 or 4 rollers. We send them [catalog] *sheets, also list of second hand machines, without prices.*

You had better go and see them, also Thomas Phillips & Co. of Akron, Ohio, who want a fast press for printing paper bags. Please attend to these when you have time, and let us hear how you are getting on. . . .

[20 *June*] . . . *We sent by yesterday's Express your Books* [*and*] *stationery. Enclosed are directions for keeping the books. Any of the smaller articles we do not send you can buy in Chicago, & take out in trade.*

We will send the saws ordered for Morgan & Beach invoiced direct to you and you can communicate with them, render bill, &c.

Marder, Luse & Co. are credited $100-- and same charged to you. Please get a list of stock on hand at the Type Founders and our other consignment accounts and send us a copy.

The Cleveland Leader Co. have ordered a 36 x 54 in. Double Cylinder, so you need not see them.

We shall prepay freight on all goods shipped to you, and charge no boxing. Whenever possible charge both these items in making sales. Shall we send [*type?*] *cases to you Simonton's length or regular? Enclosed* [*not present*] *are dimensions and weight of Hand Stop and Large Hand Cylinder Presses. A schedule containing this information respecting all our presses will be sent to you shortly. On Monday we will send the samples from our Gold St. Works. Perhaps, the regular made goods will look better in the windows and for hanging up. You must use your own judgment about this.*

Please send a description of store that we may take out proper insurance.

When in Cincinnati we offered the Methodist Book Concern, Mess. Hitchcock & Waldren, a N^{o} *9 Double Cylinder Press for their old Double Cylinder and $5000--. Also 2 Single Small Cylinders, 3 Rollers, for their old Dbl. Cyl. and $6250--, and stereotype apparatus $1700--. We to pay man's time and they his expenses. If necessary to secure order we could do a little better. You had better give this personal attention and go and see them as it is important that we should have this order.*

We were sorry to hear of the Rand & McNally matter and hope you will be able to get Cottrell order countermanded, without making an unwise bargain. Do your best. . . .

Presently Kerkhoff returned correspondence in kind:

[3 *October* 1874] . . . *Enclosed are checks* [*from*] *Chas. Councilman & Co.* $\$15^{70}$ *and Barnhart Bros. & Co.* $\$20^{63}$.

I have just returned from St. Louis and Springfield. The St. Louis Fdry continue to make [*type*] *cases, and for the present could not be induced to place an order. The Branch of Boston Type Fdry will take cases when their present stock of Simon's are run out. They have a full stock consigned to them of these cases. Trade is very quiet and no presses are enquired for.*

We have had the foreman of the Springfield Journal to look at the Stop Cyl. presses, and I called there hoping to take an order but they are still undecided. I shall keep them well looked up.

The discrepancy in the accounts arises from the fact that one or two of the parcels of goods invoiced late in August were in transit. I am at work on the list & shall forward them by next mail. . . .

P.S. I have seen Mr. Jones in relation to the Coupon press. He knew of the press and has tried to buy it to keep it out of the hands of one of his workmen. Will write further in a day or so. W.H.K.

[From Louisville, 14 May 1875] . . . I was in St. Louis Tuesday & Wednesday and saw Mr. [Daniel] Houser the day of the consolidation of Democrat & Globe. They have decided to retain the Globe form (quarto) and their Four Cyl. will not be equal to their combined Editions. Mr. Houser has arranged with me to telegraph him in about 10 or 12 days from now to meet me in Cincinnatti† *to look at the Commercial Presses. He will order a press the first thing after money [comes] into the Democrat Office and expects to turn in the four Cyl. in trade, but not the Six Cyl. as I thought you would prefer the quarto to the folio press and would not be willing to take both. Next Monday they move the press into the new office & after that Mr. Houser will be able to leave.*

*Dr. J. H. McLean [St. Louis] will put in a Litho. press soon as some alteration is made in the building. He prefers to deal direct with you. Please write him and send a catalogue as he may be induced to do all his own printing one of these days.**

Brown Bros. of Columbus, Ind. are not ready to buy [a] press and will not put in steam for some two or three months. They like the Hand Stop cyl. and have now sheets & prices and all information.

The Courier-Journal building of this city is being roofed in and Mr. Haldermann† *[Walter N. Haldeman] said last night that he should decide and place the order for new press in three week's*† *certain. I am convinced from a long conversation with him that as far as he is concerned the order is safe, though Mr. Medill has spoken most favorably of the Bullock with the Chicago improvements, and there is a strong effort to let in a Bullock press. Mr. Wateson*† *[Henry Watterson, Courier-Journal] is now in Cincinnatti & I expect to meet him . . . & he will look up the Commercial presses thoroughly. A great deal depends on his opinion and I expect to send you some information to morrow night, after seeing him. Their Chief Clerk [Meredith] and foreman is*† *now in N.Y. Campbell sent a man down here to take on these men to N.Y., paying their expenses, so that they might see his new press at work. He offers to put one in on any terms and claims 30 or 40,000 per hour with adjustable folder attached. Mr. Aldermann*† *agrees that Campbell is visionary and that he cannot afford to try experiments in his new building. Unless these men who have with them Mr. Rowell of this city, whom you know, bring back a very startling report Campbell's proposition will not I think be further considered.*

Mr. Aldermann would have gone up to Cincinnati but he is the only one of the firm in the city now, and will wait Mr. Waterson's report before seeing the presses.

I expect to be in Columbus, O. on Tuesday and return to Chicago same night. . . .

* See *Age of Great Guns.*

[From Memphis, 31 May] . . . On my way down I stopped off at Columbus, O. "The Catholic Columbian" will need a press for their paper shortly, and would like to hear from you with a list of the sec-hand Gordon presses you are to get from Washington.

Glenn & Co. who were looking for a press to do proof book work are stopped by the death of Mr. Glenn. They will put one in this year if the business is continued by the remaining partners.

Mr. [Murat] Halstead in speaking of the Louisville Courier Journal, said "the fact is, just as soon as the Bullock folks saw that they were turning towards the Hoe press they got scared and Williams [of Bullock] made great concessions to close the contract at once."

I have put the matter of cases right with Allison Ltd. and sold them 100 pairs to be sent from Chicago. They showed me an invoice from Roberts & Co. for 100 pairs at $175--. I sold them at regular price.

Mr. Halderman[†] talked very freely with me at his office and showed me a letter received from Mr. Williams speaking of the two presses in the [New York] Sun Office and claiming high speed for them though they are old style. I find also that [R. M.] Pulsifer of Boston [The Herald] wrote a letter strongly recommending them and I am reminded that Mr. Story[†] of Chicago [W. F. Storey, Chicago Times] received such a letter from Pulsifer and from N.Y. Herald which shows Mr. Williams' plan of working.

They take a Double Bullock and the contract is, to run at 10,000 sheets per hour, if necessary (20,000 papers) but regular speed of 8000. I tried to show them that this had not been accomplished. Mr. Haldermann did not intend to order before seeing your presses at the Commercial Office, but the report brought by Meredith & foreman from N.Y. which was that the Campbell was a fizzle and that there was some trouble with the "Tribune Press" and that the Herald double was working well & spoken of highly and this coupled with Mr. Williams' presence here with the Pulsifer and Medill letters influenced him to place the order against his original intentions.

I have sold Rowell of Louisville a small lot of 40 pairs cases & some galleys.

At Nashville, Tenn. I found that large drum Cyl. of Roberts & Pervis has been breaking. It is now being repaired and new cut bevel gears made to drive the inking cylinder. They have lost the job for which the press was bought & it is now an Elephant on their hands and they would be glad to find occasion against you. They have nothing between this large press and a half med. Gordon to do all their work.

I leave for St. Louis tonight. Annexed is order [for a press]. . . .

Kerkhoff was working tirelessly, and although he seems to have enjoyed the brutal conditions of travel, professional calls, gossip collection, and nightly financial reports, he missed a certain amount of appreciation from the home office. Hoe's patronizing letters, with the usual stilted phrases of formal correspondence, seldom included recognition of work well done. Although times were lean, taking the initiative, the Midwest representative asked for a raise.

> [From Hoe, 19 February 1876] . . . *Your application for an increase of Salary was received & has been considered with a view to meet your wishes as far as we can consistently do it in view of the present condition of business.*
>
> *We are sending off some of the clerks & more of the hands as we have no work for them and so far as indications now go we do not see that the future is so promising either here or in the west as would now warrant an increase in the expenses of the Chicago Branch. At the same time we are prepared to say that if the business there shows an increase over last year to warrant it we will accede to your request. There are quite a number of Drum Cylinder presses new & second hand, two rollers, that we think could be placed in offices in the west, and if necessary at prices even lower than we have named, and we would like to have you take up this matter especially & try what you can do and if you think it will pay us for you to go on such an expedition we will send a new list & name the lowest prices. We are very desirous of disposing of this stock on hand, much of which however good is not so saleable here, in view of the more recent improvements. For country papers & for job work nothing could be better.* . . .

All in all, a very disappointing letter, Kerkhoff thought—not really different from all the other lengthy tirades and exhortations. Could the Colonel not spare more than a fleeting moment to acknowledge his humbling request? Tired and weary late at night in some shabby hotel, he wrote again and got, at least, a prompt reply:

> [From Hoe, 24 February] . . . *In reply to yours of 23rd inst. we have to say that we do not see why you should take what we have said or done to be any indication of any dissatisfaction with your work. The branch at Chicago has been quite as successful as we could expect in these times, but you know that so far there has not been much profit realized and we only wish to be prudent about it and not count on too much in the future before it is realized. We wish to show our appreciation of your services and we expect that a reasonable request to wait and see how matters go would satisfy you. If you are in need of money you may draw the $500-- and let the matter stand a while & until you come in to New York when we doubt not in a personal interview all can be well understood and disposed of in a satisfactory way. When you have disposed of the pressing applications now in hand & secured the orders it would be well to come & have an interview also in regard to the disposition of the stock we have here about which we wrote to you.* . . .

Evidently Kerkhoff had asked for a $500 yearly raise; as usual the company response simply parried the request and thrust more work at the unhappy clerk. Soon after this exchange the Colonel seemed to remember that Kerkhoff never complained while on the road, and set him going again:

> [*From Hoe, 27 November 1876*] . . . *We think it would be well for you to go to Detroit again and see Mr.* [*James E.*] *Scripps, from there to Toledo, thence to Toronto, Ont. Please see all the printers, and especially the Rev. S. Rose, "Guardian" office. He has an old press we think you can arrange to take in trade for a new one. Leaving Toronto go to Montreal, Buffalo, Rochester, Syracuse and Utica and any other intervening places where you think there is any possibility of getting orders. . . . Stop at all intermediate places. . . .*

Other assignments continued:

> [*From Hoe, 29 March 1877*] . . . *Yesterday there was a gentleman here (we forgot his name) who owns "The Field" in Chicago. You can ascertain who it was when we say he is a large manufacturer of Reaping Machines and has now a suit with McCormick for some infringement. He has gone on to Phila., but will return to Chicago in a few days, when we want you to see him.*
>
> *We showed him* [*it was Charles W. Marsh—The Field was a sports paper*] *our different presses and offered him the second hand 41 x 57″ Column pattern 4 Roller Single Large Cylinder, named on list, at $2250--. The 37 x 51 in. Campbell Press would answer his purpose, but we think there is nothing better for him, to do his newspaper and other work, than our Two Revolution Machine, and we would like you to call his attention to this.*
>
> *Yours of the 26th enclosing Draft for $1275.-- and order for N° 1 press, is just received.*
>
> *The press sold F. A. Klatt Aug. 31st/76, is a N° 4, not 5, Washington Hand Press—N° 4 frame, inside, 29 in., N° 5—32¼ ins. . . .*
>
> [*P.S.*] *He* [*Marsh*] *saw the new four roller drum Cylinder press, with Cyl. & table distribution & seemed to prefer it. Unfortunately we had no 2 Rev. Mach. to show running. Barnhard & Bro. & Spindler*† *are after him sharp and you had better get them on your side by giving some commission, if necessary.*

And so it might have gone, on and on, perhaps with no raise at all. But somewhere about the beginning of the 1880's Kerkhoff had had enough. He resigned his position, perhaps acknowledging to himself that the years with Hoe had certainly been instructive, if not lucrative. His frequent brushes with Campbell and Bullock colleagues, sometimes with William H. Williams himself, showed him there were other places to go. For William H. Kerkhoff it was Bullock; he joined that struggling company and a few years later was one of its officers.

A clerk named Crowther may have been the lad Kerkhoff engaged to tend the Chicago office. In 1878 he directed the expansion of the branch the Hoes had been almost too pessimistic to foresee. In larger quarters the company even had space to sublease:

> [*From Crowther, 16 January 1878*] . . . *Your telegram to hand saying —"Rent the Loft to April to #89* [*Dearborn Street?*] *at the best terms you can." We saw Mr. Hanlon and the best we could do was as follows—The Rent to be $1580 him to take the lofts just as they stand and, the rent to commence to day—This will save white-washing. . . . We hope to forward the Lease & ½ Months rent tomorrow of course we have to connect the pipes with heating apparatus & Elevator Running. This we are doing today since the frost is rather bad, as our man when he shut down some months ago did not do it properly, and the water pipes connected with boiler are frozen, and one is too broken. We have the Steam Fitter working now, and they*[†] *will have to work late to night to finish the job, as we have faithfully promised to have the Elevator running tomorrow, and we are taking every precaution agst: pipes bursting when we light the fires. . . .*

Startling news followed just two days later:

> [*From Crowther, 18 January*] . . . *We have had several enquiries respecting* [*our*] *building. A fire occurred and burnt out Campbell Ptg. Press Co.* [*a branch office*] *and a lot of 2nd hand Machinery, Cushing Ptg. Co. and a few small Job Ptrs. We went on and saw these people. Campbell said it was not suitable for them. Cushing Ptg. Co. liked the 1st Loft, but they would want Power and a longer lease. J. J. Hanlon, 304 Dearborn, said they could use 2 lofts, and if we could arrange with Cushing to take one of the top rooms, they might take the 2. We saw Mr. Cushing and after he settles with the Ins. Co. will see about it and also let us know about presses. Mess. C. Goldsmith & Co., Dealer in Paints &c. offered $1000 for 1st loft with Power, from the 1st of May, and will give an offer this week for the 3 Lofts. We shall look after these people, and if possible let the whole building. Mr. Petrie promised that he would extend the Lease to any good Tenant.*
>
> *Mr O'Donoghue says he can't send for Press this month; Rowe & Huxfud charges for removing him are $559^68 so you will see what a large amt of money it must have cost him. He said that if he had the money he would have given me the order to ship right away.*
>
> *Rowe & Huxfud are dissolving Partnership, Huxfud going out; he said today that for every $1½ he put in, he will get $1, and glad to get out at that. . . .*

Although Crowther did not travel as much as Kerkhoff (telegraph and the mails were perhaps workable alternatives), he faced a similar wave of competition—now from other large newspaper press builders. He reported an imported German lithographic press printing "beautiful" halftones from glass plates. Inquiries for big machines were now coming in from Denver newspapers. And one local prospect was especially tantalizing:

> [*From Crowther, 24 October 1878*] . . . *Henry O. Shepard & Co. "Inland Printer" 183 Monroe St want our lowest figures on the following Press:*
>
> *#6 Pat. Stop Cyl. 32" x 47" #5 Imp. 2 Roller 29" x 42"*
>
> *set up in their office complete—for Cash—Also price ½ cash balance on time. State time when you can deliver, which must be inside 2 months.*
>
> *We hardly expect to get the order, as we expect they will favor the people who advertise largely with them, but we should like to have a good try, and Mr Shepard said that every Press builder should have the same chance. We drew his attention to the fact that we were figuring on our best Press, and should not expect him to compare our prices with Presses of inferior make. He said that everything would be considered, and the cheapest and best would get the order. This would be grand. . . .*

The following week Crowther's fears were realized:

> [30 *October*] . . . *We saw Mr Shepard this morning (H. O. Shepard & Co) to see if we could do anything with him. He said we were a far way out in prices; we expected him to say this, and if we had cut them in two it would have been the same. He dare not give us the order because the people who pay him a lot of money for advertising would not like it. . . .*

Although Crowther remained with the company well into the 1900's, he was to be only second in command to a family member who had been groomed for the position. Chicago and its wide environs was to be the territory of a young Hoe just as Robert II had claimed England.

Stephen Smith Hoe (1847-1887), Peter's first son, was educated in public schools "after the manner of all his family," [27] said the *New York Times*. Taking up his first duties as agent in the sales department, he was an unusual salesman indeed: quiet and thoughtful, with a kind disposition and "generous to a fault," he nevertheless became so occupied with business matters he "shunned society," never married, and was "idolized" by friends and employees.

In 1885 Stephen took charge of the Chicago office. The following year, upon the death of his uncle, he returned to New York for reorganizational discussions; there was no doubt Chicago and the Midwest would play a significant part in the company's expansion. But suddenly taken ill, he died July 1887 while at his uncle's Tarrytown home. Almost the entire Hoe establishment appeared at his funeral. While calling little attention to himself personally, Stephen had nevertheless gradually built the inland business to twice its volume, using the very publicity and advertising techniques in Chicago his cousin Robert would soon consider abhorrent and unnecessary in New York.

Stephen had sent a steady stream of puffery to *Inland Printer* just as soon as he arrived in Chicago; he wrote hints to pressmen on packing and bearers (March 1884), and proclaimed in a full-page advertisement in each issue that R. Hoe & Company "At their Western Branch [now 180-182 Monroe Street] have on hand every article required for printing. . . ." Within six months he revealed that the firm would build a press manufactory on Chicago's North Side—199-201 East Van Buren Street, taking occupancy September 1884. On the premises were a full stock of machines and other printers' materials and a complete repair shop. The Illinois office, like the British branch, almost became an autonomous organization. And the Midwest had grown sufficiently to command and tax all its resources.

The Detroit *Free Press* began 1831, Sheldon McNight printer and proprietor. Six years later the office was destroyed by fire, but by luck a steamboat frozen in the nearby Detroit River had a complete printshop in her hold. McNight bought it on the spot. In 1846 the paper was issued by Messrs. Baggs & Harmon, who had just won a contract to print Michigan's statutes. They needed a power press and somehow recalled the name Hoe (Columbus, Ohio, already had such a machine). Harmon went to New York but couldn't decide on a model, whereupon Baggs joined him in selecting a "Medium."

> Its owners were anxious concerning their purchase, almost fearing that it would shake itself to pieces when started; but one day the press was got in running order, and away she went, two men turning, another one feeding, and the office full of spectators. Everybody was delighted and astonished, and it was weeks and weeks before the sight-seers ceased to climb the stairs. Publishers came here from St. Louis, Chicago, Cincinnati and other western cities to see the press work and find out about it, and within a year Hoe had more western orders than he could fill. . . .[28]

This machine, affectionately named "Peggy," was soon the first power press in the Northwest; in 1886 it was still workable and on its way to a country shop when a new one went into Baggs & Harmon. The Detroit *Free Press* has long since become a long and prosperous journal.

Cincinnati got its first power press 1834 when the *Gazette* bought an Adams. Soon book printers too had enough of them to make the Ohio town a "literary emporium of the West." In 1840 Calvin W. Starbuck founded the Cincinnati *Times* and long remained its editor. He had the misfortune to take an 1861 Bullock which failed, and in dealing for a "superior article" apparently over-extended himself seriously: in January 1863 Hoe won a suit against him in the Ohio Supreme Court for $2953.70, and actually realized $2977.01 (interest included).

Murat Halstead became editor in 1865 of the Cincinnati *Com-*

mercial (established 1843) and made it one of the great Western papers. Despite what Mott called "a genius for news enterprise," [29] Halstead seemed unsure of his costly new Hoe web press:

OFFICE OF THE CINCINNATI COMMERCIAL
M. Halstead & Co., Proprietors

Cincinnati, May 17, 1875

R. Hoe & Co.
Gentlemen:

We are under obligations for your consideration. We will pay ten thousand dollars on or before June 15th. I will make the payment as much earlier than that date as I find practicable. Just now we find a great deal of paper wasted. We cannot have exactly the quality we want. The quality is good, however.

I hear that the Bullocks† Co. is making great concessions. The Louisville Courier-Journal has ordered a Bullock, I hear. They had thought of taking your press. The Cleveland Leader is also on the point of ordering a Bullock. By the way has the Chicago Tribune one of your presses in operation? . . .

M. Halstead

Halstead's competition, the Cincinnati *Gazette* (it merged in 1883 with the *Commercial*) just a few months before had bought a Hoe, and with more conviction. Founded 1827, the *Gazette,* with Stephen S. L'Hommedieu one of its owners, had been printing by steam just seven years later—barely two decades after the London *Times* itself. But in the westward movement of presses and publishers, another city was destined to shine—St. Louis.

"We love the North and the South, the East and the West, and the Union above all things," [30] said a St. Louis compositor in 1860, referring waggishly both to America's impending rebellion and to struggling trade unionism that had started in Missouri seven years before. Despite the vastness of the territories opening up across the country, and the shortages of manpower as well as commodities, the prospect of growth and steady employment became strangely critical. The union was an unsure ally.

At first represented indifferently in various compositors' unions, pressmen in the East had steadily fought for organized recognition. By 1865 they had formed the "A. and C."—Adams and Cylinder Press Printers' Association—which later became New York Printing Pressmen's Union No. 51. (Although a platen press, the Adams' heaviness and power drive discouraged its classification as a jobber.) In this way large numbers of book printers became union members long before newspaper pressmen, and except for a constant fight against the "rat," a worker who accepted below-union wages ("As a rat undermines foundations, so the rat undermines the trade" [31]), the printing industry in America came to grips with mechanization far more amicably than that of Europe. In the West, however, there were problems.

The Missouri *Republican* in St. Louis dated to 1808; it later came into the hands of George and John Knapp and N. Paschall. The Knapps had done power press business as early as 1846 with R. Hoe & Company, and were early running Lightning machines. They tried everything in between, too:

> The press used in printing the Missouri *Gazette* from 1808 to 1822 was a Ramage, the first to be set up west of the Mississippi River. . . . The next press was the Stansberry† patent level with a capacity of 240 impressions per hour. It was followed by the Washington hand press, . . . and it in 1837 by an Adams power press with a capacity of 800 per hour. Then came in succession the Hoe presses: in 1843 a single-cylinder; in 1853 a double-cylinder; in 1858 a four-cylinder, and in 1864 an eight-cylinder with a capacity of 20,000 copies per hour. In 1872 the *Republican* installed two new presses, a Bullock and a Walter, and in May 1879, these were supplanted by a new Hoe machine capable of printing, folding, pasting the backs and cutting the ends of 30,000 eight-page papers in one hour.[32]

This last, the Hoe "Double Web Perfector and Folder," was for its time the largest press in the world. Inveterate gamblers on St. Louis' long-range future, the Knapps made their paper one of the most influential in the state. Traditionally a bitter foe of the St. Louis *Globe-Democrat,* the *Republican* engaged in long circulation battles in which pressmen were sworn to secrecy over actual runs. The Knapps would find themselves fighting printing unions, too, in a frantic attempt to keep staffed, but that was a losing battle for the whole community.

By the 1870's St. Louis and the Midwest seemed to some as tame as the East and the sleepy South. The scarcity of pressmen in California was tempting restless men beyond their endurance:

> The printers of the middle 'fifties were an adventurous race, with a contempt for employment and employers that was almost magnificent. This led to the easy throwing-up of "sits" [situations] and much wandering. The West was calling loudly. . . .[33]

But many of these happy-go-lucky fellows—knowing they had rare skills for which the pay was always good—used their profession only as a ticket to see the wonders of the country. Others, however, did not travel empty-handed. A London correspondent in 1849 had noted, "The Yankee has an admirable trick of carrying a printing press upon his shoulder wherever he goes—he cannot live without his paper. . . ."[34]

One of the earliest printed newspapers in California (founded 1846) was the Monterey *Californian,* variously issued on coarse

Spanish writing paper and cigarette paper, at times the only kinds available. Its proprietor later wrote of its inception:

> The press was old enough to be preserved as a curiosity; the mice had burrowed in the "balls"; there were no rules, no leads, and the types were rusty and all in pi; it was only by scouring that the letters could be made to show their faces; a sheet or two of tin was procured, and these, with a jack knife, were cut into rules and leads. Luckily, we found with the press the greater part of a keg of ink, and now came the great scratch for paper. . . .[35]

The *California Star,* San Francisco, started a year later with an old Hoe hand press and similar ingenuity to find supplies.

News took seven weeks to reach the California coast, and quantities of newsprint even longer. Upon the sudden discovery of gold many printers abandoned work to pan the streams; in the ensuing fury there were duels, assassinations, losses at sea of types and paper, and other agonies to make the publication of newspapers nearly impossible. Not many survived.

In 1848 the *Californian* and the *Star* merged to become the *Alta Californian* under new owners, Edward C. Kemble & Company. Early in 1850 Edward Conner, formerly a mail clerk at the New York *Herald* and now agent for R. Hoe & Company, arrived in San Francisco. He quickly sold a steam press to the *Alta Californian*—the first power machine in the state, and in June of that year became an employee of the paper himself. (In 1857, after attempting to start another San Francisco paper, Conner went to Valparaiso, Chile, to establish a job shop and newspaper.) Manuel M. Noah, the Major's first son, also settled in San Francisco, becoming for 20 years the *Alta Californian's* editor.

San Francisco was clearly the metropolis of the Pacific far West, and merchants and agents soon followed adventurers and prospectors with equal determination. George A. Van Bokkelen ran a paper and printer's warehouse there in the 1850's, and was both friend and agent for the Hoe's:

> [Van Bokkelen to Peter Hoe, 19 June 1858] *. . . I find it impossible to get any quail at present, as the game law is in force, but as soon as it expires, I will get them for you and send them. . . .* [I want you to send me a Washington hand press] *and if you have not shipped any to other parties here in the last two months and have no orders for them now, please ship me another two weeks after this and another two weeks after that, in all 3 presses. . . . Regret to hear of sickness in your family. We are all well. . . .*

A decade later Jerome B. Painter handled Hoe affairs in San Francisco, but it was apparently a minor portion of his activities.

Painter, with a prospering type foundry and other commercial ventures was doing better than most miners. After some advertising in his *California Farmer* had amounted to $300 he took it in trade, and asked Hoe to honor a demand for that sum from Mess. Siberia Ott (726 Broadway, New York) for shipping a grand piano to California. Painter added (3 January 1865), "We have some thoughts of investing some money [$10,000 or more] in U.S. Bonds . . . and we would like to know if you would be willing to make the purchase for us? . . . We dislike troubling you with the business, but our limited acquaintance in your city is our reason. . . ." Painter obviously became too successful for hawking saws and type cases.

H. L. Tatum, who apparently replaced Van Bokkelen, had known very little about power saws, and less about steam presses. His technical education via long and patient letters proceeded immediately. Tatum soon visited some shops, like the Oakland *Tribune*, and at another, also in Oakland, sold a casting machine for Siamese type. His instructions continued:

> [*To Tatum, 2 May 1878*] . . . *We are pleased that you are making so thorough a canvass among the printers, and trust you will acquaint yourselves*† *with every want they may experience for goods in our line. There are however two things we wish you particularly to guard against in your efforts.*
>
> *1st Do not hold out any inducements of extra long credit to the printer. We wish to avoid this as much as possible, for our trade is large and very extended and to put our capital out on such endless credit is simply ruinous. It is better that the printer wait until he absolutely needs a press, than to be hurried into buying by offers of long credit. In the mean time you can see them often and learn just when is the right moment to talk business.*
>
> *2nd Be careful not to come into contact unpleasantly with the Type Founders. We desire, so far as is for our interests, to retain their friendship and good will, but wish to manage things so that eventually the trade shall be exclusively in our hands.*
>
> *Mr. Palmer* [*Palmer & Rey, San Francisco typefounders*] *was in to see us and we made arrangements with him to sell our wood work. Please quietly ascertain what Painter & Co. are doing in their Foundry—how their trade is looking. . . .*

When Tatum proposed to take second-hand presses in trade the Colonel was extraordinarily lucid, but firm:

> [*To Tatum, 13 May 1878*] . . . *We cannot do so without, in a short time, filling our works with second hand machinery which is slow of sale and cannot be sold with our guarantee without thorough overhauling and complete repairs. Although we do not guarantee presses of other manufacturers still, before sending them out, we always repair them, often adding improvements which make them better than when new. . . .*

These smaller Adams Presses named in our list [of used machinery] have been in stock for years and there is little hope of ever selling them. The prices affixed are simply the cost of repairs put upon them, in expectation of disposing of them, but they are now superseded by other machines. . . .

You can readily see why we are unable to take so many second hand presses, putting in them more than our profit, repairing them, holding them a long time, and, if sold, giving credit, running risks of collections and finally, ending up with a net loss, instead of gain after all our trouble. . . . You will find that many manufacturers are willing to sell on almost any time, to introduce their machines, but we cannot accept this ruinous policy and supply first class machines & we furnish no other. . . .

Our facilities for turning out accurate and durable machinery are unequalled by any other machine shop in the country and this, coupled with our extended experience, enables us to make presses intrinsically worth from 60 to 77% more than those of other manufacturers.

The first cost in purchasing is perhaps greater, but when the superiority and durability of the machines is[†] *taken into consideration, the economy of the purchase is at once apparent. . . . We are preparing a book [of shipping data] for your own use, containing weights, etc. and will send it shortly. . . . We are pleased to notice your manner of advertising and believe it will do good. The article in the Mining Journal was excellent. . . .*

With this encouragement Tatum went back to his onerous canvassing.

The San Francisco *Call* (founded 1856) was long a conservative morning family paper, owned by the proprietors of the *Bulletin* (1855), a much more radical, crusading evening paper. J. W. Simonton, at the Associated Press office in New York had retained his connections with the *Call,* and in 1877 when the paper bought a web press was asked to have Hoe explain certain installation charges. Hoe complied:

[To Simonton, 15 May 1878] . . . [Homell is one of our best mechanics], a man in whom we have every confidence, created by long and faithful service.

He says when going to a strange city, at first it is necessary to spend more money than when settled. As soon as possible he took a room at $1-- per day and had his meals at a Restaurant. Often these were irregular and, you know, that men when working nights, as he did to accommodate and hurry the erection of press, take an extra meal in the evening. A portion of the time he shared Mr. Walters' room, but though he compensated for this in various ways, treating to meals, etc. at one time paying half his monthly rent $12--, Homell should have gone to a regular boarding house as we tell all our men to do, because when eating at Restaurants often the ma-

chinist will take out the Foreman to dinner and treat around, all of which adds to the expense.

We believe his report, that his expenses were as stated, and that he spent in addition while there a great part of $50-- of his own money that he took with him.

He has given an accurate daily account of his time and we have charged other firms for the work he did at their offices. We pay him more than half the price charged and he is worth more to us in the shop than out of it. . . .

The American frontier was a vast wilderness where neither Hoe agent or Hoe catalog penetrated with any celerity. Yet the newspapermen there, from Canada to Mexico, put out editions with the same vitality and dedication as their Eastern counterparts. In the Rockies one enterprising printer hauled a complete shop on the backs of 45 mules to found the *Idaho Sun,* 1884. The *Rocky Mountain News* (founded 1859) in Denver was washed away by a disastrous flood in 1864—the press wasn't found until 35 years later, buried in sand.

A Campbell drum cylinder shipped to Fort Macleod, Canada, for its *Gazette* employed a dozen brawny Indian braves to crank the press, "their faces bedaubed with red and yellow paint."[36] In Idaho Indians reportedly discovered that molasses was the main ingredient in rollers—they stole them and ate them like candy. Indians also ran off with frontier printing types and inlayed them on their fancy smoking pipes. Job printing and barter were essential: one Nebraska publisher accepted a bronco from a subscriber and promptly turned it over to a compositor for wages.

The presence of a printing shop was nevertheless considered vital to the dignity of an ambitious settlement, its proprietor often doubling in offices: humorist Edgar Wilson (Bill) Nye founded a Wyoming paper in 1881 and accepted the postmastership (which paid almost nothing). "In my opinion," he said, "my being selected for the office is a triumph of eternal right over error and wrong. . . ."[37]

The mystery and comedy of evolution also penetrated this rough and raucous territory, coupled to the usual editorial skirmishes Easterners had long enjoyed. In Prescott John H. Marion's *Arizona Miner* battled a rival: "[Berry's] first great care was to fill himself with whiskey, after which it was his custom to walk, like the swine that he is, on all fours to his den. . . ." William J. Berry of the Yuma *Sentinel* replied with only slightly more elegance:

> We had the extreme mortification of seeing the editor of the *Miner* . . . laid out in the refreshment room, dead drunk, with candles placed at his head and feet, and a regular "wake" held over him. It was then for the first time that we discovered Darwin's connecting link between the fish and the quadruped. . . .[38]

The taming of the west would take a little longer.

For R. Hoe & Company, meanwhile, the volume of more serious —but no less spirited—business by mail again increased (after the Civil War) across the growing nation. But problems continued to be human in scale:

> [*January 1865 ("†" omitted)*] . . . *I im work one of your Cilinder Press for about 8 Years. I allwase been thinkin there ot to be a cutter ateadged to it. I wend to work and got a cutter up and it works well.* . . . [Y]*ou can have it of you wish befor einy Boddy als.* . . .

> [*November 1872*] . . . *I have such confidence in your integrity* . . . *I will send* [*a used press*] *to you & let you fix the price afterwards.* . . .

> [*November 1872*] . . . *Send* [*collection agents Dun, Barlow & Company*] *instructions to sue. There is not good feeling between* [*these Ohio partners, one of whom*] *has got a home.* . . .

> [*February 1874*] . . . *We use our double cylinder night & day. With two setts*† *of pressmen. One of them has been in the habit (unknown to us) of loosing his springs and driving at maximum speed.* . . . [T]*wo* [*main gear*] *teeth have gone.* . . .

> [*February 1874*] . . . *I anticipate a break in the Shaft of the "universal joint" of the No. 3 Press you re-built for me at the time of the* [*Chicago*] *fire.* . . . *Send me out one.* . . .

> [*February 1874*] . . . [*I want a good used newspaper press*]*—but having started with $15.75 in my pocket am not yet in condition to buy a new ten cylinder Hoe.* . . . [*Added in another hand:*] *Get* [*financial*] *report.* . . .

> [*February 1874*] . . . *Enclosed please find draft for amount of your bill for Press &c purchased. The press works like a charm.* . . .

With a network of hardware and typefoundry agencies and a few exceptional traveling clerks, as well as a brisk exchange of correspondence, R. Hoe & Company was nearly satisfied the country was pretty well covered. But back in the northern reaches of New York state a situation developed that proved to be most distasteful. The Colonel for once had been too successful.

John Lovell, a prominent Montreal printer and publisher (acquiring a newspaper, *L'Ami du Peuple*—"Friend of the People"—in 1835), became a steady Hoe customer when he purchased a cylinder through Toronto agent Charles Palgrave:

> [*To Hoe, 17 May 1856*] . . . *We have much pleasure in saying that Mr. Van Allen arrived here on Sunday morning. He commenced putting up the Press early on Monday morning and by Tuesday evening he had it ready.* . . . [*It*] *is now working as well as we could wish.* . . .

It seemed the start of a perfect relationship.

About 1873 Lovell decided to erect a book manufacturing plant in the United States to furnish employment for several sons who were coming of age to help manage their father's operations. The 1870's were flush years of expansion for book publishers—into paperbacks and cheap volumes in series of all kinds. The Montreal printer hoped to exploit this new book market with some very ambitious publishing schemes.

Lovell selected Rouses Point, New York (on Lake Champlain, a mile and a half from the Canadian border) as a spot close to home but convenient to the New York traffic. He proudly sent a prospectus along with his requests for quotations, and made a good impression. "You will certainly have the finest printing house in your section of the country," Hoe responded (24 November 1873), probably not even sure where it was, "and we wish you all possible success."

Curiously, the U.S. Book Trade Association, meeting in Philadelphia in 1874 and equally interested in Lovell, quickly suspected the foreigner of intending by some borderline ruse "to defraud the revenue of both countries and to undersell all the traders who pay their honest dues."[39] They insisted upon mobilizing to "stop his fun," but there was little they could do—directly. They watched, nervously.

Within six months R. Hoe & Company was engaged to furnish both pressroom and bindery equipment. "If you send us a plan of the floor or floors on which you are to place the presses," said the company (10 June 1874), "showing the position of main shaft and Engine from which they are to be driven, also the lights & points of compass, we can probably lay out the rooms as well here if not better, than by sending on a man. This will save you his expenses and time." Shipments of equipment began in September.

Following the presses, Hoe naturally recommended (11 July 1874) a line of bookbinding equipment: "we manufacture everything that is used, and made in the most superior manner. . . . We have a great many of the tools &c in stock and can supply the others at short notice." Hoe had easily scooped the whole Lovell order and almost singlehandedly equipped a complete book manufacturing plant of tremendous capacity.

By mid 1874 John Lovell & Sons, General Printers, Stereotypers, Electrotypers, and Bookbinders, capitalizing on cheap rural labor, had a thriving enterprise in the tiny wilderness village. The Canadians had spent almost $200,000 building and equipping the shop, at first christened "Lake Shore Press," then "Lake Champlain Press." Brothers Joseph, Robert, and John W. Lovell were promptly dispatched there to govern the work, and two—John and Charles—

soon married local girls. Rumors notwithstanding, there seemed to be good omens about the whole project.

By 1876, however, the plant was in financial difficulty, a result of poor initial financing:

LOVELL PRINTING AND PUBLISHING COMPANY.
(Limited)
Printers, Publishers & Blank Book Manufacturers. Works:

23 & 25 St. Nicholas Street, Montreal — Lake Shore Press, Rouses Point, N.Y.

Montreal, 10th May, 1875

Messrs. R. Hoe & Co.
New York

Gentlemen,

My son, John W. Lovell, returned here on Saturday, and, I am sorry to know, without effecting the loan so often referred to. He says that it is still open with some hope of getting it.

He tells me that you desire to get a mortgage upon the machinery, &c, which you furnished. Let me beg of you not to urge the matter for a few weeks. If the loan cannot be obtained upon the property at Rouses Point, money must be raised here. A few days more will decide as to the prospects in New York. If unfavorable I shall make another attempt here to raise sufficient money to pay off our liabilities, and you shall be among the first to be paid. Let me say, that with the exception of your claim, our liabilities are not large, and our prospects are very encouraging. I state this to assure you that there is not the slightest danger of any one having a claim on your machinery except yourselves.

Before commencing to build at Rouses Point I was promised a loan of $60,000. Had I received this sum, our money matters would be now easy.

If a satisfactory payment be not made to you by the early part of next month, I will see that the mortgage be transmitted to you.

Yours truly,
John Lovell

Strangely, it appears that Hoe and Hoe alone was pressing Lovell, and probably for an amount quite insignificant to the New York machinery maker. A week later John W. Lovell elaborated upon his urgent mission:

[*To Hoe, 18 May 1875*] . . . *The loan the writer endeavored to obtain while in New York, especially to enable us to pay off our indebtedness to you is still a matter of negociation. Mr. McDayall in whose hands it is informs us he has every prospect of carrying it through this time and we are in daily expectation of receiving the amount, when a settlement would at once be made with you. In the meantime we are endeavoring to obtain the money here* [*Montreal*] *and have also called a meeting of the Directors of the Company for*

> *2nd June next when an effort will be made to sell the balance of stock, \$59,000^{00} and to make the two calls of 10% each still due on the subscribed stock.*
>
> *We deeply regret the great inconvenience you have been put to by our not meeting note due on 17th ult. and still more the necessity of having to ask you to wait a short time for the one due tomorrow.*
>
> *Our property at Rouses Point has cost us almost \$100,000^{00} and is unencumbered and the plant contained in it has cost nearly \$100,000 more and while we feel quite sure your account will be paid in full we have still to ask your kind indulgence. . . .*

Hoe nevertheless insisted—an act it had rarely taken in its long history—and indeed filed a mortgage in November 1875. Not content, however, with such reluctant ownership in a remote manufacturing plant, Hoe began threatening court action as subsequent notes fell due. Although the book factory by end of 1875 was almost in full operation and heralded in the trade press as an awesome success, Hoe would give no more credit or time:

> [*To Lovell, Rouses Point, 18 January 1876*] . . . *Referring to your favor of the 14th inst. when navigation opens, or now if preferred, please send on the Baxter Engine and we can undoubtedly sell it for you, and will credit your account with all we receive for it.*
>
> *We would like to accommodate* [you] *by taking back the stereotype apparatus, but as it was made especially for you and we have so little call for this style machinery, fear we could not find ready sale for it. We will, however, put on our list a memorandum of the different articles, and if we have orders where they can be made available will send for them, instead of making new. . . .*
>
> *We have no objections to your removing two of the Standing Presses charged at \$150 . . . each from Rouses Point to Montreal. This we understand is what you desire. . . .*

In March Hoe used a favorite scheme to help recoup money. Lovell was asked (9 March 1876) to "execute all orders which may be sent to you by Mess. Dodd, Mead & Co., charging and rendering bills for the same to us directly." Without captive manufacturing facilities, Hoe nephews Dodd and Mead, struggling as tyro publishers, rejoiced at unlimited manufacturing credit. It also seemed the Colonel intended to be helpful to all.

The Lovells, meanwhile, were frantically entreating Montreal financiers and friends to assume some of the indebtedness. Several agreed to buy \$10,000 parcels of stock, and in 1877 Hoe also received \$12,000 in securities. The Colonel, however, was still not satisfied. As the elder Lovell began to see his cause slipping he hoped at least to bring some of the larger Rouses Point equipment to Canada where it could be better employed. The response was tough:

[To Lovell, 29 March 1877] . . . We have no objections to the Eight Cylinder Rotary Press being removed to Montreal provided you will pay us the $900 now owing, & the note of $1000 due April 6th. The remaining three thousand dollars we should wish secured in some manner, as satisfactory as at present & paid when due by Feby 4th 1878. We consider the press worth what you agreed to pay for it, less deterioration. . . .

Hoe pushed his quarry harder and harder. This operation was unlike any other transaction in the company's long record.

There is no doubt the American book publishing industry was behind the Colonel's effort to drive the Lovells out of Rouses Point and out of book manufacturing in the United States. It is possible that Harper & Brothers, Hoe's friends and colleagues, sometime creditors and relatives (by marriage), had been spokesmen for the agitated industry. The pressure for this disagreeable work must have been enormous. As late as 1891 the mention of activity at Lake Champlain was still inflammatory: "The trade have been consoling themselves for some time with the thought that operations at Rouse's Point had ceased," said *American Bookmaker*. "This is untrue. . . ."[40] But it was certainly true as far as the Lovells were concerned.

In August 1877 the Montreal firm was obliged to auction its plant equipment, and then the property itself. There were few buyers. Hoe sent veteran agent John H. Lockwood to supervise:

Rouses Point, Aug 5, 1877

Mess. R. Hoe & Co.
Gentlemen,

Your favor of 3rd inst., with enclosure has been received.

We are getting along nicely with packing of presses &c. and shall commence loading some time on tuesday.† I have made all arrangements with Agt. here for boat who is to have it ready tuesday morning.

Mr. John W. Lovell left for Boston last evening, and proposes to call on you early tuesday A.M. I judge from what he said that his business in Boston is to make arrangements for a loan on Building so that he can take up the mortgage which you now hold on Building.

There seems to be considerable uneasiness as to what you will do with the property after the Machy is removed. Mr. L says that he expects the Buildings here will be the next after the Montreal property; this latter troubles him more than any thing else.

Mr. J. W. Lovell will take with him a list of Stereo. & Electro. Machy that they would like to retain. I endeavored to ascertain from him what he could do as to paying for it but he could give no positive answer. If he makes any arrangement with you for the machy described in list which I enclose [not present], please telegraph me *as early on tuesday as possible, as we will be ready to commence on it by tuesday P.M. Address telegram to me at "Lovell Printing Office" as I shall be waiting for it. Mr. Lovell Sr. was here yesterday and was a little severe; he said that he could not blame me for looking out [for] your interest but that I should have put prices on the property which would have been "respectable." He further said that it would still leave the Co. in debt some $25,000, and that he did not know where the next move would be made; he endeavored to ascertain from me what you intended to do, but I could give no information.*

Before he went away he became quite reconciled, and made a very kind offer to allow us to run Engine & use the elevator when we were ready to lower the Bindery & other machy from upper floors.

Most every one here has an interest in the Lovell Ptg. establishment, i.e., a promise to pay, & they would be willing to take most any thing for their claims.

I will telegraph you as soon as we commence loading boat, also give name so that you can insure. I hardly think that we can get all in one Boat altho the Agt. claims that Boat will hold 125 tons. I hope you may be able to make arrangements with Mr. Lovell for mailing as per list enclosed, as some are very much worn, especially Battery, Cases, Galleys, Cabinets, Stands & Compo sticks.

Some few things must have been taken to Montreal as I cannot find them. I shall however scour the place & get all that is to be found.

Your obt. Svt,
John H. Lockwood

Two days later the search was completed:

[*To Hoe, 7 August*] . . . *We have everything now ready & shall commence loading as soon as the Boat is ready & hope to have all on board by saturday*[†] *night. . . .*

The standing press bought by Mess. Perrault & Co., Montreal, has not been removed. I have notified him that every thing will be removed this week & presume a check will be sent for it. If not, shall put on board with other goods.

As soon as I can get name of Canal Boat will telegraph you. . . .

[*To Hoe, 9 August*] . . . *I will endeavor to get order from The Lovell Ptg. Co. for return of premium when I go to Montreal. None of the Mr. Lovells here can give an order for it.*

I telegraphed you this A.M. that we were shipping Machy on Canal Boat P. P. Gaylord and that she would take all of it & leave here Saturday night. Everything is now on board except six Iron pieces, which will be put on board by 12 N to morrow. . . .

There is nothing special to communicate. Will take out B/L for goods to be del^{d} *at foot Broome St. N.Y. The boat which I have is one of the best & should insure at a very low rate—price 15¢ per 100*^{lbs} *or* $\$3^{00}$ *per ton of 2000*^{lbs} *loading & unloading to be done by Canal Boat Captain. . . .*

Later for another tiny sum John W. Lovell had to beg more time:

> [*To Hoe, 1 August 1878*] . . . *My note for $250 in your favor falls due on the 4th inst. I have tried to obtain the means to retire it or pay part of it, and hoped to have got money here today to remit, but have failed in* [so] *doing and so have to ask you to grant me renewal.*
>
> *I enclose new note for $250 & $6 for interest. I will endeavor to make payment on the notes falling due and you may be assured will do my best to discharge the whole. . . .*

The young publishing entrepreneur would not forget this humiliating experience, and it was perfectly clear to him that the smug American book establishment had been behind it all.

Lovell soon appeared in New York defiantly operating as a publishing "pirate" who would exasperate all the great book publishing houses with a network of cheap series. Revenge would be sweet: as his audacious schemes succeeded he aspired to monopolize the entire book reprint industry. His brother Charles similarly went into business as a New York book manufacturer and almost accidentally participated in the development of new bookbinding methods that all would find useful. Like the Hoes, surprisingly enough, the Lovells made an indelible mark on the American graphic arts industry.*

But R. Hoe & Company indulged in no other sharp practices its partners could not live with. It never really had to. With the best talent and the finest tools, with watchful eyes on both foreign developments and patent infringement, with a national sales force and three generations of the family in the business, the company could not be beaten. Yet, for a variety of reasons demonstrated perennially in any hectic marketplace, many customers were free to invest in the products of dozens of other manufacturers—with predictably mixed results. In the wake of countless machinery salesmen crossing and recrossing the United States to sell their wares, a Rhode Island printer bewailed the confusion:

> In the first place, we don't want so much poor machinery, engines, presses, etc. In the next place, whenever a press breaks down because of the poor material and workmanship used in its construction, we don't want the builder to fall back upon his dignity and swear that it has been broken through the carelessness or incompetency of the printer. It is a fact, as plan as it is undeniable, that more than one half of all our printing machinery—especially in kinds of job and country newspaper presses—are a disgrace to our American mechanics. Why is it? Not that they are imperfect in principle or design, as they come from the inventor's hands, but because of the poor material and worse workmanship used in their construction by the

* See *Books for the Millions.*

> builder, to say nothing of the many presses built so light that they literally "shake themselves to pieces." . . .[41]

Uniquely, perhaps, at R. Hoe & Company inventors were one and the same with manufacturers; each supported the other's reputation. *Scientific American,* aware of the problem from both printing and engineering standpoints, had no hesitation to comment:

> We never had or heard of any work from R. Hoe & Co., for instance, that did not give satisfaction in quality *and especially in cost,* when practically contrasted with machinery from professedly cheaper shops. . . .

Around the country, too, year after year printers almost always had a good word for Hoe products and service. The *New York Tribune* elevated the firm with classic elegance, in 1869 ending its tour of the factory loftily applying to the brothers a line from the Roman poet Horace: *"Serus in caelum redeani"*—"May you return late to heaven."[42] The plea was answered thunderingly.

Part II

JOURNALISM AND DEMOCRACY: THE TECHNOLOGY OF NEWS

It is generally acknowledged that Gutenberg developed movable type and printing press only to reproduce books, and in this endeavor achieved simply an economy of labor which gradually cheapened the printed word—and learning—everywhere. Although the technology of printing moved slowly from the fifteenth to the eighteenth century, for students the book was (and is) a superlative communication medium: durable, reliable, unchanging. At the opposite end of the spectrum is the newspaper: ephemeral, inflammatory, shallow. For a short period, however, intellectual thinkers in England came very close to functioning in and controlling the public press, such as it was, when the cheap daily became a mass reality. The newspaper almost became a literary document.

Applegath's vertical type-revolver at 1851 Exhibition / The Walter Rotary / Robert Hoe letter of 27 June 1856

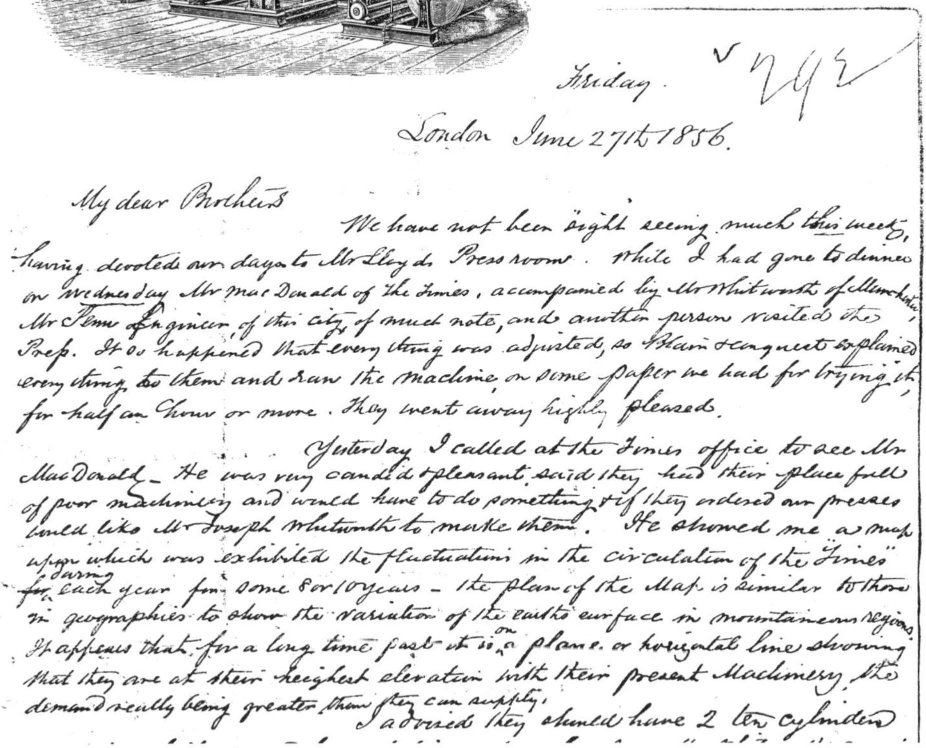

Friday. √ 292

London June 27th 1856.

My dear Brothers

We have not been sight seeing much this week, having devoted our days to Mr Lloyds Press room. While I had gone to dinner on Wednesday Mr MacDonald of the Times, accompanied by Mr Whitworth of Manchester, Mr Penn Engineer of this city of much note, and another person visited the Press. It so happened that every thing was adjusted, so Bain & Conquest explained every thing to them and ran the machine on some paper we had for trying it, for half an hour or more. They went away highly pleased.

Yesterday I called at the Times office to see Mr MacDonald – He was very candid & pleasant, said they had their place full of poor machinery and would have to do something & if they ordered our presses would like Mr Joseph Whitworth to make them. He showed me a map upon which was exhibited the fluctuations in the circulation of the "Times" during each year for some 8 or 10 years – the plan of the Map is similar to those in geographies to show the variation of the earths surface in mountainous regions. It appears that for a long time past it is on a plane or horizontal line showing that they are at their heighest elevation with their present Machinery, the demand really being greater than they can supply.

I advised they should have 2 ten cylinders

The popular paper was a rare child of the Industrial Revolution. After three centuries of printing by hand, a marriage of two vital forces—the mechanical press and the steam engine—produced the power-operated cylinder printing machine. When Koenig at the London *Times* brought the concept to fruition in 1814, the printing press as a servant and journalism as an institution seemed established. The distinguished historian Sir Stanley Morison observed:

> Books there were in the Middle Ages, books there will always be because men who can write will write; but printing is a process optional to literature and to booksellers. On the other hand, newspapers are the invention of the printer, and printing is absolutely essential to newspapers and newsvendors. The business of newspapers is to

> give the news of the moment at the moment. Hence their entire existence and progress depend upon speedy multiplication and transport.[1]

The monthly, bimonthly, weekly, thrice-weekly, then daily newspaper followed through the early 1830's with larger and larger circulations made possible by improved technologies. Before this phenomenon succeeded, however, the complexion of literary thought in Europe was quite a different affair.

A century earlier Joseph Addison had announced in the *Spectator* his intention of "bringing philosophy out of the closets and libraries, schools and colleges, to dwell in clubs and assemblies, at tea-tables and in coffeehouses. . . ."[2] Along with Pope, Steele, and others in England, Addison created an essay form which developed middle-class morality and aesthetics. It was limited in appeal only because that class of citizenry itself was small. In 1704 there had been less than one weekly newspaper buyer per 100; in the first half of the eighteenth century readership tripled but still remained insignificant numerically. Addison claimed, nevertheless, that 20 people read each *Spectator* issue, giving it an appreciable circulation of one-half million, or 1 Englishman in 11.

Within a short time the effects of the Industrial Revolution—and the French Revolution—began to appear in England. The slow-paced, protected life of British society was jolted as never before, and quite unprepared for what followed:

> In the first quarter of the nineteenth century it was becoming increasingly clear that what was politically, socially, intellectually, and spiritually a new society was growing up in England for which neither the institutions, nor the ideas, that had been inherited from the eighteenth century would suffice.[3]

By Marshall McLuhan's controversial dictum—"Societies have been shaped more by the nature of the media by which men communicate than by the content of the communication"[4]—so drastic a change must have been associated with a failure of old communications and the creation of new. In those altercations the old medium (the quarterly) died and the new (the newspaper) began.

Lewis Coser listed eight institutions in the history of Western man which stimulated intellectual activity: the salon, the coffee house, the scientific society, the monthly or quarterly review, the literary marketplace (book publishing), political sects, the "Bohemian" milieu, and the "little" magazine.[5] The presence of censorship he considered a negative influence. Three of these categories are unusual in that they furnish thinkers an opportunity to appeal not only to peers, but to the general public as well: the public house, the periodical review, and the political rostrum—all of which depend on reaction from the masses. The newspaper is not mentioned at all.

The coffee house, almost unique to England, enjoyed immense popularity as a meeting place for the exchange of thought, news, gossip. Unlike the French salon (for which an invitation was necessary), the public house welcomed all levels of people and all kinds of argument. It nurtured a tolerance for peculiar views and prejudices. Those who excelled in verbal persuasion became not only the best entertainers, but they achieved considerable social power as well.

Although not immediately a profound literary device, caricature in England paralleled coffee-house discussion in importance. Reading was still far from universal. The political or social cartoon, a one-sheet drawing printed in black and often colored by hand, was usually very cleverly contrived and sometimes cruelly biting. As vehicles of criticism and comment for illiterate thousands, eighteenth-century cartoons were long instrumental in shaping British outlook on European events.

Apprehension and hatred of the French, first over irregularities of the French Revolution, and then over economic competition and outright war, gave the British press and the local meeting places plenty of ammunition for nationalistic commentary. As Napoleon grew in power he was mercilessly lampooned; with the added tension of his possible attack of England interest in continental affairs was feverish. During the time of his expected invasion (1803-1804) 250 caricatures were issued on that subject alone. Similarly, dispatches of victory and defeat were the "stuff of life" to average Englishmen, none so excruciatingly awaited as that of Wellington's final victory at Waterloo (18 June 1815), which took three days to reach London over only 70 miles of land and sea.

Long after Waterloo a reaction against French democracy postponed (until 1832) the possibility of a new liberal society anywhere in Europe. Until then, too, any real freedom of the press would have to wait. Although the caricature declined abruptly after the Napoleonic Wars, its artists and techniques would find definite places for themselves when the popular press emerged. Meanwhile, however, the more literary periodical reviews would enjoy their finest hours—disturbed, paradoxically, by profound questions of social values and goals.

The so-called literary or intellectual reviews were largely modeled on *Gentleman's Magazine,* a fairly exclusive (circulation 4000-5000) quarterly dating to mid-eighteenth century. It contained strong erudite and classical material, passing gradually, laboriously, through discursions on giant fungi and Greek inscriptions to the newly discovered mysteries of gothic architecture and other elements of romanticism. Read at home, the monthlies and quarterlies seemed (at first) only a step removed from the actual confrontations afforded by coffee houses and cartoons. With a general subscription audience less easy to define, and publication policies for very catholic tastes, they promised to furnish a compromise between perishable oratory and interminably late, costly, or inaccessible books. Upon the conclusion of long illustrious careers Coser admired their intentions:

> The great English reviews of the nineteenth century and the intellectuals associated with them made a deliberate effort to shape the taste of the new middle-class audience, which eagerly turned to them in search of standards for behavior and opinion, of guidance, and of enlightenment.[6]

The public, bewildered by industrialization, poverty, class inequalities, and other nagging social problems, certainly needed help. But from the viewpoint of the intellectuals themselves, writing for or editing these great publications, it was quite a different story.

England approached the nineteenth century with an unhappy collection of antiquated laws, none more so than those associated with the right to read and the privilege to publish. An incredible domicile window tax (which existed until 1851) was calculated, said Charles Dickens, by the resulting darkness of homes, to be more damaging to mass literacy than the Stamp Act itself. The stamp duty, however, was also a truly vexing arrangement.

The British Stamp Act was first promulgated in 1712 to control the publishing activities of politically minded thinkers. Originally a tax of 1 penny on every "whole" sheet, it was increased in 1776, 1789, and 1797—retaliating, obviously, to increased agitation—so that, with a per copy tax of 3½ pence by the end of the century, a newspaper invariably priced 6 pence was more than workers could afford. The legislation seemed well designed for its purpose.

The stamp tax, however, applied only to papers reporting news; even so, its application by 1780 brought income from 14 million annual stamps, double what sales had been in 1753. Usually a colored mark imprinted at the corner or center of the front page, it could at times be a proud badge. The *Manchester Guardian* commended the social impact of a stamped paper:

> Its most striking physical feature was the crimson Revenue stamp impressed in the top right-hand corner of its front page. . . . The stamped newspaper was a means of social ceremony and obligation. Not many people had it, and the man who owned it, owned it to some extent for the benefit of his neighbours. . . . People presented their compliments to one another and begged to be favoured with a few moments' loan of the *Gazette* or the *Dispatch*. Being borrowed, it was duly returned, or it was passed on by consent to another applicant. . . . It was property.[7]

The *Guardian's* first issue (5 May 1821—the day Napoleon died) was printed at 200 per hour on a Stanhope all through Friday night, for Saturday distribution. Newspapers were perforce published only in the largest towns and read very little outside them, as what news (gossip) they contained was local and usually overpowered by merchants' advertisements (on which there was also a 1 shilling tax per insertion.) In London, of course, the newspaper filled many additional needs.

Coffee houses—with journal reading privileges—and newspaper "societies" (6 to 12 families jointly obtaining a single subscription) both prospered despite the vicissitudes of the papers from an economic point of view. In 1821 5000 such societies reportedly accommodated 50,000 families, and even as late as 1840 there were between 1600 and 1800 coffee houses in England. W. H. Smith, famous today as stationers and booksellers, started in the 1820's as a public reading room boasting 150 newspapers and periodicals on hand. Despite the inconveniences, the public seemed determined to read.

Newspapers consisting of opinion only, or of material previously published in book form, heretofore exempted, were newly regulated by one of the so-called "Six Acts of 1819." This law sought to silence a radical and anti-religious press by taxing periodicals (now 4 pence) containing news or commentary on news (with the exception of religious papers). Ironically this fostered the reprinting of cheap, insipid gothic novels in newspaper form (for which few critics saw any socially redeeming value), as well as the more explosive works —the target of this legislation—of Thomas Paine and other radicals who had indeed first appeared in books. Nevertheless the anti-government papers continued to publish against growing odds (many editors were silenced), led by a most extraordinary newspaperman, William Cobbett.

A self-taught radical thinker, William Cobbett (1762-1835) became exceedingly clever in his ability to arouse and persuade through the press. Fearing arrest after exposing the graft of some Army officers, he fled to the United States at the end of the eighteenth century where (as "Peter Porcupine") he issued newspapers with a flair and daring that paradoxically characterized the American press for decades. He returned to England and established the *Political Register* in 1802, editing and writing with a combination of controversial polemic and business acumen the rest of his life (except for another short flight in 1818). His paper cost as much as 1 shilling ½ penny at one point, but it was uniquely in demand:

> Here was a new kind of journalism, which trenchantly commented on domestic events and prescribed remedies for the desperate state in which the workers found themselves. It was a journalism which

> struck home, as none had done since the days of Paine, even to village-bound men who had remained indifferent to all papers as long as they merely reported news from faraway places, whether London or Austerlitz, in which they had no slightest interest.[8]

When Cobbett saw his paper jeopardizing the licenses of the coffee houses, he developed an alternate home edition—a 2 penny pamphlet without news but full of philosophy. Its circulation jumped to over 40,000 in two months, eclipsing anything of similar nature. He organized an efficient distribution system utilizing public coaches to send bundles to agents in rural towns. To circumvent the 1819 regulations he rechristened a series of his political treatises as "Cobbett's Monthly Sermons," so as to appear as a nontaxable religious tract. With *Political Register* Cobbett did much to encourage other editors to challenge the government's authority.

In 1831 a movement began to support a cheap press in defiance of the Stamp Act. Henry Hetherington, another radical printer, issued *Poor Man's Guardian,* an unstamped penny weekly, and others soon followed. Hetherington once sent a corps of newsboys to the streets groaning under bundles of waste paper to engage the police—while the real edition went out the back door and over rooftops. Hetherington also claimed at one point he was only lending the paper to readers for 1 penny, as the Act dealt only with works put on sale. When officials began to watch him he was obliged to take circuitous routes to work, "generally in the disguise of a drab-coated quaker."[9] Although Hetherington was jailed for about a year (during which time he continued to issue the *Guardian*), a jury in 1834 finally decided the controversial weekly was not a newspaper according to the Act, thereby strengthening the untaxed periodical press considerably.

After lengthy discussion in Parliament the stamp duty was reduced in 1836 to 1 penny per sheet, but this slight relief did not mitigate the violent feeling against the principle of a tax on literacy. Gradually many reformers argued against the Stamp Act, the London *Times* with justifiable sarcasm:

> [It was] a tax on knowledge . . . a tax on light, a tax on education, a tax on truth, a tax on public opinion, a tax on good order and good government, a tax on society, a tax on the progress of human affairs, and on the working of human institutions.[10]

That same year (1855) the stamp tax was fully and finally repealed, and a tax on bulk printing paper was similarly stopped in 1861. The effect on journalism in England was electric.

With the sudden removal of government duty the popular London *Daily Telegraph* reduced its newsstand price to 1 penny, forcing all the others to do likewise. With a new, workable price formula, dailies also sprang up in Manchester, Sheffield, Liverpool, and other cities.

The news was now available to all, and, with the establishment of syndicates furnishing stereotypes of fiction and other feature material to provincial proprietors, the most rural of them was at no disadvantage. Since the average man worked a six-day week, with Saturday often his longest, Sunday newspapers tended to be most in demand for leisure activity. (Puritanical laws prohibited even the museums from opening on Sundays.)

On the European continent itself factors much the same as in England mitigated against a free press. In France some unique solutions would strangely be proffered from the world of art.

"Three hostile newspapers are more to be feared than a thousand bayonets,"[11] Napoleon had said, undoubtedly having battled both. The French kings before and after him had apparently agreed; they kept an iron hand, exercised through a system of licenses, on printing and publishing of all description. A decree of 1810 limited the number of printers (only 60 were permitted in Paris), and each needed approval from the Ministry of the Interior, gave an oath of allegiance to the Emperor, and submitted five copies of every publication for the censor's examination. Although Louis Philippe abolished censorship in 1835, the practice of licensing lasted until 1870, its ramifications sufficiently harassing. Of course, those publications hewing close to governmental lines and policies would naturally enjoy favor. And the journal closest to official approbation was Armand Bertin's *Journal des Débats.*

Originating during the French Revolution, *Débats* had become the most influential paper in Paris (15,000-20,000 subscribers) during the Restoration; it tolerated no radical or liberal political thought. In the arts, however, there was latitude for the most violent criticism: musicians and artists, after the Revolution, had become public and social figures often in open competition for official stipends and public approval. Hector Berlioz, for example, accepted the coveted position of concert critic on Bertin's paper in 1835—and remained for 28 years despite the hazards a busy impresario might not be expected to endure. In 1846 he wrote:

> Thanks to the ill-will provoked by my criticism in the *Débats,* and to the raging anger of envious natures, I can make my way in Paris only like a red-hot cannon ball that hisses, burns, shatters. I have noticed that this hostility increased during my absence in Germany. . . .

By the time Berlioz retired from the paper in 1864 the climate for serious exchanges of arts and sciences had deteriorated. "The intellectual press he had been bred to and had helped make famous," wrote Jacques Barzun, "was already an anachronism."[12] Berlioz and his contemporaries had been among France's first literary figures to experience public attention through the press. For those, like

Berlioz, without political ambition, the government-supported newspaper was a refuge; for others, of course, the ideologically unacceptable situation was intolerable.

Charles Philipon (1800-1862) settled in Paris in 1825 and took up the new fad called lithography. Described by Henry James as a "little bilious, bristling, ingenious, insistent man," [13] he started *La Caricature* (1830) as an illustrated comic weekly, its first issue with text entirely written by Balzac, and successively featuring artwork by Daumier, Monnier, and Grandville, among others. Two years later he founded *Le Charivari*—"Chatter," a daily comic paper, with Louis Desnoyers as editor. These publications so lambasted the government—Daumier's cartoon of Louis Philippe's metamorphosis into a pear is classic—the whole publishing industry fell under close scrutiny. In 1833 alone there were 411 official "stoppages" of the French press, with 143 judgments resulting in 65 years of prison sentences and 350,000 francs in fines.

New laws controlling the publication of drawings in 1835 were almost specifically directed against Philipon. After 22 confiscations in one year and three staggering fines (considered "lenient" government treatment) *Caricature* was completely suppressed by the law, and *Charivari* was forced to turn to social satire exclusively—a formula which, in spite of its limitations, worked so delightfully the paper continues to modern times, having inspired the equally enduring *Punch, or the London Charivari.* In 1847 Philipon started *Le Journal pour Rire*—"Journal in Jest," containing copious artwork by Gustav Dore. Philipon, said Randall Davies, "raised caricature from the precarious issue of occasional prints to the regular position of an indispensable auxiliary to journalism." [14] This achievement was perhaps all too successfully accomplished within just a few years, but these were revolutionary times.

Honore Daumier suffered a major share of government harassment and imprisonment on Philipon's behalf, but he apparently felt his comic expression crucially important. In reflecting on the caricaturist's work, Henry James defined the artist's mission in concert with the press: "Journalism is the criticism of the moment *at* the moment, and caricature is that criticism at once simplified and intensified by a plastic form." It seemed natural that editor and artist work together. James added:

> Daumier became more and more the political spirit of the *Charivari,* or at least the political pencil; for M. Philipon, the breath of whose nostrils was opposition, . . . is to be credited with a suggestive share in any enterprise in which he had a hand.[15]

Philipon and Daumier, fortunately, found some support among

other journalists who also cried for an independent and responsible public press.

Emile de Girardin (1806-1881) is recognized in France as "Father of Modern Journalism." The illegitimate son of General Count Alexandre de Girardin, he acquired a handsome fortune—mostly by stock speculation—which, like Brougham in England, he dutifully used for "educating the masses," although he was often accused of mere charlatanry. As a young, ambitious thinker—writing short editorials in "bullet-like" sentences—he made the press a balance of idealism and commercialism. His wife, Madame Delphine Gay, conducted salons at which she collected "society" news (and once predicted that when tobacco reduced men smokers to somnolence women would take control of society). Girardin's concern for the poorer classes of France was multifaceted: he is also credited with establishing a network of small-town savings banks.

Girardin's first paper was *Le Voleur*—"The Thief" (1828), which consisted of nothing but material lifted from other publications. The sheet, and others like it, prospered until the Revolution of 1830. Girardin also started *La Mode* (1829), for which Gavarni did many exquisite fashion plates. He also tried an illustrated comic weekly, *Silhouette* (1830), but it too failed a year later. Also in 1830 he began *La Garde Nationale* as an organ for France's National Guard: its enthusiastic reception became his first major success. As he began to understand the popular function of the press Girardin campaigned for the abandonment of *"caution"* (money deposited with the government from which fines were deducted) and stamp duty, so that papers could be launched and sold more cheaply.

In 1836 without waiting for stamp tax abolition, Girardin founded *La Presse,* France's first important popular daily, charging one-half the price of the older Paris dailies. Lacking any political affiliation and sold singly instead of by subscription, *La Presse* introduced heavy advertising to help pay costs. With a policy of enlightenment—and immensely appreciated serialized romantic fiction—Girardin alone made newspapers "the universal habit" in France they were becoming throughout the civilized world. He employed Balzac, Dumas, Gautier, Hugo, Sue, and many others who went on to literary fame. An English journalist, Henry Vizetelly, meeting the elderly Girardin, found him with a Napoleonic lock of hair on his forehead and a weakness for "extravagant" neckties. Daumier immortalized him in his cartoon depicting the press mightier than the sword—showing Girardin driving a plume to the heart of France's president.

The success of the "penny" press in France, however, soon engendered a "yellow" journalism which destroyed much of the enlightened thinking Girardin had created. Under the Second Empire

Paris newspapers managed openly to support candidates unfriendly to the regime and to maintain some measure of liberty. But it was sly work. An editor who could command insult and innuendo, said Roger Williams, "might publish in confidence and know that his readers understood his point of view." [16] Under such conditions, however, France—like most of continental Europe—was a poor example of journalistic freedom.

Repression of the press throughout Europe brought with it the inevitable opposition by radical segments which could not be stifled at any cost. In Germany, with the profusion of some 56 petty principalities constituting the German Confederation of States, the babel of social and economic criticism was perhaps correspondingly loudest.

Karl Marx got an early position on the *Rheinische Zeitung*—"Rhineland News," a mildy liberal Cologne newspaper dating to 1842. Within a few months he became editor and turned the paper into a radical journal full of practical politics. Thanks to his tremendous powers of persuasion and satire, circulation zoomed. One of his anti-Russian editorials, however, came to the attention of the Czar, who protested to the King of Prussia—and Marx was obliged to resign. The paper collapsed.

In Paris Marx next helped to launch a subterranean German-French journal, ingloriously crushed when the first issue was confiscated by French authorities. In 1848, after the failure of several revolutions sweeping Europe (in which he had high hopes), Marx returned to Cologne to revive his old paper. With him was Frederick Engels, whose prolific and compelling writing style endeared him to the German. Again arrested, this time for protesting local taxes, Marx spoke so eloquently at his trial he was acquitted—but once more expelled from the country.

Landing in London, Marx tried again in 1849 to revive the *Zeitung* as a German-language review, but it survived only five issues. He agreed as European correspondent to write a series for Greeley's *New York Tribune,* although at first Engels handled some of the assignments as Marx feared his English was too poor. Not surprisingly, these dispatches often covered in great detail the state of European cannon technology (as today's reporters cover rocketry and missiles), since international politics have always hinged on the latest tools of war. Marx, however, was nearing the end of his career in journalism; his apprenticeship editorializing and propagandizing to a mass audience—piqued by harassments of censorship and economic embarrassment—nevertheless became austere foundation for the central cause of his life.

For years Marx had been writing and rewriting *Das Kapital,* the first volume of which he laboriously brought to publication in 1867 (its two other volumes issued posthumously by Engels). Marx was quickly established as the intellectual leader of an international socialist movement which inspired masses throughout the world. Without the experience and frustrations of his newspaper years it is possible Marx might not have determined to persevere. His ideological success, however, never brought his followers—even today—any degree of editorial freedom in which his coveted dialectics could transpire.

With the adoption of Marxism after the 1917 Revolution, Russia with its totalitarian Communism has permitted only a strict authoritarian press—government control even more tyrannical than conditions in Germany had been for the theory's founder. "A newspaper," Lenin had said, "is not only a collective propagandist and collective agitator; it is also a collective organizer." Stalin called it "the most powerful weapon of our Party." Khrushchev branded it "our chief ideological weapon." [17] These concepts are not what journalism has come to mean for the rest of the world.

Before the (Hoe) high-speed type-revolving press, the important English reviews had ample opportunity, despite their relative infrequency, to appeal to the general public. With the increase of national literacy the common Englishman, at the beginning of the nineteenth century, was only temporarily handicapped by the flowery language and lofty objectives of educated writers whose own comprehension of science, nature, and government was just emerging. For both there were new experiences and rewards.

The publishing dynasty of John Murray was vitally connected with the new forces of discovery and reportage. Lieutenant John MacMurray resigned from the British Marines at the age of 23 to become a bookseller. To a friend he wrote: "Many blockheads in the trade are making fortunes, and did we not succeed as well as they, I think it must be imputed only to ourselves. . . ." He dropped "Mac" (anti-Scottish feeling was high), bought a shop on Fleet Street and specialized in medical books. Isaac Disraeli, Benjamin's father, became a Murray author.

One two occasions Murray attempted to launch gentleman's magazines, but they failed. In 1783 he tried once more, starting *English Review* that year. The book publishing situation was so placid the new periodical intended to

> . . . give some account of *every* book and pamphlet issued in Great Britain and Ireland and in America, with notices of the chief publications of the Continent.[18]

With the prospect of a literary revival, Murray's son John II (1778-1843) newly educated, soon determined also to become the London agent for Archibald Constable's *Edinburgh Review,* which had just begun.

With impeccable credentials and appealing to the great Whig families, the Scottish journal, founded 1802, had been organized by Sydney Smith, Henry (Lord) Brougham, and Francis Jeffrey. In 1807 Murray added his imprint to Constable's, but, not yet satisfied—having lost his father's *English Review,* he formed his own magazine, *Quarterly Review* in 1809. Catering to Tories and aristocracy, the Murray review soon rivaled the Scottish one almost on its own ground: the first issue contained three articles by Sir Walter Scott and was so well received it was reprinted after selling out. John Gibson Lockhart (Scott's son-in-law) was long its capable editor. Murray started another venture, a daily newspaper called *Representative* with Benjamin Disraeli its editor; although meant to rival the London *Times* it quickly failed.

Additional literary publications soon followed these crusaders: *The Examiner,* an anti-establishment journal, was edited after 1808 by Leigh Hunt, who also contributed heavily to it; *Westminster Review* (founded 1824) served the middle class with radical Benthamite philosophy; *Fraser's Magazine* (1830) was founded by William Maginn and published Carlyle and Thackeray. The quarterlies were carefully managed and deservedly popular. *Edinburgh Review* and *Quarterly Review,* both addressed to aristocracy, took opposite sides in government questions and often sought to discredit each other—but discretely, of course.

The editorial content of the gentleman's quarterlies included speeches, poetry, "entertainments," and natural sciences. They were far from politically oriented. There were frequent topical articles on "the spirit of the age," its social manifestations, solutions, and consequences. Book reviews, sometimes years late, were often excuses to introduce unrelated dissertations and occasionally ran as long as books themselves. They were tastemakers: Macaulay's *Edinburgh Review* piece on Milton in 1825 was instrumental in making him an immediate literary success. History of all kinds was particularly welcomed, as was modern philosophy.

James Mill, father of John Stuart Mill, wrote prolifically to furnish purpose and contentment to contemporary British life. In an 1824 *Review* he first explained the novel problem of the literary periodicals if they were to survive, for, unlike books, they must be appreciated (bought) instantly upon publication. To catch immediate attention, then, each issue had to be a bit sensational—often by patronizing the opinions of those currently in power. Mill hoped, however, that *Edinburgh Review* would not have to resort to subterfuges he recognized in other magazines, and issued a policy statement:

> We [the publication] have no claim to be trusted, any more than any one among our contemporaries: but we have a claim to be

tried. . . . We may be sanguine enough, or silly enough, or clear-sighted enough, to believe, that intellectual and moral qualities have made a great progress among the people of this country; and that the class who will really approve [our contents is] sufficiently numerous to reward our endeavours.[19]

Competition would soon be an additional factor in the pragmatic publication of these works.

Mill's (and Jeremy Bentham's) philosophy of Utilitarianism—what is useful is good; what furnishes the greatest happiness is right—gradually permeated English society through the journals, the newspapers, and other popular tracts designed, largely, to bring about "the diffusion of useful knowledge"—meaning technical training to cope with the Industrial Revolution, and economic and social instruction to effect a measure of common prosperity. In his *Review* articles Mill reported to his aristocratic friends much of his methodology and expectations.

Since middle-class readers avoided crass pamphleteering and yet seldom read books, Mill rated the newspaper high as a propaganda and educational tool. Lord Brougham, hard at work along similar lines, in an 1835 *Review* enthusiastically agreed:

Here [in the press] . . . is a channel through which, alongst with political intelligence and the occurrences of the day, the friends of human improvement, the judicious promoters of general education, may diffuse the best information, and may easily allure all classes, even the humblest, into the paths of general knowledge.[20]

Brougham had founded the Society for the Diffusion of Useful Knowledge in 1826, and in the following decade apparently learned much about media selection for his program; he also organized mechanics' institutes as well, to give laborers an appropriate place to do their reading.

For *Edinburgh Review* Thomas Carlyle similarly demonstrated how that intellectual magazine, now almost 30 years old, was keeping abreast of the times:

Were we required to characterise this age of ours by any single epithet, we should be tempted to call it, not an Historical, Devotional, Philosophical, or Moral Age, but, above all others, the Mechanical Age. It is the Age of Machinery, in every outward and inward sense of that word; the age which, with its whole undivided might, forwards, teaches, and practises the great art of adapting means to ends. . . .[21]

Quickly, too, others rushed to interpret the effects of incessant whirring, clattering gadgetry. William Hazlitt, parliamentary reporter and drama critic, was credited with giving his pieces "permanent" interest which raised them from everyday journalism to literature. He correctly saw in the dawning mechanical era the possibility that dissemination of education and literature could become a mass phenomenon:

> There is no doubt . . . that the press (as it has existed in modern times) is the greatest organ of intellectual improvement and civilisation. . . . From the moment that the press opens the eyes of the community beyond the actual sphere in which each moves, there is from that time inevitably formed the germ of a body of opinion directly at variance with the selfish and servile code that before reigned paramount. . . .[22]

Hazlitt was optimistic that the quality of journalism he perceived in the gentleman's magazines would simply be printed faster and distributed more widely than before. (Television enthusiasts a century later—finding the task of "intellectual improvement" still unaccomplished—would see the same potential in that new medium.) But a few less sanguine observers realized that the masses were losing touch with the intellectual even as they strove to communicate.

In an 1825 *Examiner* Leigh Hunt had warned that the magic of the machine was no substitute for creative thinking:

> A business of screws and iron wheels [that is, a printing press] is, or appears to be, a very common-place matter; but not so the will of the hand that sets them in motion, not so the operations of the mind that directs them what to utter. We are satisfied respecting the one by science. But what renders us sensible of the wonders of the other, and their connection with the great and hidden mysteries of nature? Thought—Fancy—*Imagination.*[23]

Like his father, John Stuart Mill in *Examiner* (1831) was also concerned with current journalism. He was not fooled by mass media whose output was far from educational:

> The grand achievement of the present age is the *diffusion* of *superficial* knowledge; and that surely is no trifle, to have been accomplished by a single generation.

Although now better informed, the public, according to Mill (who also edited *London Review,* 1834-1840), was still easily duped by charlatans—many of whom, he admitted, had earlier fooled wiser men. Mill recommended that authorities familiar with "moral and social philosophy" guide public opinion carefully, although he feared that such authorities did not exist, "except, indeed, to a certain limited extent in the United States of America." Meanwhile, he concluded, "as the old doctrines have gone out, and the new ones have not yet come in,"[24] everyone must judge world matters for himself.

Five years later in *Westminster Review* Mill revealed a suspicion that the moral and social philosophy he advocated was indeed being dispensed, by—of all things—the newspaper. Just as civilized peoples in concert had conquered the savages (who had not learned to "cooperate" among themselves), Mill saw isolated British laborers, on the threshold of a better society, in need only of similar cooperative spirit. Along with new organizations like trade unions and publicly owned corporations, he saw the press hard at work:

> A more powerful, though not so ostensible, instrument of combination than any of these, has but lately become universally accessible—the newspaper. The newspaper carries the voice of the many home to every individual among them; by the newspaper, each learns that all others are feeling as he feels, and that if he is ready, he will find them also prepared to act upon what they feel. The newspaper is the telegraph which carries the signal throughout the country, and the flag round which it rallies. . . .[25]

The power of the press to consolidate and channel the thinking of the masses, and its potential speed for doing so (Mill's reference to the telegraph—not yet a working electrical system—perhaps more prophetic than he realized) was thereby described in simple terms possibly for the first time in "modern" England. The concept of every man intuitively readying himself for community action was exactly what monarchies (with the most cumbersome—but effective—legal apparatus) had long managed to avoid.

Mill himself, however, although he saw more "humaineness" and a decline in bigotry and aristocratic ostentation, was disappointed that in spite of their new found cooperation the masses had not materially improved their lives through the newspaper. Yet he judged Parliament and the London press the only remaining respectable professions in which an individual could work, for better or worse, to influence his countrymen. Years later (upon publication of his *On Liberty*), when 1858 government persecutions contradicted his observation that England was enjoying "liberty of the press," his once inviolable principles had so mellowed he overlooked the stern official action because it wasn't "persisted in" nor was it "political" in nature. A genius almost personally at a loss for having no worthy audience among the people he so wanted to reach, Mill found the contemporaneous situation morally and intellectually desolate. The Industrial Revolution had failed him.

Carlyle, too, clearly saw what the machine age was doing to the unsophisticated public and its fickle new media. He warned his aristocratic readers that the more popular sister branch of journalism had not only kept pace with progress, it was enjoying a growing strength of startling—idolatrous—proportions:

> At no former era has literature, the printed communication of Thought, been of such importance as it is now. . . . The true Church of England, at this moment [1829], lies in the Editors of its Newspapers. These preach to the people daily, weekly; admonishing kings themselves; advising peace or war, with an authority which only the first Reformers, and a long-past class of Popes, were possessed of; inflicting moral censure; imparting moral encouragement, consolation, edification; in all ways, diligently "administering the Discipline of the Church."[26]

The literary quarterlies, patently aware of mass media trends, were in fact charting their own decline even as they marveled at the power of the press.

In that first third of the nineteenth century, so prolific in the generation of democratizing and analytical thought in government and the sciences, the reviews had indeed managed to issue perishable material somewhat faster than the publishers of books, but not fast enough. The void was quickly filled. The Industrial Revolution had touched the printing press, and like a howling dervish it was throwing off weeklies and dailies everywhere on the British Isles. *Gentleman's Magazine* itself in 1825 noted there were 35 weekly newspapers published in London alone, all selling for 3 or 4 pence, all competing for lower class attention. And the family press had not yet entered the scene, nor the socialists or other special interest groups.

The penny press was a major exertion, especially for the lower classes, in the increased dissemination of knowledge—gratuitously called a "March of Mind." Directed to those just beginning to read, it was fostered by a few wealthy patrons to counter the cheap radical (political) press. William Chambers' *Edinburgh Journal* (founded 1832) and Charles Knight's *Penny Magazine* (also 1832) were the outstanding successes among family papers, the latter issued under the aegis of Brougham's Society for the Diffusion of Useful Knowledge.

Knight (1791-1873), an extraordinary writer, editor, publisher, educator, assumed the superintendency of SDUK in 1828, recognizing the London *Times'* steam-driven press as a perfect tool to reach the common people. He thoroughly planned and executed the details of printing and distribution to give the public, in *Penny Magazine,* a "quality" product despite the technical difficulties of untried methods in an untested program. He modified woodcut techniques—appreciating the value of illustration to semiliterate readers—so as to retain pictures despite poor ink coverage (without makeready) of the fast power machines. Where the hand press had produced only 1000 copies of a newspaper daily, Knight's rotaries with two sets of stereotypes turned out 16,000, the last issue as sharp as the first.

Thanks to steamboat, railway, and "fast" overland van, the SDUK magazine could be available throughout England within three or four days, and sold for a penny everywhere. Where there was Cobbett and Hetherington, there also was Knight. In all, said a preface to the collected first volume of *Penny Magazine,* "ready and cheap communication breaks down the obstacles of time and space," an achievement Knight judged possible only in "a high state of civilization."[27]

The enthusiastic reception of *Edinburgh Journal* (which also carried fiction but no illustrations) and *Penny Magazine* (without fiction) was dazzling, the latter reaching 200,000 within a few years (it ceased in 1845). Countless imitators in the 1840's, thanks to Knight's experimentation, made a feature of heavy illustration. The *Illustrated London News* (1842) was the first to succeed at subordinating text to picture: Herbert Ingram, a Nottingham printer and bookseller, reached a 66,000 circulation the first year. (When Knight noticed an Ingram artist sketching at a London trial, he realized that the problem of rapid woodcutting had been solved.)

Lloyd's Weekly News (also 1842) with tremendous advertising (Edward Lloyd's main contribution to journalism) reached 350,000 in 1863, requiring the proprietor to install the first American high-speed rotary presses—R. Hoe & Company's—in England. Lloyd almost singlehandedly gave Hoe the civilized world as a market, and the grateful pressmaker reciprocated by giving the United Kingdom the literary freedom it so laboriously sought. For with high-speed machines censorship and control proved futile.

With heritage harking to Milton's *Aeropagitica* and beyond, England in its own way had long fought a steady battle for freedom of the press as a principal aspect of individual liberty. But the Fleet Street conflict—within the shadow of Buckingham Palace—was always slow and costly; the complex strata of new urban society and the Industrial Revolution's profit motive made objectives unclear and suspect. When city journalists at last spoke up they too were more a babel than a war cry: the concentration of commerce and finance in and around London was more calculated to pit newspapers against each other (or splinter them hopelessly) than unite them against the excesses of official control.

Despite the colorful activities of editors like William Cobbett in tweaking the censors' noses, during the first quarter of the nineteenth century the British press had (like the American) been predominantly political in character. Much of what we consider important in a paper today was missing:

> The idea of news as information played a correspondingly small part, and such important events as the Battle of Waterloo might be dismissed in a few lines, swamped by editorial comment and scurrilous gossip. . . .[28]

The paucity of news gathering techniques (versus the abundance of opinion and rumor) and the fact that the public, genuinely interested in the war, was enjoying its coverage not in newspapers but in caricatures and other nonliterary media, were of course mitigating. The public press was, in fact, a small and exclusive entity in bustling London life. In 1829 the total circulation of seven morning papers was only 28,000 and six evening sheets tallied 11,000; these accommodated the business and government man who needed daily intelligence. Sunday papers, however, totalled 110,000, indicating that leisure-time reading by the working-class family would encourage these great trumpets of radical—socialist—thought to effect more change than the sober weekday press.

The Englishman's taste for political matters was whetted, nevertheless, and with the 1832 Parliamentry Reform Bill one barrier after another fell to the democratizing of British government. The newspaper after that year, however, took surprisingly little credit for these important accomplishments.

Workers, whose original quest for technical knowledge was purely a defense against the creeping sophistication of the machine age, were now irretrievably exposed to reading. In over 700 mechanics' organizations (by 1850) the newspaper reading room became more popular than any library, and an aggregate 100,000 membership soon changed its reading diet from technical matters to national and world affairs. With the removal of the Stamp Tax in 1855 these newly habitual consumers had the wherewithall to afford their own papers, and buy them they did.

Daily British newspaper circulation trebled between 1855 and 1860, and then doubled between 1860-1870. The *Star* was liberal, popular; the *Standard,* conservative, candid; the *Morning Telegraph* was pungent, independent (circulation 90,000 in 1859). The *Times,* imperious as ever, was actually losing ground (circulation almost 59,000 during the Crimean War, down to 35,000 at the end of the decade). Nevertheless there was something for everyone:

> The poor man now has his newspaper and reads it at home along with his wife and children, while formerly he went to the public house and paid two-pence for a glass of bad liquor or a cup of villainous coffee to get a "sight" of the *Times,* the *Chronicle,* the *Morning Post,* or the *Advertiser.*[29]

But so vast and hungry a public was begging for intensive exploitation. With remarkable speed and charity moralists and teachers instituted a didactic press (*Penny Magazine* and *Edinburgh Journal* were followed by dozens ranging from religion to gothic fiction) to give countless families wholesome, topical instruction; their cheapness was commendable, of course, but they had to be credited only cautiously for any effort toward making the man in the street a capable voter. The penny press was a distinct disappointment to the intellectual, its condescending nature especially infuriating. Said an 1833 critic:

> The people wish to know what is the cause they are reduced to live, in many cases, in wretched dwellings devoid of comforts; and by way of answer they are [in *Penny Magazine*] treated to a dissertation on the pyramids. . . .[30]

More purely educational efforts—with popular appeal—never materialized.

With the development of woodcut techniques, the great illuminated press also skimmed off a sizable readership. These gazettes eventually disappeared as other papers similarly borrowed the devices of pictoral and photographic reporting. Comic papers and scandalous guides to London low life also had unfathomable circulations. But curiously still missing was a widespread competitive popular press which might operate for the good of the people from a political, national, point of view. Apparently there was no point or profit appealing to a public denied the strength of suffrage.

The tightly knit haughty Fleet Street papers were joined by others in Manchester, Liverpool, and elsewhere, but in the tiny British Isles there seemed little room for wide divergence of opinion based on economic conditions. Political discussion was usually theoretical, pluralistic, parochial. Each paper within its social or geographical domain almost deliberately avoided overriding popular support, even for noble public purposes. Altick wrote:

> Whether or not [the working class] really wanted daily papers, as had been alleged during the campaign against the tax, workers did not buy them for some thirty or forty years after the penny daily came into being. The papers themselves made no great effort to attract lower-class readers. They remained what they had always been, papers for the upper and the substantial middle classes, giving most of their space to weightily reported political news and devoting relatively little attention to such topics of mass interest as sport and crime.[31]

Heavy political fare was unquestionably dry for the middle-class reader as well as the working man, to whom the franchise was again extended in 1867. It may be that the democratization of England proceeded rationally through the century largely out of government fear of overt revolution; it may also be conceded that British newspapers kept the fear percolating—but scarcely more.

Continental European journalism, like other institutions that ought to have been liberal, enjoyed little freedom. The French papers, said James Bryce, contained the best and the worst of all big countries; their pluralistic outputs effected a constant murmur, to be sure, but with absolute censorship throughout the country they were poor models of press enlightenment. In Germany Bismarck was the first to suborn the press, choosing not to stifle it, but, according to Bryce, "poisoning the well of truth at its source."[32] Of middle European and Russian conditions McLuhan wrote, "Their journalism exudes the private point of view of the literary Mandarin."[33] The European press was a constant disappointment to American observers.

Not until about 1883-1896 did a wave of "New Journalism" overtake England—even then largely following the 1884 Franchise Act which had again dramatically multiplied the numbers of British voters. The movement—accompanied by new yankee mechanical equipment—actually imitated the editorial vitality of American journalism of the 1840's. Instrumental were George Newnes (*Tit-bits,* 1880; *Strand Magazine,* 1891), Cyril Pearson (*Pearson's Weekly,* 1890), and Alfred Harmsworth, Lord Northcliffe (*Answers to Correspondents; Daily Mail,* 1896), who demonstrated the fecundity of great untapped populations by founding newspaper dynasties that were truly massive in scope and common—almost vulgar—in appeal.

By the turn of the century the democratization of England was complete for all practical purposes, although the domestic press could take little credit. Steinberg noted:

> The transformation of the newspaper into an instrument of mass-information and mass-education, into the voice—often shrill, but always unfettered—of democracy, is the major contribution of the United States of America. . . .[34]

It is true the yankee model was principally a technological one: professional journalists for fast stories, fast presses for late news. But the American penny press editorial formulas had to go along to make the system work. It did so admirably: in both hemispheres it bred popular strength and national solidarity that would permit modern democracies to withstand crushing world wars as well as countless social conflicts. But for both America and England events soon revealed that socio-political newspaper coverage alone was not enough.

In the second half of the nineteenth century the march of science and technology began to replace the abstract progress of earlier thinkers. But the technical innovators no longer had access to the newspaper—the fast medium they would need—for their communications. The telegraph with its world outlook left philosophers (and scientists) little quiet for rumination as popular events, whether petty or stupendous, vied for limited newspaper space. Literary intellectuals like Carlyle, Emerson, Thoreau, Marx, Nietzsche, Zola, Dreiser continued to editorialize, but their erudition was lost in the scramble for more entertaining copy. As political institutions settled into electioneering routines the public forgot it still needed intellectual guidance for uncharted waters.

The satirical wits similarly lost their impact; they had flourished enigmatically when they were most abused. Hence *Punch,* founded 1840, never reached great political importance, as had Philipon's comic papers. "Certainly," Henry James admitted, "the pages of *Punch* do not reek with pessimism," a condition he equated with harassed French caricature. This explained to his satisfaction why the caricaturist as interpretor of news had not achieved a "high destiny" in America. "The newspaper thrives in the United States," he said, "but journalism perhaps does not; for the lively propagation of news is one thing and the close interpretation of it is another." [35]

Although *Punch* gave its middle-class patrons an occasional glimpse of intellectual humor along with its satire and slapstick, the comic magazines displayed less and less ingenuity, and assumed for themselves almost no role in serious politicking. Their wit, in the absence of real challenge, dissipated. More disappointing, perhaps, was the increasingly genteel nature of protest in England, devoid of Cobbett's flair or Marx's passion.

William Morris (1834-1896) was long intellectually unhappy with the established institutions of the British government, although after an Oxford education he had turned to escapist poetry to occupy himself. In 1876 he began to voice an interest in politics, opposing Disraeli's foreign policy and producing a manifesto, "To the Working-Men of England," the following year. He so committed himself to the Socialist cause over the next few years—Marx had influenced him profoundly—he was actually selling working-class publications (especially the Social Democratic Federation's periodical *Justice*) on street corners.

Like no one else, Morris appreciated the importance of the printed word in winning support. While he contributed steadily to *Commonweal,* a Socialist League journal of which he was editor in 1885, he also wrote frequent letters to general newspapers where he knew numbers of readers, as yet uncommitted to his quarrel with the Industrial Revolution, could be reached:

> [To *The Athenaeum,* 10 March 1877] . . . Your paper has so steadily and courageously opposed itself to those acts of barbarism [indiscriminate repairwork on old buildings] which the modern architect, parson, and squire call "restoration," that it would be a waste of words to enlarge here on the ruin that has been wrought by their hands. . . .

> [To *The Daily News,* 20 November 1885] . . . I have just read your too true article on the vulgarization of Oxford, and I wish to ask if it is too late . . . to spare the few specimens of ancient town architecture. . . .

But in the sanctity of his Socialist world—and with the obsession of his growing feeling for good printing (he was to start his Kelmscott Press in 1890)—Morris lambasted the public press despite its propaganda value:

> [From *Commonweal,* June 1888] . . . Then there is the enormous mass of printed paper which is not books or literature, but which the public pays for every day, since I suppose a faculty once acquired produces a habit and must be exercised even when it is the mechanical one of reading print. . . . I believe, indeed, it is thought by some that this habit of the consumption of newspapers is the first step in education. Good! The second step, I take it, will be the end of that habit.[36]

Morris as aesthete and intellectual was too idealistic for his Socialist coworkers (in 1890 he was ejected from *Commonweal*), his disdain of mass newspapers possibly a contributing factor. In *News from Nowhere* (1890), his futuristic description of a Utopian Social state, popular journalism played little part. On the day of the "revolution":

> The ordinary newspapers gave up the struggle that morning, and only one very violent reactionary paper (called the *Daily Telegraph*) attempted an appearance, and rated "the rebels" in good set terms for their folly and ingratitude in tearing out the bowels of their "common mother," the English Nation, for the benefit of a few greedy paid agitators, and the fools whom they were deluding. On the other hand, the Socialist papers . . . came out full to the throat of well-printed matter. They were greedily bought by the whole public, who, of course, like the Government, expected a manifesto in them. But they found no word of reference to the subject. . . .[37]

Morris' Utopian revolution was calculated to take place in the year 2000.

The quarterly gentleman's magazines, consequently, soon gave up any pretense of following the world of popular taste and current opinion. Admittedly the discoveries of science were increasingly technical and time-consuming to appraise. In 1859 the publishing house of John Murray issued naturalist Charles Darwin's *On the Origin of Species by Means of Natural Selection,* which theorized that the variety of life on earth was explained by species adapting to differing world conditions. It was, said Sir Julian Huxley, "a book that shook the world and . . . has altered the substance and the direction of human thought more profoundly than any other publication of the age of print."[38] Darwin (1809-1882) had to spend much of the remainder of his life defending himself alternately against irate scientists and befuddled laymen.

Soon after the treatise appeared it was attacked in *Quarterly Review* (Murray's own magazine!) and *Westminster Review,* Samuel

Wilberforce, Bishop of Oxford, in the former spotting all its threats and countering them thunderously:

> . . . a dishonouring view of nature. . . .
> . . . a tendency to limit God's glory in creation. . . .
> . . . absolutely incompatible with the word of God. . . .[39]

At a famous 1860 debate, when Wilberforce with mock politeness "begged to know, was it through his grandfather or his grandmother that [he] claimed his descent from a monkey?" Thomas Huxley devastated the Bishop with his own sarcasm. It was unfortunate, however, that the quarterlies had become camps of conservatism incapable of the speed of riposte and delivery required by science's accelerated pace. But the general public, thanks to inflammatory and taunting press coverage, went to the opposite extreme. Darwin was amazed the book was popular at all, writing to a friend:

> . . . What a grand, immense benefit you conferred on me by geting Murray to publish my book. I never till to-day realised that it was getting widely distributed; for in a letter from a lady to-day to E. [wife Emma], she says she heard a man enquiring for it at the *Railway Station*! ! ![40]

What was more, the clerk said he'd soon have it. The public deserved to be acquainted with the concept of evolution, but *The Origin of Species* is not easy reading, and popular expositions would have been helpful. The general reader was instead titilated with gibes wherever they could be found—like Sir John Herschel calling evolution "the law of higgledy-piggledy." Misunderstandings consequently multiplied.

Darwin himself had scrupulously maintained that descent with "change" did not necessarily imply "improvement" of species; he avoided such words as "evolution," "higher life," "lower life." A biographer admitted:

> Had we heeded Darwin's warning [to use correct terminology], we would have been spared much of the confusion and misunderstanding that exists between scientists and laymen today. . . .[41]

Among misinterpretations engendered by this unenlightened situation were manipulations of a deliberate and ominous nature. Darwin soon saw political thinkers apply his work to some very peculiar uses which—thanks to journalism's misapplications, succeeded all too well. Marx had written to Engels:

> Although it is developed in the crude English style, this is the book [*The Origin of Species*] which contains the basis in natural history for our view. . . .[42]

The dismayed naturalist understandably declined Marx's offer to dedicate a volume of *Das Kapital* to him. But he also watched popular thought about evolution everywhere degenerate into ridicule and simian humor, popping up in all the current literature. Newspaper people, more than any other group, had so much fun at Darwin's expense they failed utterly to notice they had contributed to—perhaps caused—the public's alienation from all science. Journalists never recognized their responsibility to report science (until the atomic bomb made its point) as thoroughly as they covered politics. Today national policy is often dictated on laboratory blackboards, the public still as mystified and uninformed about it as it had been a century earlier by secret treaties among kings.

Seldom able to entertain themselves in other creative channels (as did Morris with his furniture, wallpaper design, and artistic bookwork), most scientists and philosophers, from Darwin on, kept well to themselves. By publishing their thoughts deliberately, carefully, they had no particular fear of bishops or general readers. Thinking people increasingly found the task of proselytizing painful and largely unnecessary, as public support for universal truths is seldom needed. Despite compunctions, they preserved their identity:

> Yet the intellectuals remained intellectuals, proud of it, and even in the more competitive ranges of, say, journalism, always conscious of some separateness of outlook from those who bought and sold material things.[43]

At the same time, however, as the purse strings of most ambitious research eventually revert to the public at large, no thinker can afford to disregard popular attitudes which themselves find effective voice through the media.

Some scientists of course successfully popularize their undertakings, although even the most patronizing have observed, as Brougham did not, that the less educated classes were not generally demanding of explanations:

> The mass of mankind in the Western World were, it is true, profoundly though indirectly affected by the progress of nineteenth-century science. They were provided with all kinds of wonderful devices and conveniences, which vitally altered their lives. . . . In popular magazines and newspapers they read superficially about the wonders that modern science had uncovered. . . .[44]

But man's serious thinking about life and society, alas, had scarcely begun. The popular mass media at the end of the century was as superficial as Mill had found it at its beginning.

Lynn Thorndike described the magnitude of the rift between intellectual man and his brother in the street:

> If the system of public education were more truly disciplinary, if the newspapers largely replaced crime, sensation, and sentimental matter

> by straightforward statement of political, social, and economic happenings, and the progress of knowledge, if the magazines minimized love stories, adventure, and personality in favor of matters of more moment to civilization, if advertisements gave sound advice as to good manners in public thoroughfares and conveyances, or sensible medical, legal, and financial counsel, if the moving pictures were employed more for purposes of instruction—perhaps the average man could keep up. . . .[45]

In Europe the intellectual, for a few fleeting decades of the nineteenth century, could almost have closed this gap, but the newly installed mass newspaper, with its peculiarities, frustrated his chances. In America there was such enthusiasm over its technological progress these profound questions never even came to light.

RICHARD MARCH HOE

1812–1886

Courtesy Library of Congress

5.

The Incredible Lightning Presses

TYPE REVOLVER LEGENDS AND SCHEMES / SWAIN AND THE PHILADELPHIA LEDGER / THE FAST NEWSPAPER FRATERNITY / THE ERA OF HOE'S "LAST FAST —" / PATENT EXTENSION

About 1841—many years earlier than is generally recognized, Richard March Hoe, then 29, was ready and anxious to undertake his most ambitious challenge in the art of press manufacture. Perhaps his youth and impetuousity were responsible, considering the sheer improbability of it. Despite the recent success of his father's cylinder presses—indeed, with his own improvements built into them, little attempt had been made toward increasing the speed of printing beyond about 1000 impressions per hour. For newspaper work, especially, this rate was woefully slow: papers with circulations around 30,000 would, even with a battery of machines, need 24 hours to print the day's news.

Yet, thinking about greater speeds was far different from attaining them. Richard remembered clearly how his father's shop only within a decade or so had replaced wood-cutting tools with simple metal lathes. Neither Americans nor British could well measure the minute dimensions which made complex machinery go or not go—machinists had not yet even decided how screw threads should be pitched, nor how castings might best be seasoned, machined, honed, mated, fastened, or lubricated. These matters, each little and yet nagging, would have to be swept along with considerations of press design, bed dimensions, and steam power available. The ramifications were almost endless, and around him Richard had few experts with whom to confer.

The Harper foreman who wept to see steam presses installed did not cry alone; many feared the trend toward mechanization and speed. Although America's growth during the early nineteenth century had been hampered by a nation-wide shortage of manpower—hence the frenzy to invent labor-saving machinery—those whose jobs were jeopardized did not think well of fast presses.

Throughout the country pressmen's unions were understandably alarmed by the intrusion of cylinder and jobbing machines. The Washington D.C. chapel resolved to undercut "the price of press-work and the introduction of rollers and roller boys as will enable employers to have their work done as cheap, better, and with greater certainty by hand than by the use of steam or power presses."[1] By 1835 most unions similarly conspired to make steam printing uneconomical, but gradually the sheer novelty and ingenuity of the machine coaxed admiration and challenged men's competence to make it work. The "music of the presses" (originally all the sounds of a busy hand printery—"the creaking of the heads, the thumping of the balls, the noise of the running in and out of the carriage"[2]) was now coupled to the hiss of the steam engine. The result soon gave employment, directly or indirectly, to thousands in a growing American book industry, and tens of thousands in the world of newspapers.

Despite his many hours at the shop on routine problems, the Colonel kept returning to thoughts of breaking the speed barrier. James Collins reconstructed the problem:

> The type for a newspaper page was locked up in a "chase," or iron frame, and laid flat on the bed of the press, where it rolled back and forth, and a cylinder rolled the paper over it sheet by sheet. Suppose the type were fastened around the cylinder instead, and the paper rolled against it—would not that mean twice as many copies an hour?[3]

Since there would be no unproductive return motion of the cylinder to reach its starting point, the concept seemed doubly efficient. Hoe heard frequent rumors about such machines here and there, and kept discussing them with his brothers. They were not easily won. They reminded him that in 1828, when their father was still puzzling the construction of Napier's machines, three New Yorkers had developed just such a revolutionary press.

In 1825 the Methodist Book Concern, nearby, possessed four iron hand presses, and a decade later, 30. But its workload was increasing even faster. Charles G. Williams, an inventor, Barker Badger, an editor at the religious publishers, and a pressman named Short had constructed a type-revolving machine with four feeding stations which they were anxious to demonstrate. Azor Hoyt, then shop superintendent, thought the odd machine might print the *Christian Advocate,* the organization's popular (8000 circulation) weekly. Although a senior director of the firm disapproved, Hoyt rashly tried it for one issue: for this insubordination he was promptly fired, and the paper reverted for a long time to its battery of slow hand presses.

Although understandably poorly printed in comparison to hand work, the trial presswork involved tapered column rules and the lock-up system Hoe himself contemplated. Thomas McElrath, then a proofreader and clerk at the Concern, remembered the press, as did Peter Hoe, who believed its trouble was imperfectly machined parts, permitting type to fall out during its operation. Constructed by Worrall & Company, the novelty was subsequently sold to Duff Green and shipped to Washington. The Hoes could ill afford a similar fiasco.

There were also rumors of Wilkinson's press at the New York *Sun,* irritating with promises of success that would eclipse R. Hoe & Company. And there was news from abroad, where Rowland Hill's patent attracted frequent attention and Applegath-Cowper-Morison revolving machines seemed already to be at work. Patenting his saw-making machinery in England in 1837-1838, Richard himself had examined the British state of the printing art and returned unimpressed by developments there since Koenig's departure. He was convinced his "dream of a madman" (Hassard's words) could be made to succeed. But settling on technical details was not so easy. Richard agonized:

> From 1841 to 1846 he thought and studied and experimented in vain. He was sanguine in temperament and happy in disposition, but little by little he became distant and restless, caring nothing for society and less for self-indulgence. Despondency fastened on him, and his family became alarmed for his health. While in this moody and abstracted condition, urged all the time by the publishers of the dailies fast growing to be "great," he roused himself from lethargy and renewed his experiments. . . .

This Richard Hoe was no longer the lad whom his partners just a few years before feared might not apply himself to business. But still needed was a stroke of genius:

> One night in 1846 he tossed from side to side in his bed and, unable to sleep, thought he would get up and walk about the room. As he rose, as with a flash the vista opened before him and he saw as clearly as if with the physical eye a solution of the long vexing problem. It was a plan for securing type on a horizontal cylinder—the panacea for the trouble of his patrons. Rushing to his desk he drew the diagram, which, after the details were perfected, resulted in the lightning press. . . .[4]

These "inventive" anecdotes from the New York *Herald* were preceded by others in the *National Intelligencer.* They illustrate, if nothing else, the public's desire to understand—to visualize—the romantic moments of genius and good fortune which propelled inventors to fame and wealth. Richard's path was clear:

> On the next morning, in connexion with the foreman of their works, Mr. Bowen, he began his successful experiments. That gentleman, on hearing the idea suggested, immediately declared that an eight-cylinder press, to work off 20,000 sheets within the hour, could be made upon the principle as easily as one of but four, which Mr. H. desired then to produce. . . . On his way home [from Washington, having filed a caveat] he met Mr. Swaine,† of the Philadelphia Ledger and Baltimore Sun, to whom he explained his invention. Mr. S. immediately ordered one. . . .[5]

At Gold Street nevertheless the Colonel received full support from his partners and the project was launched.

In the Experiment Room Tucker was similarly briefed in confidence; he too was ready for the challenge. He pondered Hoe's proposal, using wedge-shaped column rules parallel to the cylinder's axis, thickest at top where lateral pressure hopefully would lock the typematter in place, column by column, despite the considerable centrifugal force of fast revolution. Tucker saw no problem:

> This seemed so feasible that we fitted up a type bed, filled it with type, put it on a cylinder about five feet in diameter and revolved it at one hundred revolutions per minute; and to the surprise of many, and gratification of all, the type showed no signs of flying off the type bed. . . .[6]

Upon securing the type on the central cylinder and mounting four impression cylinders against it, the Colonel's rotary machine would print at every revolution as many copies as impression cylinders. The prototype (very similar in general appearance to the Applegath-Cowper-Morison), with one boy feeding at each impression cylinder, was calculated to produce an astronomical 8000 to 10,000 impressions per hour. The company with trepidation committed itself to design and construction of all particulars. If it were only a matter of thoroughness and precision, Hoe care would make it succeed; but its novelty went beyond that.

Richard, more confident from the start than his apocryphal 1846 insomnia would indicate, had lost no time acquainting friends with the unusual engineering concept. One was indeed the Philadelphia *Ledger's* William Swain:

> [*To Swain, 1 September 1843*] . . . *I shall be in your city in about 2 weeks when I expect to be able to tell you something definite about the 4 Cylinder machine I spoke to you about. I wish you would consider a little & tell me if you will desire to have one and fix on the*

size you will require if you should wish one. My best respects to Mr. Simmons. . . .

Swain's long and close association with the Hoes made him a steady confidant and a faithful customer. A few weeks later another remark about the machine slipped into brother Robert's routine correspondence:

[23 *November*] . . . *We are in immediate want of a barrel of Varnish in all respects like the last you furnished us from some one of your town folks. If you do us the favor to have a barrel sent out immediately it will much oblige us. Please draw on us at sight for the amount of the Bill. Richard has his new press advancing finely—more of this anon. . . .*

By the end of March 1844 it appears that, despite representations in New York to the major newspaper proprietors, Richard Hoe had staked the Philadelphia *Ledger* to take one of his new type-revolvers—if not in fact the very first of them. No other printer was similarly bombarded with reminders. With Swain pleading he could not come to New York, the Colonel was free to recount recent events as he saw them:

New York March 26th 1844

Friend Swain,
We have been expecting to see you on here and hope to have that pleasure soon. Our new velocity or "planetarium press" has recently been the object of no little interest to the fast newspaper fraternity. We have had some four or five of the principal ones to see the operating model and they are unanimous as to the fact that it will operate to entire satisfaction. At this time we might have orders for three or four if we chose to take them but as it is a new thing we preferred rather to make only one at the outset, the order for which has been given us by our friends Messrs. Greely[†] *& McElrath. Mr. Beach says we may go on & build one for him also. The day of double cylinders we apprehend for such papers as "the Ledger" is pretty much your past. We hope you will come on and see for yourself.*

Very truly yours
R. Hoe & Co.

To Mr. Wm Swain Esq.
Phila.

Considering the financial difficulties of the young *Tribune,* it is hard to expect Greeley and McElrath (who had seen the Williams machine fail) to have seriously ordered so unorthodox and so expensive—$6000—a press, except to capitalize on the publicity and arouse the envy of colleagues. More likely it was another impetuous act of Greeley's which McElrath ignored.

An encouraging note from Benjamin Brown suggests that the press even later was still not fully operable:

> [*To Robert Hoe, 17 February 1845*] . . . *Please give my compliments to the Col. of Mount Pitt*[?], *& say I hope he will loose*† *no time in getting his Planetarian*† *press in full operation & I hope with such a great success as his Iron Fly Boy.* . . .

Perhaps too the enthusiasm of the "fast fraternity" was only the Colonel's imagination at work, for in fact as journalism historians tell the story, Hoe was having considerable difficulty getting newspapermen to endorse the press. They were more than skeptical:

> Publishers argued that the notion was an insane one because, disregarding the difficulty of printing from a curved surface, centrifugal force would fling the type all over the place as soon as the cylinder attained any speed. . . .[7]

There would have been no better introduction of so sensational a press than a big metropolitan daily, itself sensationalizing the day's news. On the other hand, of course, if the machine failed—as had the others—perhaps a quiet death in Philadelphia would be preferable.

Diversions lurked everywhere. The company at this moment (1844-1845) was also constructing a rotary newspaper press with a stationary flat type bed; although a quarter-size model had worked well, the big machine developed such concussion it had to be abandoned. The Colonel wisely knew where and when to quit.

Somehow the inventor would not let this most dramatic printing machine fail. He had made the type cylinder so large the arc of the curved surface was only slightly bowed; with special quoins and the V-shaped column rules the type could be locked on the removable curved bed (soon called a "turtle") so securely that centrifugal action, instead of flinging the types outward, actually forced the quoins and rules to bear more tightly against the typematter. It neared perfection, but it was still far from ready. Another year went by as work progressed.

One printer privileged to see the remarkable prototype many years later recalled his awe to Richard:

> Perhaps you may remember, some twenty-three years ago, in the year 1845, when you were engaged with me in fitting up a model printing office at the [American Tract Society] Tract House in New York you took me into a secret apartment, partitioned off from the rest of the shop, at your up-town establishment, and there showed me the model of a new invention of a Type-Revolving Cylinder Press, upon which you were working with much patience and expending your capital most lavishly, in order to make the machine attain your ideas of cheap and rapid printing of newspaper publications. At that time the sight of your new machine, together with your explanation of what you desired to accomplish, filled my mind with wonder and interest. . . .[8]

Building a steam engine for Swain in 1845 furnished Robert an occasion to ramble with the Philadelphia friend who had not forgotten New York politics. James Harper, senior of the four brothers, had been elected reform mayor in 1844; his integrity in office, as in business, was a credit to publishing, although his humor and that of his cronies were often a trial (George Kendall rescued him one morning from a women's temperance group by interrupting, "Come, Harper, let's go out and have *another* drink . . ."[9]). But friends could not let James Harper accept another debilitating term:

> [*Robert Hoe to Swain, 4 April 1845*] . . . *The Colonel has not time this morning to write particulars but you will probably receive a long winded communication from him in a few days touching matters and things in general—reasons why every man ought to vote the native ticket . . . and that Harper is not elected next Tuesday, etc. All of which you must wait for patiently. . . .*

Needling references to the planetary press nevertheless no longer punctuated the letters to Swain. The company had won a firm commitment from Philadelphia. For his part, Swain may have expected this to be an opportunity for nationwide attention which he himself would generate. The custom of buying a bigger press to "confirm" rising circulation would here have its most stupendous application. Although a long wait was still necessary, Swain was neither impatient for delivery nor worried about cost:

> *Philad. Jan'y 3, 1846*
>
> *Dr Colonel—*
>
> *Yours of the 31st ult. came duly to hand, in which you give a pretty account of the New printing machine. I am indeed glad it works to your satisfaction, and hope as I doubt not it may be made to to*† *mine. I suppose of course you have it made adapted to the* [*weekly*] *Dollar Newspaper as well as the Ledger size. There is one thing I fear will trouble us, viz the necessity of using thick rules—this will be a serious objection in point of appearance. I will enclose you a copy of the Dollar Newspaper that you may see its present size. We may want to bring the Ledger up to the same size in the Spring, possibly. Hope every thing will be as perfect as can be made. It is as you know very important to you as well as to us; to you while so many fast presses are being invented, one for the Sun in N. York, and others in Europe. . . . In haste, yours as ever*
>
> *Wm M. Swain*

Fully ten months later a formal contract was drawn up:

> *Phila. Oct 1st 1846*
>
> *I agree to make for Mess. Swain Abell & Simmons one of my four cylinder Rotary Planetary Presses with two beds or chases & column* [*dimensions undecipherable*] *size width as the Daily Ledger, Phil-*

adelphia, and put said machine up in their office cor. Third & Chesnut[†] *Sts Phila complete, for the sum of five thousand Doll. This amount to be paid after the said Machine has been put up & in complete successful & satisfactory operation for one month. If in the course of Two Years they desire to exchange for a Six Cylinder Machine I agree to give them one for a sum not to exceed twenty five hundred dollars difference or to make them one of such size & capacity of six cylinders as they may require for the sum of Five thousand dollars. In case within Two Years any reduction should be made below these prices I hold myself bound to them for the difference.*

Rich[d] *M. Hoe*

[P.S.] *The above machine can be run without risk twelve thousand impressions per hour. RMH*

Swain, Abell & Simmons and all Philadelphia were ready.

At last, with well concealed apprehension, Hoe shipped off his first rotary press about March 1847, sending William Conquest to erect it, and came along himself to supervise construction. Swain was immediately delighted enough to coin an epithet that would continue many years:

Ledger Office
Phil[a] *March 19/47*

Dear Bro:

The machine is now running with the fly strings on, and is working well & beautifully; the boys are begining[†] *to feed better almost as well left handed as right & the fly strings & rollers work well. Swain says that is the "Chicken."*

I send you two Dollar Newspapers printed first side. The men in the composing room & all others seem to abandon all prejudices & acknowledge it the perfection of printing presses. Hoe's Last Fast Press is the name Swain has given it—for several reasons, as he says, beneficial to us in a business way. It (the name) saying that Hoe had made fast presses before & that he considered this the ne plus ultra of speed & in principle, &c, &c.

I shall endeavor to have the Foolscap press put in proper trim to morrow. Another thing I have concluded that it is best to say $7500-- for machines of this size & $9000-- for double size.

We shall put on the Ledger to[†] *Sunday evening & then I will give you some more news. . . .*

Mary is quite well again. How are you all at your houses?

Yours affy
Rich[d] *M. Hoe*

To/R. Hoe & Co.
New York

While there was no need for secrecy paralleling the London *Times'* introduction of the Koenig just three decades before, Swain on the contrary had no intention of missing the traditional fanfare of an exclusive story. The first rotary issue appeared 22 March 1847, and the details of its presswork were among the paper's first columns:

> We have now got fairly at work, running off the immense nightly edition of the Ledger upon the newly invented Cylindrical Rotary Printing Machine which a few days since we announced to our readers as having been manufactured for the Ledger, by the Messrs. R. Hoe & Co., Patentees and Printing Press Manufacturers, of New York. . . . The one we have has been in course of construction for us months past, and is the first and only one yet built, commenced or ordered, except by ourselves. The manufacturers already have our order for a second one of the same kind. . . .[10]

The *Ledger* described the unusual press as a breakthrough for large papers that formerly could not print enough copies through the night for the morning's sales. Its revolutionary principle proved once and for all "not only whether any machine could be invented by the most ingenious mechanic, capable of applying this principle to printing, but, even if there could be, whether the principle itself could be adapted to letter press printing." [11] To a layman audience, the *Ledger* necessarily introduced some curious new terminology:

> The essential principle of difference between this last invention and all other kinds of presses or machines hitherto used, consists in the fact that while upon those the types are locked up with "quoins" and a "mallet" and "shooting stick" [quoin-tightening tool], in a "chase," and laid upon a "bed" of iron, the surface of which it flat, with this one the types are screwed up with a wrench, in what our compositors are pleased to denominate a "turtle," constituting both bed and chase, and placed upon a large cylinder, four and a half feet in diameter, more or less, which revolves upon its own axle within four other smaller cylinders, one fourth part the size of the larger one, these revolving also upon their own axles in an opposite direction.

Each of the small cylinders, the *Ledger* explained, received a sheet of paper which, at every fourth revolution, met the turtle of type. There were two inking rollers to each impression cylinder, fed—through a system of rollers and vibrators which distributed ink to the large type cylinder (thus acting as its own ink feeder and distributor)—by a fountain placed underneath.

> With two of the cylinders the white paper is fed in above, and the printed sheets are thrown out below, and with the other two it is fed in below and thrown out above. The room taken up by the machine, paperboards and all included, is about sixteen feet in length, and nine feet high by about five feet in breadth. The type used upon it are the ordinary type hitherto used on the Ledger, which are made to assume a circular form in the "make up" by the use of bevelled column rules.
>
> For the attendance of this press ten persons are required, viz., a

> superintendent and an assistant, and four to feed sheets in, and four to receive them as they come out. This is the same number only that have been required to attend the two presses heretofore used in printing the Ledger.[12]

The paper found the machine's "great simplicity" to be another asset, and its operation so quiet "it can hardly be heard in an adjoining room." The whole Philadelphia printing community soon assembled to give the proud inventor a gala party seldom seen since the Waygoose feasts of Moxon's time:

> *Phil. April 16th, 1847*
>
> *Dear Bro:*
>
> *I will send you an account of the meeting to-night to see the "Last Fast Press" run & to take a little of the creature comforts, at D. K. Minor's Hotel, opposite the Ledger office.*
>
> *The Press does work beautifully & the edition of the Ledger is rising—31,900 papers worked off on the machine including stoppages for changing rollers, &c, &c, in 4½ hours is doing pretty well. It would astonish you to hear the compliments I receive—some of them are too strong, too much pork for a shilling. Both Messrs Swain & Simmons have concluded to have another machine & have given me the order to make one. But as Mr. Abell is to be here on Monday they wish me to wait in order to consult him whether it shall be a 6 cylinder or a 4. . . .*
>
> *We had a fair company to the jollification, some toasts & good jokes—say a fair proportion for Editors, & Printers you know are ready in such matters. About 30 persons were present at the supper but many more were down to look at the machine & I believe all were pleased. . . .*
>
> *Mr. Steele told me you remember last Spring or Summer that he would advise Messrs Sanderson to take one half* [*interest in the patent?*]*—to jump at it. Now ascertain how their pulses beat, & we can have a conversation together on the subject.*
>
> *Your aff. Bro.*
>
> *To/Robert Hoe* *Richd M. Hoe*

Two years later Swain had occasion to reveal just how important the unique press had become in the paper's production routine:

> [*To Richard Hoe, Paris, 27 June 1849*] . . . *We had a very bad break of one of our Hoes Last Fast two or three weeks since, which required about two weeks to have repaired. You can imagine my anxiety during the time depending upon the other alone to do the work of both sides of between thirty eight and forty thousand copies of the Ledger daily and about the same number of the Dollar Newspaper, weekly. Had this press given out, we should have been awkwardly situated for there is no other office in Philadelphia that could have printed half our edition.*
>
> *I do wish M'Elrath & Greeley would get you to make them a six*

cylinder press, for we want one, double size, if not a pair, and situated as we are at a distance from your establishment, we would rather not have the first one made. . . .

Curiously, the Lightning's success at Philadelphia (and Paris) must have suggested to Charles Williams' 1828 investors that Hoe's "infringement" of his rotary might bring them something in damages. The Colonel would hear none of it:

[*Hoe, Paris, to New York, 19 September 1849*] . . . *As to the Hoyts doing anything or making any trouble, they would find themselves very much mistaken & they would only be making fools of themselves to undertake it* [*a suit?*], *for three reasons, 1st. their machine failed, no matter from what cause, second, it was not like mine, third, their patent has long ago expired & cannot be resuscitated, so they could gain nothing in any event. I presume you have entered the caveatt*† *as I requested & if so I have no fear of any body, for I am quite sure of success. . . .*

This confidence was long the foundation of Richard's drive and determination. Those customers who met him and learned to know him became as steadfast as he.

William Swain, perhaps Hoe's closest and most loyal friend, had not only accepted the first Lightning and ordered a second on the spot, but five years later, when the pair of four-cylinders became inadequate, he replaced them with two eight-cylinder machines costing $20,000 each—a mammoth Hoe investment. By this time (1853) *Scientific American* repeated a ready pun for the Lightning: "no other that we have seen, has yet been able to 'hoe it out,' but it 'has hoed all others out' in fast printing."[13] Swain continued tirelessly on the *Ledger* until December 1864 when he sold the paper to George W. Childs. He died 1867.

Swain, Abell, and Simmons had met in New York as compositors in Major Noah's employ and formed a strong and trusting friendship. Together they set up the Philadelphia *Public Ledger,* almost amazed at how successful the penny paper could be. They were sorely tempted to repeat the performance: in 1837 with blond, bewhiskered 31-year-old Abell (1806-1888) as manager and the third partner "silent" (Simmons died 1855), they set up the Baltimore *Sun,* in format exactly like the *Ledger.*

At first 15,000 copies were printed daily on a "little Napier press." After a few years a Hoe double cylinder was acquired:

New York Decr 16, 1842

Dear Sir

RMH returned last evening and as he is engaged writing to somebody else till mail closes, I have volunteered my services in your behalf. The press will be shipped to morrow on board of a vessel which will *sail next day. We shall send Wm Colby on, Monday evening. Richard was detained in Phil^{a} till Wednesday evening by a subpoena from Mr. Alexander to attend court to testify in the matter of a libel suit against you on Tuesday next—to prove your proprietorship of the Ledger, but after considerable persuasion they let him go. We think it proper to inform you also that they seemed to be very busy about the matter. Why didn't you give more for that press?*

Yours very truly
Robert Hoe

To Wm M. Swain Esq.
"Baltimore Sun"

The plaintiff was possibly Charles Alexander, a Philadelphia publisher who established *Saturday Evening Post, Godey's Lady's Book,* and the unsuccessful *Daily Chronicle.* True to his word, Hoe got the machine off to Baltimore the next day:

New York December 17, 1842

Gents
We have pleasure to hand you invoice & Bill of Lading of the cylinder press this day shipped by the schooner Maria which sailed this day for your city. As mentioned in our letter yesterday Wm Colby will leave here on Monday for your city. When the press is put up and running please remit $2750-- and we will give you a paper acknowledgment for the same. With much regard we are yours truly for the proprietor

Robert Hoe

To/A. S. Abell & Co.
Baltimore, Md.

P.S. The crack in casting is nothing. RH

[*Invoice*]

Messrs. A. S. Abell & Co.
To the "HOE" printing press & S[aw] Man[ufactor]y

1 double cylinder printing machine "HOE" patent	$3500
Less cash discount agreed upon	750
	$2750

If young Robert was unsure of his addressee he was also imprecise about the firm's own name: variations of "'Hoe' Printing Press, Machine & Saw Manufacturers" appeared until about 1844. Response was quick and gratifying: "The Hoe Press works first rate," Abell wrote (28 January 1843), "and the way we make *her* travel is a caution to the natives. It is a first rate machine, just such a one as we expected from your unrivalled establishment. Glad to hear you [Richard] are about to take yourself a wife. . . ."

In 1853 the Baltimore *Sun* installed the first Lightning ever made —discarded by its sister paper in Philadelphia.

Similarly exploiting fast train and telegraph communication—as a friend of Morse Abell heartily endorsed the telegraph, the Baltimore *Sun* became one of the four principal American newspapers (with the New York *Sun*, New York *Herald*, and Philadelphia *Ledger*) of the 1830's. Situated in the third largest city in America and covering Washington, D.C. as well, the jaunty paper was ready for dramatic expansion and unprecedented architectural innovation.

Abell determined to build monumentally: for a new Baltimore headquarters he chose a plot where no less than five buildings had to be torn down. His grandiose intentions naturally interested Richard Hoe, who—with avowed competency in the matter—entertained some expectations of orchestrating the entire project.

Not at all reluctant to plan a newspaper office from the vaults up, the Colonel had assimilated the principles of construction engineering as well as the canons of architectural embellishment. His proposals were authoritative and convincing:

New York, April 26th 1850

Dear Sir

We send by Express a front view or elevation for your inspection. The first story to accommodate your P[ost] office; the columns of cast iron. Between the front and the entrance to the stories is a lobby or open space 6 ft wide, extending the whole width, which is calculated for cegar smokers, idlers, or persons waiting for delivery of letters, as this plan is expressly for the accommodation of the Post office. The 5th story is lighted by small windows under the cornice, but principally by [artificial] light.

The front, above the first story, is intended for brown stone; the statuetts† to be solid blocks in the press and sculptured afterward, as done in Paris.

This design is suitable for either brown stone or marble.

We shall send in a few days a design for a front in granite & also in cast iron.

Yours truly
R. Hoe & Co.

A.S. Abell, Esq.
Balt. Sun

P.S. The sketch sent being in pencil will be easily effaced if not handled carefully.

But Abell had already met a more compelling genius. Originally a New York watchmaker and engraver, James Bogardus (1800-1874) invented an eight-day clock, cotton-spinning and sugar-mill equipment, and metal engraving machinery, among many ideas which rewarded him handsomely. In 1847 he designed for himself a prefabricated metal and brick factory at the corner of Center and Duane Streets, New York, its chief novelty a system of precast iron columns and cross members (in what he called "the rich architectural designs of antiquity" [14]—that is, fluted shafts with traditional Greek capitals, many of which may still be seen above Greenwich Village). These were simply bolted together at the site. Bogardus' factory would turn out such columns by the thousands, some going to Harper's new plant after their 1853 fire.

Aside from a natural attraction for the mechanical interconnection of structural members, it is easy to see why Richard Hoe himself endorsed cast-iron construction: it reduced the volume of masonry walls and other dead areas and permitted heavier floor loads and larger unrestricted floor space—just what his growing presses would require.

Under Bogardus' supervision a "Sun Iron Building" designed by New York architect R. G. Hatfield would rise five stories to startle Baltimore and set a pace for commercial architecture as the "Sunpaper" would set for journalism. According to Hatfield's blueprint, the counting (business) office occupied the first floor, composition and editorial departments had the third, telegraph companies leased the second, and other tenants had upper stories. With the mail room, two new eight-cylinder Hoe type-revolving presses and their steam engines occupied the basement. From September 1851, when it was first opened, the building served the *Sun* perfectly for 53 years, with one possible exception:

> [*Richard Hoe to Abell, 12 September 1851*] . . . *The pump for water closets has been a very unfortunate affair, as the one I found the party went to fix to suit me & broke it all to bits so I concluded to let him slide altogether & we are now making one ourselves that we shall be sure will do the business. Patience & box up your water closely* . . . .

No matter what it was, the Colonel always preferred to make a gadget in his shop rather than buy one.

It is likely that, in proffering an iron building to Abell, Hoe had only been promoting the work of Bogardus and Hatfield, for he knew them both. Bogardus had won early admiration. Tucker noted:

> During 1842 the firm made a number of grinding mills for Mr. James Bogardus. These mills were extremely simple and have been used quite extensively for grinding paint, ink, grain, etc. They were . . . soon in such demand that Mr. Bogardus started a shop for their manufacture.

For Hatfield the Hoes would soon render similar service:

> In 1854 the firm made a hydraulic testing press for Mr. R. G. Hatfield, an architect. The cylinder was 4 inches bore, and the press was furnished with the necessary fixtures for holding the materials tested to ascertain their lateral, tensile and crushing strengths. . . . The tests of the strength, &c., of the various metals given in Mr. Hatfield's work on 'House Carpentry' were made with this press.[15]

Excited by the apparition of his new edifice, Abell was understandably more anxious to get all his equipment installed and running than attending to bills. "Please find Canal receipt for 'Super

Royal Cylinder Press' shipped this day," wrote Hoe (15 August 1851). The note was penned by "Seymour," probably Henry, a son of Mary Hoe, the oldest of the brothers' six sisters. (Another son, George, would have interests elsewhere: Stephen S. Hoe (19 April 1877) to Gordon L. Ford, *New York Tribune*, introducing his cousin, hoping to enter the coal trade.)

A few days later (19 August) shafting and pulleys went to Abell with the routine warning, "We propose drawing on you at sight for say $3000 unless we hear from you to the contrary. . . ." On 22 August a paper cutter and other small articles were shipped down with the patient acknowledgment, "We have your favor of yesterday directing us not to draw. . . ." But correspondence escalated into blustery—but slyly insistent—senior hands:

> [*Richard Hoe to Abell, 23 August 1851*] . . . *How are our men getting on; what is the state of things about the corner of Balt*[e] *and South Sts*[?] *I think I shall come on to see after things a little, the last of next week & I shall endeavor to take Mr. Swain along. I wish you would ask Mr. Conquest to either write me a letter stating how he is getting on, or tell you; also say if he wants some more help so as to have the shafting going up at the same time he is getting up the engine. . . .*

The following week (28 August) R. Hoe & Company billed Abell for $1207.01 for the shafting, gently requesting, "[P]lease let Mr. Conquest have what [expense] money he wants," followed by a small single-cylinder press, a double-medium wetting trough ($1800), and steam valves. Through October Hoe was working feverishly on still another Abell press (although this order was put aside to repair a New York *Sun* machine). At last Abell's large poster press was ready to ship and the balance was $2946.06; the obligatory "[W]e propose drawing on you at sight" was reinstated—and A. S. Abell & Company soon payed all obligations without further incident.

Twenty years later the old friends were negotiating once more:

> [*Hoe to Abell, 25 June 1870*] . . . *The offer you make is too low, as we are sure that our R. M. Hoe would say and we have good reason to think that he would find fault if we accepted it. We would agree to give you at*† *over 12,500$ that we may realise for the 2 old presses, & agree to take them and $27,500, and as that is so near your figures we trust that you will be satisfied with it & direct us to go on with the new machines. They shall be superior & we will lay our selves out to do our very best as to their quality & character in the construction of the presses. The more we investigate & learn the more we are satisfied that these machines are every way best suited to your use. . . .*

Abell took almost no time to consider the counter offer and accepted.

The Baltimore *Sun* had indeed made a lasting place for itself in newspaper history. Within fourteen years it proudly participated instrumentally in three great inventions: telegraph, iron-frame building, and high-speed printing (in a fourth, the Linotype, Abell had no interest). Others could hope to fare as well.

Swain's successor in Philadelphia had no trouble keeping up the pace. Although Childs promptly doubled the *Ledger's* price to 2 cents when he bought the paper in 1864, his circulation continued to grow. Inevitably, other enterprising journalists watched such prosperity, but similar growth, artificially stimulated, was not always lasting. William J. Swain, son of the *Ledger's* founder, in 1870 started the Philadelphia *Public Record* to compete with Childs, offering free maps as premiums. He wrote to Hoe (20 November 1872) "[I]f my circulation continues to increase as it is now doing, in less than sixty days my press will be unable to perform the work required, in time. . . ." Five years later, however, heavily in debt, Swain sold the paper at a sad loss.

For Childs' *Ledger,* however, the earlier Hoe Lightnings were scarcely suitable for 80,000 daily editions. In 1867 the paper invested heavily in the "planetarium" concept with innovations making it even more spectacular:

> Mr. George W. Childs now ordered three Four Cylinder Type-Revolving Presses to print the Philadelphia *Public Ledger,* two copies on a sheet, one press to have folding machines attached to fold the sheets to half-page size. The folding machines were made of the usual striking blade style, placed one below each fly-table, and as the sheet-flier laid a sheet on the table it was cut in two by the machine and each copy folded to half-page size. This arrangement was so satisfactory that folders were soon attached to the other two presses, and also to a Six Cylinder Press ordered shortly thereafter.[16]

For the *Ledger* 10 separate folding machines (9 Chambers and 1 Buckley), originally installed to back up the presses, obviously had declined in importance, as would the first successful Bullock press, so costly to its inventor. Childs (1829-1894), starting modestly as a bookseller and publisher with a strangely prophetic life-long desire to own the prestigious *Ledger,* built around these bright new Lightnings a headquarters that was to be a newspaper showplace.

For the gala proceedings to inaugurate the new *Ledger* building in June, Childs had invited a host of Philadelphia dignitaries and other personalities from around the country. Although he found it necessary (as usual) to decline, the Colonel fondly recalled his long association with the paper:

> New York, June 24, 1867.
>
> My Dear Mr. Childs:
>
> As I telegraphed you, I regretted very much that I was compelled

to forego the pleasure of being present at your opening of the NEW LEDGER BUILDING, as I had fully intended to be. Every advance in the usefulness and prosperity of the LEDGER cannot fail to interest me, and I would have liked very much indeed to join with your many friends in wishing success to the LEDGER and its present proprietor and manager.

What should be the duration of a public newspaper, the world is not yet old enough to tell; but if one may judge of a tree by its fruits, the LEDGER, for all I can see, is likely to be perennial. At any rate, in its present hands, it will lack for nothing.

I well remember the patient industry and firm resolve which characterized the establishment of the LEDGER, to make it in all respects a perfectly reliable newspaper, and on all subjects to consider the public welfare, treating all subjects and persons simply upon merit; always aiming to express, on important national events and topics, the best sense of the American people. I have been familiar with its whole history during the thirty-one years of its existence, and was very intimate with its former proprietors, especially with my old friend, Mr. Swain, whom I always regarded as one of the soundest thinkers connected with the newspaper press. I know it was his and his partners', Messrs. Abell and Simmons, constant study and aim to give the LEDGER a sterling character and individuality of its own, lifting it far above all party or personal influences, and, regarding only the public welfare, associated the best talent in its editorials and conduct—the LEDGER alone responsible, the writers always anonymous. His (Mr. Swain's) sound views, and steadfastness in maintaining them, and able financial management, established the LEDGER on a most enviable and durable basis. These facts are well known to you, and no doubt induced you to become its proprietor.

I am certain that the many improvements you are making are but the legitimate growth of the paper in your hands, and doubt not that the resources, energy, and ability you are so bountifully bestowing upon it will increase its usefulness, power, and prominence.

There is no position in life, it has always seemed to me, more ennobling than that of a conductor of a well-established and prominent public journal—none that gives opportunity for greater usefulness, nor one that involves more grave responsibility.

I trust you may live long to enjoy the high position acquired by your own ability and industry.

Yours, very truly,
Richard M. Hoe.

Geo. W. Childs, Philadelphia.[17]

Childs found time and resources to be philanthropic as well, later with his banker partner, Anthony J. Drexel, establishing the Childs-Drexel Union Printers Home at Colorado Springs, Colorado.

About 1847 the New York *Sun,* despite (or because of) its debilitating investment in its own Wilkinson-Beach machinery, had also been anxious to speed up its presswork. Its four-cylinder Lightning—the second machine made—went noisily to work in November of that year with hearty avuncular encomiums from *Scientific American.* The scientific weekly was already boasting recognition as an institution in its own right.

Rufus Porter (1792-1884) had worked as an electrotyper, among other skills. He edited the *New York Mechanic* in 1840, bought it the following year and renamed it *American Mechanic;* it ceased in 1842. He then began *Scientific American* (1845), his belief in the superiority of America through invention and science a trust in the future that hardly seemed at first justified—although R. Hoe & Company furnished him frequent inspiration. After one early call at Gold Street he lauded Hoe's cylinder machines: "Printing by steam is an operation of much curiosity, and should be seen by all visiting the city." [18] (At about the same time he saw Clay and Rosenberg's patent steam-powered type-setting machine which he understood was going into the *Sun.*)

The four-page science newspaper had cost Porter $100 to start; two years later he sold it to schoolmates Alfred E. Beach and Orson D. Munn for eight times that. Its weekly circulation would shoot up from 200 to 10,000 in 1848, and 30,000 in 1860, but its founder, an artist as well as tinkerer, was restless to pursue many chimeras. He intended to send a hydrogen-filled dirigible (first proposed in 1834) carrying 50 persons to the California goldrush areas in 3½ days; he foresaw the steam-driven automobile, the machine gun, and an elevated railroad. He reportedly sold to Samuel Colt the idea for a revolving firearm.

Porter's "color printing machine" for wallpaper was a huge impression cylinder surrounded by (and geared to) a number of design cylinders each with its own inking apparatus—very similar in concept to Hoe's planetary press then (1846) still in secret construction. Porter nonchalantly gave away his version: "any person who may be disposed to construct and enjoy the exclusive use of this invention may have the most favorable terms. . . ." [19] He may only have adapted what he, like others, saw in the Hoe Experiment Room; admittedly he never actually built such a machine.

Hoping perpetually to communicate his inventive enthusiasm to others, Porter was often militant:

Men of thought! be up and stirring—Night and day;
Sow the seed—withdraw the curtain—Clear the way!
Men of action, aid and cheer them—As ye may.
Aid the dawning, tongue and pen;
Aid it, hopes of honest men;
Aid it, paper—aid it type—
Aid it, for the hour is ripe,
And our earnest must not slacken—Into play.
Men of thought, and men of action!—Clear the way! [20]

His prestigious name remained on the paper's masthead long after he had left.

Alfred Beach nevertheless carried on the scientific tradition. He invented a raised-character typewriter for the blind, and followed with his pneumatic-tube mail system and passenger subway—squelched by the nefarious Tweed Ring's own transit plans. He remained with *Scientific American* until 1896. His partner Orson Munn too was extraordinarily active in disseminating patent information via his agency: when the Commissioner of Patents in 1857 admitted that more than one-fourth of all applications came through Munn & Company, Munn suggested it was more like one-third.

Yearly inventors' fairs in New York and elsewhere, and the fervid public interest in mechanical novelties which characterized the nineteenth century, gave *Scientific American* an official and unofficial mandate to assume responsibility for American genius of all kind. It cajoled inventors to bring forth their labors, fought for them (especially during the Civil War) against government bureaucracy,* championed their rights and properties against "socialized" invention without compensation, and broadcast American ingenuity around the English-speaking world.

With its peculiar concern for machinery's progress *Scientific American* "examined with no little pride and satisfaction" the *Sun's* new press, "a noble monument of mechanical genius and skill." [21] A reporter clocked it at 12,000 impressions hourly, but supposed it would do 50,000 if it could be fed that fast. The paper continued for decades to give Richard Hoe considerable flattering editorial coverage (as well as some confidential advertising discounts); it was a way of sitting in admiration at a master's feet.

After a month's operation a copy of the *Sun* came to the Colonel's attention, and he was somewhat more critical of its faulty inking:

> [*Richard Hoe, Paris, to New York, 12 December 1847*] . . . *I think that by the looks of the Sun that something is not quite right in the distribution of ink; ask Tucker if the fact that on the distributing surface the ink rollers move slower than on the type & that when they strike the type they are obliged to increase in speed & may wipe or slur the ink over the type instead of putting it on fairly has not something to do with it. . . .*

This led to long discussions of improved inking mechanisms for all Lightning machines. The problem was corrected.

Long before the Philadelphia *Ledger* felt the need for expansion, the *Sun* in 1849 had already contracted for a new, larger Lightning. "It sent its check to Col. Hoe for $20,000, and in a year a press with eight cylinders, capable of printing 20,000 copies an hour, was

* See *Age of Great Guns.*

completed,"[22] said the paper's historian. Despite Wilkinson's expectations of eclipsing Hoe with his own *Sun* machine, the management knew how to appreciate—without enmity—a press that worked. And indeed it functioned with surprising versatility, said *Scientific American*:

> In this one press, it may be said, "there are eight combined," that is, in respect to its effective power. One, two, three, or more tympan cylinders can be detached with the rest left free to work. This makes it very convenient, for it requires but a moment's labor to set the press so as to work with any number less than the eight attendants.[23]

This feature had not escaped the feeders themselves, who in the confusion and hubbub of getting out an issue could, for a few minutes at least, sneak away from their stations almost without detection. "One old warrior who was a 'taker-off' in 1870," wrote George Isaacs, "reminded me that the pictures of these presses are not complete in all details, for 'tin cans' are not on show. The 'tin cans' were tied to the frame adjacent to each feeder on the 'long run' machines to avoid the feeder leaving the press."[24]

The *Sun* gave a banquet—another great printers' waygoose—at New York's Astor House upon the installation of its mammoth press: "Col. Hoe was the hero," said Munn, "upon whom the honors of the occasion were showered." This splendid affair on 29 January 1851 was engineered by Moses S. Beach with his brother Alfred assisting. Novelist George P. R. James made a "few happy remarks," as did Major Noah, Henry Ward Beecher, and other Hoe friends. The Colonel's apparently modest reply was not recorded, but Benjamin Day substantiated the *Sun's* urgent need for the extraordinary machine; circulation, of course, had been responsible:

> Constant and vexatious complaints of the late delivery of the paper could not be avoided up to the time that I left the establishment, and I suppose these complaints continued until my friend here, the worthy guest in honor of whom we have assembled this evening, came forward with his great invention of setting types upon a cylinder, the success of which, I doubt not, has brought gladness to the hearts of a multitude of newspaper men. Mr. President [Dr. Chapin], it is really not in my power to express sufficient admiration of the energy and genius which has brought about so stupendous an improvement, and I trust some more able voice than mine will do more justice to the merits of Colonel Hoe's wonderful invention, from which, I doubt not, will date an important era in the art of printing.[25]

Dr. Edwin H. Chapin, pastor of the Universalist Church of the Divine Paternity—Horace Greeley and P. T. Barnum's church—was a true friend of printing and publishing: at a dinner a few years later he responded to the toast, "Editors of the Newspaper Press!" with a "masterpiece of oratory" which electrified his audience. He had not prepared himself until the very day of the affair, when he visited Harper's plant "to see a power-press and gather inspiration by looking at the machinery." [26]

Orson D. Munn of *Scientific American* as a constant admirer of Hoe's wizardry considered himself fortunate indeed to secure an invitation; he saved his wit, however, for a much wider audience:

> Col. Hoe . . . comes along with his Great Rotary, and *hoes* out *Suns* at the rate of 20,000 an hour. No wonder all the guests at the Messrs. Beach Dinner were more than planet struck. . . .[27]

The new Hoe Lightning at the *Sun* was indeed the largest and fastest press in the world. But R. Hoe & Company least of all rested for contemplation, as dozens of inquiries came in for planetary machines.

With several Lightning orders in their possession, the Gold Street brothers breathed a great collective sigh; the entire project would be financially sound after all. *Graham's* reporter in 1852 revealed the Colonel's hedged—but very vulnerable—gamble:

> [T]heir utmost calculation, as to the probable number which they should ever be called upon to manufacture, was five and twenty, but so marvelously has the demand exceeded their wildest imaginations that they have already built sixteen. . . .[28]

And they had scarcely polled all the large establishments nor touched the European market; neither did they count on prestige sales—what might only be called "Lightning fever."

Each application, perhaps for the first time in press erection, required meticulous care:

> In all contracts for type-revolving presses the most minute details were necessarily specified, and exact measurements taken of everything. Not alone the size and shape of the paper, but length and width and number of columns, depth of headings, rules, dashes and margins. Drawings of the location which the machine was to occupy were made, and every precaution which prudence could suggest was taken that no error might occur. Long before the machine was completed tracings were forwarded that the preliminary work for its reception might begin; and then, down in the subterranean depths of a cellar or under the adjacent street, as a precaution against the ravages of fire, excavations for the foundation and pit were made, and a solid stone wall was laid, lined with a tank of boiler iron, on which the machine was to rest. A skillful mechanic accompanied the machine to erect and start it. This duty was frequently assigned to Mr. Richard Taylor, whose familiarity with its various parts was surprising, and won for him many friends.[29]

The *New York Tribune,* at first so enthusiastic about the type revolver, became a hard customer to close. Although the Colonel was still tinkering with the ink distribution system, the company was bombarding McElrath constantly—for years. Every development seemed directed at Greeley's paper:

> [*Richard Hoe, Paris, to New York, 5 April 1849*] . . . *I am very sorry you cannot persuade Mr. M^c^Elrath to order a six cylinder machine. I know he could do his work on it to his entire satisfaction, & am quite sure he will order one at some future day, even should he have a couple of presses made of the old kind. Please say to him from me that this* [*Paris*] *machine works better than any before it, & that the flys are "charmant" as the Frenchmen say. We have not the slightest difficulty with them. I wish he would not decide to order any machines until my return. I should like very much to have a long talk with him.* . . .

The Colonel's indefatigable campaign nevertheless plodded another two years, but when the *Tribune* finally took a four-feeder in its vaults it was all praise. Staff writer John Hassard spoke for the paper as if he had discovered the machine:

> The four-cylinder press built for THE TRIBUNE in 1850 was the ninth machine of the class that Colonel Hoe's firm had made, and it did admirable service here for many years. It threw off about 8,000 impressions an hour, printed, of course, on one side of the sheet only, and as the process had to be repeated for the other side, its actual capacity was 4,000 complete papers an hour, which at that day was something unprecedented and marvellous. But having once established the principle of the rotary or "lightning" press, Colonel Hoe perceived that the development of a greater capacity was a simple matter of detail. . . . THE TRIBUNE in less than two years required additional facilities, and the first six-cylinder press ever made was constructed for us in 1852. To this we were soon compelled to add an eight-cylinder machine, and finally, in 1855, the great ten-cylinder lightning press, the largest style which this class ever reached, and for nearly twenty years the champion printing press of the world. . . .

Hassard's reportorial skills nevertheless breathed life into cold machinery:

> Deep under ground, and extending far beyond the lines of the sidewalk—generally to the middle of the street—were the lofty vaults occupied by the complicated machines. THE TRIBUNE used three of Hoe's lightning presses. The ten-cylinder machine towered aloft like a house. Eighteen feet in height, thirty-one feet in length, and nine feet in width, it required the space of two entire floors; it was approached by staircases and galleries; and platforms were built around it for the workmen. . . .
>
> During the early part of the night the pressmen prepared the apparatus with the greatest care. The bearings were oiled; the tapes and

> blankets were examined and, if necessary, renewed; the proper consistency of the ink was tested; the ink-reservoir was filled; the inking-rollers were removed, cleaned and replaced. . . . Meanwhile huge stacks of paper had been moistened in the adjoining vaults, and a supply had been placed on the feeding shelves. About midnight the plates of the "first side"—that is, the four pages which come on the inside of the sheet when the paper is folded—were lowered by a hoisting shaft from the stereotype-room at the top of the building. Instantly the pressroom started into activity. The four-and-twenty feeders threw off coats, rolled up sleeves, ran and clambered to their places beside the tiers of shelves. Scores of gas jets began to blaze. The metal plates—four for each of the three presses—were quickly lifted into place and secured by clamps screwed to their beveled edges. The pressman stood with his hand on the starting rope. The distant engine, which had been breathing impatiently for the past half hour, got in motion. The shafting under the vaulted ceiling began to turn. There was a cry, "All ready!" and then, as the rope was pulled, the three great machines, slowly at first, but with accelerating speed, set off on their tremendous revolutions.[30]

The *Tribune's* working-class readership could comprehend and visualize these details as enthralled as today's public following a rocket to the moon:

> It was a smooth and sure and comfortable sort of movement, and starting the big five-story [that is, its own levels] press was as easy as starting a clock; but the whir and clank of myriad wheels, the rattle of the wooden flies, the regular pulsation of metallic strokes, made, nevertheless, a deafening noise. Now the boys seized the fast accumulating piles of printed sheets, and ran with them from the receiving tables to a platform, where they were piled up. . . . So for three hours, without pause, the work went on. . . . Carriers, newsboys, and the agents of the great news companies crowded the vaults as daylight approached; and during the last hour especially, when every part of the machinery was revolving at its greatest speed, and men were running hither and thither under the gaslight with immense piles of the printed sheets, and customers were clamoring to be served, the spectacle of a newspaper press-room was one that never failed to impress the imagination and fix itself upon the memory.

Thomas McElrath, who so long resisted the Lightning, fell hard for other schemes more difficult to justify. Having invested heavily in a railroad which failed in the 1857 panic, he lost $120,000, resigned as president of Nassau Bank, and had to sell his *Tribune* holdings. "I am very sorry for M^{c}," the Colonel wrote (18 August 1857), "but he has no one to blame but himself. . . . It will make no difference in our operations with the Tribune folks. . . ."

With typical lack of agreement, perhaps, the New York *Herald* was prepared to dispute a concern of Hassard's over escalating costs of Lightning presses. Bennett was all praise for the economy of the Hoe machinery. The *Herald* in 1848 had scrapped its earlier presses, confined as they were to now intolerable flat-bed speeds. It had taken Lightning machines numbered 6 and 7. With figures to back his claim, Bennett in 1861 described the transformation of his shop to the Commissioner of Patents:

> [I]n the year 1848, after obtaining the type-revolving machines from R. Hoe & Co., the Herald was printed at the rate of 8000 impressions per hour; that these machines have been replaced by others capable of giving off 18,000 impressions per hour; that in the meantime the size of the Herald has been doubled, so that if now printed on the old double cylinder machines only 3000 impressions per hour could be obtained, so that the type-revolving machines on which the Herald is now printed have *six times* the speed of the fastest presses of any other kind in use; that if the double cylinder presses were still used it would require six of them, each giving 3000 impressions per hour, to work off one side, or twelve presses to work off the complete edition in six hours and a quarter; that it would require the setting up of six forms of type, and as the composition of a single form of the Herald costs $1,000 per week, it would entail an extra expense of $5,000 per week, or upwards of $260,000 per annum, and six times the room would be required for compositors, proof readers, &c.
>
> Deponent further says, that for the last five years the cost of repairs on the type-revolving machines during the same period was $4,289.99 or about one and five-eighth cents per thousand impressions, which, according to deponent's experience with the double cylinder machines, is a saving of one hundred per cent on repairs.[31]

The Colonel too was ecstatic over Bennett's product. "The Herald astonishes all at the [*La Patrie*] office," he wrote (31 May 1849). "It really is a God send to me to be able to shew it so beautifully printed. . . ."

The controversial paper contracted for its ten-cylinder in the summer of 1856 ($10,000 down, $5000 weekly for three weeks). The installation was quite uneventful, and many years later visiting the paper's new Broadway and Ann Street headquarters *Scientific American* was more impressed watching Hoe's new 60-horsepower steam engines activate a battery of printing machines. With income through the 1860's of over $1 million annually, the New York *Herald* continued invincible and incomparable.

Early in 1851 Henry J. Raymond, George Jones, and Edward B. Wesley—all with considerable penny press experience—joined to form yet another newspaper, the *New York Daily Times*. The *Sun*, they observed, was enjoying immense success but it was full of trivia; at the opposite end of the spectrum, wrote Elmer Davis, was uncompromising tradition:

> There still survived some excellent examples of the type of newspaper which had prevailed in the earlier decades of the century—the so-called blanket sheets, literally big enough to be slept under, especially by those who had tried to read them. They were massive, expensive, and dull; dignified if not respectable; content with a small circulation among gentlemen who had plenty of time, if not much inclination, for reading.[32]

Although Greeley's *Tribune* had been financed a decade before for only $3000 (and was returning a generous profit of $60,000 annually after ten years), the *Times* partners could not manage to start with less than $100,000, half of it, claimed Augustus Maverick, for the mechanical plant. Indeed, the four-cylinder press alone was $12,500. Raymond would be editor and Jones business manager (Wesley, a banker, was not at first actively involved). The paper's editorial offices, however, would not for a long time share the costly care bestowed on the pressroom.

For the *Times* ("Daily" was discarded in 1857) a half-finished brownstone on Nassau Street was hastily furbished. By September, when the paper was to appear, the editorial department was still a shambles. "All was raw and dismal," said Maverick. "[Floors were] open lofts, destitute of windows, gas, speaking tubes, dumb waiters, and general conveniences." The staff, furthermore, travailed eerily:

> [Raymond's assistant wrote] at a rickety table at the end of the barren garret; his only light a flaring candle, held upright by three nails in a block of wood; . . . the city editor, and the news-men, and the reporters, all eagerly scratching pens over paper, their countenances half-lighted, half-shaded by other guttering candles; . . . Raymond, writing rapidly and calmly, as he always wrote, but under similar disadvantages; and all the night the soft summery air blew where it listed, and sometimes blew out the feeble lights; and grimy little "devils" came down at intervals from the printing-room, and cried for copy. . . .[33]

The copy boys nevertheless rushed to serve no less than 18 waiting compositors, who, in turn, perforce fed the appetite of the great press itself. "A dozen employes of the mechanical departments" had all been seduced by Raymond from Greeley's *Tribune,* and a few, like James W. Simonton, had come along from the *Courier.*

Although scheduled to commence on Tuesday, 16 September 1851, the Hoe machine, capable of 10,000 hourly, was not ready until the 18th. The first issue revealed some of the problem:

> [Pointing fist] *The New-York Daily Times* is printed upon one of Hoe's celebrated *Lightning Presses,* built expressly for this establishment and just put up in the Press room in the basement of No. 113 Nassau street. The Press is one of the finest ever built, but as it is perfectly new, never having been worked until this morning, its work

> is not as good as it will be after a little time. The greatest part of the impression this morning was imperfectly printed,—part being clearer and more distinct than the rest. This is unavoidable from the operations of a new Press, but a few days will effectually remedy it.[34]

Issued as a large four-page paper and sold for 1 cent, the well edited *Times* in ten weeks jumped to 20,000 circulation. With enormous paper costs, however, the partners raised the price after 12 months to 2 cents, at the same time doubling its number of pages to placate newly acquired readers. By the first year's end too the Hoe press counter had tabulated 7,550,000 copies of 312 daily issues. The *Times* was a solid enterprise and grew quickly.

To turn out an 8-page issue every morning, half of the *Times* had to be made up by 10 p.m. the night before; the other four pages were held open until 3 a.m. for late news. At that time the press was started, finishing its run in three hours. Between 5000 and 6000 copies had to be in the mail, and other bundles delivered to railroad stations, before the average businessman arrived at his office. All had to be folded by hand—the chore of newsboys, some of whom boasted they could fold 1400 copies in 62 minutes.

After its first anniversary Jones became ill and resigned; first Wesley and then (in 1853) Fletcher Harper became business manager. Fletcher too left after a few years, and Jones eventually returned. Under Raymond's capable and steady leadership, however, the paper prospered; modeled after the London *Times,* it kept a high moral and conservative level, balanced in news and well mannered socially.

In the spring of 1854 the *New York Times* moved to new quarters on the corner of Nassau and Beekman Streets. With reporters evidently pledged to the paper's high ethical plane, Raymond's night editor once had to turn to the less principled mechanical department for (in Maverick's words) "beating Bennett at his own game." In September 1854 when two vessels, the sidewheel steamer *Arctic*—of the Collins Line—and a troop ship, collided at sea in dense fog, only Bennett had the details. Lost among 300 lives were the Maya archaeologist-artist Frederick Catherwood and most of Collins' own family; a tragedy. A *Times* pressman was sent to Bennett's office to steal an early copy of the *Herald;* with the purloined disaster snugly in its columns, the *Times* rushed to the street an hour before the *Herald* and outsold Bennett's paper all day.

Circulation that year was about 28,000, and new presses were indicated. Despite their exorbitant price, the *Times* in 1855 managed to issue a dividend—its first. A year later it was planning another new office, this time resolving to abandon its enforced tradition of dark and ratty accommodations. A five-story palatial Times Building on Park Row was the solution. Maverick described the prestigious neighborhood:

> The building occupies the northern end of the block bounded on the east by Nassau Street, and on the west by Park Row; abutting upon the open space formed by the junction of six streets, and known as Printing House Square. This square is sacred to the Press; for within a stone's-throw of each other are situated the offices of four of the leading daily papers of New York,—the *Times, Tribune, World,* and *Sun,*—and those of a dozen weekly journals and the Sunday papers; besides the great printing establishment of the American Tract Society, the warehouses of paper manufacturers, the shops of book-dealers, job-printing houses, and a countless variety of places in which print is in one way or another coined into ready cash. . . .[35]

The structure—and the staff—seemed especially intimidated by Hoe's large and complex machines: "The press-room vaults are of extraordinary dimensions, extending around the three fronts of the building," Maverick added.

The Colonel's substantial bills for Lightning presses must have prompted Raymond—ordinarily favorable to industry's progress—to opine in an 1852 editorial that mechanical creation should devolve free to the public. Ever the inventor's champion, *Scientific American* retorted:

> Had Messrs. Hoe not possessed the means and ability to construct their improved printing presses, the daily papers of this and other cities, in all likelihood, would still be jogging along with their old slow Napiers. . . .[36]

Raymond conceded. A new six-cylinder type revolver was installed in 1855, but a few years later the *Times* got a redeeming bargain when a ten-cylinder went on sale.

Like many other money men, Edward Wesley was hurt badly by the 1857 panic, but he fortunately held his newspaper stock. Two years later again readying another new headquarters, the *Times* ordered a ten-cylinder on which Hoe commenced work. But from Boston came the news that a similar machine would be returned almost unused. Wesley would consider it.

The Boston *Traveller* of Worthington, Flanders & Company dated to 1825, and was an early candidate for the four-cylinder revolver:

> [*Richard Hoe, Boston, to New York, 21 December 1849*] . . . *I had a long talk with Messrs. Flanders & C° yesterday. They are obliged to run their press too fast* [*which*] *is the reason why it breaks. I am to meet them this P.M. to discuss still farther the 4 cylinder. I proposed to charge $8000 & take their press for $2500.--. They are hardly prepared for this. They seem to think 6 or 7000 was enough or that I should allow more for their press. I hope to bring them up to my way of thinking this P.M. I have seen most of the other printers & all seem to be doing well & many talk of wanting machines in the spring. I think we have only commenced to do business in Boston. . . .*

Within a decade the *Traveller* had taken possession of its ten-cylinder. But finding its new format too expensive under depression conditions, it ran the big machine only two weeks before returning to its old four-cylinder.

> [*Richard Hoe to Robert, London, 29 September 1857*] . . . *They now wish us to take back the ten and make them a six or eight. I told Mr. Worthington that the only way we could do* [*that*] *would be to let the* [*New York*] *Times have that press & he would have to pay all expenses, & also* [*all*] *that we had so far expended on the Times Ten Cylinder. I am going on to Boston next week to arrange the matter with him. It will be a loss to the Traveller folks of about $7000, i.e. $4000 that we have spent on the Times press according to the cost book & about $3000 expenses bringing back the press & fitting it for the New York Times.* . . .

In December the Boston press arrived at the Hoe shop for some "trifling alterations" while the *Times* tidied its vault for it. In May 1858 the ten-cylinder printed "to the satisfaction of us all," the Colonel wrote, "including Messrs Wesley, Raymond & a number of amateur lookers on. . . ." The paper prospered.

In 1861 the *Times'* George Jones explained its unique way of using its two Hoes:

> At the present time the daily edition of said paper, which is ninety thousand (90,000) copies, and also the weekly and semi-weekly editions, are all printed on a six and a ten-cylinder press, with one complete form of types; one side being printed on the six-cylinder press and the other side, which contains the latest news, on the ten-cylinder press. This course is pursued for the reason that the ten-cylinder press can print the one side of the whole daily edition in less time than the six-cylinder, so that the completion of that part of the form can be delayed for the latest news whilst the other side is being printed. That there is no other kind of press known to him [Jones] by which the daily edition of said paper could be printed without the extra expense of duplicating the forms, and that said type-revolving printing press is of great value to the public, and indispensable to the success of newspapers of large daily circulation; and that the inventor is entitled to a highly liberal remuneration for the production of an invention which has been, and continues to be, of such great benefit to the public.[37]

The *Times* installed stereotyping in July 1861 and bought additional presses during the Civil War. By 1865, with format enlarged to 7 columns, it became the largest paper in America and indeed equal in reputation to its London idol. It had won its right to pontificate—reviewing *The Origin of Species* in four long columns, the paper concluded: "The fundamental limitation of Mr. Darwin's theory springs from a fact in his own mental structure. He is but a Naturalist. . . ." [38] Raymond, like many newspaper people of his time, vacilated between journalism and a political career, regarding them as two facets of the same jewel (and so they were). He died in 1869, age 49.

The roster of Lightning presses quickly grew to make that style of machine a major item for R. Hoe & Company. The first 23 were sold by 1855:

Philadelphia *Ledger**	2–8 cylinders	
New York *Sun*	1–8 "	and 1–4 cylinder
New York *Herald*	2–4 "	1–6 "
New York Times	1–6 "	1–4 "
New York Tribune	1–6 "	1–4 "
Baltimore *Sun*	2–4 "	
New York *Sunday Dispatch*	1–4 "	
New York *Commercial Advertiser*	1–4 "	
New York *Evening Post*	1–4 "	
New York *Staats-Zeitung*	1–4 "	
La Patrie, Paris	1–4 "	
Boston *Times*	1–4 "	
Boston *Traveller*	1–4 "	
Boston *Journal*	1–4 "	
Philadelphia *Evening Bulletin*	1–4 "	
Cincinnati *Commercial*	1–4 "	

* Not in chronological order.

The *New York Times* added:

> It will be observed that the use of these Presses is general, in all offices whence circulate large editions of the daily or weekly journals. The peculiarity of the applications of mechanical ingenuity—the nicety of operation—the multitude of parts which require to be smoothly combined,—the wilderness of tapes, bolts, cylinders, rollers, "turtles," rules, tables, wheels, cranks, and levers, that have to work exactly in shape, or not work at all,—all these facts render the putting up and the practical operation of the great Press a delicate matter. The cost of an eight-cylinder Press is in the vicinity of twenty-five thousand dollars. The time occupied in the construction of one of them is about four months.[39]

William W. Harding took over his father's old and conservative paper just before the Civil War and gave the Philadelphia *Inquirer* new life; with large war maps and good battlefield coverage its circulation reached 60,000 in 1862. Harding was ready for a Lightning although a little late in ordering:

> [*Harding to S. D. Tucker, 7 January 1863*] . . . *We have concluded to have Six Cylinder Press ordered to suit the six page plan, that is, with Six Turtles on the Typo Cylinder and will feel obliged to you if you can visit Philadelphia as purposed by you. You can get the*

full details and directions. The favor you can do me is by managing matters so that there will be no delay. . . .

[P.S.] Telegraph on receipt of this, marked "Free," when we may expect you.

The success of the machine made Harding an enthusiastic friend:

PHILADELPHIA INQUIRER

June 2, 1868.

Hon. A. G. Cattell, Washington:

Dear Sir:—This will introduce to you Colonel R. M. Hoe, the inventor of the renowned Fast Printing Press.

Col. Hoe's invention has done an incalculable service to mankind, and revolutionized entirely the art of printing and publishing; but the debt of gratitude the civilized world is now under to him, seems not to have been repaid to him, and he is now about asking for an extension of his valuable patent. When the sacrifices, the incessant toil, labor and expense, incident to the introduction and perfecting of his press, are taken into account, the desire for such an extension is not an unreasonable one, and I, among hundreds of the friends of Col. Hoe, in and out of the newspaper business, most cordially hope that his wish may be gratified.

The Colonel will fully explain to you his views on the subject, personally, and I trust that your influence may be interested on behalf of the proposed extension.

Yours truly,
Wm. W. Harding[40]

Such allies would soon be needed desperately.

In 1858 the New York *Express* and the Cincinnati *Enquirer* desired machines; the former settled for the Boston *Herald's* old four-cylinder when the *Herald* stepped up to a six. The St. Louis *Missouri Republican* had a four-cylinder in 1859, and five years later noted that its new eight-feeder had to be shipped via four independent railroads: New York to Salamanca on the New York & Erie, south to Cleveland on the Atlantic & Great Western, Cleveland to Cincinnati on the Cleveland, Columbus & Cincinnati and the Little Miami railroads, and finally to St. Louis on the Ohio & Mississippi.

Oswald Ottendorfer's New York *Staats-Zeitung* (founded 1834) would reach a circulation of 49,000 daily in 1876, making it the largest German language paper in the United States; it had ordered a six-cylinder in 1857 and twenty years later had two of them. The *Brooklyn Eagle's* two eight-feeders in 1876 were the only Lightnings printing an evening newspaper, and later when its proprietor, Isaac Van Anden, discarded a four-feeder to make way for two tens, he regretted losing what many printers had come to consider the "prettiest bit of mechanism" ever made: "He wished he had held on to the four cylinder and would gladly have sacrificed what he received for it for the 'simple pleasure of seeing it work.' "[41] A European visitor as early as 1851 had said it was worth crossing the Atlantic just to see a Lightning run for ten minutes.

The Methodist Book Concern had modernized fitfully. On one occasion its new A. B. Taylor feeding equipment in an 1857 trial automatically sent 65,000 sheets through a press without error. At last it consented to improve its newspaper work with a Hoe Lightning. The austere organization, while specifying certain innovations, perhaps little realized what an auspicious purchase was theirs. Methodist conservatism had been supreme: having rejected the first American rotary (Williams') they accepted the very last of that style to be manufactured:

> The last Type-revolving press built was a four cylinder machine, ordered by the Methodist Book Concern in 1876, to print the *Christian Advocate* in sixteen-page form. It was arranged with a combined collecting and folding cylinder and a pasting apparatus at each end, by which the sheets, after their first sides were printed, were collected on the cylinder and flown, and after their second sides were printed they were pasted in the proper margins, folded to page size and piled on a table. This was the first press built by R. Hoe & Co. which pasted the sheets for folding, and it was continued in use until superseded by a web machine in 1890.[42]

The Concern paid $15,000 for the machine; it was installed in December and ready for the first issue of 1877.

Long before Methodist Book reaped the rewards of the Lightning another religious paper had seen some holiness in the Colonel's work:

> Colonel Hoe, who invented this press, has earned for himself an imperishable name, and so popular has it become, that several have already been sent to foreign countries, and others are ordered, besides those which are used in this. . . . Truly man is but little lower than the angels, and crowned with glory and honor. His inventive and creative genius proved him indeed to be a Son of God, and worthy [of] all the effects which have been made to extricate him from the bondage of error and iniquity, and make him indeed joint heir with Christ, and a glad participant in the glorious liberty of the children of God.[43]

The ten-cylinder machine, first appearing in 1856, was 38 feet long, 15 feet wide, 16 feet high, and weighed 21 tons. It had 14,730 separate parts and cost $30,000; it printed 24,000 sheets per hour, all of which the *Gazette* duly noted.

Scientific American of course was similarly fascinated by the giant's components: 6000 bolts and screws, 1200 wheels, 202 wooden rollers, 400 pullies, 400 tape guides, 500 yards of belting, and cogs, arms, braces, etc. too numerous to specify. "[A]nd," said the journal, "it operates with astonishing precision, . . . a smoothness and regularity that commands admiration."[44] The ten-feeder was offered in four sizes: 28 x 38 inches, 31 x 42 inches, 35 x 47 inches, and 38 x 52 inches (paper dimensions).

"Precision, . . . smoothness and regularity" were sales features R. Hoe & Company might not have expected the public to notice. But they were, as usual, unmistakably built in. McNamara noted:

> Daily service in competent hands had but little effect; its motions, with the exception of the fly and nippers, being rotary, little wear was visible, and as bushings were placed wherever needed, readjustment was inexpensive and easily made. Under this state of facts depreciation was trivial, notwithstanding the enormous original cost. Standing alone, without a competitor in the world, recognized as the champion newspaper press, publishers were forced to buy it, and thus the Hoes had a monopoly.

The Chicago historian went on to describe the exquisite beauty of the mechanical phenomenon:

> At midnight, when all but printers are wrapt in slumber, feeders and pressmen are congregating below preparing for their nightly toil. The pressmen oil every bearing, scrutinize carefully the tapes, see to the condition of and set every roller, and charge the fountain. Meanwhile the brawny feeders have each put up their "lifts," and† rolled down to shape and topped off with a number of spoiled sheets. As the "turtles" for the first side are lowered from above they are placed on trucks and wheeled in front of the press. Two ring bolts are screwed in, three feeders heave away on the tackle, a sheave pulley running on a shaft directly above is run into place, the "turtle" is lowered on to dowels set in the rings of the main cylinder, the ring bolts are withdrawn, and a hand below hooks on the bar to a loop on the under side and tightens up the screw.
>
> The press is then moved ahead to receive the other pages, the top rollers are put in, and at a signal from the foreman one feeder places his foot on the starting bar, and as the central orb of the system begins to slowly revolve, all the rollers and cylinders move in unison, the flys drop down with a bang, and soon, the belt being on full, with accelerating speed the sheets shoot in around each cylinder, up the inclined tapes like sparks from a rocket, and are caught by the flys and laid down flat, and are turned and evened up by the fly boys ready for the second side.
>
> Here, in the pressroom of a morning paper, amidst the glare and flash of gas jets and the hum of revolving wheels, are we enabled to comprehend the vast strides made in our beloved art. To the mechanic the sight and sound of perfectly balanced machinery is harmonious; to the casual spectator, mounting up to the surrounding galleries, it is inspiring, the "poetry of motion" exemplified, and a proud monument to the matchless skill of its inventor.[45]

A few years later, when the occasion demanded some cost figures, the Colonel rather haphazardly jotted down the cost-book information based largely on the *New York Tribune* and *Herald* machines:

[*Richard Hoe to Robert, Paris, 19 April 1858*] . . . *The expense, or rather our charge for the drawings of the first two* [*10-cylinder*] *machines including the tracings sent to Messrs Whitworths,*[†] *amounts as per charge book to the sum of*		$ 967.50
Patterns for Press with platform		1,211.72
The 2 Presses		29,875.18
Putting up with some changes in foundations &c&c		2,578.00
The Tribune press on which are all (& some more) of the improvements sent over to Messrs W&Co. for Times, all complete platforms &c &c		13,660.85
Putting up do		451.90
The putting up 10 cyl. for Boston Traveller which was returned		760.88
From the charge for drawings to ascertain the cost, deduct 20% leaving actual cost		744.00*
From the charge for presses, patterns & putting up, deduct say 22½% = to	33,665.90*	
less percentage	7,474.82	
	26,191.08	
I think you might take off 25% & not be much more out of the way than the above. This would make		$33,665.90
[*less*] *25%*		8,416.47
		25,249.43
If you take from the cost of Tribune press		$13,660.85
25%		3,415.21
You have		$10,245.64
You will see if you take the amount of Herald two presses with patterns, & without the charge for putting up, you will have the gross sum of		$31,086.90
deduct 25%		7,771.75*
		23,315.15
Add drawings, say		560.00
Allow 224$ for cost of tracing		224.00
		$23,875.15* . . .

* Arithmetic errors.

Robert also asked for additional data regarding the machines' endurance:

Feby 14th 1859

Messrs. Raymond, Wesley & Co.
New York
Gentlemen,

We desire to send to our Mr. Robert Hoe in Paris answers to the following questions; will you do the favor to give them as far as you can.

How many Printing Machines have you for your Newspaper Work, and what are they, and how long have they been in use[?]

Were your 10 Cylinders the first that were built[?]

What number of revolutions of the main cylinder do you regularly run the 10 Cylinder Machine per minute, and how much faster have you at times run them?

What is the average number of impressions per hour including stoppages[?]

What have been the breakages and what have been the delays consequent on them[?]

How do the finger motions of the 10 Cylinders compare in point of durability with the 6 & 4 Cylinders which you have used[?]

Do you find more fingers broken on the 10 Cylinder in proportion to the number of Cylinders, than in a 4 or 6 Cylinder[?]

About how often do fingers or grippers break on your ten Cylinder machine[?]

Have you any more difficulty with composition Rollers on the 10 Cylinder than on the 4 or 6 cylinder[?]

What occasions you the most delays[?]

An early reply would much oblige

Yours truly
R. Hoe & Co.

Identical letters were sent to Bennett and Greeley. The three firms responded promptly.

Raymond Wesley & Co.
New York February 22nd 1859

Gentlemen:—

In reply to your inquiries of 14th inst. we beg to state:—

1. We have two of your Printing Machines for our Newspaper work, one a Ten Cylinder and the other a six Cylinder Press. The six Cylinder Press has been in constant use for nearly six years; the ten Cylinder has been used ten months.

2. We run our ten cylinder Press on an average speed of 28 revolutions per minute and have had it 35 at times.

3. The average number of impressions per hour is about 15,000.

4. We have had but one break since the ten cylinder Press was put up. That was caused by the breaking of a tape which caught between a pinion and intermediate cog wheel and broke the latter. The delay which this occasioned was merely the time required to [*install?*] *the cylinder and did not exceed ten minutes.*

5. We have broken but one finger on the ten cylinder press since it has been in operation and that caused no delay. On the six cylinder press we have broken not more than an average of one finger a year.

6 & 7. These questions have been already answered.

8. Our rollers last quite as long on the ten cylinder as on the six cylinder press and work quite as well.

9. The breaking of tape causes more trouble and delay than any thing else connected with the working of the ten cylinder press. The time lost from this cause are,† an average per night of seven hours working time, is about thirty minutes.

These answers are given on the information of Mr. Clausen the foreman in our Machine Department and are believed to be entirely reliable.

Very Truly Your obt servt
Raymond, Wesley & Co.
Proprietor of "The Times"

Mess. R. Hoe & Co.

The *New York Tribune* told a similar story:

[*To R. Hoe & Company, 22 February*] . . . *The ten cylinder rotary press was put to work in May 1858 since which time it has printed over seven hundred thousand (700,000) impressions per week. The first break down was occasioned by lack of metal in the actuating pinion. The second by a shrinkage flaw existing where one of the arms attached to the rim of main wheel. The third and last by a small tool accidentally passing through the tape gears of the tenth cylinder.*

No gripper motion has ever been out of working order. . . .

The main drum of Ten Cylinder press makes 24 revolutions per minute (almost without variation) the average production is about fifteen thousand per hour. [M]*ost of the delay is occasioned by imperfect paper. . . .*

On the subject of rollers I beg to remark: we make our own, from the very best glue and molasses the market affords, paying strict attention first to the material used in cleaning; second to the state of the atmosphere in which we keep them, and lastly to the fact of allowing no set of rollers to work more than 3 hours at one time. . . .

Similarly, the *Herald* maintained a speed of 28 to 30 revolutions, suffered little breakage and few delays, and complained more of tapes stretching due to atmospheric changes.

There were, of course, occasional minor mishaps, for which replacement parts were usually at hand. A potentially disastrous episode only confirmed the soundness of the system:

[*Richard Hoe to Robert, London, 27 June 1859*] . . . *As to the speed of the Ten cylinder machines. When they are pushed for time you will find them, both at the* [*New York*] *Times & Herald, running up the speed seldom over 32 or 33 revolutions, but I have seen the Herald at work as high as 35 revolutions; but it is my opinion that at the speed of 30 revolutions per minute, with a proper organisation of the feeders & some extra persons to put up the paper on the feed boards, that speed would produce more work or the best results. . . .*

Some little time ago at the Ledger Office, Phil[a]*, the belt ran off the governor pulley & away went the engine with both presses on. The engineer was in the vault eating his supper & a tremendous speed had got up before he started. Every feeder & the assistant press man, who was in charge, all ran out of doors instead of throwing off the belt & stopping the presses. The engineer who shut of*[f] *a stop steam valve in the boiler room, to which he ran, not knowing as he said but* [*that*] *the pieces would begin to fly off the presses. He asserts that the machines were running at the rate of seventy to eighty revolutions per min. Not a p*[*iece*] *was broken & no damage done except blacking the blankets pretty thoroughly. . . .*

Throughout the lives of these machines there were few complaints from printers. Henry Bullen admitted, "We never heard of any accident to a curved type form on a Hoe type-revolving press; on the other hand, Hoe's predecessors never got so far as actual printing. . . ."[46]

By mid century, of course, the Lightning had proved itself the unqualified triumph Richard Hoe expected it to be. It established his reputation. Said McNamara:

> No wonder his temerity won the admiration and plaudits of the world; and the resultant effects soon proved how well he deserved them, for, coming hand in hand, at a time when the influence of the telegraph and the railroad were being first felt, its influence upon journalism cannot be overestimated.[47]

The Colonel had done for the newspaper what Gutenberg had done for printing in general, giving employment and cheap daily information to millions. Even pressmen were now too boggled to protest: at the 1858 Chicago convention of national typographic unions, one officer arose:

> When I look around me, Mr. President, and see the great strides which the art preservative has taken since Dr. Faust took his first impressions upon a few wooden blocks, and compare it with "Hoe's eight-cylinder," which gives us twenty thousand impressions in an hour, I open my eyes in amazement and say, "What next?" . . .[48]

This euphoric condition persisted for the following twenty years (brought to an end only by machinery still more breath-taking). McNamara judged the transition at close hand:

> For a quarter of a century this [Lightning] press was regarded as a grand success. That it is so no longer is no fault of its inventor. . . . So long as types alone were used this press was a necessity. The audacity of its designer prompted him to despise the flat bed, nor wait until some other gifted genius should solve the problem of multiply-

ing forms by stereotyping. The tissue-paper mold, like the tissue ballot, was potential! The papier-mache matrix was its doom![49]

In 1861 the Colonel, now quite fond of the extraordinary invention that had linked him indelibly with fame (if not with God and the angels), was anxious to retain its possession by having the patent extended.

It seemed a fair thing to do. The Patent Commissioner's own examiner furnished the following endorsement:

UNITED STATES PATENT OFFICE,

June 25, 1861.

To the Hon. D. P. Holloway,
Commissioner of Patents:

Sir:—After carefully examining the case of Richard M. Hoe, who has applied for the extension of a patent granted to him on the 24th day of July, 1847, for an improvement in Rotary Printing Presses, I have the honor to report:

First—I believe the invention was new and patentable when the patent was granted.

Second—It belongs to a class of inventions requiring great skill and forethought, and worthy of high reward when successful.

Thirdly—The sum of seventy-eight thousand six hundred and eighty dollars and ninety-four cents ($78,680.94) profits estimated to have been derived from the invention, is small in comparison with the millions of dollars shown by the testimony in this case to be on the side of the public. There has been no opposition to the extension of this patent, and the most extensive printing establishments in the country have derived benefits which have induced the most prominent publishers to give their affidavits in Mr. Hoe's favor.

The papers are full, clear, and satisfactory, and I recommend the extension to your favorable consideration.

Respectfully submitted,
T. R. Peale, Examiner[50]

On 8 July 1861 Commissioner Holloway read a statement from Stephen Tucker on the invention's "novelty," and lauditory affidavits from George Jones of the *New York Times,* John Lawler of the Philadelphia *Sunday Dispatch,* Charles O. Rogers of the Boston *Journal,* James G. Bennett of the New York *Herald,* and Horace Greeley of the *New York Tribune,* all attesting to their complete inability to issue mass newspapers without the machine. Colleague Isaac Adams too paid strong compliments.

Despite a national capital nervously tooling up for impending rebellion, Hoe's petition for extension was granted for another seven years. The Colonel went home to New York satisfied, although he would soon wonder whether war conditions could permit the projected exploitation of his revived property, especially through the South. But the magnitude of that American struggle and the part played in it by Northern newspapers nevertheless meant continued long production hours for R. Hoe & Company. For the busy firm the time ran all too quickly.

About 1868 Richard Hoe began once more to fret the coming expiration date of his patent for the Lightning. Its novelty would become public property in July 1868 unless Congress interceded; the Commissioner of Patents could do no more for him. Galvanized to action, the Colonel used every means available to him to present his case, imploring every printing friend to acquaint congressmen and other Washington officials with the situation. The company also revealed that it had earned exactly $1,237,845.58 from U.S. sales (of which the Colonel's 3/7ths share was $208,833.74) and $300,000 in England.

Testimonials from friends and customers came in unstintingly, many letters of introduction to Washington dignitaries among them. The Colonel bundled the correspondence and introductions along with some data on the firm and its presses and published them in pamphlet, *Richard M. Hoe's Memorial to Congress for an Extension of Letters Patent.* All the important names in American printing were there, all agreeing the Colonel to be the creator of machinery indispensable to their very business success.

But sheer number—over 300 endorsements—was not enough. Congress put off the matter uneasily, reluctant to grant concessions (as in 1826 with Walker's request) where it had no particular technological comprehension.

Richard M. Hoe tried again the following year, still hopeful his innocent cause would be honored. Confrontation with the 1868 Senate Patent Committee had shown him where Washington's concern lay. He admitted:

> When asked by the Committee how much I had made out of the invention abroad, I stated about one hundred thousand dollars. On examination I find the sum was less than sixty thousand dollars, and a large portion of that sum is in tools and machinery in our shop in London. This I think should not be taken into consideration here. . . .[51]

Congress' interest in actual remuneration to the inventor Hoe saw as the paramount obstacle to his success. He felt obliged to display publicly all the expenses and profits related to the Lightning presses, and reiterate that his tiny personal reward, "which includes manufacturer's profit," was only $208,833.74. To many former friends and sympathizers, however, this did not look like a trifling figure. Hoe wisely tried to marshall additional forces:

[To Ellis H. Roberts, Utica, 24 February 1869] . . . What we now desire is a very strong letter from you introducing our Rich[d] M. Hoe to Senator Conkling asking that he will interest himself personally to promote & secure the renewal of my brother's patent in the senate during this present session. . . .

Roscoe Conkling, a brilliant speaker, skillful pleader, and shrewd lawyer who was beginning to champion the constitutional rights of newspapers, could have been a powerful ally.

[To the Honorable T. A. Jenckes, Washington, 5 April 1869] . . . I trust you will be able to call up my Bill and pass it in the House to-day or to-morrow. . . .

Jenckes headed Congress' Patent Committee. But there was no quick action on the thorny bill. More delays. Yet the Lightning was not easily lost, and the Colonel fought on. He sent Robert to Washington to witness the moment of truth:

Willards Hotel, Washington
Jany 20, 1870

Dear Brothers

We arrived here last evening. This morning I called on young Mr. Stanton with whom I had a pleasant interview. After reading what I had written & the paper made by [Hoe lobbyist] Mr. Stevens he proposed to make a petition for Richard to sign to be presented to the Senate and he would accompany it with a statement as counsel explanatory &c, to which I assented. In alluding to Mr. Stevens he answered that he [Stanton] was not in favor of any lobbying but would feel safer to let the application stand on its merits. At the same time [he] would be glad of any influence that Mr. Stevens might exert. Nothing will be done before early in the week, after Mr. Stevens arrival.

Mr. Wells[t] received me very pleasantly and I am to meet him to morrow at 10 A.M. Please send me the Phila Inquirer of 18th containing the Editorials for Mr. Stanton. I shall draw on you for my expenses here. On Saturday we may go to Richmond to return here Monday or Tuesday.

Your affe Brother
Rob[t] Hoe

Mr. R.M. Hoe
& P.S. Hoe

"Young" Stanton was Edwin L., son of Edwin Stanton, controversial Secretary of War, whom the Hoes knew. Gideon Welles had been Secretary of the Navy until 1869; he was a Hartford newspaperman. The passive profusion of names, visits, appointments—so unlike other Hoe correspondence—testifies to a certain helplessness and frustration:

[Robert Hoe, Washington, to R. Hoe & Company, 21 January] . . . This morning I was in the Hall of Representatives waiting for Mr. Wells[t] to get the paper, early, when in came Judge Niblock who

heard what I had to say and seemed pleased. I am persuaded that the method our Richard has pursued here to have the petition rest on its merits with intelligent members to advocate it is the most sure way to get it passed in both houses.

Mr. Jenks† told me this afternoon that it was no use for me to see any members of the house, that if it passed the Senate it was easy to get it through the House. I shall see Mr. Willey I hope to morrow. The Senate was so thronged to day & absorbed with the Virginia bill that I did not disturb Mr. Willey as yet.

Mr. Stanton said to me to day that he knows Mr. Carpenter of the Senate Committee well. I see that he fears that Mr. Stephens† may do us harm unless very discreet. As I see things now I would not employ Mr. Stevens. Be sure to see him & get him to see me here on his arrival before saying or doing any thing for us.

The feeling that we are wealthy Mr. Stanton says is the principal obstacle in our way. . . .

Now that we have engaged Mr. Stevens it is of course best to use him wherever we can but I have grave doubts whether his paying any money will benefit us. . . .

Have not heard from you or home since we left Nyoryork.†

Mr. Stanton wants a copying press as pr memo annexed [not present]. *Send no bill. . . .*

Despite the Colonel's many printing friends, some of whom were indeed members of Congress, and despite the lobbying efforts of Stevens and Stanton, the Hoe appeal had no enthusiastic support. Letters continued:

[*To McKee & Fishback, St. Louis, 29 April 1870*] . . . *Will you be so kind, if consistent with your view, to favor me with a letter of introduction to Senator Carl Schurz of your state asking him to aid me in the Senate to get the bill, now reported passed. . . .*

Messrs. McKee and Fishback were proprietors of the St. Louis *Missouri Democrat.* Schurz too would be a powerful ally.

Here may have been one crucial affair where the Colonel wished for more political connections, more clout, more friends in better places. Such associations had never before seemed important to him; his strength had been self-made; his successes based on performance. As time went on many factors worked against him, forces he could not control. The patent had already enjoyed 21 years of life (even the Civil War had not interfered with his ability to manufacture the machines); the novelty—and personal awe for the inventor—was wearing off. The company's phenomenal growth and prestige in the industry did not permit much sympathy, and the contemporaneous development of web presses—by several builders—began to render the Lightning anachronistic, evolutionary, academic. Trade opinion had weighed these arguments and found, surprisingly enough, sudden stern opposition to the Colonel's pleas.

"[Like others, we] unthinkingly signed the petition," said Robert S. Menamin, editor of *Printers' Circular*. When signatures were first collected the matter was one of "compliment," he continued, instead of remuneration. "Patentee" was urged as a "graceful balance" to the prefix "Colonel." But suddenly Hoe talked of money, and his old friends began to consider the appeal more commercial than honorary. The Senate Committee on Patents in its July 1868 report, having already determined that Hoe himself earned about $300,000 from the patent, concluded: "Considering this case in the light of the usual applications for the extension of patents, the committee would have to decide *that the profits aforesaid should be regarded as a fair and reasonable compensation to the inventor,* and would not feel authorized to recommend another extension. . . ."

Menamin agreed that a gross profit of $700,000 on one invention was "no mean offering," and opined: "if Mr. Hoe's application again fails, as we trust it will, he can still console himself that not one in ten thousand of his fellow-inventors has been as liberally rewarded for brain-labor and expenditures as himself." Hoe's application claimed the war had hurt him—that "by reason of the condition of the country from that time [1861] to the present, the demand for such printing machines has been very limited." "This argument is itself a stroke of inventive genius of the highest order," said Menamin, who knew that, because of the war, the expansion of journalism had been tremendous; sales lost in the South would rapidly return.

"Of all the anxieties by which the Republic is actually or theoretically convulsed," Menamin concluded, "none is less pressing or less worthy of the attention of the representatives of the people than the imaginery fear that Richard M. Hoe's name will swell the roll of unrequited genius."[52] The many years of R. Hoe & Company's quiet good will had not sufficed.

Still more scathing was an anonymous letter to the editor of *Printers' Circular:*

> . . . Let us start fair . . . by asking who is *Richard M. Hoe*? I answer promptly (from my own experience as well as the testimony of others), one of the most accomplished and successful machinists in the world. But the Patent Laws are intended primarily and mainly for the use and protection of *inventors*—a class not often identified, and never to be confounded with, manufacturers of machinery. Is Mr. Hoe an "inventor"?

This critic found the Smith press to be the invention of Wells, the Hoe double-cylinder perfector to be Napier's, and "the most complete platen press known," the work of Isaac Adams. The rotary or type-revolving press, "the great *mastodon* of the tribe," was similarly little more than the creation of Morison in England, six years earlier. Yet Hoe's contribution to printing was nevertheless deemed substantial:

> He turns out the most carefully finished work, as a whole, in the Republic, if not in the world. I never saw a botched article from the establishment. More than forty years ago my old shopmates chose the Smith hand-press because of the finish of all its parts. Not an ounce of weight in its frisket-frame nor a shaving of wood on its tympan frames was left in our way. The polished brass hooks and buttons, the light neat spring points, the sure grip of the wrenches and aptness of the screw drivers—each appealed to tasteful workmen; while all the important points showed the same harmony; and showed the sweating pressman that some one or more, far off in the blacksmith's shop, thought of his troubles and wished to lessen them. . . .

And for such quality products the Hoes had not labored in vain. "With fifty years' apprenticeship in one line, and twenty-one years of special protection," said this critic, "who dare enter the lists with them? As well attempt a foot race with a locomotive." But the consensus was that Hoe receive no further concession. "He is asking of Congress what he does not need, what he does not deserve, and which we must take care that he does not get. . . ."[53]

James Moran offered three points to consider when dealing with Hoe's critics at this time. First, many press inventions—indeed the processes of inventing—were often not as truly original as their creators and friends believed; second, Hoe acquired honestly from others many useful patents (and some less so) which he could technically call "his" property; and third, he certainly lavished more care and servicing on his equipment than any of his competition. "That is not to say that Hoe was an idealist," Moran concluded, "and never adopted another man's ideas without recompense, but often those who complained had themselves adapted earlier inventions."[54]

Scarcely a simple adaptation, however, the Lightning was perhaps the one great conceptual project from his shops which the Colonel could truly call his own; he had a proud, almost smug, satisfaction with it as no other—he had persevered where so many had failed, succeeded where so many had scoffed. It was the machine that made R. Hoe & Company world famous and—not coincidentally—created a modern world of journalism. As early as 1847, upon the completion of Swain's four-cylinder, Richard had insisted that an engraving of it be sent to the *Franklin Institute Journal:* it was possibly the only instance of his seeking professional recognition. Unfortunately his old friend Dr. Jones died in March 1848 and the respected journal, edited for about two years by a committee, did not see fit to notice the marvel.

Although continued possession of the patent may have been important in British sales, where the machine was introduced fully a decade behind its domestic appearance, the Hoes could never have plead poverty or obscurity. Indeed, it was probable his pride only wanted "patentee" affixed to his name just a little longer—but how in commercial America to phrase such a sentimental request? An 1859 *Printer* had already commended the Colonel's modesty:

> The name of "Hoe" is patriotically identified with "the Colonel's last fast," (whatever the number of cylinders may happen to be) and if he were a fame-seeking man, surely he must have about enough of it by this time, for wherever printing or civilized language are known, his name flies, wafted on the manifold wings of intelligence. . . .[55]

But in a quiet way Hoe indeed yearned for the coveted "patentee"—a word of rank among inventors—just as he treasured his military title. Nevertheless, as time slipped away and other problems materialized, the Colonel abandoned his desire. The Hoe rotary printing machine fell into the public domain, superseded by Hoe's own improved machines. The patent was of no further interest.

With the manufacture of about 175 Hoe type-revolving presses an auspicious era in American journalism—truly mass circulation newspapers—became a reality. But the technology did not stop; modifications would follow rapidly. "Each press is employed only for a short time," Pasko explained paradoxically of all printing machinery, "but is absolutely required then." Of the planetary press he went further, describing its zenith and nadar: "The Hoe lightning press was for many years regarded as one of the wonders of the world . . . and the only value now . . . is as old iron. . . ."[56] Surprisingly, not one was ever preserved for posterity.

Newspaperman Will Irwin, recapitulating the history of nineteenth-century journalism, saw Richard M. Hoe's Lightning press loom large, not as multistoried machinery, but as an important example in the intellectual progress of mankind. He put its quite unexpected technical success into perspective with similarly startling associations:

> [Benjamin] Day had scarcely installed his plant before he replaced his Washington hand presses with a flat-bed steam press. . . . Then, in the late forties, and when Bennett was ready to conquer new worlds, Richard Hoe discovered that the way to get speed out of a press was to put the type on the cylinder. . . . Last of all, in 1859, Darwin crowned his century by publishing his "Origin of Species." From now on men were ready to accept ready-made theories less and less. . . .[57]

HOE DISPLAY, PHILADELPHIA CENTENNIAL

From Harper's Weekly, 9 December 1876

6.

Stereotypes and the First Web Machines

THE REFRACTORY PRINTING PLATE / CRASKE AND THE PAPER MACHE MATRIX / BULLOCK AND THE PUBLIC PRINTER / HIGH SPEED PAPER PROBLEMS / THE PHILADELPHIA CENTENNIAL

Throughout the eighteenth century many efforts were made to convert pages of loose typematter into durable printing plates. Two important advantages would accrue: (1) once corrected, a (solid) plated page presented no further worry as to subsequent typographic error, or worse, wholesale pieing; (2) once duplicated, expensive foundry types could be returned to their cases for reuse. William Ged, a Scottish goldsmith, made plaster of paris casts of typeset pages and on several occasions attempted to organize businesses—but with unreliable partners—to exploit his method. Although he was, after many discouragements, appointed printer to the University of Cambridge in 1731 with license to print Bibles and prayer books, his technique survived only fitfully. Few commercial books enjoyed such large editions as to require their publishers to adopt the process.

Even with the increasing demand for literature, other objections to Ged's method remained: (1) the plaster cast, destroyed during removal of the plate, could therefore be used only once; (2) types so employed were usually fouled and abraded by traces of plaster. But the idea was still sound. A Frenchman, Gabriel Valleyre, in 1730 had developed a process using clay, and others in France soon tried many mixtures, including sand, talc, and gypsum. Firmin Didot in 1795 coined the name "stereotype" (Greek stereos, fixed + type) for this kind of platemaking, and soon managed to print many books by it.

Gradually, well established classics in England and France were still the only form of printed product justifying, by their long life, the expense of stereotyping. For book publishers two additional savings emerged: from plates fairly easy to store a number of small printings could be pulled, thereby obviating (1) expensive, time-consuming resetting of type, and (2) costly inventories or sheets held from an earlier large printing.

Charles Mahon, Earl of Stanhope, about 1800 revived the stereotype process with a method wherein the matrix was baked like pottery. Contemporary inventors, including Cowper and Applegath, made other improvements. Nevertheless, few of England's small publishers had sufficiently standard works meriting stereotyping, and few British printers cared to expose their types to such deleterious treatment. Yet, by 1824 Johnson's *Typographia* lament for displaced compositors suggests that the practice had in fact grown substantial:

> We conceive that the inventor of stereotyping is not worth the pains of our tracing; and more particularly when we reflect that so many of our brethren who well deserve . . . a comfortable subsistence . . . are now through this process, reduced to a very humble pittance; . . . at one season of the year, nearly one-half of the valuable body of men alluded to may be considered as destitute of employ on account of [publishers not resetting] the standard works, which is the summer's stock work."[1]

Jean Baptiste Genoux, a French printer, in 1829 received a patent for a new "wet mat" technique. Several thicknesses of dampened brown tissue paper were "brushed" or "beaten" down over a page of type until the pulp filled every crevice around the letters. On this "flan" (so called for its resemblance to French custard tarts; transliterated as "flong" in British use) heavier paper was then pasted as a backing. After drying, the entire matrix, surprisingly enough, withstood molten typemetal ladeled over it to cast a plate. Joseph Kronheim, a London wood engraver, patented a similar process in 1844. The process was faster than "dry" (clay or plaster) methods, several plates could be cast from a single paper matrix, and the cheap, light, flexible matrices could be stored or shipped easily and safely. The typematter was easily cleaned. The wet mat or papier maché ("mashed" paper) method became the standard everywhere, although —still—until the 1850's only long-run bookwork was worthy of stereotyping at all.

Swiss printer James Dellagana learned Genoux's process in 1848 at the Paris newspaper *Constitutionnel,* and with a brother set up a stereotyping shop in London. In 1855 he patented equipment for casting type-high plates (stereotypes before that had always been only about ⅙th of an inch thick) and for casting them hollow to save weight and metal. These improvements quickly made stereotyping practical for newspapers which had been reluctant to work with plates that had to be mounted (usually on wood) as yet another step in the hectic delivery of pages to the press. In 1861 Dellagana invented a molding press known as a roller or calender, which was universally adopted. He quickly found a most prestigious customer.

Between 1856 and 1859 the London *Times* experimented with wet process stereotyping, hoping to speed the mechanical steps of newspaper publishing as pressures of urgent news—the Crimean War—demonstrated the crucial timing problems dailies everywhere would soon face. Cowper had tried to curve flat stereotypes by heating them, but they tended to crack or break. In 1860 the *Times* boldly installed Dellagana's complete wet mat (papier maché) system, especially tailored for its Applegath presses. The first plates were cast type-high in single columns; soon entire pages in one piece were ready to print on rotary presses. John Walter openly shared these trials with other papers, and some of them were thereby encouraged to install similar (Walter) stereotype departments. A new era dawned in newspaper printing.

Stereotyping had been introduced to America about 1810 through the work of John Watts, a printer apparently familiar with the Didot and Stanhope systems. About the same time typefounders George and David Bruce also brought their skills to the problem, adding improvements. The Bruces had started as printers in 1806 (with borrowed type and press), learned the secret of stereotyping in 1813 (printing the first American stereotyped book—*The New Testament*—a year later), and in 1816 began stereotyping for the American Bible Society. They quit the printing business; David continued the stereotype shop while George began a typefoundry (1822), both continuing merrily for decades. They were long Hoe friends.

In Philadelphia Jedediah Howe and Lawrence Johnson both worked at stereotyping and then formed a partnership (maintaining their separate foundries). Upon Howe's death in 1834 an employee, John Fagan, bought and enlarged Howe's plant. As in England, however, only occasionally did book—and periodical—publishers resort to stereotyping, encouraged but gradually by the spread of literacy and its clamor for reading materials.

Joseph Adams, in addition to his woodcutting skill and his advanced makeready techniques, possessed a strong inventor's streak as well. To protect the delicate linework of his engravings, so easily damaged in handling and press, he developed a process for electroplating the surface with a copper solution—in effect making the first electrotypes. The extravagant 1846 Adams Bible, of which Harpers printed 50,000 in serial form and 25,000 complete in hand-tooled morocco, were splendid masterpieces attesting to the virtues—and long life—of electrotyped plates.

No doubt with a strong recommendation from the Harpers, Adams (whose electrotyping experience actually dated to 1839) commissioned R. Hoe & Company to make some machinery of his own design for performing electroplating on a commercial scale. Tucker explained:

> Electrotype plates were now coming into use in place of stereotype plates for the better class of printing. The coating of the face of the wax molds or matrices with powdered blacklead was a very disagreeable process and had hitherto been done by a hand brush, but Mr. Joseph A. Adams, an expert wood engraver and a noted pioneer in electrotyping, patented in 1855 a machine for this purpose to be turned by hand. The firm built a few machines from Mr. Adams' design, but the mechanism proved somewhat slow and noisy, nevertheless it was a great improvement on the old hand process. The firm afterwards bought this patent.

Much as Robert Hoe had accidentally been introduced to the value of cylinder presses, so also was the company's fortuitous involvement with Adams:

> The introduction of electrotyping led to the bringing out of many other implements and machines to facilitate its various processes, such as the wax molding press, the electrotype battery, melting furnace, backing pans, shaving machines, saw tables, routing machines and a host of other articles, the whole forming a substantial addition to the firm's business.[2]

William Filmer, a practicing "electro-stereotyper," set up a Boston shop in 1851 and moved it to New York two years later. He described the craft in *The Printer* (1858-1859). With the Adams-Hoe blackleading machine he was more satisfied than Tucker, finding it thorough, quick, economical, and clean. Others thought so too. Filmer wrote:

> I am sorry to have to record that several electrotypers have constructed machines and have them in use, which are obvious infringements on this patent, as they use a vibrating or reciprocating brush, etc. Mr. Adams has spent his time and money in making various improvements for the benefit of electrotypers, and the least acknowledgment that they could make would be to pay him the small amount which he asks for the right. . . .[3]

Filmer also worked with R. Hoe & Company to develop electrotyping equipment, and was naturally interested in inventors getting their royalties.

Nevertheless, so advanced was the science of electroplating that by 1869—within a decade—"second generation" systems were on the market. Tucker described them:

> Electrotyping was now fast superseding stereotyping for most kinds of book and job work, and new and improved devices and machines for more quickly and cheaply producing the plates were being gradu-

> ally introduced, such as the dynamo for precipitating, the hydraulic molding press, power planing, shaving, sawing, beveling, routing and similar machines. The Blackleading Machine the firm had been making was slow and noisy, so it was modified and simplified into the form in which it is being used at the present time.[4]

Of many wet cell batteries, Ringwalt opined: "The latest and probably the best battery is that devised by Joseph A. Adams, and manufactured by Hoe & Co. It is lined with glass to prevent leakage. . . ."[5] Adams made additional improvements to his equipment in 1870 and the company also nurtured the technique with a series of elaborate machines. For fine bookwork and long runs electrotypes remained in use well after World War II.

Charles Craske (1822-1905) had come to America from London in 1837. In 1850, then a 28-year old "engineer," he introduced papier maché stereotyping to the United States. Four years later he was able to cast curved full-page plates for the New York *Herald*—long before Dellagana in London—in a technique that would remain standard for decades. Bennett's acceptance of the new process, however, did little to commend it to others. Between 1857 and 1858 Craske had considerable difficulty acquainting newspapers with it, and even the *Herald* lost interest. While the inventor doubled his efforts, new impetus suddenly came from Europe.

Despite a retainer from the *Times,* the Dellaganas had not been enjoined from working with other customers. Thomas Catling of *Lloyd's Illustrated Newspaper* was most anxious to print from "stereo.," but its prize Hoe Lightnings, so laboriously acquired, could not accommodate stereotype plates. Catling explained:

> The "turtles" of Hoe's machines, designed specially to hold type, required that any substitute, in the shape of stereo., should be type high, that is, nearly an inch; twenty-nine thirty-seconds of the inch, to be exact. A solid column of metal, with the requisite curve, was secured, but the weight proved too great for the revolving cylinders of the machine.

The Dellaganas tried their hollow casting process for Catling, but because of the pressure under which the Hoe cylinders met, "the printing caused such a depression as to make it appear with a white space down the centre of each column."[6] Even when the turtles themselves were abandoned in favor of a special curved stereotype bed, the 3/16-inch stereotype plate would continually develop flaws, cracks, or bend unevenly. The situation seemed hopeless. To all appearances Richard Hoe was avoiding the problem, for he was, at that time (1861) still petitioning the British government for an extension of his patent for the basic Lightning presses utilizing type.

R. Hoe & Company surprisingly had not seemed particularly interested in stereotyping. As early as 1819 Matthew Smith patented a stereotype block (a "patent" base, like scores of others over more than a century, for mounting the thin plates at type height) containing hooks to secure plates of varying sizes quickly and conveniently. But developments in this area fell more appropriately into the domain of the type founders, with whom the Hoes usually did not care to compete. As late as 1859 the Colonel offered no stereotype system for either newspaper or book work, despite a powerful challenge from his brother:

> In 1859 Mr. Robert Hoe, who was then in Europe, sent to the firm here a curved stereotype page of the London *Times,* which had been worked on their [Applegath] press, and stated that this process of stereotyping was in successful use in that office. . . .[7]

But with other pressures closer to home the Colonel would not be stampeded; was he not the sole master designer of a vast printing industry?

The New York *Sun* (again) did not think so. When its old celebrated eight-feed type revolver would no longer suffice, the paper in 1862 went back to tinkering for its own solution. Moses S. Beach resumed control in January and soon installed complete stereotype equipment, including his own roll-feed and paper-wetting apparatuses and machinery for cutting the web after printing. These developments put the *Sun* first with a perfecting press and an accompanying "bindery" line. The newly prospering paper in 1868 was bought by Charles A. Dana, who ran it for the next 29 years. Another paper too tried to make its own way.

As the *New York Tribune* approached the Civil War (during which its circulation would be the nation's highest—160,000 daily) it possessed the largest mechanical facilities. Thomas N. Rooker, composing room foreman, was a man of considerable ingenuity: one of his inventions was a type case with adjustable bottom (always "full" for composing, always "empty" for distributing); another, manufactured by Hoe, had smaller compartments to reduce compositors' hand travel. Wishing to avoid the double setting of copy and yet have two presses simultaneously printing identical pages, he realized that stereotyping was the answer.

Rooker dutifully consulted the Colonel. The latter rose magnificently to the question (perhaps to stall, perhaps in a practical joke taken seriously) with an elephantine scheme described by Henry Bullen:

> He advocated the continuation of the use of one [type] form, and proposed to build for the *Tribune* a twenty-feeder type-revolving press, with an output of 50,000 impressions per hour. This press would have been 36 feet high, and would have necessitated a new building. . . .[8]

It is possible, as Neipp suggested, that "despite his genius, Richard Hoe did not see, at that moment, the printing of newspapers by means of stereotypes."[9] Indeed, friends were still trying to persuade him that the curving of cast plates could be done at all. Brother Robert had slyly enlisted Conquest's support:

> [*To Conquest, 3 May 1862*] . . . *We have the* [*New York*] *Evening Post 8 cyl about ready & a 4 cyln for Harpers Illustrated Weekly to print cut electrotype forms is underway. We are at a loss whether to bend the copper sheets in the curved mould & so cast it, or to cast it in a flat mould & then bend after it is backed up. I intend to write about this to RMH, but forgot to do so. Have you seen any thing done in the former way*[?] *We want all the information you can send us on the subject. . . .*

The following month Tucker, writing to the Colonel in London, would emphatically disagree that electrotypes could not be successfully curved, urging him to patent his system in England—and possibly glean an order or two from the *Illustrated London News* in the bargain.

At the same time, however, the indecisive company was forcing major customers to make their own choice. *Harper's Weekly,* begun 1857, reached 100,000 before the end of the Civil War; for this growing volume the publishers in April 1862 wanted not only speed but precision, obliging Hoe to promise (12 April 1862) that its four-feed Lightning, printing from stereotypes, "will print 5000 impressions an hour and register surely within ⅛th of an inch, & work the cut forms of your weekly newspaper as good as any single large cylr press. . . ." For this $10,000 purchase Harper's would get, free of charge, either a press for bending the flat plates or a curved mold for casting them to that configuration. The Colonel could not be more equivocal.

Rooker, meanwhile, with no time for fanticizing, had politely rejected Hoe's grandiose solution. At stake was the very life of his paper. He had no choice but to explore stereotyping for himself. Encouraged by Greeley, Rooker asked New York book stereotypers to devise a fast system for newspapers which would also permit repeated castings of a type form. There was no response. From London, however, he received the Dellaganas' offer, for $30,000 yearly, to furnish their system as used on the London *Times.* He was on the verge of accepting when he heard from Charles Craske.

Having no manufacturing facilities of his own, Craske had brought his improved specifications to R. Hoe & Company. The Craske process was sound, and the Colonel must have known he could benefit more from business with hundreds of little stereotype shops than from the sale of an immense 20-feeder to one or two large newspapers. Since the equipment's development was essentially completed by Craske, the company found itself in the stereotype line quite without costly trials in Tucker's sequestered Experiment Room:

> We made the [Craske] stereotype mold, shaving machine, beds, etc., for this purpose, and after some little experimenting the result was so favorable that stereotyping was soon adopted by the *New York Herald, Sun, New-York Times,* and other papers. The success of this quick process of stereotyping was of immense benefit to newspapers of large circulation, for it enabled them to duplicate their forms and use additional presses, and it was also of much importance to Messrs. R. Hoe & Co., for they were called on to supply the additional presses required.[10]

Charles Craske's 1860 contract with the *New York Tribune*—and machinery furnished by Hoe—started a new industry. On 31 August 1861 the *Tribune* began printing from curved stereotypes, the first newspaper in America to do so with a reliable process. "The type-revolving presses printed from curved plates as easily as from type forms," Henry Bullen grudgingly admitted, "and their sale went merrily on."[11]

Hoe quickly approached the New York *Herald* with the new system:

> *New York, April 9, 1861*
>
> *J.G. Bennett Esqr*
> *Dear Sir*
> *The tools and fixtures necessary to stereotype the "Herald" forms will cost about fifteen thousand dollars, certainly not over two thousand—the beds requisite to work the plates can be made for one hundred dollars each.*
>
> *Yours very truly*
> *R. Hoe & Co.*

Bennett had not been especially interested at first. When the same offer was made to the *Tribune,* however, he promptly installed a stereotyping department, as did other large metropolitan papers.

Craske's method of furnishing curved plates for Lightning presses did not reach England nearly so quickly. Catling was impatient:

> As the demands of newspaper proprietors became more and more urgent, stereotypers found their opportunity, and the now familiar process was developed. Taking a cast from a page of type on the flat [stone], the papier-maché mould or flong was made flexible enough to be bent, and when this was placed in a curved casting-box, plates were readily made to fit any machine. Before this amazingly simple discovery could be adapted to *Lloyd's* it was necessary for Colonel Hoe to provide fresh beds for the stereo. plates, in place of the discarded "turtles." These were designed and made in America. Thus it was that *Lloyd's* had to wait till February 2nd, 1862, before it could be printed from stereotype. The *Times,* with the assistance of the

> Dellaganas, had been enabled to secure the great advantage of the new method some time previously.[12]

Not all Britishers eagerly awaited Hoe stereotype equipment. A visitor to a large London shop noticed two Hoe shaving machines in the firm's yard, "exposed to the elements, crumbling to dust and ruin under complete neglect." He asked why. His host answered that "such was the disinclination of his stereotype-finishers to adopt new improvements of any kind, and such also his own reluctance to urge his workmen against their will. . . ."[13] The visitor had known them to be excellent machines.

As late as mid 1870's R. Hoe & Company offered equipment for plaster cast stereotyping, but the paper maché process would triumph. For a Canadian printer a quotation could be attractive enough:

Estimate for Stereotype Apparatus Paper Process
for plate 16 x 20 ins.

[*6 June 1870*]

for Mess. Jno. Cameron & Bros. London, Ont.

1 N° 3 Steam Drying Press for form 16 x 20 ins.	250.00
1 N° 3 Melting Furnace 20 ins. diam.	150.00
1 N° 3 Casting Mld for form 16 x 20 ins.	250.00
1 Mould for Casting border Rule for form	20.00
1 Skimmer Wrot Iron	4.50
1 Oil brush Goats hair	2.50
1 Chalk " "	2.50
1 Beating " Bristles	6.00
1 Paste " Iron bound	2.50
1 Circ. " For Serving	1.50
1 Wooden Seive	3.00
1 Pr Shears	2.50
Soft unsized Paper / Tissue } say	10.00
Box[ing]	15.00

For a larger installation, however, a complete stereotype shop could cost well over $12,000, but no one questioned the investment. The process quickly became a standard operation throughout the printing industry. The original wet mat process was eventually replaced by several dry mat techniques which proved faster and more convenient. And with the successful application of printing from stereotype plates other problems—and solutions—quickly materialized.

In 1861, after almost fifteen years of commercial sales of the famous type-revolving press, R. Hoe & Company received one of its most prestigious orders:

> During this year the Superintendent of Public Printing ordered a

Four Cylinder Type-Revolving Press and the necessary outfit for making curved stereotype plates, and these were erected in the Government Printing Office, in Washington, early in 1863.[14]

Before this time the company—perhaps understandably—had gotten only a small trickle of government work. In 1837 it sold to Edward Stubbs, agent for the Department of State, seal presses (stamping "U.S. Consulate" with city), along with a large quantity of red, white, and blue bunting. Six years later Stubbs bought ten more presses, for which order Robert Hoe was thankful:

> [*18 January 1843*] . . . *It will afford us pleasure to receive your order for any thing you may require in our line & you may rely upon our prompt and careful attention in their execution.* . . .

This kindly thought paid off: a request followed for a copying press, another seal press, and blotting paper, the last item, like the bunting, duly furnished without comment. The company waited patiently, however, for substantial printing machinery orders for a government shop the Hoes must have trusted would soon be needed.

Until 1861 the nation's printing in the city of Washington had been handled by privately owned plants, many of them old Hoe customers—Peter Force, Thomas Allen, Blair & Rives. Charles Van Benthuysen, Hoe's old Albany friend, about 1846 underbid the Washington printers and set up his own shop in the capital. Another, Gales & Seaton—having forfeited the 1829 Napier but who, like Van Benthuysen, bought one of the earliest Hoes, apparently believed the brothers capable of everything. They asked:

> [*To R. Hoe & Company, 11 March 1855*] . . . *You will oblige us by informing us at your earliest convenience whether it be in your line of business to undertake the reprint of a certain number of the volumes of the "Annual of Congress" (four in all) on bourgeois* [*9 point*] *type, neither new nor worn out, making in all about three thousand large octavo pages, the index and some interspersed extracts or documents to be in brevier* [*8 point*] *type; and at what price per 1000 ems you could procure the execution of this work, line for line, from printed copy to be furnished.* . . .

The company had to decline.

Contracts with such commercial plants, however, were expensive for the government, and potentially embarrassing. Among others, Fletcher Harper testified he thought the United States could save 50 percent of its printing and binding expenses with its own shop. Yet talk of forming a government printing office had gone on for decades. Finally in 1861 the four-story printing plant of Cornelius Wendell, the largest in Washington, was purchased to become the Government Printing Office. Wendell, formerly a Van Benthuysen foreman, had succeeded to the latter's business. Its pressroom housed 23 Adams presses which, one year, turned out a 225,000-copy edition of the *Annual Report of the Commissioner of Agriculture,* and whose 3 Hoe cylinders also did mountains of work.

The entire Washington shop employed 350 hands, and although it was deplored as an ugly building situated in seasonal swamps, its mission and prestige would justify its existence handsomely. The burden of government wartime work would require all the mechanical assistance the printing industry could muster for it, and a two-year wait for a Hoe type-revolver must have been something like expectations of a messiah—or a monster.

The new Hoe four-feeder at the Government Printing Office was especially built to print tomes like the popular but controversial agriculture annual. (Some lauded the book as a great treasure for farmers, others found it "a bundle of trash.") Over 150,000 copies were planned for 1863. Recent advances in stereotyping had nevertheless made it possible to take much of the drudgery out of this long job:

> The method of making the curved stereotype plates for this machine was entirely new. A thin sheet of tempered steel, whose normal form was the required curve of the stereotype plates, was drawn down flat on an iron plate, and in that position was coated with clay into which the form of type was impressed. The steel plate was then released and resumed its natural curve, the clay was dried and put in a curved mold and the stereotype plate cast in the ordinary manner. This method produced excellent plates, and it is now being used with wax instead of clay for making curved electrotype plates for the Rotary presses now in use for the finer class of printing. . . .

Surprisingly enough, the long awaited $15,715 press, engine, and fixtures (shipped by special permission via government carrier) arrived at the printing office at a tense moment.

Despite what were thought to be fair working conditions, government bindery workers had struck, in the fall of 1862, for more pay and fewer hours. They won the country's first eight-hour day (for $24 weekly) in the printing industry. But Hoe's mammoth machine, its requirement for feeders as well as pressmen notwithstanding, was evidently seen by still hostile workers as a new and continued tack to put good men out of employment. Mysteriously, wrote Tucker, "when it was put to work it met with the most determined opposition from the printers in the office":

> The blankets and compo rollers were found cut and accidents happened to the machine continually and no one seemed able or willing to master it. The obstacles and delays became so serious that the use of the machine was finally abandoned. . . .[15]

Such industrial sabotage—one of the inevitable reactions of troubled men watching the Industrial Revolution everywhere threaten the job security so vital to their lives—was hard to combat. John D. Defrees, the first Superintendent of Public Printing, in his annual report for 1863 nevertheless blamed a shortage of presses for delays in printing, coyly avoiding the Hoe disaster:

> One edition of 25,000 copies of the *Report of the Commissioner of Agriculture* has been printed, bound, and delivered, and the remaining number of copies will be printed as rapidly as the number of presses which can be withdrawn from the more pressing demands of the War Department can accomplish. . . .[16]

An unwary Congress the following year obligingly appropriated $61,000 for expansion. Defrees built an additional structure to the plant—with space for nearly 1000 workers—and also purchased "12 first-class Adams presses, manufactured by Hoe & Co., of New York" at a cost of $36,000. The special Hoe four-feeder, dismantled and crated, must have gone back on the returning train. The shop now had 35 Adams and 9 cylinder presses, plus several smaller machines, when Defrees resigned in 1866.

Wendell returned to superintend operations for one year. He was faced with the same agriculture yearbook, again a gigantic job even by today's standards. No less enterprising than Defrees and no less observant of the latest printing equipment on the market, Wendell of necessity risked the same sabotage. But only a new press, specially designed, could do the yearbook without tying up the whole shop. In 1866 he bought, for $25,000, the sensational new Bullock. It required even less labor than its predecessor:

> Like the Hoe Press, it carries the forms upon the cylinder, but it differs from that press in requiring no attendants to feed it, and in delivering the sheets printed on both sides. It is a great improvement also, realized in this press, that the sheets are delivered silently, the noisy racks of the Hoe Press being wholly dispensed with.

This 1866 observer continued:

> The substitution of an automatic system of feeding for hand-feeding, which is one of the greatest economical advantages of this press, has been effected by introducing the paper into the machine, after it has been subjected to a moistening operation by passing through a shower of fine spray, in the form of an endless roll. A single roll will contain several thousand sheets, and the printing operation, including the cutting of the paper into proper lengths, will proceed uninterruptedly until the roll is exhausted.[17]

Few midcentury machines could boast so high a degree of automation. Yet, mysteriously, there was no further sabotage. The agricultural report for 1866 was the first book scheduled on the Bullock, and the printing and binding of a colossal 189,550 copies of the 692-page tome (at a manufacturing cost of 84 cents per copy) proceeded without incident. The Government Printing Office had weathered a potential crisis and was ready to tackle effectively whatever work came into the shop. Among 26 Adamses, 14 Hoe cylinders, and 13 Gordons and other job presses (1870), the Bullock had made a home for itself and promised to do as well in commercial plants wherever the latest and fastest was wanted.

William H. Bullock (1813-1867) was deprived of the very fundamental assets Robert Hoe and his sons had found of inestimable value in business: a family to help with the pioneering inventor's work. Born in Greenville, Greene County, New York, and orphaned at the age of 8, Bullock was cared for by an older brother in nearby Catskill. The lad was put to work learning the trade of iron founder and machinist; considering the sleepy industry of upstate New York this proved a commitment more foresighted than Robert Hoe's apprenticeship to a carpenter in industrial England. But thereafter Bullock could count on little fraternal help of any professional value.

The young machinist nevertheless took to his calling enthusiastically, becoming expert even before finishing his contract. He taught himself the art of pattern making, and also whatever engineering theory he could glean from off-time reading. When about 20 he left his brother's home to start a machine shop in Prattsville, New York, where he devised several contraptions, among them a shingle-cutting machine. After two years perfecting this device—apparently demonstrating a unique determination to closet himself until assured of absolute success—he removed to Savannah, Georgia, to engage in shingle manufacture. This venture failed. Sereno Newton had innocently found him there running a newspaper, but Bullock soon returned to New York.

In New York city the determined machinist next set up shop to fabricate hay and cotton presses of his own design (and also a type of artificial leg). Again he gave up, in 1849 moving to Philadelphia to start a machine shop and patent agency. He developed a grain drill, a seed planter, and a lath-cutting machine, patenting them between 1850-1854. In connection with his patent agency, however, he began to issue another daily newspaper, *The Banner of the Union* (1849-1853), the operation of which refocussed his attention on the problems of printing. He would devote the remainder of his life to solving them.

Bullock worked simultaneously on three aspects of printing neither Hoe nor anyone else had seen fit to tackle from scratch: (1) mechanical feeding of paper in rolls, (2) printing on both sides of the sheet in one pass through the press, and (3) fast, efficient on-press cutting into sheets. Taking no methodological guidance from R. Hoe & Company, which characteristically rushed to the patent office with every caveat, every improvement, however slight or speculative—resulting, to be sure, in a good record of prior claim in disputes, Bullock quietly spent fifteen long years improving his mechanisms, working some of that time both in Philadelphia and in New York. In Catskill he is said to have built the wooden cylinder press, operated by hand crank, which became the prototype of his masterpiece.

Like Hoe (and others) before him, Bullock nevertheless had not hesitated to borrow what he could from earlier workers. As a Philadelphia agent he had secured for Arsene Legat, of Paris, France, an American patent for a platen press to print rolls of wallpaper. Bullock freely acknowledged this device as giving him inspiration for his own machine. He did not mention many other forerunners whose work might have been instructive.

Thomas Trench, an Ithaca inventor, had successfully printed (with stereotypes) on a web press of his own design in 1836, and although he had been praised by both the American Institute and the New York Mechanics' Institute, his work was apparently too advanced for commercial application: he could print whole books—including the Bible—at rates like 1 per minute. Similarly, Thomas Nelson, Jr. in 1851 demonstrated a rotary machine printing two sides from a small roll at 10,000 per hour, but European printers scarcely thought it more than a "mechanical toy." An 1851 bed and platen press of Thomas H. Dodge, New Hampshire, fed from a reel, was bought by a cloth manufacturer for textile printing. These apparently passed under Bullock's gaze but were of no particular interest.

When curved plate stereotyping appeared in 1861 Bullock perceived that with it he would not have to solve anew the preliminary problem of mounting type on a cylinder—thus escaping the pitfall that had trapped Rowland Hill (and others) in London and Wilkinson (and others) in the United States.

William Bullock finally settled in Pittsburgh, 1859, when he was satisfied that his prototype was ready for exploitation. One early machine was erected at the Cincinnati *Times*, and two others were installed 1863 in the *New York Tribune*, the first of the large papers to try them; not entirely successful, Bullock had them scrapped. He went back to work. Manufacture of the final version, patented 14 April 1863, commenced in Pittsburgh and it went on sale in 1865. Despite the disruptive war economy, the Bullock was recognized immediately as a significant advance in the press builder's art.

Bullock's machine could do all he set out to accomplish. It was first to print from a continuous roll of paper, to print both sides of the sheet, and to cut the paper either before (avoiding web tension problems) or after printing. Too, its speed was phenomenal—20,000 perfected sheets an hour. The press installed in 1865 at the New York *Sun* was the first in America especially built for curved stereotypes. Six others there soon followed.

For the next two years Bullock devoted himself tirelessly to the fabrication and installation of machines for which the demand, thanks largely to post-war commercial activity, was considerable. In 1866 the Philadelphia *Inquirer* had two single and one double Bullock, the single doing 8000 (perfected) per hour. Bullock's orders grew. And, wrote Carl Mitman, "as might be expected of a mechanical genius, he was constantly working on improvements which were to be added as they were perfected."[18]

But before American family-owned businesses converted, one by one, to corporate structure to attract qualified personnel, such firms increasingly faced difficulties with one-man one-expert leadership. Bullock, alas, with only one daughter, had no family from which to groom coworkers. To some of his faithful employees he imparted his mechanical ideas and techniques, but—as the Harpers and Hoes could testify—there was still nothing like blood and marriage for business longevity.

In April 1867 Bullock was superintending the installation of a press for the Philadelphia *Public Ledger,* the very paper that had taken Hoe's first Lightning. No wonder the Colonel could not attend the festivities! Swain's loyalty to Hoe was of little avail when Childs took the paper; coming from a quiet book publishing background he had had little exposure to beguiling—if unschooled—Hoe press salesmen. But he wanted the latest and best, a Bullock. The inventor himself supervised the necessary tests to true the machine; like Hoe twenty years before this was now a crucial moment for him—all the more so as his triumph would be Gold Street's loss. The checks continued.

Suddenly, as the last printing trial transpired, Bullock's foot became caught in the driving belt of the press and was badly crushed. The damage and shock were so severe he could not recover; Bullock died nine days later, age 53. His shaky organization would have to carry on without him. Despite the dull sadness of the unfortunate death of a great press maker, the *Ledger*'s commemoration of its new headquarters proceeded on schedule. The new Bullock's exact mission is not clear, but it may have been meant for Sunday supplements and other specialties.

The Bullock Printing Press Company's Pittsburgh sales staff bravely carried on, not without considerable success. John W. Kelberg, a Swedish-born machinist, had helped with many improvements to the original machine, and after Bullock's death set up nearly all the presses installed. The firm's product had indeed been fully successful, and the patient inventor might have begun at last to reap the rewards of his long seclusion.

In the New York *Herald* plant a Bullock, printing and cutting two copies of the paper with each revolution, with three attendants produced 30,000 impressions an hour:

OFFICE OF "THE HERALD"

New York, January 13, 1869

Sir:—

In reference to the Bullock press, we are pleased to say that we have had two of them at work in the press room of the Herald, since July last, and so far, they have given entire satisfaction. The quality of the work turned out from them, is equal, if not superior to any other news press, and the undisputed fact of their great economy in the saving of labor, and in the trifling and unexpensive nature of the repairs necessary to keep them in order, make them particularly desirable.

Yours, etc.

To Wm H. Williams, *J. Gordon Bennett, Jr.*
President Bullock Printing Press Co.[19]

At the New York *Sun* the paper's proprietors suggested that Bullock incorporate Beach's pet folding mechanisms into their new presses, which was accomplished. After about two years the paper had four of them:

> [W]here the Hoe press requires that you shall print but one side at a time, and thus, if you have but one press running, that the first side of the paper shall be put to press long before the second is made up, the Bullock press gives you full control of all the pages up to a much later hour. . . . While the Hoe press requires one man to feed each cylinder, and half that number more of fly-boys to remove the sheets, the Bullock press feeds itself from rolls of paper, from which about 2,500 perfect copies of the Sun are printed automatically, and delivered ready to cut up, so that *four men can do the work on this paper which twelve men used to be required to do. . . .*[20]

The Hoes, distinctly an old-fashioned style, suffered in such comparisons. Shortly thereafter the *Sun* had a battery: "When our seven Bullock presses are working, we can turn off, without extravagant assertion, 210,000 copies an hour," said someone in the front office. Back in the pressroom itself the drama of engaging these whirring monsters was even more exciting:

> The Press-Room is situated in the basement of the edifice, and is a most capacious apartment. When we arrive at the press-room, at half past one o'clock in the morning, matters are in no very lively trim. Everything is quiet. There is not yet even a hiss of steam. Stalwart men are stretched out on huge piles of paper, fast asleep. Some of them lie face downward, with their arms stretched out at full length, and sleeping as though they would never again wake. Others are lying all in a heap, others flat on their backs, showing grimy but honest faces; and all are sleeping soundly. Other men are bringing in huge rolls of paper from the dampening room and arranging them conveniently at hand for the pressman.
>
> The Bullock press, on which THE SUN is printed, prints from a

continuous sheet, which is wound up in the form of a huge cylinder. The machine for wetting down or dampening the paper is so constructed that it unwinds it from one roll and at the same time winds it up into another roll; and as the paper thus passes from one roll to another it is subjected to a uniform shower of the finest spray, which dampens it in every fibre to just exactly the degree which is requisite for it to print to the best advantage—the construction of the press, the rapidity of the motion, and the force of the pressure, all considered. . . .

And now, as the pressmen have begun to wake up, and are beginning to put the stereotype plates on the cylinders, and the steam begins to give tokens of its coming, let us go up to the composition room again, where over the first page of THE SUN the final struggle of matter against space is to begin. "This," says the Night Editor, pointing at it as he speaks, "is the costliest page on this planet." It is now half-past one o'clock A.M. The form must be in the stereotyper's room in fifteen minutes. There is matter enough on the make-up table to fill four pages, and every line of it is important. What's to be done? especially as a fresh batch of copy has just come up marked "MUST," from the Managing Editor, who is still hard at work below. Now is seen the value of understanding *every part of one's business,* especially the mechanical part.

The Night Editor is a practical printer, copy-cutter, proof-reader, anything and everything that may be needed. He looks over the type—does not have to resort to the proofs—and orders out this and cuts down that, and reads the proof of new articles from the type—and finally "Good Night" comes from the telegraph offices—and the page is completed, and the form is locked up (that is, fastened so the type cannot fall out) and trundled into the stereotyper's room exactly at 15 minutes to 2 o'clock A.M. . . .

At 7 minutes to 2 the first press starts and delivers 200 papers a minute. In a few minutes the counters begin to count off ["There is one man who counts 300 a minute, and another who can count 400. . . ."], and get the papers ready for delivery to the newsboys and newsmen. At 2 minutes after 2 the second press begins to throw off its 200 SUNS a minute. At 9 minutes after 2 the third press starts; and so they keep on until all the presses are running and throwing off fourteen hundred SUNS a minute, two of the presses printing 300 papers a minute each. . . .[21]

Gradually the Bullock was made even more useful by the addition of reliable folding apparatus. It went into the New York *Democrat,* the Philadelphia *German Democrat,* Philadelphia *Evening Star,* and the London *Daily Mail.* But the competition was keen.

In England the Hoes watched the Bullock salesmen at work, fully aware of the various ploys but aghast at their success:

[*Conquest to New York, 14 August 1869*] . . . *The agent of the Bullock Machine has been to every office in London & offered to put a machine up on approval & take it down if it does not suit. Of course this is an inducement & the [London] Telegraph folks are just the people to grasp such an offer. . . .*

Doggedly selling and servicing type revolvers, William Conquest was suddenly sharing his British clientele with Bullock's Louis Simon. Joseph Levy had bought Hoe Lightnings through the 1860's for his *Daily Telegraph* (and still owed £1500 on them), but was indeed lured to the new siren. George Smith, *Pall Mall Gazette,* similarly had his Hoe equipment removed to accommodate the "approval" Bullock:

> [*From Conquest,* 1 *January* 1870] . . . *We made the change for the Pall Mall last night, every thing went off very well considering the short space of time we had to do the work in. Mr. Smith is as full of the Bullock Machine as ever & is waiting the result of the start before deciding to order however.*
>
> *There is a little incident happened in our favor of this order & which Mr. Smith is quite aware of. Mr. Walter of the "Times" is so disgusted with the apearance† of the papers which come off their New [Walter] Machines that they have stoped† the last new machine so that they may have an oportunity† of experimenting on that machine. However you know such things never loose† any thing in rumors; [they] are throughout London that the Machine is condemned. If this get† into print so much the better for we have been humbuged† enough lately with them and the Bullock, & now the old "Wilkinson" Machine has been taken out of a Cellar & is now going up at Drydens† as we suppose to have new life put into it. . . .*

Messrs. Dreyden, formerly Edward Lloyd's manager, were engineers and press builders. Other news was even more delicious:

> [15 *January*] . . . *The Bullock has been started at the Daily Telegraph & is so far a failure. Mr. Collins the [Telegraph] Machine Man says that so far the feeding & delivery of sheet seems pretty fair, but the distribution is shocking, in fact the Colour Runs off entirely in about ten minutes after every start. The man who has come out with the machine from Philadelphia is a Swiss & dont seem to know any thing about the Machine. He will make any alterations that any one suggests & he has just got into the hands of a party [Levy?] who will humbug him nicely. The agent who made arrangements with Mr. Levy is a German Jew who certainly only understands money matters so between them they will make some of the Bullock Co. dollars fly easily away. . . .*

> [22 *January*] . . . *I only returned home last night & have just heard from Mr. Lloyds Manager Mr. Hance that the Bullock Machine has worked the supplement of the Telegraph with perfect satisfaction and at the rate of 4,500 sheets per hour. When I asked him where he got his information from he said that the agent who sold the machine told him. However I shall know all about it to night when I see*

Colley† the [Telegraph] Machine Minder & will send you word by the next mail. . . .

[29 January] . . . Notwithstanding that there has† been so many new printing machines in the London Market lately, there is another made its apearance† from Jules Derriey of Paris, & at first sight I thought it must be one of your own prospectuses. The speed as stated for the machines where the sheets are laid on by hand is Rather amusing. It is very clear that the french† people dont mind throughing† the "hatchet." . . .

The Bullock Machine has had what the agent calls a fair trial, that is, it worked one edition of the advt Sheet at the rate of 5,000 per hour. The machine is now stoped† & I do not think that the Levy[8] care much about working it. However they are willing to keep [it] & from what I can learn they are to have the machine gratis & I suppose this is the reason why they will be willing to keep it. . . .

[5 February] . . . The Bullock machine has only worked one edition since I wrote last week. It done† about 1600 of the ¼ sheet supplement & from what I can hear the set off troubled them. Coincidentally the number produced was between 5 & 6,000 per hour during the time they was† running. The great complaint is that there is so much time lost in getting a start.

The agent Louis Simond† is troubleing† the Newspaper proprietors so much that some of them are not willing to give him any further interviews. The Daily News folks are a little old womanish & are almost going into fits from fear that they have done the wrong thing by ordering the last [Hoe] 8 Cylinder. This bit of weakness is more attributable to Mr. [J. R.] Robinson than the other part of the Company. There is a probability that Simond may get an order or two but this is a great question.

If the Telegraph folks would work the [Bullock] machine constantly & had none of our machines to fall back on we should now have this question settled, but as it is, the matter is an open question. The machine man at the Telegraph says that finally the Machine will be used for the odd numbers during the day so that every thing will be in its favor & so keep the pot boiling all the time, untill† some one else has one where the machine will have more work to do. . . .

[12 February] . . . The Bullock Machine has not been working this week & it is not likely that it will be used again only as a curiosity. There is another new feature. . . . It appears that the reason why the paper for newspapers is so low in price in London is partly on account that the sheets are made at the mills in double and in some cases in tribble† width & cut. However they find that the paper for the Bullock Machine can only be made in the one width only, just the size of the sheet wich† makes it cost too much making. & Another thing is that it is impossible to make this common paper without having a good number of breaks in each Roll; this causes so many stopages† that almost destroys the whole machine. . . .

Finally:

> [26 *February*] . . . *The man who came out here with the Bullock press has left for America & the matter now rests with the agent & the Levy*[s]. *The press has not be*[*en*] *in use since we wrote last & is not likely to be. They are now using our last eight feeder wich*[†] *stands along side of the Bullock Machine. . . .*
>
> *We have nothing to fear in respect to this machine. There are so many drawbacks to it when you go thoroughly into the matter. Notwithstanding that these rival machines have come in and realy*[†] *done themselves no good, they have had their detrimental effect with some folks in respect to our machines. People speak with some matter of doubt as to wich*[†] *story is right. However I have never had any difficulty in clearing up the matter with any one. . . .*

These Bullock vicissitudes abroad were only of academic interest locally, as most American newspaper publishers were quite unaware of them. As the Bullock patent approached reissue in 1870, C. A. Durgin, Hoe's New York patent lawyer, assured him the company would not be inconvenienced by it.

In the United States nevertheless Bullock sales went on. By 1872 the Chicago *Times* had one and ordered two more; the Chicago *Tribune* was directing Hoe to refurbish their eight-feed Lightnings—but two years later gave them up altogether. Robert Hoe II in Chicago, upon the *Tribune's* taking a Bullock, wrote his son (8 May 1873), "Tell your Uncle Richard he must hurry up on ours in this country. . . ." The Boston *Herald* took a Bullock on trial. When Tucker, visiting Vienna's 1873 exposition, heard the New York *Herald* was returning a ten-cylinder for a double Bullock, he cut short his vacation and came home. The partners were mobilizing to action.

Despite generally slow business conditions, Hoe developmental work was progressing fitfully. There would be labor problems:

> [*Richard Hoe to Robert, London, 7 November 1873*] . . . *The time to make up stock for future sales is when everything is dull, just such a time as the present; now we have had no difficulty in returning to 10 hours & besides reducing wages 10%. I suppose if the men find things very brisk next spring we shall have to add on the wages, but we shall by that time have them all working by the p*[*iece*] *or so nearly all that they will not talk of 9 hours for a day except to get more wages. . . . If it were not for the Web machines I do not know what we would do for work. . . .*

Through 1874 no less than five web machines were nearing completion, but not fast enough. Occasional letters of inquiry were not as friendly as they seemed:

THE TIMES OFFICE
Washington St., and Fifth Ave.

Chicago, May 11, 1874

Gentlemen:

I have your telegram of this date. It is not quite convenient for me to go to New York just now, but I will certainly see your press before I purchase another machine. I hope it is what is wanted. The chief fault of the Bullock is its want of speed. It will go no more than 6000. We have three of them, and they are not adequate to our edition, which range† *from 35,000 to 45,000. We must have more speed in future presses.*

Please write me about your press. What will it do at a fair average speed? Not what can it be forced to. I have seen one of our Bullocks run for a few minutes at the rate of 10,000, and yet we cannot rely upon it for more than 6,000. At a fair working speed, what will your press do? What space does it occupy, in height, length, and width? The Bullock is now made double, as well as single, but there is too much of the double concern. What is wanted is rapid single machines.

I have great faith in you, and confidently expect that you will furnish the desideratum.

Yours very truly,
Mess. R. Hoe & Co., New York *W. F. Storey*

Wilbur F. Storey had bought the Chicago *Times* (dating to 1854) in 1861 and ran it with considerable gusto—his sensationalism required him at one point to defend himself simultaneously in two dozen libel suits. Men like this would not wait.

In 1873 when the *New York Times* installed its Walter press the paper was sued by Bullock for paper-cutting and wetting infringements. The Colonel decided to help the Joneses, George and his son Gilbert, fight back. His own paper-cutting patent was violated by both the Walter and the Bullock (in the New York *Herald* office), and he was miffed: "I would like to give young Bennet† a lesson but would also like his business," he admitted (12 June 1874). Litigation continued for five years.

An impetuous Robert Hoe III had another approach. He recommended buying out the Bullock company, or getting control of it, if only to obtain the patents. He began locating parties willing to sell stock when William H. Williams, its president, himself approached the Colonel to offer three-quarters of his holdings. Hoe wasn't sure such a major investment would really give him any advantage even for a few years, but he listened to Williams' proposal.

After some bragging about its successes, Williams in October 1874 offered the entire firm (except cash, bills receivable, and unexecuted contracts) for $150,000—which the Colonel thought three times too high. A year later they met again, by which time Hoe learned that certain Bullock patents in England had expired. Williams soon came down to $60,000, but the Colonel instead suggested—perhaps another of his jests—that Williams place all the Bullock resources in Hoe's hands, Hoe to build both Bullock and Hoe machines, they to share royalties. Both agreed to think about it, but no agreement ensued. To the Colonel's surprise Bullock won its suit against the *Times* in 1878, and he found himself reluctant to tangle further with his competitor. "Patent suits are the most uncertain that I know of," he had once said earlier (18 September 1874).

Despite the rather nervous delight of Hoe salesmen out-foxing Bullock agents from town to town, R. Hoe & Company—as with earlier brushes with the Adams brothers—chose instead to fight on the drawing board. When Hoe and Bullock in 1878 disputed ownership of sheet-wetting designs, both backed off: Bullock surrendered its claim and Hoe developed a concept which used no offending spray whatever. The Colonel had ordered the spray apparatus not put on the Philadelphia *Daily Times'* Centennial press; the newspaper was invited to return the wetting equipment and take the new, which, the company said, was better anyway.

Two Bullocks at the Exposition would nevertheless print the New York *Sun* and the New York *Herald* before a thrilled public. Williams was probably so happy about this, Robert Hoe III gossiped, he was "very full of liquor and talked rather wildly." Williams divulged he was holding onto the firm only to turn it over to his son, whereupon the young Hoe slyly suggested to his uncle that R. Hoe & Company offer the boy employment.

Victory in law courts would be small reward for Bullock (or Hoe), and because of several infringement cases the Pittsburgh firm, without its inventor-head, faced growing money problems. In 1887 it reorganized in Chicago (cash capital $100,000) with J. G. Knapp, president, Conrad Kahler, vice-president and superintendent, Robert Tarrant, secretary and treasurer, and—of all people—William H. Kerkhoff, general manager—"a combination which is hard to beat,"[22] said *Inland Printer* ingenuously.

R. Hoe & Company eventually bought some Bullock patents, and, according to Henry Bullen, profited handsomely with them. Characteristically, the Hoes professed warm, almost paternal relations with displaced Bullock people. The Colonel in 1877 actually had assisted Bullock's daughter, Mrs. Watters, who left a note:

> [4 *August 1877*] . . . *I was over to see Mr. Hoe about* [*employment for*] *a young man. . . . I am Mr. Bilocks*[†] *Daughter of the Bulocks*[†] *Printing Press. . . .*

Hoe agreed to take on the boy, from a destitute New York family.

At first, of course, during the competitive postwar years the Hoes were threatened as never before. While Bullock had been only one of several press builders whose products threatened to invade Hoe's newspaper territory, the sophisticated Pittsburgh machine was indeed well designed and had received just attention. The Colonel and his crew, however, had not been idle.

Unlike Bullock, whose fifteen years of experimentation were shrouded in secrecy, the Colonel left a conspicuous engineering chronicle of the same period. Following the initial introduction of the type-revolving press he applied himself diligently to the difficulties of getting good web performance. During some twenty years the Colonel and Tucker, alone and together, worked constantly to turn the Lightning into something even more potentially stunning—a web perfector with versatile cutting options at the delivery end. Before starting anew with design concepts that would originate in the drafting room (a later approach), the inventors chose first to utilize a workable press mechanism, putting Hoe well ahead into new experimentation with a minimum of tooling up.

Within a year of the installation of his first type revolver, Richard had an improvement for it:

> In 1849 Mr. R. M. Hoe filed a caveat for a method of printing one or more rolls or webs of paper on a Type-revolving press, cutting the webs into sheets and flying them. . . . As a form approaches the two cylinders at one side the paper for them is drawn from their rolls by drawing rollers, is printed by the form and then stops until the opposite form approaches when it is again drawn forward, printed and stops, and so on until the paper is entirely printed on one side and rewound into rolls. . . .

The following year saw another major development:

> In 1850-1 there was made in the experiment room a part of a type-revolving press consisting of an impression cylinder with feedboard and delivery apparatus, full size. . . . The [printed] sheet then ran into delivery tapes, and in the operation of printing and delivery it was naturally turned upside down. . . . It was the intention to cut the sheets . . . as well as to perfect [them].

Tucker in 1852-1853 made two large custom-built Lightning-type rotary perfectors; one (destroyed by fire after a few years) was for Thompson's *Bank Note Reporter*—circulation 100,000, the other for George F. Cooledge, printing Appleton's spellers from electrotypes hand-bent to the impression cylinder, still running at century's end. Foster's "Prestonian" machine of the 1870's, with web feeding, was very similar to these early Hoes.

Tucker was promptly put back on the type-revolving work:

> Mr. R. M. Hoe thought that if both outside and inside forms were put on a cylinder at the same time its diameter would be doubled,

> and on that account it would be more convenient to put the type on, and also, its increased diameter would give space around it to print two or more webs on it at once. . . .

With these early attempts R. Hoe & Company still had not produced a workable web-perfecting paper feed. Tucker and the Colonel must have nearly exhausted themselves on the possibilities. They considered a new approach:

> Just about this time the firm bought Mr. Victor Beaumont's patent, dated September 6th, 1853, for a method for printing a continuously moving web of paper on a Type-Revolving Press, cutting it into sheets and flying the sheets. . . . This method of printing a web is very ingenious and is precisely the same as the web is printed on the firm's present Type web presses. But the scheme as a whole did not seem very practicable. . . .

In 1854 the Colonel returned to the nagging problem himself:

> During this year Mr. R. M. Hoe filed a caveat for three plans for flying sheets cut from a continuously moving web of paper. The first . . . [involved] switches or an air blast, [in] four separate channels. . . . The second method, as an outcome of the first, was to make the four channels or paths of such relative lengths to each other that the four successive sheets would all arrive at a given junction point simultaneously. . . . This plan has been called 'long and short paths.' The third method was to receive on a cylinder, say, four successive sheets, one upon the other, and then direct them off in a body, by switches or an air blast, to a sheet-flier to be flown together. This plan is known as the 'collecting cylinder.' . . . This caveat has been of great benefit. . . .

The 1854 caveat, like the Beaumont patent, was particularly useful in forthcoming litigation, but was not the full answer, mechanically.

Richard Hoe in 1858 explored still another idea for the handling of web paper on a type-revolver:

> Hoe filed a caveat for a Web printing machine to print four webs from two type cylinders carrying wedge shaped type. . . . Each web [after perfecting] then passes between a pair of cutting cylinders furnished with a serrated knife, but having here and there a tooth removed so that the sheets are still slightly joined together. The sheets may be flown by a rotary sheet-flier, . . . [or] by an oscillating sheet flier, . . . [or] two horizontal inverted sheet-fliers, . . . [or] the web may be folded uncut in a zig-zag pile. . . . None of the above devices, however, have been patented. . . .

A final inspiration, developed before the Civil War put an end to this exasperating spate of paper-handling trials, had a frustrating conclusion:

> During 1858 we built in our experiment room a full size web cutting

> and folding machine for sheets the size of an eight page *New York Herald,* with cutting cylinders which cut the sheets from a web or roll of paper. As the sheets descended from the cutting cylinders the first folding blade, which vibrated horizontally, struck the sheets centrally, . . . a second folding blade descended vertically through a slit in the [folding] plates . . . [and] a third blade struck it horizontally through a central slit in the plates. . . . [But] it was discovered that a patent for folding between plates had been granted to Mr. George K. Snow, in 1850, so nothing further was done. . . .[23]

During the war Tucker was occupied with everyday problems, and the Experiment Room probably grew dusty. He did not forget the recent decade's frantic work, however, and was still at work (on paper) with a variety of schemes:

> *New York June 16th 1862*
>
> *Dear Colonel,*
>
> *Your favor of 30th ult. is to hand & contents noted.*
>
> *You seem to have the impression from my letter to Mr. Conquest that I do not wish to make the 2 Cylinder rotary perfecting press but this is not so. I feel that our reputation is much interested in every machine we ship abroad & should prefer to send none but perfect & tried machines & therefore would rather the first one of its kind should be put in operation here. Nevertheless you will see by the enclosed tracing* [not present] *that I have not been idle in the matter, but have succeeded in dispensing with feeding cylinders, drop rollers & tapes, & feed from both boards onto one cylinder & into fingers direct. This I think a great improvement both in simplicity & accuracy.*
>
> *I intend to enter a caveat for it as also for the apparatus on the other tracing* [not present] *for cutting a roll of paper into sheets of any required length as it passes on to be printed. The inking apparatus I think will answer as it is drawn but I am not so sure about the double acting sheet flyer. I think it probable we may require a light downward blast on each side of it to keep it straight or possibly 2 sheet flyers set back to back with downward leading tapes between them.*
>
> *I have thought a great deal about "the great question, of getting more speed with if possible less complications" and am strongly inclined to the opinion that the proper course is to feed from a roll, work both sides, cut either before or after being printed as found most convenient, the latter if possible but instead of one type cylinder & four surrounding imp. cyl. to have probably not more than three type & three imp. cyl. . . . & thus print two continuing sheets. Of course for every type & imp. cyl. added another sheet can be printed but owing to the facilities of making stereo plates I think it would be better to have smaller machines & more of them than a many cylindered machine in which when you wish to stop one cyl. you must stop so many.*

17th. Peter has just showed yours of the 31st ult. in which you countermand the two perfecting machines in favor of a 4 Cyl. I can hardly say I am glad for I have got fairly interested in the perfecting machines.

However I shall have the drawings finished ready for a future occasion.

I disagree with you in thinking we shall not succeed in backing the electro shell on a curve, but think it will be done & so effectual & the press work so well that the London Ill. News[†] may find it to their advantage to order 2 of them for their cut side.

I recommend you to patent in England if possible:

1st Backing up electrotype shell in a curved mould or backing pan.
2nd Cutting a roll of paper into sheets as it passes in to be printed.
3rd Apparatus for cutting the roll into sheets of any required length as it passes in to be printed, or out after being printed.
4th The arrangement above sketched for printing 2 or more continuous sheets ad libitum.
5th The arrangement (caveated here) for printing 2 or more continuous sheets from 2 type cyl. by surrounding their imp. cyl.
6th Feeding 2 sheets of paper direct into the fingers of one & the same cylinder.

Should you think it worth while to patent the above (I suppose they can all be included in one patent) I will send you drawings & Specifications.

Very truly yours
Stephen D. Tucker

To/R. M. Hoe Esq.
London

After the Civil War the two were back at work:

A patent was granted to Mr. R. M. Hoe and myself, June 29th, 1869, for a web press in which type and impression cylinders were arranged in two vertical series facing each other, one series consisting of one type cylinder between two impression cylinders, the other series of one impression cylinder between two type cylinders. Thus a web passing between the upper pair . . . would be printed on both sides, and a second web simultaneously passing in the opposite direction . . . would also be printed on both sides. . . . This seemed a very simple arrangement, but . . . nothing further was done in the matter.[24]

Despite all these expensive prototypes, R. Hoe & Company had not solved the web-perfector feed problem, nor other technicalities lurking in the delivery end. Yet the work of others, especially in Europe, became a nagging reminder:

[*Robert Hoe to Richard, Geneva, 13 September 1870*] . . . *George Jones of the N.Y. Times has also returned* [*from England*], *and in an interview he told me of a visit to the London Times office; that he and his son saw each of the perfecting presses at work satisfactorily in every respect, printing 10 to 11 thousand perfected sheets in an hour, and is sure they will have to introduce them here if we do not—or supply something equally good or better. He wishes me to urge you to go and see the Times machine at work.*

We hope that you will, if you approve them and consider it worth our while to be interested, find what Mr. MacDonald would be willing to do with us to introduce them here. Mr. Jones saw the Maranoni[†] *machine at work with Mr. [Benjamin] Wood but thinks little of that plan. . . .*

George and Gilbert Jones were in fact themselves so sold on the *Times'* Walter press they would take no less than four of them by 1872. But their warning was welcome. It did not often happen that a competitor's product was superior, and it is instructive that the Hoes had no compunction to act only as agents for a foreign machine. Meanwhile, however, their research continued. They tried enlarging the Lloyd two-feeder perfector into a four:

> It had been decided to build, as fast as orders in the factory would permit, a Four-Feeder Rotary Perfecting Press, on the same plan as the Two-Feeder Perfecting Presses previously made for *Lloyd's Weekly Newspaper,* of London, and most of the drawings and patterns for it had been made during 1869 and 1870. In April, 1871, an order was received from [B. Wood of] the *New York Daily News,* for a Four-Feeder Perfecting Press to print the *News* four copies on a sheet, the sheets to be flown open. . . .

Because of space limitations—four feed tables facing each other, and the sheets all passing through the machine in a common channel—Tucker reverted to the 1853 "long and short path" flying apparatus. The machine ultimately produced 8000 sheets per hour (the speed of the feeders) and was considered a success. But again the Colonel reviewed his progress and returned to some central problems:

> The firm had now turned its attention again to a continuous sheet or web printing machine. The success of the flexible matrix for curved stereotyping made the printing part of a web machine perfectly feasible, and cutting the sheets from the web presented no difficulty, but the great obstacle to complete success in web printing was in disposing properly of large size sheets printed at great speed. . . .[25]

Tucker's chronicle had indeed been a disappointment. The firm's progress looked bleak:

1849	Printing one side of roll on Lightning press; rewinding printed pages, reversing the roll direction, then cutting. Not used.
1850	Printing one side of sheet on Lightning, flopping the web. Not used.
1852-53	Custom rotary perfectors/type revolvers. Workable.
1853	Printing one side of roll and flopping. Not used.

1853	Beaumont's patent: printing one side of roll, flopping, and cutting into sheets. Better, but not used.
1854	Three caveats for flying sheets from web, including "long and short paths" and "collecting cylinder." Usable.
1856-57	Type-revolving cylinder improvements not connected with web work.
1858	Caveat for printing four webs from two cylinders. Not used.
1858	Web cutting and folding machine; workable only for cylinders too small for newspapers (Snow patent). Not used.
1859	Cutting sheets from roll to feed a perfector press. Not used.
1861-65	War work; two-feed rotary perfector; four-cylinder color press.
1869	Printing two webs in two vertical banks and perfecting. Not used.
1869-70	Wood's Four-Feed Rotary Perfecting press. Successful.[26]

The Colonel, now 59 in 1871, had supervised some of the most intensive engineering toward solving the many facets of web presswork, and was still not rewarded with success. The Bullock, though not perfect, to be sure, had appeared five years earlier and captured the public's fancy as a symbol of journalism's progress. Glowing accounts of the Walter press were also coming in. These were constant reminders of unfinished business.

Significantly, however, the force which would affect the Colonel personally—the factor which usually resulted in Hoe success—was another request from his old British friend, Edward Lloyd. Touring Europe in 1871, Hoe received an order for a press to print Lloyd's newest weekly. It involved a dilemma Hoe could not bring himself to reveal: usually including a large suburban subscription, the large British weeklies had to be folded (for mailing) on the press—introducing to the beleaguered inventor a demand for trouble-free folding when he had not quite solved simpler web feeding problems. And now Hoe agreed to execute the order in England—under Lloyd's eyes.

In the United States, meanwhile, there was some comfort in web handling minus the versatile delivery options Hoe believed publishers would insist upon. The great metropolitan newspapers simply wanted fast printing speeds so as to make street sales as early as possible. They could be persuaded to take a Hoe web press without folding apparatus, and the company by 1874 could reliably make such equipment. As he had done many times before, the Colonel apparently singled out a paper—the *New York Tribune*—as test vehicle for this program. It was a crucial campaign, as the prestigious paper was in new hands.

Whitelaw Reid (1837-1912) was "tall and slender, with a drooping moustache and 'intelligent eyes' under a sweeping mass of dark hair,"[27] wrote Baehr. He had reported the Civil War and gained other newspaper experience in Ohio. Among several offers in 1868 from big Eastern papers, Greeley's was most alluring: the old eccentric claimed he was quitting his active direction of the paper. Reid, 31, took the job; his father-in-law, millionaire D. O. Mills (and others) put up money for his share in the business. Reid was in full charge by December 1868 as managing editor. In 1876 he would write a friend, "You'll be glad to know that the office does well. We made $85,000 last year & got everything into beautiful shape. . . ."[28] Reid labored diligently on the paper until 1889, when he was appointed Minister to France, after which service he found little time for further newspaper work.

The 34-year-old *Tribune* had left its crumbling "old rookery" in April 1875 and moved into "Tall Tower," an extravagant new headquarters in the style of a Florentine campanile, on Nassau and Spruce Streets facing City Hall Park. Composition (soon aided by Mergenthaler's Linotype, which Reid helped finance) took place on a light and airy ninth floor; editorial offices and Reid's suite were on the eighth—all well appointed. Of this building, the tallest in Manhattan, Baehr wrote:

> In keeping with the progressive character of the establishment the presses in the basement were reinforced by a new Hoe web affair that tossed off 16,000 to 18,000 *Tribunes* in an hour. This speed enabled George Pearce, the night editor, to keep the forms open until 3 A.M., instead of locking half of them at midnight. In the basement were other attractions besides those which drew the connoisseur of newspaper technology. Breaking with the Greeley bonedry tradition, a beer salon was installed there. . . .[29]

Earlier notes between Hoe and Reid had revealed a few new problems before the impressive machine could roll:

> [*R. Hoe & Company to N.Y. Tribune, 23 December 1874*] . . . *Your paper maker is correct about the size of the hole in the spool. . . .*
>
> *We learn that they wish to send you rolls of 18 in. diameter; this will cause more trouble and . . . waste of paper in the rerolling, when uniting two rolls in wetting. It certainly will be less trouble to the paper maker, but it is evident that whatever labor or trouble he saves, you must take . . . ; and you know they do make rolls at the mills in a well united continuity of paper three feet in diameter with a circle of boards at either end, so preserving the paper during transportation. . . .*

As discussions crystallized, "Tall Tower" talk and other puffery would not distract the Colonel:

> [*R. Hoe & Company to Mess. Dun, Barlow, 5 January 1875*] . . . *Will you please obtain for us, confidentially, a full and detailed report of the financial standing of the N. Y. Tribune Co.?* . . .

The firm was sound, and the work undertaken.

The *New York Tribune* was pleased with its dazzling new web perfector. At a trial run in the Hoe shop visitors had their appetites whetted: "While we stood looking at it, in less than 20 seconds over 70 copies of *The Tribune*, six at a time, printed on both sides, were laid down by its magic fingers." [30] With the machine in the plant basement the paper commented:

> A constant blurring of the side first printed on each sheet seems to be an inseparable feature of the work on the Walter press, while the Hoe press, besides avoiding this difficulty, is less complicated than anything previously manufactured in the way of a fast printing machine. American industry in this department has long been without a rival, and it is a peculiar pleasure to find that Col. Hoe is able to maintain its preeminence. . . .[31]

Six months later it reported how the press' phenomenal speed had dramatically benefited the reporting of some news (alas, not an urgent matter in world affairs):

> Last night it was desirable for the accommodation of friends in Saratoga to issue a midnight edition of THE TRIBUNE, containing an account of the boat race, for delivery at Albany, Troy, Saratoga and elsewhere, by daylight this morning. Rapid work, of course, was required. Within precisely one minute after the last stereotype plate was given to the pressman, the perfecting press began moving, and twenty seconds later was producing eight-page papers, beautifully printed on both sides, at the rate of 12,480 copies per hour. Not a break nor an instant's delay of any sort occurred. . . .[32]

The following year (February 1876) the *Tribune* took another web machine, replacing its last six-cylinder Lightning:

> [*R. Hoe & Company to Whitelaw Reid, 24 January 1876*] . . . *We have made a careful examination in your press room and find that the present driving shaft is amply strong enough to drive the Web machine. . . .*
>
> *If you can allow us to commence on Monday, after your edition is off, we can have the Web Machine ready in time for your edition of the following Monday, and if possible for Saturday's edition. . . .*
>
> *We trust this will be satisfactory to you, and think the saving in wages in your press room alone will be far greater than the expenses attending the change. . . .*

Reid still had an eight-feeder Lightning. For this second perfector Hoe rendered bills (23 March 1876) requesting payment of $6000 or $7000 "as proposed," adding "Our wants just at this time are much greater than we expected. Would therefore if convenient be glad to receive check as early as Monday or Tuesday next. . . ." The correspondence included a personal request (10 April 1877) from Stephen Hoe: "Can you [Gordon Ford, *Tribune* Manager] recommend a Stenographer who has the time and ability to give lessons[?] Father is desirous my brother [Frederic] should learn. . . ." Frederic was 19.

The company proudly referred customers to the *Tribune* installation, claiming modestly of the new perfectors:

> [*To S. S. Foss, Woonsocket, R.I., 20 April 1878*] . . . *We use every effort and spare no expense to have them thoroughly made, and one of our presses, properly handled, should run ten years with scarcely any outlay for repairs. The new web machine at the Tribune may be seen running every Wednesday morning*. . . .

For a prestigious Canadian publisher the Colonel made special professional—and avocational—overtures to corral an order. The results were poignant.

George Brown started the Toronto *Globe* in 1844, when he was 26. Outgrowing the facilities of his father (who also ran a newspaper), Brown that same year came to New York to buy from R. Hoe & Company the first cylinder press to be used in Canada. So enthusiastic was he for its 1250 impressions-per-hour speed (and so remote must Toronto have seemed to Hoe) that Brown on the spot also became Hoe's Canadian agent. But as Canada's journalism developed, its practitioners kept one eye on British developments as well as those of their southern neighbors.

In 1853 the *Globe* with 6000 circulation needed a new press and a steam engine to run it; Brown strangely chose a 3000 impression Taylor and bought an engine locally in Kingston. The paper became a daily. Again, in 1860 Brown got new copper-faced type from James Conner & Sons, New York, and a second large double Taylor, which must have vexed the Colonel. The paper also took "a remarkable new [Chambers] folding machine from Philadelphia." Its historian wrote:

> The *Globe* could now print, fold, and mail 3,000 papers an hour with only six employees in its pressroom, most of them boys. The new machinery, Brown calculated, should save enough to pay its cost within a year. He was really ushering in the age of the big mechanized press in Canada. . . . [33]

Brown's brother Gordon in 1850 had joined the paper, two years later becoming managing editor. As (Sir) George Brown increasingly involved himself in politics and government Gordon handled the *Globe*'s everyday operations. Together they began to report—and make—Canadian history.

In 1868 the Toronto paper added a four-story wing to house a new rotary press, and in the fall a four-feeder Hoe Lightning was installed. At the same time George Brown announced the purchase of a second and even larger four-feeder, especially made for a weekly edition. At a moment when Toronto's population was just over 50,000, the *Globe*'s daily circulation was about 48,000. With such unprecedented popularity the proprietor could relax and attend to other affairs.

At the age of 43 in 1861 the confirmed bachelor married Anne Nelson, daughter of British publisher Thomas Nelson. Her two brothers, William and Thomas Jr. (developer of the 1851 rotary press) had been Brown's schoolmates. Brown brought Anne to Bow Park, his family home where farm, village properties, and timber interests (his mills in 1860 alone sawed over 4 million board feet) were almost an industry by themselves. Brown's particular interest, however, was a breed of short-horn cattle of which he had the largest herd in the world. The industrious family prospered.

The *Globe*'s second Lightning, double the size of the first, was rolling in June 1870. For the big machines Brown in 1876 bought British stereotyping equipment; he and Lady Anne nevertheless came down to Philadelphia to see the Centennial Exposition, but, despite Hoe's magnificent display, he was apparently unimpressed. While he let it be known he was in the market for yet another machine, the Colonel seized the moment to reinforce his man in the field:

Dec. 22, 1876

My Dear Sir George,

I learn from our Mr. Kerkhoff that you have in consideration the subject of adopting the system of printing from a roll of paper. Now I wish to inform you that our Web machine, so named because it prints from a roll of paper, is in my opinion very much superior to any other yet invented, because First, it will do better printing with the same paper and ink, for there is much better distribution of ink. Second, it is more accessible in all its parts to the pressman, the stereotype plates and rollers are under his eye and readily changed.

Third, the delivery by collecting a number of sheets together and laying them on a table is vastly superior to that of any other method, for the papers are laid in an even pile.

Fourth—there is no machine working at, or near, the same speed which runs with so little noise, showing greater perfection in the construction and workmanship.

We have made also three of these machines with folders attached. One was in the Centennial Exhibition and is now in the Philadelphia Times office; the other two are printing and folding the Glasgow Herald—they print and fold the Glasgow Herald at the rate of 12,000 papers per hour (4 folds). The Phila. Times, running the short way of the paper, is now running, printing and folding at the rate of 15,000 per hour double sheets, making at the rate of 30,000 of that paper per hour. We have now two machines approaching completion for the London Daily Chronicle and Clerkenwell Gazette, which has been purchased by Mr. Edward Lloyd. These two machines will print and fold, or print and deliver flat, whichever is desired, and the change from folding to laying the sheets flat can be made by the pressman in less than two minutes of time, and these machines will attain a speed of 12,000 per hour by either delivery.

I am informed by our agent in London that since our machines were erected in the Daily Telegraph office in London the manufacturers of the Walter press have not received an order. Now if you wish to examine this matter for yourself come on [down], and we will show you the machines working in the New York Tribune laying the papers flat and regular on a fly board and go with you to Phila. and see them folding the papers at the Phila. Times office. I would be very glad to hear from you just what you wish to do and will be pleased to give you figures which shall be low enough to meet any reasonable demand.

How is your farm getting on[?] Does your short horn stock grow in value as you expected[?] I have a very extraordinary breed of small white Yorkshire Pigs on my place & some good Jersey's[†] Cattle.

Yours very truly

[Sir George Brown, Toronto] *Rich[d] M. Hoe*

P.S. By mail please receive copy of Scientific American.

Brown apparently postponed his plans for new presses as adverse economic conditions began to accumulate.

While the Canadian's $450,000 cattle farm (about 300 head on 900 acres) attracted world-wide attention for modern and efficient veterinary practices, and while he proceeded to incorporate the ranch as a joint-stock venture, an economic depression in 1878 made his plans excruciatingly difficult to finance. Suddenly in December 1879 fire struck the farm—a jealous farmhand was blamed—destroying buildings, sheep, pigs, and horses (the cattle were saved). Brown rushed to the scene to direct the reconstruction, but the arsonist struck again at Christmastime, and he sadly gave up the effort.

In 1879 Brown had announced that the *Globe* would soon undergo a change of face: its blanket sheet would give way to a new eight-page format (16 when required)—the shape of a modern multi-page newspaper. The new press, alas, was to be a Bullock which could print, cut, and fold 28,000 per hour. For the *Weekly Globe* another Bullock was ordered. But tragedy struck again. In March 1880 Sir George was shot in the leg by a disgruntled pressman; the wound became gangrenous and he died a few months later. In April, almost as he expired, the Bullock began to roll. The Toronto paper (today the *Globe and Mail*) began its era of revitalized journalism sadly without its creator, and without the enthusiastic participation of his fellow rancher at Brightside.

The San Francisco *Chronicle* in 1879 took the same press as the *New York Tribune.* Founded 1865 by Charles and Michael De Young, brothers who had come from St. Louis, the paper long crusaded for good government and other California necessities. Upon the installation of its new sophisticated equipment the proud newspaper was as eloquent as a reporter could be over mere machinery:

The apparatus, which works with the most perfect smoothness, and with all its varied movements adjusted with the nicety of the most accurate chronometer, is, nevertheless, formed in all its parts with a view to substantial and lasting construction. Everything about it is solid and durable, though in no respect clumsy. The heaviest single casting weighs 2200 pounds, and the total weight of the machine is 65,000 pounds, or thirty-two and a half tons, resting upon a massive foundation of masonry. The large impression-cylinder weighs nearly five tons.

Some parts of the apparatus had to be very carefully packed for transportation, and six [railroad] cars were required to bring the numerous castings and cases from New York to San Francisco. It may interest printers to know that some of the ink-rollers, of which there are fourteen, weigh 118 pounds each, being six inches in diameter and seventy-five inches in length. There are four "form-rollers," eight vibrators and two fountain-rollers.

Some idea of the labor required to set the press up ready for use may be gained from the fact that in the pulleys alone there are 1800 separate pieces. The entire machine is only 25 feet long, 7 feet wide and 7 feet high, the highest part being the folding attachment. The space occupied is only half that needed for the old six-cylinder press. The principal portion of the press proper, however, is only 4 feet high, which brings it about up to the attendant's breast, so that its entire working is conveniently under his eye.

Suspended over the lower portion of the machine is the web or continuous roll of paper, manufactured expressly for the proprietors of this journal. It is of the precise width of two copies of the daily CHRONICLE laid side by side: if unwound, it would stretch from here to San Bruno, and each roll contains enough paper for fifteen thousand copies. . . .

Lusty Californians who just a few decades before needed little more than a shovel and a perforated pan—and itinerant printers whose hand presses arrived on mule back, were now treated to the rarest mechanical wonders:

To drop these open [printed] sheets without their being wrinkled by the air catching them in the descent was a difficulty which has been overcome by the addition of a blower. Close beside the press is a noiseless turbine or rotary blower, operated by a small belt from the main shaft. To the mouth of the blower a pipe very much like an ordinary stovepipe is attached. This rises to the hight† of the top of the fly, crosses underneath the press, and through four smaller pipes directs a constant current against the sheets, by which means they maintain an upright position until deposited by the fly. The blower can be used for ventilating purposes also, if ever so needed. . . .

[T]he Hoe press must be seen at work and not at play. A crank is touched, the wheels revolve, the tapes fly, the eyes become dazzled

> and the mind bewildered at the sheer speed with which the virgin white paper spins off the reel and shoots past the stereotype and impression cylinders. It is not until after it has been printed, cut and twice folded that the eye can distinguish individual sheets, and even then with difficulty. To the uninitiated beholder the sight of thin, fragile paper passing between so many cylinders, twitched this way and that by so many knives, and at last coming out neatly-folded newspapers, devoid of rent or wrinkle, seems little short of miraculous. . . .[34]

Even for Eastern metropolises, however, this sophisticated "Fast Printing Machine," which represented Hoe's ultimate mastery (for the early 1880's) over mechanical limitations and competitive encroachment, was a rare sight in newspaper vaults.

A larger part of the Hoe output at the end of the 1870's was little more than bits and pieces of earlier evolutionary work. Conversions and modifications of type revolvers were frequent—from type to stereotype, from sheet to roll, from flat to folded delivery. Every alteration was another "model." In 1879-1880 a "hybrid" type-revolving web press was built for the New York *Commercial Advertiser.* Hugh J. Hastings assumed the paper's editorship in 1868, at which time he had found the Colonel's work inspirational:

> The man who makes two blades of grass grow where only one grew before, is called a public benefactor—what shall we call the man who furnishes twenty "sentinels on the watch-tower of liberty," where only one existed before? Up to the perfection of your [Lightning] patent, the circulation of our daily papers were[†] confined to less than ten thousand, now they are counted by the hundred thousand, and Old Europe acknowledges the greatness of American genius in adopting the Hoe Press. . . .[35]

Hastings may be excused for his indebtedness to Jonathan Swift: "Whoever can make two Ears of Corn, or two Blades of Grass grow upon a spot of ground where only one grew before, deserves better of mankind and does a more essential service to his Country, than the whole race of Politicians put together."[36] The fast presses were indeed proving, for the cause of democracy, to be as productive as corn or grass wherever they were permitted to sprout. But by the end of the 1870's the Hoe product was not so cohesive an inspiration. Tucker described the *Commercial Advertiser* press:

> This machine printed two rolls or webs of paper simultaneously. The type forms for the outside pages and inside pages were placed respectively on opposite sides of a large revolving central cylinder which was surrounded by eight impression cylinders, a set of four at each side, and each set printed a separate web. One side of each web was printed . . . in the precise manner shown in Beaumont's patent of

> 1853. . . . The webs were then turned over by a [Crowell] device . . . and their opposite sides printed. . . .

Thus a hodge-podge of old mechanisms and other "pet" features, thrown together in what Tucker euphemistically called "various styles," seemed the best the company could do for varying newspaper requirements:

> Quite a number of type web machines of various styles were built during the next few years [early 1880's], some of which printed and folded four, eight and even twelve page sheets, but they were mostly one roll machines. Several two roll machines were built to print and fold four, eight and twelve page sheets at the rate of twenty-four thousand per hour, for evening papers of large circulation, but their complication caused such stoppages and annoyance that their use was abandoned. During these years the stereotyping process was being steadily perfected and time shortened; . . . the evening papers [were gradually induced] to adopt stereotyping and the ordinary web press.[37]

To a large extent the 1870's were faltering years for R. Hoe & Company—just when the country's great centennial celebration demanded their latest and most brilliant achievement. Yet there was almost nothing really spectacular on the drawing board.

Although the company had perhaps not really made a public name for itself until it introduced the Lightning, its proprietors, despite Sereno Newton's early advice, had exhibited only sporadically. The annual fairs of the American Institute, long the chief public display of inventors' wares, were full of puffery and quackery; Hoe was neither seeking backers nor interested in dazzling casual onlookers. The Colonel, furthermore, had declined an appointment as Institute manager, an honor he may have considered only a ploy toward his firm's yearly participation.

At the 1845 Institute fair Hoe's card press had nevertheless been on display, its sponsorship unclear. Shortly thereafter the august body commended the Colonel's saw grinding machinery, and he proudly published its encomiums in advertising. But, as with the Erie Canal and Transatlantic Cable celebrations, the Hoes generally preferred more modest "institutional" publicity to blatant promotion schemes.

With a dozen Lightning machines in daily service by 1850, the brothers had something truly phenomenal and awe-inspiring to crow about. But the firm still had ugly recollections of its fruitless endeavors both in France and England, and it avoided the great European international trade fair. Hoe's appearance at the 1851 London Exposition had been quite incidental, and furnished no compelling precedent to exhibit at the New York Crystal Palace two years later. *Scientific American* queried the company in 1853 and reported:

> We have been informed that Messrs. Hoe are not going to have one of their large lightning presses on exhibition. The reason given is, they could not get the requisite quantity of room to erect and work it. We regret this, because we are sure that this press would command the admiration of all. . . .[38]

The fair would include a Taylor press and an Adams, printing the *Weekly Journal of the* [Crystal Palace] *Association*. Applegath's had been featured at the British exposition; why should not Hoe's at the American? "In justice to our country," *Scientific American* shouted, "we want to see the biggest and fastest press in America at the Exhibition."

The intrepid inventors' spokesman then checked to find that Hoe had not even made application for space, which was not at all in limited supply. "We are sorry that there are none of Hoe's Mammoth presses in the Exhibition," the paper concluded, ruefully, "They would have given foreigners an idea of the *power* of the Press in our free country, as well as a very good idea of what American ingenuity can do. . . ."[39] Perhaps the Colonel might some day have admitted that a Lightning at the London Exhibit might have opened European doors five years earlier, but he would get that market all the same. Five more years of experience could not hurt.

By the 1870's, however, many more competitors were yapping at his heels, and the advertising arts—thanks to newspapers—had more brass than the machinery they touted. As Eugene Smith, his friend and lawyer, advised, it was increasingly necessary to fight rivals with publicity and advertising as well as caveats and lawsuits.

More than any other domestic area the American midwest revealed a great potential to grow, taking its printers and newspapers with it. For a simple state fair in Ohio the Colonel mounted a first-rate campaign:

> [*To Mess. Russel,*† *Morgan & Co., Cincinnati, 28 July 1874*] . . . *We are just about finishing a N° 6 & 7 size Large Cylinder 4 Roller Machine which will be superior to anything of the kind heretofore made. We have added great improvements in the driving apparatus and appliances for overcoming the unequal motion of the universal joint.*
>
> *The N° 7 will be done, and we can finish the N° 6, in time for the opening of your Exposition in Sept., where we should like to place one on exhibition.*
>
> *Would you be will*† *to take charge of and run it there, with the understanding that you would in the end purchase it, provided it gave satisfaction in every respect, and we are certain it will do so. We will deliver the press in Cincinnati and make the terms of payment easy.*

Please let us hear from you by return mail, as we have no time to lose. . . .

[P.S.] Nº 6—*Large Cylinder Press—4 Roller* 32 x 47 *in.*, $3600.--.
Nº 7 " " " " 35 x 51 *in.*, $4100.--.

Despite the rationale of such an offer, however, this often repeated technique was more a desire to sell than a yearning to exhibit, as the firm still believed professional testimonials far more efficacious than public displays. Even so, it was just the thing to do again in 1876, and the company began its ambitious plans early.

For the United States Centennial Exposition at Philadelphia the Colonel had apparently conquered his reluctance to exhibit. With a company history not much less than the country itself, the Hoes had at last shaken off the self-consciousness and modesty they brought to such undertakings. Although sales of expensive, technical equipment could scarcely be influenced by an ogling public, still, newspaper proprietors attended expositions too.

The three brothers, Richard 64, Robert 61, and Peter 55 in 1876, together represented some 120 years of business experience, much of it personally involved with sales and service. They themselves had attended enough international exhibits to know that good results were based on good planning. Many practical matters were quickly settled:

> [*To Pennsylvania Rail Road Company, 23 December 1874*] . . . *In our negotiation for placing machinery in the approaching Centennial Exposition we shall have occasion often to go ourselves and send our employees on to Philadelphia. We therefore, if possible, desire to buy one of your yearly commutation tickets, which shall be good for the members of our firm or any of our men, not confining it to three persons, as is your custom. Can we do this, you having the assurance that no one except those in our immediate employment shall use the ticket*[?] . . .

The company would want to show its latest presses, each gleaming and in running order. But the type revolver—the Hoe trademark and show piece for 30 years—although always a mechanical marvel and really never before publicly displayed at such a world-wide event, was no longer a grand achievement. It was, in fact, suddenly as extinct as a dinosaur, its impressive size now an anachronism. Wrote Hassard:

> It is a curious fact that for 350 years, in spite of all the ingenuity brought to the improvement of printing-presses, they steadily increased in size, in complication, and in price. . . . So long as it was possible to use in printing only one set of type form, cylinder was piled upon cylinder, and still the press grew bigger and bigger, and the pressmen went on clambering higher and higher, one mounting

over another's head, as the "Bedouin Brothers" at the circus build themselves into a pyramid. All at once this expansion process came to a full stop.[40]

Largeness itself was no longer in high regard; it was unimportant, unwelcome. By 1876 the Hoes realized fatefully that the Lightning days were ended; orders had evaporated and, more importantly, performances were eclipsed. A new symbol of excellence was needed.

Another marvel would highlight the Hoe display; the whole exhibit would be built around a new web perfector with folder attachment. *Paper World* in 1888 still extolled its virtues:

> The development of the fast newspaper press, with its numerous excellences, has been the work of the Hoe firm. The eight and ten-cylinder press looked like a finality, but it was not, and was comparatively shortlived. The web perfecting press, which prints both sides at once from stereotype plates that can be duplicated to any extent desired, thus making it possible to set several, and even many, presses at work at once on the same edition, and on a continuous roll (or web) of paper requiring no feeders, cutting the paper so that it will open like a book, pasting the loose parts, printing from two to twelve pages, as desired, or from four to sixteen, and inserting and pasting the supplements, and finally folding and counting the papers—that is the King of printing machines.[41]

As Hoe's usual policy, a local newspaper would be sought to exhibit the press at the Centennial and remove it to its offices upon the fair's close. As before with the introduction of the Lightning—only by coincidence, perhaps—the Colonel had again thought of the Philadelphia *Public Ledger*:

> OFFICE OF THE PUBLIC LEDGER
> S.W. Cor. Sixth and Chestnut Streets
>
> George W. Childs, Proprietor
>
> *Philadelphia May 31, 1875*
>
> *R. Hoe & Co.*
> *Gentlemen*
> *In answer to your letter, Mr. Childs tells me this afternoon, that he has not applied to the Commissioner for room—in which to print a daily paper during the Centenniel*† *Exhibition. So your application stands alone. He has promised to see Mr. Goshorne*† *in the morning, but too late I expect to hear the result of the interview, as I have promised to start with Philadelphia Commandery No. 2 at 12 o'clock to visit your city. I will endeavor to call and see you on Thursday next.*
>
> *Yours respectfully*
> *Wm L. Drane*

Childs liked the idea very much:

> [*Drane to Richard Hoe, 15 July 1875*] . . . *Neither* Mr *Childs or myself have given up the idea of publishing a Centennial Paper. Since our plans for space was*† *submitted to Mr. Goshorn, an idea seems to have got into his head, or the Committees, that there are "Millions" to be made in various ways, out of a paper, printed, as we propose to do it, in sight of the visitors. Mr. Goshorn would gradly*† *grant our request, but so valuable a privildge*† *could not conscientiously be given, without possibly incurring the censure of other publishers. To help him, we proposed that Circulars be sent to such publishers as he thought would buy, asking them to bid for the exclusive priviledge*† *of publishing a Daily Paper on the Centennial Ground, allowing a limited time to return an answer. This suggestion was approved by him, but I suppose that nothing has been done, as we have not yet received a Circular.* . . .

Meanwhile, a veritable network of subsidiary displays was organized:

> *Nov.* 5th *1875*
>
> *Mess. J. B. Lippencott*† *& Co.*
> *Phila.*
>
> *Gentlemen,*
>
> *We have agreed to let Appleton & Co. have a Stop Cylinder of* N^{o} *6 Size to run at the fair, & which they purchased of us, as we explained to you.*
>
> *The Steam Lithographic Press, & a Newspaper Press, we propose to give to the Graphic Co., in accordance with the understanding with you a few days since, as they agree to take them from us or pay for the use of them.*
>
> *The other machines which we would like to place in your hands to run are as follows, viz.;*
>
> *1 Single Small Cyl or 3 revolution Press, bed 35 x 51 in.*
> *1 Two-Revolution Cyl Press for fine work, bed 35 x 51 in.*
> *1 No. 6 Single Large Cylinder Press, 2 rollers* [*bed 32 x 47 in.*]
> *1 No. 5, 6 or 7 ditto 4 rollers* [*bed 29 x 42 in.*]
> *1 Stop Cyl Press of any size ditto 6 rollers*
> *1 Bed & platen power Book Press 29 x 43 in. platen, 4 rollers*
> *1 Single* [*Large*] *Cyl Job Press, bed 17 x 21 in. 2 rollers*
> *1 Hydraulic* [*standing*] *Press, say with 12 in. ram*
> *An assortment of small Machinery, to be used or not as required.*
>
> *Please say whether the sizes here named will suit, & if not which you would prefer. Also make any suggestions that may occur to you & oblige*
>
> *Yours very truly*
> *R. Hoe & Co.*

The New York *Daily Graphic* was founded 1873 by a firm of Canadian engravers anxious to capitalize on newly developed photographic newspaper illustrations; it ceased in 1889. Months later (22 May 1876), as the Lippincott display materialized, Hoe was most fussy: "We understand you have an Imposing Table not of our make, in our enclosure. This will have to be taken out. . . ."

As 1876 approached, there was sadly not the most fertile economic situation to build upon. Both Boston and Chicago were recuperating from catastrophic fires; individual businesses old and new around the country were grumbling. Hoe's regular customers knew these facts, but a client in China got a terse summary:

> [*To P. R. Hunt, Peking, 13 June 1876*] . . . [*W*]*e have sent the plate and cut for Washington Press to L. S. Ward, Congregational House, Boston, as desired.*
>
> *You have doubtless learned by the American newspapers of the dullness of trade in its various branches in this country. We really seem to be down to "bed-rock" in many industries. As this is the great centennial anniversary of American Independence, the nations have all expressed their kindly feeling toward Columbia, by representing themselves in a most substantial manner at our great exhibition in Philadelphia, of which we trust you learn the particulars. . . .*

It is perhaps characteristic that the Hoes would not plunge into wild centennial reveries without sober regard to the lean times. The company nevertheless envisioned a very ambitious display; it knew no other course. In February it sent John S. Albert the application forms. "Captain" Albert was chief of the Bureau of Machinery. Appended was a list showing, in addition to the eleven presses described in the Lippincott letter, a hand lithographic press, a [hand press] self inker, a Washington, another stop cylinder, and the main attraction, a "Pat. Perfecting Web Printing Machine with Folder attached." After listing this equipment many times and calculating its floor space, Hoe—so astute at putting giant presses into unseen basements without striking a pilaster, arch, or steampipe—was now embarrassed:

> [*To Lippincott, 10 February 1876*] . . . *We did not expect to trouble you again about the Exposition machinery until the time of opening, but upon layout of our space, which has been much restricted, find we shall be rather short of room.*
>
> *Lately we have been experimenting with the bed and platen book press (Adams), with a view of producing a much faster and better machine. We have been very successful, but shall not, for some months, yet, be able to show it in its complete and remodeled form. In view of this we would rather not put one in of the present style, which is so well known.*
>
> *Of course if you have work prepared for it, and must use it in the exhibition we will put it in, as agreed. If our other presses are run all the time, they will turn out a large amount of work and there will be a*

> *better space in which to run them if the Adams Press is omitted. . . . We shall be glad of any early reply. . . .*

There was room, nevertheless, for Benjamin Franklin's old hand press, belonging to John Murray of New York; Hoe agreed to put it on show, but, although invited, declined to buy it.

Edward S. Hoe, Peter's son, 20 years old, was dispatched to Philadelphia to set up the machinery and needed ready cash:

> *[To Edward Hoe from Theodore Mead, 5 May 1876] . . . Your telegram just recd. We have just written asking Mr. McLaughlin to furnish you what money you need for wages. I need hardly caution you to be very careful what you say to him & others. It was for instance distinctly agreed upon between him & Robert that no wetting machine was to be sent. . . .*

Frank McLaughlin was proprietor of the Philadelphia *Daily Times,* the paper ultimately honored to run the big web machine. Sadly, in a list for exhibitor's passes, however, Edward was not included. Members of the firm—Richard M. Hoe, Robert Hoe, Peter S. Hoe, Robert Hoe, Jr., Stephen Tucker, Theodore Mead, William Colby; and Employees—Stephen S. Hoe, Henry Cavin, Richard W. Taylor, Richard Dunn got them, more to visit than to work.

Alexander De Puy was sent to do the decorating. The firm specified "Indian red" for woodwork and shafting, but Captain Albert apparently objected to that color. Unfortunately, his Machinery Hall had a major flaw common to much industrial-age construction:

> *May 12th 1876*
>
> *Dear Sir,*
>
> *The roof of the Machinery Hall leaks very badly directly over where we were to put up some of our presses. We cannot therefore erect them until this is repaired for when raining the water comes down in a stream and would utterly ruin the machines.*
>
> *Will you kindly give this matter immediate attention, and much oblige*
>
> *Yours very truly*
> *R. Hoe & Co.*
>
> *To Chief of Bureau*
> *Machinery Dept.*
> *Phila. Pa.*

Things began to move fast:

> *[To De Puy, 15 May 1876] . . . The desk goes to-day. Enclosed are the keys. The Register Book and a lot of circulars will be sent this afternoon or to-morrow morning.*
>
> *Please have the painting done with all possible haste, and we desire the presses to be running during all hours each day of the exhibition. Harris' press is shipped this day, and we shall probably telegraph Kugler to-morrow—Let him go down at once on receiving dispatch. . . .*

[To De Puy, 18 May] . . . Enclosed is a check for $125. . . . Have the hangers and woodwork painted as Capt. Albert directs, and let Carlisle & Jay paint the presses, doing the work as quickly as possible.

We have no objections to Appleton & Co. putting the sign proposed over the press, and we have written them regarding the Counter and Case.

Please do not sign any permits until they are first submitted to us. Carlisle to get through and return as soon as possible. We are having signs printed to be placed over the presses and will send them on, with the necessary copper wire for hanging them. . . .

More passes were needed.

May 31, 1876

Dear Sir,

As it will be necessary for us to have at least one member of our firm constantly attending to our interests at the Centennial Exhibition, we would apply for passes for following members of our concern.

R. Hoe	*P. S. Hoe*
T. H. Mead	*S. D. Tucker*
Wm L. Colby	

also for the following named employees,

Jno H. Lockwood	*142 Penn St. Brooklyn N.Y.*
W. W. Hanna	*163 Hayward St. " "*
Wm Wainman	*41 Carmine St. New York*
Jules P. Vienat	*42½ Columbia St. " "*
C. Howell	*272 Madison Street*
R. Scott	*241 Davison St.*
W. H. Kerkhoff	*Chicago, Illinois*

Yours very truly
R. Hoe & Co.

Chief of Bureau, Machy Hall
Centennial Exhibition, Philad.

An old customer-supplier made a difficult request:

June 7 1876

Dear Sir,

Your favor of the 6th inst. is received. Our force at the "Centennial" is hardly sufficient for our personal uses, therefore [we] cannot promise to keep your [book sewing] machine "wiped off" as requested.

If our men were not fully engaged would of course be quite willing to arrange it as you desire.

Very truly yours
R. Hoe & Co.

Henry G. Thompson Esq.
Treasurer Automatic Smch Co.
Milford, Conn.

Yet, Hoe employees had sufficient time to roam about. "[O]ur foreman examined your machine in order to avoid infringement," Hoe wrote defensively (22 November 1876) to Harrisburg bindery-machine manufacturer W. O. Hickok, "not to make one, which we would be very sorry to do. . . ." The two firms had had earlier patent conflicts but were on the whole amicable.

The Hoe exhibit itself featured the large web machine printing the Philadelphia *Times* every day. The No. 5 stop cylinder issued the "Prattler" daily for Lippincott. Other machines were running miscellaneous work for other book publishers. But the rattle of Bullocks and Walters nearby could be heard too.

With printing presses at the show from several leading manufacturers, it was perhaps inevitable that an official competition would materialize. Tucker was on hand for the trials and competition presswork, although he was not in top health:

June 27th 1876

Dear Sir,

Confirming our despatch of this morning, we will prepare and print the points of merit of the presses for the Judges and will bring them Wednesday. We have thought it best to send Vienat on to help you during the trial, owing to the great heat, and your indisposition, together with the importance of the trial, and have telegraphed him at Washington to report to you tonight without fail at the Exhibition Grounds. We also telegraphed you to get H. J. Toudy & Co. or some other good Lithographer, to put a really good job on the Litho. press. You will recollect that Sir Sidney Waterlow expressed a willingness to wait a few days until this press was ready, and it is of course unnecessary for us to advise your hurrying it as much as possible.

Very truly yours
Peter S. Hoe

Stephen D. Tucker Esq.
Philada Penn.

Toudy, "Steam Power Lithographers," 529 Chestnut Street, may have been good, but, after a long indebtedness to Hoe, possibly owed a favor or too, as well. The commercial importance of the trial had not escaped the brothers, who put emerging data immediately to use:

[*To George Childs, Philadelphia, 1 July 1876*] . . . *At the Centennial Exhibition we have a Web printing Machine with a folder attachment and on Thursday last we ran it for the Judges of Printing machinery at a speed of 14,000 double sheets or 28,000 single sheets of the Phila Times in the hour & during all this time the folders did their part or duty without a moment's detention. We lost 6 minutes by bad spots in the paper. Including every delay we netted 22,000 complete papers Folded & laid neatly down during the said hour. This is double the production by folding delivery ever before accomplished & by an entirely original invention of much simplicity. We wish to show it to you in operation, and our R. M. Hoe, who is expected to arrive in str* [*steamer*] *Russia due here on 4th next, will come on to see you. . . .*

Foreign journalists came from as far as Australia, although the coveted business of that continent, more closely allied to its mother country, developed via Hoe's London office. The company sent a warning note:

> *[To William Conquest, R. Hoe & Company, London, 5 July 1876] . . . Mr. L. C. MacKinnon who is the business manager of the Melbourne Argus has been here and at the Exhibition at Phil^a looking especially to the Web machines with a view of ordering [for] that paper. We gave him a line to you & to our Mr. Robt Hoe Jr. in London. We offered to make a quarto web machine for the Argus newspaper with a cumulator delivery for flat sheets and attachment also of our new folder, so that either can be worked at their pleasure. We offered to make such a model in 6 months delivered on board here in New York with the necessary Stereotype apparatus for £4500. They have determined that a machine from us should be made here in N. York. You must be careful that your quotations shall agree with ours.*
>
> *We expect to put the 2nd Web for Herald† Glasgow Herald on board the direct steamer that leaves on Saturday next 8th inst. It ran to perfection here.*
>
> *We had trial at the Exhibition of the Phil^a Times Machine with Folder delivery & in speed exceeded during the hour, the Walter machine. The one with cumulator delivery is to be tried on Tuesday 11th inst. The Walter Press abandoned their Folder and threw them† [to] one side after two or three days experimenting and they ran the machines with flys only. The [Hoe] Folding delivery worked one hour & we had not a slip on it. . . .*

Tucker's undisclosed malady was much improved by the company's success:

> At the Centennial Exhibition in Philadelphia in 1876, the firm made a display of some of its more important products, such as a web printing and folding machine, a web machine that delivered its sheets unfolded, a double cylinder, two-revolution cylinder, stop cylinder and lithographic printing machines, large circular saws with inserted teeth and a variety of minor articles. The web printing and folding machine was built for the *Philadelphia Times,* and was in operation daily, printing and folding a four-page sheet of the *Times* at a speed of four hundred sheets per minute, and was always a center of attraction in the Exhibition. In the competitive trials of web machines, it ran at nearly five hundred sheets per minute and far surpassed, in both performance and speed, all its competitors. The whole exhibit was much admired and received the highest award of its class, a medal and diploma, and added to the prestige of R. Hoe & Co.[42]

"I did not doubt that the award . . . would be to us," said the usually cocky Robert Hoe III (1 August 1876). But who could blame him; the company had done well.

To supervise the routine matters of the Centennial's final days John Lockwood, a long-standing employee, was left in charge. Conditions surrounding the close of the fair became as hectic as the opening:

[*Lockwood to New York, 25 October 1876*] . . . *Since writing Mr. McLaughlin has been in & is very anxious to get out Web* [*No.*] *23 on Monday next. He desires us to make application* [*to remove the press early*] *as above. I have telegraphed you and await reply. Am sure that the application would be refused.*

The Iron Standards sent on were so small in holes, that we had to have them bored out to fit pipe railing. The crowd was so great to day that it was broken down again. Shall endeavor to make it hold out and not send for any more standards. . . .

[*26 October*] . . . *I have seen Captn Albert to day in relation to Web 23. His answer is no, under any circumstances until the morning of Nov. 11th. I will so report to Mr. McLaughlin.*

Ex Gov. Bross of [*Chicago*] *Tribune Co. called to see press in operation. It worked well (to use a slang phrase he was tickled to death). He took off his hat and cheered, was anxious to know if folder could be applied to their press.*

Have had a great many visitors to day, will send list tomorrow. . . .

[*26† October*] . . . *We have had notice from the Supt. . . . to have all our pulleys removed from main shaft the first thing after evening of 10th Novr, presume you will have the necessary men sent on for that & other purposes.*

Did you propose to have boxes made here by outside parties? or send on men & make them in our space?

If the former, I would suggest that they should be commenced as quickly as possible. There will be a large demand for packing boxes after Nov 10th & these people (carpenters) here will have very little idea of low prices.

I should like to devote most of next week in the city and make an effort to dispose of our wares. The time is near at hand when we will have to pack them up if not disposed of here. There are several matters which I should like to see you about. Shall I wait at office Monday morning & see you? before coming on here?

Dr. Nelson of Metho. Book Concern called this morning. He says that Mr. [*Cyrus*] *Chambers has been to see him & tells him if he buys your folder that he will put an injunction on it. The Dr. told him that he did not imagine that Hoe & Co. would infringe on any of his rights &c.*

Mr. Short of Cleveland left for N.Y. at 1 P.M. to day. He told me last evening that he should take a coupon press and would go to office to morrow (Thursday) and arrange terms &c. This Gent. has our presses in use, was formerly Sandford & Co. They are perfectly good, and will doubtless pay cash for any thing they purchase.

Mr. Henderson of [*Saturday?*] *Eve. Post was in to see Web 23. & was much pleased. Said that some time since Mr. Hoe told him that he intended to build a perfecting press to print from the Type form, but as he had heard nothing further concluded that it had been abandoned.*

Dr. Nelson called the second time to see press in operation & expressed himself amazed at the rapidity & accuracy with which papers were delivered. . . .

[27 October 1876] . . . *I have already taken inventory of what we have here, and shall endeavor to dispose of the small articles, cabinets &c. There will be no difficulty in having one of the Type Foundry*[†] *take No. 5 Wash. press if we do not dispose of it. The Inker should be ret*^{d} *to N.Y. as the Improved Inkers are seldom called for.*

I annex memo. of such boxes [not reproduced] *as we will want (not counting those for presses which you have) I shall get 2 or 3 good boxes that D. Appleton & Co. had paper in which will answer for small articles.*

Have been to see Mess. Gebbie & Barrie about Adams press. They have concluded to do nothing further at present, promise to notify us when they are ready. Saw Col. [S. S.] *McClure and reported to him about refusal to deliver press. He said that he expected as much, and seemed to think it was all right.*

Mr. Reynolds, of Reynolds McKimm & Trumpler, "Sunday Dispatch," Providence R.I. called and examined new style Dble Cyl. & 3 Revolution Small Cylinder presses. They require a Dble. Cylinder and he will call on you early next week.

Have just learned at Director Generals Office that report of Judges &c will be furnished on application in writing giving number of exhibit (no blanks are furnished or required), application to be signed by firm and addressed to A. T. Goshorn, Director Genl.

The paper ordered from Jessup Throm will be del^{d} *at Times office to day, we have had no difficulty with what we have had from Times office for past 2 weeks.*

One of the new rolls will be wet & sent up to morrow.

I have sold the old metal plates which were sent from N.Y. to try Web 23, price rec^{d} *was 5½c per pound. Shall I dispose of those for 8 page circular*[?] *They are pretty much used up, hardly think you would advise printing any more from them, & as it is very cool here it is difficult to work the ink.*

Mr. S. S. Hoe had promised to remain here to morrow PM so that I can go home in early train, and attend to getting some affidavits which will be necessary before I can vote as you will see by the enclosed telegram [not present] *sent to me after I had, as I supposed, registered. . . .*

[28 October] . . . *Last eve. I saw F. Heppeheimer of North W*^{m} *St. N.Y. He will want either a N° 2 or 3 Steam Litho. press and if you have it ready now could take it in, but will not have his work ready before Jany 1st. The object in putting it in now would be on account of mild weather, as he will have to tear away a portion of window casings. Please send word to Mr. H. when you could deliver. It would be well for some one to see him as to terms. . . .*

I presume you will ship boxes by rail so that they will be delivered in Machy. Hall. Will ascertain from Supt. of Transportation here, about how he desires them marked & will notify you. . . .

[31 October] . . . Will have N° 5 Imp[l] Inker boxed this PM & hope to ship to morrow. Will notify you as soon as it is sent. The Inker is complete with exception of under compo. roller which was returned to N.Y. for some reason (in early part of Exh[n]). Please have one forwarded to Mess. Davenport & James.

Mr. W. D. Evans of Westchester, Pa. called to day & examined Dbl Cyl. press. He seemed to think that it required a change to make it complete, i.e., a sheet iron guard around the Finger motion, while I promised to call your attention to this.

Mr. A. T. Sands of Times, Cinn. Ohio has been in several times. He will go to N.Y. tomorrow & will call on you for some information as to cost of Perfecting press with folder. He was in hope to have seen Mr. R. Hoe before he left. . . .

There is nothing special to report. Web 23 works nicely, and attracts the usual number of visitors.

I presume you are aware that we can ship all our goods to N.Y. free of charge. Have the usual blanks and can make necessary applications when all is ready. . . .

[1 November] . . . We have no news from you to day. Mr. R. Hoe & family called and were here when Web 23 was in operation.

Some time since I wrote you that I had some conversation with Mr. Mitchell of Lippincott & Co. in relation to purchase of presses on Exhibition.

He promised to consider the matter & let us know. Today he called, and the subject was again mentioned. His reply was that he personally should like to have presses that had been identified with the Exhibition but he did not know what Mr. Lippincott would say, would have to see him. I told him that Mr. R. Hoe Jr. would be in town in a day or two & would doubtless give him a call. Thinking perhaps that Mr. Hoe could do more with Mr. Mitchell than any one else, I will leave the matter, till Mr. R.H. Jr. arrives.

The pressman (Mr. Ryals) of Observer Office, Utica, N.Y. called to day. He was instructed particularly to examine newspaper Presses. This firm (Mess. Grove & Bailey) need a new Double Cyl. Press. Would suggest that a new price list be sent [to them], at same time, & call their attention to press, of which their pressman has full information. . . .

With the 10th of November the grand exposition came to a close. All who participated must have been utterly exhausted.

Harper's Weekly nevertheless found the Colonel's "powers . . . at their full vigor," upon having just completed his 64th year. And having seen the Hoe mechanical wonder at the Centennial the paper had similar superlatives:

It is hard to conceive of any advance beyond this "Web Printing Machine" in the special purpose for which it is designed. . . . [G]reater

> rapidity of execution seems beyond the limits of possibility. . . . Twenty-three of these presses are now in actual and completely successful operation. Of this number, nine are in use in different newspaper offices in this country. Twelve are running in four newspaper offices in Great Britain—the London *Telegraph*, the London *Standard*, *Lloyd's Weekly* (London), the Glasgow *Herald*—and two are in Australia.[43]

R. Hoe & Company was very pleased with itself over the historic year's activities. The brothers sent copies of the *Harper's* article to friends, and to newspaper chums in England shipped 38 canvasback ducks (2 for Conquest himself) and a barrel of apples. It was a good year, except for one growing worry.

When Richard Hoe developed the Lightning he had started a revolution just as pervasive as Benjamin Day's creation of a penny paper. Printing historian James Moran described the added implications of highly mechanized presswork:

> The rotary machine has thus revolutionized printing. Five hundred years after the invention of printing by Gutenberg, including four hundred years of printing at the rate of 250 single sheets an hour, rotary presses were turning out 70,000 newspapers an hour and handling up to 144 pages. By its development the rotary machine upset the careful labour structure of those four hundred years. Even the cylinder press did not disturb the skilled printer, who continued to perform his time-honoured mysteries of making ready. The rotary, which became a piece of precision engineering, was so different from the old machines that, at first, the man in charge was often a mechanic or engineer, who had perhaps helped to install it, and not a member of a printing craft union.[44]

Ironically, the genius who engineered this revolution was increasingly apprehensive about retaining control of it.

The firm's national—and international—prestige notwithstanding, R. Hoe & Company now found itself goaded by competition with Bullock (as well as a few distant threats from England). The Colonel who so much enjoyed ushering in, singlehandedly, the era he had established was now quite annoyed to share its honors with others. He was all the more piqued to note that competitors with few resources of their own had learned so much of their technology at his expense. The absolutely baffling problems of web press feeding and folding delivery, through the 1870's, had significantly dulled the Colonel's customary cheerfulness.

But this debilitating situation was only temporary. For R. Hoe & Company another mechanical genius, a Boston yankee, was soon on his way to New York to restore the balance for a long time to come.

Engd. by Geo E. Perine, N. York.

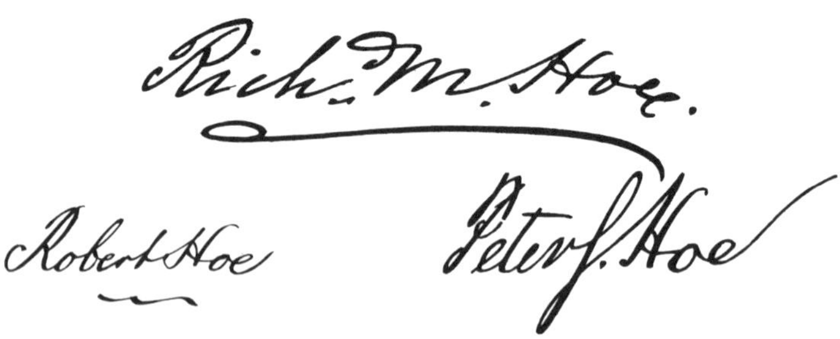

THE HOE BROTHERS

From Lossing

7.
The Boston Factory and Luther Crowell

ACQUIRING THE ADAMS PRESS AND SHOP / EMPLOYEE TENSION AND A CALL TO WAR / PRINTERS AND THE BOSTON FIRE / CROWELL AND THE MECHANICAL FOLDING OF NEWSPAPERS

Luther Childs Crowell was a precocious lad thrust into quite another sphere from the industrious Hoes and their compelling New York manufacturing activities. Son of a ship's captain whose descendants had all known Cape Cod and Boston as the hub of the world, Luther was born 1840 in West Dennis, Massachusetts. The sea would surround him as he grew, and present a strong challenge to his exceptional talents.

Educated in local schools with a short stay at a private academy, Crowell also studied medicine privately with a neighboring physician. But the sea won out—about the age of 18 he entered merchant marine service and after four years was offered a captain's commission. The prospects of a ship to command at so early an age was some cognizance of hard work and superior ability—and in the fine tradition of son excelling at his father's profession. Too, the clatter and din of the world's busy and mysterious ports were sounds of which most lads could only dream. But the boy declined; he would sail no more.

Luther Crowell's new interest was sailing of a more fanciful kind. Long days and nights on watch, studying the soaring albatrosses behind his ship, had evidently focussed his attention on the feasibility of aeronautics; he had, in fact, developed an airplane and was anxious to test it. Shortly after moving to Boston he patented his design (June 1862), encouraged somewhat by sages at Boston's Academy of Arts and Sciences that it would work. It was a vertical take-off and landing device, with wings and propellers turning from horizontal to vertical for ascent and descent. Unfortunately Crowell's financial backing failed, and the etherial project crashed forever.

Crowell's youth at least kept him indefatigable; something else—perhaps uncertain health—kept him from participating in the war. He had nevertheless chosen to make invention his career, and he had a few ready ideas to exploit. Crowell next turned to a project suggested to him during his airplane work: a machine to make paper bags. The demand for such a mundane item was real, and he received a patent for his apparatus in 1867. Five years later he patented an improved machine to make square-bottom bags similar to those universally used today. But patents do not always insure income. How Crowell supported himself is little known; his modest life he shared with (Mrs.) Margaret Doane Howard of Boston, whom he married in Wellfleet in August 1863 and who gave him two sons, Edgard D. and Luther F. Crowell. For what was to follow it is plausible he worked somewhere within Boston's tiny engineering community.

While he was developing yet another type of machine-made bag with a side seam, Crowell became involved in infringement proceedings over his earlier bag devices. He found it opportune to sell partial rights to all his bag patents to the offending party, and presumably received royalties from them. The business of professional inventor, in an age of incredible mechanical activity—as Richard Hoe himself was learning, could only sustain the most clever and determined, both at engineering wizardry and at practical commerce. Crowell, whom Collins characterized as "an odd genius," was nevertheless established and available to the trade on demand:

> He was a sea captain, with no mechanical training, who, after he retired, invented a machine to fold paper bags. Being told that folding the newspapers from fast rotary presses was a problem, he looked into it and found methods of folding and taking them away as fast as they were printed. . . .[1]

Early retirement from sea life was common, but the rest was also true. By 1873 Crowell had devised and patented a sheet delivery and folding mechanism so novel it waited two years for a practical application. Sleepy Boston was partly to blame.

The commercial capital of Massachusetts was not a lively newspaper town, as penny press and scandal sheets were accustomed to measure success. A local newspaperman as late as 1886 admitted:

> The journalistic profession here includes a large number of liberally educated people; and the "Bohemian," thanks to the better influences prevailing in American journalism, is now a rarity in Boston. Boston has reason to plume herself a trifle on the cleanliness and tone of her periodical journalism. . . .[2]

The city was indeed a bastion of scholarly thought as well as maritime activity, although its papers nevertheless enjoyed their role in public causes, incidents, and domestic tragicomedy. As Treadwell and Ruggles testified not long before, Boston had more than its share of mechanical problems as it faced the industrialization of its business world. Thanks to aging Isaac Adams, the Hoes would soon have strong New England connections and a Boston office from which to offer help.

From his earliest years the Colonel remained a respectful distance from the Adams enterprise. The printing community had generally decided that Gold Street had no machine to compete with the Adams, and that Adams in Boston had only a superb press for bookwork. Their activities, however, could not long remain separate. The Hoes were often asked to accept an Adams in trade; they were usually reluctant to do so but unwilling to offend a customer:

> [*To E. R. Jewett, Buffalo, 29 October 1844*] . . . *We trust that you will be able to dispose of the Adams press to the publisher in Ohio as you mention, for we do not like to deal in presses that are not of our own manufacture. If you think proper however you may drop us a line and state the size of the platen of your press—how old it is—its present condition—the price which you wish to be allowed for it. We cannot sell it here unless you put it a very low price. . . .*

Occasionally, too, Hoe was asked to deal with the Adams brothers as professional colleagues. At such times he was only as insistent as any impatient customer:

> *New York March 1st 1844*
>
> *Dear Sir*
>
> *Mr. Craighead wishes me to request you to call and see I. Adams & Co. and ascertain for him how soon they will engage to furnish him one of their improved presses, bed 26 x 40 inches. He thinks that by your personal application you may induce them to furnish him one sooner than if he merely wrote himself. See what you can do. . . .*
>
> *The only news in the city is the horrible explosion of Stockton gun of which the papers will give you particulars. . . .*
>
> *Yours truly*
> *Rob[t] Hoe*
>
> *To/A.S. Bowen, Esq.*
> *Care of S. N. Dickinson Esq.*
> *Boston, Mass.*

The Colonel, of course, had not failed—in his own way—to acknowledge the superiority of the Adams; he twice earnestly tried to imitate it. While in England (a good place to look around) he was urged by his brother:

> [*1 July 1838*] . . . *The improved Adams presses are getting very much into use & we think if you could find a superior bed & platten† press it would be very valuable to compete with Adams presses. . . .*

Finding nothing abroad, Hoe returned to Samuel Fairlamb, who managed to put something together:

New York January 13th 1840

Dear Sir

I am very anxious to hear from you respecting the Application of Mr. Fairlamb for a patent which I left in your hands. If you recollect the specification was defective & you were to redraw it & send it on to me to have it signed & returned to you. We have the machine nearly finished & wish very much to have the Patent issued as it will be a very important acquisition to our business. In about a fortnight I shall send you a model & drawings for a patent for the Saw Grinding machine.

Please write soon & let me know if odd volumes of the Franklin Journal could be supplied to complete a set as I find some of my friends have been so kind as to borrow a few volumes for an indefinite period.

Yours truly

To/Dr. Thos P. Jones Esq. *Rich*d *M. Hoe*
Washington City D.C.

The Company built two machines in 1839-1840 which Tucker described:

> These presses, which were very compact and occupied but small floor space, were intended to compete with the Adams press. Both bed and platen rose and fell so that they balanced each other. . . . The machine was quite simple, but was altogether deficient in both speed and power in comparison with the Adams. . . .[3]

But the Colonel was most anxious to add the bookwork market—Adams' territory—to its line of jobbing and newspaper presses; the 1837 depression for a long time pinched him unmercifully. He tried again: early in 1846 Tucker in his Experiment Room was at work on a model bed and platen power press for book and cut (illustration) printing. The prototype functioned well and was patented July 1847. That same year a number of the machines were made for the American Bible Society, Nassau Street, New York—a last-chance hope for the society's harried pressroom manager.

The American Bible Society (founded 1816) had 11 older hand presses desperately turning out literature for its ambitious tract programs. It then looked to Daniel Fanshaw for production, but even this professional printer could not keep up. In 1844 the society resolved to equip its own shop (to Fanshaw's intense irritation—he had refused to procure larger presses at the society's insistence), as reported by Robert Hoe:

> [*To Benjamin Brown, 8 February 1844*] . . . *The Bible Society have decided to do their own printing. I think we shall receive the order to go ahead today. Their old printer dies hard.* . . .

Hoe evidently furnished machines for the society, but they too were quickly inadequate. In 1846 the religious organization was planning to print 750,000 copies of the Bible by spring of the following year, and a million copies the year after. By March it was 100,000 shy of its goal—the Nassau Street facilities just too small for such a task. Then the Hoe bed and platen equipment was installed, and although still insufficient, it functioned satisfactorily until 1852, when the society found larger quarters on Astor Place. (So superior were these new facilities, however, the Society managed to furnish a Bible to every soldier, North and—by special truce—South, in the Civil War.)

The Bible House problems seemed partly managerial—plus a curious disinclination to accept the cylinder press in its shop. The Colonel reacted to both—his remarks on nepotism somewhat anomalous:

> [Richard Hoe, Paris, to New York, 2 May 1849] . . . *With respect to the affairs at the Bible House, I am very sorry to hear of the difficulties there. I do not know why the Hoyts have not been more wide awake. I fancied for some little time before I left home that trouble would come from Azor's having so many of his relatives in the house, for as a general thing a man cannot be served with the same faithfullness*[†] *or rather industry by a realative*[†] *as by a stranger, & one does not feel like obliging a relative to correct mistakes, &c, &c. . . .*
>
> *If I was home I might possibly find a young man for the foreman of the Bible House, but capable men for such situations are very scarce, I can assure you. As to working cylinder presses, any good pressman can work them or can be taught to work them, and if you could see that the great bulk of all the Book press work done in England & France is done on cylinder presses (much worse ones than any we ever made) you would never give up cylinder presses to suit the whims of others. . . .*

The Society's continued use of bed and platen machines, however, soon created new problems for the company.

About this time the Bible Society's bed and platen Hoe came to Isaac Adams' attention: he quickly claimed its features to be an infringement on his brother's press, and also claimed that Hoe's sheet flyer on cylindrical machines also infringed on Adams patents. Such charges, more theatrics (as events proved) than legal surety, nevertheless accumulated through the 1850's and intimated that an eventual confrontation between the two prestigious press makers would ensue. Neither the Colonel nor the Adamses really welcomed such conflict, and both tempered their positions. By 1859, when a Hoe lawyer reported receiving an assignment from Adams, the Colonel wrote:

> [To George Campbell, Philadelphia, 26 January 1859] . . . *I do not expect or wish to have any difficulty with Mr. Adams as I have had*

his assurance many years since that we should not come in collission.† *If the matter he wishes you to take in hand does not affect my interests of course I should have no objection to your working for him. . . .*

Adams also did a small amount of machine work, from time to time, for R. Hoe & Company.

The Boston brothers were on somewhat shakier ground with regard to Hoe's "infringing" sheet flyer, although Tucker had to recollect it hazily himself:

> It is more than probable that our present sheet-flier was evolved from the tympan of the original hand press; they are very similar in form, are both applied to laying down open sheets of paper, and both perform the operation in the same manner. Tympans were used on the old two-feeder Adams Press, which had two traveling friskets. . . . These [Hoe] tympans [however] had no friskets, the sheets being held in place by the pressure of the atmosphere while being laid down, as with the present sheet-flier. When the improved Adams Press was brought out about 1836 this tympan was modified to receive the printed sheets and lay them down in a pile automatically, and worked very successfully on these slow running presses.[4]

Richard Hoe had then developed machinery for adapting the automatic sheet fly to fast-running cylinder presses, for which the need was crucial. His device worked so well it was attached to all new Hoe machines and to a great many older ones. The Adamses were understandably reluctant to abandon their claims on a $250 attachment that every printer in America needed and wanted.

Upon the company's failure to duplicate the Adams press (without risking infringement), the young and gifted inventor's success with the sheet flyer and subsequent devices—the Lightning a dazzling "convincer"—may indeed have endeared him to the aging Adamses. Their platen business did not suffer from the New Yorkers' cylinder operations, certainly. Isaac Adams, 57 in 1859, was perhaps already contemplating retirement. Possibly using the old infringements as an excuse to feel out his competitors, he journeyed to New York to explore the situation. Correspondence followed.

Feby 25th 1859

My Dear Sir,

Yours of the 19th was duly recd. The document enclosed [not present] does not propose to accomplish all you first proposed to me, viz. to settle by reference [referee] the past, present & future, & that the affidavits of the parties interested were to be taken by the reference on any point on which either wishes to testify. Am I not correct that such was our conversation[?] Another thing I wish to have understood if you are willing which† *is that after the award of the reference is made that if I elect I may accept your proposition for a sale of your entire interest. Are you willing to do this[?]*

I have been giving the subject of a purchase more thought. Should I conclude to do so, would you throw in the patterns & leave it optional with me to take such tools only as I might wish at your valuation, & rent me the shop & other tools for one year at a fair rate[?]

Further, will you take good improved real estate in New York at fair prices for a great portion of the purchase[?] *I have a horror of being in debt.*

Yours very truly
Rich[d] M. Hoe

To Isaac Adams Esq.
(South) Boston, Mass.

The thought of buying the Adams enterprise became more and more flattering, apparently, although ramifications of the negotiations in a depression climate were genuinely perplexing. Meanwhile he brought Robert up to date:

[*To Robert Hoe, London, March 1859 (no day)*] . . . *Some little time since, Mr. Isaac Adams came to New York & made a claim for infringement of his patent in using the fly. I told him that his brother Seth had voluntarily proposed that if we would abandon the manufacture of Bed & Platen presses they would waive any claim they might have for the use of the fly on cylinder machines. He said his brother did not reccollect*[†] *any such promise & at any rate it would not apply to the renewed patents. I asked him if he remembered that he had got my affidavit to aid him in his first renewal; he had not forgotten it but as he sees a chance to make some money I think his moral perception is a little blunted. He proposed first to sell his patents as renewed & his whole press business without the patterns & tools for the sum of one hundred thousand dollars payable as might be agreed or he would leave the whole matter to referees one or three as we might elect.* . . .

We had I think much better abide the result of a suit than submit only a part to a reference as a suit would last nearly to the end of the patent—Five years yet to run (both his old & new patents were extended by Congress). . . . *The matter has given me many a headache for I really fear he would beat us in a suit & it is impossible to determine on what view a jury would assess the damages.* . . .

He stated that notwithstanding the dull times of the past years they cleared about thirty thousand dollars on their press business & had cleared nearly Eighty thousand in one year. . . .

At first I could not bring my mind to think we ought to buy, but the more I think of it the more strongly am I inclined to think it the only course for us to pursue. The possible results of a suit are frightful & [*by purchasing*] *we could certainly more than get back our money.* . . .

Robert was apprehensive, but England was far away.

[*Richard Hoe to Isaac Adams, 3 March 1859*] . . . *Yours of 1st inst. is rec[d]. It is not what we expected. Can you make it convenient to come on to New York as we should like another personal interview so that*

> *should we not conclude to purchase at your price we may talk over some other propositions which we may make if we shall not be able to agree as to a reference. . . .*

The outcome of another conference was an agreement to purchase Adams' patent rights, drawings, patterns, stocks, and tools, "in fact," said Tucker, "his whole interest in printing machinery." Adams posed no insurmountable stipulations; he had wished to withdraw from business and undoubtedly considered no other buyer. Although the company was bogged down with demoralizing affairs in England (over the London *Times*' ten-cylinder), the Colonel ran off to Boston like a boy with a new toy:

> *[Richard Hoe, Boston, to Robert, London, 22 March 1859] . . . Here I am looking after our heavy purchase which I hope will turn out as well as it promises. The orders now on hand for Adams presses are twenty & they cannot be all completed under four months if so soon. Mr. Bowen is here with me and is engaged . . . making the appraisal of the tools and stock. I shall endeavor to settle with Mr. Adams to-morrow about the rent of the shop &c &c for one year as within that time I hope to bring the whole thing to New York. . . . I should not be surprised if we could clear on the Adams presses and the other work brought in by this purchase, forty to fifty thousand dollars this year. . . .*
>
> *I feared you might have objected to so large an operation, but I know if you were here you would see . . . it was the only thing we could do. . . .*

The company's holdings after the sale would be formidable: they totaled 35 city lots (2 acres) and almost 600 employees. The Colonel got more than he expected: he took Adams' embossing machines, standing presses, and paper cutters, giving him entrance into the bookbinding field, such as it was, as well as complete domination of the printing industry.

The New England press maker was pleased to release a letter to the trade:

> BOSTON, *April 1st*, 1859.
>
> MY FRIENDS—The printers, publishers, book-binders, and type-founders, who have, for many years, been kind and generous patrons in the purchase and use of my inventions and manufactures—I have now, with much regret, to perform the duty of taking a business leave of you.
>
> Owing to a state of circumstances, which I need not explain, I am induced to relinquish the place, which I have heretofore occupied, in the business of furnishing presses of my invention, and other apparatus peculiar to your several vocations, to Messrs. R. Hoe & Co., a firm sufficiently well and favorably known, to afford assurance that they will fill my place with satisfaction to my friends and patrons,

and with honor, as well as profit to themselves; and towards whom, therefore, I desire to bespeak an extension of that confidence and patronage with which you have so long favored your grateful friend,

Isaac Adams [5]

R. Hoe & Company was equally anxious to make a statement to the public, but in addition to felicitations it wished to warn (in the Colonel's words, 10 May 1859) "those cheap men, who do nothing but make poor machines & sell cheap," who might hope to benefit from patent confusion. This requirement took another month to formulate:

New York, *May* 1, 1859.

The Printers and Publishers of the United States are respectfully informed that the undersigned have purchased of Isaac Adams, esq., his entire interest in the patents peculiar to the "Adams Printing Press," and other kindred machinery, together with his patterns and tools used for building the same: and that the business of making and repairing as prosecuted by him, will be carried on with renewed vigor and vigilance by them, on the old premises, situated in Foundry street, South Boston, to which place orders may be addressed as heretofore.

The Adams machines have been largely used, and held in great favor. They will not be allowed to decline in our hands, either in style or staunchness. We may fairly appeal to our established reputation as pressmakers, as a sufficient guarantee of undiminished excellence in every particular, and if increased dispatch is possible, it will be imparted to the business.

We also embrace this opportunity to notify those who are manufacturing machines that contravene the patents in question, that measures will be promptly taken to protect our legal rights. All persons using such machines, without license from us, are liable to suit as *particeps criminis* [accomplice]: and those who purchase such machines after this date will have no cause or complaint if dealt with rigorously.

Among the many points covered by these patents, we beg leave to call attention to two which have been frequently infringed. The thirteenth claim of the patent of 1830 reads, "I also claim the mode herein described, of laying the ink upon the types by *passing the rollers between the bed and the platten*."† The sixteenth claim of the patent of 1836 reads, "I also claim the employment of the *fly*, or its equivalent, for receiving and piling the sheets."

Communications or orders may be addressed to the Boston Factory, or to our principal office in New York.

R. HOE & CO.,
Printing Press, Machine, and Saw Manufactory,
29 and 31 Gold Street, New York.[6]

Yorkston strangely credited two unlikely parties with making this sale possible: "This important consolidation was conceived and carried out by 'John Henry's *Printer*,' aided by George P. Gordon."[7] He may have been hinting that Gordon had expressed no interest in the Adams property, and that Henry's long and generous Hoe coverage had won for the firm the approbation of the trade.

In its full-page article *The Printer* acknowledged that "In a general way, the Adams presses may be said to have won their high reputation with book work, and that of the Hoe presses by newspaper work," adding, ". . . yet, these two courses of usefulness may now be fairly considered as practically blended in the everyday operations of large establishments, where the orders bring in every variety of work. . . ."[8] While many important papers, like the New York *Sun,* had indeed started in job shops, and while many newspapers long maintained thriving captive job departments of their own, there were nevertheless few printing plants of any size whose salesmen might as likely bring in a book order as one for a newspaper. The industry, forced by the increasingly large expenses involved with specialized equipment, was already obliging shops to choose among the specialties of the art.

R. Hoe & Company's competence, nevertheless, now spanned four distinct trades—job printing, bookwork, newspapers, and bookbinding, and the brothers' enduring solidarity gave them an invincible commercial position. Lehmann-Haupt much later observed this phenomenon:

> The activities of the house of Hoe have formed one of the chief contributions of the United States to the development of modern typography and have placed that name on the honor roll of the great printers and type founders. The striking thing is that this has occurred despite the fact that the Hoe inventions have not been fundamental in the sense that the Nicholson patent for the introduction of the cylinder in impressing and inking was fundamental and subversive of earlier methods. But by a century of persistence in the further application of fundamental principles, the successive generations of that house have done more for the improvement of machine printing than any other agency of their times.[9]

The company would long continue the manufacture of Adams presses, making, in all, over a thousand of them in 57 sizes, many still in use at century's end.

The Colonel had performed considerable juggling of real estate properties to meet Adams' requirement for short-term payment, although at the same time he reserved the right to anticipate instalments where possible. While he told an old telegraphy friend, Zenus Barnum, he was paying Adams $125,000, a letter to his lawyers seems to reveal the final price:

Oct. 28th 1859

Gentlemen,

Your favor of 27th inst. is received. In consultation we have concluded to state the exact position to you, and take your advice before acting.

We purchased from Mr. Adams his Patent Rights for One Hundred Thousand Dollars. The Tools, Stock & Patterns, say Fifty One Thousand.

The Tools, Patterns & Stock we think is all on which we can be taxed in Boston, and as we yet owe Mr. Adams Seventy Thousand Dollars we can swear that our indebtedness in Boston is greater than our property there.

Under these circumstances what do you advise[?]

Yours truly
R. Hoe & Co.

Messrs Hubbard & Houghton
Atty^s at Law, Boston, Mass.

One payment was just a little tardy:

Nov 2nd 1859

Dear Sir,

In reply to your favor of 28th ulto. we beg to say that we would rather not give our Note for the amount of Steam Engine &c. as we are not in the habit of giving Notes. Indeed you are the only person who holds our paper.

In a week or two it would be quite convenient for us to remit the amount, but if you require it we can make immediate payment.

We are, dear sir

Yours truly
R. Hoe & Co.

Isaac Adams Esqre
Boston, Mass.

By the Fall of 1859 only two $36,000 notes remained, one to be paid March 1860, the other, 1861. "I should like to know how it would seem to be entirely free from debt of any & all descriptions," Richard wrote Robert (16 August 1859). "It must be a very independant† sort of feeling."

Adams retired contentedly to the Sandwich, New Hampshire, he had known as a youth—a town to which, he claimed, he would not return unless he could buy it. He now was able to do so; he had amassed a fortune reckoned as high as $3 million. He acquired many farms and pasture land around Sandwich and planted great stands of white pine. In 1868 he graciously gave the Colonel a hearty endorsement for his type-revolving press, calling it "an invention of great value to the public, and of such value that it is difficult to estimate it by a sum of money."[10]

The sale quickly became history as only minor adjustments appeared noticeable to the general public. Occasionally distant printers were unclear about the change of events. To a Montreal customer the company wrote (14 May 1878), ". . . we beg to state that we own all of Isaac Adams' patents, and that no one manufactures the Adams Press except ourselves. All those manufactured since 1860 should have our name upon them, and previous to that date, the name of

Isaac Adams. . . ." The Hoe management quietly took possession, but it would become a wartime tinder box.

The more Richard returned to Boston the more persuaded he was to leave the shop where it was. Telling Robert it would soon be an excellent spot for one of his sons, the Colonel placed Alfred S. Bowen in charge and was very satisfied with his control. "Mr. Bowen keeps things very snug & orderly & confines the business to the specialties particularly," he reported to his brother (16 August 1859), although by the end of the year the Colonel accused him of thinking, like the natives, that "Boston is the world." Bowen had been sent there even before his personal affairs were settled: "We have rented your house on Henry Street to a Mexican," Peter Hoe reported (13 June 1859), "Mr. S. J. Migil for 500 dollars balance of year." Relieved of domestic worries, Bowen began to put the shop in order, working from a stream of missives from Gold Street.

New identification for the Boston factory was needed immediately. Peter discussed it with his brothers and told Bowen (29 April), "In regard to the Sign, one to designate the Office should of course be required. The Colonel says have the name painted on the end of the building in large letters. Same or similar to our uptown works which you will remember as 'R. Hoe & Co. Printing Press Machine and Saw Manufacturers.'" Also covered were wages ("when paying in paper money . . . use only such as can be readily cashed"), shipping presses by schooner ("it is our usual custom to insure"), repairs by landlord Adams ("get him to put in those windows . . . without our paying extra"), and so on, including directions for making some old machine tools work better.

Tucker, about to win a partnership in the firm, now wrote a little pompously of his new proprietory interests:

> [*To Robert Hoe, London, 20 June 1859*] . . . *The Boston shop appears to give no trouble in the managing & why should it? It is an old established & well known shop & the printers are satisfied their interests will not suffer by its having fallen in our hands. Certainly our being the sole proprietors of the sheet flyer must tell favorably on our business, besides extricating us from an ambiguous & ugly position & putting us into an honorable one. Another very important point is that we are now complete in every branch of the press business & have no occasion to persuade a customer that a cylinder press is better than an Adams press but can let him take his choice & we are sure to profit by it & if he is dissatisfied afterwards we are not to blame. This is a very agreeable feature, for it is very unpleasant to traduce the quality of your neighbors goods in order to sell your own. . . . I hope you will excuse my thus intruding my views of the matter upon your notice. . . .*

Robert had expressed some reservations about the Adams purchase, but through the year heavy correspondence from Westchester County (the Colonel had broken his leg on 8 November) smothered his protests.

In 1856 R. Hoe & Company had received a large order—"a source of much gratification"—for a complete government printery at Madras, India. Now the Adams reputation loomed as large: in 1862 the Turkish government ordered 10 Adams power presses which were shipped from Boston 30 April. Bowen sent a description of the $5990.59 purchase:

> [*To New York, 29 April 1862*] . . . *Each press is packed by itself in ten (10) boxes and there are seven iron pieces of each press sent without boxing, making seventeen (17) numbers to each press. N°s 1 @† 17 comprise one press complete. N°s 18 @ 34 another,* . . . [*and so on, until box No. 170*]. *The box N° 171 contains extras which belong to no particular press. They are duplicates of parts most likely to wear or break. The registered numbers of these presses are 696 to 705 inclusive.* . . .
>
> *My folks at home are as well as usual. Richmond* [*a son*] *is at Port Royal, S.C. He is Gr Master 1st Mass. Cavalry. Nicholas* [*another*], *1st Lieut. Top. Engineers, is at York Town on Genl McClellan's staff.*
>
> *My respectful regards to Mrs. Hoe.* . . .

Alfred J. Bowen, a third son employed as accountant in the Boston shop, in August 1863 left for Kentucky, possibly to look after family property.

Sales of Adams presses for a long while seemed slow, which the Colonel blamed on unsettled national conditions:

> [*To Robert, 17 January 1860*] . . . *The political excitement in the South on the Slavery question is very great, & will I think affect general business some for a while. As to secession it seems nonsense to me for it cannot be carried out without a civil war resulting.* . . .

But trouble loomed ahead, as the impact of rebellion gradually bore down on newspapers, printers, and inevitably pressmakers. For the Hoes it struck in Boston rather than New York. When orders fell off there was no choice but to let men go. Through the spring and summer of 1862 the weekly payroll dropped from $825-900 to $500, not without repercussions:

> [*Bowen to New York, 28 May 1862*] . . . *Enclosed I send a petition signed by a large majority of the men employed in this shop, to be allowed to quit work at five o'clock on Saturdays. Please let me know what you think of it.* . . .

The response next day was negative.

Bowen found a slack summer a good time for refurbishing:

> [*Bowen to New York, 11 June*] . . . *We shall devote next week to repairing our steam engine and water pipes. The city authorities have*

attached a meter to our water pipe so that our bill for water will be in proportion to the quantity used; therefore it is necessary to stop every leak on the premises. . . .

When the lease expired that year Bowen pleaded with Adams (24 July), "in the present depressed state of business, we are now paying too much, and we desire to know your lowest terms in case of a renewal." Adams obligingly gave the shop a reasonable rent.

In the fall Bowen was still reporting (17 September) "everything is stale & flat and unprofitable," although conditions also gave him a chance to look around:

[17 *October* 1862] . . . [Y]*esterday I went to see a new card and job press patented by N. Ames, solicitor of patents in Boston. It is for sale and I think worth looking at. I wish Mr. Tucker would come on and see it. . . .*

But if the paucity of new orders was a disappointment, the repair and service work would thrive. Customers were holding off on purchases, trusting their machines to inexperienced help, and pushing them too fast. For one printery after another the results were often costly.

The Boston *Transcript* had been founded 1830 by Lynde M. Walter, a young man captivated by the idea of a small daily. Measuring about 10 x 15 inches, his paper featured literature, theater, advertisements (to be sure), and other cultural events. Walter's sister Cornelia assumed control after his death, and although the format was enlarged somewhat, it continued to be the organ of Boston's elite. Printers Dutton & Wentworth had been instrumental in the *Transcript's* formation, and the Dutton family retained its interest for a century. "Two excellent printers and pleasant gentlemen,"[11] said Frederic Hudson of the partnership. They were also the state printers.

Even when they were complaining they could be pleasant:

Boston, Jan 9, 1854

Messrs R. Hoe & Co.

Gent. Yours of the 6th came duly to hand this morning, containing your bill for $140.26. We make no objection to any item in the bill with the exception of that of Mr. [*Charles*] *Yetman's ($76.13) which we think high. We do not wish to find any fault with him, but all he did on the press for the 17½ days, was of no avail; it did not work any better after all his experiments; besides we think he should not have spent so many days after he discovered he could not remedy the defects. It was only after your foreman came on that a remedy for it was discovered. . . .*

We believe Mr. Y. was satisfied himself in a week that he could not effect his object. There was[†] *many days he was not in the office, and others when he did nothing. Then there was a loss of time in waiting for your order, whether to return or not. We told him at the end of a week, if he could do no more, he had better return and report its condition to you. These facts, and knowing that the press never did any good work from the start, is† the reason we object. . . .*

If you think this item will bear an abatement please make it. We leave it at your discretion. . . . We wanted a first rate press and paid your price. It was an unlucky machine. . . . It destroyed the best font of type ever put on a newspaper in America in a month. We will say no more. . . .

With sentiments of respect and esteem

*Your obt ser^ts^*
Dutton & Wentworth

The Boston *Transcript's* Lightning (a later press) met with an accident Bowen described:

[17 *October* 1862] . . . *The press was started before the form was secured and the arms and brasses which carry the compo. roller intermediate between the distributing surface and distributing cylinder* [*were damaged*]. *We are repairing as best we can for temp. use, but send new parts.* . . .

For the Colonel Henry W. Dutton & Son, successors to Dutton & Wentworth, said of their type revolver:

We look upon the invention of this press as among the greatest mechanical achievements of this prolific age. It has caused a complete revolution in the printing business, as it is capable of producing from twenty to fifty times more impressions than any other press, and consequently has now become an indispensable machine in every newspaper office where large numbers of papers are issued.

Although it may be financially against our interest to have the patent extended, yet had we a vote on the question, we should most cheerfully give it for a renewal. The whole world owes a debt of gratitude to Col. R. M. Hoe for the invention, and as he has passed the meridian of his usefulness, he is justly entitled to the fruits of his inventive genius to the end of his days. . . .[12]

The Colonel, of course, scarcely considered himself beyond usefulness.

The Boston *Journal* started 1833 as the *Mercantile Journal,* one of several cheap papers ($4 per year by subscription, 6 issues weekly) which predated the popular penny press sold individually on the street. Guided by its editor, Charles O. Rogers, its whig coverage melded into vigorous Civil War trumpeting—later turning, under Colonel William W. Clapp, to rank sensationalism.

Rogers had constant problems with his Lightning, Bowen reported (6 December 1862): "The Boston Journal six cylinder Press was badly broken this morning. . . ." One cylinder had to be returned to New York for repair. Within two weeks there was trouble again:

[Bowen to New York, 16 December] . . . I have just been over to the Journal office. The same plate that you fitted to the N°. 4 cylinder of their six cyl. Press last is broken again. Please send me a casting as quick as possible and we will fit it up here. This is the fourth time this cylinder has broken down at the same point and I think I have found the cause at least of this last break.

The gear attached to the movable cam turns the way to unscrew the stud from the nut that holds it to the plate. The head of the stud is sunk into the gear; this was dry and rough and had evidently been hot and stuck fast so as to unscrew and loosen the nut and allow the cam to strike the finger rod when it should have passed clear. Fortunately every thing fell outside the press so that nothing was broken but the cam plate and one or two teeth from the pinion that you put in last week. If this stud had been fitted with left-hand screws they would have been more safe. . . .

Rogers nevertheless also gave the Colonel a fine testimonial in appreciation for such close care and service:

If there is one man living whose ideas deserve the protection of this Government in the way of a patent, you are the man. No invention of the age has had or will ever have such an effect on the world as your "Last Fast Press." . . .[13]

The Boston *Herald* was the most popular paper in all New England. Founded by a group of printers in 1846, it first saw light of day on a double cylinder Napier, replaced in 1850 by another Hoe. Robert apparently gave his young brother Peter the careful instructions a new employee would need to deal with the paper:

New York, April 18th, 1850

Dear Brother,

We have just received your letter from Albany. Enclosed is bill of press for Mr. French [of Boston Herald] which we have sent to day by Fall River Line. If he wants a man to come on to put it up he can telegraph us. Mr. S. Clarke Esq., attorney, State St., drew up the mortgage before. Please have them† do the same this time & have the policy assigned. I hope you will not extend the time beyond 3, 6, & 9 months. He may as well borrow the money as we. See that it is all done up without delay & get the notes. We have merely written to Mr. French to day that we have sent the press & that you will call on him. We drew on him on the 16th for $590.00 at sight. We are all well. They are putting up the last [Gold Street] iron girder this morning. Nothing new.

Truly & Affy
R. Hoe

Jonas French had specifically asked that Charles Yetman put the new press in order and look over his other machines as well.

The *Herald* next bought one of Hoe's four-cylinder revolvers in

1856, replacing it two years later with a six. Edward C. Bailey, new editor and proprietor, claimed (for Hoe's *Memorial*) the latter printed daily an average of 70,000 and occasionally as many as 100,000: he further affirmed that his run could not be printed in time by any other machine nor without prohibitive costs being passed to the consumer.

Upon purchasing the six-cylinder Bailey had been billed for $17,-500, "$1,000 higher than I anticipated," he retorted (7 April 1858), "as I understood Col. Hoe to say the price would be $16,500. . . . The charge for boxing & carting $136, appears high, though I will not object to it. . . ." His undiminished good nature and sincere enthusiasm prompted Bailey to buy a second six-cylinder. Bowen reported:

> [*To New York, 29 July 1863*] . . . *We are putting the Herald six cyl. into Bailey's press room to day and as soon as we get it under cover will commence to take down and pack the old four cylinder.*
>
> *We have a large number of our men out jobbing for printers who are moving, making changes in their pressrooms, and having their old presses repaired so that we are not getting on with the* [*new*] *cylinder presses as fast as we expect to. . . .*

Bailey, 58 in 1869, sold the paper and expected to retire.

If the war had not directly accounted for Bowen's heavy and slipping work schedule, he would soon have some stultifying side effects of the nation's tension. Hoe suddenly had a nasty crisis in Boston:

> [*From Bowen, 26 June 1863*] . . . *We have a strike in the shop this morning. I have had many complaints from men working on the large lathe that other men in the shop interfered with and troubled them in various ways so much that it was impossible for them to stay in the shop. Three men had left in this way but none of them would name the parties who attacked them. This morning another man who has been at work on the lathe for two or three weeks said that he could not stay with safety in the shop any longer, that billets of wood had been thrown at him, that he had been threatened and hindered so as to render his position a dangerous one.*
>
> *A Mr. Wilcox, a man from New York, has of late spent a large share of his time running about the shop talking and hindering other men and who when spoken to by Mr.* [*George*] *Mitchell had said that he felt no interest in the shop because his wages were not so high as he thought they should be, was found to be one of the most active assailants of the men at the lathe and as we had already decided to discharge him at the end of the week, when I heard this I determined to discharge him at once and told him to put on his coat, come in to the office & get his wages and leave but he refused to leave. I then sent for the police and he was taken out of the shop. When I returned to the shop I found the men assembled and Mr. Guest was haranging them on the subject of Wilcox's discharge and the necessity there was for sustaining Wilcox by quitting work in a body unless he was set to work again.*

I said very well if such is your determination you must go. I should much rather have you go to your work but Mr. Wilcox will not come back while I am here.

2 o'clock P.M. about half of the men have come to work and I think the balance will be in to morrow morning. . . .

The last sentence seems to have been added after a hopeful watch of the situation, but more than a few hours was necessary. The initial harassment of the lathe men resembled premeditated sabotage, but Hoe's government work was quite minimal. Only the general wartime tension and fatigue (next to erupt a few days later in the July draft riots of New York) can be blamed.

[*From Bowen, 29 June*] . . . *The men have nearly all returned to their work; convinced, I think, that the position they had assumed was untenable.*

Some good men have left—not because Wilcox was discharged, but rather in disgust with the conduct of the other men. I believe there will be no difficulty in filling their places in a few days.

I have not discharged Guest yet. I think it better to let the affair settle a while and after a week or two to let him go. He is not a profitable hand on the work we are doing at present—his wages are too high for the amt. of work that he does.

During its first two years the war had gone badly for the North, and the nation's press (thanks largely to R. Hoe & Company) was quick to report battlefield disasters almost as they happened, and to demand imperiously that Lincoln and his generals act one way or another. The president himself was worried: after Gettysburg he announced a large draft to muster sufficient force to defend Washington. Rioting erupted in New York:

[*Robert Hoe III, to his father, 15 July 1863*] . . . *There is much excitement & a great many rumors flying about, but I think matters are very much exaggerated & we shall be quiet by to-morrow. It seems to be nothing but a mob of thieves & pickpockets set on to make trouble by the N.Y. so called "Peace Men"—secessionists.*

The shop was threatened yesterday morning & Mr. Tucker closed at once. None of our men are engaged in the riot, which is made up of Irishmen & women bent on plunder. . . . We are not molested here altho [*printers*] *Mr.* [*C.A.*] *Alvord, Harpers & Jno. A. Gray have been warned to stop this A.M. . . .*

Despite hostile activity so close to home, the Colonel perceived Boston to be his main problem. He disliked leaving Bowen (not a company officer, and somewhat regretfully passed over in that regard) in charge of the unsettled shop.

While the New York factory turned out a number of special ordnance machines, the Boston branch had devoted itself to rebuilding, repairing, and erecting presses, especially Adamses, for the New England area. Its work was strictly homefront business: "I have been called upon by a committee of Gentlemen to subscribe, in the name of R. Hoe & Co., to raise a fund for bounties to volunteers in Boston," Bowen had written (27 August 1862). "What shall I say?" The management task now seemed demanding enough for a trustworthy lieutenant—from the Colonel's own family ranks. He had three lads from whom to choose.

Robert's two sons, Robert (III) and Arthur, were 24 and 22 respectively, both enthusiastic about the business. But Theodore Mead was two years older and very deserving of an assignment.

Theodore Hoe Mead (1837-1909) was the first son of sister Elizabeth, who had married Enoch M. Mead. Descended from Mead families in Massachusetts since 1631, Enoch had been a prominent New York dry goods merchant who fortuitously retired just before the 1837 depression. Theodore, born that year, grew up with his Hoe cousins, dashing in and around New York with them in their smart buggies, wary only of the Bowery gangs whose toughs enjoyed baiting uptown youths. Mead was educated at a private New York school and at 16 entered his uncles' growing firm. After working four years he journeyed abroad, spending three years in Europe, where he became fluent in Spanish, French, and German. Back with Hoe from 1860 on, Theodore was polished and ready.

Alfred S. Bowen, after a lifetime of service with the firm, was persuaded to retire as superintendent of the South Boston works about the first week in July, 1863. He resided for many years in Bergen, New Jersey. Peter Hoe and his promising nephew were both dispatched to Boston to see at first hand how serious was the shop's altercation. The Hoes had never before experienced such difficulty with employees. A careful analysis was made:

> [*Mead to Stephen Tucker, 23 July 1863*] . . . *I found matters here in some respects in a worse condition than I expected; on the other hand some difficulties had been exaggerated. For instance the disaffection among the men was entirely useless and gave me no trouble after the first few days. They are a very good set to take them altogether*[†] *but had always had their own way, I imagine. Uncle Peter can tell you what changes I am now making. There are two which I am happy to say, are already completed: one, the change from dirt to cleanliness and from idleness to industry—the next thing is to introduce system into the industry—an element that has been often wanting here, for Mr. Mitchell does not know what it is, and Mr.* [*Joseph*] *Bond's feelings on the subject have not yet had a chance to develop into very clear ideas. I proceed carefully, however, acting mostly from behind the scenes.*
>
> *I hope you will take a run on here before long. . . .*

Despite his youth, it would appear that Mead possessed sound judgment and almost military efficiency in reorganizing the shop. Striking at slovenliness, idleness, and chaos could not have been more appropriate for restoring morale, discipline, and sense of mission.

Like many a good commander Mead demonstrated too that loyalty works both ways. There were things he could do for his men. He wrote to Isaac Adams, in retirement in New Hampshire, complaining that the South Boston factory, still on lease, was frequently inundated with flood waters and mud because there was no drainage. Adams again graciously admitted that "he supposed the water always stood within six inches of the floor of main building," and promptly commissioned his brother Seth to install drains:

> [*Mead to Tucker, 31 August 1863*] . . . *We are about reduced to order again but shall be annoyed a good deal by Mr. Adams draining operations which are to commence immediately. The men are very unsettled. Every body tells me that there never has been so much trouble with hands before.* . . .

About the same time, with morale and efficiency as justification, Mead trimmed the staff from 80 at the end of July to 73 a month later—no doubt discarding some grumblers and latrine lawyers whom he suspected of earlier complicity. In New York his uncles had been so distressed they were guardedly hopeful:

> [*To S. P. Rounds, Chicago, 28 July*] . . . *In our Boston works we have encountered so many obstacles that we have almost discouraged but having recently made there a complete change in many respects, we will in future have no difficulty in supplying the demand for Presses manufactured in Boston.* . . .

The Massachusetts works would in fact soon bristle with sharpness. Mead next found time to venture to Boston's prestigious printing establishments just across the Charles River. He recounted his confrontations with the moguls of American book publishing:

> [*Mead to New York, 27 August 1863*] . . . *I this morning made the tour of the Cambridge printing offices with that package of cyl. press work* [*samples sent from New York*] *which made quite a sensation. I think there is an order or two to be got for small sized presses before long—if they were New Yorkers I should be certain of it. If you can send me another assortment* . . . *I am to see Mr.* [*Henry O.*] *Houghton again on Monday & should like to have them before that time.* . . .

But Mead's cockiness soon vanished:

> . . . *I was astonished to see my name among the conscripts; would rather send a substitute (an exempt) than pay the $300. & will see if they are to be had here & at what price. Will you please see at what price they can be got in N.Y. as I presume they are plentier & cheaper*

there & there would not be the trouble & risk of sending him on by R.R. How much time have I to make these arrangements & is it necessary in N.Y. (as it is here) to report in person to the military authorities? . . .

This had not been Mead's first brush with the military. The year before he—and his cousins—had successfully met the problem, or so Robert III thought:

31 Gold Street
NY Aug 26/[18]62

My Dear Father & Mother,

We did not reach New York on Monday until after Eleven o'clock, owing to stoppages on the road, so I almost wished I had remained [at Cross River] to take a fish—or two. . . .

With regard to the Drafting we have completed the list, with 12 names [indispensable Hoe workers?]. Mr. Van Vechten [company lawyer] was in & wanted $100 each, down, which I gave for Theodore, Arthur & myself, as the rest have all done or promise to do. I think we shall have no difficulty in finding substitutes & engaging them. There is an alien in the saw shop who thinks of enlisting & whom I shall try & engage in the morning. The boys Theodore referred to I think are both unsound as they tried once to enlist & were not received as their wind was not good for a "run." . . .

I hope we shall have volunteers enough to avoid draft or at least to lessen our chance some what. Two men from the shop have been in to-day for the bounty offered of 25$†. . . .

With love to [sister] Laura & all,

Your aff. son
Robert Hoe Jr.

To Mr. & Mrs. R. Hoe
Cross River, N.Y.

Mead nevertheless lost no time checking the manpower situation in Boston. Returning a Spanish translation the next day to New York, he also reported the distressing effect of bounty subscriptions (such as that to which Hoe had probably contributed):

[28 August 1863] . . . *Substitutes here [are] about $400 owing to the high state & town bounties in some parts of New England, which are paid to drafted men whether serving by substitute or in person. Brokers in the article tell me that substitutes can be had in N.Y. for little more than half what they get here. The best place they say is the neighborhood of the Battery. . . . I would not have a deserter. . . .*

[P.S.] I would give more for a stout able bodied man than for a poorer one.

The frustrations of business added to his own discomfort. Mead had to explain that the company's railway presses would be delayed two months, as all the local foundries were doing "dry sand work for government—propeller blades and the like." His research continued:

> [31 *August*] . . . *I am told by brokers here that by taking a little pains substitutes can be found in New York without difficulty at about $200. That is all that was paid here until the Eastern bounties raised the price. If necessary however I should prefer paying a little more than the exemption fee to send a stout foreigner in my place. . . .*

This paragraph was Mead's last mention of his draft problem; unquestionably a husky (German) immigrant was found to replace him in battle. But it may have come up again the following year, as it did for Robert, who in August 1864 reported his new substitute was named Leonhardt.

On the home front Theodore Mead continued to discharge the multifarious duties of company officer. Tucker complimented him for some mechanical innovations, and his uncles thanked him for several translations, approving his occasional sharp replies where called for: "Mr. Santo's letter certainly seemed to me of the last degree of impudence. I will write him. . . ." Mead promised (3 October 1863). Employee problems continued:

> [*Mead to New York, 5 September 1863*] . . . *Our best apprentice who was running a planer at $3.50 a week has left us & gets $1.00 a day. Another who has been here 6 mos. is very discontented with $2.50 a week & is really worth more as he is strong & quick. Other shops here are giving more—do you recommend our doing so too?* . . .

> [24 *September*] . . . *The Herald folks have managed to ruin one of the cylinders on their 6 feed press.* . . .

Mead also calculated that the Adams presses were costing 17 percent more to manufacture than their list price.

If Mead seemed unusually reluctant to go off to war, another mitigating factor had some bearing: on 3 December 1863 he married Anna Rebecca Johnson of Philadelphia, daughter of Hoe's old business colleague. The groom was back at work in January:

> [*To New York, 8 January 1864*] . . . *I have today hired a pattern maker who will be I think very valuable. He is thoroughly acquainted with all our patterns & for fifteen years did all the Adams press woodwork.* . . .

The home office objected, however, and Mead was obliged to let the man go. He threw himself back into the Cambridge sales area:

> [9 *January*] . . . *Mr. H. O. Houghton today orders two more 27 x 41—4 roller presses at $2265 each. He has a large contract & will have to fit up a bindery. He talks of giving us the order but is going to look about first.* . . .

Houghton had won a boggling contract from G. & C. Merriam to print and bind thousands of dictionaries without even possessing the facilities to do so; with this impetus his shop soon became the best in the country. His incidental publishing ventures would stabilize in 1880 as Houghton Mifflin (of which the Riverside Press became a subsidiary), and increasingly he performed his fine presswork on Hoes.*

Mead also tried to reach A. K. P. Welsh, another fine Boston printer:

> [*13 January*] . . . *Welch*† *Bigelow & Co. I imagine are very short* [*of cash*]. . . . *Mr. Welch is hard to find—I begin to consider him a sort of myth.* . . .

Mead nevertheless called attention the next day to $2317.29 outstanding, and somehow managed to mention that Welsh, Bigelow also owed money to cousin Frank Dodd. When reached, Welsh paid only $1000 on account, and tossed back the discomforting intelligence that he was going to claim some reward for a Hoe invention in which he had participated. Welsh had bought a new press and specified special inking apparatus which, although executed by George Mitchell of the Boston shop, he considered his own idea. Mead reported:

> [*16 January 1864*] . . . *Mr. Welch*† *it seems has applied for a patent. Is there anything on which he can base his claim even supposing him to have suggested the improvement to Mr. Mitchell?* . . .

This matter was solved amicably enough, and Welsh was thereafter engaged, for several years, in testing similar press improvements:

> *Dec 23rd 1868*
>
> *A. R. P. Welch,*† *Esq.*
> *Cambridge, Mass.*
> *Dear Sir,*
>
> *Some one suggested putting adjustable iron bearers on the perfecting press. Would you be willing for the sake of having them to dispense with the leather which now covers the bearers?*
>
> *With regard to the shifting tympan cloth do you not think on the whole, considering the difficulty in making it slip over the blanket, especially when overlays are pasted on it, and the possibility that the marks showing the edge of the successive impressions may show on the final work, that it would be better to put, say, five covers on a roll in the cylinder and stop the press for three or four minutes to roll an entirely clean one on than to have a ratchet working all the time? The self acting apparatus for winding up the cloth would give the pressman more trouble than stopping a few minutes now & then.* . . .
>
> *For the other press if you will try it I will have a distributing cylinder made to take ink from the fountain and distribute* [*it*] *before it is laid on the table.* . . .
>
> *I had better remark that in order to diminish the friction of the*

* See *Books for the Millions.*

shifting surface cloth we were obliged to cover the blanket and overlays with a smooth sheet of paper.

Yours very truly,
Rich[d] M. Hoe

This test work was fairly complicated:

[*Richard Hoe to Welsh, 5 January 1869*] . . . *We have commenced putting in the reeling apparatus & I desire you would make a trial of it. . . . You will have enough blanket so as to change every hour. There will be enough on the reel for 12 to 14 surface cloths. We shall sew a rod in both ends of the blanket so that you can very quickly take it out to get at your overlays. . . .*

[W]*e think we can arrange the sheet flyer to work as you wish; but wish to know if you prefer that we shall arrange the fly that the offset sheet will lay on top; or if it will make no difference to you. . . .*

I enclose you a sketch of our proposed addition to the cut press. You will see that before the ink is put on the tables, it will have been more fully distributed than now. . . . I propose to make this apparatus so that you can easily attach it to your machine & take it off if you like; thus with both systems you will be able to tell me which is best. . . .

These adjustments to a new perfector continued another year, at which time the test equipment was effectively written off:

[*Peter Hoe to Theodore Mead, 20 July 1870*] . . . *I wrote Mr. Welch*[†] *this morning asking for a bill of the presses he proposes to return, and size of platens. It would be just as well to let the presses remain in his shop until wanted if he would like to use them. Nothing new. . . .*

Much of the experimental work done in Boston would end, however, with the closing of the branch itself.

Theodore Hoe Mead had been made a member of the firm in 1868 and continued to exercise many legal duties until he retired in 1900. Like his uncle, Mead was active in social reform: he was a manager of the New York House of Refuge on Ward's Island, a member of the New York Prison Association, and concerned with other philanthropic work. He had been an active sportsman but ill health forced him to reduce his regimen in 1885 and he began to travel again.

With a "natural taste for *belles lettres*," Mead wrote copiously: on horsemanship for women, on a free public library system for New York city, on maintaining health without medicines, on restoring "the original melody" of the English language (lost by Americans' tendency to clip and obscure vowels). With his wife Anna he had six children, one of whom, Lawrence, worked six years with the company before going on to a railroad engineering career.

By the end of the 1860's Boston conditions had deteriorated somewhat. Inflation (machinists' wages were 75 percent higher in 1869 than 1860), the fragmentation of work, the distance, and the gradual reduction in Adams orders, were factors in the shop's decline. A final event was almost welcome:

> In 1868 a new, wide street, called 'Broadway' was opened in South Boston, and ran directly through our factory, completely destroying it. As there was now ample room in the new factory building just completed here, it was decided that instead of taking other premises in Boston it would be more economical and convenient to have the works in New York, and they were accordingly brought here early in 1869.[14]

Stephen Hoe was sent to close the shop ceremonially as Seth Adams presented the deed of the property to the city.

Hoe agent William H. Kerkhoff thereafter made frequent calls throughout the New England area. On one 1872 trip he arrived at Boston's Tremont House and quickly telegraphed some orders to New York. Then—hastily in pencil—bedlam:

> *Tremont St. Boston*
> *Oct 14th/*[18]72
>
> *Dear Sir*
>
> *I am almost compelled to write to you and having but a few moments before mail time, have to write quickly, please excuse all that may follow.*
>
> *This is the case—*
>
> *Pat Donohoe*† [*Patrick Donahue, Boston*] *Pilot office*
>
> *Campbell is here trying to sell him a press. Howlnett*† [*Hartnett & Salomons*] *is trying to sell him a Double Cylinder No. 5 for $3500 put up second hand—the thing is to be settled within 2 hours. Donahoe*† *intends to get to work in a temporary building with temporary press—Campbell offers to lend him a press until he can build one of course with the provision that he builds his permanent press.*
>
> *Hawlnett*† *is on the same move. He offers to take away the cylinder if Donohoe orders his next press through him. You will see that you are wanted here, for I am in a false position with my limited responsibility but this time I must make a plunge and burst if I must, my reward will lay in my notice—*
>
> *I have got hold of "Blue"* [*Maturin Ballou, Boston Globe*] *and this is the result*
>
> *We give Mr. Donohoe a new No. 5 Double Cylinder with 2 rollers back of each cyl like I.S.M. Smith's & Co. which is four usual cylinders for all I know, put up in Boston for $6250.00 dollars*† *and we take the same back for 75 per cent that amount in exchange for a larger 4 cylinder or other press any time within one year, breakages excepted. The press to be up here in 14 days (your list says ready in 1 week).*
>
> *Now for the reason for this: Campbell is making a great effort here*

claiming that his press (new) is just the thing for papers having "cuts" in and is willing to make any effort to get his press in—"Blue" wants 2 Adams presses which he (Campbell) wants Donohoe to replace with his book press taking all risks if they don't suit.

Hawlnett wants his press in so that he may secure the trade for all that follows including the 2 Adams presses.

Donohoe's paper must be printed at once & he decides this as I said in 2 hours. I have taken the order as mentioned. Please ease my mind by a telegram as soon as you get this, and say soonest day said cylinder can be sent.

You still have the controlling power of all this for if you telegraph me that press can't be supplied why it shuts up the thing at once.

One thing I would say that your communication must be decisive—and though you will kindly give me hints and instructions and I shall bend every nerve to accomplish this business in your way. Yet you must say yes or no to my arrangement, so that in case I cant do what you want I can fall back upon doing what I know I can do.

Everything else is well & I hope to send many orders. I have seen Mr. Dow again & received a written order. Campbell has tried very hard in this quarter.

I shall have something to say to you about Blue when I may see you.

You must please imagine the weight of my apology in send^t you such an epistle.

Yours faithfully
W. H. Kerkhoff

Soon there would be bedlam for all of Boston. An insurance report had warned that Boston (like Chicago) was dangerously vulnerable to conflagration; its thriving business district, furthermore, housing large numbers of paper merchants as well as printers, publishers, and booksellers (Lee & Shepard's bookshop had just suffered a serious blaze) was poorly served by inadequate water mains. To make matters worse in 1872, an equine epidemic had incapacitated Boston's fire horses. Saturday morning, 9 November, fire broke out and raged for 18 hours, putting over 700 buildings—more than 900 businesses—out of commission.

Hoe's Boston friends and customers had nobly demonstrated their allegiance when the firm asked for testimonials on behalf of the Colonel's request for patent extension. Now it was Hoe's turn to reciprocate. In refined Boston, much more than Chicago, the substantial graphic arts industry was absolutely stunned by the destruction: the city was second only to New York in printing and publishing activity. Some 23 printers in town were wiped out; all clamored for new presses and other equipment to start anew. Dozens of telegrams and letters bombarded Gold Street.

Thomas, Carroll & Company, machinists and press erectors "temporarily at 390 Federal Street" (where they remained), wrote on Rand, Avery stationery:

> [*To R. Hoe & Company, 12 November 1872*] . . . *Will you please furnish us with a large cast gear and shaft and stud fitted and connecting rod crank unbrazed* [*for an Adams*]. . . . *We had the broken gear taken to our shop last Saturday afternoon and it is now beneath the ruins as we burned compleatly*† *out by the greatest fire ever in Boston.*
>
> *We expect our insurance is good as we are in the Royal of Liverpool, England. We feel that all the stock and tools are* [*an*] *entire loss.* . . .

The press awaiting the lost gear was probably gone too; in any event Thomas, Carroll would have done much better ordering whole presses (by telegram) instead of replacement parts. The Boston printers were themselves far more compelling:

> [*J. H. Bufford, telegram 12 November*] . . . *Can you fill my order for a Number two 2*† *Press, Answer immediately.* . . .

But Bufford sent an anticlimactic letter the same day: "The fire did not touch us & we want more presses at once. How many & what sizes have you on hand? Answer J. H. Bufford." His greedy demand did not seem as urgent as others':

> [*Moses A. Dow, 11 November*] . . . *My presses are all destroyed, and I wish to know how soon you can furnish me with four of the largest Adams presses, 34½ by 49½ platen, sockets all brass as is usual with your best presses.* . . . *I can do no business till I get them all.* . . . *None but Adams presses will do.* . . .

Dow's *Waverly Magazine,* a compendium of amateur poetry, seemed low in priority, but its proprietor was insistent. He telegraphed the next day, repeating his plea: "Let me have four large Adams presses. . . ." Mail continued:

> [*H. W. Dutton & Son, 11 November*] . . . *What have you for a press suitable for our use?* . . .

To help understand the actual situation in Boston, Theodore Mead was dispatched there for the emergency. "Please send Mr. Kirk-Hoff† as soon as possible. Transcript presses saved," he telegraphed on 12 November, countermanding Dutton's earlier query. When Kerkhoff arrived the next day he was promptly corralled by Dow, who must have hunted him down. "Make for M A Dow four Adams presses thirty five by forty nine four roller at sixty seven hundred dollars each. Cash on delivery," he telegraphed on the 13th. The poetry would not wait.

After a few days looking around the ruins, Kerkhoff found much woe among printers whose insurance companies were not paying. "[Lithographers] Armstrong & Co. have tried hard . . . [but can raise no credit] owing to the backwardness of Insurance men & the fact that their best friends are also sufferers by the fire. . . ." "[Charles] Crosby & Co. . . . get *none* of their insurance, and it looks as if you [Hoe] had the bulk of their capital." Although Bufford was not burned out he could not raise money from accounts that were. "The difficulty of getting suitable rooms [for shops] will keep orders back," Kerkhoff further warned. Even the most optimistic scarcely knew where to begin.

In 1872 the Boston *Transcript* had just moved into splendid new quarters—"one of the ornaments of Boston," said Frederic Hudson—on Washington Street, the city's "newspaper row." These new premises were agonizingly destroyed in the blaze that ironically spared the *Boston Globe,* which had started that very year by taking the *Transcript's* old office. At the *Transcript* Kerkhoff had actually found its "double-six" machine salvageable, and Dutton wrote Hoe that they were returning it for rebuilding:

> [*To Hoe, 20 November 1872*] . . . *We hope and trust that you will put it in hand at once, and work* <u>*day*</u> *and* <u>*night*</u> *till rebuilt, and return it to us at the earliest possible time. You know our situation . . . in our present calamity. Our entire establishment is in ruins. . . .*

The *Transcript,* whose circulation was only about 17,000, threw itself on the *Globe's* charity for press time. For two months this cooperation continued (the *Globe's* own budding circulation was only 5000), a condition always uncomfortable for competing papers. Dutton soon acknowledged Hoe's assurance that he would do all he could, adding,

> [*To Hoe, 26 November*] . . . *We are at the mercy of Mr. Ballou, and he may at any day say to us, we must leave. . . . The* [*press*] *beds can be repaired here. Mr. Hartnett at a great risk removed them from the cellar filled with battered type, and says he can restore them here, to fit the rings as originally constructed. . . .*

The *Transcript* survived and continued (until 1914) so conservatively it refrained from using photographs long after the Levy screen had made them possible. Its generous host was quite another operator.

Maturin Ballou (whose surname so puzzled Kerkhoff he put it in quotes) was more a theatrical impresario than publisher. From his stationery:

> [Letterhead]
> Office of
> Ballou's Pictorial Drawing-Room Companion,
> Flag of the Union, and Ballou's Dollar Monthly.
> [Back of stationery]
>
> [The Drawing-Room Companion is] a record of the beautiful and useful in art. This paper presents in the most elegant and available form, a weekly literary melange of notable events of the day. . . . In

> every respect it is an illustrated record of the times. . . . [No advertisements.]
>
> [The Flag of the Union is] an elegant, moral and refined Miscellaneous Family Journal, . . . a paper ["of the mammoth size"] for the million. . . . It contains the foreign and domestic news of the day, so condensed as to present the greatest possible amount of intelligence. . . . [No advertisements.]
>
> [Ballou's Dollar Weekly is] a miracle of cheapness [containing in 100 pages of reading matter in each number] news, tales, poems, stories of the sea, sketches, miscellany, adventures, new discoveries, biographies, wit and humor, . . . also spiced with a record of the notable events of the times, of peace and war. . . . This work is intended for the million, north or south, east or west. . . .
>
> The Weekly Novelette [is also published], . . . a rich casket of jewels, [quarto, 16 pages].[15]

With such purple prose and obviously wishing to be all things to all readers, Maturin M. Ballou founded his daily newspaper, the Boston *Globe,* in 1872. He had dabbled in many publishing and printing schemes. But he was not so familiar with crass journalism as with his literary pursuits, and the paper lost money at an alarming rate. The enterprise was saved, however, when General Charles H. Taylor (1846-1921) became editor within the first year. Having learned printing as a boy, Taylor distinguished himself in the Civil War and gained wide newspaper and political experence. In 1869 he had started *American Homes,* a successful 10-cent monthly which was destroyed in the Boston fire.

For the *Globe* Taylor had to use every resource he could to keep solvent; in 1877 he daringly cut the price from 4 to 2 cents and announced his support of the Democratic Party. Circulation jumped to 30,000 in three weeks and to 150,000 in three years. He made many improvements to his mechanical department, and at the time of his death was considered the dean of American journalists. When Edward C. Bailey sold his Boston *Herald* in 1869 Taylor got him for the *Globe*; when the bankrupt Boston *Post* could not take its new 1876 press (winner of a gold medal at the Centennial), Taylor offered $15,000 and got it. The Boston *Globe* celebrated its own Centennial in 1972, long a Hoe customer.

Rand, Avery & Company, whose shop had escaped the November fire, was working around the clock to accommodate Boston's urgent printing requirements almost singlehandedly. With a reputation as America's finest printers, the firm had corralled the country's foremost customers, particularly Boston's scholarly book publishers. George C. Rand in 1852, with eight power presses, had once before worked night and day to turn out 300,000 copies of Stowe, *Uncle Tom's Cabin* for a public eager to learn of slavery's miseries.

Just eleven days after the great Boston conflagration Rand, Avery itself caught fire, a stunning blow to the partners' sensibilities and a staggering loss to many publishers with plates and other materials stored there. Kerkhoff, who had at first raced from one distraught printer to another, telegraphing and writing well into the night to keep Gold Street informed daily, now had time to devote all his energies to Rand, Avery reconstruction:

> [*Letter, 20 November 1872*] . . . *It is almost impossible to present the position of their presses to you, as the fire men will not allow any one to enter the Cylinder press room, but 10 Adams presses have fallen thro. two and three floors, and the balance appear to suffer but slightly by water. The heat appears to have been considerable.* . . .

> [*Telegram, 21 November*] . . . *Send Rand & Avery four Adams press machinists one of them a good man to assess damages ten large Adams have fallen through floors but little heat remaining presses injured but slightly.*

Rand had considered himself responsible for several press improvements, and in this period of great loss his inventiveness seemed to console him. A week after the fire he was understandably "at home, sick"—but studying his plans for special universal joints which would "connect and run two lines of shafting of different angles and shafting on different levels." Despite a fully refurbished shop, Rand, Avery did not long survive the eroding decline of nineteenth-century printing standards.

With the final spray of water on Boston's ashes the fire turned into a morass of claims and settlements. R. Hoe & Company was creditor to many shops which could no longer produce without further advancements. The messages were plain: "Messrs. Rockwell & Churchill want the presses but ask you to waive one hundred dollars per quarter," Kerkhoff wrote (20 November 1872). Undoubtedly the company did so, and probably even helped old press-erector colleagues.

R. W. Hartnett and M. A. Salomons ("Manufacturers and Dealers in Printing Presses and Bookbinders' Machinery, . . . Special Attention given to Repairing Printing Presses") wrote on 19 November: "Many of the printers now purchasing presses &c are depending upon their insurance money to pay for their presses & when that is received they will pay Cash in full. We are furnishing them from our stock, on liberal terms of payment. . . ." They hoped Hoe would likewise give them liberal terms too. The cooperation which must have followed unfortunately led to reprehensible excesses:

> [*Robert Hoe III to Robert, England, 26 December 1873*] . . . *Hartnett & Salomons have swindled us dreadfully. After a regular fight we settled the a/c & everybody now says "have no further dealings with them" so that matter is settled.* . . .

Boston's widespread destruction enabled the Hoes once again to practice their skills in complete shop design. Charles G. Greene had founded the *Morning Post* in 1831 as a $4 yearly daily. It prospered as a Democratic organ, becoming notorious at the end of the century for yellow journalism. But the proprietors knew how to get a bargain when they shopped for a press:

Boston, Feby 3, 1874

Messr. R. Hoe & Co.
Gentlemen,

You sometime since kindly offered to lay out our cellar for us. Our present cellar you arranged so acceptably that we cannot refuse to seek your advice and accordingly [send] a plan of cellar in new building; the blue marks indicate erasures.

In the basement we must get, besides mail & folding rooms, two presses, one or two boilers and our double engines. The latter take about 12′ x 10′. The boilers and engines, we think, would be advantageously placed as indicated, considering the flue & passageway. Please favor us sometime this week with your views returning the plan and oblige

Yours Respectf
Beals, Greene & Co.

The effects of a miscalculation in the planning of such a pressroom could be serious. This unlucky event happened to the Boston *Traveler,* which in 1857 had traded a ten-cylinder Lightning for a new six feeder, and in 1874 proudly ordered another new machine. Thomas, Carroll & Company were called in to help erect the machine and applied themselves so assiduously Hoe was inclined to disallow the bill:

THOMAS, CARROLL & CO.
Printers' and Bookbinders' Machinists,
No. 390 Federal Street

Charles H. Thomas
Joseph F. Carroll

James T. Wadsworth
M. Wagner Brush

Boston May 25 1874

Messr. R. Hoe & Co.
New York

Sirs,

Yours of the 20th inst. is in hand and we are rather supprised† at the reckoning you have made of our time on the Traveler work. We have our time sheet to refer to, and if you will reckon that part of the time from Saturday 5 p.m. till Monday morning you will find that part alone is nearly double your figures; it is 1½ time till 12 midnight and double time till Monday morning, which will be for one man only, and we had three men there during that time and on saturday† and saturday night we had other men there besides. And we had to alter the shipper bracket, Platform rest on No. 5 feed rail, *[and on] No. 4 Roller sockets make a stud for Drop roller feed gauge to take the place of one lost. Cut of[f] platform collumns,† which were ordered to be done by your forman.† Does your forman know anything about unboxing and getting the turtles up in the compossing† room and old ones down and boxing them?*

In regards to helpers we would say that we received orders and got seven riggers and your forman kept them till sunday† night, and you have a copy of Riggers bill as we have it and that tells their time.

We had men there from the unboxing till your men left Boston.

We enclose you another bill showing you the number of hours each day.

We have done considerable adjusting to the press since [for] which we have not charged you.

Hoping the new bill will meet your approval we remain

Yours &c

Thomas, Carroll & Co.

Aside from the high time charges, Hoe should have found most disturbing the extra work occasioned by faulty and missing parts. Correspondence the following fall reveals the situation:

[Worthington, Flanders to R. Hoe & Company, 10 October 1874] . . . Owing to your mistake in making our press several inches higher than in the plan furnished you, we were put to a large expense and great inconvenience; in building a new drain, over one hundred feet long into an adjoining street, in order to drain our press room. Even now the drain does not drain off the water entirely; we are obliged to use a pump to draw the water from under the press.

We have been at fully $1200 to $1500 actual expense rendered necessary solely by your mistake, and we claim that one thousand dollars at least should be deducted on that account from your statement. . . .

Planning care and attention would obviously not be misspent for such installations.

By the end of the hectic decade Boston was back to normal. Hoe sent a new agent, J. C. Brocklebank, to New England, pointing out that among printing jobbers only Phelps, Dalton & Company received goods on consignment, but all the others must be visited regularly. His instructions were classic:

[To Brocklebank, 23 April 1878] . . . We want you to take a thorough canvas† of the trade and learn just about what proportion of orders for woodwork, cases, cabinets, stands, etc., we receive, and why we do not have them all. Learn what prices they arrange with them to sell our goods exclusively. We do not want to open consignment accounts, but intend if possible to fix prices so we shall control the Eastern market in this line.

Of course you will be careful not to urge parties not† to buy elsewhere, who already send to us for everything they want. . . . You will understand how to handle this matter and can undoubtedly send orders for these small goods. Never overlook them. They do not amount to so much as a press, but attention to these small matters often develops good trades, and helps to keep our men busy. . . .

Sales of Hoe equipment, of course, were brisk when agent and home office worked together. John Wilson, Jr. wanted quality equipment for his Cambridge Press. He was considering a Hoe two-revolution press:

[*Theodore Mead, New York, to Wilson, 17 May 1878*] . . . *They are running here in N.Y. as he* [*Brocklebank*] *told you, on very fine work at a very high speed. For instance the letter press of Scribner's Monthly and of St. Nicholas is printed regularly at a speed of sixteen hundred & eighty per hour and the wood cut side of these magazines —a part of the edition only—at twelve hundred & fifty. . . .*

It appears he bought the machine. Among newspapers, however, it was not so easy to please and keep restless customers.

As late as 1878 the *Herald* was the only Boston newspaper large enough to require stereotyping. Its circulation had reached 90,000 in 1861, settled to 65,000 momentarily, but after a decade was at a steady 100,000, its Sunday edition (begun 1861) among the most prosperous in the country. Hoe was particularly anxious to have proprietor R. M. Pulsifer see his new web machines, but correspondence was curiously discouraging:

[*To Pulsifer, Boston, 17 July 1874*] . . . *To make a Web Press of the size you require, we should have to get out new patterns, and our price is $25,000--, put up* [*at*] *your office.*

The press will run as rapidly as the paper will admit. We think you can safely rely upon a net production of 10,000 per hour, or 20,000 of the "Herald" per hour.

The dimensions of press are as follows, Length 21 ft., width over all, including driving shaft, 10½ ft., height 7 ft. . . .

We hardly know what sum to name in trade for your old Six Cylinder, as we already have two of them in stock, and the prospects are that quite a number of these presses will be thrown into the market. . . .

Having vetoed the Hoe proposition, however, the *Herald* a year later was obliged to discard the two six-feed Lightnings unceremoniously. They were apparently exhausted:

Boston May 17/[*18*]*75*

R. Hoe & Co.
Gentlemen

My object in coming here this morning was to learn what shape the two six cylinders that came out of the Herald office was† *in. As I knew that one of them was offered at a low price to the Providence parties.*

The[*y*] *have not been repaired nor are the*[*y*] *likely to be for some time. And Hartnot*† *is not in a position to offer them only in the shape the[y] are in. Now I am prepared to talk to them understanding about the Herald* *Presses.*

The parties whose card I enclose [not present] have come here lately from Rochester and will want a steam Lithographic press in the fall. I met Armstrong to day on the street; he says the work is coming in lively and will want another press soon. . . . Mr. [Louis] Prang has called on me at the Parker House; canot† say what the result will be at this writing.

Respect
T. Cochran†

For their recent perfidy Hartnett & Salomons were apparently not allowed to pursue a bargain without some harassment. But what must have intrigued both Cochrane and the Colonel was what the *Herald's* new press was going to be like.

Pulsifer had shopped at home for a new, modern machine; it was designed by one of Massachusett's own mechanical wizards. Thomas Jefferson Mayall (1826-1888) worked as a lad in a Roxbury paper mill; he invented a rubber belting system, and a machine for printing wallpaper and calico. He had a coffee hulling machine and a slew of armaments (armor-piercing shell, automatic cannon) to his credit among 200 U.S. patents and 70 in England. He was, said a later prospectus (almost posthumously) for Mayall's cable, electric, and pneumatic elevated railways—"The systems of the future,"—"One of the leading and most prolific inventors of this country of prolific inventors." [16]

Such genius seemed to have no trouble putting together a high-speed press (with his 1867 "accumulating cylinder") much to *Herald* specifications, but his own ancillary folding equipment on the machine would not work. Mayall discussed the perplexing problem with Luther Crowell, and the yankee friend obligingly took a look at the installation. Crowell with genuine surprise must have seen that his 1873 mechanism for folding a web of paper in a rotary motion was exactly right, and he applied it to the press with consummate success. Now he realized too that he had a valuable commodity.

Crowell had heard of R. Hoe & Company, manufacturers of newspaper printing presses; perhaps he repeatedly passed their well identified South Boston shop. Why not offer them the folder? He dashed off a letter almost as he finished fitting the mechanism to Mayall's press, and got an answer the very next day. The reply was not encouraging, but Crowell could not know that so rapid a response alone indicated fervid Hoe interest:

July 24, 1874

Dear Sir,

Your favor of the 23[d] inst. is received and contents noted.

We cannot say whether your invention would be of value to us without knowing more about it. If you send us a model, description or information whereby we may judge of its utility, we shall be pleased to make examination and report to you.

Very truly yours
R. Hoe & Co.

L. C. Crowell Esq.
Boston, Mass.

Crowell unfortunately was probably completely stymied by such an arrogant requirement. "An ingenious Yankee skipper who could not even read a drawing and had to work from patterns or models,"[17] said *Scientific American,* Crowell could scarcely put his thoughts cogently to paper and was unlikely to build a working model when the machine itself could easily be seen. Nothing transpired. Crowell had quite innocently solved a problem baffling to paper engineers and press manufacturers alike. He continued to improve the mechanism over the next few years, but, having no means to exploit it commercially, did little with it.

In England Edwyn Anthony was at work on a similar device, and intimations of Crowell's success were at least disheartening. E. Anthony & Sons, New Bedford, Massachusetts, had bought Hoes for their *Evening Standard,* endorsing them happily for the Colonel's 1868 *Memorial.* Edwyn, one of the sons, settled in England, where he became part owner of a Hereford newspaper; with William W. Taylor he developed press folding equipment about 1875, one device, for example, turning a web to print the second side, patented the next year. The Colonel admired this apparatus and (in 1877) got their American patents assigned to R. Hoe & Company.

Some anxious inquiries from Anthony necessitated a long reply—which the Colonel attempted without even remembering Crowell's introductory correspondence two years before.

Nov 24th 1876

Mr. Edwyn Anthony

. . . I do not think you seem at all to appreciate the situation here. Although Mr. Crowels† patent is only for trimming the paper for the purpose of laying ½ the paper on to the other half making what may be called the first fold & he evidently only contemplated folding & nothing farther—in fact we know that was all he was called upon to do, for we have satisfactory evidence that he was at work on a web machine of a competitor of ours in Boston; [*he was*] *called in to make a success of what seemed to be a failure & we are informed that to a certain extent he has succeeded. But the difficulty seems to be this, that he has done just enough in trimming the paper over a smooth bar or smooth fixed roller at an angle with the direction of the papers track in his folding device to make it impossible to work your patent except subject to him unless you can obtain your* [*U.S.*] *patent by proving that <u>not your invention</u> but your <u>English</u> patent was granted before he perfected his invention.*

From what we are told we fear as he knows exactly [*to*] *what time he must be able to date his invention; that it is barely possible if they or he is obliged to draw some on his imagination to antedate your patent. As to priority it might be done. We none of us know Mr. Crowel personally or any of his friends & he may be above thinking of doing a dishonest or dishonorable act & I trust it is so & he may be able to show clearly that this idea was conceived by him before the sealing of your English patent & if he does so we cannot see that anything can be done by you or us until this obstruction can be removed. Of course he cannot use what is claimed in your patents beyond that shown in his, but although his patent here is of date later than yours he can go before the Comr. of Patents & prove the date of his invention & so prevent our having a broad foundation claim & we must be in a measure subject to him. I send enclosed a note from our Patent Solicitors for you to see what they think. We have directed them to see Mr. Hamson [Anthony's agent?] & if they can do so to aid him.*

The Mister R.M.H. desires to be kindly remembered to your Brother & family & self & wife.

Yours very truly
R. Hoe & Co.

[8 Florence Terrace, Shepherd's Bush Road, London]

[Added in another hand] P.S. If Mr. Crowel can prove to the satisfaction of the Com^r of Patents that he invented all that is contained in your invention or Patents in England on his applying for a reissue of his American Patent he would secure the right here for all in your Patents. I do not think he can do this but we must wait to see. I have had a drawing made for a machine & if this was satisfactorily settled would at once go forward & make a machine for one of our newspapers. Yours, RMH

The Hoe brothers still had not met Crowell, handling the situation mostly by mail. But recollections of basic matters were growing cold: to M. A. Salomons & Company, Boston, asking (6 December 1876) when the "Mayale†" press was installed, and whether it had a gathering cylinder on it. (It did.)

Within these two years, furthermore, the Boston *Journal* was again in need of new equipment:

BOSTON JOURNAL
Established 1833

Morning Journal	Evening Journal
Weekly Journal	Semi Weekly Journal

Published by the Journal Newspaper Company
120 Washington Street.

Boston March 8 1877

Mess. Hoe & Co.
Gentlemen

I had a conversation yesterday with your Mr. Hoe regarding a new press. I find that the only serious objection to the new press which he proposed is the speed. We work on our 8 cylinder every afternoon in one hour 17,280 papers. Could some improvement be made to obtain greater speed[?] It would answer very well for a Morning Edition, but I want a press of the description given me by Mr. Hoe, which will enable us to print both sides (and fold or fly) from type that will have at least 15,000 pr hour and can be run with entire safety.

Truly
W. Clapp

[*R. Hoe & Company to Clapp, 16 March 1877*] . . . *It is easy to promise to make a machine to give the speed you speak of, but we wish to be on the safe side, and, while we think our press will print at the rate of 12,000 per hour* [*we*] *do not like to agree to make one to do over 10,000.*

Possibly we shall attain even a higher speed than 12,000 per hour. We are now making a machine of the kind, & when it is completed shall be glad to show it to you.

If in the mean time you will give us an order for one suited for your paper, we will commence it immediately & guarantee 10,000 per hour. The price to be $15,000, & your old six Cylinder Machine. . . .

[*To Clapp, 20 March*] . . . *For some time past we have had under consideration an entirely new plan for a Web Perfecting Press, to print from a form of type.*

It was not a machine of this character we at first proposed making for you, but since receiving your letter we have given the subject more careful attention, and are now prepared to take your order, guaranteeing it shall print your paper at at least 15,000 per hour, & perhaps faster.

The press will go within the space now occupied by your six Cylinder Rotary, but will not stand as high.

If you decide to give us the order we will go on with the New Machine at once. . . .

[*To Clapp, 22 March*] . . . *The machine we propose to make for you will print from an endless sheet or roll of paper, from type forms, locked up in the same manner as in your Rotary presses. The beds, however, will be of a different curve from those on your Eight Cylinder Machine, but the type may be changed from one to the other with the same facility as from your six to your Eight Cylinder beds.*

The speed named, 15,000, would be that number of perfect papers per hour. We did not propose to put a folding attachment to this machine, understanding you did not require it.

One proposition is to make you this machine, as above, for $15,000. and your old six Cylinder Rotary Press.

We guarantee it to print at a running speed of fifteen thousand perfect papers per hour in a satisfactory manner.

In addition to the regular pressman, you will require probably two men and a boy to run the press. . . .

This rambling barrage was clearly calculated to prevent another Boston coup by "a competitor of ours," a feat that did not often happen, to be sure, but was costly enough when it did. Matters concerning Crowell had made all Boston affairs somewhat unsettling.

Anthony in England too was a pesky genius to work with. In anticipation of getting his patent rights the Colonel began to build his devices. Conquest was kept informed:

[To Conquest, 14 April 1877] . . . Last week we received an order from Boston for a machine of this kind [web fed], which we purpose making after the Anthony plan of turning the sheet & putting on two rolls. This will give a speed of about 16000 per hour without difficulty provided the paper turns, of which the Col. is confident. We are now testing the plan in the works. Mr. Anthony has not yet sent us the U.S. patent & probably has been unable to secure it thus far. Under the circumstances, we think nothing had better be said to him of our having any order. . . .

Because of a discrepancy in the plans, however, Hoe was obliged to reveal to Anthony that there was indeed an order on hand:

April 25th 1877

Dear Sir,

You will be glad to hear that we have been experimenting satisfactorily with some of the details of the New Machine. We have an order for one & are going on with it immediately.

From the directions you sent, we have been unable to make the tapes, in connection with the apparatus for turning the paper, work properly, & will be glad if you can send us a drawing or working model of this.

Also please send us any instructions or suggestions you think may be of value.

Yours very truly,
R. Hoe & Co.

Edwyn Anthony Esq.
"The Elms" Hereford, Eng.

But Anthony was not entertained by the challenge of building a single rival folder to match Crowell's on a rival Boston paper. He was not content with half-hearted measures and refused to let his property slip away. Within the next few months, however, as Anthony's patent registration bogged, the Hoes were closing their arrangements with Crowell. If not a clear draftsman, Anthony was a biting correspondent:

8, Florence Terrace
Shepherd's Bush Road, Nov. 9, 1878

Dear Sir

We beg to acknowledge the receipt of your letters of the 18th of October last, and of various documents and other papers.

We are very much surprised and annoyed at the decision of the Examiner. Having fought only on the express condition that in your opinion and in that of your legal advisers we had a really strong case, and having had from you since its final completion but before the decision the most positive expression of belief in our success, we naturally are astonished at the result being otherwise. And a perusal of the case has greatly intensified that feeling, because it is perfectly clear that the decision does not rest on reason, justice, or law. An appeal would beyond question effect its reversal. We agree with Messrs. Munson and Philipp that the point of law they raise in their letter to you is quite sufficient to upset the decree; and we are further of opinion that the decision may be set on one side on several other grounds. The Examiner decided against Crowell on every issue except with regard to "Exhibit No. 4." Among other things, we think it can be conclusively shown that Crowell utterly failed to establish the genuineness of that document, and that it does not constitute a sufficient explanation of the alleged invention. We shall send in a day or two some notes on the case, and we shall esteem it a favor if you will kindly forward them to Messrs. Munson & Philipp.

Our natural annoyance must not prevent our tendering you most cordial and hearty thanks for the pains you have taken. And we desire most especially and emphatically to thank Colonel Hoe for his exertions on our behalf. We also wish (and we should be glad if you would undertake the office for us) to convey to Mr. Philipp our appreciation of his able argument and of the happy way in which, in re-examination, he set right [a witness], who in one point had become so hopelessly and completely confused. And likewise to Mr. Tucker for his evidence, so clear and untouchable under cross-examination. Most ungrudgingly do we pay our tribute of satisfaction and thanks to all concerned on our side, notwithstanding that by the purchase of his patents, in place of Crowell, you yourselves have become our opponents. We will not complain that, under our agreement and with the full power we had given you, our application to you, about our Interference case, was equivalent to one to ourselves, and that we ought to have been informed of it. Nor do we think it necessary to go into the question whether on such a purchase taking place, you can hold our interference case on any other terms than as trustees for ourselves. We say this because we believe that, though the transaction in question has apparently rendered our interest no longer identical, it is not really so. We will endeavor to make this fact plain.

We ourselves, with regard to appeal, are pulled in opposite directions. On the one hand, we are eager and longing to appeal, because we do not think we have had justice dealt out to us and because we feel the most profound conviction that we can get the decision reversed. On the other hand, we wish to avoid personal inconvenience and expense. If we appeal, we have decided to go into the case most thoroughly, to come over to the States ourselves, and to carry the matter if necessary into the Supreme Court of Appeal. Thus our wishes on the point are like an evenly balanced scales. We are, therefore, ready to make a very considerable concession, and to undertake not to appeal on the basis of <u>*either*</u> *of the two following arrangements.*

1. That on all machines in which the paper is reversed but not transferred, you pay us the full royalties reserved to us in our agreement with you dated the 11th day of October, 1876; and that on all machines for the United States in which the paper is reversed and transferred, you pay us, so long as Crowell's patent for the same shall remain in force, three fourths of the said royalties.

2. That you pay us the sum of three thousand pounds sterling for all our rights in the United States, we undertaking to assign to you all American patents we may obtain, and the agreement of the 11th of October, 1876, to become void in so far as it refers to the U.S. royalties and to remain in force in regard to our British patents.

As to appeal from your point of view, there can be no doubt that you have decided rightly in preferring to go on without it. I will not stop to speak of the awkwardness and unpleasantness, in view of all the circumstances, of your forcing us to appeal, of your opposing as applicative [that] which until now you have conducted on the other side, which we are making in accordance with an agreement entered into with you, and the merits of which you have so strongly and repeatedly urged to be entered, on our side. But I will pause a moment to point out the folly of haggling over our paltry royalties and of spending time, money, and energy in the Law Courts which would be infinitely more profitably spent in bringing out the invention. In a few months, we solved the question in all its generality, flat surface, type, or stereotype; we provided the most copious drawings and explanations; we made several models of machines including a full sized one of the very machine in issue (figs 22 and 23, sheet 5, British specifications); and we constructed full size the two apparatus[es] (figs 5 and 6, sheet 2, of said specifications) and drove them with perfect success both in connection with a web printing press and with a wetting machine at a rate far greater than paper has ever yet been printed on. Had we (as we originally intended) worked the thing ourselves, our machines would by this time have been in at least a dozen offices. It is now <u>*more than two years*</u> *since our agreement with you was signed and not a single machine is yet finished. We do not speak thus with the view of charging you with any remissness in the matter. We presume the "Interference" case has not gainsayed its name and that it has* <u>*interfered*</u> *with your progress. Our object is to bring into strong relief the wisdom of not posing over our unusually small royalties and of their wasting in the Law Courts our strength and most precious time, but of buckling to [face] the rich and profitable market which awaits you. If we don't appeal, you have under Crowell the exclusive use of the reversing and transferring machines. There is no doubt of our obtaining the two methods alluded to by the Examiner, for none of the evidence addressed by Crowell affects that part, and since you now stand in his shoes, there would be no opposition. Therefore you have at the present moment, if we forego our appeal a broad untouchable foundation claim which will enable you alone to produce two roll machines and which will give you a practical monopoly of fast newspaper machines. If you produce a two roll machine, the great newspapers over here will be obliged to have it. [If] one of the three Dailies (Telegraph, Standard, News) lay one down, the others must do the same. Even the Times could be compelled to follow suit. Its proprietors shewed great folly when, not content with the unparalleled position of their journal, they turned themselves into second rate printing machine manufacturers. But I am quite sure they would never allow other papers to possess swifter machinery than their own and that they will prefer rather to purchase from you. Probably within 12 months of the introduction of a satisfactory two roll machine, you would have upwards of thirty orders by the above four great papers alone.*

Independently of these considerations, an appeal is for you in any case an evil. Be the issue what it may, it cannot be otherwise than not so satisfactory to you as if we had not appealed. If we win, in addition to the expense, worry, and delay entailed by it, you will have to pay us the full royalties instead of the three fourths which we are now willing to accept. The last evil is little enough, but both together are not half so serious as another drawback not at first sight so obvious. If you work the transferring or reversing under Crowell's patent, your monopoly will last till <u>May, 1894</u>; if under us (as of course you must do if we appeal successfully) the invention will become common property in <u>March, 1889</u>, i.e. <u>more than five years previously</u>. If we fail in our appeal, you would certainly save, of a portion of our invention, three fourths of our pigmy royalties—a saving not even sufficient as a set-off merely to the vexation and delay of law proceedings. But that is not the whole case. It appears to us that, win or lose, you will, by opposing us, make <u>your agreement with us void at our option</u>. We do not think there is any need at present, and we hope and believe there will not be in the future, for us to enlarge on this aspect of the question.

We should be much obliged if you would kindly telegraph a reply to this letter. You need merely say "Appeal," or "No Appeal" (as the case may be); and the address "Anthony, Hereford" is sufficient. When you write us next, we shall be glad to know

1. What progress has been made with the machine which has been in hand now more than twelve months?

2. Is it a two, three, or four roll machine? For type, or stereotype only? and is it a reversing or transferring machine or one which reverses without transferring?

3. Have you received our letter dated the 14th of September last, and the drawings, &c, sent at the same time?

4. And, if so, what is your opinion about the automatic tension? and what are your views about our remarks anent folding by cutting into fractional sheets?

In conclusion, we trust you will accept or reject our offers as they stand. Any suggestion of a further statement of our royalty, or of a lesser one than that named for our U.S. rights would merely serve to fruitlessly annoy us.

We beg to remain, Dear Sirs,

Yours very truly,
Anthony and Taylor
(per Edwyn Anthony)

Messrs. Hoe and Co.
New York

A machine with one roll of newsprint produced a 4-page paper; a "two roll" press gave 8. The "one roll double [width]" yielding 8 pages became the standard "unit" in modern printing press classification. Providently for Hoe, the inventive works of both Crowell and Anthony and Taylor eventually devolved to its possession, solving confusion and infringement problems that might have developed:

> In 1880 the firm bought of Messrs. Anthony & Taylor of England, a certain interest in their two British patents dated respectively March 27th, 1875, and August 31st, 1875, and also their entire rights to all United States patents that might be obtained containing the inventions and devices covered by the British patents, and finally bought their remaining interests in said patents. These British patents described (1) bars placed obliquely, and a combination of such bars with rollers, for changing the course of a web of paper, for turning it upside down, for shifting it sideways, for slitting it lengthways and bringing one part under the other, etc. and (2) utilizing these manipulations of the web to accomplish certain results in its printing and delivery. Many of these results had previously been attained in the United States but had not been patented in Great Britain, hence in view of the firm's increasing business with that country, these patents were desirable.[18]

While the Colonel palavered endlessly with Anthony over matters of doubtful value, Crowell had not been idle. Upon receipt of his third folder patent (1877) he promptly engaged the Atlantic Works, East Boston (Abishen Miller, President), to build improved machinery as orders were received. The first, of course, would be for the Boston *Herald*. These activities apparently reminded Hoe that his folder problems were far from settled.

As late as April 1878 Richard M. Hoe was still ruminating over a possible impending legal encounter with Crowell. In M. B. Philipp's Washington office he had met Mr. Livermore, Crowell's attorney, who advised a meeting between Hoe and Crowell as the most amicable approach to discussions. Livermore "generally spoke of the great ability he considered Mr. C to have, in the simplicity of his contrivances, &c, &c," the Colonel recorded (24 April 1878). "I replied that from what we had learned he (Mr. C) wanted too great a share for his inventions and services. Mr. L replied that he thought if we entertained the matter we would find he would not be expensive in his demands. . . ."

By springtime, finally, Hoe had just about made up his mind to offer Crowell something substantial for his inventions, and engage him for further work as well. He offered the elusive yankee inventor $8800 to free himself from his Atlantic contract, $15,000 additional for all Crowell's patents, and $2500 as a year's salary to work at R. Hoe & Company. Crowell would be permitted to remain two months at Atlantic to finish his folding machine "and attach it to a Bullock Press in the Boston Herald office." The Colonel waited tensely for a reply. Puzzlingly long incommunicado, Crowell at last explained his silence:

20 Walnut St
Chelsea
Boston July 29" 1878

R. Hoe & Co.
Gents.

You must excuse me for not writing at an earlier date. My physician threatened me with all sorts of horrid diseases and even death if I did not drop every thing and take a vacation a month ago. I have tried to take the vacation. . . . I see nothing in the way of the machine being put up in the Boston Herald office and finally purchased by them.

Hoping to hear from you soon, I am

Very respectfully Yours
Luther C. Crowell

The Boston inventor, then 38, was unfortunately almost as delicate in health as his letter suggested, but he agreed to the company's terms. Tucker, uncomfortable at becoming Crowell's supervisor, apparently avoided the man but duly described his accomplishments:

> In 1878 the firm bought of Mr. Luther C. Crowell all his patents and applications for patents for inventions in printing and folding machines, and also engaged his services for a term of years to aid in developing such machinery. In Mr. Crowell's patent of August 22nd, 1876 for folding paper, the printed web is slit longitudinally into two narrow webs, and these can be cut transversely into, say, four-page sheets and folded separately, or one part of the web may receive a line of paste and, by passing around a triangular device, be transferred under the under part, and the two associated webs be cut transversely and the sheets folded up together, forming eight-page sheets of four open leaves. . . .[19]

Robert Hoe III divulged that Crowell's early efforts with the firm were not without costly trial and error problems—perhaps a reflection of the yankee's difficulty with blueprints. The bag-making machinery had contained folding apparatuses which seemed useful in newspaper work, and Crowell was instructed to exploit them. These were indeed important devices: (1) a "V" former or "cow catcher" which made the first and second folds while the paper was still in web form; (2) a rotating cylinder folder (an improvement over Campbell's and Tucker's); (3) a high-speed paper cutter which also stacked newspapers in 25's or 50's. By these inventions Crowell increased press speed from 10,000—the Colonel and Tucker's maximum—to 24,000. "After many efforts," Robert Hoe nevertheless wrote, sourly, "and the failure and destruction of several machines which had been constructed at great expense, the Hoe 'Double Supplement' machine was produced, . . . for a third time [effecting] a complete revolution of the methods of fast newspaper printing."[20] The Double Supplement, which indeed took 10 years to perfect, was commendably fully credited to Crowell, and the company sold many of them.

From one assignment Crowell plunged into the next. The *National Cyclopedia* bravely kept up with him:

> As the demand arose for a newspaper of six, ten, twelve, and even fourteen pages, he solved this problem by the aggregation of two presses set at right angles together, each printing a web four pages wide, which were assembled in a place common to both machines and laid one upon the other, double ply, and he adopted a series of angled bars, which enabled him after having split both webs longitudinally into two-pages wide sections, to collate these sections which previously were side by side one upon the other, thus making a four-ply web. This was the beginning of the famous quadruple presses. . . .[21]

Furthermore, having offered Crowell a substantial price for his inventions, and having—at a propitious moment in 1878—bought his patents, R. Hoe & Company became clear owner of some crucially valuable properties. Several infringement claims by others working exactly in this area were frustrated, one of them reportedly among the largest interference cases ever brought before the patent office: settled in 1879 after four years, it obliged several New York papers, using folders made by other manufacturers (especially Bullock), to come unconditionally to Hoe's terms. Another in 1884 was equally decisive. Despite the nightmare prospects of web presses smashing themselves to smithereens in the Experiment Room, Hoe's investment in Crowell was—or should have seemed—justified. At the end of the century Crowell was fifth in a list of 39 inventors with over 100 U.S. patents—he had 293 to his credit.

But although a tireless and compulsive worker, Crowell was still a moody New Englander, unhappy with his good fortune as the Hoes would have it. Within his first year in New York he was grumbling about salary—a disagreeable matter the Colonel left to Tucker to solve. In 1882 Crowell got $5000 retroactively and his pay was doubled, the firm begrudgingly acknowledging "the arduous labors performed." But Crowell soon rebelled again:

Brooklyn June 2 1887

S. D. Tucker
Sir

As it is evidently the wish of the firm that you consider the matter of arriving upon terms for a continuation of my services I direct this letter to you! Your manner and words convince me that you are disposed to give full credit for such services as I have given. I will add that you and the firm collectively and individually have always treated me honorably and in many instances with more consideration than my conduct deserved. I do not propose to go with argument or debate but will say that I am aware that every valuable method or device which I have been instrumental in bringing about has been attended with and largely dependent upon large expenditures of money; that the opportunities and facilities offered are to be equally considered; that the legal talent retained by your firm and always exerted in the interest of those efforts also claim consideration, but I submit for certain consideration a review of my efforts to take advantage of all these agencies for the benefit of your firm. You as well as I are cognisant of the fine thread upon which many valuable matters hang! and to discuss the merits of a race-horse or a yacht with the expectation of arriving at a satisfactory conclusion as to whether their success depends upon their build or their management would involve more time than you or I would care to give. . . .

It has been some time since I notified your firm that I had terminated our old agreement and therewith gave proper notice. I have not acted up to the letter because I looked to the spirit of our contract. Therefore as is always the case when one party to a contract is guided by the spirit and the other by the letter one or the other naturally gets [*hurt*]*! To get to the matter in hand . . . I will say that the following are my propositions*

1. *That I be made a member of the firm by assignment—without cost—of an interest equal to or of the value of ($10,000) ten thousand doll on the basis of the purchase of Col. R. M. Hoe's interest as made by the firm. This for reasons of policy is with restrictions as will be given.*

[2.] *I to receive a salary of eight thousand doll per year, commencing with the present year.*

[3.] *That I receive fifty prct of the profits derived from any of my inventions relating to the art of printing which I may introduce or which I have allready*† *Patented and which have not been introduced. . . .*

[4.] *I agree not to take any dividend for my share in the firm for five years, and also that all profits derived from inventions as described shall be paid to me in stocks or shares in the firm stock at their valuation.*

I hope the propositions will not strike you as unreasonable and that you will also bear in mind that however it may result I shall give your firm my best efforts and services. . . .

I am obliged to be in Boston for the remainder of the week and as I am considering offers from a firm there pertaining to the future you will oblige me if you will kindly acknowledge receipt of this crazy letter written between 12 o'clock midnight and finished at twilight this morning.

With best wishes &c. I am yours very truly

L. C. Crowell

P.S. I have taken steps to finish all matters relating to your firm which I have incomplete, and pending applications will also be finished and assigned to your firm as agreed upon &c. LCC

Crowell may have observed with some professional jealousy how easily Tucker, Colby, and Carpenter had found their overtures for firm membership warmly received, and expected his own genius to be similarly rewarded. Tucker could not consider such a rash ultimatum, and said so. Crowell meekly returned from a Boston visit and accepted a more modest appeasement, but such confrontations continued as each work contract expired. By 1891 the yankee inventor's salary nevertheless was $10,000, but he was worth it—and he knew it.

Like Tucker himself, Crowell continued frantically to remain loyal to the important engineering challenges and his prestigious affiliation with the industry's foremost manufacturer—as far as his temperament and health would allow. He worked "to his heart's content,"[22] said *Dictionary of American Biography,* but he yearned for the professional recognition he could not achieve:

> He was modest and retiring, and always shrank from publicity, so that very few, even among newspaper publishers, ever knew that the inventions by which they so largely benefited were due to him. . . .[23]

Although he delighted in being called a "Cape Codder" (and died there in 1903), he and his family had settled willingly enough in New York. He remained with the company about 25 years, almost to his death, his final years crushed by the conflict of his difficult personality with the imperiousness of a third generation of Hoes:

CROWELL HOMESTEAD
South Wellfleet

Oct-10-1900

Dear Mr. Hoe,

. . . I regret to say that my health is very poor. Since my illness of last winter my heart has troubled me very much, and often I am unable to sleep for 3 to 4 consecutive days & nights, and this is terrible. My physician forbids any mental exertion, and very limited physical exercise—he assures me that in a few weeks perhaps I will

be able to take more exercise and also work out the various plans and improvements which I have been so anxious to complete.

I do so wish that you could have given me, this spring, the encouragement to finish up the "Wrapping Machine" matter which you give in your letter—is it possible, that there is no one of all the many in your employ capable of completing the last machine constructed—when it is so nearly completed, that a few revolutions would teach a young novice as to what is needed to complete it. I often feel that I must go to the works and complete that machine. . . .

The last time I was at the office I had been without sleep for 4 days and nights, and as my physician had prescribed—but just—the "Digitalis" to regulate the heart action, I was strangely affected by it. . . . At this time I asked Mr. Chew your cashier for the sum due me when he replied that it was your orders that he should not give me anything whatever. You cannot imagine my feelings. I tried to appear unconcerned but I feared that my strength would not take me out of the office. That you should instruct him—the cashier—to do and say that which I felt to be a disgrace and insult—coming from him—in fact I was so completely affected that I went home and took my bed. . . .

I will give full particulars of my circumstances in my next letter. . . . Hoping you are in good health and that business is all you desire, I am yours very truly

Luther C. Crowell
Oct-10-1900

How sad! Crowell's genius with folding equipment had not only saved the company its large domestic newspaper sales; it had helped R. Hoe & Company conquer everywhere—even England and the world.

TEN-CYLINDER TYPE REVOLVING PRESS

From Richard Hoe's Memorial

8.

England and World-wide Success

THE FRENCH CONNECTION / LONDON SUPER SALESMAN LLOYD / OLD THUNDERER'S AMERICAN LIGHTNING / HUMBUGGED BY A MANCHESTER IDOL / THE BRITISH SHOP AND EUROPEAN RIVALS

Enthusiastic reception of Richard Hoe's type-revolving press in the United States brought the company to foremost attention. America had need of such success stories. Despite its concerted attack, through science and invention, on the shortage of manpower by providing machinery of every description, the new nation, so well removed from the fashionable centers of Europe, seemed unable to stand alone. Innovations could not survive unless introduced from England or France. Many inventors were so frustrated by this attitude they all too often rushed to Europe for encomiums they could not get in America.

R. Hoe & Company had early received the cheers, the laudits, the testimonials few contemporaneous American firms enjoyed. The unique "planetary" press created a sensation in journalism: "A revolution in newspaper printing took place," wrote Robert Hoe, simply. "Journals which before had been limited in their circulation by their inability to furnish the papers rapidly increased their issues, and many new ones were started."[1] But Colonel Hoe was immediately aware that the London newspapers were capable of circulations as high as those in New York—some of them, like the dignified *Times,* already in need of Lightning presswork. Others in the British Isles and Paris would want it soon. Anticipating the westward expansion of America, which he was well sited and eager to serve, however, Hoe was hopeful only of selling his patents and letting European manufacturers operate as they pleased. The much needed cash would settle old loans, if nothing else. His 1838 experience had reminded him he could expect little more; few Americans had ever done well on such missions abroad.

Upon the success in 1846 of Swain's Lightning in Philadelphia the Colonel had thought of selling its rights in England—to the *Times,* of course. He wrote in 1847 to the Newtons, asking whether Old Thunderer was still using its 1837 presses, and would not be interested in something new. The response seemed favorable for a British visit. William Swain also encouraged him to go, and agreed to accompany the young inventor and his wife Mary on the trip.

In October 1847 the party left East Boston on the steamship *Britannia,* the Colonel dragging with him dozens of unfinished shop problems to be handled by transatlantic mail. Despite considerable pitching that made Mary and Swain ill, Richard poked about (2 October): "This is a very good strong ship, & the Engines which are the regular English Marine Engines made by Mr. Robert Napier of Glasgow, are fine, good, substantial ones. . . ." Upon arrival in London Hoe and William Newton, his patent agent, assembled drawings and specifications and began to make newspaper office visits, for which he had several letters of introduction. The young (35) inventor's enthusiasm and frankness were not always assets:

No. 2 Ely Place, Holborn
London, Nov 14th, 1847

Dear Bros:

It is now ten days since I filed my specification & although I have been constantly engaged in efforts to obtain an order for a machine I have not yet succeeded. I delivered my letter of introduction from F. Hall Esq. [*New York Commercial Advertiser*] *to Mr. Sampson who was very polite & kindly gave me an introduction to Mr.* [F.H.P.] *Wetherell*[†] *who is one of the business men engaged in the Times. I was delayed several days in waiting for him as he was not in town & when I did see him he said they were under such obligations to Mr. Applegath that although they would like me to leave with them a proposition, still they could not entertain it until the expiration of their contract with Mr. A. Our minister Mr.* [George] *Bancroft gave me a letter to Mr. Walter the main proprietor of the Times which I left with Mr. Wetherell. I intend to see Mr. Walter himself when I leave the proposition which Mr. Swain & myself have concocted. I have an appointment with the proprietor* [*Herbert T. Ingram*] *of the Illustrated* [*London*] *News who is about starting a daily paper. I have found it somewhat difficult to reach the right parties here. . . .*

Nov. 18th. Nothing new as yet from any of the London Newspapers—they all feel very poor here & I do not believe they all think I tell the truth, and are frightened by any Yankee Doodle, expecting to be Diddled by him. . . . Give my love to all. I hope [*daughters*]

Emily & Addy are good girls. Remember me to Mr. Colby & any who make enquiries.

Yours very affecty

To/R. Hoe & Co., New York *Rich[d] M. Hoe*

Wetherall would be appointed Printer of *The Times* in 1852; he was obviously already functioning in that capacity. Sobered somewhat by the formalities of British business, the still innocent yankees crossed to France to make similar calls.

A French friend, Frederic Gaillardet, guided the party around the capital's newspaper offices. Originally a lawyer, Gaillardet traveled extensively in America and in New York had founded *Le Courrier des Etats-Unis* (1839) to unite the French of the New World, during which time he must have met the Hoes. In 1848 he had returned to France and long collaborated with Emile de Girardin on *La Presse*. Newton too accompanied them about Paris, as he spoke fluent French, and by a stroke of good fortune offers and counteroffers were quickly flying.

[Richard Hoe to New York, 12 December 1847] . . . *After we had got settled in our hotel which Mary has described as also our manner of life* [that is, modest?], *we delivered our letters. One to Mr.* [Richard] *Rush our Minister given me by Mr. Bancroft, one to Delamarre, Leroy, Chabrol & Co. from Messrs S*[anderson]. . . .

Mr. Newton then had some business with a Mr. Perot[†] *a Machinist & Engineer just outside the barrier. In the course of conversation Mr. N. spoke of the press & Mr. Perot said he had the day before dined with a gent who would be likely to take hold at once of the machine; he made an appointment with him & called for us. We went with him & to my surprise he took us directly to Mr. Delamarre, the Banker to whom we had had letters of introduction & Mr. Swain a letter of Credit. After opening the drawings we had a laugh over the singularity of the circumstance that his house had been giving me letters to parties whom they supposed wanted my machine, when he himself was ready to treat for it. We made a bargain. . . .*

Perrot too was a lucky find; one of the few capable machinists in Paris, he was later rewarded with some engineering work. Delamarre (his first name perhaps lost forever) was an impresario only P. T. Barnum could match. Immediately he boggled his guests with a grandiose plan:

. . . *He made a proposal to enter into an association* [corporation] *to put up several* [Lightning] *machines & charge the cost of the machines to the association, then charging for the Patent or the use of it two francs on every thousand printed, this to be divided into two shares one for the association the other for me. He calculated to print all the principal journals of Paris & with the three principal journals of Paris* [La Presse, 35,000; Le Siecle, 30,000; Constitutionnel,

25,000] *made out a statement of income of six thousand pounds per annum, one half for me. This would be very nice & I do not know but it would be the best arrangement to make. . . .*

The Colonel nevertheless thought he could bargain in the future with Delamarre and his exploitation scheme; he still hoped for a quick $30,000 to $50,000 in cash by selling the patent. But the Frenchman gave the inventor no opportunity to look further.

. . . I have entered into an arrangement with Mr. Vannard (who is the printer or publisher put in front as is usual here while the real owners of the papers are in the background) of the "Patrie" & some other papers; to furnish a 4 cylinder machine & put it up in his office for 30,000 francs [$6000] in about six months. The proprietor & owner of the building where the machine is to be put is Mr. Delamarre. . . .

During the second French republic the Paris daily *La Patrie* was vigorously enjoying its new freedom from censorship. Founded 1814 in a flurry of nationalistic expression, the paper had changed hands several times until 1844, when Delamarre acquired it.

A former officer in the Garde du Corps of Charles X, Delamarre had married into a prominent banking house where his financial perspicacity quickly earned him a Chevalier of the Legion of Honor in 1841. The newspaper apparently came to his attention as a property for sale, and he began to visualize himself grandly as one of the new imperialist journalists who would guide and accompany France once again through glorious times. Although a consistent supporter of Napoleon III, for his paper's position on political matters he was often censured (in 1859 escaping complete and final suppression only by a general amnesty) but his drive remained undiminished. Delamarre was a foresighted but impetuous reformer—for conservation of natural resources, for decent wages and pure foods, and for mesmerism and the occult.

Like his newspaper colleagues in America, Delamarre took active personal interest in reporting city life—accidents, crime, catastrophes—although his paper, said Joel Munsell (who judged its coverage "able, false and unfair"), was "a thorough aristocrat in feeling." [2] But it was extremely popular: with a circulation over 35,000 in the mid 1840's and heading toward 70,000 by 1849, it had obvious mechanical problems that conventional presses could not solve. The Colonel began to see two Lightnings—numbers 8 and 9—marked "sold."

. . . You will see that Mr. Delamarre or Mr. Vannard has the privilege of taking another machine at the same price. This machine will be made here. . . . Mr. D. says that when the first machine is put up he will be ready to make an arrangement to work or make & sell the machines or to charge a tax or duty on each thousand printed on the

machines. . . . Mr. Swain & Mr. Newton consider the agreement as good a one as I could have made. . . .

Awaiting settlement of many matters, the Americans had ample time to sightsee, to visit typefoundries and printing shops (20 years behind America, Richard judged), to complain (Newton had bungled some details), to ruminate ("The profits [Delamarre] makes up seem to be enormous but I do not see but it must be so. . ."), and to hope:

[To New York, 27 January 1848] . . . I feel sure that [by the time Delamarre's press is running] the Times people will be through with their experiments & ready to treat. I understand they are expending the money necessary to perfect the machine of Mr. Applegath, he not having the ability. All the hands employed about the Times office with whom I have conversed think the machine will never work which they are building & so "thinks I" you cannot conceive how glad I shall be to get home & should like never to leave again for foreign parts. We are all well, Mr. Swain & Mary dreading the sea voyage. . . .

The Colonel had instructed Gold Street to commence work immediately on the *La Patrie* machine, with beds to accommodate, in addition, Delamarre's other papers, *Constitutionnel* and *Commerce.* He also told Tucker to learn French—fast. In a final letter from Sheffield he reported good final news from the skittish *Times*: if Applegath failed, Hoe would be summoned. There were fine prospects of selling other machines—especially job (card) presses—in London, and rumors that the stamp duty would soon be repealed. The elated group returned to America eager to discharge its obligations.

In October 1848 Richard returned to Europe, the crated four-cylinder for Delamarre still at sea on its own way. Accompanying the Colonel was Stephen Tucker, then 30, well briefed and ready for this prestigious assignment, although far from fluent in French. R. Hoe & Company was not unaware that this Old World press installation would rank in importance with Koenig's historic contribution just decades before. Many incidental matters needed attention:

Hotel Chatham, Paris
November 11th, 1848

Dear Brothers,

I sent day before yesterday to you a short letter as I had but little to say & I have little more to say now except that I feel rather home-sick. We are now engaged in making a plan for the [French] Customs of the machine complete which seems to be necessary as well as the peice† drawings.

We have examined the place & fixed on a place in which to place the machine. I have a line from Quesnel Freres & C°, but the ship had not arrived but is expected every hour. They say "she is over due & they hope has been delayed only by adverse winds as she is a good *ship & has a good commander." They will inform me of her arrival instantly on her entering the harbor. Also that it will cost 68 francs per ton freight by railroad to Paris & take 3 to 4 days & by River Boat 15 francs per ton & 12 days. Of course, the difference being about $100 for a week I shall order it by River Boat & in the mean time be preparing for it. We are to make an application to-morrow, or rather Monday, to have it sent "sur*† *[sous] plomb" or under lead seals to Paris. Should we not succeed in this we may be obliged to go to Havre & take the most of the machine out of the boxes at the Custom House there. . . . Saw Mr. Kendall of the Picayune to day. He is engaged here writing letters home & getting Lithographs made for his history of the war with Mexico. . . .*

The Colonel was also surprised to meet Joseph Adams and his wife on the Rue de la Paix; the wood engraver strangely would soon be immensely helpful.

. . . Wednesday, 15. . . . I have very little else to write about except the state of France & I will say as regards business it seems generally about the same as it was with us at [18]37 when our Banks all broke. I do not know as it is quite as bad here. But there is one fact here which is too apparent, notwithstanding the great amount of suffering stated in the English papers as existing in Paris, that during one week we have been here we have both noticed the absence of beggars & if you compare Paris with London I would state in the past week I have been only 8 times asked for alms, whereas in princely London I counted one morning in a walk of an hour & a half 13 demands. . . .

We shall have much emigration I think from France at any rate, & particularly if a monarchy should be again forced on the people which I think will not be the case, for I believe the masses are for a Republic. I shall write a letter to Swain by next steamer as well as one to Bennet† *& to McElrath & let them publish my speculations if they think fit, but without name. . . .*

Tucker was a little more cheerful:

[To New York, 29 November 1848] . . . Well, here we are exercising that greatest of all virtues, Patience, awaiting the arrival of the Press from Havre, which we expect in about 10 days. However we have not been altogether idle in the meantime, as I have made a drawing of a side view of the machine as set up for the purpose of passing it through the Custom House, also a drawing of the Proof Press which is now being made. The counter shafts, cone pulleys &c for driving the machine are likewise under way. I suppose the Col. has informed you that he got permission from the Director of the Customs to bring the machine through to Paris without its being unpacked at Havre. . . .

There is no doubt but that 8 or 10 of these machines could be profitably used by the newspaper offices here, as "La Press"† *& other* *papers are obliged to set up duplicate forms. . . . About business I do not [feel] we can say much as yet. The machine being something new it will be necessary to get it up & in operation before any thing can be done. . . . However nous verrons [we shall see]. . . .*

Hoe and Tucker had now been abroad two months, desperately trying to keep themselves busy until the machinery arrived. The Colonel finally admitted to such "homesickness" he instructed his brothers to send Mary over to him; she would arrive early in February 1849. His enforced study of the French political picture, in which he would ordinarily have little interest, was unfortunately most sanguine:

[Hoe to New York, 13 December] . . . You will of course see by the papers that Louis Napoleon is elected, by an overwhelming majority, President of the Republique Francaise & I can tell you it was done more quietly than most, if not all our elections & I feel sure that matters are about to take a steady improving course here. The Republic I think, is a "fixed fact" & I do not believe that Prince or rather President Napoleon with all his popularity can subvert it for any length of time. . . .

Girardin of the rival *La Presse* had repeatedly entertained Hoe's solicitations, and by December 1848 appeared to be fully ready to buy a Lightning. But the Colonel repeatedly put him (and others) off, believing the imminent running of the *Patrie* machine would absolutely overwhelm French (and British) publishers and put all of Europe nicely into his pocket. But another report from Tucker indicated that progress would be excruciatingly slow:

[Tucker to New York, 28 December 1848] . . . Our Press is here at last & in the office & the masons are now building the foundations, but such a† slow mortals I never saw before. The dirt is all carried out in baskets which are straped† on a mans back. The man backs up, stands untill† his basket is filled with a wooden shovel, & then the man walks off & dumps it out over his head. Oh! for a couple of yankee irishmen† with a yankee wheelbarrow & shovel. I am in hopes that by monday† I shall be able to donn† my overhauls† & go to work in earnest. We have the shafting, pulleys, &c, all ready waiting for the masons to finish, for they have a way here of putting the shafts under the floor. . . .

Tucker tried to see some benefit in their long Paris stay:

. . . I am strongly of opinion & very much in hopes that we shall make something out of it yet, but it must necessarily take time or it will be sacrifised† as the french† people are a shrewd set & will not buy a thing of this kind without seeing a probability of its yielding something & besides it looks very suspicious to see a person want to sell out a thing of this kind in a hurry & it also gives them an advantage over the seller which they know how to improve. The only

way to do [it] is to disguise our intentions, by having the appearance of remaining here. . . .

Richard nevertheless complained to all correspondents of the exasperating delays. All consoled him sympathetically, knowing his propensity for work, but his British agent scarcely realized how ominous had been his humor about it:

Office for Patents
66 Chancery Lane
Novr 21st 1848

Dear Sir,

I enclose a letter received for you by the last steamer—you will no doubt learn by it that Zachary Taylor has been elected president of the United States.

I fully calculated upon your wasting your time in Paris but I can tell you that you have only just had a taste at present—and before you have finished your matters you will have had a most beautiful specimen of French business affairs. Dont forget in making your drawings to colour all the cast iron one colour all the brass another all the wrought iron another and so on and you must give the weights of all the pieces.

We have received notice from your brother that the hand presses [jobbers for London] are under way but they have not yet arrived here. When they do we will give you due notice. I remain dear Sir

Yours very Obediently
R. M. Hoe Esq. Paris *W. E. Newton*

P.S. I ought to have told you before leaving London that in my interviews with Perrot & Delamarre I told them that we had made improvements conjointly on the machine. You will know my reason.

Newton, necessarily represented as co-inventor to obtain the British patent, nevertheless revealed some capable participation in the machine's promotion:

[Newton to Richard Hoe, 29 December 1848] . . . About a week ago there was a lecture and description at the Society of Arts about printing and printing machinery and at that time a great blaze was made about the wonders of Applegaths machine of which large drawings were exhibited. The person who undertook to explain it said it was a most wonderful thing and bestowed on it the highest encomiums but as he was no mechanic he was continually puzzled to answer questions as to its operations and after getting into most inextricable confusion he was obliged to admit that its movements were so complicated that altho' he had seen it he could not clearly understand them all and therefore could not explain them. . . .

Here the salesman in Newton took command:

. . . So, when he had finished I explained the construction and operation of the machine and pointed out the disadvantages and objec-

tions to which it was liable and shewed to all who had any mechanical knowledge that the machine must necessarily be an extremely expensive one and one that owing to its peculiar construction is extremely liable to derangement. I then by means of a piece of chalk & black board explained the construction of your machine, pointed out its extreme simplicity and stated that its operations were limited only by the ability to feed it and I hope I made the machine perfectly understood as I satisfactorily stood a pretty stiff oral examination and explained every thing & answered every question that was asked. People seemed very much pleased with the machine and I was rewarded with two or three rounds of applause upon finishing. . . .

The smug London *Times* was unconcerned.

It was now 1849. The Colonel had met Armand Bertin of *Journal des Débats*, the proprietors of *Le Siècle*—"The Century," and representatives of *Constitutionnel, National,* and *Moniteur*—all of whom now anxiously awaited the trial of Delamarre's machine. Some were ready to buy, purely on the yankee reputation for ingenuity, but Hoe insisted on the clinching demonstration. While Tucker at last sank the bed plate at the *La Patrie* office the Colonel returned to England to confer with his advisors and creditors, the Sandersons, and fetch Mary at Liverpool. The *Times*' continuing difficulties were a rhapsody to him:

[*Hoe to New York, 26 January 1849*] . . . *I have seen the* Times *Press in operation. It is a very complicated large machine & I do not believe they will ever be able to work it any faster than they do now, 8000 per hour with 16 persons. Mr. Wetherell,*† *the manager, has said to me that if I will let him know when I am ready, he will come over to Paris to see my machine. . . . You cannot imagine how jealous & envious Mr. Applegath is, but it is all sport to me. I met him at the "Times" & was introduced to him by Mr. Wetherell. I will merely state that I am satisfied that if I had brought my press out as much later than his, as his has been brought out later than mine, he would call me a Pirate, &c, &c, but when I look at his machine I feel that I can afford to be magnanimous. . . .*

While in London the inventor made extensive arrangements with printing equipment manufacturers and salesmen Hopkinson & Cope to sell Hoe job presses, inking machines, and other accoutrements; he also sent to New York a small table-size printing apparatus (later called the "Harrild") which he believed would sell "like hot cakes" to American merchants. He was delighted and relieved to welcome Mary, who had, alas, been seasick again and looked "thin & pale." With new credit (and advice) from the Sandersons and Mary to accompany him, he was ready once more to face the elusive Delamarre:

[*Hoe to New York, 8 February*] . . . *I have this course in view in Paris: after the machine is put in operation & is well going, I shall tell Mr. Delamarre that I cannot go into the exploitation, but that I would like him to make me an offer for the French Patent & as he seems to feel so certain of the success of the exploitation, I have hopes he will do so. If he does not, I shall then take the next best course, that is, to take orders from the various journals & after that shall strive to sell those orders with the Patent right. . . .*

At last progress could be measured (7 March): "Last evening we put steam on the press & had it going round up to 9000 per hour without forms on. The people about the office opened their eyes wide & every one except some of the compositors, who will lose their situations, looked delighted & some of them told me that it was beautiful but it would cut off their arms as the paper would only require to be set up once. . . ." Success! Events moved faster:

Paris, Hotel Chatham—
Mar 16th 1849

Dear Bros:

I have only time to say that we have started & tried the machine & send you the two papers under date 3rd Mars which we printed yesterday (first essay with ink); every thing about the press works to a charm but I fear we shall lose two or three weeks ere we get the feed boys up to a good speed. We worked yesterday at 4300 & they had as much as they could do at first; but toward the last of 1000 sheets they did better. I have offered them each a premium soon as they arrive at 8000 per hour, & they seem determined to reach it. . . .

The flys work admirably & are the delight of all beholders; nothing of the kind has ever been seen here before. In fact the machine made all the lookers on converts, for many have been under the impression that it would not go well, & many that it would not go at all.

The tapes [best London linen] work to a charm & I should think would run 6 months without any breakages or alterations. . . .

How I wish I had now here 4 of the Feeders from the Herald or Sun offices. I would wake up these sleepy foreigners quick to the tune of Yankee Doodle, march! quick step! run! but I must wait until I can force up these slow & easy people who in every thing else but work are quick enough; but when it comes to working quick in our business they know nothing about it. Excuse this letter; I will endeavor to give you a much better & more interesting one next week.

Remember me to all.

Yours very affectionately,
To/ R. Hoe & Co., NY *Richd M. Hoe*

True to his promise, the following week's news was delicious:

Paris Hotel Chatham
Mar 23rd 1849

My Dear Bros/

I have much satisfaction in being able to inform you that last evening we commenced to print, regularly, the "Patrie" of which I send you some copies taken without care from the heap. In fact they are as bad as the worst & the best were much better.

The rollers made here are not the best & ours had lain so long that they were not in prime order. However Mr. Delamarre is having a casting kettle put in & we shall teach them to make rollers. Mr. Adams (Engraver) has volunteered to help me in teaching them how to make rollers.

The press works as you say it should—like an "old Stager," & the "sheet flyers" are the admiration of all beholders, & your humble servant feels well pleased with himself that he persisted in putting them on. It is 12 o'clock, Tucker is in bed preparing for to-night. At various times during our progress I have seen indications of opposition from some of the employes in the office of the Patrie but we have treated them all with extreme politeness & I have been very familiar in talking with the workmen which is not at all the Fashion here for "Les Ouvriers sont Cannaille"† [*canaille; "The workers are trash"*] *is the opinion of nearly all the employers. You will see that they are very much flattered by kindness. The working of the press last night removed all doubts, & each seems to vie with the other in praising it. Of course the compositors who will be affected by it are not so well pleased, but they say it is beautiful, charming, & must "break their arms." There are 70 compositors in the office & they make 2 compositions each day; of course some 30 will soon be discharged. . . .*

As the feeders slowly approached their target speed (by May they reached 8400 for a few minutes—that is, 2100 each) the Colonel parried and thrust with Delamarre, frequently out of town for days at a time. The boys not only reached their mark, they learned to do even more. The Colonel later recollected:

[*Richard Hoe to Robert, London, 27 June 1859*] *. . . Now in our difficulties with feeders in Paris . . . we arranged to put up the paper without stopping the press. The extra hands (as a* [*feed*] *board began to shew a small heap of paper) would take a lift or smaller quantity & roll it up, slip it under the upper part of the paper on the feed board & work it under by degrees so as not to interfere with the feeding. I have often known them, after they were au fait* [*under way*] *to feed steadily on & never stop the press for two hours & sometimes more at a time. . . .*

The feeders in New York—ten years later—were not working this efficiently.

Tucker was obliged to spend every night in the pressroom lest "accidents" to the press sabotage the Colonel's negotiations. Such meticulous attention to the machinery, the workmen, and their operating conditions finally earned the approval of their French hosts:

[*Hoe to New York, 31 May 1849*] *. . . I must dig through all the troubles here. We have to teach every body their*† *trades; they have a* *fashion & it is very difficult to get them out of it. So many of the workmen printers know so much more about printing than we mechanics do that they will not do as we tell them, but we have convinced them that we can get on without them by doing ourselves in half the time they take to accomplish the same thing. I find talking is not of any use, acting does the business, & the men, having found out that we can do any thing either about the printing or press, make good rollers, &c, begin to pay us some respect. In fact as some of our yankee friends would say they conclude we are some pumpkins. . . .*

The Colonel's continuing political analysis, however, was again off the mark:

[10 May] . . . *I wish much to be home & feel that Europe is tumbling to peices.*† *In fact to my mind a general war in Europe is almost inevitable, & the only thing* [that] *will prevent it is the fact that Royalty has not a superabundance of cash, & are not exactly sure of their men. I do not believe that there will be any more trouble in France at present, but every thing around here is in a distressed state, Germany, Prussia, Austria, Italy, &c. . . . I have come to the conclusion that even with our much laughed at Republic we have a firmer, stronger, & more durable government than ever has been seen in the world before. . . .*

Country after country in Europe indeed suffered through fitful political and social adjustments to the Industrial Revolution which (thanks to innovators like the Hoes) was reforming work habits everywhere. To his intense surprise and consternation (and secret delight) Richard was almost swept up in a serious imbroglio:

Paris, June 14th 1849
Hotel Chatham
rue Neuve St. Augustin, 67

Dear Friend,

I have only a few minutes to write & say that although we had a demonstration yesterday as you will see by the papers you receive, still to day everything appears as quiet this morning as if nothing had happened. I send you a chart [not present] *with this of the representations of France, which I thought might be interesting to you.*

The sight yesterday was worth seeing but at one time I was rather too near when two or three platoons fired in among the crowd wounding many & killing some. I should like very much to see you out here—suppose you take a trip over. Paris would please you. Should you do so, in London as you pass through you can enquire for me at Newton & Sons, 66 Chancery Lane, London, who would be able to tell you where I am should I happen to be in London or elsewhere (than here).

I am getting on very well here. The press is working beautifully & is the Lion in the Printing way. G. W. Kendall is still here & a number of other Americans. I wish I was a good hand at writing letters I would give you a description of affairs on the continent of

Europe. They are in a situation sufficiently complicated at present & what the end will be no one knows.

Drop me a line or come yourself.

Yours very truly,
Rich[d] M. Hoe

George Roberts Esq.
Boston

To his brothers Hoe also described this shocking event where, by his own estimate, 150,000 Frenchmen shouting "Vive la Constitution" were dispersed by waves of infantry and cavalry. The immediacy of the news was prophetically obvious to him—as was his own value to journalism: "The necessity for a fast press was shown last night to a great extent," he told his brothers (14 June 1849), adding that the Paris Lightning through the night had printed 69,000 copies.

Ebulliently the Colonel invited dignitaries, friends, acquaintances, and newspaper people to see the planetary press in operation, including President Napoleon, the *Times* London proprietors, press builder R. W. Cope, machinist Joseph Whitworth, and the curious Dr. Dionysius Lardner (then London *News* correspondent). At the same time, however, he sadly began to feel that Delamarre had all the while been "humbugging" him.

After interminable meetings, the wily French banker in July 1849 paid 10,000 francs ($2000) on account, confirmed his order for a six-cylinder, and intimated a third press as well would be wanted. His check was rushed to Sanderson Brothers & Company for the delinquent Hoe account. Tooling up for the manufacture of the six-feeder in France was commenced; Tucker was directed to begin drawings and specifications at once.

The Colonel was furious with "these heathen of the old world," obliged to spend weeks more collecting the balance from Delamarre and arranging payments for the second press. At the end of September he was at last able to leave:

[*Hoe, London, to New York, 28 September 1849*] . . . *If you knew how I have striven to get away from France, & the numberless vexations & trying occurrences I have been obliged to overcome, & how much my patience has been tried, you would think that all had not been smooth water with me here, & that had I not possessed more patience & perseverence than Job, I should have left in disgust long since. . . . I am very sorry I have not 50,000$*[†] *in cash to bring home, but I have worked as hard for it as any one ever did. Of that I am sure. . . .*

He and Mary returned to New York by mid October, understandably exhausted. One friend commiserated:

OFFICE OF THE PUBLIC LEDGER,
S.W. Corner of Third and Chesnut† Sts.

Philadelphia, Pa. Oct. 21, 1849

Messrs. R. Hoe & Co., New-York
Gents—Allow me to congratulate the the† "Col." and his lady upon their safe return to America, and also your house upon his arrival. I am anxious to see and talk with him and anticipate the pleasure ere long. We are to have a special meeting of the Directors of the Magnetic Telegraph Company at Washington on the 16th of next month where I shall expect to see him certainly if not before.

You mention that you have forwarded to us "La Patrie" with an article of interest about "Hoe's Last Fast." I have not yet received it. Nor did I ever receive one which the "Col." stated he should send me from Paris. This one may probably arrive however tomorrow. I am anxious to see it. Best respects to the Col. and his lady. If I could get away I would slip on to N. York and give 'em both a shake—a welcome shake of the hand.

In haste,

Respectfy Yours,
Wm M. Swain

[P.S.] *The Col. mentioned in his last letter to me that he had concluded a bargain for his French patent—For how much did he sell—Did Tucker return? Swain*

Tucker remained in Paris to superintend the assembly and trial of the second Delamarre press. He performed some social duties as well:

[*Robert Hoe to Tucker, 15 April 1850*] . . . *This will be handed to you by our friend Mr. Wm Geo. Stewart who is also a good customer of ours. Having said this much we know that we need not say more to have you give him a warm reception in Paris. He is vice consul to the US from Mexico. You may remember our old friend Mr. Jno Granger. Mr. Stewart succeeds to his business. We have had very flaming accounts of the "New—Paris—Rotary Continuous-Sheet press," in the New York papers.* . . .

In September 1849 final signed documents of the Hoe-Delamarre transactions had been forwarded to New York and the Colonel anxiously said "finished" to his most enervating experience. He never again attempted such negotiations and it was fortunate his brother had the temperament for them. But the end was nowhere in sight.

The turbulent political situation in Europe produced an oppressive reaction. Tucker explained:

But the stamp duty of one sou [1 cent] per copy, which had been taken off at the Revolution of 1848, was restored after Louis Napoleon's election as president in 1849. This at once cut down the circulation of the one cent papers to such an extent that the six cylinder press was no longer needed, and it was countermanded. This led to a law-

> suit which kept me in France for another year, and which resulted in *La Patrie* being condemned to pay the firm a good round sum in damages. . . .

It is hardly strange the exasperated Colonel would not let Delamarre cancel his order without pressing for handsome recompensation. He had earlier been "not a little astonished" that production costs tabulated in New York had made the first machine—Swain's—much more than he expected, and now the additional promotional expenses and delays—his time in Europe, and the commencement of the second machine were almost incalculable. He saw the French publisher a devious ally, his installation unlikely to continue serving as a display for potential British or any other customers. The litigation and exact damages are not recorded (riots in 1871 would destroy the French archives), but Hoe evidence suggests that, with suits and countersuits, final settlement between the two parties did not occur until January 1859—at which time Delamarre was ready to order (from Marinoni) another six-cylinder machine.

Yet Europe—especially England—seemed so promising for the Lightning. The Paris machine had gotten a fair degree of publicity, but even the best encomiums quickly fade with time. Announcements of the Great Exhibition of 1851 suddenly revived R. Hoe & Company's prospects—at least *Scientific American* thought so:

> Mr. Hoe will no doubt attend the great Industrial Exhibition in London, in 1851, and come off victorious in competition with any other. This press will be the *flying heavy artillery* of the literary world. . . .[3]

Strangely enough, the Colonel made no attempt to put the machine on display; having relied on the Paris installation to impress newspapermen, he may have counted on their natural curiosity to cross the channel for a good look at it. Others, however, also slighted the big international fair, prompting one patriotic American to act decisively for his countrymen.

George Peabody, an American financier and philanthropist whose London business dated to 1837, was chagrinned, as the exposition materialized, to learn that the United States government had made no appropriation for representation—only 519 yankee exhibits, mostly tawdry items, would be displayed. He personally advanced $15,000 to remedy the situation, and, according to a congressman in 1856, "[D]ay after day, as some new product of American ingenuity and taste was added to the list,—McCormick's reaper, Colt's revolver, . . . Hobbs' unpickable lock, Hoe's wonderful printing-presses, and Bond's more wonderful spring governor,—it began to be suspected that Brother Jonathan was not quite so much of a simpleton as had been thought. . . ."[4] Peabody (whom Congress later reimbursed) had obviously collected these prestigious items from friends in or around

London, and unfortunately would not have found a Lightning among them. For that particular piece of American ingenuity, nevertheless, British credulity was once again taxed by another old friend.

Horace Greeley visited Europe in 1851 to find himself involved in an official discussion of the newspaper stamp duty, then undergoing reexamination. In gracious testimony before a parliamentary Select Committee on Newspaper Stamps the *Tribune* chief jumped to the chairman's opening question:

> Are you able to give the committee any information with respect to the press in the United States, and whether the absence of a stamp in the United States is productive of any bad effects, in your opinion?
>
> I will readily answer your question. We could not comprehend the operation of the stamp; it would be impracticable, under our institutions, every way. . . . [I]t would require an entire revolution in the habits and feelings of the people generally. . . .

After descriptions of the freedom of the press as Greeley knew it to devolve and exist in America, the committee took the opportunity to inquire about another aspect in which there was avid—but guarded—interest:

> At what rate can you print the *New York Tribune*?
>
> Our press will work but 10,000 an hour; there is one faster, an eight-cylinder, which will work 20,000, they say; I call it 18,000. . . .

With the London *Times* straining to produce 12,000 hourly this intelligence appears to have been understood incredulously as a declaration of faith instead of fact. The question was rephrased:

> Do you believe that there is a press at work in New York which will print 18,000 copies an hour?
>
> Yes; I have seen it at work, and it will throw off as fast as men can feed sheets upon it.
>
> Have you seen the press that is used for the [London] *Times?*
>
> Yes.
>
> Do you consider that the press at New York prints with more rapidity than the one at the *Times* office?
>
> Yes; the press at the *Times* is fed on what I would call a drum or vertical cylinder; we feed on horizontal cylinders, directly on and around our greater or type cylinder. One man stands over the top, and another below, and so on, feeding, and the four cylinders are fed on each side of the large cylinder, on which the form revolves, and they feed the four paper cylinders, one above the other, on each side of the large cylinder; the sheets are laid over and over on the face as fast as the great cylinder can be turned, and every time it makes a revolution it has printed eight journals.[5]

The committee had much to ponder. Many of its questions suggested that the political and legal aspects of government press regulation could be seriously influenced, not by lofty principles of freedom and justice, but by the mere mechanical ability of newspapers to reach large masses. R. Hoe & Company's growing record for speed would be a tantalizing model for British journalism, if not an innovative threat to British self esteem.

Although Old Thunderer was a private enterprise, there was understandable national pride in its world-wide reputation. Its historic position as the most progressive paper in a mechanical sense, thanks to the Walters' employment of Koenig (whom few remembered as German) and the ingenuity of Applegath, was another world record of which the paper was proud. Perhaps the *Times* management itself was behind a rather coy inquiry by Hoe's British agent:

> *[W. E. Newton to Richard Hoe, 1 July 1853] . . . You are no doubt aware from the newspapers that some alterations are contemplated in our stamp & newspaper laws. These alterations are likely to very greatly increase the sale & many offices are now looking out and preparing to make extensive alterations. . . .*

While for many years handling Hoe's patent applications, W. E. Newton had gotten perhaps the most tantalizing glimpse of all into advancing American printing press developments. He had secured the type-revolving patent in England for Hoe, May 1847. Figuring to participate in the company's success, Newton suggested that he might function as British agent for sales as well. He professed to have a printer client who would take a Lightning and who would function as a demonstration to all of London that the machine was sound. He added, "I don't doubt with this gentleman's assistance getting a machine into the Times office even and many others . . . ," urging the Colonel to send a machine immediately.

R. Hoe & Company were not accustomed to ship on consignment, and even though their relationship with the Newtons was long and close, the brothers in New York must have sensed that these patent agents—despite certain technical competence—were not the salesmen they wanted for so large and lucrative a market. Robert Hoe himself, even more than Richard, anticipated frequent visits abroad and became increasingly disposed to handle European sales. As time went on the company's indebtedness to Sanderson Brothers & Company—reaching about $23,000 in 1857—obliged their appointment to the coveted position. But it was not mere courtesy:

> *[Robert Hoe, London, to New York, 10 September 1856] . . . Mr. W. E. Newton is absent in Scotland but his brother has written to him and he may meet me at Manchester, as he is anxious to be our agent here. I think it far better to leave the business all together in the hands of Mr. [Edward] Sanderson. The house of SB & Co. is one that is so respectable & well known that it is a great advantage*

to be associated with such people. It inspires confidence in us at once. . . .

Yet not until almost a decade after the admittedly impetuous French installation did the British become acquainted with Hoe's marvel—and no thanks to Hoe salesmanship, either. Another wild and furious journalist was the catalyst.

"The English, always slow to adopt improvements," wrote Robert Hoe III, "did not appreciate the value of these [Lightning] presses until the year 1856, when Edward Lloyd of 'Lloyd's Weekly Newspaper' in London, having seen the one in the office of 'La Patrie,' ordered a 'six-Cylinder' machine."[6] Like military geniuses who instinctively recognize tactical superiority through new weapons of war, Lloyd keenly sensed this Hoe invention would insure his own success in journalism.

A printer by trade, Edward Lloyd (1815-1890) had sold books and newspapers on the London streets while still a boy. At the age of 24 he formed the idea of an independent liberal newspaper, bringing one out within a few years and rankling over its oppressive stamp taxation. In 1842 he started *Lloyd's Penny Illustrated Newspaper,* containing literary items and reviews for which no stamp was necessary. The government, however, found a few lines to be "newsworthy" and threatened action. Lloyd quickly changed its title and got a stamp for it. Meanwhile experiencing considerable difficulty with newsdealers (the 3-pence price was half that of similar papers), he stepped up its advertising: at one point he embossed thousands of pennies with the phrase "Lloyd's Weekly London News, 3p, post-free," until parliament, in order to stop him, hastily passed legislation prohibiting the mutilation of currency. According to Thomas Catling, long his editor, Lloyd trimmed his beard and hair several times a day so as to pick up gossip and entreat barbers to post his handbills. These stunts all worked to perfection.

As Knight and other penny paper pioneers had demonstrated, the London public was hungry for illustrated fare. Lloyd's weekly soared to 33,000 in 1848, on its way to 90,000 in 1853, and headed still higher two years later when the tax was removed. What must have been a foreign newspaperman's courtesy call at the *La Patrie* office to see an American novelty turned for Lloyd into a driving desire for possession. He placed an order and bombarded R. Hoe & Company with carelessly scribbled letters impatient for the machine's delivery:

12 Salisbury Square
May 2nd/[18]56

Dear Sirs,

I have just received your favor of the 15th of April in which you state the machine will be ready to sail on the 17th of April *only two days later than date of your letter. I thought it might probably be a mistake that you meant May 17 but I have put everything in hand from your plans. I cannot take any thing from the [basement] pier &† as it belongs & forms part of the next building. With regard to the shafting the motion is got by cog wheels for the four feeder [Applegath?] so that I shall have the direct motion for your machine & cog wheels for the four feeder. Your plans [are] plain and explicit and I shall do every thing to carry them properly out and I hope wether† a fortnight after the arrival of the machine, to write for the second which I hope you will let me have as soon as possible as I do not know what I shall do with only one as I must have both forms on together. I must use a four feeder for one pde [period] which will be most awkward.*

In haste, yrs truly
Edw Lloyd

[*Lloyd to New York, 9 May 1856*] . . . *Since I wrote to you I have conversed with my pressman on the subject of the layers on. He does not expect the least difficulty about it and thinks the hands you send will be required but a very short time. In fact he thinks our hands being used to our size sheet will be able to lay on as fast as the hands you will send. We have the fastest and best layers on that are to be had here. I am anxious to hear from you as to when the machine is likely to arrive here. I suppose you send it by sailing not steamship. I am getting on with all you require, as also with the room for the second machine which I shall order the moment the one coming over succeeds and which I must beg you will let me have as soon as you can after receiving my order, as one machine will be comparatively useless.* . . .

Lloyd's enthusiasm and eagerness to buy a second press even before the first was installed was hauntingly like the Colonel's grand reception in Paris. But there was some method in this British madness.

[*Lloyd, 16 May*] . . . *I have received yrs wherein you put my doubts at an end by explaining the time* [as] *the 17th of May not April that your press is to leave New York for here (where I hope it will supersede all others and lead to something that will repay you for yr indefatigable perseverence and industriousness in carrying out a most important invention which all others have failed to do). I need not tell you I shall be most happy to see any gentleman connected with yr firm but I would rather that that gentleman should be here at the time he could be a witness to the success of yr machine, which I have the greatest interest and faith in. You have done well in working the machine before you sent it away.*

How is yr perfecting machine going on[?—]*that's the machine that would sell here by wholesale but I fear the price would stand somewhat in its way. You ought to have one at my office* [for display] *altho it would not be of much use to me. I should like to see one here to show but that will be an after consideration. I will send the plan for the second machine room for second six feeder next week. I fear*

I shall be much cramped [in the pressroom] for the lenght† *for second machine, while I have super abundance of Breath.*† . . .

At last Hoe notified Lloyd that the precious machine was on the steamship *Ericsson,* due in London on the last day of May. Dispatched to England at the same time were two veteran Hoe press erectors, William Conquest and James Blair. Had he really thought this first type-revolving press on the British Isles would be the prestigious event he hoped for, the Colonel would have come himself. But two crack mechanics (and his brother) were almost as good. Remembering the trouble with attendants in Paris, Richard sent along with Conquest and Blair six American boys experienced at feeding the hungry machine. "[T]heir appearance in quaint coloured blouses created quite a sensation in Fleet Street when they turned out during the dinner hour,"[7] Catling wrote. The lads also helped install the press. Soon all was in readiness:

[*Lloyd to New York, 30 May 1856*] . . . *You took me by surprise. I did not not*† *expect the machine would arrive here for a month, but the moment I received your information I went to work in real ernest*† *and am quite ready and expect it here tomorrow Saturday or on Monday. Mr. Conquest and Mr. Blair together with the hands have arrived here safely. The tables are ready for the composing room; when I had my last new type I saved six pages of the old matter for the purpose of imposing for your machine to try it & every facility that I can properly give I shall be glad to do.*

Messrs Conquest and Blair appear to fully bear out the high character you give, and are evidently men of business. They seem most anxious for the success of the machine. I am only waiting the success of it, to send the order for the second and shall do so the instant it starts off and trust you will use every exertion to get it done as early as possible as I shall never feel safe with only one. There is a French one making a great fuss here, but I hope to shew the printing world here THE News Press. . . .

[*13 June*] . . . *The press is progressing. I hope it will be ready to try by Monday next, the 16th, and I trust start fairly off by the following friday*† *the 20th when I hope to send you the order for the next, which I am sure you will use all speed possible to let me have. I had applications from all quarters to see it in operation, among others the "Times," all of whom seem anxious about it, but not more so that*† *I shall be to introduce it. Messrs Conquest & Blair are doing their utmost and you could not show a greater interest in the matter than they do.* . . .

[P.S.] *The main ring which was broken* [*in transit*] *& mended is in the oppinion*† *of Messrs Conquest & Blair just as strong & sound as can be wished. I have not had the pleasure of seeing Mr. R. Hoe yet but suppose I shall soon.* . . .

Robert Hoe (with his family) had indeed timed a visit to Europe with the heartening music of the first Lightning in England—hopefully to be followed by a confrontation at Printing House Square. Understandably preoccupied with that consuming vision, he had little interest in the ancient treasures of his itinerary:

> [*Robert Hoe, London, to New York, 19 June 1856*] . . . *On our way from Sheffield to London we visited Warwick Castle, Kenelworth† Castle, & Stratford on Avon and was much delighted with them; & all the country from Birmingham to London which is one great landscape garden now in its glory. What with all the old ruins, castles, monasteries, cathedrals & abbeys, including Westminster Abbey the greatest of all, and the moslems* [*mausoleums*] *and tombs in them we are quite delighted to turn our backs on them toward living London. . . .*

Robert was most anxious to meet Lloyd. Apparently so unlike the elusive, aloof, slippery Delamarre, Edward Lloyd had been hard to visualize in New York. But Robert found him genuine enough:

> *. . . I find Mr. Lloyd is delighted with the machine. He is a plain clever man, not unlike our old Mr. Colby at 50 in appearance and size. I am persuaded he will do all he can to advance the Machine. It is bringing him in contact with newspaper men here with whom he has had no acquaintance and will be a great thing for him in every respect. . . .*

Conquest and Blair, for their part, had worked like demons. At Liverpool Blair got the press through customs without opening the boxes. When Conquest was sent home in July for a visit (typically, via Birmingham, Sheffield, Manchester, and Glasgow for shop inspections), Blair, working alone, became so nervous and insomniac he was ready to collapse. His work finished, he too returned to America.

The first copies of *Lloyd's Weekly* from the new press were dated 6 July 1856. Circulation—140,000—was handled the most expeditious way the world had ever seen. All had transpired exactly to the publisher's most ambitious plans and expectations. Catling, who delighted in 1878 to interview an African gorilla proclaimed to be Darwin's "missing link" (by a "playful rubbing of its chin . . . we became quite friendly"), and who later enjoyed a "right royal time" in 1893 visiting the "great [New York] works of Hoe and Co.," revealed his employer's wily machinations:

> The necessity of having a second machine being quickly recognised, Mr. Lloyd brought his keen business faculty to bear on the negotiation. Mr. Hoe had been so eager to introduce the machine into England that he sold the first one at half-price, but Mr. Lloyd's offer to take a second at the same figure was rejected. While making arrangements for going out to New York to discuss the matter, Mr.

> Lloyd wrote the inventor that he thought it was greatly to his interest a second machine should be supplied. If there should be any failure with the first, the repute of the Hoe press would be seriously damaged; whereas a second machine would ensure success, and thus be certain to lead to a demand from other papers. Mr. Hoe saw the matter in this new light, and a second machine was quickly supplied on the same easy terms.[8]

Lloyd may literally have saved his own life with his spritely letters; the compelling argument for two machines had, of course, often been used by Hoe himself. While a delightful diversion for those at leisure, a trip to the United States at that moment would have been a burden to Lloyd despite the urgency of the mission. Lloyd reminded a reporter not long afterward:

> [T]he object of my going was to see Hoe, and arrange for two machines on certain revised terms, so that if one broke down, I should have another to fall back on. Just before the time for sailing I received a letter from Hoe telling me that I could have just all I wanted. In consequence of that letter, I did not go. The ship I was booked for went to the bottom. . . .[9]

Lloyd had almost sailed on Collins' ill-fated *Pacific,* which—just 16 months after the *Arctic* disaster—apparently struck an iceberg and sank; Hoe's old friend Edward Collins was ruined.

Although he had made similar concessions years before when introducing the Lightning in Philadelphia, and although enthusiasm for a second press was painfully identical to the Delamarre debacle—both sales fiascos, the Colonel at last relented. After all, Lloyd had at a moment's glance in Paris seen its worth, and his infectious scribblings and exuberant personality showed the Hoes that here was an avid entrepreneur—more than Swain or Delamarre—who would not let the novelty go unnoticed.

According to Robert Hoe III the Salisbury Square order was not fully discharged until 1857, but that included the controversial spare as well. The first press, of course, with its initial "music" had already done what both Hoe and Lloyd wanted. Wrote Tucker:

> [W]hen started on the edition at a speed of 12,000 sheets per hour [the machine] ran so successfully that Mr. Lloyd said this one trial was sufficient, and giving the feeders a handsome present, said they could return to New York. Mr. Lloyd at once gave an order for a second press, and this, together with the successful running of the machine so established its reputation, and also that of the firm, that orders were quickly received from the London *Telegraph, Illustrated London News, Manchester Guardian,* Edinburgh *Scotsman, Glasgow Herald,* and others.[10]

The receipt of these prestigious British orders paved the way for Hoe around the world, without significant competition for the next twenty years. *Scientific American* had again been prescient: "The time was when we used to import our printing presses from London," it had said in May 1856, upon the *Ericsson's* departure with Lloyd's Lightning, "but the tables have turned in our favor, and we are paying back our debt with compound interest." [11]

Almost immediately upon his arrival in June 1856 Robert Hoe had also met Herbert Ingram, proprietor of the immensely successful *Illustrated London News* (150,000 weekly). He was prepared to order an eight-cylinder for the type side of the paper and described his problems with the other side:

> [*Robert Hoe, London, to New York, 19 June 1856*] . . . *He wants to know if we can make a Machine to print the cuts. He says that they have tried to engrave upon curved blocks but they find it difficult to draw straight lines on them and get a proper perspective & the blocks twist also. They dont want to Electrotype the cuts if they can avoid it.* . . .

Although the busy publisher and member of parliament returned with Thomas Middleton, his press builder, to admire the Lloyd machine, Hoe had to admit the New York shop was so occupied he could not promise reasonable delivery. The Ingram sale seemed lost.

Providently, the abolition of the compulsory stamp duty in 1855 had made it possible for both old and new British papers to multiply without restriction; no other press builder could meet their needs. But the most prestigious paper of all, Old Thunderer, was still not among the happy converts. Robert Hoe II, whose training was admittedly "mercantile" rather than mechanical, acknowledged his brother's engineering triumph and now sensed the crucial importance of a special mission to Printing House Square, his most urgent challenge.

As circulation for the London *Times* doubled in the 1840's John Walter II was again searching for solutions. David Morison in 1839 had patented a planetary-type stereotype rotary press in England, but nothing came of it. Applegath's 1827 four-feeder, despite its speed of 6000 hourly, could not satisfy the paper's growth. Wrote Ellic Howe:

> For the first time in connection with printing *machinery* we meet with difficulties related to factors human as well as mechanical. The trouble lay in the feeding, for the attendants were unable to lay on more than two sheets in five seconds (1,500 sheets per hour each, or 6,000 for the team of four), with the required precision. This was due to the large size of the sheet used for THE TIMES, which was difficult to handle and bring to the gripping apparatus from the feed board.[12]

Although Tucker had heard reports that the *Times* was making inquiries about the Hoe type-revolver as early as 1846, Walter—with what Hassard so aptly called "characteristic contempt for American smartness"[13]—had long balked at stepping off the British Isles for help. That year he again asked Applegath to consider an eight-cylinder machine. Applegath remembered:

> In the summer of 1846 I learnt by a mere accident that the demand for THE TIMES was limited by the want of a quicker machine, and knowing by the experience of the DAILY NEWS that 6,000 per hour was about the maximum rate of their four-feeder . . . determined upon changing the reciprocating motions of the forme into a circular one.[14]

Applegath called upon his old associate, Cowper, and they began to draw up new specifications. Again they "consulted" Koenig, whose 1811 patent had described just such a press. But work was slow.

The eight-feeder was expected to double the capacity of the four, but it was fraught with complications. The 13-inch impression cylinders were repositioned vertically around a vertical type "cylinder" almost 17 feet in circumference. It was not really cylindrical, but actually a polygon with as many sides as columns of type. Nevertheless, by October 1848 it was running, capable of 8000 at first, 10,000 to 12,000 if its sixteen boys were really sharp at feeding and taking up the sheets. Although "fruitful in mishaps,"[15] said P. M. Handover, it went to work. On 29 December of that year, as it once again introduced a new machine to its faithful public, *The Times*—goaded perhaps by Newton's chalk talk—made oblique reference to Hoe's Lightning presses:

> [N]o art of packing could make the type adhere to a cylinder revolving round a horizontal axis and therefore aggravating centrifugal impulse by the intrinsic weight of the metal. . . .[16]

This comment, of course, was received at Gold Street with peals of laughter—by 1848 a half dozen Hoe type-revolvers in the United States were proving daily that type could indeed be locked on a cylinder, and well printed.

Still occupied in Paris and fuming about it, the Colonel felt more indignation than humor over the *Times'* obstinacy. His friend Edward Sanderson had revealed other knowledgeable Britons to be similarly puzzled by the intractable paper:

> *Sheffield 6th Dec 1848*
>
> *My dear friend,*
>
> *I returned from London last evening & find my occupation will not admit of my paying a visit to Paris, which I much regret. By this time I hope the "last fast" has reached you & in progress of putting up. I feel confident of success both in Paris & in London also, for W. Braminall told me that Applegaths machine did not work at all right & it certainly would be worth while to get some [of] the Times*

folk over when you have put the Press in working condition. . . .
Wishing you every success I remain

Yours Very Truly
Colonel R. M. Hoe *E. F. Sanderson*

Although the paper purported to be bound by Applegath's contract, the British inventor, financially ailing, was in no position to fight, and in fact might have welcomed an opportunity to escape. Increasingly intimidated by Walter's stubbornness—the decision to use this old idea of Koenig's, after all, was his—Applegath worked hard but was genuinely baffled by some aspects of the problem:

> . . . Why the New York layers-on can place 2500 sheets *larger than the* [London] *Times,* while our hands have only attained to 1500, I cannot pretend to determine; but *the rate of motion* being slower in the *Times'* machine than in Mr. Hoe's (at the same number of revolutions), the superior production of the New York machine must be explained on other principles than an inferiority of mechanical means. . . .[17]

While Applegath needed takers-off as well as feeders, the Hoe had automatic flyers which deposited the sheets neatly in piles. Applegath attributed to "superior slight of hand of the New York men" what were advances in rotary press technology. Nevertheless, the 1851 Select Committee, to which Greeley had addressed wasted words, was advised definitively that "Applegath's machinery . . . is superior to anything in use in any part of the world."[18] The British government seemed in collusion with Old Thunderer.

The highly contrived Applegath machines were built by Messrs. Thomas Middleton & Company, Southwark. Walter took two eight-feeders, and a year later, a nine-station model. A four-feed vertical for Ingram's *Illustrated London News* was shown at the 1851 exposition with considerable popular interest. As the Crimean War began to electrify all of England with telegraphed news the *Times* approached a circulation of 55,000, a staggering load to deliver daily —20,000 of which had to be ready by 6:15 a.m., when morning trains took them to the provincial towns. Advances in stereotyping, fortunately, had alleviated the problem somewhat, as typematter could now be plated for both the older flat-bed four-feeders (which printed the "inner forme"), and also for the nine-feeder, which had its own curved casting box.

Applegath, however, had not a minute's relaxation even with stereotyping. Having satisfied John Walter for the time being, he determined to adapt a new rotary concept to smaller presses, for which there might be some demand. His "Victorian" system employed two parallel type cylinders printing alternate columns of a page, say, columns 1, 3, 5, and 7 on one cylinder, and 2, 4, 6, and 8 on the other. The greater curvature of a small impression cylinder aggravated the problem of handling a polygon as if it were cylindrical, however, and a return to tapered types was again considered. Two years later, having built two machines on this principle, the British inventor had nowhere to go with his idea but R. Hoe & Company:

> [*Applegath to Hoe, 24 March 1855*] . . . *I am engaged just now in working this columnar principle out to its limits.*
>
> *Taking the Times rate of motion as a datum, viz. 1500 revolutions per hour × 200 inches circumference = 300,000 inches. I can make a double action "Victoria" of 50 inches* [*diameter*] *(by means of dealing with the type in single columns) or four revolutions for one of the Times . . . or 12,000* [*copies*] *per hour. A single action Victoria may be only 40 inches dia. (1/5 of Times)* [*yielding*] *7500, manufactured with a set of chases for 250£.*† [*As*] *4 feeders now cost 800£,* [*even*] *600 would be cheap.*
>
> *I dare not assume the rate of motion which Mr. Hoe gives for his four feeder, 10,000 per hour—or 2500 revolutions (his cylinder is rather larger than the Times I believe). Neither have I any right to doubt what is stated both in letters & in print. The Times rate is very quick for the composition rollers and for laying on.*
>
> *Hoe 2500 × 200* [*in.*] *= 500,000* [*in. per hr*] *÷ 50* [*in. paper width*]
> *= 10,000 or 20,000 double action*
>
> *This machine is well worth Mr.* [*R.M.*] *Hoe's notice as it is both economic and durable, and I would show how cylinders of such small diameter can be used with the common type of a newspaper office. The register or falling of the columns into their exact place was a fear which the first machines working dispelled and we have never detected the variation of a hair line.*
>
> *I should like to make a proposition, if you would not deem it too great a liberty, which might be beneficial to Mr. H. & myself. I suppose the 1*[d] *stamp will be removed &* [*soon effect*] *an opening for rapid machines* [*here*] *as well as in the "States."* . . .

The scheme was not likely to attract the Colonel, of course, as his own type-revolving system, now represented in dozens of newspaper shops, was well beyond the developmental stage, and simpler. To some extent Applegath began to recognize the superiority of a horizontal cylinder, and was reportedly talking to John Walter III about it. The vertical machines had indeed reached their limit: Applegath's 1853 eight-cylinder was capable of 12,000—but no more. "[H]is best efforts," wrote Ringwalt, "could not prevent his presses from being supplanted in the office of the London *Times,* by those built by Hoe."[19] American technology would at last triumph.

Unlike the various London and provincial papers which would promptly import Hoe equipment without qualm, Old Thunderer had more to think about than out-performing a rival down the lane—although competition and restricted production were indeed at the root of its desire for faster presswork and earlier street sales. Walter had engaged John Cameron MacDonald as chief engineer (replacing Wetherall) in 1856, and instructed him almost immediately to get new presses. The newspaper's own three-volume history (Morison), with two factual errors in one line, portrayed Walter as curiously unconcerned as to where the purchase could be made:

> The Chief Proprietor, determined not to allow office pride to stand in the way of efficiency, was prepared to buy machines from outside. Acting promptly, Printing House Square ordered two of the latest rotary and horizontal cylinder machines of the type originated in 1846 for the *Philadelphia*† *Public Ledger* by William† Hoe. The machines were ten-feeders. . . .[20]

This official history, however, did not even mention a trip by Walter to the United States to see these presses for himself, although a later (1953) account—with another error—did:

> The People at Printing House Square had been slightly shocked in 1856 when Lloyd, the Sunday newspaper proprietor, ordered one of Hoe's flat-bed† six-cylinder machines from its inventor. When this was known John Walter made a thorough investigation of his machine room, and visited Philadelphia to examine the machine room of the *Public Ledger.* He decided that the Americans were ahead of Printing House Square and all London, and ordered two ten-cylinder Hoes. . . .[21]

In light of subsequent developments, the excruciatingly painful decision to "buy American"—reflected by conflicting stories and typographic blunders—apparently distressed Old Thunderer staff and historians alike for a long time.

After celebrating the success of the Lloyd press, Robert Hoe had the confidence he needed for his confrontation. It was surprisingly easy: the titans were expecting him.

Friday
London, June 27th 1856

My dear Brothers

We have not been sight seeing much this week, having devoted our days to Mr. Lloyds Press room. While I had gone to dinner on Wednesday Mr. MacDonald of the Times, accompanied by Mr. Whitworth of Manchester, Mr. [*John*] *Penn, Engineer of this city of much note, and another person visited the Press. It so happened that every thing was adjusted, so Blair & Conquest explained every thing to them and ran the machine on some paper we had for trying it, for half an hour or more. They went away highly pleased.*

Yesterday I called at the Times office to see Mr. MacDonald. He was very candid & pleasant; said they had their place full of poor machinery and would have to do something & if they ordered our presses would like Mr. Joseph Whitworth to make them. He showed me a map upon which was exhibited the fluctuations in the circulation of the "Times" during each year for some 8 or 10 years. The plan of the map is similar to those in geographies to show the variation of the earth's surface in mountainous regions. It appears that for a long time past it is on a plane or horizontal line showing that they are at their heighest† elevation with their present Machinery, the demand really being greater than they can supply.

I advised they should have 2 ten cylinder Machines which seemed to meet his approval. Soon Mr. Whitworth came in. He said he would undertake to build them any number of Presses like the one at Lloyds but was rather not disposed to undertake a 10 cylinder from the drawings, particularly as we have none running in America. When I told him that we should propose to send out a competent machinist who could superintend the whole affair he thought there would be no difficulty.

We had some private conversation together. Then I soon found his ideas were to take the Lion's share but he moderated considerably when I told him we had thought some smaller establishment would answer our purpose & may be satisfied with less. He said he would do the work much better than any other establishment and the prices should not be more than 10% pr ct† higher. In addition to the manufacturer's profit he would be satisfied with 10% on our portion of the receipts. I promised so soon as we received orders to come to Manchester and see him. It appears to me best not to bind ourselves positively to any one concern. Mr. Whitworth said he would do the fair thing. Mr. MacDonald said the work done by the London Machinists was uniformly bad & cannot be depended upon.

Yesterday Mr. MacDonald called at Lloyd's in the afternoon with Mr. Applegath's brother[-in-law] and several others of the Employees of "The Times" to see the Press start. They came just as we were about to start on this week's edition and of course we had some stoppings. In addition to the bad shape of the paper the side margins were decreased very much in consequence of our wide column rules. The margin left was only about 5/8 on each side so we got some sheets on the Rollers. They seemed to make all proper allowances, and on leaving Mr. McDonald† appointed Saturday at 11 a.m. for an interview with me. He remarked that the matter had occupied their serious attention & Mr. Walter had been consulted & that he presumed the introduction of our machines in their office was simply a matter of time. In the winter they expected to come into possession of some property adjoining when they would prepare a new press room. That I should know their determination on Saturday.

You have now all the news about the Times to this date.

This morning I called to see Mr. Little [Ingram's printer] and Mr. Ingram. They severally went with me to see the machine at work. Previous to going however Ingram offered 3000£ for a six feeder machine for his type side of "The Illustrated London News." He wishes a machine with 12 beds, that is, a duplicate set, and the rings of the main cylinder & the beds so adjusted that he can print 8 pages or 12, the 4 additional pages to embrace his present supplement. When he prints the 8 pages only, which he will have to do until he has a Machine of sufficient capacity or size to print the 12 pages of cuts for the other side, he wishes the turtles to be in their places at equal distance from the extremity of the cylinders. For this I asked 3500£ delivered on board vessel in New York. He protested against the price. On our return he said he wished I would on Monday hand him a proposition of particulars in writing, & proposed payment as follows: 300£ in advance, 500£ on delivery, balance 3, 6, 9 & 12 months. Whether I shall get the order or not on Monday remains to be seen but I mean to be firm.

This morning I rec^d a letter from Mr. Sanderson that he expected to be in London either to day or to morrow. Mr. Newton being sick & not yet out to business I have had to get on upon my own hook. . . . It is perhaps better that we should have stood upon our own reputation and the performance of the Press. I remarked to Mr. McDonald that I had been offered letters to Mr. Mobray Morris and others but I had not considered them necessary. He remarked they were unnecessary.

I am in hopes we shall get away next week, as it is very hot & disagreeable in London at present. It is time for the Mail & I must close with love to you all in which we all join. Hoping to hear from you in a day or two I am

Your affectionate brother
Robert Hoe

To Mr. Richd M. Hoe
& Mr. Peter S. Hoe, New York

A letter to Morris unnecessary—what understatement! An interview with this *Times* officer would have been disastrous. More xenophobic than all the others, Mobray Morris distrusted everything foreign, America more so (he called the Atlantic cable a "great bore"); it was he who told the 1851 Select Committee that the Applegath was the world's best. He was probably the least equipped to know.

The happy Colonel at home thought his brother's progress so far to be all they desired, and enthusiastically sent him on a summer-long continental tour which his family, especially son Robert, greatly enjoyed. The owners of the London *Times*, by scarcely more than an auction nod, had at last agreed to do business with yankees, but they had not lost all their senses.

The prestigious order for two ten-feeders was officially given in September 1856. The machines were to cost £6000 apiece, but MacDonald got the price reduced to £10,000 ($50,000) for both, promising to order a third if the first pair worked well. (The Colonel had calculated that two such presses would cost the firm at most $30,000.) Edward Sanderson, nominally Hoe's agent in England, was asked to consent to the extraordinary discount:

> [*Sanderson to Richard Hoe, 1856 (no date)*] . . . *Now £1000 per press is a heavy deduction, but as Robert can bear testimony every endeavour has been made to prejudice Mr. Walter against you, & I feel satisfied if you can nail the bargain for these presses at the price named every obstacle is removed in the clear course of a most successful career for you in Europe. The "Times" Printing house is one of the shew places of the world & when it is seen that your Machinery does the work, it will be acknowledged that "Hoe & Co." are "Some pumpkins" after all. . . .*

Robert Hoe III justly considered the *Times* order "a gratifying tribute to American ingenuity," for the paper's belated recognition of Hoe superiority had certainly the element of admitting its earlier obstinacy. But Old Thunderer's proprietors (not unlike Delamarre a few years before), well aware of this concession, made sure to include in the purchase a thorny face-saving stipulation: domestic manufacture! The difficulties were compounded enormously.

As leader of an elite corps of newsmen and a national example of the highest principles of newspaper conduct, the London *Times* was ever sensitive of its position in British society, in the European balance of powers—indeed, in world opinion. Perhaps piqued by encroaching yankee domination of industry in many trades and markets, the paper's management acted chauvinistically for its countrymen in denying these slick Americans a complete victory. The *Times* owners probably felt, said Tucker, that "their interests would in some way be compromised if it were known that they were using presses imported from the United States."[22] McNamara suggested "more selfish reasons" but did not specify them, although he intimated the paper may have wanted repair facilities and spare parts close by—thinking exactly akin to that of Hoe's Napier customers only 25 years before.

John Walter nevertheless specified that his new Hoe presses be built in England by British workers with domestic materials. Such arbitrary stipulations were not really strange, of course. Earlier the Walters had similarly demanded that, in selling them two printing machines, another foreigner—Koenig—was not to release other presses within 10 miles of the City of London during the lifetime of the patent. Restrictions like these resembled the schemes Treadwell and others had relied upon, although by midcentury the steam-powered rotary printing machine was hardly a secret or monopolistic device. Mass production and vigorous advertising had not yet shown that these petty restraints would prove quite useless and unnecessary.

R. Hoe & Company had scarcely entertained thoughts of a branch sales office—no less a manufactory—in England, considering the long lull in Europe since the Paris machine had been installed. But growing indications that the *Times* would insist upon domestic fabrication —either by Hoe or by a British machinist—revived the idea:

[Richard Hoe to Robert, London, 12 August 1856] . . . Mr. Swain thinks there is money enough to be made in England to start a shop over there. I have made an estimate of the cost & cannot make it exceed $20,000. dollars† for tools & fixings for a shop. I have thought over the matter in every way & think that although there is surely a great deal of money to be made by starting a shop, & that that is much better in all aspects than contracting to have others build the machine; still a sale of the entire Patent right is the best thing you can by any possibility do. We already do here a great deal of the work by tools & I think with as little manual labor as Mr. Whitworth or anyone else would, & if you engage with any one to make the first machine they should in a great measure be confined to making tools so that all that were charged to us or the making [of] the machine should be ours, & that no tools should be made that did not meet the approbation of the person who superintended. . . .

When the Sandersons disapproved of a Hoe London factory the prospect was shelved. Too, there seemed no time to go about it efficiently; but a dilemma remained.

The *Times'* impeccable reputation in the international newspaper world would admit of no clear glory for the company. Despite the loss of engineering control Hoe would be blamed for any failure; similarly, the local manufacturer—not the Colonel—would be credited with any successes. The construction of so elaborate a machine, demanding careful selection of materials, soundness of castings, excellence of machinework, and patience in fitting and erecting—all had to be meticulous. "I am not sure that it would not be better to make an arrangement with Mr. Whitworth rather than some weaker concern," the Colonel told his brother in England (14 July 1856), "if he will say for what prices he will do the work. . . ." Robert agreed that Joseph Whitworth of Manchester was the only firm in England with the capacity and ability to handle the important assignment.

Many years of service from Whitworth machine tools had been a natural inducement to engage the firm. Whitworth could unquestionably perform as well as his tools (the defective wheel cutter and other irregularities had not yet turned up), having recently distinguished himself anew with his celebrated quest for precision. At the invitation of the British government, in 1854-1855, with a small committee, he had just visited the United States to inspect manufacturing shops, especially to see methods of mass production.

Whitworth had not failed to recognize the importance of printing to the growth of America's industrial strength:

In every State in the Union, and particularly in the north, education is, by means of the commonest schools, placed within the reach of each individual, and all classes avail themselves of the opportunities afforded. The desire of knowledge so early implanted is greatly increased, while the facilities for diffusing it are amply provided through the instrumentality of an almost universal press. No taxation of any kind has been suffered to interfere with the free development of this powerful agent for promoting the intelligence of the people, and the consequence is, that where the humblest labourer can indulge in the luxury of his daily paper, everybody reads, and thought and intelligence penetrate through the lowest grades of society. The benefits which thus result from a liberal system of education and a cheap press to the working classes of the United States can hardly be overestimated in a national point of view; but it is to the co-operation of both that they must undoubtedly be ascribed. . . .[23]

Delicately avoiding England's own unflattering stamp tax situation, Whitworth pointed out that Prussia, with high literacy but a restricted press, had done poorly in giving its people direction, purpose, and contentment.

A contemporaneous New York *Herald* furthermore contended that Whitworth and fellow engineer John Penn had long before (1849) met the Colonel in Paris:

They went with him to the *Patrie* office to examine the new press. After they had examined it Mr. Whitworth said that he had seen the Applegath press, which was being built for the *Times,* and which was not then quite finished, and that if he were one of the *Times* proprietors, or could have his way in the matter, he would let the whole of that thing go, and get in a press built on the American system, as he termed it.[24]

In New York Whitworth toured the Hoe shop and later expressed his great admiration for American workmanship to the *Times'* management. He seemed to be an enthusiastic Hoe ally.

Whitworth's minor brush as a youth with the manufacture of toy platen presses had been little help for this project, and his ongoing armament work, in any case, would soon grow to such proportions he would find little opportunity to supervise any routine shop matters. (Ironically, 10 years after his death the Whitworth firm merged with another armament king—Armstrong—from which a strong printing press division would emerge.) It is nevertheless strange, however, that the huge factory and resources of this master of efficiency and precision would find the Hoe order an almost insurmountable task.

Robert Hoe rushed to Manchester and quickly drew up a contract with Whitworth. Significantly, the British firm was to work from drawings furnished by Hoe. A record of the meeting was sent to Gold Street:

Manchester, Sept. 13, 1856

Rough Memorandum

That Jos. W. & Co. make the Machine for the "Times" and any other machines that may be hereafter ordered. The Machine to be made at the regular prices charged [by] Jos. W. & Co. The price at which the Machines are sold be mutually agreed upon. The difference between Jos. W. & Co.'s invoice and the price charged to consumers be allotted as follows—15% to J.W. & Co. & 85% to R.H. & Co.

Jos. W. & Co. [to] take entire charge and the responsibility for the business, and guarantee all sales for the above commission of 15%.

If Messrs R.H. & Co. employ any other party in Great Britain to make the Machines Messrs W. & Co. are to receive the 15% as patent right, the same as if they have made the Machines themselves. Messrs. R.H. & Co. to be at liberty to supply any Machines from America to any part of Great Britain, without paying any commission to Jos. W. & Co. as this agreement is not intended to deter them from supplying Machines from America. Should any difference of opinion arise the matter to be referred to two Engineers in this Country.

From London, meanwhile, a clamor of newspaper orders had accumulated, unlike anything the company had experienced at home.

Upon the abolition of the newspaper stamp (although advertising and paper taxes remained), there was "an immediate and remarkable" impetus in the British press of the provinces where there had been none before—Manchester, Liverpool, Sheffield, Edinburgh, Birmingham. And each metropolis had at least two large rival papers. In view of an electrifying number of rotary orders, William Conquest (after his short trip home) was instructed to remain in England to erect and service all the Hoe equipment. He settled comfortably into British life, raised a fine family, and soon became known to the Hoe establishment somewhat deferentially as "Mr. Conquest of England." His harried calls upon industrial—newspaper—centers throughout the British Isles made him as elusive and peripatetic as Robert Hoe himself.

"Playhouse Yard," Whitworth's vast factory, tooled up for the big Hoe assignment—over which Conquest was given complete supervisory authority, although the 36-year-old yankee first had to satisfy imported press customers:

Guardian Office, Manchester
Oct. 6, 1857

Dear Sirs,

In consequence of a reduction which we have made in the price of our journal & a consequent large increase of our circulation we find ourselves much baffled[?] by want of means of printing sufficiently rapidly. Under these circumstances I write to urge upon you to hasten as far as you possibly can the making of our four-feeder machine.

When [in August] I gave the order for it to your Mr. Robt Hoe he told me that he would urge you to use your best endeavour in pushing on the making of it but I have since heard nothing of it.

I now write to request you will do your utmost to let us have the machine at the earliest possible moment. Should it be in your power by working day and night or by over hours to hasten its completion, I should be glad if you would do so even though it should increase slightly the cost of the press.

I am gent[s] yours faithfully,
John Edw[d] Taylor

John Edward Taylor II (his father had founded the *Manchester Guardian*) edited the paper 1861-1871, although he was on the staff long before and was proprietor until 1905.

In 1855, upon removal of the stamp tax, the *Guardian* became a 2-penny daily (from a twice-weekly at 5 pence) and two years later, upon reducing its price to 1 penny, it had indeed found itself in a circulation dilemma only the Hoes could solve. The paper's New York correspondent had dutifully—pointedly—described the eight-cylinder version in 1856, but the staff, to its later annoyance, preferred the smaller model, for which it waited nervously.

In 1857 Robert Hoe and family returned to England, this time to find Manchester matters in a sorry state. At Whitworth's shop the young, cocky foreman assigned to the Hoe project, W. Hulse, refused to let Conquest oversee the work, claiming any problems would best be solved by themselves. Hulse and others apparently formed a bad impression of yankees when A. M. Freeland had earlier visited the shop, pilfering ideas. The first press was to be finished by November, but Conquest thought otherwise:

[*Conquest to New York, 16 October 1857*] . . . *The affairs at Mr. Whitworth's shop are much about the same as usual. The end framing is just about finished. A number of the pulley shafts and tape rollers are fitted. The cross stretchers I have had removed since I have been here this time. Mistakes are plentiful yet the drawings are correct; when I point out the mistakes and refer to the drawings or my letters as the case ma*[y] *be, they look at each other and say, whose fault is this, of course nobody's.*

I have free access to the shop but not to take my coat off or do any work more than point out the difficulty. I have felt a good deal like taking my coat off this week to see a man finishing off the first finger motion; he has been about a month on it or more. They have tried various shaped cams, & have got a cast Iron friction roller on the opening stud 1⅜ diam. and say that no other size will work. I have had the whole of it thrown away & it is now nearly completed. I have told Mr. Hulse very plainly to keep to the drawings and we shall be satisfied right or wrong. The first cylinder is near finished. The beds are not commenced to be turned, the counterballance[t] not cast, the surface being fitted on, all the small work well under way. I still think the machine will be done about New Years day. I have seen Mr. Whitworth; this time he ask[t] me how I thought they were getting along. I told him just how the things were for he is scarcely ever in the shop. He says the next one will be finished in one half the time this one has taken.

I was at the office of the Manchester Guardian yesterday. Mr. Garnet[†] *[h]as not settled where the new machine is to be placed but will let me know as soon as he can, so I can not give you any info more than the width of the columns, at present.*

The Times foundation [h]as been suspended on account of their deciding to make a basement under the machine room. This will make the fdn some ten feet high. I have made a plan and elevation. The brickwork is three feet thick built in cement with 13½ in. party walls from top to bottom. I have made it heavy but it seems to suit them very well.

Please remember me to all. I hope the money matters [a depression] in N.Y. will soon be all over—It must have suspended business very much. I forgoten[†] *to say that the [London] Bells Life People proceed very cautiously; they are still making enquiries. Myself and family are quite well. Please to accept my regards. . . .*

Visiting Playhouse Yard again in January 1858, Conquest discovered so many discrepancies even Hulse became concerned enough to permit the yankee freer access:

[19 *January* 1858] . . . *After a thorough understanding he told me to come in the shop at six o'clock in the morning and stay until nine at night if I chose. I then told him that he had now placed me in the position that he should have done some months ago. We had a warm time for a few moments strictly in a business point of view and parted quite on friendly terms. . . .*

To Robert Hoe Conquest reported (11 February 1858), "They can now see that they are *dead beat* and *don't hesitate to own it, & I feel ashamed of their folly.*" But now the press would not be ready until March. Conquest looked after his other charges:

[*Conquest, Manchester, to New York,* 25 *February*] . . . *I am glad to hear that you are all quite well and business improving. Since writing to you last I had a very good start with Mr. Tallis Press and the Paper looked very well the second week* [so] *that I had confidence enough to leave the machine in charge of their pressman. . . . I received a letter yesterday from Mr. Tallis, and another one to day respecting some beds or turtles he want*[†] *making*[†] *to work the Standard newspaper. . . .*

John Tallis, London Publishing Company, had ordered a six-cylinder in August 1857, demanding it be ready in five months (against a £400 penalty); at first he wanted beds for a paper like the *Illustrated London News,* but although the Hoes met their deadline Tallis would have other vexing difficulties.

. . . *I commenced to put up the Guardian Press last Monday week and shall run it to morrow morning, and print the edition on it on*

Monday next when I hope we shall have a good start so as to be able to send you an order for a six cylinder for the same office by the next mail. We have got a very good room for it. The foundation is built of bricks with a stone copeing† and is a very good job. I have commenced to make rollers for this machine. . . . I wish to make a short job of this machine on account of the Times machine which is near completion.

There are six sets of the tapes on the [Times] machine and the others will be on next week. The beds are not yet finished; they were so badly done that I had them turned over again. [T]his affair has so much displeased Mr. Whitworth to see the blunders of his manager and foreman that he told me he would take the matter in his own hands, which he evidently has done—the work is progressing rapidly and the Times first machine, and the [Manchester] Examiners machine, will be done together in about three or four weeks at the furthest.

In respect to having some of Mr. Lloyd's feeders to start the Guardian machines I scarcely think it would be worth while for one reason: the press proprietors[?] of newspapers in London and Manchester are well aware of the speede† of these machines and all of them very naturally expect more or less trouble with the feeders at the first editions and will not confine the trouble to the machine at all; as well as this, they say their men have got to learn and they may as well commence at first as at last, so I think or hope this will meet your views of the matter. . . .

Please to tell Mr. Tucker that every thing is all right about this machine—the Guardian—and look† very neat and convenient also. The engraving you sent with the machine [h]as got a prominent place in the Guardian office.

Feb. 25th†/I have had the Press running this morning; every thing seems to be quite right, it is settled to print the first edition on it on Monday next. We are all quite well. . . .

Conquest was now spending almost all his time in Manchester.

In Paris Robert, distraught that the *Times* triumph was turning to disaster, quickly developed a special plaque for the contracted machines:

Invented by Richard M. Hoe
New York, United States of America
Manufactured by Joseph Whitworth & Co.
Manchester

"This perhaps is but a small matter in a business point of view," he wrote the Colonel (11 March 1858), "but as yet not having realized much profit I should think that we are entitled to some of the *glory*."

Conquest nevertheless felt Whitworth had gained respect for the plucky Americans, whose machine demonstrated some "mechanical excellence" after all. The Colonel already thought the press six months overdue, and deplored the cost of his colleague's false pride in time and money. Conquest's reports continued to distress all:

> [2 *April 1858*] . . . [*I*]*n reference to having a pressman from New York to run the Times presses. I think it a good idea and I will use all the influence I can at the Times office for them to have a man that is thoroughly acquainted with the Machine which I am satisfied would be to their interest. . . .*
>
> *I have got the Times Machine in full running order and will print with it to morrow when it will remain for Mr. Macdonald*† *to see it running on Tuesday next, then it will be taken down and sent to London as quick*† *as possible. The machine works very well but no better than one of ours. Either*† *is the work any better done in any respect after so much talk about the accuracy of their work.*
>
> *I have a good deal to do with Mr. Whit*[*worth*] *lately; he has given me full power to give the men their work and have* [*it*] *done to suit myself. This has given me a good opportunity to point out the ridiculous alterations that Mr. Hulse has made in the machines at no small cost. Without either having apearance*† *or convenience, I am satisfied it will have a good effect eventually, or I hope so.*
>
> *I do not point out the alterations with a view of finding fault or making any thing unpleasant, but I feel as though I should explain how the money has been foolishly squandered away without the slightest benefit, either to the machine or its makers. I received a letter from Mr. Tucker on Wednesday in which he says he is pleased to hear that Whitworths have been obliged to yield at last. So am I in one sense of the word but on the other hand I am not satisfied to see the waste of capital to bring them to a bearing. The second Times machine is up in the shop and about six men working on it. If I were in the shop for the next three months I could finish it.*
>
> *The Examiner Machine will . . . not be done for some time. The Main Shaft has been thrown aside on account of a bad place in the journal. There is*† *many parts have been thrown away both by bad metal and bad castings. There is no dependence in their castings. There has been a man boreing*† *roller mandrels for three weeks and has only one good one yet. I am working in the shop from six in the morning until nine at night, the same hours as the men. Having full power I thought it would be best to be in the shop all the time the men were there, so that they shall have no excuse on that account. Mr. Hulse was a little surprised when I told him I was coming at six in the morning but he has but little to say in the matter. We are on very friendly terms to all apearances but I think it only goes skin deep with him. . . .*
>
> *The Examiner people, and the London Star, are very anxious about their machines but I fear it will be some time. . . .*

Six-feeders for the Manchester *Examiner* and the *London Morning Star* were also Playhouse Yard productions, and Whitworth was actively soliciting others, such as the indecisive *Guardian*—he was especially loathe to see another American machine brought to his home town. Taylor, Garnett & Company was indeed in a quandary.

Despite frequent references to *Manchester Guardian* people being pleased with their four-feeder—and likely to take a six if payment could be eased, Taylor's staff had done some bad calculating and were not happy at all. Long in anxious competition with the *Examiner*, the *Guardian* had projected its 1857-1858 circulation very conservatively. Larger subscriber increases and delays getting the Lightning into production were a bittersweet surprise. The press had still been at sea when Peter Allen, business manager (and Taylor's brother-in-law), had the sinking feeling it was too small—"American machinery," wrote the paper's historian, "already unsuitable before it was introduced."[25]

At the end of its trial run Allen agreed the press was working well, but maintained its purchase was a mistake and that a six- or eight-feeder was "looming in the distance." Even with a new machine apparently installed within the year the paper's circulation was still a problem. In December 1858 Taylor was negotiating for two eight-cylinder machines, insisting the Colonel take back a six-feeder at full price (less 10 percent depreciation) and that the new equipment be discounted 15 percent. The company took these intimidations in stride, determined to keep the *Guardian* out of Whitworth's grasp.

After eighteen exhausting months Joseph Whitworth & Company had at last cleared the floor for new work:

Manchester
April 23rd 1858

Messrs. R. Hoe & Co. [*New York*]
Gentlemen,

. . . As we may shortly receive an order for a four cylinder Printing Machine we should be glad if you could send us drawings of that size by an early mail. Your previous letters intimated that you have a set of drawings completed.

The first "Times Machine" is dispatched & Mr. Conquest & a staff of workmen are in London attending to its erection. The other machines are progressing & in two months all will be cleared out of our workshops.

Now that we have acquired a knowledge of these Machines we hope to get along more briskly.

We are, Gentlemen

Yours very obediently,
Joseph Whitworth & Co.
W. Hulse

More briskly indeed! Although Whitworth was still the best name in machine tools, the Colonel had lost all respect for the fumbling master of precision:

[*Richard Hoe to Conquest, 7 February 1859*] . . . *Can you give me a description of Whitworth's universal horizontal boring machine or table, about its size, usefulness & cost[?] You need not tell W&Co. that we have taken the advice of one of your letters & got up patterns for a Universal shaping machine. I expect by the time you return home we shall have a number of them in use. . . .*

The extra Tumblers which you wrote Mr. Tucker W&Co. had put on the other end of the finger rod will I should think cause the fingers to break more frequently—a contrary effect from that intended, I suppose. Are they using their own make of finger[?] I suppose you have rec[d] *those we sent to Mr. Lloyd for the Times press. If you see Mr. Macdonald*[†] *you might say to him that I should very much like to make a ten feeder for them & would send them one that would give them entire and perfect satisfaction & that we would ship it within six months after receiving the order here.*

I suppose that Messrs. W. & Co. have not added much to their reputations by their failure, I may call it, in making these machines. I also see that W's guns have burst as well as his presses. Do any of the parties you talk with about presses want him to build them[?] From the patent & drawings which bro. Robert sent me of Applegath's new press I do not think it will even amount to anything & the sooner he gets it done the better. . . .

Hoe-Whitworth relations declined steadily.

In July 1858 Whitworth submitted his bill for four machines, in which the two ten-feeders were listed at £5744 and £2924, or about $43,345 together—leaving the American firm less than $7000 from which to deduct its own related expenses and salaries. There would be no profit. When Whitworth "directed" Hoe to deposit £5000 to his account the Colonel cried "impudent"; he accused the "Manchester humbug" of "unpardonable" stalling to consume the life of the patent and padding the bill with unnecessary tools, drawings, and patterns. He had already instructed Conquest to record every factory error—and its cost, and to weigh the second machine's parts during installation to see if casting costs had also been inflated.

Conditions were exacerbated when the first machine was delivered in London. While it is easy to understand that machining errors might have slipped past Conquest, one difficulty reported by McNamara could not have been foreseen:

> It is probably not too much to say no machinery of such magnitude was ever constructed [in New York] with greater care or nicety. Every part was not only fitted accurately, but every contingency was provided for. No better evidence of this can be shown than the fact that the first press constructed in England according to the plans furnished could not be started. It is related that the English workmen had not calculated the great strain of the various tapes, upon which the atmosphere had no little effect. The spring of the tape shafts from this cause necessitated barrel bearings, which it would seem they had overlooked.[26]

The successful trial in Manchester suggests that London's wretched climate was entirely at fault. Another problem was employee reaction:

> [*Robert Hoe, London, to New York, 25 June 1858*] . . . *It is now a week that we have printed the inside form of the 56,000 Edition of the* [Times] *supplement and have every night done better. The feeders at the speed of 27 revolutions we run perform very well and the machine is now getting in thorough condition. The Applegath faction have acted very ugly at times but Conquest got along with all of them pretty well until last night, when he discovered the color suddenly to fail & the form to fill up. On examining he found the fountain had been filled with the dirt, paper, grit & refuse from their old machines which delayed nearly two hours. Mr. Macdonald*† *was present most of the night as he was aware there was some underhandled influence at work & he told me this morning he was never was*† *so angry before & that some of them will probably have to walk for it.* . . .

Robert, so much in love with "living London," followed this sickening report of sabotage with a retaliatory outburst:

> . . . *London is rendered shockingly disgusting about the river Thames. The whole body of water is black with corruption. On the bridges the stench is intolerable from it.* . . .

After all these frustrations the brothers knew only one way to set things right. The Colonel passionately insisted that Walter as a "privilege" let him manufacture the third press in New York to show how well it could be done. But no order came; instead there was some new high-level perfidy in London as well as Manchester:

> The Hoe machine was first used in Printing House Square in August, 1858, and, after some accidents, settled down to print so rapidly that it necessarily stimulated Walter. Hoe's machine was a sheet-fed rotary horizontal. The Applegath was a vertical rotary. Walter now broke it up [and began thinking about modifying the Lightning]. . . .[27]

Hoe's most prestigious British customer would not likely remain so. In Moxon's land there was no great waygoose feast for the successful inventor—and now press "joiner" to the world—as had occurred in Philadelphia and New York.

The ten-feeder Lightning at the London *Times* was nevertheless a tribute to Hoe after all, despite its careless fabrication. Rated at 20,000 impressions, it would outstrip anything of Applegath's—the company's only real competition in this area. It required 24 men to operate (at a weekly cost of £60), even though there were no takers off. Without the additional boys and without extra tapes and guides so common in British shops, the "Mammoth Printing Machine" was a formidable marvel nevertheless. *Printers' Register* in 1867 was still urging members of the fraternity to judge for themselves:

> We should like some of the old school of slow-coach printers to go and see these [Hoe] machines at full swing! It would astonish them beyond measure to see the host of pale-faced men and lads at work in the midst of the wilderness of wheels, cylinders, and straps, and to hear the deafening thunder that accompanies the storm of sheets pouring down on all sides faster than the eye can follow them, or the mind realize their numbers.[28]

Although the domestic ten-feeders gave considerable trouble—MacDonald found the safe sustained maximum to be only 13,000, the name "Hoe" had caught on. Between 1860 and 1870 Hoe machines, with four, six, eight, and ten stations, would be more widely used by the principal London papers than any other, some, like the *Daily Telegraph* and the *Standard,* running as many as five each. Applegath himself finally developed a ten-feeder stereotype press for the *Morning Herald,* but he was fast losing the race. The *Times* installation had repercussions for the whole United Kingdom.

As penny papers multiplied upon the elimination of the stamp tax they threatened to swamp the long established, better journals; but the Hoe rotaries permitted the old papers to reduce running expenses and remain competitive. The rewards for all were even greater. Hassard wrote:

> [T]he example of *The Times* and the other great London journals, in introducing the American invention, was quickly followed by all papers of importance in England, Scotland, Ireland and Australia. The result was an immediate and unexpected increase in circulation, bringing with it a corresponding augmentation in advertising, and followed, of course, by a considerable reduction in the subscription price. This again increased the circulation, so that it is, perhaps, not too much to say that this invention was largely instrumental in preparing the people for the extension of the suffrage accorded to them not long afterward. . . .[29]

The contribution of press technology to the democracy of an entire nation—an empire!—was indeed not too great a conceit. The Hoes bore it without affection.

But while the brothers could accept praise with equanimity, they could not tolerate rank effrontery. They refused to pay Whitworth's enormous bill. As with their interminable quarrels with Delamarre and Dyer they would not abide what they considered flagrant business injustice. They consoled themselves that any competent judge—indeed, the whole engineering profession—would concur with their every complaint and contention. On both sides of the Atlantic the Hoes busied themselves with affidavits and testimony, while awaiting Whitworth's justification of his expenses and methods. Whitworth instead criticized the drawings.

Most of the disagreement centered around the machine's paper grippers or fingers, which Hulse had repeatedly executed poorly and attempted to rectify by modifying the design. The British firm blamed faulty draftsmanship:

> [*Conquest to Robert Hoe, 9 September 1859*] . . . *I cannot conceive how Mr. Whitworth can concientiously*† *say that thier*† *mistakes can in any way be caused by the inaccuracies of the drawings sent to them. I can solve the matter in no other way than they were inexperienced in the matter and deficient of a pure good feeling to be taught anything in the matter.*
>
> *The finger motion in particular* [*has*] *given them a great deal of trouble; for this reason the corner of the impression cylinder being cut out to allow the tail of the finger rod to make a revolution. . . . I was told by some of the men that some of the finger cams had been broken three times but Mr. Hulse said in the office in your presance*† *that they had been broken five times. On my next visit to Manchester Mr Whitaker* [*a Whitworth machinist*] *drawn*† *my attention to this matter wich*† *was at No. 7 Cylin^{r}. I inspected the finger motion and apearantly*† *the motion looked all right yet the corner of the cylinder had escaped my notice. I then told the man to turn at a good speed wich broken*† *the cam instantly. I then detected the difficulty at once; Mr Whitaker had seen so many broken previous to this that his mind was perfectly made up to see the* [*failure*]. . . .
>
> *I know of no drawings of the finger motion that are wrong in any respects or that would lead a stranger astray with a moderate amount of care. The revolving finger cam cannot be planed to its proper shape in any other way than by fitting of it to its own position. . . . This whole matter I pointed out to Mr Whitworth before the machine left the shop. It is of no use for anyone to say that the finger motions will not work to the drawings. . . .*

Conquest located Whitworth employees who admitted that Hulse had told them to pay no attention to his directions and advice. There were other intrigues:

> [*22 November*] . . . *In reference to secretly altering the drawings* [*—this*] *is quite a new feature of the affair to me. If such a thing has been done of course I am not aware or I should of*† *told you. Some are tracings of the mach^{e} and some, drawings. If the drawings have been altered it would be a very hard matter for me to say that I could detect it, . . . especially the drawings of the Finger Motion. . . .*
>
> *In reference to the character of the mistakes, I given*† *to Mr. Macdonald*† *about 33 in number wich*† *I know to be facts and wich I have a copy of. . . . The only answer I can make as to why the mistakes did occur is as I have previously stated, for want of diligence and experience. . . .*

The Hoes demonstrated with diagrams and models that if the proper steel—Swedish, not English—had been used, and the drawings followed, the fingers would have been fine. Surreptitiously, however, in

New York the finger motion was subtly redesigned to reduce future trouble. MacDonald, torn between British allegiance and sincere admiration for the yankees, for months did his best to mediate, but neither side would budge. At last a formal referee was engaged.

Testimony before the arbitrator, J. E. McConnell, a respected engineer, began November 1859. Both Conquest and Robert Hoe made effective presentations, but neither as surprising as another witness:

> [*Robert Hoe to New York, 9 December 1859*] . . . *Mr. Farthing* [*Whitworth's draftsman*] *at first . . . demurred at giving his testimony,* [*but*] . . . *his scruples gave way & he gave his testimony on every point in our favor. Mr. MacD. said the umpire was quite staggered to hear it all. He (Farthing) said that in his opinion the drawings were complete & sufficient. . . .*

Meanwhile, curious and interested parties would make frequent trips to Joseph Levy's office to see the latest New York-manufactured ten-feeder erected and run. Somehow MacDonald knew it had improved fingers although, according to the Colonel, a micrometer was needed to spot the difference:

> [*Conquest to Robert Hoe, 1 March 1860*] . . . *It seems very singular how Mr. Macdonald*[†] *should say that the finger motion should be so very different to that of their own machine. I have never measured any part of the motion therefore could not say to any one that there was any difference; . . . therefore if they have got any dimentions*[†] *of this mac*[e] *they have got it by stealth. . . .*

Despite this privileged information (probably given him by the Hoes), or more likely because of it, MacDonald played Hoe's game. He brought Whitworth personally to see the *Telegraph* machine, he told them (26 March 1860), "& was particularly careful in pointing out to him how beautifully smooth & easy the action of the finger motion was. . . ."

But these peripheral antics had no influence on the referee, who handed down his decision on 5 April 1860. Essentially he upheld the Whitworth invoice and furthermore judged the Manchester firm entitled to keep all drawings and patterns in its possession and to continue making and selling type-revolving machines, if it wanted to, persuant to the original Hoe-Whitworth contract. It was a sad defeat, mitigated only by the incredible commercial success of the invention throughout the British Isles. But there was no time for tears. Almost immediately the Colonel had to reorganize his materials and reshape his arguments for yet another defense of the coveted property on foreign soil.

With Robert Hoe himself hustling about England in search of press orders, Whitworth's own interest in the fussy work, like Napier's, soon evaporated. But now the revolving-type patent would soon expire, and unless an extension could be obtained other wily Britishers would rise quickly enough to challenge their monopoly.

The Colonel now resolved to hold onto the precious rights, extend them if possible, and manufacture presses in England as well. Because he was an alien, the Lightning had not been patented in his name; Catling himself had been puzzled to find the documents under "Newton." (Hoe's British patents had become so extensive by 1859 he asked Newton to make copies of all of them, and Rowland Hill's seminal 1835-1836 work as well.) Nevertheless in 1861 the inventor was duly prepared for the auspicious hearings on the matter. Catling was there:

> It came before the Privy Council on February 1st, Lord Cranworth, Lord Chelmsford, Lord Justice Knight Bruce, and Lord Turner being present. The scene comes vividly to mind, as I was called upon to exhibit to their lordships the technical method of securing the type in the "turtle" that was before them. With Colonel Richard Hoe was Mr. Edward Lloyd, Mr. Edward Levy (now Lord Burnham) and the legal gentlemen engaged.

Hoe's petition stated "the machines were very expensive, and consequently for many years the patentee was unable to induce newspaper proprietors to give orders for them; that he obtained no orders in England till after the alteration of the Stamp Duty, in June, 1855; that the first order was in 1856, for *Lloyd's Weekly Newspaper,* and that up till that time the patentee had only received orders for fourteen machines." After a boggling glance at the drawings and specifications Lord Cranworth spoke for the Council:

> "Looking at this patent, we are all satisfied on the score of its merits, that it is a most useful invention, possessing the great merit and charm of simplicity as far as such a subject can be made simple. . . ."[30]

The Privy Council granted an extension of five years beyond the 14 of the original (1847) patent. This was at least enough to encourage R. Hoe & Company to establish a London factory. No one knew better than the Colonel that in five years the technology of newspaper presswork would undergo further revolution.

In 1862 the company opened a proper sales office (and repair shop) in London. Not surprisingly, it was located in Crown Court at 13 Salisbury Square—immediately adjacent to Lloyd's own offices and pressroom. No doubt William Conquest was expected to escort prospective customers next door to see Lloyd's machines in motion and receive some enthusiastic comments from their ebullient owner. He seldom failed them. Throughout the British Empire the Hoe salesrooms were a mandatory stop for newspaper people. For the Hoes it represented a return—after only 60 years—of the family name in a manner their father would greatly have enjoyed.

Lloyd's weekly newspaper grew steadily: 347,000 in 1863, 412,000 two years later, his Lightnings taxed to their limit. But R. Hoe & Company was ready with new equipment. Possibly because of Lloyd's insistance on printing with stereotypes, the Colonel and Tucker applied themselves to the "Two-feeder Rotary Perfector," an assignment that interrupted the company's development of a more general web perfector. It was, however, an easy task, except for one trouble area:

> The machine was of double width and printed two copies on a sheet, side by side, from duplicate forms of stereotype plates. The two impression and two type cylinders were all two feet in diameter, and placed in a straight horizontal line, the impression cylinders in the center and the type cylinders, with their inking apparatus, on the outside. . . . The sheets from both feed-tables followed the same path through the press, and as they issued, a vibrating switch directed them to the right and left alternately to two sheet-fliers, one at each end of the machine. The cylinders printed a sheet at each revolution, and as it was printed on both sides the product of the press was equal to the product of an Eight Cylinder Type-Revolving Press. But in operating the machine, the 'set-off' gave great trouble and various devices were tried, such as shifting the tympan often, heating the impression cylinder, coating it with oil, etc., but the best remedy was found in finally obtaining a quick-drying low-priced ink. . . .[31]

Mr. Lloyd was again fully satisfied; two years later he ordered another. Capable of 90,000 impressions hourly, these machines were also taken by the *Daily Telegraph* and *The Standard*—and two additional ones for Lloyd.

Lloyd's pressroom—presumably containing these historic presses—was destroyed by fire in 1873 (oily rags were blamed). In 1876 Lloyd bought the *Daily Chronicle* and a year later built his own paper mill, "Bow on the River Lea," the largest in the world for its time. It pioneered in the making of papers from esparto and wood pulp.

Joseph Hatton, veteran London correspondent for the *New York Times,* visited the remarkable publisher:

> I was directed to 12 Salisbury Court, and there, in an unpretentious little room, I found Mr. Edward Lloyd, a hale, hearty, middle-aged, florid-complexioned, white-haired gentleman. He introduced me to his son, a stalwart young fellow. . . .[32]

When Lloyd bought the *Chronicle* he knew exactly what to do. Hatton continued: "Immediately on the conclusion of his purchase Mr. Lloyd called to Messrs. Hoe, of New York, to make him eight thousand pounds' worth of machines. . . ." Each of the new presses was to have a folder and cutter (to sever all folds for the reader, who was heretofore expected to tear them). These were the first machines in England with these advantages; the first issue emerged May 1877, and a year later circulation doubled. Lloyd had to call on Hoe again, asking,

> . . . his New York friends to make a double machine that should print two complete *Chronicles* at once, cutting, folding, counting as before, but using up a web of paper double the previous width and weight, and capable of printing 25,000 per hour. . . .[33]

It took Hoe more than a year to accomplish this feat, but the company did it:

> R. HOE & CO.
> Printing Machine Makers and Engineers
> Tudor Street, Whitefriars, Grand Street, New York,
> and Dearborn Street, Chicago
>
> *London E.C. Oct. 19th 1877*
>
> *To Messrs R. Hoe & Co.*
> *504 Grand Street*
> *New York*
>
> *Gentlemen*
>
> *Your favor of the 5th inst. to hand, & with thanks for the number of the "Christian Advocate." The Article on the Web Machine is the most comprehensive one that I have seen & must do your Firm much good.*
>
> *The Evening Times Machine at Glasgow is working very well. We have their Bill already made out—but as our man has only left the Machine in their hands for two weeks we did not think it well to present it yet; however we shall do our best to send you some cash. [H]owever, business being so bad even people who are supposed to have plenty, do not care to part with it in these times.*
>
> *I fully expect to have got an order from the Daily Telegraph this week for a Web but now the great Murder Case is over Mr. Levy does not seem to care about ordering & says that his paper may go down to nothing for all he knows. What they realy† want is a reduction of £205 as stated in My last weeks letter, that is to say, they want the machine with the folder & also to lay down the Sheets openly for the Sum of £2975.0.0 otherwise they will not order. . . .*
>
> *The Chronicle Machine is working very well & no sign of creasing the sheet. Where we have any orders pending I do not fail to follow them up [but] even the Chronicle folks of Newcastle will not order the Machines until there is only sufficient time to have them finished by about this time next year. They say that there is a probability that something new may come in between now & the time they require the machines.*
>
> *We regret to hear that the drawing paper [we sent] did not suit & that the price was no less than you have bought it for in New York. . . . With kind Regards*
>
> *Yours Very Truly*
> *William Conquest*

For Lloyd's unflagging enthusiasm over Hoe machinery the Colonel for a long time reciprocated by buying quantities of printing paper manufactured at Lloyd's mill. Like steel importations, incoming paper had many problems:

> [*Richard Hoe to Lloyd,* 10 *December* 1866] . . . *I have been waiting in the hope of being able to write you that we have sold the paper, only the first lot of which has yet arrived—did you insure it or not*[?]
>
> *I have today received an order from my friends of the* [*New York*] *Tribune for you to make for them one thousand* 1000 *Reams thirty seven inches by forty seven & three quarters inches,* 37 *in. x* 47¾ *in., say Sixty four pounds to the ream* (64 *lbs*) *of* 500 *sheets at* 4½¢ *per pound* . . . *packed in the same manner as the lot rec*[d] *free on board of vessel in London. Please get freight as low as possible. If the paper suits them you will I think have a steady customer & a a*[t] *good one. I should have succeeded sooner had they & also some of our other friends not been under contracts.*
>
> *We have not found any one who wants the size of the lot we have* [*on hand*] *& I fear it will not bring cost, but we may yet meet a purchaser. We find there was* 60 *odd reams badly damaged with water on the voyage.*
>
> *You may send this lot for the Tribune direct to us, that is, ship to R. Hoe & Co. New York, with bills & bill of Lading & consular certificate of printing paper for Newspaper unsized so as to enter it on the lowest duty, as you will remember sized paper is charged* 10% *more.* . . . *How do you get on at the mill*[?] . . .

Hatton surmised the strong synergistic relationship between the two creative men: "The very latest inventions in the generation and use of steam, the newest ideas of Hoe in the way of printing, are pressed into the service of these two papers [*Lloyd's* and *Daily Chronicle*]. Colonel Hoe is Mr. Lloyd's ideal machinist; Mr. Lloyd is Colonel Hoe's ideal newspaper proprietor."[34]

Between 1856 and 1862, 35 Hoe type-revolving presses in all had been erected in England. Wilson and Grey in 1888 said of them:

> The machines were admirably built, and the small amount of trouble they gave, coupled with the splendid results, soon rendered them popular, not to say universal, with newspaper proprietors who were fortunate in having to cater for an increasing demand.[35]

As the firm's business in Great Britain grew in importance R. Hoe & Company was obliged to reconsider its thoughts on manufacture in England. It was not an easy decision.

> *14 Torrington Square*
> *London, May 8*[th] *1866*
>
> *My dear Bro's:*
>
> *I this morning rec*[d] *yours of 24th ult. I cannot say that I am very much surprised at your aversion to making a manufacturing concern of our business out here. The small* [*Salisbury Square*] *shop, as it is now arranged, can be managed by Mr. Conquest & a visit of examina-*

tion once or twice at most each year would enable us to keep things from getting wrong over here, whereas a manufactory would require some one on whom entire confidence could be placed to be here all the time. I mean business confidence, for Mr. Conquest could not do more than see well & thoroughly to the manufacturing department, but it seems a very great pity to see so fine a business in all your hands pass away, as I think must inevitably be the case, unless we conclude to manufacture here.

It appears to me that it will be many years ere wages will fall with us at home to a point low enough for us to make & ship profitably to this market. I know that we have already here a fine business & should we commence making here we could have just as large a concern as we might desire. No doubt it would be necessary to invest all the profits over & above a good support for one of our partners here in improving the facilities of the business for, say, the next three to five years.

I have looked for a suitable place for the business but did not intend to enter into any lease or purchase whatever until we had talked over the matter face to face. The little jobbing shop Conquest will make pay without any difficulty, & he will manage it well. It is desirable that we should decide either to let the matter end as it may & only keep our trade here just so far as we can do without any addition to our shop here—making all large machines in N.Y. & getting as much profit as possible.

One thing is certain, I take it, a rise in prices would finish our business here in the shortest possible time so that if we cannot make at home & sell out here for the same prices we have been doing we must then either make up our minds to do without this business or make our machines out here. I have many times said to you both that having no sons to follow me I had not for myself that interest in the establishment of a concern here. I would otherwise feel so that after giving you when I get home all the pro's & cons of the subject as I see it. I intend to allow the decision to be made without any force on my side either way.

I am convinced that it is very important we should be able to do our work at home very much more economically with more room so that we might have more tools to be worked by cheaper men & boys as is done by all the most successful shops here. The saving is very great in having tools of a character that lads cannot do bad work & must do it rapidly. . . .

I hope we shall experience no difficulty from Quarantine on arriving in our bay. It would be extremely unpleasant to remain there any length of time. I presume the Cunard ships will not be detained unless there is disease on board & as they do not carry emigrants it is most likely they will have clean bills. We anticipate very much pleasure from receiving our acquaintances with our home relatives and friends.

Your very affectionate brother,
Richd M. Hoe

To Messrs R. & P.S. Hoe

Even as the partners deliberated, the "little" shop was becoming a sizable enterprise:

June 9th 1866

To Robt Hoe Jn^{r} [III]
New York
Dear Sir,

. . . I will make enquiries about a saw maker from some of my friends in Sheffield. I never meet with any one in London in that line of business. In Reference to me† taking a trip to N.Y. is† entirely out of the question. I cannot see the Phantom of a Chance, so therefore do not entertain the idea, althought† I hope the time to arrive some day but certainly not just now.

Our little Shop business is fairly under way; we have now 7 Machinists & 2 Labourers at work, & we have plenty to do for some time to come. We work to a disadvantage at presant,† for we ask all the work we do, we have some new tools to make wich† is rather a draw back to us. There is one difficulty that we shall have to contend with. Wich must be over come by our doing a better job than our neighbours & at the same time we must make a charge high enough to pay a fair profit. There are some jobs where thier† very apearance† will command a large profit, others where we cannot clear over 15 or 20% so that we must pay due attention to the looks of the work done, & charge in accordance.

Most new jobs here are done by contract & as soon as we can bring some men up to our class of work, we shall be able to do work as cheap as any one here. As soon as we can get the books in in† order we intend sending a Monthly Statement of items wich will just show what we are doing. With kind Regards

Yours truly
William Conquest

Robert Hoe III was growing interested in the British office (and bibliographic London in general), and indeed the partner most likely to be entrusted with its operation.

A year later the little shop was capable of making type revolvers—and as cheaply as the Colonel predicted, one instance most unpleasant. A man named Weston apparently contracted to buy two eight-cylinders with secret intentions to ship them to the New York *Herald*. Suspicious, Hoe would not sell, and had to face court action to do so. Of course, many legitimate orders in addition to the "Weston affair" came in.

The Salisbury Square location quickly became cramped and in 1869 a factory was leased at Dorset and Tudor Streets, Whitefriars, large enough to accommodate up to 100 workmen. The young partner was on hand:

[*To Robert Hoe, Liverpool, 11 June 1869*] *. . . As we were not all ready to go to London I went up for a couple of days to take a look at the Shop, & matters generally.*

The new premises are most satisfactory in all respects. They are in excellent order, well built, with a fine light on three sides. I was surprised to find so nice & convenient a place. There is ample room for further enlargement when necessary & a comparatively small sum laid out in alterations would give not only about double the floor room, but make it more convenient. The main front is facing the Thames. . . .

In September Robert brought John MacDonald to see the new factory; on a warm day he was more likely impressed than he would have been a few months later:

[*Conquest to Robert Hoe, London, 13 November 1869*] . . . *We have had some cold weather & 7 days of frost. We have no means of Boarding our shop so have now commenced to turn the exhaust steam from the engine through the bottom shop. Then I am in hopes that enough will find its way through the floor to warm the upper floors to that we shall not take any harm from the cold. It is to our interest to have the shops moderately comfortable. . . .*

By the end of the year three large presses were nearing completion, one of them Lloyd's perfector, the other two rotaries which Conquest was painting "Slate color of Rather a bluish tint" similar to that of Whitworth's tools. The shop's profit that first year was over $12,000. Delighted, the brothers sent over a large quantity of the Colonel's own pippin apples, but they suffered badly at sea and only about five barrels were saved.

Under hard times in 1870 the workforce was reduced to 50 men and boys, the latter being trained as distinctly "Hoe" apprentices. Conquest had now added stop cylinders to his inventory, but Robert soon had doubts that creeping diversification was wise:

[*Robert Hoe to Richard, London, 16 January 1872*] . . . *I have written to Mr. Conquest to say that instead of commencing to introduce and make new and various printing machines in England we ought to do nothing to interfere with the Rotary press business. . . . To make one perfecting press in England must be at loss unless we can follow it up by numerous subsequent orders to be got for them. You have already undertaken a single stop, country cylinder, and* [*now*] *propose a wood-cut stop cylinder press also. It seems to me that if you introduce examples of these machines and others that we have in hand into England in detail, if we cannot supply the demand there for them, other manufacturers will be persuaded to do so. . . .*

My own opinion is decidedly to confine the London shop in its present size and capacity to the Rotaries. . . .

The Chicago fire and the resulting desperate need for printing equipment dramatically turned the company's attention away from England. Robert reiterated his position a few months later, proposing that if the existing London branch could not command sufficient business it should be shipped back to New York. He added:

> *[To Richard Hoe, 2 April 1872] . . . There is a desire on the part of Mr. Tucker and others of the firm to establish a shop instead in Chicago, the great western trade center, whenever the London shop can be transferred over. . . . This is not seriously proposed at present. My own judgment is in favor of increasing here [New York], as business demands concentrating all hands and means where we can be together and supervise, improve, and control. . . .*

And still later went on:

> *[To Richard Hoe, 19 August] . . . What we have to fear most of all is foreign competition here. All that you do to teach John Bull our trade he will be quick to use to his benefit and incidentally to our personal business disadvantage. This may be a narrow view but national results must be expected to follow national causes. Ideas and wares circulate very freely now a days between the old and new world. . . . The weather here is so confounded hot. . . .*

A few years later, back in England and busy, Robert seemed to have forgotten his objections:

> *[To Robert Hoe III, New York, 11 December 1874] . . . On Saturday, you know, we only run the shop a half day and noon comes very soon these dark mornings. I have come to the office through a disagreeable rain. . . . Scarce a day passes without our being able to see or write to some of our customers who are on the lookout for Web presses to fold the papers, and to learn about our machines and what we are doing. . . .*

He was surprised that Conquest's weekly salary of £9 ($40) had not been increased in fifteen years, and recommended a £50 bonus for 1873. At the same time there was so much routine administration at the London shop Robert regretted that nephew Richard March Hoe (Peter's son) "cannot settle down here," to the potentially large and lucrative business. This young Hoe had other interests, however, and the aging senior partners were all too conscious of the few eligible men in their families. He gave Conquest a £100 check as a wedding present for his daughter Lucie as well as the £50 for himself. There was no longer any discussion about closing up operations in Britain.

Conquest gradually confined himself to the management of the shop and Robert functioned as salesman to all of England (although living a major part of each year in Paris). The British branch prospered. "We have some first class workmen—the best to be had, steady fine young men," Robert recorded (26 June 1879). In the early 1880's the Tudor shop, like that at Salisbury Square, was judged too small, and larger quarters were found "lying between the Borough Road, Mansfield Street, Earl Street, and Dantzig Street." It was a four-acre block, but actual floor space amounted to about seven acres;

almost 1000 employees were at work. Conquest fulfilled his duties "with great ability and fidelity," wrote his admiring son-in-law Tucker, "until his death in 1887." Conquest's passing seemed to his anglicized family the lonely end of a glorious era, and hoped it would not be overlooked by old friends in the United States. An 1887 London *Times* carried only a tiny notice:

> On the 7th inst. at his residence, 2, Cople-villas, Thurlow-park-road, S.E. WILLIAM CONQUEST, aged 66; for 30 years London Manager to R. Hoe & Co., New York and London. American papers, please copy.[36]

Henry R. Vizetelly (1820-1894), a wood engraver, printer, and publisher, had started *Pictorial Times* (1843) and *Illustrated Times* (1855), running the latter to 200,000 when Herbert Ingram decided its competition with his *Illustrated London News* was intolerable. (Vizetelly's paper featured a series by caricaturist C. H. Bennet, "The Origin of Species," wherein human "types" emerged from animal forms in "Darwinesque development.") Ingram bought the paper in 1859 and squelched it.

Vizetelly was deliberating another sheet when suddenly Ingram and his son, traveling in America, perished in a steamship accident on Lake Michigan. Mrs. Ingram asked the accomplished rival to help run her vast enterprise. It was a big job.

Herbert Ingram (1811-1860) had been apprenticed to a printer at 14; he then established himself in London as printer, bookseller, and news agent. Early in his career he observed that whenever issues of the metropolitan newspapers contained illustrations, they sold better. He thought of publishing a paper full of pictures, but soon realized that wood engraving was so slow a process even the most casual deadlines could not be met. He solved this problem brilliantly by parceling work to several engravers, even dividing single large figures into six or eight blocks (glued or bolted together) among as many workers. The life of these wood engravings was remarkable: the *Illustrated London News* Christmas issue, 1882, was 425,000 impressions, after which the cuts were still in good condition.

Ingram's novel paper appeared 1842, its first 16-page issue containing 32 cuts. It was a sensation on Fleet Street. Mason Jackson, a contemporary, wrote of Ingram:

> His enterprise helped to change the character of public taste, and allured it into channels which were previously open only to the wealthy and the refined. His practical knowledge as a printer and newsagent were of infinite value in organizing and conducting the varied details of newspaper business. He was ever on the watch, and made opportunities where other men would have been indifferent and inactive. . . .[37]

In its edition of 2 December 1843 the *Illustrated London News* had described its new Middleton "steam printing machine" with two feeders and two takers off. The following year it bought a Napier perfector. The paper was produced at the 1851 London Exposition by Applegath and Cowper on their vertical press, attracting much attention. Ingram's stunning approach to cheap journalism invited many imitators—all situating themselves noisily in the "line of literature," as Fleet Street and the Strand were called.

Along with Vizetelly's own contributions were *Penny Illustrated Paper* and *Graphic* (among others), and, although Ingram in 1848 did start his cheap daily, the *Morning Telegraph*, it was premature. Dozens of pictorial papers followed in other capitals—Berlin (*Deutsche Illustrirte Zeitung*), Paris (*L'Illustration, Le Monde Illustré, L'Universe Illustré, La Republique Illustrée, La France Illustrée*), and in America (*Harper's Weekly, Harper's Bazaar, Frank Leslie's Illustrated News, Spirit of the Times, Daily Graphic, Gleason's Pictorial*). Although there was room for all, the rank imitators were inevitably less successful, some poignantly so.

John Tallis II (1818-1876), like his father, was originally a London bookseller. Visiting America in 1849, he had been fascinated by the New York *Herald's* four-cylinder Lightning—a demonstration of speed and efficiency he never forgot. A few years later his urge to engage in illustrated newspaper publishing—with a Hoe—had prompted him to order the six-feeder:

> [*Conquest to New York, 16 October 1857*] . . . *At my first interview with Mr. Tallis about the foundation for his press he said I could put it up just to suit ourselves, as Mr.* [*Robert*] *Hoe had allowance in the agreement to pay for it. I told him that it was an unusual thing for us to do unless there was an agreement to that effect; he said there was. . . . I answered to this effect, that there was no time to loose*^{t} *and with his permission I would proceed with the fdn & leave the affair as a matter of argument for himself & Mr. Rob*^{t} *Hoe to settle. . . .*

While the press was hurriedly being installed (against a penalty clause) in a nearby factory Tallis rushed over to the Strand to see Ingram. He offered £150,000 for the *Illustrated London News*, hoping to purchase all his dreams at once, but—at a last moment—Ingram changed his mind and refused to sell. Infuriated, the "hotheaded Tallis" (Vizetelly's words) vowed revenged by starting a rival paper which would sweep him under. For Tallis' *Illustrated News of the World*, woodcutters, paper, and a complete shop were ready at the end of 1857, but the January financial panic in London hit him hard. His backers withdrew.

> It was enough to daunt any man but Tallis refused to be beaten. He still had one great asset—the Hoe Printing Press which he had just installed in his Water Street printing office. He managed to persuade the proprietors of the *Daily Telegraph* of the efficiency of his machine and arranged to print their newspaper for the sum of £100 per week. With this security he was able to go ahead with his own paper. . . .[38]

Tallis had quickly ordered new beds London *Times* size, and promptly began printing for Joseph Levy. The *Illustrated News of the World* finally appeared 6 February 1858, indeed looking very much like Ingram's paper. But Vizetelly was merciless:

> Having, by the payment of some preposterous sum, secured publishing offices within a door or two of the "Illustrated London News," he entered on his doubtful campaign with insane confidence. The paper, a commonplace affair, without a spark of talent either in its engravings or its articles, proved a ludicrous failure. . . . In the course of a year or two Tallis had lost nearly all he possessed without doing the "Illustrated London News" the smallest damage, whereupon his newspaper venture came to an end.[39]

In 1860, when the *Daily Telegraph* got its own Hoe, Tallis needed new schemes. He invented the partial printing of metropolitan news for the provincial papers but was plagued by creditors who saw no prospects for such syndication. He then commenced furnishing stereotypes of national news—The London "Type-High" Stereotype Company—which, again, a creditor begrudgingly took over, only to find it a thriving business. Tallis could not survive these losses and harassments, and died a poor man.

Having brought his multifaceted experience to Mrs. Ingram's aid, Vizetelly was apparently exploring anew the type-revolving press as a solution for circulation problems on an undisclosed paper:

> *15 & 16 Gough Square*
> *Fleet Street, London*
> *Jan. 26, 1865*
>
> *Gentlemen,*
>
> *I should feel obliged by your informing me whether you have one of your 6, 8, or 10 cylinder printing machines in an advanced state so that you could be in a position to supply the same within a much shorter period than you usually require. . . .*
>
> *Supposing that I am enabled to procure you an order for two of your machines please to let me know what amount of commission you can allow me. I may mention that I have had no communication whatever with your representative in this country. With regard to the stability of the parties for whom I am acting I may state that on giving the order they will be prepared to pay with Barings*† *or any other*

respectable bankers hand whatever amount you consider necessary to ensure you against the slightest risk.

Your reply by first mail will oblige

Your obed[t] servant

[*R. Hoe & Co., New York*] *Henry Vizetelly*

The company was not likely to encourage such clandestine transactions, and probably said so. In any event Mrs. Ingram sent Vizetelly that year to Paris as correspondent, where he remained seven years, dramatically covering the German siege.

As for itself, *Illustrated London News*—despite its earlier wood-engraving innovations—seemed, like the London *Times,* reluctant to deal with Hoe; Ingram may indeed have resented competitor Lloyd's success in that direction. After first discussing an eight-feeder (and taxing Hoe ingenuity for printing woodcuts at high speed—actually an insoluble problem), he switched to thoughts of a six-cylinder machine (but balked at the price) although his prosperous paper was large enough to justify a ten. According to Mason Jackson Ingram took the six-station Lightning, adding very little to the city's clatter:

> In London, and in all our great centres of population, the newspaper has become the unfailing accompaniment of the City man's journey to business. At the railway stations journals of every kind tempt the loitering passengers, while the illustrated papers appeal to him in a language of their own. . . . Day by day the demand is made, and the supply is ready. The printing-machine never sleeps and is never tired. Its voice is one of the voices of the night—most unmusical, yet with a mysterious meaning. . . .[40]

The development of the photographic halftone screen eventually forced the exquisite wood engravings off the illustrated (and daily) newspapers everywhere.

Londoner Joseph M. Levy (1812-1888) as a youth bought a printing shop in Shoe Lane and was soon destined to manage great newspapers. He acquired the *Telegraph & Courier* in 1855 as a bad debt and reissued it at 1 penny—the first London paper at that price. At year's end its circulation was 27,000, about half that of the *Times,* but growing steadily. Always well aware of mechanical developments in America, Levy particularly admired the racy format of the New York *Herald,* and his imitation of that paper's style, called catering to "the million," brought him phenomenal circulation. At any rate, Levy ordered a ten-cylinder in June 1859, abandoning his four-feeder Middletons, and innocently walked into Hoe's international dispute.

Completely frustrated at getting an American-made machine into the London *Times,* the Colonel saw Levy's order as a perfect lesson—for Old Thunderer and all Europe—of Hoe's superior workmanship. To guarantee its welcome he suggested Robert give Levy "a good long rope"—that is, as before, charge only half price for it. Ready in December and containing all the latest improvements, it was in Tucker's opinion "far the best 10 Cylinder we have made."

> [*Conquest to Robert Hoe, London, 26 February 1860*] . . . *The work progresses favorably at the Telegraph office. The machine is all up with platforms, also eight of the Cylinders regulated by hand, the remaining two cyl will be regulated by to morrow at noon, when the machine will be ready for a trial to morrow afternoon. I had a little alteration in the End platforms wich† I explained to you. The cost was but trifling; the hand rails have been made bright* [*plated*] *and add much to the apearance† of the machine.*
>
> *There is yet more to be done in the apearance of the machine; first the salt water has taken the paint off the side frames; however this* [*repainting*] *can be done after the machine has done its first weeks work. I would suggest that there be a brass name plate on the front lower platform wich will just meet the eye of a visitor. . . . The name plate that has come with the machine is of Iron and does not show itself. . . . Mr. Levy is a daily visitor and he seems to be well pleased with the progress. . . .*

This little piece of brass, costing £2, worked to perfection:

> [*Conquest to Robert Hoe, 1 March*] . . . *The name plate is put on the machine. It is not very large, but it seemed to have the necessary effect, as a gentleman in company with Mr. Levy give† vent to his feelings at the sight of it, when Mr. Levy reply† was that he believed in free trade. . . .*

By 1866 Levy had four ten-feeders; two years later his circulation reached 200,000—thrice surpassing the London *Times.*

Through these years the company's operations in England settled down to a hectic "business as usual" atmosphere, with Levy frequently on the docket:

> *Jan 7th, 1864*
>
> *To Messrs R. Hoe & Co., New York*
> *Gentlemen,*
>
> *I writen† to you some two weeks ago saying that we had been promised an order for a Six Cylinder Machine. The Gentleman* [*James Smith, publisher of The Sportsman*] *has this morning closed the order; the machine it† is to be a No 8, same width as Mr. Lloyds furnished with 8 type beds, 2 sets of roller stocks, two roller moulds, pulley blocks, wrenches & e&e* [*equipped and erected?*] *in our usual way, for the Sum of Three Thousand One Hundred & Seventy Five Pounds Sterling, payable £600 at the receipt of the order, £1000 at the receipt of the Bill of Lading for the Machine, in London, the balance when the machine is at work.*

The machine is promised to be at work by the 1st of July next, without fail. 2½ % off for all Cash paid. I am in hopes that you will consider this a fair bargain; first he pays the full price for the machine, & also £25 for putting up. We are one set of stocks [rollers] less, & he pays for all land carriage. I have also charged £5 for each Stereo type bed extra, on account of large size. The agreement is that we furnish the articles on the 3rd page [of this letter—not reproduced] at the prices stated.

[I]n addition to these articles Mr. Smith required a complete Stereo Apparatus, but he has not yet concluded to order this untill† we can give him an estimate as to the cost. The articles required are one mould of large size, one shaving table, with furnace fittings & melting pot & ladle. . . . The price paid for them here would be 175£ sterling. The hot plate & press is included for this sum.

I will send the particulars for the size of sheet to be worked on this machine by the next steamer. The ship with the 8 Cyl for Mr. Levy has arrived this morning; it will be stored for the next three months at least.

It is likely that the Courier folks at Manchester will purchase the Birmingham old four Cyl.; if so, I shall be able to commence business putting up their six cylinder machine. [I]t makes me feel bad to hear of these arriving and no place ready for them or even likely to be for some time to come. How is business in the shops[?] I suppose we may now begin to promise the Machines for here on short time, say 5-6 or 7 months. The putting up of the machine for Mr. James Smith is included with the machine for £3175 which is £25 more than catalogue price. With kind regards

Yours very truly
William Conquest

P.S. We have recd the first instalment of the cash for this machine to day, less 2½ % for cash. . . .

A decade later Levy's *Daily Telegram* was a client of considerable importance, having in 1874 ordered eight Hoe web perfectors. The proprietors still knew how to drive a bargain:

[Conquest to New York, 13 October 1877] . . . We have not yet come to terms with the Daily Telegraph folks in reference to the order for a Web Machine with folders also to lay down the Sheets their full size also to perforate the unfolded sheet at the Heading so that it can be separated by the Reader. The price we quoted was £3500, that is to say, £3000 pounds for the machine & £500 for the folder. These prices are subject to 7% discount for cash. These terms are precisely the same as they paid for all their Webs. . . . They want what they now Call a substantial deduction of £205.

I have fully explained to their Manager Mr. Lloyd that they have already got what may be called two Substantial reductions firstly by quoting the machine at £3000 in place of £3150. Secondly by only charging them £500 for the folder in place of £600 & even thirdly by allowing them 7% discount for cash which we do not allow to others

& which was done on consideration of their first large order. . . . Mr. Lloyd told me that he had the power to give me the order providing we should take off what he calls the odd money—the modest sum of £205. . . .

We have sent you by this mail 16 copies of the Glasgow Herald which contains the article concerning the Evening Times Machine which was started by General Grant & christened after him. . . .

Mr. J. Johnstone [London Standard] has left London for "Spain" where he intends to remain during the Winter. We have a Cash Balance of £814.3.6 with Bills in the House to pay amounting to £368.19.3. With Kind Regards

Yours Very Truly
William Conquest

Levy's manager (apparently not related to Edward Lloyd) had shrewdly bargained for a host of discounts on a machine that wasn't even experimental in nature. But the Hoes had long before gotten a small personal favor from Levy that would have satisfied them.

Edward Seymour (third son of sister Mary) at Yale had earned some money as correspondent for the *New York Times.* (In 1858 he gleefully described the indictment of fellow student Eugene Smith, son of Matthew Smith III, for shooting a volunteer fireman in one of the ongoing scrapes—common in many towns—with pugnacious fire gangs.) In July 1859 his uncle Robert gave him a letter to Levy—without stating their relationship—enabling him to become New York correspondent for the *Telegraph.* With his first piece appearing 1 September, Edward became a practicing journalist, and like many Hoe nephews escaped the family press business.

The press-building industry, especially in Europe, had changed dramatically since the cylinder machine was introduced. Although responsible during the first third of the century for dazzling progress in printing press technology, David M. Napier & Sons, Lambeth, had gradually avoided the immense and increasing costs of cylinder press construction. The design, tooling, and production of precision machines were growing beyond the capabilities of the frugal Napier shop as father and son conceived it, the latter, James M. Napier, with "customary icy calm" willingly forfeiting the business. Nevertheless he did not overlook what he considered a colossal misappropriation of his father's inspiration.

In 1848 when the British *Mechanics' Magazine* featured Hoe's four-cylinder Lightning, James wrote the editor to claim that his father had designed just such a machine ten years before, with the "inevitable" shabby consequences:

. . . Mr. Hoe was in London after the [Napier] patent was obtained, and the invention was then fully explained to him, and many details which were not described in the specification, to which, of course, he had free access, were liberally laid before him, by the inventor himself in a friendly way. . . .

It is hard to picture the canny Scotsman quite so frank with a rival press builder—especially the yankee who had destroyed the entire American market for Napiers just two decades before. Reducing his charges ad hominem—like old feuding penny editors, the son quite forgot his icy reserve:

> . . . [A]fter wading knee deep in the polluted soil of historical exaggeration and corrupted facts, Messrs. Hoe, (for who can suppose that the article has emanated from any other quarter?) blinded by a desire of renown, plunge overhead into the slough of culpable self "appropriation," which first begets barefaced "repudiation," and ends in composed and settled "annexation." . . .[41]

Never before had so responsible an individual made such damaging insinuations about Hoe integrity. The Colonel was shocked.

Richard denied the charge vehemently to his brothers and immediately asked J. C. Robertson, editor of *Mechanics' Magazine,* to compare patents and render an opinion. His response was prompt and thorough:

> [*Robertson to Richard Hoe, 7 February 1849*] . . . *I have agreeably to your request examined the copies you have sent me of the specification and drawings of the two Patents for Printing Machinery, taken out on your behalf, 4th. May 1847 and 18th. July 1848, in the names respectively of Mr. Wm. Newton and Mr. W. E. Newton and carefully compared them with the specification of a Patent also relating to Letter Press Printing, granted to Mr. David Napier 18th. April 1837. . . .*
>
> *I have no hesitation in stating it to be my firm conviction that Mr. James M. Napier's charge against you of having appropriated to yourself the inventions of his relation Mr. D. Napier, must have originated in error and misconception and that they are in fact without any just or reasonable foundation. . . .*

Curiously, although Richard was actually in England in 1837, it seems possible that young Napier may have confused Hoe with the earlier visiting Daniel Treadwell. Fortunately, no one in England or America took the charges seriously.

Napier & Sons, nevertheless, had played a crucial but passing part in the history of the fast newspaper press. As British printers after 1856 perforce looked to America for mechanical developments, only the French browsed in the showrooms of British manufacturers. They were so traditionally behind the times they seemed incapable of catching up.

In France an awakening machine industry had dimly perceived the novelty and value of rapid printing machines, although, paradoxically, the nation's illiteracy and a heavily censored press with monumental newspaper taxes (until 1871) had little need of them. In the 1820's paper had been so expensive journals were reduced to small formats with room only for racy expositions of opinion, to which the French reading public obligingly became accustomed. The famous printing house of Pierre Firmin-Didot began using a Stanhope in 1818 and four years later brought to France its first steam press. In 1823 a two-cylinder Cowper and Applegath was imported for *Bulletin des Lois,* Paris, while Paul Dupont, a large job printer, built his own. Napiers and Koenig and Bauers (from Oberzell, Germany) soon poured into France, nevertheless, much to the consternation of French labor.

Printers during the 1830 July Revolution saw their opportunity to stop the frightening mechanization of pressrooms. They smashed machines at the government's Imprimerie Royale, at the *Débats, Courier Français,* and *Constitutionnel,* and at other large shops, and expected the monarchy to support their demand for the return of hand presses. The king ignored them.

Parisians saw their first perfecting press in 1838, but not until about 1850 did a domestic French machine emerge—with a "grande flourash," *Scientific American* reported, ". . . to throw the American one of Messrs. Hoe entirely in the shade." It consisted of a series of lateral cylinders a little more than half the size of Hoe's *La Patrie* press, needed no cords or tapes, used only three men, printed from stereotypes at 15,000 per hour on undampened paper, folded the sheets, and was cheap to purchase. Stereotyping (an agonizing failure for newspapers elsewhere) took only 15 minutes per plate:

> A few sheets of tissue paper are placed together and passed upon the form containing the types. Thus the mould is formed, the metal is poured upon it, and as soon as it is cold the stereotypes are ready for the cylinder. . . .[42]

With its usual sarcasm over foreign claims—of course, often garbled or exaggerated, *Scientific American* supposed: "The next thing we will hear about, from Paris, is a machine which will take in the manuscript at one end and trundle it out at the other (never wearing the type) into printed books, papers, and radical speeches." But this 1850 invention of a young man named Marinoni was no joke to Hoe.

At the age of 12 Auguste Hippolyte Marinoni (1823-1904) apprenticed in Paris as a compositor and a little later was put into the shop of Gaveaux, a maker of hand presses. These were still the only kind in France, printing for their newspaper proprietors only 4000 or 5000 impressions daily. At 15 Marinoni was annoyed to notice that even the Imprimerie Royale imported its presses from England; he resolved one day to make French printing machines for his countrymen. By the time he was 30 Marinoni would actually be selling them to England.

At 20, married, with two children, and working in a foundry, Marinoni learned to cast characters as well as set them, and in several shops came into contact with important newspapermen, above all, Emile Girardin. The editor often invited him to dinner, for which he had to rent a dress suit, as he could not afford to buy one. Fortunately, Marinoni had one great quality—love of work; it sustained him and made him tough. He had one other—he needed only four hours of sleep, so that he could work indefatigably in endless stretches.

Marinoni was soon employed as a mechanic in the *La Presse* pressroom. In 1847 Girardin commissioned him to construct the first French four-feed machine, inspired by Applegath's in England. Only 24 that year, Marinoni's genius was all the more charming in contrast to his social self-consciousness. He himself recollected:

> In 1848 I was at *La Presse,* with Girardin; the [political] conditions were such that I went two and a half months without sleeping in a bed, passing day and night in the shop, napping a few hours at a time on tables. For food I had only some bouillon, not having time to go to a restaurant. Sometimes Girardin, finding me greatly fatigued, took me with him in his coach despite my blue overalls covered with ink, and made me sleep in his hotel suite. Here was where he received all those celebrated personages, politicians, writers, artists. I was shy, and Madame Girardin often tried to engage me in conversation. "Well, Hippolyte, what do you say?" In such moments I wanted to be swallowed underground!" [43]

In 1854 Marinoni began to manufacture presses as a business. Although *La Presse* had a multiple-feed machine, such giants were so rare and little appreciated progress would be exasperatingly slow. At an international exposition in the 1860's one of the Firmin-Didots, commissioned to buy printing equipment, almost collapsed when Marinoni told him that his new machine could replace as many as 30 presses.

"Thirty hand presses?" Firmin-Didot asked, amazed.

"*Mais non;* not hand presses, mechanical presses!" [44]

Here, Marinoni remarked, was a practicing printer who still thought in terms of ancient hand operations.

One of Marinoni's early machines was jointly produced with Hopkinson & Cope of London. Based on a Napier, it was called the "Anglo-French" perfector. Marinoni soon sold presses to the British *Sporting Life* (4 machines), *Weekly Times* (3), *Weekly Despatch* and *Referee* (6). The Colonel soon met him, and brother Robert also dropped by to chat.

> [*Robert Hoe, Paris, to New York, 4 March 1858*] . . . *This P.M. I passed several hours with Mr. Marononi.*[†] *We visited La Patrie office.*

The machine he made was at work printing 10,000 impressions pr hour by the watch. The old [Hoe] machine is removed for repairs. The new machine runs very well but is not quite as thoroughly finished as our work but is very creditable to him. We then went to the office of "La Presse" and then to the office of [printer] Napoleon [Chaix].

Mr. Marononi resembles Mr. Lloyd in appearance and manners. He is a wide awake, energetic business-like man & says he would very much like to make an arrangement with us to furnish machines for England. He says they could be made as well and cheaper by him than at Manchester. He certainly is not above his business and I have no doubt would be very energetic in the execution of orders.

All the reply I made to him was that I would consider his remarks, but at present could not see our way clear to make any change in our arrangements. No letters received from you to day nor any papers. . . .

Marinoni had indeed obtained the French rights to the Lightning presses, and—of all things—Delamarre had been his very first customer for one.

Continuing on very good terms with R. Hoe & Company (to whom he assigned some patents), Marinoni soon built Hoe type-revolvers as a vast improvement over the Applegaths. One for *La Liberté* in 1866 was a six-feeder using stereotypes which he could afford to sell for 30,000 francs ($6000) versus Hoe's own version for 140,000. The Frenchman capitalized on his good connections with Hoe, not so much on ethical scruples, but because he found the American name known and respected. Neipp reported: "Marinoni, announcing his new machines through the press, never failed to say, with the greatest loyalty, that his co-inventor was Richard Hoe, the constructor of the immense American Newspaper machines so common in America and England. . . ." [45]

One clever New Yorker found Marinoni's claims (and his prices) too appealing to resist. Congressman Benjamin Wood, brother of Mayor Fernando Wood, had acquired the New York *News* (founded 1855) in 1861 and—sharing his brother's provocative pro-South, anti-Lincoln attitudes—made it extremely popular in the city's poorer tenement areas. With circulation up to 100,000 by 1870, he needed fast presswork—and gave a substantial order to Marinoni. The Hoes were understandably surprised.

[*Robert Hoe to Richard, Geneva, 13 September 1870*] . . . *I went to see Mr. Wood (Ben) as you suggested. He was quite frank to inform me that he had ordered a six feeder from Maranoni*[†] *of Paris, and had paid him $2000 in gold upon it. That it was to be put up here and when satisfactory he was to pay the balance $9000. He said a stereotype apparatus is to be sent with it, that was to complete a plate in 8 minutes. He had an agreement to become agent for the introduction and sale of the presses here.*

> *I showed him [a proof of] the cut . . . upon which Maranoni had written in French "Invention of Mr. Hoe." Mr. Wood promised to examine into the matter and then to act justly toward us, and make an arrangement with us if possible to use the French machine, and to have us make him a second machine. He seems to think he would have to depend upon us and work in our interests; [he] promised to do nothing without a good understanding with us. Mr. Wood thinks this machine will be the most valuable for general work here. . . .*

The following year, when the *New York Tribune* (because of construction delays) could not accept delivery of a new Hoe perfecting four-feeder, Robert had an inspiration.

> *[Robert Hoe to Richard, London, 29 April 1871] . . . I therefore went to see B. Wood and made, after a good deal of negotiation, a bargain with him to take it for his "Daily News" for the sum of $4000 and his four cylinder press. He agrees to countermand the Maranoni*[†] *6 cylinder and not receive it on any consideration. We engage to make him a six feeder hereafter if he wants it for the sum of $13,500. Of course he does not want any folders attached. He had made an arrangement with Maranoni to sell as well as to manufacture the machines, on paying [a] Royalty. Had we not anticipated this agreement as we have done, a law suit would have been very likely to arise between us and Wood. . . .*

To his own surprise, perhaps, Wood had been pleased with the Hoe equipment, and Stephen Tucker had gotten some web feeding ideas while working on it. The Hoe manoeuver had at least precipitated no difficulty with Marinoni.

Because of antiquated thinking among French printers—little changed since the Colonel's experience at *La Patrie*—however, Marinoni soon had considerable resistance at home. Like Treadwell not too long before, he established a printing plant to demonstrate his machines and contract work for them. One paper accepting his offer was *Petit Journal,* which in 1877 had bought four Marinonis at 40,000 francs each (again, versus 120,000 for Hoes)—and unlike the Hoes they were perfectors. A few years later *Petit Journal* gave up the race for new equipment and placed its mechanical work in Marinoni's shop. The paper unfortunately fell so far behind in payments Marinoni had no choice but to take over the publication. By 1884, with Marinoni in full charge, the Paris sheet had a circulation of 750,000—largest in France—and the master press-maker was hoping to reach one million.

Soon other Parisians with mechanical flair entered the field. Pierre Alauzet worked in several machine shops, including Gaveaux's, and set up for himself in 1846 to make printing equipment. Inspired by machinery at the London *Graphic* office, he made a large cylinder press for *L'Illustration.* His son Leon expanded the firm, but it merged with Marinoni's after World War I. Jules Derriey started fabricating printing equipment 1854, and in 1867 displayed a large stereotype newspaper press destined to replace—finally—the Hoe Lightning at *La Patrie.* He and Marinoni soon furnished rotaries to all the French papers and color printers. His son entered the business about 1900, but the firm was, like Alauzet's, swallowed by "Etablissements Marinoni" after the war. Henri Voirin, also a pressmaker, similarly merged with Marinoni in 1920.

Despite his increasingly successful monopoly, the modest Marinoni kept in touch with his old friends. Richard Hoe too felt it necessary to watch press developments in France, and his brother Robert enjoyed long buying trips to Paris, perhaps using Marinoni and Perrot visits as an excuse.

The Colonel had been particularly interested in Marinoni's flat-bed press drive mechanism, and negotiated for U.S. rights for the device. It enabled him to bring out an 1862 "railway" (bed mounted on "trucks" resembling railroad wheels) newspaper machine. *Scientific American* was impressed with it for country shops: "The press runs remarkably easy; we took hold of the fly wheel and readily drove the apparatus with one arm." [46] Tucker named it a steady company asset.

Marinoni next developed an 1867 high-speed rotary perfecting press, again (with stereotypes) simple and cheap. When he brought out a lithographic machine in 1869 the Colonel liked it so much he proposed immediately that Marinoni patent it in the United States and let Hoe manufacture it. Although the French company to this day makes printing and bindery equipment, the firm, like France itself, does not greatly influence the world market.

In Great Britain the Lightning had struck far more fertile territory with its revolutionary technology, a factor which contributed to its own quick obsolescence. "But the day was now over," wrote Howe, "when the newspaper printing machinery could be installed and left in place for a quarter of a century." [47] Although the *Times* would not rise much beyond 60,000 or 70,000 in the 1860's, early delivery of the sheet on the street was a constantly nagging problem. It meant thinking about a new press almost all the time.

John Walter III (1818-1894) was, like his father and grandfather, passionately interested in the mechanical perfection of newspaper printing methods. He assumed control of the *Times* in 1847, when his father died. Although he had indeed bought two ten-cylinders for a temporarily embarrassing situation, he did not think of himself as a steady yankee customer by any means. The Walters once proudly inventoried their impressive involvement with newspaper technology: the logotype system, Koenig's steam press, the Applegath rotary, paper maché stereotyping, the (1868) Walter rotary, and Kastenbein's machine type composer—Hoe Lightning conspicuously absent. Walter nevertheless had sufficient mechanical expertise to attempt his own improvements, and his paper had adequate wealth to explore them.

Old Thunderer's proprietor brought new requirements and specifications to his veteran manager John MacDonald (who was also the *Times* reporter on science and mechanics), and his chief engineer, J. C. Calverley. Both would labor for six years on the project, variously named the "Walpole" and the "Walter" (for John Walter IV). This independent research was a great surprise and disappointment to the Hoes, contemplating "second-generation" equipment for the prestigious paper. They were long very cordial with MacDonald, whom they believed likely momentarily to sign the order for it. Robert Hoe considered Walter's unabashed copying of his brother's apparatuses perverse and spiteful:

> [*To Conquest, 29 April 1862*] . . . *The last course of "the Times" is consistent with their policy to be exclusively English. I think personally that Mr. MacDonald is above such meanness, for so I must characterize their readiness to use our invention without making us a due consideration, but he is no doubt overruled by the proprietors. I hope you are sound again in body. Wish often that I could pop in and see what is going on as of old.* . . .

It was a very different story when John Bull learned from yankees! Hoe mentioned the matter again:

> [*18 May*] . . . *We are glad to hear that you have got the Star Machine to work so satisfactorily. As to the London Times, the course taken by it in reference to another press is very inconsistent with Mr. McDonalds*[†] *statements to us. We can only infer that he does not concur with his employers in the matter. Perhaps they are waiting for RMH to return. There is nothing very new in the shop.* . . .

R. Hoe & Company had grown very fond indeed of what Moran saw as "an interplay of ideas between publishers and their pressmen and manufacturers and their technicians"[48]—when the company was in control of it. MacDonald, although a Hoe friend, could not help. The paper's historian said of him:

> He was a man of unusual ability, having a firm grasp of the engineering basis of modern printing, an appreciation of foreign politics, and a loyalty to the traditions of Printing House Square and the Walter family so unquestioning that he bore the brunt of the litigation over the Parnell libel case that brought the circulation and finance of the paper to disaster.[49]

MacDonald died in 1889.

Unlike most of Hoe's newspaper friends who knew nothing of mechanics, Walter was a born tinkerer. In 1862 he and his crew began experimenting with Hoe's Lightning in order to turn it into a perfector printing from paper rolls; four years later his web machine was capable of 12,000 8-page papers hourly. Although the newsprint had to be dampened (at first), a reel could be changed in one minute. Walter directed that three be built, ready in 1869, with a fourth to follow.

Young Robert Hoe (III) was on hand to watch the development of these machines:

> *[To Robert Hoe, Liverpool, 11 June 1869] . . . I had quite a talk with Mr. Macdonald.† He is full of his new machine. . . . He says they have it now so nearly right that they see clearly the way to an "entirely perfect machine" of the kind. According to his own statement they use the greatest care in the making of the paper, & he thinks their power of controling† the paper manufacturer was a great feature. . . .*

Over the weeks the conversation continued:

> *[To Robert Hoe, Liverpool, 17 June] . . . He [MacDonald] gave me to understand that they were prepared to make some arrangement with us to manufacture the machine for the United States. I think their intention is to make [it] for this market [England] themselves. . . . It is not a mere matter of pride with them to have a machine of their own, but they mean to make some money from the sale of them.*
>
> *I stated that we were making a machine, the plans for which were matured, but at the same time there might be something of value to us in theirs, and we likely had improvements that would be of service to them. Mr. McD seemed to imagine that we had never given the matter of a fast perfecting press attention & John Bull-like, were sticking to the one thing, "the old rotary press." I rather insinuated that we could not afford to be idle, as making presses was our only business, and we did not have a newspaper business to fall back on. . . .*

Surprisingly, the Britishers had taken seriously the fanciful possibility of transatlantic cooperation:

> *[To Robert Hoe, 24 June] . . . He [MacDonald] had conferred with Mr. Walter as to the propriety of showing us the machines or of submitting the plans to us for inspection, with a view of our making the machines for America, but, as we were making a new machine of our own, they thought we had better go on with it independently, and they with theirs. . . . The machines are made in the "Times" office but the heaviest pieces are probably partly done out, & by Whitworth & Co. . . .*

The prospect of a British-American alliance had quickly passed.

> *[To Robert Hoe, New York, 18 September 1869] . . . Night before last I saw the "Times" Machines in operation. Two of them were at work, and I saw one of them work off a roll of paper, without stoppage, at a uniform speed of 175 per minute. They put on & started a new roll in about two minutes without shifting the tympan.*
>
> *The whole thing is very simple. I suppose the patent office draw-*

ings show it pretty well. A great feature is that it is capitally well made, at the same time the noise is terrific, making it impossible to speak so as to be heard in the same room. There are from ten to twelve thousand papers on a roll. . . .

Mr. McDonald† has a Machine Shop across the Square opposite the Times office. I think their intention is to go into the manufacture & sale of their machine, probably as a Company. They will no doubt soon offer our ten feeders, or one of them at least, for sale. . . .

I showed Mr. McDonald our shop here & the new machine made for Mr. Lloyd; he seemed surprised. . . . I saw [John] Ruskin the other day; a funny looking & rather dried up little man. . . .

As "endless-sheet press fever" (in Robert's words) swept through the metropolitan papers of America and England, the Hoes, working frantically and admittedly late in the field, watched nervously as reports came in.

At first the Walter machine seemed a failure: "All the newsmen say that one side is so white, and the other so black and dirty, that it is impossible to read either,"[50] *Printers' Register* noted. Nevertheless the work continued and eventually met success. It was in fact a good, simple machine—"the beau idéal of the mechanical engineer," said Grant, telling why:

While each of the ten-feeder Hoe machines, a long and lofty row, requires eighteen men to feed and work it, the new Walter machine occupies a space of only fourteen feet by five, or less than any newspaper machine yet introduced, and requires only three lads to take away, with half the attention of an overseer, who easily superintends two of the machines while at work. The Hoe machine turns out seven thousand impressions printed on both sides in the hour, whereas the Walter machine turns out twelve thousand. . . .[51]

The *Times* with these Walter presses had to be hand folded for several years, until George A. Wilson's "Victory" press with folder (first used 1870—five years before any Hoe folder—on the *North British* [Glasgow] *Daily Mail* and *Glasgow Weekly Mail*) came into the shop. At any rate, for the Walters, at least, British supremacy seemed restored.

Interestingly, the Walter press, with curved stereotypes, high-speed web feeding, and automatic folding, was exactly right for the *Illustrated London News,* although elaborate modifications were involved. In 1877 W. J. Ingram, a son, and James Brister, manager for 17 years of the machine department, modified a Walter press especially for illustration work, the engineering of which took Middleton two years to perfect:

Then [after Koenig and Applegath] came Hoe's American machines, and finally the Walter Press, the principle of which last invention has, in the Ingram Rotary Machine, been successfully applied to the

> printing of cheap illustrated newspapers. By the old "two-feeder" machines the engravings were printed on one side of the sheet, and, by a second printing, the type on the other side. They turned out 1500 impressions of the engravings in an hour, while the type side was printed (by a six-feeder American [Hoe] machine) at the rate of 12,000 impressions an hour. The *Penny Illustrated Paper* is printed by the Ingram Rotary Machine at the rate of 6500 an hour. It prints both sides of the sheet at once, cuts each number to its proper size, folds it, and turns it out complete. It occupies no more space than an ordinary perfecting machine, and only requires four men to attend it, while thirty men and five "two-feeders" would be required to do the same amount of work by the old system.[52]

"[A]nd little more needs to be said," *Engineer* added, "to prove that it bids fair, to effect as great a revolution in the printing of illustrated newspapers as the latest Walter and Hoe Rotary machines have effected in the printing of daily newspapers."[53]

But the Walter press itself, made (like "everything in the place, except the paper,"[54] said Hatton) by the *Times'* machine shop in quantity, proved versatile and adequate for many papers. It went into the London *Daily News*, the Edinburgh *Scotsman*, the Liverpool *Post*, and Birmingham *Daily Post*, all papers which gave up Hoes.

By 1872 the Walter's snobbish fame reached America: the press that had replaced Hoe ten-feeders must have been considered something miraculous indeed. The *Missouri Republican,* St. Louis, quickly took one, and the *New York Times,* basking in the publicity, was printed on it at the 1876 Centennial, "Surrounded with crowds of visitors intently watching its perfect and regular action, 'like a thing of life.' " Although similar to the locally available Bullock, the Walter was selected by the New York paper perhaps only because of its association with its British namesake. It proclaimed:

> The Walter press is the most perfect printing-press yet known to man; invented by the most powerful journal of the Old World, and adopted as the very best press to be had for its purposes by the most influential journal of the New World. . . .[55]

John Walter had visited the *New York Times'* new Park Row headquarters in 1866, pleased to find the structure similar to the King's Printing House in London; in 1874 he reciprocated by designing his new Queen Victoria Street façade like the *Times'*.

In the New York plant the Walter equipment evoked considerable admiration, but Bullock insisted there was infringement and won its case 1878. The *Times* nevertheless ultimately took four machines printing from stereotypes, their operation further simplified somewhat by the paper's engineer, Gilbert Jones (George's son), who eliminated half the attendants originally designated. The *Missouri Republican's* experiences were far different. "The 'Walter' Machines are a failure here," Robert Hoe III reported (2 June 1874) from Chicago, "& so considered by the newspaper people. Mr. Knapp said his cost him over $30,000.00." That paper gave them up that year.

In 1873 two British newspapers, the *Telegraph* and the *Standard*, had ordered from Hoe in London modified "Walter" presses, hoping for the best of both systems—"a kind of 'Hoe-Walter,'"[56] said *Printers' Register*. The London *Daily News* also placed an order, but Hoe had so many machines to complete no delivery date could be agreed upon, and the paper ultimately went to Walter for 8 machines.

Lasting until the early 1890's, the Walter was just about as far as a British press had been able to go in domestic and international sales, with R. Hoe & Company all the while smarting over English success purloined from basic Hoe designs. When Walter died in 1894 his second son, Arthur, took command, but legal complications forced the estate to dissolve the partnership. The new *Times* proprietors, interested pragmatically in getting out a newspaper which had indeed become the journalistic standard of the world, were understandably no longer disposed to experiment with—no less manufacture for sale—printing equipment which had become enormously complex and costly. They promptly bought from Hoe—London, to be sure—but admired them unashamedly:

> If the linotypes fascinate by their ingenuity, the huge printing machines [Hoe's latest three-roll balcony type] impress by their size and power. Each can print some 26,000 copies of *The Times* in the hour. . . .[57]

By 1912 nearly all the London newspapers were printed on Hoes, but Old Thunderer, as always, had been the prize catch; Hoe not only (again) won a prestigious customer, it had now also eliminated a powerful competitor. For the *Times,* ended was the midnight excitement of secretly printing the news; gone were private experiments to make type stick to great cylinders; absent was the house-made web perfector preferred because it was British. "When the *Times* replaced the Walter presses by Hoe presses in 1895," said David Pottinger, "the victory was complete."[58]

Hoe Adams, parts numbered for assembly / Early Hoe drum cylinder
Swain, Abell & Simmons account, 29 June 1841

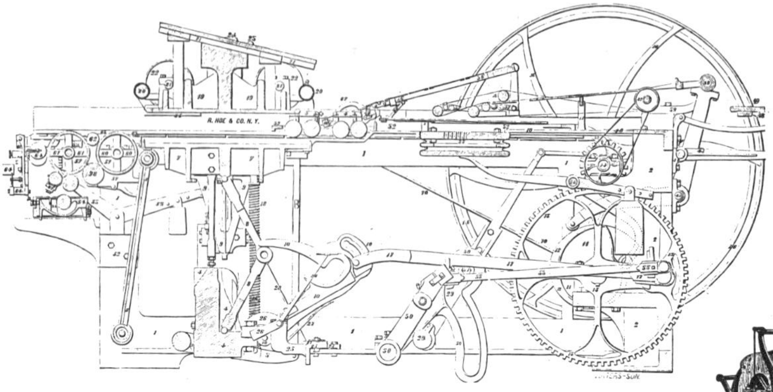

Part III

DEVELOPMENT OF MASS MEDIA: THE GIANT NEWSPAPER PRESSES

The history of journalism in pre-Revolutionary America, punctuated by the wit and industry of Benjamin Franklin, the 1735 "Freedom of the Press" trial of John Peter Zenger, and the 1765 Stamp Act (which directly precipitated American independence), recalls only distant—Colonial—episodes in the long chronicle of resistance to censorship in Europe.

Protestations to the stamp duty and the proclamation, by American Colonies, of their independence from Great Britain became the unrivaled mandates for a new journalism as well as a new government. Never before had writers and printers won the latitude and immunity they suddenly possessed. The earlier reporting of more or less derelict maritime news and cloaked political opinion was almost no sufficient

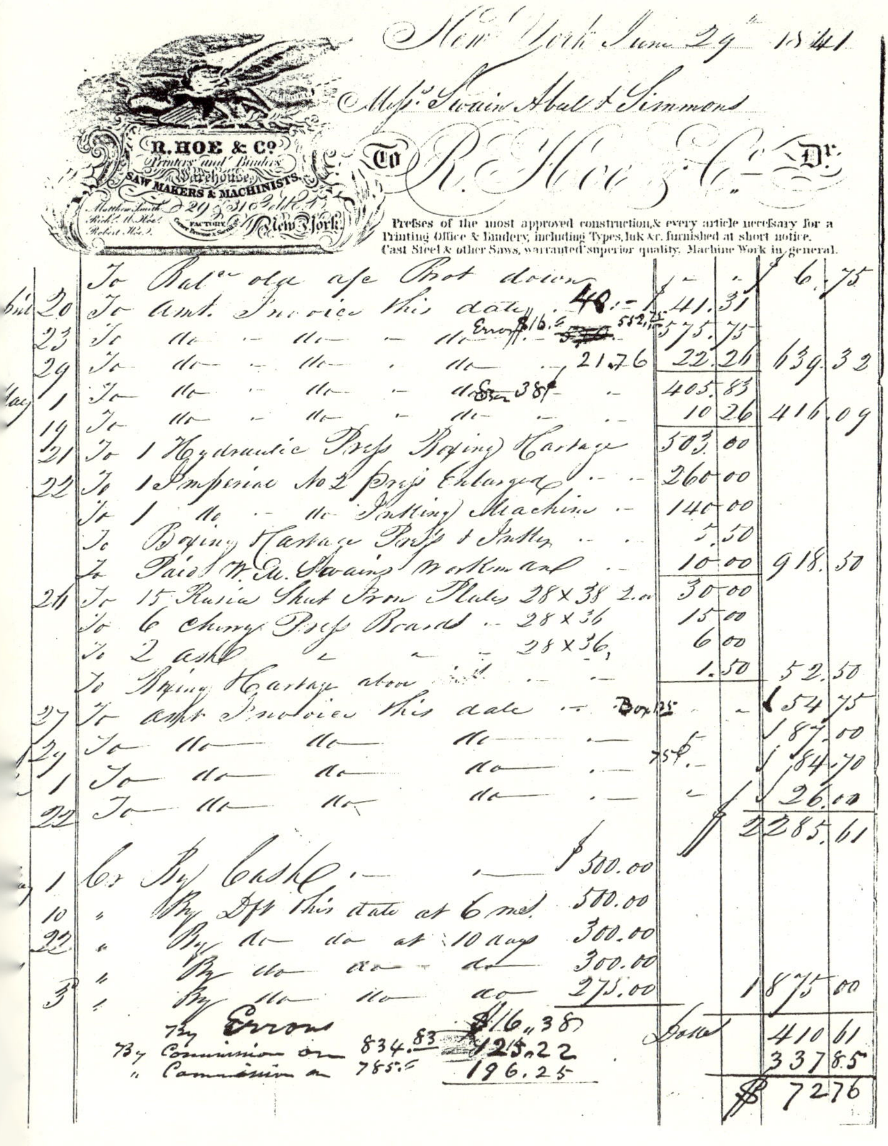

precedent for either newspaper editor or his trusty readers—now jointly and uniquely charged with the guidance of a great new ship of state. Democracy had dire need of friends at home and around the world as well.

At the beginning of the Revolutionary War about fifty print shops were at work in America, most turning out an occasional news sheet only as events occurred to fill it; the paper, oddly enough, was only the creation of printers usually drumming for other business. Although wartime activities obviously generated more news—and more demand for news—labor and material costs rose steadily, as did competition from coffee-house gossipers and other news mongers. Local items were trivial, often printed only to confirm or deny rumors already known;

much intelligence was exchanged belatedly by letter-writing "correspondents"—when not lifted whole from neighboring and foreign journals. Advertising and subscription fees were seldom satisfactorily collected; even promises of barter were disregarded when crops failed. Struggling newspapers under these conditions had to seek substantial advantages—fresh, compelling news, perhaps—if they were to succeed.

Upon the establishment of a United States government there was considerable agitation in the realm of politics. Unpaid veterans were bitter and vociferous; taxes were high and inequitably collected; unsure direction in government threatened to stalemate or disrupt the national machinery altogether. Fighting for individual rights and for political causes was quickly recognized to be a matter of winning public support, and the printed word would be instrumental in both small and great issues.

After the armistice an extraordinary spread of newspapers testified to a freedom of the press in the United States not appreciated anywhere else in the world. Nevertheless, still curtailed by prohibitive price (6 cents—which also bought a pound of cheese, sugar, or tobacco) because of scarce materials and high printing costs, the American newspaper remained a medium for the rich and influential. Yet, from 1800, when the U.S. capital moved from Philadelphia to Washington (displacing violently the center of political activity and creating another rash of newspaper founding), the number of papers increased from about 200 to 1200 in the 1830's—65 of which were dailies. And, as visitors to this continent recorded, the American sheets were immensely popular:

> The influence and circulation of newspapers is great beyond anything ever known in Europe. In truth, nine tenths of the population read nothing else. . . . Every village, nay, almost every hamlet, has its press. . . .[1]

America proved, despite its rustic appearance, to be far more literate than the countries of Europe; the public (hungry for news, especially of the Old World) took to reading with little inducement from philanthropic educational societies. The almost absolute freedom of the press, however, was difficult to control; extremists rushed in with politicking and political philosophies which have beguiled and shaped American election campaigns ever since.

In 1789 the *Gazette of the United States* was founded in New York, financed by Alexander Hamilton to disseminate Federalist doctrines for promoting a strong Constitution. Schoolmaster John Fenno, its editor, announced it would be a

> . . . National paper, published at the seat of Government, [designed to supply] Early and Authentic Accounts of the Proceedings of Congress—its Laws, Acts and Resolutions; communicated so as to

> form an history of the Transactions of the Federal Legislature under the new Constitution.[2]

That such a self-appointed critic of government, commercially oriented, should so promptly be deemed desirable, after centuries in which no previous regime had tolerated exposure (no less criticism), was a curious comment on democratic procedures which theorists, talking of checks and balances, might have considered above tyranny.

Yet it was just another manifestation of yankee pragmatism at work. The early American printer, said Rollo Silver, "saw the output of his printing press as a primary tool in the building of a new country." [3] But others also noticed he possessed the chief means of communication: when George Washington desired to complete his presidency with a final message to the people, he had only the newspaper to turn to. He selected the Philadelphia *Daily Advertiser* in September 1796 for his famous isolationist message.

Hamilton's *Gazette* inevitably attacked and provoked Thomas Jefferson and his anti-Federalist friends. They started the *National Gazette* (1791) to respond, and a political war soon followed. Other purely political papers added to the flutter, some to champion the activities of a state legislature, others to influence a single national election, a few to promote an isolated candidate. Of the 1801-1833 period Frank Luther Mott observed: "Politics were paramount in most papers; they dominated the news, furnished the subject matter of the essays and letters to the editor, and pointed the editorial paragraphs." The period was characterized by tremendous growth, unsettled international conditions, unprecedented political and bureaucratic construction, unclear libel and censorship laws, and unquenchable libertarians—labeled by Mott as the "Dark Ages" of American journalism:

> Few papers were ably edited; they reflected the crassness of the American society of the times. Scurrility, assaults, corruption, blatancy were commonplace. Journalism had grown too fast. . . .

These were certainly not the tactics of British journalism even under extreme conditions; in praising the integrity of the American editor John Stuart Mill obviously had theory more than practice in mind.

In these newspaper squabbles American statesmen generally declined to participate. Jefferson, although a firm believer in public education and free press as instruments of liberalism, "had never quite understood the rough-and-tumble journalism of the times," [4] Mott had said. Others, like Hamilton (whose "Federalist" papers ran serially in the one paper and who controlled editorial policy on others) and Andrew Jackson (who during elections handled the press very effectively), had no trouble recognizing what journalism could do. Outside the government, however, few educated persons appreciated the power of print as newspapermen wielded it. The common man, of course, loved it.

Alexander Hamilton started the New York *Evening Post* in 1801, persuading rich merchants and Federalist thinkers to lend $10,000 (and to cancel the notes) for the establishment of a political paper. William Coleman was editor, although Hamilton guided the paper personally, along with two others, the *Gazette* and *American Minerva.* Strongly democratic, the *Post* had good commercial coverage and some literary distinction, although its circulation remained small.

William Cullen Bryant, when he became the *Post's* editor in 1829, brought a poet's romanticism to the paper that almost impeded its growth. While he admired the steamboat, telegraph, railroad, and other inventions, he did not use those devices as ingeniously as other editors, and failed to modernize his pressroom until 1850 (although the paper had an eight-cylinder Lightning by 1862). "Though it spoke for the common man in flaming editorials," said Calder Pickett, "[the *Post*] failed to publish a newspaper for him." [5]

Philip Hone called Bryant a "black-hearted misanthrope," having seen him scuffle with the *Commercial Advertiser's* William Stone (Hone was amazed such a brute could write beautiful poetry). Bryant's humor was certainly uncivil: his half century as editor, chronicling man's foibles and follies, suggested to him that a reverse Darwinism was at work:

> If we hold to Darwin's theory—as I do not—how are we to know that the vast multitudes of men and women on the earth are not the ruins, so to speak, of some nobler species, with more elevated and perfect faculties, mental, physical, and moral, but are now extinct? [6]

Bryant, James Gordon Bennett, and Horace Greeley personally shaped the colossal American newspaper industry that shortly followed.

Alexis de Tocqueville upon arriving in America (1831) perceived the intricate world of journalism almost at a glance:

> All the political journals of the United States are, indeed, arrayed on the side of the administration or against it; but they attack and defend it in a thousand different ways. . . . The facility with which newspapers can be established produces a multitude of them; but as the competition prevents any considerable profit, persons of much capacity are rarely led to engage in these undertakings. . . . The journalists of the United States are generally in a very humble position, with a scanty education and a vulgar turn of mind. . . .[7]

Ignorance and vulgarity could not long enjoy public approbation, but their effects for a time were loud and irritating. De Tocqueville himself was startled by at least one example:

> Ambition is [Andrew Jackson's] crime, and it will be his punishment, too: intrigue is his native element, and intrigue will confound his

> tricks, and deprive him of his power. He governs by means of corruption, and his immoral practices will redound to his shame and confusion. . . .[8]

This kind of language was far from exceptional, and statesmen like Jefferson and John Adams were beginning to warn that, although the paper's contribution—like the political party, quite unpredicted, unlegislated—to the cause of democracy was unmistakable, regulation of the press might be a growing necessity. Such offensive reportage, of course, would inevitably produce a counterforce which manifested itself, surprisingly enough, in still another journalistic phenomenon with similarly vulgar motives—the "penny" press.

The penny or one-cent newspaper emerged a dynamic but unsure force at the end of the first third of the century. In Philadelphia *The Cent* (1830) and in New York Greeley's *The Morning Post* (1833) were the first, but short-lived without a simple, consistent scheme for making cheap, popular journalism work. The missing formula was provided by printer Benjamin Day, who, despite a raging cholera epidemic, started *The Sun* (1833) in New York with immediate success. In its first issue Day had written:

> The object of this paper is to lay before the public, at a price within the means of every one, all the news of the day, and at the same time afford an advantageous medium for advertising. . . .[9]

Printer Greeley (who similarly began his *Tribune* in 1841) later explained more clearly to an editor friend what "news of the day" really was: "[T]he subject of deepest interest to an average human being is himself;" he wrote, "next to that he is most concerned about his neighbors. . . ." [10] But means to apply the unique formula remained with a privileged few. One of them, Frederic Hudson, acclaimed the "newspaperman's newspaperman," wrote:

> In the origin of the Cheap Press its owners were mostly printers. They were practical men. They had worked in the old party newspaper offices, and had had their joke and their moral over the articles they had put in type. This was their journalistic experience. They had no very comprehensive ideas of newspapers. New ideas did not often appear where they worked. They knew that local news was always interesting. Village gossip had taught them this. . . . So the Cheap Press commenced with local matters. . . .[11]

As printers first, they understood fully the mechanics of producing a publication and were therefore in an advanced position to devote their main energies to the collection of news and the interrelationship of events with their private opinions or ambitions. News for them became a personal affair, sometimes driving them into government office when editorializing did not seem sufficient, sometimes surreptitiously creating it when times were dull.

These intrepid editors, nevertheless, did not fill the one-cent papers with penny ideas; competition made them scurry to give the public everything they could find. As soon as product and price generated a daily habit, the formula was confirmed. "A business which had fed too much at the back door of politics," said the *Sun's* centennial editor of the printery, "became one which thrived on its own enterprise." [12] Of course, this renewed industriousness continued to have its critics. Philip Hone carped at it continually:

> [5 September 1843] . . . These are glorious times for the penny papers. The yellow fever at Rondout, a prolific source of lying and vituperation, the quarrel among the Episcopal clergy, the squabbles between the Calhounities and the Van Burenites for the Presidential nomination, a great murder trial in New Jersey, robbery, forgery, and defalcation—all these things furnish food for the taste of our "reading public" to repletion. . . .[13]

But even Hone probably read at least one of them daily. Others in the profession, with similar sensibilities about the penny formula sordidness, changed their publishing policies to literary or religious miscellanies. They prospered too.

Among the great newspapers launched about this time were the New York *Herald* (1835), The Philadelphia *Public Ledger* (1836), and the *Baltimore Sun* (1837). Within five years of Day's commencement 19 other penny dailies appeared in New York, as well as tri-weeklies, semi-weeklies, weeklies, semi-monthlies, and monthlies. The occupational hazards of this phrenetic new profession were promptly described:

> The life of an editor is comparatively short. He wears out before his time. The exacting toil he pursues, which is rarely, or never broken by a solitary day of relaxation, shatters his nerves, exhausts his vital energies, and makes him gray-haired almost in middle age. To him the course of nature is reversed, and night is turned into day. He labors when other men sleep. Nothing tells sooner on the constitution than this. The close room in which he usually sits, the stifling odors of damp newspapers from the mail, and the blinding glare of the gas-lights increase the wear and tear upon his system. . . .[14]

Yet there was no arguing with success. The aforementioned newspapers, all founded by printers, survived—as did many others—well over a century.

The *Herald's* James Gordon Bennett, starting ingloriously with borrowed money, further directed his sheet's "important" news paradoxically to the "unimportant" citizen, the commonest laborer and tradesman, to whom he promised lively human interest and strict political freedom. To be sure, his promise of independence reflected not a belief in honesty, integrity, or balance, but a renunciation of his long submission to political cliques and parties, favors and debts, job seekers and other harassments. His paper in fact pioneered in sensational crime and sex—and the public thrilled to read it. Because of his exhaustive coverage, furthermore, the *Herald* was considered the first real *news*paper.

At any rate, the penny press, more than the political sheet, created habitual readers out of potential idlers:

> Americans are a nation of newspaper readers. There are papers for girls and boys, for teachers and taught, for trades, arts and sciences, for the lovers of the wonderful, the religious, the agriculturists, the metallurgists—in a word, the physician, as he rides to the expected birthplace of some young patient, and the undertaker, as he drives to the house of morning, must each have his paper to while away the time. . . . [15]

The public soon enough would want and need somewhat more intellectual fare with its breakfast toast and coffee.

A century later Marshall McLuhan, reinforcing the universal observation that man's interest is man, further identified the newspaper's unique methodology: the "confessional" nature of presenting one intimate item after another in "mosaic" form gives the paper a powerful vitality:

> It is for this reason [that is, ability to furnish "inside" stories] that the press seems to be performing its function most when revealing the seamy side. Real news is bad news—bad news *about* somebody, or bad news *for* somebody. . . .[16]

All the good news in newspapers, according to McLuhan, is in their advertisements: new products and services, irresistible sales and premiums—making a perfect balance in the healthy paper. As long as its finances remained stable the newspaper had surprisingly little to fear.

Aside from de Tocqueville and other European visitors (whose awareness of European suppression and censorship gave them a vivid basis for comparison), there were few critics of American journalism as an institution. Newspapers of any reasonable persuasion survived, expanded, prospered. Of course there were conflicts: politicians frequently chaffed under editorial attack and (when public caning failed) sued for libel; collaborators in organized crimes sometimes retaliated with mayhem and coercion (not always brought to justice). Those readers whose propriety or privacy was invaded boycotted a paper or cancelled their subscriptions. Joseph Pulitzer nevertheless regarded such conflict a noble mission for the whole profession:

> What is everybody's business is nobody's business—except the journalist's. It is his by adoption. But for his care almost every reform would be stillborn. He holds officials to their duty. He exposes secret schemes

> of plunder. He promotes every hopeful plan of progress. Without him public opinion would be shapeless and dumb. Our republic and its press will rise or fall together. . . .[17]

Coincidentally, perhaps, like politics itself in the United States, journalism in America was often and largely a matter of personalities rather than issues. The physical product, generically, had become so well established by the end of the nineteenth century no one questioned its right to exist unencumbered by harassing regulations, legal or social. George Henry Payne found this freedom—and license—to predate the Revolution:

> From now on [after Zenger's trial] the history of journalism is not of its own struggles, but of the struggles of the ideas for which it stands. The thing itself is established. Henceforth it is a story of development; development in close connection with the idea of democracy, from which it sprang, to whose influence it owed its quick growth, and to which in turn it contributed as no other single factor in civilization, except Christianity, has contributed.[18]

Americans were consequently at a loss to account for European methods of control. Occasionally, like Greeley at the 1851 British inquiry, they found opportunity to tell friends how annoying a stamp tax could be. By midcentury, however, with the appearance of Hoe's fast type-revolving presses, such official regulations would not only have been nagging, they would have been stifling if mass circulation papers had been required to submit to pre-publication licensing, per-issue censorship or taxation, or penalization for petty infractions.

Within a few years of their inception, nevertheless, the penny papers in America had demonstrated much more than their British namesakes. They possessed a vitality and impetus that engendered admiration and wonder. Samuel Bowles of the Springfield *Republican,* one of the profession's most influential practitioners, in 1851 himself marveled at the phenomenon:

> The brilliant mission of the newspaper is not yet, and perhaps never will be, perfectly understood. It is, and it is to be, the High Priest of History, the vitalizer of Society, the world's high censor, the medium of public thought and opinion, the circulating life-blood of the human mind. It is the great enemy of tyrants, the right arm of Liberty and is destined, more than any other agency, to melt and mold the jarring and contending nations of the world into one great brotherhood, which, through the long centuries, has been the ideal of Christian and philanthropist. Its mission has just commenced. . . .[19]

Ironically, at this moment, not far away, the newspaper was unready for such praise.

Midcentury New York was developing the perfect opportunity for journalists, in Pulitzer's words, to make ongoing plunder and corruption their business, to expose, to hold officials to their duty. Indeed, the press of that city, largest corps in the country, could not fail to notice irregularities in the municipal government—at every level. Its conspicuous silence against the notorious Tweed Ring, unfortunately, was, said Mott mildly, "far from creditable to journalism." [20]

Willam M. Tweed was born 1823 on Cherry Street, in the heart of Hoe's lower East Side. He quit school at 11 and apprenticed in his father's chair factory, but, restless and hungry for attention, found the volunteer fire company a gallant outlet for his bravado. He distinguished himself in the holocaust of 1835, at Taylor's explosion, and at many other Manhattan fires. Tweed named the "Americus Engine Company No. 6," and gave it the tiger as symbol. From foreman of the crew he jumped into city politics—barely winning as alderman in 1851 despite some shabby campaign tricks. Two years later—with his father's Franklin Square shop smoking from Harper's great fire (Tweed's quarrel with the Harpers would grow)—he was well entrenched and using his uncanny talents effortlessly to bilk the city treasury. In 1857 he was elected one of twelve supervisors with unconstrained management of the city's money, and soon practiced various extortions from big and small businesses and the state of New York as well.

Tweed had not developed his enormous network of schemers without help—and precedent. Fernando Wood (1812-1881) at 20 had opened a "wine and segar shop" which failed because he spent so much time politicking. With a sailor's groggery and lotteries he soon grew so rich from gambling he retired in 1849 to devote himself entirely to politics. His earliest ambitions elicited Philip Hone's most damning comment: "Fernando Wood, Mayor!! . . . [He] ought to be on the rolls of the State Prison. . . ." [21] Wood succeeded nevertheless and was mayor of the city 1855-1858, 1861, and 1862. He displayed a gift for audacity and chicanery an ambitious young politician could only admire:

> Wood demonstrated how power could be centralized. As the model of craftiness in government, he was a masterful exponent of applying corrupt practices. . . . [He was] in a very real sense, Tweed's mentor.[22]

Wood and Tweed's personal dislike for each other kept them fighting for years for the control of Tammany Hall, the Democratic party headquarters. Wood, a master of duplicity (of which Tweed had no need), failed, however, to unite the party in his support, a feat the younger man soon accomplished handsomely. Tweed's success was begrudgingly described by an 1869 *New York Times* in metaphores the Hoes would have appreciated:

> [Tammany is] a wonderfully and admirably-constructed and conducted machine. . . . The wheels work smoothly, the pulleys run

> without a jar, the cogs slip into one another perfectly, while everything is kept well oiled and greased from the public funds of our wealthy citizens. . . .[23]

At election time the party apparatus fairly hummed with bribery, rowdyism, blackmail, and, of course, unconcealed repeat voting.

The city was blanketed with a network of Tammany clubs which enlisted boys' gangs, immigrants, professional criminals, and greedy officials for its tasks. One of the most sumptuous halls was the 7th ward's "Young Men's Democratic William M. Tweed Club," 105 East Broadway, near Tweed's home and just a few steps from the Hoe factory. Early in 1871 it was lavishly appointed with parlor, bar, billiard, card, and meeting rooms, and office. Strangely, except to complain about deteriorating pavements (Tweed was also New York Street Commissioner) the Hoes appeared oblivious to the nefarious regime. In their efforts to be scrupulously honest tradesmen the brothers had almost neglected their responsibilities to the community and to their consciences.

To develop his power without interference from the public, "Boss" Tweed had not overlooked the necessity of suborning the metropolitan press. Various New York papers received heavy patronage and city advertising which kept them quiet; on others reporters were bribed with gifts and appointments to sinecures. Although official municipal notices were limited by law to 17 newspapers, Tweed was making payments to 87 (37 of which—without other support—perished with the ring's downfall). Editorial criticism was consequently long sporadic and ineffective until the evidence became overwhelming. Finally the *New York Times* and *Harper's Weekly* resolved to fight the giant machine.

The *Times*' George Jones and *Harper's* Thomas Nast were both offered enormous bribes to quit their accusing, taunting, and incriminating exposures (Harpers also lost its lucrative New York textbook contracts). Tweed's most explicit wrath, against Nast's exquisite cartoons, recalled the power of political caricature in antimonarchic France just a few decades earlier. "I don't care a straw for your newspaper articles," he wailed to cronies, "my constituents don't know how to read, but they can't help seeing them damned pictures." [24] Within months a great reform meeting at Cooper Union demanded that the "wisest and best citizens" save the city. An executive "Committee of 70" foremost New Yorkers was formed, with William Havemeyer, a former mayor, heading the group. Tweed was soon arrested and the ring destroyed.

After several trials Tweed was found guilty of 51 of 55 offenses (each offense consisting of four counts), given 12 years, and fined $12,750. His consecutive sentence was contested, however, and in December 1875, while visiting his family, he eluded his guards and escaped. It was a monstrous embarrassment for the county sheriff, William C. Conner. Tweed himself had made typefounder James Conner, William's father, a Grand Sachem of Tammany, and the son had imitated his father's career closely. William joined the volunteer fire department, the National Guard, and Tammany Hall; he was elected Supervisor of New York County, County Clerk, and in 1873 Sheriff. Technically responsible for Tweed's imprisonment, Conner exonerated himself to the complete satisfaction of his biographer, Henry Bullen:

> By great energy and a liberal outlay of his own funds, Sheriff Conner succeeded in capturing the fugitive in Spain, thus erasing a stigma placed on his official conduct by the guilt or the carelessness of a subordinate.[25]

Ironically, Spanish authorities had recognized Tweed from a Nast cartoon. His health failed and he died in prison 1878. Although estimates varied widely, the Tweed Ring reportedly had cost the city about $200 million. It was a crime that boggled even the hardened big city journalists.

George Rowell, who championed American journalism even more than his own newspaper advertising agency, perhaps had Tweed's career in mind:

> Good men are not afraid of criticism by the press. It makes statesmen and unmasks hypocrites; it incites to great deeds; it brings to every man's door the record of progress made in every department of learning and activity; it develops science, and whiles away the tedium of heavily hanging hours. It fits out exploring expeditions to discover new fields for civilizations; it lays before its readers the doings of the world's busy multitudes, the fall of empires, the uprisings of nationalities, the record of the storm-king's progress around the world, the decisions of the forum, the acts of governors and legislators; it tells the farmer what to plant and when; it develops the latest inventions; it weighs in the nice balance of purity of motive the deeds of the ruling powers; before it the unjust tremble, and on its pages, as on the walls of the ancient banqueting hall, the wicked rulers may read "weighted in the balance and found wanting"; it stirs the heart of benevolence to greater deeds of charity, it inspires the desponding, deters the plotters, and from the ruler to the humblest citizen, it throws its Minervian aegis around all alike. . . .[26]

But in this exhaustive mandate Rowell and his cohorts persistently overlooked one important area perhaps more flagrantly ignored in America than in Europe.

Like the British press, the American journalistic community, so devoted to partisan political and governmental activities, had also failed to demonstrate similar consuming notice of another growing, tyrannical institution: science. To be sure, nineteenth-century experimentation and technology did not at first seem likely to threaten the public quite so much as unchecked political ambition. Science—the sublime understanding of Nature—was (like politics) nearly within the comprehension of all, anyway, and often within the realm of personal experience. This was especially true in the important domain of agriculture and animal husbandry.

The gentleman farmer in America was a vestige of the eighteenth-century "Age of Enlightenment," an admirable philosophy holding that an educated man could control his environment. As the sciences made dramatic progress—most pragmatically affecting crops and veterinary sciences—many affluent civic leaders, merchants, and industrialists (unlikely farmers indeed!) treasured their country estates where nature's order and tranquility prevailed. The farming enthusiasm of William Cullen Bryant was typical: extremely proud of his Roslyn, Long Island, property, he sent detailed planting and pruning instructions to his caretakers even when traveling around the world. P. T. Barnum was so avid a "farmer" he was president of Connecticut's Fairfield County Agricultural Society. There were almost as many such country gentlemen as those whose living depended upon the fields.

Although, wrote an agricultural historian, "Gentleman-farmers with fat steers, superior produce, and walnut-paneled barns that were show places" [27] were much resented by professional farmers, they nevertheless gave needed support to struggling American agricultural societies, fairs, and journals. They became a knowledgeable and helpful force in shaping the country's rural standard of living. And, like the moderately schooled general public, they professed to be equally enlightened in their understanding of the more abstract sciences just emerging.

In the United States Swiss-born naturalist Louis Agassiz (1807-1873) enjoyed enormous popularity. He authored countless abstruse works on prehistoric glaciers, fossil identification, and animal classification. (Boston *Transcript*: "Prof. Agassiz, in his treatise on turtles of North America, leaves out one curious species entirely—the hard-shell 'turtles' of Hoe's printing press." [28]) In an era when modern man —even those confined to office work—prided themselves on comprehending nature's laws, all could appreciate Agassiz. By describing the orderliness of God's universe he brought the citizen and scientist together as friends.

After enthusiastically received Boston lectures Agassiz in 1847 came to New York. Horace Greeley, who himself had an expensive "model" vegetable farm at Chappaqua, New York, welcomed him and published his talks verbatim in the *Tribune* and reissued them in pamphlet form. Agassiz's glacier studies showed him that entire species had been eradicated by the intrusion of the ice ages, while new forms appeared after each of them, forcing him to conclude that God created life not once (as the Bible said) but over and over again. Too, he determined that glacial activity—"God's great plough"—was responsible for much of what the Bible attributed to the Flood. Charles Darwin wrote to compliment him for his work, but the appreciation was not reciprocal.

While many American scientists had recognized isolated examples of evolution before Darwin brought them together, Agassiz, trying desperately to reconcile his own theories with the Bible, was repelled by the idea. Through the 1850's (innocently abetting the South's position on slavery) he lectured that Blacks and whites were distinct peoples, created separately by God. In Charleston, the Southern locus of academic thought, disregarding the Scriptures, according to an incredulous *Scientific American,* he "avowed his disbelief in the unity of the human race!" [29] Agassiz even claimed the world's languages to be "created" separately, bearing no relation to each other.

Darwin welcomed his fellow naturalist's comments on evolution. "I have ventured to send you a copy of my Book . . . on the origin of species," he wrote, modestly. But Agassiz studied it with growing irritation. "This is truly monstrous!" he thundered in the margin of one offending passage; "What is the great difference between supposing that God makes variable species or that he makes laws by which species vary?" [30] he asked in another. Agassiz soon began lecturing against Darwinism, his remarks published 1862 in the scholarly *Atlantic Monthly* (placing itself squarely alongside the implacable British quarterlies)—and reprinted nineteen times in book form.

Although Horace Greeley refused to publish growing attacks on Agassiz, the wise old newspaperman quietly slipped away from him:

> Visitor—"Why, Mr. G., you are a Darwinian."
>
> Mr. Greeley—"Am I? That was what Agassiz said, and [although disdaining fishes evolving into monkeys] he added, 'It is only at the highest stage of the whole development that we find man,' and I think so, too. I suppose [sniping at Secretary of State Hamilton Fish] that accounts for some men calling each other curious Fish. . . ." [31]

While Agassiz could not abandon his faith (such as it was), another anthropologist enjoyed jabbing the Bible's Fundamentalist tenets.

Dr. Josiah C. Nott, a Mobile, Alabama, physician (cousin of Dr. Eliphalet Nott), wrote profusely and irreverently on the enigmas of race, including a giant 1854 treatise on the "types" of mankind. After an illness he wrote a friend, "I have been well enough to skim Darwin's book—the man is clearly crazy, but it is a capital dig into the parsons—it stirs up Creation & much good comes out [of] such thorough discussions. . . ." [32]

Asa Gray, another American naturalist, had quietly managed to reconcile scientist and believer: "Mr. Darwin . . . we trust implies that all was done wisely," he wrote, "in the largest sense designedly, and by an intelligent first cause." [33] But quiet reconciliation was never good news. When Fundamentalists at the 1925 Scopes evolution trial in Dayton, Tennessee, remained unconvinced, a journalist gleefully reported the event's ultimate outcome: "Darwin made monkeys out of them," [34] he quipped. Science would fare no better in twentieth-century America. "Most of the newspapers treated the whole case as a farce instead of a tragedy," [35] commented Clarence Darrow, John Scopes' counsel.

After the Civil War, as public tastes widened, many newspapers threw off their political mantles and labeled themselves "neutral" or "independent." Whitelaw Reid heralded this new formula:

> Independent journalism! that is the watchword of the future in the profession. An end of concealments because it would hurt the party; an end of one-sided expositions . . . ; an end of assaults that are not believed fully just but must be made because the exegency of party warfare demands them; an end of slanders that are known to be slanders . . . of hesitation to print the news because it may hurt the party . . . of doctoring the reports of public opinion . . . of half truths . . . that is the end to which every perplexed, conscientious journalist a new and beneficient Declaration of Independence affords.[36]

Partly responsible for broadening public interest in world matters, the New York *Herald* continued to dazzle its readers. The paper employed a variety of stunts (notably Stanley's search for Livingstone, 1871; an 1872 hoax that New York's zoo animals had escaped) and other devices to make—not merely report—news. Symptomatically, these were really manifestations of its new owner's caprices:

> James Gordon Bennett, Jr., though much of the time an absentee editor, never relinquished his control. He was brilliant and alert, and often daring. At the same time he was an insolent egotist, a tyrannical employer, self-indulgent, and vain. . . . He was contemptuous of men, and often hired and fired by whimsy. . . .[37]

Yet through the 1870's the paper's circulation hovered at 100,000, often higher.

Other New York papers also maintained a quick pace. Charles Dana's independent *Sun* was considered the "newspaperman's newspaper," its circulation also over 100,000. When Greeley died in 1872 Whitelaw Reid, 35 and ready to step up, made the *Tribune* respectable and lively. For the *Evening Post* Bryant relied on his son-in-law Parke Godwin until the paper was sold (1881) to Henry Villard; Carl Schurz was chief editor for a while. The *New York Times* remained small but prestigious. This "New Journalism" prospered somewhat in direct proportion with its independence, but nowhere more conspicuously than Joseph Pulitzer's *World*.

Six foot two inches in height, 37, and politically ambitious, Pulitzer (1847-1911) snapped up the failing New York *World* in 1883 and (like many others before him) issued a bold salutation:

> [The *World*] will from this day be under different management—different in men, measures, and methods—different in purpose, policy and principle—different in objects and interests—different in sympathies and convictions—different in head and heart. . . .
>
> There is room in this great and growing city for a journal that is not only cheap but bright, not only bright but large, not only large but truly democratic—dedicated to the cause of the people rather than to that of the purse potentates—devoted more to the news of the New than the Old World—that will expose all fraud and sham, fight all public evils and abuses—that will battle for the people with earnest sincerity. . . .[38]

World circulation was soon on its way to 250,000, then—over a million.

Mott attributed Pulitzer's fantastic success to six factors: exhaustive news coverage, well written; newspaper stunts (like its fund for the Statue of Liberty pedestal and other community projects); quality editorials—which most occupied Pulitzer himself; size—boasting "the only two-cent eight-page paper," it often went to 12 and 16 pages; copious illustrations, if only crime diagrams or cartoons; promotion via coupons, contests, and advertising. Another, color comics and banner heads, would follow.

The New Journalism also capitalized on sensationalism—sordid crime, scandalous sex, disasters, and monstrosities—to provoke greater emotional involvement with the day's events. These were aided by increased use of illustration, particularly photo-engraving. Stephen Horgan applied Frederic Ives' halftone process to several papers, the New York *Daily Graphic,* New York *Herald,* and *New York Tribune,* and the *World* too soon picked up the novelty, especially for its Sunday editions. Pressmen had generally discouraged heavy illustration ("cutwork" always required fussy inking) but, said Mott, "the Boston *Journal* did the trick for its edition of Sunday, May 6, 1894, using its Hoe quadruple press to print no less than twenty-one halftones. . . ." [39] Other papers had to try it too.

William Randolph Hearst (1863-1951) got his first experience on his father's San Francisco *Examiner* and applied the New Journalism with fiendish gusto. But, like Pulitzer, he yearned for a New York paper as the supreme symbol of success. He bought Albert Pulitzer's *Morning Journal* in 1895, staffed it with crack newspapermen, and—with all of the *World's* tricks and techniques —squarely faced Pulitzer for a contest:

> In 1895 the *World* had the largest, cleverest, and most popular Sunday paper in the country. Hearst, whose San Francisco paper had

> long rented rooms in the *World* building as its New York office, established himself there and in January, 1896, began secret negotiations designed to hire for the *Journal* the entire staff of the Sunday *World*—editors, writers, and artists. . . . The Hearst checkbook was a strong argument, and soon Pulitzer was surprised and staggered by the departure of the whole staff. His editor in charge soon pursued the deserters, however, and, adopting the Hearst language, brought them back. Hearst thereupon again raised the bid, and twenty-four hours later had them all once more. . . .[40]

The Sunday papers in particular, with batteries of new color presses and "iridescent" inks, threw spectacles steadily at the bedazzled public, culminating in the giant supplements and comic sections. These shinnanigans so escalated that, by 1898, the circulation battles of the two titans led unmistakably to the very real Spanish American War:

> Certainly the most powerful and persistent jingo propaganda ever carried on by newspapers was led by the New York *Journal* and *World* in 1896-98, and the result was an irresistible popular fervor for war which at length overcame the long unwillingness of President McKinley and even swept blindly over the last-minute capitulation by Spain on all the points at issue.[41]

The two dailies now both had circulation over a million.

The big New York papers at the turn of the century had also become big businesses as well. Colossal investments made the newspaper more a corporate institution than the personal organ it had been, and brought a degree of (financial) conservatism along with professional compartmentalization. Nevertheless, where in 1892 only 10 papers in four cities had circulations over 100,000, twenty years later there were over 30 in a dozen towns, and, on the average, circulations of all American daily papers had doubled.

Frank A. Munsey (1854-1925) watched the success of these "million dollar" newspapers and wanted to join them. With a yearly income of $500,000 by 1900, from his shrewd magazine schemes, he began to think not of single papers, but chains of them. Running the fine old institutions on a ruthless trial and error basis, he literally destroyed many of them, like the New York *Daily News* (bought 1901), the Boston *Journal* (1902), and the Philadelphia *Evening Times* (founded 1908). Although some of his papers made money, Munsey was universally hated as "the great executioner of newspapers" (Mott's words), despite increasing indication that such consolidations were inevitable.

With men like Bennett, Pulitzer, Hearst, and Munsey as its best customers, it is perhaps no wonder that R. Hoe & Company would begin to show symptoms of its own eccentricities in business. The firm advertised less and less, and became very bureaucratic in its administration. Robert Hoe III, its head, withdrew increasingly to his Grolier Club activities in New York or his private residence in London, where he could not be reached by daily crises. But if the Hoe organization for whatever reason was no longer the simple press-maker it once was, all the other newspaper impresarios were similarly no longer the printers their predecessors had been. They had to make a special effort to understand the various printing processes but were only incidentally involved with them anyway. With somewhat better educations and a sense of America's direction and potentialities, they addressed themselves to growing responsibilities on behalf of the democracy which gave the papers life a century before. This called for editorial house cleaning of profound proportions.

The "Dark Ages" of the first third of the nineteenth century, if not honorable and rational by gentlemen's standards of the day, had nevertheless introduced a rhetorical violence which, like the slapping of a newborn baby, sent the country on its way screaming. It is possible a yapping press spared any real bloodshed while it drove sluggards and malcontents irately to the polls—if not into office. The more enlightened penny press that followed, on the other hand, might by its mildness have produced a conservatism or aloofness easily fatal to the spirit of democracy. Nevertheless it did furnish the seed of one idea that redeemed all faith in American newspapers—objectivity.

James Gordon Bennett's pledge of political independence, unclearly understood, perhaps, had been the touchstone of the New Journalism. But its power was more salesmanship than performance. Political neutrality in the hectic, formative years of a struggling government was, in reality, as extraordinarily naive as the political sheets had been raucously over-critical.

But for the general public, these were hardly sophisticated times: the average reader, poorly educated and limited in national and international perspectives, could hardly vote knowledgeably if guided only by unbiased human-interest news. The inducement of political neutrality, whether sincere or not, met wide professional incredulity at the time, and perhaps for that reason functioned largely as a lure—a grand appeal to the intellect—to increased circulation. Few readers had any commensurate need for scrupulous reportage, but all were now drawn to the pageantry of party platforms, candidate integrity, and election probabilities. The opinionated newspaper held its lead:

> But whatever may be said of the enterprise and success of the penny press, and however important its part may have been in the news revolution, the fact remains that up to the time of the Civil War it was not the independent penny press but the partisan political press that dominated American journalism.[42]

Thus, during the first seventy-five years of American democratic experience, political protocol and tradition were strongly recorded and shaped by the partisan press. With growing agrarian-urbanization and slavery-industrialization polarization, at no other time in American history were party issues and the fate of the nation so critical. It was almost no time to be neutral.

> And by this time [1852], the influence of the public press, as the educator of the masses of the body politic, had become palpably manifest. . . . The journalist has a greater immediate audience [than the orator], but his words reach afar off, as weighty in the uttermost parts of the land or of the earth, as in the deep basement where they issue from that marvel of modern machinery, the Hoe Press. . . .[43]

As new territories joined the Union newspapers played back and forth to report and consolidate opinion—and, even before the telegraph lines were strung, to augment news—across considerable distances. De Tocqueville had noted:

> [The press] rallies the interests of the community round certain principles and draws up the creed of every party; for it affords a means of intercourse between those who hear and address each other without ever coming into immediate contact.[44]

Such exchanges were crucial to the success of a sprawling country: and soon enough not just community interests and party creeds were at stake. Ringwalt observed:

> In this country, especially, newspapers have exerted a boundless influence. They have made and destroyed countless reputations, elevated and deposed innumerable officials, furnished an indispensible prerequisite to genuine popular government, raised immeasurably the standard of civilization, diffusing far and wide its blessings, and in view of their expanding power, he would be a bold man who would venture to affix a limit to their future achievements.[45]

Ironically, as a handful of responsible journalists had once found it their conscientious duty to attack the tyranny of monarchies, hundreds of "irresponsible" newspapers were now—almost unaware of their sacred mission—by their penetrating exposes and inflammatory opinions—defending the existence and freedom of democracies. At first, perhaps, the public had not minded occasional manipulation (when it was astutely spotted):

> The personal opinions of the editors have no weight in the eyes of the public. What they seek in a newspaper is a knowledge of facts, and it is only by altering or distorting those facts that a journalist can contribute to the support of his own views.[46]

But only with significantly improved literacy would this contention become convincing.

With unyielding logic, nevertheless, press "independence" and "objectivity" became a shibboleth, through the nineteenth and twentieth centuries, to identify obvious distortions and editorializing. To be sure, as the reader and his children accumulated more education, the more blatant political organs—and the "independents" with more insidious biases—became recognized for what they were. Too, the sensational "yellow journalism" of the 1890's was thereby—almost incidentally—dealt a serious blow.

But no newspaper ambitious to exercise its influence and hold its grip in the marketplace really refrained from editorializing, supporting parties and candidates, commenting on municipal and state operations, and crusading for public and private causes as they found them. The hallowed quest for objectivity was appreciated professionally to have only the most specific application: internally the press praises objectivity partly because monopolistic wire services necessarily disseminate only a single version of a story, and because it serves to improve the harmony among the many positions a newspaper must take on issues in its columns. Yet among other libertarian countries objectivity has totally failed to achieve similar importance.

Ostensibly to furnish readers a steady arsenal of pure data with which to form their own opinions, the newspaper world drills its cub reporters with a kind of Hippocratic denunciation of slant:

> The journalistic ideal of fair play demands not only that the reporter in gathering and selecting his news shall seek to tell both sides of the story, but also that the phrasing of his account shall not be colored by his opinions.[47]

Thus what the editor-in-chief wants reported to him—today as always—are indeed the facts only; he will himself supply the cant before an item goes to press.

Often exacerbated by pressures of vested interests as much by honest limitations, incoming news is altered, puffed up, minimized, or discarded altogether. Newspaper objectivity is scarcely more than a matter of insuring internal fidelity of worker for employer, for at the bottom of even this simple relationship is the cognizance that the reporter is often middle-class liberal, while his boss is business-oriented conservative.

The newspaper's success in bringing news of any description to the reader with dispatch and vitality is fraught with difficulties. The modern editor, said Ben H. Bagdikian, discards four incoming stories for every one he prints; on large metropolitan dailies he sees up to ten times as much as he can use. His selections furthermore are restricted by the dispatcher, who decides how many pages may be devoted to editorial matter after the advertisements are cared for. But even the most scrupulously "independent" editor faces pitfalls of content. Bagdikian continued:

> It is not always a dispute over the accuracy of facts that is at stake, but a dispute over which facts are most important. And here the doctrine of objectivity in its most stringent form usually fails. . . .[48]

Political writers long ago had already dismissed the drive for objectivity as utopian. Bryce commented:

> If every newspaper gave only the news and without bias, thus enabling the formation of a sound and sober public opinion, it would be an unmixed good. [But] this virtue is not to be looked for in such a world as the present. . . .[49]

The dilemma of more mundane newspaper independence is solved, strangely enough, by the plurality of that financial monster, advertising. Even the earliest papers appreciated revenue from advertisers, and the public (in spite of itself) enjoyed the copy as much as the columns. Newspaper proprietors recognized that the client's freedom to place announcements—in journals he judged (by his own standards, to be sure) decent, sincere, impartial—hopefully reaching his widest public, discouraged radicalism of every sort. Ironically, the more advertisers a paper has, the less it fears the commercial pressures of a few. Walter Lippmann's conclusion, "A newspaper that can really depend upon the loyalty of its readers is as independent as a newspaper can be, given the economics of modern journalism,"[50] does not really ignore advertisers' loyalty as well.

In 1947 the Commission on Freedom of the Press was formed by Robert M. Hutchins, Chancellor of the University of Chicago, upon the quixotic suggestion and with financial support of Henry R. Luce of Time, Incorporated. When the group made its report mere objectivity had become something quite unwanted: it was "no longer a goal of the press; it was a fetish."[51]

The newspaper after World War II was thought by the Hutchins Committee to be in danger from several directions. It was immensely powerful, of course, but at the same time the tool of a few "lords"; it was very subjective in its coverage and services. The report—unfortunately premature to the sudden introduction of television—was made within the memory of newspaper circulation wars and the puzzling observation that total numbers of American papers, from a peak in 1909 of 2600 dailies, were down to 1750 only 40 years later. The Commission's principal recommendation was that society needed "a truthful, comprehensive, and intelligent account of the day's events in a context which gives them meaning." It had noted the increased complexity of veracity: "It is no longer enough to report *the fact* truthfully. It is now necessary to report *the truth about the fact.*"[52]

The power of the press to cleanse, by exposure of public and private prestidigitation, has always been appreciated, of course, although individual papers themselves have sometimes not been above reproach. Their crusading work on behalf of democracy in general is far more commendable—a noble assignment perhaps never quite appreciated by their readers.

A purely partisan newspaper nowadays would be unnecessarily myopic, and would have struggled painfully for the last century. Since the Civil War there have been few radically divisive issues in American government on which competing papers could capitalize (the doctrine of isolationism a possible exception). A political press today could scarcely support itself; largely American papers are mildly republican, democratic, or socialist, although essentially quite independent from party control. Indeed, there is so little ideological animosity among journalists there were in 1947, according to Hutchins, 174 localities in the United States where, by joint use of a single printing plant, "a Republican and a Democrat newspaper run peacefully through the same press but at different times of the day." William Allen White believed the press between the world wars had changed from leader of public opinion to a lazy business, "mere peddlers and purveyors of news,"[53] and McLuhan agreed in 1964 by asserting that no one any more read editorials at all.

Much of these developments are a reflection of public apathy about most political issues: voters today seriously lack an ideological framework for understanding their government, and generally leave the more profound issues to elites. Television has abetted this indulgence by catering to the public's infatuation with personalities. The affluent masses nevertheless are enthusiastic enough about preserving the economic status of American democracy (by whatever government administration), and here the modern newspaper has found rich ground for exploitation.

The paper is always expected, of course, to function foremost as critic of the manifold processes of government; this is its mandate and its responsibility. Although the early press had, in Charles Beard's words, "little or nothing to do with truth telling,"[54] its duty in providing a medium of exchange to elicit the will of the people was as important to the government's success as sessions of legislators manifesting that will. Incredibly, no official device exists for this purpose except, perhaps, ephemeral election literature itself. Without the newspaper, Bryce noted, "there could be no democracy in areas larger than were the city communities of the ancient world." He made his point even clearer:

> In modern civilized countries . . . nearly all knowledge comes, not by talk, but from the printed page. The voter who cannot read a newspaper or the election address of a candidate is ill-equipped for voting. . . .[55]

The random and boisterous conduct of the partisan press, to be sure, concealed its redeeming service to democracy. It was a check of government provided by no other institution. Thus the social responsibility usually overlooked until the Hutchins Commission enumerated the press' functions really existed throughout the life of newspapers in America. Still, as with events 200 years ago, no single paper has any specific responsibility today, but by free enterprise and free press all papers continue to express themselves, as de Tocqueville had noted, on both (all) sides of the question. The press would often be wrong, but the public could be trusted to digest, appraise, reject, in what has been called a well-known self-righting process.

In today's world, however, multivariegated national coverage of political issues is not sufficient for the preservation of American democratic government. In the interest of national security the press must often be silent—the reporter's most difficult dilemma. The reader has a burden of similar proportion: he must forever accept slanted reportage as the real price he pays for his pennies-a-day information. At work in both cases is the teleological preservation of the United States itself under varying conditions of war, peace, and a spectrum of diplomatic crises between the two.

The coverage of world news by the newspapers—which Bagdikian defined as "information about distant events transmitted speedily to a popular audience"[56]—is a curious anomaly. For much of man's recorded history he has done well enough without such news. That today's citizen has a legitimate claim to "all" distant news, in an "objective" or contemporary history sense—and for only a few cents—is a reflection only that electronic communications can indeed effect such coverage efficiently.

The assimilation of world news gives the American voter an additional buffer, an ambience, which helps him see his country in better perspective and helps guide his elected representatives through both internal and external problems. "Leaders have always understood that knowledge is power and that to share current political and social news is to share the power to govern," Bryce wrote, "which is why full public information and democracy are inseparable partners."[57]

Thus the essential responsibility of the newspaper is its instrumental part in perpetuating American—and world—democracy. The government itself, however, has demanded more of the press in recent years as its ideological existence becomes more and more threatened. Precedents go back almost to the American revolution.

The United States government, beginning in 1792 with a series of postal acts which favored newspaper exchanges, proudly acknowledged its concession to journalism. The Postmaster General in 1848 explained:

> [Newspapers have] always been esteemed of so much importance to the public, as the best means of disseminating intelligence among the

> people, that the lowest rate has always been afforded for the purpose of encouraging their circulation.[58]

Today, of course, that concession and the same conditions still obtain. While modern critics of the newspaper are again shouting "lazy and superficial," "tied to worn-out formulas," more perceptive analysts find the newspaper, on behalf of government, now working hard for those early concessions. In today's popular press the administration "leaks" unpalatable news in small parcels, samples public opinion through "off-the-cuff" remarks, and tests foreign powers with vague threats and arbitrary policy statements it can later claim were misquoted. Journalism, said Hohenberg, has literally replaced diplomacy, and James Aronson maintains that "the press to a large degree has become a voluntary arm of government."[59] This new role of the press is not to be taken lightly.

Walter Lippmann long ago suspected that newspapers could willingly be subverted to the interests of national administrations:

> The press, in other words, has come to be regarded [by some] as an organ of direct democracy, charged on a much wider scale, and from day to day, with the function often attributed to the initiative, referendum, and recall. . . . the Court of Public Opinion.[60]

But, he quickly added, government by newspaper was not workable, it was not even thinkable. The press, he claimed, was only a "searchlight," a medium, by McLuhan's definition of electric light, simply radiating energy totally without message. Yet, by 1947 the Hutchins Commission urged the government to intrude in the world of journalism for the very purpose of using it as a tool. Of its thirteen recommendations, the committee stated:

> We recommend that the government, through the media of mass communication, inform the public of the facts with respect to its policies and of the purposes underlying those policies and that, to the extent that private agencies of mass communication are unable or unwilling to supply such media to the government, the government itself may employ media of its own.
>
> We also recommend that, where the private agencies of mass communication are unable or unwilling to supply information about this country to a particular foreign country or countries, the government employ mass communication media of its own to supplement the deficiency.[61]

Considerable manufactured news from government agencies has indeed rolled from mimeograph machines since 1947, seldom properly identified as to source and ulterior motive.

The effects of television on the newspaper have been a profound influence in these developments. That fluorescent medium has demonstrated the public definitely wants today's news today (as before), but finds the perishable screen image (and infinitesimal allocations of news time) to be no handicap. Newspaper readers, tossing the sheets away daily, never really needed a durable record—tons of such scrap are now a great ecological nuisance, in fact. McLuhan attributes to television the immense popularity of weekly news magazines which have (unlike newspapers) doubled their circulations in the last two decades, apparently for those few who do crave news in depth and permanency.

Like the authoritarian presses of the world, American newspapers run the risk of supporting a form of government, right or wrong, and becoming purely propaganda tools for special administrations. While manipulative handling of news in Russia and China, for example, may be rationalized as pragmatic necessities for large masses of illiterate or poorly educated people, no such claim can be proffered in the United States. American papers nevertheless need not lose everything to stubborn principles. They can restore their usefulness immensely if they broaden their coverage of opinion, criticism, the arts, and the sciences—subjects which today hardly reach the public in newspaper form.

The newspaper as a vital, democratic tool has, where popularity is concerned, largely been replaced by television. While no one now asserts confidently that electronic journalism will continue this democratizing process, it is certain that the press in new social roles will find it necessary and desirable to referee television's conduct, lest it—or any other futuristic medium—overwhelm democracy itself. That specter has already been described.

George Orwell, as a newspaperman himself, curiously visualized many journalistic travesties in his world of 1984. He described past issues of the London *Times* repeatedly altered to make "fact" follow fiction; he foresaw "rubbishy newspapers containing almost nothing except sport, crime, and astrology."[62] In his horror of bureaucratic government in which the television set (which cannot be turned off) plays a major part, Orwell's hero realized that ultimate freedom relied (again) on the reestablishment of free government. But this would be an electronic age revolution; the newspaper was already an anachronism.

In its new, present-day socially recognized role the press is a reliable ally on untested ground. "It is still too early," said Siebert, Peterson, and Schramm, "to discern what the full-blown form of the theory will be. . . ."[63] John Hohenberg was more troubled:

> After an existence of nearly three centuries, therefore, the relationship between free peoples and a free press has been placed under an increasing strain. It is scarcely an exaggeration to say that the free press is on trial in every open society, and its partnership with free peoples is being called into question. . . .

Whatever the method and however difficult it may prove to be, the revitalizing of the press is a matter of first priority for the cause of representative government and the health of democratic societies. . . .[64]

STEPHEN DAVIS TUCKER

1818–1902

Courtesy Michigan Historical Collections, Bentley Historical Library, University of Michigan

9.

Genius at Work in the Experiment Room

THE SECRET APARTMENT / FOLDER SUCCESS AT THE GLASGOW HERALD / THE TRIBUNE FAST PRINTING MACHINE / DOUBLE SUPPLEMENT AT PULITZER'S WORLD / DE VINNE'S PRESSES

"I began my apprenticeship with Messrs. R. Hoe & Co. in June 1834," wrote Stephen Davis Tucker, determined not to incorporate in his memoirs too much of his personal affairs. "Mr. Robert Hoe, Sr. died in 1833 and the firm then consisted of Mr. Matthew Smith, Mr. Richard M. Hoe, and Mr. Sereno Newton. I was indentured to Mr. Matthew Smith, and a very kind master he was. . . ."

So began Tucker's modest manuscript of 147 typewritten pages, composed about 1890 when he was in his 72nd year. Although its author called it a history of the company, and although his position in the firm was immensely knowledgeable and influential, Tucker had been far too sequestered—and far too self-conscious—to chronicle the company's progress broadly or leave behind something of his thoughts, his feelings, his views of life and the people he met. He compiled, instead, scarcely more than the annals of experimental projects assigned to his care. With modest education and ambition Tucker let himself be swept by the dynamic Hoes into their world, their problems, their families. Like one of the Colonel's orphans, Tucker needed that security, and for it he labored unselfishly throughout his life.

Even strict apprenticeship seemed like parental guidance to him:

> Mr. John Colby [head of hand press and jobbing rooms, and forge] was my first foreman and he was pretty testy. Apprentices were not pampered in those days. Laborers were scarce in the factory and apprentices took their places. I recollect that one of my duties while

> youngest apprentice was to sweep out my department thoroughly every Saturday afternoon.

Tucker moved quickly through several of the press building rooms, but his talent seemed to lay not in the performance of repetitious mechanical chores, but in the more unusual jobbing assignments. From the start he participated in the making of a barrel-stave saw and joiner, a coining press, a brick-maker, a cotton gin, an embossing machine, wood planers and lathes, and grinding mills, all in addition to custom-designed printing machines. Around 1842 the volume of this developmental work—both for private individuals (inventors without their own shops) and for the proprietors themselves—became sufficient to give it proper space of its own:

> About this time the firm concluded to have a private experiment and model room. The rules of the patent office at that time required that every application for a patent should be accompanied by a working model of the invention. Besides, Mr. R. M. Hoe wished to have a number of experiments tried, and it was thought that models and experiments should not be exposed to public view. Accordingly a private room was partitioned off in the fourth floor story and Mr. James Blair and myself were assigned to it.[1]

This tactic may have been suggested by the 1835-1836 work for the Howe Pin Machine Company, which had insisted that a private room be used.

Through the next few years Tucker handled dozens of privately inspired and financed machines of all description, increasing considerably his practical knowledge of materials, mechanical motions, and engineering techniques. In 1846, although he was made foreman of the "hand press and miscellaneous department," he retained charge of the Experiment Room. He was 28.

Shop duties, however, did not preclude the necessity of traveling to newspaper and jobbing offices around the country to install or repair equipment. Such journeys were a mixed pleasure: the opportunity to see new and exciting places was often mitigated by the difficulties of travel. Tucker fondly remembered what must have been his first trip, during the 1841-1842 winter, an excursion to balmy Louisiana, where the New Orleans *Bee* needed a mechanic. In a succession of such calls Tucker, still young (and—for at least a decade—testing a fancier name for himself) may have indeed worried the Colonel that he would appear too green for the task:

> *New York April 13, 1843*
>
> *Dear Sir,*
> *This will be handed you by Mr. Davis Tucker whom we send to put up your cylinder press. He is not only competent to put it up, but will instruct any one you may select how to run it properly, if neces-*

sary. We shall be glad if you can send us any orders for printing materials by Mr. Tucker or put him in the way of obtaining orders from good offices in your city. Any favors of this kind or attention bestowed on Mr. Tucker will be at any time cheerfully reciprocated by

Yours very truly
Robert Hoe

To Messrs. L'Hommedieu & Co.
Cincinnati, Ohio

To the firm's employee a letter (in care of Hommedieu) was more demanding:

[*To Tucker, 24 April 1843*] . . . *We suppose that by this time you have arrived in Cincinnati and have had time to look about some. Since you left we have made considerable reduction in the prices of our presses and printing materials generally . . . on business credit, or if cash to be paid we make 6% discount. . . . Perhaps by returning through Buffalo you can sell a press (single or double cylinder) to the Buffalo "Commercial Advertiser." They have wanted one for some time. . . . If the expense of returning by the way of Buffalo is much more than the way you went, or the least expensive*† [*meant to be deleted?*] *comfortable way of returning, then do not go to Buffalo but come directly home, for it is probable that during the summer we shall have a man travelling throughout all the principal western cities. . . .*

After promising the Boston *Atlas* a press (31 October 1844) from all new patterns, "a very superior machine—in short a *model* press," Hoe had to send Tucker after it:

[*To Boston Atlas, 2 April 1845*] . . . [Y]*our pressman did not seem to understand very readily about the press & the "sheet flyer" particularly seemed to be out of his immediate comprehension. . . . Mr. S. D. Tucker who is in our employ . . . is abundantly competent to put your press in successful operation. . . .*

And in Philadelphia:

[*Tucker to New York, 16 December 1845*] . . . *I find that the great cause of trouble with Graham & Cummings' press is the poor quality of the paper, at the same time I think that Mr. Cummings is somewhat prejudiced against them.*† . . . *I raised the roller yesterday so as to make the paper throw out more horizontal which improved it somewhat. . . .*

[*16 March 1846*] . . . [T]*here was not a sheet slured*† *during the whole night. We started at 11 o'clock & got through at 6 this morning. Neither did we get a single sheet on the rollers during the night which is a mystery to them* [*an undisclosed printer*] *as they have been used to getting on about fifty. . . .*

On one such trip to New England Tucker brought home something of value:

> I happened in a composing room in Boston just at this time and saw a novel manner of taking proofs. Two bars of square iron about type high, were lying across an ordinary imposing stone and carried a small iron roller. A galley of matter, or a small form, was placed between the bars, the paper and a blanket were laid on and the roller passed over them and back. This seemed such a simple and quick way of taking proofs that the firm decided to bring out a proof press of this style. . . .

Hundreds of these were made, some as patent medicine premiums.

Returning from his long stay in France, erecting the Paris machine and attending other matters, Tucker borrowed another idea:

> It had always been the practice in the factory to keep the small tools, taps, reamers, drills, &c., in open racks in care of no one. The men took them from the racks, if they were there, and were supposed to return them, but very often did not. If not there, an hour or more might be spent in hunting up a tool wanted. If worn out or broken, the tool would of course be returned in that condition if possible, leaving it for the next man who required to use it to put it in order. If a man broke a drill, for instance, before he had finished using it, he took it to the blacksmith shop and waited there until it was dressed, and so with all the other tools.
>
> I had observed while abroad in 1849-50 that the best workshops were provided with tool rooms, in which all small tools were made, repaired, given out and kept in charge, and after my return in 1851, I wished to have one in my department, which, after some objection on the score of expense, was agreed to. All tools were brought into the tool room, and the rule made that no man should make or repair a tool; all were to be gotten from, and returned to, the tool room. The great convenience and economy of this arrangement were quickly apparent and it was quickly adopted in the other departments.[2]

This procedure, of course, became standard shop practice almost everywhere.

Tucker had amply demonstrated his loyalty and flexibility to the industrious Hoes, but he had hardly shown any genius beyond what most sharp mechanics could do when confronted with problems. Through the 1850's he began to introduce some ideas of his own, but here too he was scarcely more extraordinary than other Hoe foremen, who were also occasionally responsible for a good idea, a patent, a whole machine. All foremen had occasion to travel widely for press repair or installation work, and a few were persuaded, from time to time, to remain for months as pressmen. They were all probably the best hands in New York for understanding the increasingly complex machines.

Young Tucker nevertheless took to his chores with special application and appreciation, coming, evidently, from a modest rural family. He was born 28 January 1818 in Bloomfield, New Jersey. His parents,

Jane and Benjamin Tucker—whose grandfather came to America 100 years before—had apprenticed him at the Hoe establishment to learn "the art, trade and mystery of finisher or whitesmith,"[3] a worker in tin or iron who finishes or polishes, as distinct from the blacksmith who forges. It was perhaps a peripheral skill he never did practice.

The lad's long stay in France gave him time to court a French lass, marrying Aimée Désirée Cherouvrier of Le Mans about 1852, whom he brought home with him. They had no children. Tucker's lifelong professional race with time and motion would find a strange, tranquil compensation in his increasingly expensive avocation: he collected those cosmic machines completely without moving parts—sundials.

As a result of his detention in Europe Tucker learned of another mechanism which prompted him to develop perhaps his first press design:

> While abroad I visited many printing offices, and saw in an office in Paris a press made by Dutarte, with which I was much pleased. It was a Stop Cylinder Press, the first I had ever seen, and it seemed to me that the plan of having the cylinder, while stationary, seize the sheet, and then to gear direct into a rack on the side of the type bed, must make perfect register. The bed was driven direct by a crank, and this required a crank arm so long that in order to keep the press low the crank ran partly in a pit in the floor. This was objectionable, but I thought it might be avoided. . . .

Discussing this with his employers, Tucker was encouraged to build a small model for job work. In 1851 he began making drawings for one with a 9 × 12 inch platen. With its positive sheet control, by fingers that obviated repositioning, and an excellent inking system, the press became practical for commercial job work; it did 3500-4000 sheets per hour and was nicknamed "Little Astonisher" by printers who thought only a hand press could do fine work. Tucker continued:

> This press worked so well that the firm decided to make a larger size, so in 1852 a 'Medium' size was built. The method of driving the bed through a slotted lever allowed a short crank, so that no part ran below the floor; the four form rollers produced excellent work, the stationary cylinders while taking up the sheet gave perfect register, the absence of tapes on the impression cylinder and no shifting of fingers were a great convenience. Hence the machine soon became popular, the only objection being that it stood rather high, and this gave it the name of the 'High Stop Cylinder.'[4]

The stop-cylinder mechanism was indeed a sound one. David Payne in 1858 developed—but did not patent—the Wharfedale (named after Yorkshire's Wharfe River valley) which became popular in England for bookwork and periodicals. Many imitations followed. Despite its former allegiance to Hoe *Scientific American* in 1870 dutifully revealed it was being printed on a newly imported two-feeder Wharfedale.

R. Hoe & Company nevertheless patented many improvements, and as late as the 1890's customers were still satisfied that the stop cylinder was, in Robert Hoe III's words, "the most perfect flat bed cylinder printing machine that has ever been devised."[5] Later models with a 36 × 54 inch bed ran at 1000-1500 impressions per hour, and an even larger version, 45 × 62 inch bed and 1700 speed, was available in 1902. Harpers' special Hoe stop cylinders in 1875 did excellent pictorial work for the *Weekly* and the *Bazar* (an 1865 Hoe two-cylinder rotary for the two magazines' type forms had already proved so good it was still in use in the 1880's). In England Newsum of Leeds brought out a press called the "Anglo-American," based on Hoe's stop cylinder.

Tucker was at the heart of these developments. He had participated in fashioning an 1858 country press, an unpopular two-color machine, and other accessories. Railroad ticket-dating presses followed. His own inventions soon increased: finger-like paper grippers (1859), an "undershot" sheet-flyer system (1864), railroad coupon press (1868), and various technical improvements to the stop cylinder (1870) and other mechanisms.

If the indefatigable experimenter seemed to be a permanent fixture in the Hoe shops, at least one old friend recognized his rare potential: A. B. Taylor in 1858 offered him a partnership. Taylor's machines had enjoyed moderate success, but their manufacturer increasingly needed fresh ideas to remain competitive. Tucker was certainly the genius Taylor needed to revitalize the Hague Street enterprise, which would sadly decline after Campbell's departure. Of course Tucker was genuinely flattered by the offer and had almost innocently reported the good news to his employers. Their reaction was a common one:

> When I mentioned the matter to Messrs. R. Hoe & Co., Col. Hoe said that the firm had already decided to admit me as a partner in the near future, but that under the circumstances I would be admitted at once, and a proper agreement would be signed on the return of Mr. Robert Hoe, who was then in Europe. . . .[6]

The Colonel had reported in February 1858 that Taylor offered to sell Tucker one half interest and give him entire control of the shop, an arrangement that would have made A. B. Taylor a formidable rival indeed.

But thorny questions were raised by a new partnership, chief of which was whether or how to share with Tucker the brothers' extensive real estate holdings, the value of which was heavily intermingled with that of the business. Robert replied:

> [*18 March 1858*] . . . *It appears to me better that we should not dispose of any interest in the real estate but lease it to the concern at a*

rent to cover the interest on its value, insurance, taxes and assessments, the buildings to be kept in repair by the concern. So many owners only complicates the title and with such a lease I don't know that it would be any advantage to Mr. Tucker to own one twelfth of it instead. His obligations with the lease would be much less & of course he would sooner realize profits in money. . . .

I suppose that Mr. Tucker knows that we have each been in the habit of discounting business paper for the firm with our private funds. He should know the extent of it and the benefit this capital is to the concern in discounting our purchases &c. . . .

This discussion soon raised—fleetingly—the matter of turning the partnership into a corporation, Richard commenting (26 July 1859), "I am not sure that it will not be to our interest to come under a sort of stock concern & have scrip to represent the interest of each. . . ." Apparently this received no encouragement, but it came up again later. (In 1881 the partnership consisted of 420 shares, of which Richard had 131, Robert 122, Peter 60, Tucker 35, Robert Hoe III 30, and Stephen Hoe and Theodore Mead 21 each.)

At any rate, it was left to Tucker to decide whether he wanted to participate in the real estate ventures. The Colonel took every precaution to make him content. "He works like a beaver," he told Robert (19 September 1859), "& does everything well & systematically, & I think is invaluable in his place. I never met his equal in every respect." But time passed and even the Colonel's broken leg did not bring Robert home from Europe; the long delay seemed inordinate. There was no point waiting any longer:

[*Richard Hoe to Robert, 3 January 1860*] . . . *I think that it will be for our interest to close up our arrangement with Mr. Tucker & let it be known. He feels as if he would rather have his interest include the real estate, & I think it more for our interest to have it so. If we leave out the real property it will give us, say not over a clear rental of 8% & the am*[t] *he will pay for the 1/12 interest will be diminished much more than his income, for instance the firm pays, say net 8% rental & clears on its capital 15 or 20%. We shall be getting for the real estate only 8% from his share, when if it were a part of the capital he would be paying his share of the extra 7 to 12%. I think he feels a little sensitive on this subject, & I hope you will look on it as I do—he is an invaluable man, I really do not know his equal.*

I think we should publish the facts of his connection soon, & if you do not agree as to the real estate we can leave that point to be settled on your return. . . .

Robert agreed quickly enough, and the papers were drawn up—ten legal sheets dated 28 May 1860.

Stephen D. Tucker for $40,000 ($11,536.72 in cash, the balance to be deducted from his future profits), with the hearty approbation of

Richard, Robert, and Peter Hoe, became the fourth partner and owner of one-twelfth of R. Hoe & Company, "engaged in the business of manufacturing saws, printing presses and other machinery." That this was the first addition to the firm in some twenty years, and that Tucker was not a member of the family (although at least 6 sons and nephews were similarly employed and perhaps hopeful of that honor), was not a little remarkable. By this provident development, nevertheless, Tucker became a recognized member of the hard-working team, his commitment as consuming as theirs. It was a prestigious promotion and honor for a lad of meager circumstances. By way of celebration Tucker promptly announced he would like a three-month vacation.

Aimée Tucker had now been in America about eight years and wanted to visit her family. The couple traveled to France but the apparently happy reunion strangely enough ended in disaster: Aimée died in Paris in September 1860. Many years later Tucker's daughter Aimée recollected:

> *. . . Poor Father. I remembering† his laboring for so many years to pay off and settle her affairs. She left everything to him until his death or remarriage, when it was to be divided between her father and her three sister,† the latter† to be used for the education of their children. . . .*[7]

In 1862 Tucker married again, this time to Sarah Ann, daughter of fellow Hoe employee William Conquest. There were several children, four of whom survived him: William C., a civil engineer, Edwin D., an attorney who handled occasional matters for Hoe, Aimée, and Florence. Tucker was now as close to the Hoes, in both their business and social families, as he could be. He tackled his assignments with unswerving application.

The success in 1861 of the curved stereotype plate and the construction by Bullock in 1865 of a complete web-fed printing machine were advances that somewhat startled the company much as had Taylor's interest in Tucker. Hoe's own work in this area—thanks to Tucker—was nevertheless substantial, if slow, at first. Admittedly, the type revolvers were selling so well the Colonel was reluctant to sabotage them:

> [*To Robert Hoe, London, 6 December 1858*] *. . . I am sure that the day is not far off when some one will succeed in printing from a continuous web, and Tucker & I have put in a caveatt† which I think covers the ground very well, & we are privately trying the only difficult portions of the problem, which are cutting the paper & delivering it after it is cut. We do not mention this to any one & do not intend to, even if we should prove successful beyond our expectations, as we do not wish to kill our present press until we think it absolutely necessary to do so, but we have come to the conclusion we must be ready before any one else. . . .*

The Colonel admitted he had assigned this project to a single workman, who would be months on the job before any apparatus was ready for tests. The plans called for a machine with two form cylinders and four impression cylinders around each of them, delivering 40,000 perfected copies hourly. He judged it simpler than a ten-feeder, and smaller—and "the finale to rapid printing." But without any real stimulation, the project faltered.

Richard usually reacted more quickly to concrete press orders than to hypothetical possibilities or rumors of competitors' capabilities. Thus it was not until 1871 that he got the assignment which obliged him to tackle this most baffling of printing and folding combinations. While in Europe the Colonel received an enthusiastically proffered order from Edward Lloyd for yet faster equipment for his weekly. Ever the innovator, Lloyd insisted it be a web press, and the manufacturer could not refuse an old friend. He resolved boldly to catch up at once with his rivals and surpass them, combining the curved plate, web feeding, perfecting of sheets (printing both sides in one pass), and cutting and folding individual papers. Still in his London hotel—thwarted somewhat by the Franco Prussian War from further travels—the Colonel pondered the whole problem anew.

Hoe's ten-cylinder Lightning had been capable of up to 20,000 impressions per hour, but on one side of the sheet only. To contemplate the printing of both sides, at similar rates of speed (too fast for human feeders), press designers were obliged right from the start to consider automatic web feeding from rolls. Robert Hoe III described the theoretical difficulties:

> First. The set-off [accidental impression from still-wet ink] of the first side. Devices were used to overcome this and the ink-makers were induced to pay special attention to the manufacture of rapid-drying or non-setting-off inks.
>
> Second. The difficulties in obtaining paper in the roll of uniform perfection and strength. The paper-makers were led to make a study of producing large rolls of paper meeting these requirements, and became much more experienced in its manufacture. . . .
>
> Third. The difficulty of the rapid severing of the sheets after printing.
>
> Fourth. A reliable and accurate delivery of the printed papers.[8]

Except for the last two requirements, prior solutions had been more or less adequate. Delivery, including on-press cutting, alone seemed a major task, but the Colonel sought more: usually enjoying large suburban subscriptions, weekly newspapers should really also be folded (for mailing) on the press. He insisted upon this complication even before the older problems were settled.

Many new press manufacturers had similarly entered the picture with a variety of delivery capabilities of varying success:

> The Bullock Co., of Pittsburgh, was trying to introduce a machine in which the printed sheets were seized by grippers on leather belts and deposited on a fly-table as they traveled over it. The London *Times* had built in its own office the Walter Press in which the sheets were partially thrown to the right and left alternately by an oscillating sheet-flier, as shown in [Victor] Beaumont's patent of 1853 (owned by R. Hoe & Co.) but which required an assistant at each table to straighten the sheets out. The Victory Co., of Liverpool, had built a machine in which the sheets were folded by the usual vibrating blades. Messrs. [Walter] Bond & [Joseph] Foster, of Preston, [England,] were bringing out a Double Type-Revolving Press, each press printing one side of a web in the precise manner claimed in the above cited Beaumont patent of 1853, but all of these machines fell far below our adopted standard of speed for a web machine, of 200 eight-page sheets per minute.[9]

Joseph Foster & Sons, press builders, had begun 1860. Tucker had even forgotten an obscure British folder patent which, owned by Hoe, forestalled an infringement suit with the "Victory folks." Percy D. Hedderwick, son of the proprietor of the Glasgow *Citizen,* in 1870 patented a rotary press which proved too complex, although its folder had merit. Tucker's concern with competitors' delivery systems illustrates that, by 1871, rotary *printing*—thanks to stereotyping—was already performing faster than any known delivery solution.

The Colonel had decided to construct the new Lloyd machine in London, and was on hand when work began. As time passed he could no longer linger, and returned to New York. Conquest kept him informed:

> [*To New York, 16 March 1872*] . . . *There is a party in Preston bringing out a Perfecting Machine for working type & from what I can see of it it is exactly the same as the machine which you have a patent for. He has sent tracings of the machine to all the principle*[†] *newspaper offices in Gt Britain. However every one is pretty well aware that we are making one & they have but little chance of success. Most of these folks spoil themselves by constructing a machine for its cheapness without a concideration*[†] *as to whether the machine will be a usefull*[†] *one or not.*
>
> *However we are so far on with our drawings that there is but little chance of any failing point. The principle*[†] *of the patterns are now at the foundry. The cylinder & fountains are nearly finished & I have no doubt but in two weeks that we shall have the Side frames up and into shape.* . . .

The original "Prestonian" had actually been designed by James Livesey, son of the owner of the Preston *Guardian;* it was not important, although a later (1876) model without folder was very popular. Similarly, two years later the Duncan machine was summarily judged: "in looking at a cut of a new Victory Press," Richard wrote his brother

(5 January 1874), "I notice one peculiarity in flying the sheets which seemed to me to be like & an infringement of my patent. . . ." On further study he was relieved to learn (11 December 1875) that the 1870 Duncan and Wilson folding patent (No. 3031) was so faulty, "nothing can save [it] from death if it should ever be contested." The Colonel was fretting so much over folder developments, he groaned, they gave him a headache.

Meanwhile, the Lloyd project had been turned over to Tucker for completion. Upon investigation in London he found that Lloyd actually preferred to have papers delivered flat, without folding, and other proprietors of dailies said the same. Since a large portion of each edition went to the provinces, it was more convenient and cheaper to send it in bulk, flat. Tucker put aside the Colonel's plan for a blast of air to start the folding operation, and resorted to an old delivery scheme he knew to be sound.

In 1853 Hoe and Tucker devised a collecting cylinder (also called an accumulator) for flying or receiving the sheets of a fast-running press. Ownership of this concept would be immensely important in dealing with rival folder developments, but it lay unused for 20 years in the Hoe files. *Scientific American,* in 1915 calling Tucker "one of the most brilliant mechanics that this country ever produced," [10] gave him entire credit for the collector; this was corroborated somewhat by correspondence in 1874 in which the Colonel confessed that although his name was on the caveat he scarcely remembered it. Tucker had brought it to his attention. The inventor described it:

> In this apparatus, six, or any predetermined number of successive sheets, are collected on a cylinder and then switched off in a body to a sheet-flyer which lays them on a table.[11]

Thus it became possible to print 300 papers a minute, while the fly was required to make only 50 journeys (carrying 6 papers) at the same time.

Tucker did not recall, too, that the revival of this device was not without legal worries. Mayall's accumulator dated only to 1867, for example, but might have spelled a contest. Proprietorship called for fancy proof:

> [*Charles Carpenter to Munson & Philipp, 15 January 1877*] . . . *There are no witnesses living who have any knowledge of the collecting cylinder of which R. M. Hoe filed a caveat, except said R. M. Hoe and S. D. Tucker. Henry Skidmore who made the drawing for the caveat died in 1873, and James Blair who was also employed in our drawing office at that time died in 1872. Do not the facts, that the caveat was filed in the Patent Office, in 1854, and that the original is their*† *still, furnish all the evidence needed on this point?* . . .

The next day another thought emerged: a Hoe old-timer could recognize Skidmore's personal drafting style "just as well as the hand writing might be known." Apparently, however, ownership of the collecting cylinder was not challenged.

The accumulator of the Lloyd press actually collected nine papers, at which point they were released and dropped onto the delivery board. Here the large sheets, according to Isaacs, had a tendency to scatter and were held in place by a stream of air. The machines were soon called "puff-puffs," after the sound of the air pumps.

In any event the problem of rapid "open" delivery was solved. Patented in Great Britain 1872 and in the United States 1877, the simple Tucker device was most important:

> This apparatus was applied to Mr. Lloyd's machine in 1873, and operated so satisfactorily, flying 8-page sheets at a speed of 12,000 per hour that the London *Telegraph* at once ordered eight web machines, and the London *Standard* six machines. The *Telegraph's* machines were built in New York, while the *Standard's* were built in the London factory, which was sufficiently large for the purpose, and this started manufacturing there in a business-like manner. . . .[12]

Orders followed from the *New York Tribune*, Philadelphia *Times*, Cincinnati *Commercial*, Sydney *Herald*, and others. Effectively solved, too, had been roll feed and curved stereotypes. Robert Hoe concluded:

> There was no limit to their capacity for printing excepting the ability of the paper to stand the strain of passing through the press, which produced, when put to its speed, 18,000 perfect papers an hour, delivered accurately on one feed-board. The average speed, however, in printing offices was 12,000. . . .[13]

Wilson and Grey were also laudatory:

> This compact and beautiful piece of mechanism is justly held to be one of the most successful printing machines in the market. As a piece of engineering skill . . . it is very highly spoken of by all who employ it.[14]

Lloyd, delighted once more, had pronounced it a triumph, and at Gold Street it was joyfully expected to "knock the spots off" the Walter machine. Through the next few years Lloyd would order the impossible again and again, and Hoe would furnish it.

Despite the consensus of British customers, the company suspected that American publishers would demand folding apparatus attached to their web presses. With feed boys and fly boys, folding boys and paper counters—and supervisors for both groups, no newspaper vault seemed large enough for the traffic. "The old morning paper pressroom was a Babel of confusion," De Vinne wrote, "for the work of printing was seriously impeded by feeders and paper-folders, who were often in the way of the pressmen." [15] Whether delivered to the public by newsstand, by carrier, or by mail (for which three sophisticated Hoe addressing machines were available, 1861-1869), in the United States some folding was invariably necessary. Hoe surmised correctly that proprietors would not only want folding equipment, but they would want it working as fast as the printing. It seemed impossible.

The crucial problem of folding newspapers on the press—at the same speed with which they were printed—eventually had to be solved. Inventors approaching this point seemed to stop, instinctively, awaiting a genius to show the way. Like others, the Colonel himself had not been certain it could be done well: "We do not make any machine for the purpose," the company wrote (19 September 1851) to an inquirer. "There is a machine invented at the east & one of them has been in use here but it has not met with favor as it is complicated and the price is large, vis. 1000$.† attached to a press." He probably referred to the machine of George K. Snow, of Boston, which involved folding between plates; others had appeared, Brown's, 1850 in Edinburgh, Livesey's, 1851 in Preston. Hoe promised to keep the writer informed, but with labor plentiful and cheap hand folding would be economical for decades to come.

Separate steam-operated folding machines were nevertheless attractive to large newspapers and book-printing plants, those of Cyrus Chambers, Jr. (1833-1911) dating to 1856.* Sheets had to be removed from the press and refed to the folder, employing long thin blades to push the fold line into rollers which creased them. Chambers obtained the necessary accuracy by using register pins (like those on hand presses) to puncture and hold the sheets in position, retracting as the blade descended. The *New York Tribune* took one of the first Chambers folders, and streams of curious visitors to its mechanical department made the inventor understandably nervous:

> [*Richard Hoe to Chambers, 11 May 1859*] . . . *Yours of 9th inst. is at hand. We have not made a folding machine or model of your machine for any person & was*† *not aware that Mr. Bacon had such an one. As to our man or one of theirs* [*Bacon's*] *measuring the machine at the Tribune office it may be although we were not aware of it.* . . .

Bacon is not identified. Chambers and his brother manufactured folders (and equally clever brick-laying machines) against constant trade resistance—the expensive devices could be very balky—and infringement, one instance of which dragged on into an eight-year suit. But Chambers won and persisted.

Similar machines nevertheless had been attached directly to presses to avoid rehandling, but no folder could function faster than 8000 per hour. This barrier seemed insurmountable: it gave Bullock trouble,

* See *Books for the Millions*.

as well as Potter. Campbell's (unpatented) folder attempts also failed and went out of use. While a radically different folding principle was sought to make the accessory keep up with increasing press speeds, the independent machine continued to be useful where much folding was done. R. Hoe & Company endorsed two makes, not without exasperating friction and complicity.

The Forsaith was one. Remarkably enough, the London *Times* late in 1869 had placed an order through Hoe, and Conquest received the machine a few months later:

> [*To New York, 26 February 1870*] . . . *We have got the* [*Forsaith*] *folding machine for the "Times" at work in the shop this morning & shall send for Mr. M*^{c}*Donald*† *to come & see it at work before we take it to their office. We are running it 35 sheets per minute. We have not had it working satisfactorily but we have no doubt but it will be all right when we have the cams & pieces properly set. There is*† *a good many crooked pieces in it & you can severely call it irregular machine fashion. However we have no doubt about it*† *working well. . . .*

Old Thunderer unfortunately could not even use the machine:

> [*R. Hoe & Company to S. C. Forsaith, 22 April*] . . . *Our Agent in London, Mr. Wm. Conquest, 62 Dorset St. Fleet St. writes that the folding machine was put up in the "Times" office, & they were much pleased with its working, but it does not fold the stamp for postage in sight. He states that he was very particular on this point in his letters to you, stating that if the paper could not be folded in five folds to show the stamp, it would answer folding it four times, & show the stamp. . . .*
>
> *Perhaps you had better communicate with him direct. . . .*

Forsaith also manufactured millsaws, lath and shingle sawing machines, and later other industrial equipment; he was for a time a Hoe agent.

> S. C. FORSAITH & CO.
> Manufacturers of
> [cut of folder] Patent Newspaper Folding Machines,
> Gage's Paper Cutters, Etc.
> Agents for the "Universal Printing Presses."
> Printers' Machinery, and Material of all Kinds, furnished at short notice
> [Fist] Send for Descriptive Circulars and price-lists.
>
> *Manchester, N.H., Aug. 18*th *1877*
>
> *Mess. R. Hoe & Co.*
> *New York City*
> *Dear Sir,*
>
> *We enclose a letter from Mr. Elder which being directed to us of course was opened, and you have probably received a letter for us in* [*an*] *envelope directed to you and if so please send it to us. Mr. Elder evidently made a mistake in making the enclosures.*

Why did not you sell him one of our new Folding Machines, such as you could warrant and satisfied† him in, as he has always used our Folders and knows what they are.

When you get a chance to place a Folding Machine it is policy to put in the best and therefore we hope in future you will remember that our Folders cannot be beat. Sorry you did not send Mr. Elder one [of] our Machines and save you some trouble.

Yours Resp't
S.C. Forsaith & Co.

Elder is not identified; his order probably went to Philadelphia.

By 1877 Cyrus Chambers had established his domination over the folder business, but he continued to defend himself vigorously:

[Hoe to Mess. Chambers, Brother & Company, 8 June 1870] . . . Mr. Nelson of Edinburgh, to whom we recommended your machine & for whom we made the enquiries, tells us that he has ordered the folder of you direct.

We have a lot of machinery to go to him by Glasgow Steamer & if you will advise when you intend to ship, will put ours on the same vessel. Under the terms of your favor of 28th ultimo. we suppose we are entitled to the commission of 5%.

If you will send a few of your circulars, we shall be glad to hand them to customers making enquiries of us for folding machinery. . . .

Chambers did not at all think this commission fair, and said so. Shortly later, when another prospect appeared, Hoe was careful to remind Chambers that work and expense were connected with such recommendations:

[To Chambers, 14 July 1870] . . . We have an order for a 12° [twelvemo] folding machine that will also fold 8° [octavo]. Your No. 5 machine we presume will answer. Please say how soon you can finish one, & whether you can give the discount of 10% as it does not pay us to take orders at a less rate.

In a former letter you referred to the Commission we make on our Presses & material. If you favor us with any order for Cyl. Presses, we shall be glad to make you 10%. . . .

The two companies continued testily to serve one another, perhaps sensing they were not, in Franklin Bailey's recent words, "congenial spirits." Another foreign order really put their civility to the test:

[Chambers to R. Hoe & Company, 20 May 1874] . . . The machine ordered by Mr. Wolfmann† of Valpairaeso† will be ready to shipp† about the last of next week. We can not see how you can expect the usual Commission from us. The machine was ordered by Mr. W. direct from us, and was Constructed, and we were waiting instructions for shipment.

The terms were Paid at works $770. Boxed $785. To be paid for by English bill on the presentation of B/L. Exceptions: may be Countermanded within two months by paying fifty dollars.

We expected the money to be in the hands of Yourselves, "McKellar Smith & Gordon," † *or some Banker, ready for us on presentation of the B/L. But since it is not, and we are quite satisfied with Mr. W.'s responsibility, we will shipp the machine as instructed through you and await his remittance or, if you choose to pay us on presentation of the B/L we will allow you a discount of 5 pr ct. . . .*

After some heated exchange R. Hoe & Company, who had furnished much other typographic materials to Guillermo Helfmann in Chile, coyly explained to the innocent customer why its services were not forthcoming:

[*To G. Helfmann, Valparaiso, 1 June 1874*] . . . *Please accept our thanks for the remittance of £100—which is placed to your credit. . . . The Hand Press with some other articles ordered will be shipped by next steamer, and the balance including the elevator, in about four weeks.*

Mess. Chambers Bros. refuse to ship without the money, and having little personal knowledge of the machine we would not care to make payment until you had received it. The Folder would probably please you, but for reasons not necessary to name, we are unable to go on their premises and examine it. Without such examination we could not of course give the recommendations desired, as to working, durability, etc.

Under these circumstances, if you wish to deal directly with them, we think they will do what is right. We much regret our inability to serve you in this instance. . . .

Much of this disharmony was born of an unfortunate accident which occurred a few years before:

A patent was granted on December 10th, 1867, to Mr. R. W. McGowan, of this city, for self-acting or tumbling register points. As this seemed a good invention the firm bought the patent for $500 and at once commenced putting the points on presses. A year or so later Mr. Cyrus Chambers notified R. Hoe & Co. that they were infringing on his patent for register points, granted April 5th, 1859. On investigation this was found to be true, and the firm settled the case by paying Mr. Chambers $10 per machine to which these points were applied during the life of his patent.[16]

Tucker's recollection that Chambers objected rather promptly seems off the mark so far as a small penalty is concerned. In May 1876 Robert Hoe III tried to meet Chambers at the Philadelphia Centennial, but did not. Shortly thereafter he received additional counsel:

[*From Munson & Philipp, 28 October 1876*] . . . *Please send to our office the models in the cases of S. D. Tucker's revolving segment to*

> *partially arrest the sheet, and that of the Phila. [Times] machine. Our Mr. Munson will examine them on Monday morning and during the day will call upon you.*
>
> *Your customers using the Chambers points are liable from the date of his reissue only.*
>
> *Please ascertain the exact date when the Duncan-Wilson [Victory folding] machine was set up and when run and how long, at the Christian Union. . . .*

About the same time Chambers' letter also came in:

> [*From Chambers, Brother & Company, 21 October 1876*] . . . *Our offer was (without prejudice) four thousand (4000) Dollars cash for all the yielding points you have heretofore built & for all you may build hereafter, or twenty (20) Dollars per pair for all you have built in the past or may build in the future during the life of the patent or any extension thereof—limited to their use on printing presses only.*
>
> *We desire that you include all, if you accept the latter proposition. We think our offer is a liberal one & one that you can accept without injustice to yourselves. . . .*

Although negotiations apparently brought the figure down to $10 per installation, the infringement was still an expensive one. On 23 December 1876 Hoe sent Chambers a check for $2045 in payment of all past usage. Two years later (12 April 1878) the company offered to manufacture for Chambers finished points with sockets and springs and round brass plates, "at $45.00 per pair, including royalty, which we pay to you." By this time, however, the practice was to avoid perforating the paper altogether.

Chambers' firm hold on folding technology in America seemed invincible. In 1875 a New York religious weekly, *Christian Union,* had been enjoined by Chambers from using its Duncan folder, and he soon told Hoe customers at the Centennial that he would stop them too. Chambers owned the only viable mechanism yet developed in the United States.

Hoe was desperate for its own folding equipment. Exploratory discussions with several inventors, among them Fred Beach, Albion, New York, Percy Hedderwick, Glasgow, and an inventor in London named Gubbins, had all been unpromising. An alliance with Chambers seemed the only way out. The relationship between the two firms changed dramatically.

CHAMBERS BROTHERS, & COMPANY

Patent book-folding and Newspaper-folding machines that paste and trim. Also tempered clay brick machines [cut of factory] Mechanical engineers, founders and machinists, bolt clippers, shafting, pulleys. Forgings and general repairs

52nd Street, below Lancaster Avenue

Philadelphia, July 21st, 1877

Messrs. R. Hoe & Co.
Grand St. N.Y.

Gentlemen,

Agreeable to our promise to your Mr. Theo. H. Mead we enclose you tracings of the leading features of our improved all rotary five fold folding machine & press combined, together with C. Chambers Jrs.' improved cutting device [not present].

There are four sheets in all, sheet one represents the cutting cylinders of a press (or they may be the last impression cylinders) with the improved cutter attached. Clearly shown in Figs. 1 & 2, the principal of the cutters is two serrated cutters one concentric with one cylinder & the other projecting slightly (see Fig. 1) with shear projection & these serrations interlock (see Fig. 2).

Fig. 3 is the first folding Cylinder with Nippers, Blade, Lever cam openers, Brakers, Strippers, Points, &c. . . .

Sheet 2 represents the 2d & 3d folding devices, blade out of contact, & the carrier bars & curved guiding bars. . . .

Sheet 3 represents the 4th folding devices together with the switching device. . . .

Sheet 4 represents the 5th folding device & the rotating packer. . . .

Shall be happy to give any information we can that you may desire relative to these improvements, so long as your negotiations with us for an interest in folding machines may be pending.

Yours &c.
Chambers Bro. & Co.
B.

P.S. Applications for the patents of this invention are being prepared for the United States & Great Britain. C Bro. & Co. B.

Collaboration with Chambers was born of desperation. But so little folding experimentation (and expertise) at the Hoe shop meant that press progress itself would literally halt if delivery problems could not be solved.

Old hostilities probably emerged to sever the discussions with Chambers, or perhaps Anthony and Taylor in England began to appear sufficiently productive and more manipulative. With the sudden and dramatic acquisition of Luther Crowell in 1878, however, R. Hoe & Company was pleased to find itself within hopeful grasp of the solution, and no further search for outside talent was made. Crowell literally carried the firm across the abyss; although substantial, in-house progress was just not fast enough.

Stephen Tucker in 1874 had solved—for the moment—the vexing problem of high-speed on-press folding around his collecting cylinder, capable of 15,000 papers per hour. This Robert Hoe III described:

> The first folders made by Hoe & Co. consisted of the combination of a "gathering cylinder" with a rotary folding cylinder and tapes conveying the printed sheets under horizontal folding blades, somewhat similar to those before described, which thrust them at the proper moment between folding rollers placed at alternate angles, finally delivering them on travelling belts by a small flier. The first of these folding machines were put upon the presses made for the Philadelphia "Times" and operated in the Centennial Exhibition, in 1876.[17]

The Colonel had also received an order from Childs' Philadelphia *Ledger* in 1874 for a web folding machine, and was nervously holding off production pending further patent searches and experimentation. But still another assignment was even more compelling.

The receipt in 1875 (after some six months of discussions) of an order from the *Glasgow Herald* for two web machines to include apparatus so as to deliver eight-page sheets folded to quarter-page size at the rate of 12,000 hourly, was a chance to give Tucker's device a prestigious international application. Robert, in Paris, was counting heavily on it:

> [*To New York, 24 December 1875*] . . . *I hope you will not let the dull times prevent you from pressing the completion of the Glasgow Herald machines; of course they must be made to work well but please press forward with them. . . . I am sure it will pay to work night and day,* [as] *we shall be out of work in London if they do not succeed very soon. . . .*

He hoped to persuade provincial papers to discard their Walters.

The first folding apparatus was shipped to Glasgow in June 1876, as another was hastily put on the Philadelphia *Times* machine for the exposition. Although the device represented only a first step in a long succession of developments in high-speed printing and folding, Robert Hoe III concluded of this era:

> With great ingenuity, added to long experience, and by the acquisition and adoption of every device which should aid them in their efforts, Hoe & Co. succeeded in providing machines of unrivalled designs, efficiency and speed. . . .[18]

Cutting the web after printing had been equally involved. Tucker explained:

> I had intended using in the cutting cylinders of these [Lloyd] machines the peculiar form of stationary knife devised for the web cutting and folding machine built in our experiment room in 1858, but found that it had since been patented in the United States by Mr. William

> Bullock. It was held in the Patent Office that, as we had never applied for a patent, nor put it into practical use, we had forfeited our rights to it, and that Bullock's patent was valid. However, I devised a jumping knife that we used in machines in the United States until Bullock's patent expired in 1880, and then substituted our own knife, which, with a modification of the springs, we are using in our machines today.

In addition to something old, the Hoe web printing equipment developed by Tucker had something strikingly fresh:

> In designing these new machines I made a radical departure from our customary styles by dispensing with all moldings, ribs, etc., as far as possible, and making the members of the framework rectangular in section with all corners rounded. The various brackets, arms of wheels, and smaller parts, were of similar form or oval. This gave the machine a massive and strong appearance, although in fact some of the parts were hollow where no great strength was required. . . .[19]

The smart new general appearance was applied to Hoe's stop-cylinder and lithographic presses, and other machines. To Tucker's delight the styling was almost universally copied by other press manufacturers.

Tucker did not mention, however, that this cosmetic attention was more a standing directive than an inspiration. Robert Hoe had written twice about it:

> [*To New York, 17 September 1858*] . . . *I like Mr. Whitworths style of rounding the corners of the parts so as to avoid angles as much as possible. His design is to study nature as shown in the details of the branches of trees, where every limb springs from its fellow, to combine simplicity and grace, strength and symmetry of form. . . . I think we should adopt the style. . . .*

Two years later and sadly disenchanted with Whitworth performance, he nevertheless persisted (27 April 1860): "I hope [Mr. Tucker] will notice particularly the style of Whitworth & Co. in the mouldings & curves. . . ." New to America, at least, this appropriate design treatment—a symbol of modernity—was welcome for post-Civil War sales.

Word soon came from Scotland that the *Glasgow Herald* was happy with its revolutionary equipment—except for one thing.

> THE GLASGOW HERALD
> THE WEEKLY HERALD and THE EVENING TIMES
> George Outram & Co., Publishers
>
> *Glasgow 4th Oct. 1876*
>
> *Messrs R. Hoe & Co.*
> *New York*
> *Dear Sirs,*
>
> *You have probably learned from the member of your Co. on this side how the new Web Printer & Folder are doing. The folder seems*

to be quite a success, & the printing is improving day by day. The speed we have not yet had tested. We are greatly disappointed however to be told by Mr. R. Hoe Jr. that the Machines cannot be adapted to print & deliver folded or unfolded our Evening folio paper (half of the Herald size), & that our only alternative is a third machine costing £4,000.

We cannot, I fear, persuade our partners to make up their mind to this course, & at such a cost, especially in view of the fact that there are several Delivery processes on the Bullock, & other machines which we expect can be adapted to the new machines, printing, cutting, & folding the full quarto sheet or the half sheet alternatively.

What others can do we cannot imagine is beyond your inventive power to accomplish; indeed the Mechanics your firm sent assure me that there is no difficulty, & that Mr. Tucker on having the matter fairly under his consideration will see his way to overcome any such difficulty. We earnestly trust that it will be so, as you can readily believe we are very averse to get any other firm to alter the machines, or bring another make machine in besides yours.

In a sentence we want the New Machine to print & fold the 8 page Herald, & (alternately) to print & fold or print & deliver the 4 page Evening Times (or Herald supplement of 4 pages). If you undertake to do this, please inform us at the earliest possible time, & the cost.

We send you this [*letter*] *direct as it may miss Mr. Robt Hoe Jr. Capt. J. M. Walters, one of our Partners, leaves shortly for the States, & may have the opportunity of waiting upon you. At all events I will try & pressure him to do so.*

Yours truly
Alex Sinclair

Such loyalty and confidence in Hoe could not go begging. Again the Colonel rose magnificently to the challenge.

Unfortunately, after much analysis, there seemed no easy way of altering the Glasgow paper's machines, and the Scotsmen were persuaded—at a 12 percent discount—to take the third machine:

[*Proposal*]

Messrs R. Hoe & Co. agree to make and erect in the premises of the Glasgow Herald proprietors, Glasgow, within four or five months from receipt of order, one of their Web Printing Machines with all the latest improvements to print & fold simultaneously at the rate of twelve thousand perfected copies per hour and not less than ten thousand five hundred copies per hour, the folding to be of not fewer than four folds. This machine [*is*] *to be supplied with two sets of iron form roller stocks, three sets of distributor stocks, two roller moulds, a full set of spanners, lifting crane and a front platform. It will also be so constructed that it may print & deliver in single sheets the Evening Times with its present quality of paper, and of its present size (or half Daily Herald) without folding, and with a motion to count the quires in lots of either 24, 26, or 27.*

It will be especially held in view that the machine is meant to print & deliver the Evening Times though fitted to print & fold the Glasgow Herald in emergency. The price to be paid for this machine shall be three thousand five hundred pounds (£3500) which shall be due in a month after the first issue of the Evening Times from it.

Messrs R. Hoe & Co. will erect the machine at the Herald buildings with own expense. If, after a month's trial, the Herald proprietors find that the machine does not perform as stipulated in this agreement, Messrs Hoe & Co. agree to take it away at their own expense. Also, should the machine not be completed in the time specified, the Herald proprietors shall be at liberty to cancel their order without any charge for damage of any kind.

After the erection of this machine, or sooner if the Herald proprietors desire, Messrs R. Hoe & Co., should the Herald proprietors then wish it, will undertake to alter one or both of the two last printing & folding machines furnished to the Herald, so that their folding apparatus may be detached if necessary and the machines simply print & deliver the Evening Times size of paper. This operation they undertake to complete on each machine within four weeks after commencing and at an expense as much lower as possible but not higher than five hundred pounds (£500) for each machine so altered.

Messrs Hoe & Co. think that no extra stereotype apparatus will be required for the new machine, but, if the Herald proprietors desire it, they will furnish it at the same price as stated in their contract of 20th May 1875.

New York, December 14th, 1876.

R. Hoe & Co.

Some slight disagreement in connection with this proposal, perhaps over the price Hoe offered for an older press, got quick, cryptic action:

HOE, TUDOR STREET, LONDON.
GENIAL GLASGOW SENATE PERMEABLE. HOE.[20]

A confirming letter (20 January 1877) to Conquest revealed: "We have just telegraphed you to go to Glasgow at once and take it at any reasonable price. . . ." Delays crept into the plans, but some august publicity from *Engineer* was welcome:

[A] considerable number of gentlemen [from a Glasgow trade convention] paid early morning visits to the printing-office of the *Glasgow Herald,* in order to see that journal printed and folded by two of Hoe's web perfecting presses. . . . Printing from a continuous web was a familiar sight to those who had visited the Exhibition at South Kensington in 1872 and witnessed the "Walter" printing press throwing off the *Mail,* and most who saw its wonderful speed and excellent work may have thought that a stride had been taken in the art of printing which would satisfy the trade for some time to come. The Walter press, however, was not long without its rivals, and foremost amongst these stands the Hoe printing and folding press. . . .[21]

The company could only advise a little procrastinating salesmanship:

> [*To Conquest, 19 March 1877*] . . . *It will be impossible to have the Glasgow Machine done at the time first named. The only way will be to try & keep Mr. Sinclair in good humor over the delay.*
>
> *We think your plan of making folding Machine drawings a good one, & shall be glad to see the tracings as soon as you have them ready. . . . Did you ever make any drawing for the Cheap Engine, we spoke of some time since? We have nothing new in a business way. . . .*
>
> [P.S.] *We would like to know how the folders put on* [*a Walter press at*] *the Scotsman work & what they are like. . . .*

Finally, only the sticky matter of collection for the entire installation remained to be settled:

> [*To Conquest, 7 April*] . . . *According to our agreement with the Glasgow Herald the money for the two web Machines was to be paid within 6 months from the time they were at work performing what we agreed they should. That time expired according to our understanding last month.*
>
> *We therefore desire that you will either see Mr. Sinclair and or write to him and ask for the money. . . . Use your best discretion in this affair so as to prevent any unpleasant complication in reference to either orders, as we have trusted to their honor very much, you know. . . .*
>
> [P.S.] *The china came safe & is very satisfactory & I thank you for the trouble taken. Ys. RH*

Tucker had been pleased to conclude of the Glasgow installation, "These machines gave great satisfaction, and from that time forth practically all web machines built by the firm folded the sheets." [22] The company had regained its foremost position.

Despite these dazzling performances in Philadelphia and Glasgow, R. Hoe & Company seemed woefully tardy at lining up a major New York customer as showplace for its new rotary equipment. But a concerted effort was now launched to convince the *New York Tribune:*

> *Jany 23rd 1877*
>
> [*Whitelaw Reid, "Tribune" Office*]
> *Dear Sir*
>
> *We would like to place in your office one of our new Web Perfecting Machines with folding apparatus, being sure its introduction will be to our mutual benefit. This press will deliver the Tribune folded four times, at a running speed of 12,000 papers per hour.*
>
> *The advantages to you will be first, the saving of all the wages paid for feeders at your present folding machines; second, the saving of the wages of the feeders on the old Eight Cylinder Rotary Press; Third, you will keep but one kind of paper and do away with two systems in your printing office.*
>
> *Besides these advantages, you will have that resulting from the*

prestige of being the first to introduce and use the machines here. At a rough calculation, we estimate it will insure a saving to you of at least one hundred dollars per week in money.

Under the circumstances, we are willing to make a very liberal arrangement, and propose to furnish a machine, with folding apparatus, as above, provided you will [pay] what appears to us will be one half of the saving in your office, or say, fifty dollars per week for three years commencing from the time the press is first in operation: also your eight cylinder Rotary Press, your present folding machines, and two old horizontal engines now stored at our works.

It is our especial wish that this proposal shall be considered as strictly confidential. The desire to have one of these machines in operation as soon as possible in New York, where we can show it to our customers, alone induces us to offer it to you, as we do now, at what is less than its absolute net cost in material & labor. As to the workmanship on our new machine, we are sure it will outlast any two of other manufacture & do nearly double the work in a given time.

Yours very truly
R. Hoe & Co.

Reid capitulated to this beguiling appeal, unaware how exasperatingly more complex was this machinery over a "conventional" ten-cylinder Lightning. Hoe was already two months late with it:

NEW YORK TRIBUNE

New York Nov. 28, 1877

Gentlemen:

You are aware that we have made all our arrangements for printing The Weekly Tribune next Wednesday in the proposed new form on the new press, which you agreed to deliver to us on or about the 1st of September last. You also know that this change in form has already been postponed beyond the limits of safety for the ensuing year's business, on account of the unexpected delay with this press.

It is of the utmost importance to us to know, now, definitely, whether we can depend on having this press in running order next Wednesday morning, or whether we must begin at once to make our arrangements for printing the weekly in the new form as best we may on other presses, and gathering a force to fold it by hand. You do not need to have your attention drawn to the inconvenience as well as actual damage this would involve.

I have directed our Mr. Fitzpatrick to wait for an answer to this letter.

Very truly yours,
Whitelaw Reid

Messrs. R. Hoe & Co.

Two weeks later the machine was ready to install, but with limited folding capacity:

[From Reid, 9 December] . . . I am obliged by the information in your favor of Saturday. It would be a great advantage to have the

press put in at once, even with only three folds, if it could be put in in a way in time to print our next Weekly, it would save us at least $150 dead loss. If, however, it cannot be put in in time for that, and [you] can have the machinery for the other two folds made to work perfectly in the time you name, it would be better to keep on at that work. . . .

Reid's impatience began to show:

[From Reid, 13 December] . . . *I am obliged by your recent favor, advising me that the additional work on the new press would only occupy in your your judgment about four days, and that we might pretty certainly count upon having it print our next Week's Weekly. I sincerely hope there may be no mistake about this, and would suggest that in case the trouble is only with the final folds that it would be best to put the press up any way and complete the work on the folds here. The damage to us of working as we are at present is more than you will imagine. . . .*

Another, that same day:

. . . *I have just received your favor of to day's date telling me that you expect to begin sending down the press this afternoon or tomorrow morning, in case a final test this afternoon on the delivery of the last fold seems to be satisfactory.*

I wish only to say that it seems to me of the utmost importance that the press be put up promptly whether that final test be satisfactory or not. I don't think you realize even yet the extent of the inconvenience and loss we are suffering from, not having the press in such shape as to do the printing and make the flat delivery of The Weekly in its new form. We can stand delay in adjusting the other parts, but we are losing heavily, from your delay in this. I hope that on no account you will delay matters so long as to suffer the least doubt about your having the press in our press room in working order by Next Tuesday. . . .

At last:

[From Reid, 18 December] . . . *I am exceedingly glad to learn that the delays are over, and if I had received your note sooner should have directed you to bring in the press at once.*

As it is, I think you had better delay it now until the afternoon of tomorrow so that the running of the presses on the weekly may not be interfered with by the chill of open doors. By one o'clock tomorrow afternoon, we shall be exceedingly glad to see the machinery coming in. . . .

And more trouble:

[From Reid, 19 December] . . . *I am sorry to advise you that, for the first time in years, The Tribune missed a large part of its morning mails, and even some of the morning express bundles, in consequence*

> *of delays by your workmen in putting in the new steam chest. Some of our people who witnessed the trouble think it came partly from bad workmanship and partly from a failure to understand the necessity for having the work done in time. The old steam chest having been removed, to make place for the new one, we were absolutely helpless. . . .*

But some good news—months later:

> [*From Reid, 18 April 1878*] . . . *We shall be glad to deliver the 3 folding machines to which you are entitled under your contract with us for the double web press, whenever you choose to send for them.*
>
> *Some of our customers complain of the Dailies printed on the new press, insisting that if they are cut they ought to be pasted. Others appear to like the new form as it is without any pasting. The experiment has hardly been tried long enough to enable us to decide yet which way the preferences of the majority will be found. . . .*

At R. Hoe & Company the staff and crew heaved a sigh of relief in unison.

Even with Luther Crowell in the shop from now on, folding problems always seemed to keep ahead of attainable press speeds. As late as 1930 B. C. Digby, of the Hoe London office, commented on the problem:

> The most intricate and important part of the press nowadays is the folder. Claims have been made that copies are printed at the rate of sixty thousand an hour, but to deliver at this speed is not an easy matter. However, this is the objective, and before long, no doubt, this will be obtained. . . . Nearly all the parts in a folder are working at a faster surface speed than the press. . . .[23]

The on-press folder at any rate was a workable reality. There was time at last for a host of other varied projects, all clamoring to be finished. But first the *Tribune* had one final thought.

John R. G. Hassard (1836-1888) was a "lank-built man with sandy hair and side whiskers . . . with real charm of style and breadth of culture." [24] Although he intended to enter the church, his poor health turned him instead to writing; he contributed to Appleton's *New American Cyclopedia* and was soon a *New York Tribune* reporter. By 1868 Hassard was editorially second in command on the paper but Greeley's frequent absences actually made him much more important.

Hassard was also the regular literary editor and music critic—in 1875 sent to Bayreuth, Germany, to cover the premier performance of Wagner's "Ring of the Nibelungs." But he was analytical too. In 1878 he obtained several hundred coded telegrams exchanged between factions of the Democratic Party; after countless hours of brutal cryptanalysis he decoded them, revealing scandalous vote-buying conditions in the 1876 elections. But the strain was too much; he traveled to Europe to recuperate, and although he contributed frequently to the paper his health declined thereafter. "Mr. Hassard all his life was a scholar, a keen lover of books, and in every sense of the word a brilliant journalist," [25] said the *New York Times.* With R. Hoe & Company he revealed still another facet, an appreciation of printing technology.

In 1878 Hassard wandered over to Grand Street for a progress report on the company's newest machinery, in which the *Tribune* had just invested so heavily. Only an incorrigible mechanical engineer could enjoy the chore—the shop was no longer a novelty that nontechnical reporters would appreciate—but Hassard turned in some credible copy, appearing in three installments. Hoe duly sent clippings to agents around the country and to England, and a note to the *Tribune:*

> [*To Whitelaw Reid, 10 April 1878*] . . . *We thank you very much for those appreciative articles and for your kind mention of us in connection with your reference to the Tribune & its machinery from the first.* . . .

Coincidentally, another letter from the paper must have crossed in the mails:

> NEW-YORK TRIBUNE
>
> *New York April 10, 1878*
>
> *My dear Colonel,*
>
> *Few men walk about in these days of wonderful achievements with cloaks of glory and wreaths of bay such as The Tribune has put on you within the past few days, and you cannot, I hope, but feel that your friends rejoice heartily in the truly merited honor. None of them, however, rejoices more sincerely than I do.*
>
> *Now let me suggest to your firm the propriety of having the three articles which have been printed in the Tribune illustrated with cuts from your catalogue—showing the various presses mentioned in the articles—and made up into a sheet like our weekly and then have 100,000 copies printed for distribution in Europe and America among printers and librarians, so that you and I at the next Centennial—don't laugh, we shall be here or in Heaven at that time—we or others can decide whether any improvement has been made in presses for rapid printing since April 10, 1878.*
>
> *God bless you*
> *Thomas N. Rooker*

If Rooker was plumbing for a *Tribune* job shop order he was sadly disappointed. The Hoes thought the reprint a capital idea, but preferred a small booklet format—and well printed, at that. Theodore Low De Vinne had just assumed the proprietorship of Hart's printing company; he had always done an outstanding job. Why not engage him for the Hassard?

De Vinne (1828-1914) was born in Stamford, Connecticut; he came to New York in 1847 and shortly joined the printing firm of Francis Hart, whose little shop had one card press and one Hoe cylinder, but unlimited capacity for infinite care and quality. De Vinne became a partner in 1858 and, despite slow growth, maintained the hard-won reputation for excellence. At one point, no doubt hoping to benefit from a master's suggestions (as with Welsh in Boston), Hoe sent a new machine on approval:

> *[Francis Hart & Company to Hoe, 27 Janaury 1863] . . . You wish to dispose of the four color press belonging to you which we have on our premises, and ask for an offer.*
>
> *We think well of the plan of the press, and not meanly of its performance. But we are of the opinion that any presses you may build of the like kind in the future will be stronger and more perfect; and we have an aversion to buying experimental machinery of any kind. Still, we feel that you have been courteous in placing your press in our hands; we think we can make it useful; and will make the purchase, if you will make the terms easy.*
>
> *We will take the press at $1,000, if not expected to pay for it under a year, and should feel better satisfied if you would allow us to work out one-half of it. We wish our business would allow us to send you a check, in princely fashion, . . . but you must know that the printing business is one of small profits. . . .*

No doubt the production of not a few Hoe catalogs (heretofore printed by William Van Norden) started in this way.

Upon Hart's death in 1877 De Vinne took over the business. His scholarly history, *The Invention of Printing* (1876), had begun to bring him considerable fame as a master typographer, and he continued to write prolifically. John Oswald later said of De Vinne:

> His contribution to quality in printing took the form of fine presswork, and he demonstrated his willingness to go to any length required to get the best results from the union of type and plates with paper and ink. During the 60's the cylinder press was coming into use and a quarter of a century later the half-tone made its appearance. Paper manufacturers were confronted with a new problem and De Vinne helped them to solve it. . . .[26]

At first, the only papers available to printers had been those finished for hand-written ink; dampening was necessary to break through the sizing, although water and oily ink were indeed incompatible and hard to control. Bruce Rogers, nevertheless, described the result:

> On those rare and happy occasions when time and expense are secondary, and where the finest possible finished product is desired, printing on dampened paper, if skilfully done, will produce a result much superior to the ordinary dry printing. The punching of the type into the softened paper raises printing almost from a two- to a three-

> dimensional medium, and the slight halo or highlight created around the individual recessed letters gives a sparkle and life to a page which cannot be obtained by dry printing. Moreover, as less ink is required, a cleaner impression is possible and the vigorous pressure of the type into the paper causes the print to become an integral part of the paper, rather than merely to lie on the surface. It should suffice to say that all fine books of the past were printed on dampened paper. . . .[27]

The dampening of newsprint, however, tolerable enough at cylinder press speeds, was becoming a considerable impediment to mass circulation papers. Yet through the 1870's in England and America there seemed no other recourse:

> In order that our morning journals, or indeed any journal, may be properly printed, it is indispensable that the paper should be first drawn through water, in eight or ten sheets at a time, and then allowed to lie for two days under the power of a pressing machine, applied to it for the purpose of making what is termed the "damping," equal. No paper would be readable were it printed on dry paper. . . .[28]

Although machine presswork had originated in London, practitioners there proffered no solution for avoiding the wetting process. The original Koenigs, Napiers, and Applegath-Cowpers, lacking provisions for makeready, necessarily functioned with dampened paper in conjunction with soft cylinder packing. Illustrations—even the best woodcuts—suffered so much with wetted paper they were discouraged wherever possible.

Working through the 1870's with S. D. Warren, a Boston paper merchant, De Vinne developed a technique for printing superbly on undampened paper calendered (machine-smoothed) or coated with fine white clay. With hard packing and careful makeready the whole wetting and drying process could be discarded. (Presswork on dry paper unfortunately introduced static electricity, often still a problem today.) De Vinne won the thanks of an entire industry.

R. Hoe & Company had repeatedly acknowledged De Vinne's passion for fine printing, commissioning him to print much (if not all) of its later catalogs, price lists, and multicolored samplers. The "scholar-printer" would soon reciprocate by performing a variety of large, high-speed quality work on Hoe machines designed for his use—jobs which, like the abandonment of wetting, the trade thought quite impossible. In 1883 he gave the Hart plant his own name, and it continued until 1926, when it merged with the Wynkoop-Hallenbeck-Crawford Company. With his typographic and style manuals and his cost analysis systems, De Vinne had begun to put some sense (and profit) into American printing practices. His aim was to change a belatedly acknowledged art into the technological science that Hoe's giant presses were making it.

Remarkably, this busy craftsman was personally consulted for the printing of Hoe's little book, now even smaller a project than earlier conceived. It was assigned to one of the newer nephews in the shop, perhaps Peter's son Stephen, for follow up:

> [*To Whitelaw Reid, 30 April 1878*] . . . *We have consulted with Mr. De Vinne of Mess. Francis Hart & Co. regarding the publication in pamphlet form of the articles upon Printing and Web Presses.*
>
> *They will occupy about 32 pages, in the same size and style as the book sent herewith. The engraving would be inserted folded. He proposes printing an edition of 1000 for $75, and we will pay one half the expense and take one half the edition, if you will do the same. Our idea is to issue it with imprint of the Tribune Association and put a price upon each copy of say ten or fifteen* [*cents*] *or whatever sum you may consider proper.*
>
> *After the first edition extra copies will be furnished by Francis Hart & Co. @ $30.00 per thousand.*
>
> *Do you wish to make any corrections in copy or shall we print from the* [*news*]*papers?* . . .
>
> *P.S. Please return sample book herewith.*

How precious to find a junior executive of the largest printing machinery works in the world addressing the proprietor of one of the largest newspaper enterprises in America about a thousand copies of a ten-cent pamphlet! With infinite patience Reid responded with complete confidence in the scheme, noting—tongue in cheek—that if it failed Hoe could use the books anyway. Delighted to proceed, the Hoe production manager next admonished the country's foremost printer:

> [*To De Vinne, 2 May*] . . . *We want you to give us a very nice edition, charging half the expense to the "Tribune Ass^{n}" and half to us.* . . .

Theodore De Vinne too must have been amused. His firm replied with utmost seriousness:

> [*To R. Hoe & Company, 3 May*] . . . *We send herewith* [*not present*] *sheet of thick paper, proposing to use it on the little pamphlet you have ordered. Will it meet your views? It will cost $7.00 more per 1000 than the thin paper of our first estimate.*
>
> *Of course we shall try to give you a neat bit of presswork, and we think we will, but we don't think the Tribune type will print as clearly as ours. We have, however, made arrangements with them to use it.* . . .

Unfortunately, this association of prestigious press-maker, giant newspaper publisher, and superb printer combined to produce an almost unreadable little treatise. It was entitled *The Fast Printing Machine: An Account of Recent Improvements in Newspaper Presses* (32 pages, trim 5 x 7 inches, type page 16 x 28 picas). Henry Bullen was probably one of the few librarians to bother with it.

Theodore De Vinne was not the only printer to experiment with new equipment and techniques. Others began to see possibilities with the exciting high-speed newspaper technology now emerging. The result brought new impetus into book and magazine publishing as well.

John B. Ford came to the United States from England in 1849. After working at Appleton's for a time, in 1867 he went into business for himself as J. B. Ford & Company, book publishers. One of his partners was his eldest son, Edward L. Ford, recently an artillery captain in the Civil War. The Fords published Henry Ward Beecher's life of Christ, Greeley's recollections, and the large *American Mechanical Dictionary,* giving them a lively trade.

In 1869 the firm started *The Christian Union,* a popular religious weekly with Beecher as editor. Circulation jumped from 72,000 in 1871 to 132,000 the following year. For this extraordinary production Ford bought a four-cylinder Hoe and four Chambers folding machines. The possibility of coupling the folders to the press became obvious to a mechanically minded fellow: "Captain Ford invented a method of running the printed sheets mechanically from press cylinders to folding-machines, instead of collecting and folding them by hand." [29] This may have been one of the first attempts in the graphic arts at "straight-through" machine coupling that led ultimately to in-line mass production.

In 1873 the firm bought a new Victory perfecting press with folder attachment, and the Captain again saw improvements. He added "combinations" to make it print a 24-page paper and a 4 or 8 page cover, fold and paste the spine, add cover to book, trim, and drop "7000 perfect covered papers per hour"—a feat that bookbindery tinkerers and book machinery inventors would have loved to claim their own. But Chambers had put a stop to this work. When Beecher became involved in scandal (1875) *Christian Union* circulation dropped precipitously, forcing Ford into bankruptcy and reorganization. (*Christian Union* continued until 1893, at which time it became *Outlook,* lasting to 1935.)

Colonel Hoe was not reluctant to use some of Ford's ideas, provided they were legally defensible:

Confidential

Washington Nov 16/[18]*77*

Messrs R. Hoe & Co.
N.Y. City
Gentlemen;
Your favor was forwarded to me here, hence the delay in replying.

Not being an expert in the matter of following the delineations of drawings, I called at the Patent Office and examined the Model of Ford's press. I could find nothing so far as I was able to see that infringes on Campbell's press. The only things that are similar are, that both are Web presses, both work with a roll at each end and both have a set of plates at each end. Campbell has no tapes, while Ford has. Ford has a reservoir with a revolving wheel to apply the paste, while Campbell has a paste reservoir exactly like one of your ink fountains on your Cylinder presses, and the paste is applied with a long blade.

The folding apparatus on Ford is operated like Forsaith's folding machine, while Campbell's folding knife is operated by revolving around. I can find no similar devices in the two aside from the general principles, unless it be the double end principle. Of course there may be devices on one press that may infringe on the other press, which a press maker could discover which I cannot. Ford's press appears to me to be just as distinct from Campbell's as the Walter, Bullock, or your presses are.

You will please consider this confidential, as I have enough trouble with Campbell without having any more.

Yours Truly
Edwin Cowles

Cowles is not identified. Hoe took some Ford patents; one was described by Tucker—it was the "Duplex Printing-Machine and Folding Apparatus Combined," which Cowles had inspected. It printed two separate webs, folded and pasted the sheets, and delivered books or pamphlets covered or uncovered. If book and magazine publishers could accept "newspaper quality" printing this machine would revolutionize (once again) an entire industry. They were not quite ready to do so.

Hoe nevertheless paid royalties to Ford (and to his family after his death in 1880—he was only 38), but the fees were a distinct disappointment:

FORDS, HOWARD, & HULBERT,
No. 27 Park Place
(Goods Entrance: 24 Murray St.)

New York, July 17th 1878

Messrs R. Hoe & Co.
Gentlemen,

Please do me the favor to inform me by bearer what amount there is now to the credit of Mrs. Ford under your agreement with me.

You informed me some time since of your intention to use the device already patented in a machine you were then building, which would of course make something due under the agreement, and I presume there may be something else beside.

Yours truly
E. L. Ford

Two days later Ford reminded the Hoes he had worked diligently with Munson and Philipp putting a pasting patent in good order, and as he was leaving for Montana for a year's stay needed a quick answer.

> [*Ford to R. Hoe & Company, 20 July 1878*] . . . *Your favor of yesterday is received. I am surprised at its contents, for last Oct. or Nov. when Mr. Robert Hoe Jr. and I were discussing the terms of our contract he distinctly gave, as a reason for putting the royalty at a low figure for machines printing from two rolls of paper, the fact that you wanted to use this feature in a press you were then building (for the Lloyds Weekly I think) and that was given as an instance of the way in which you could afford to use it if you got your own figure (which you did get).*
>
> *I supposed therefore that of course the Lloyds machine would be one, and I hoped, in accordance with your expressed opinion of probabilities, there would have been others.* . . .

It would seem the captain was well on his way to the Rockies before he could receive an explanation. The property was probably not very important to the company, Robert Hoe himself concluding, "He was unable, however, to develop his plans to any practical result; but deserves the credit of being the first to patent, if not to conceive, the idea of the association of printed sheets [that is, pasting together the products of more than one press] for this purpose."[30] Ford had been decades ahead of his time.

Yet at the same moment another customer raised a question similar to "straight-through" production. The Baltimore *Gazette,* planning a new installation, wished foresightedly for a more expeditious coupling of activities between floors. Significantly, the company discouraged the idea:

> *May 3, 1878*
>
> [*Gazette Company, Baltimore*]
> *Gentlemen,*
>
> *Enclosed is a plan of your pressroom showing the proposed location of the press, driving shaft, form elevators, etc. The ceiling is a little scant in height, but will answer, as the head of the feeder will be about 6 ins. clear of it. As your Engine is not in place the position of its belt and size of pulleys are of course imaginery, but probably not far from correct. The shaft can be extended across the room if preferred.*
>
> *We also send a mason's plan of the foundation, with all the measurements for its location, from the wall of the building given in the pressroom and not in the cellar. Our man forgot to take the thickness of the rear wall of your building and we have put it in as 12 ins., but this is of no consequence.*
>
> *In your favor of the 29th ultimo you speak of a steam elevator large enough to take in the form trucks* [*along with the forms?*]. *This is rather a dangerous thing to do, unless you have a steam ele-*

vator of a very expensive kind, for if it be stopped an inch or two too high or too low there is danger, when running the trucks on and off, of upsetting them or throwing the trucks [undecipherable]. Besides, when the Turtles come off the press they must go to the sink to be washed and as it would not be pleasant to have the trucks soiled with ink and dirty lye, the turtles would have to be lifted off to be washed and then lifted on again to go up stairs.

If the elevator is intended for turtles only, we would recommend our balanced form hoistway so extensively used for this purpose. It requires no power, but when the turtle is taken off a balance weight is put on so that it is always balanced and is easily and quickly run up and down by hand. It takes the turtle from off its table and lowers it into the pressroom, where it is received on a low two-wheeled truck and taken to the press. When it comes off the press it is on the truck, taken to the sink and washed on the truck and then taken to the [Composing room?].

We have marked in the plan the hoisting hole suitable for such a hoistway and will send you the cost of it with the balance weights all ready to put up, if desired. . . .

Should you wish changes made of any kind, please advise us and we will make plans to suit.

Your truly
R. Hoe & Co.
Carpenter

Apparently getting a favorable reaction to his hoistway, Carpenter responded (7 May 1878): "We also enclose drawing of the form passageway from the composing room floor to the pressroom. . . . In the composing room it is closed by a trap door, so that the turtle tables can be run up to have a turtle taken off or put on. If you adapt this arrangement you will require a two wheel form truck which we can supply." All such shops had their tiny elevators with chains and pulleys, counterweights and hooks; better materials handling would have to wait a long time.

Another channel of endeavor also became—at last—a lucrative addition. In 1881 the company finally perfected the "Rotary Type Endless-Sheet Perfecting Press," based on the Colonel's original planetary type-revolver—the machine he projected (6 December 1858) and assigned to a single workman for development. As a perfector it put an end to the dilemma of printing with one set of type forms. "It was introduced," said Isaacs, "specially for rapid 'getting away' on afternoon papers, where the time and expense of stereotyping were ruling factors." Hoe had described it to William Clapp (20 March 1877) and installed one (1879) at the New York *Commercial Advertiser*. But after stereotyping and superior folding mechanisms became common it declined in importance.

This press nevertheless led the way to periodical work heretofore thought impossible on fast machines. One prestigious order along these lines was an 1884 Government Printing Office press to print daily the *Congressional Record,* which, as daily copy fluctuated, could be adjusted to handle from 8 to 32 pages. About this time Theodore De Vinne returned to enlist Hoe's assistance in producing mountains of new presswork.

De Vinne was printer of *St. Nicholas,* a popular children's magazine begun in 1873, and *Century Magazine,* an eclectic monthly (formerly *Scribner's*) reaching about 200,000 circulation in the 1880's. One of his shop's largest jobs was the 9000-page *Century Dictionary and Encyclopedia,* but much other bookwork was similarly scaled. In 1886 De Vinne erected a large headquarters on Lafayette Street, New York (still there), to conduct these mammoth operations. That same year he ordered a special Hoe rotary to print, cut, and fold *Century's* 32-page advertising and text signatures, confident that speed was not of itself an enemy of quality.

De Vinne defended his gamble with almost missionary zeal:

> Web presses for newspapers are common enough, but this press has distinction as the first, and for three years the only, web press used in this country, for good book-work. . . . [It] is not so fast as the web press of daily newspapers, but it performs more operations and does more accurate work. It is not a large machine, nor is it noisy, nor does it seem to be moving fast, but the paper goes through the cylinders at the rate of nearly two hundred feet a minute. . . .

For De Vinne there remained only the problem of printing the magazine's illustrations with the same dispatch. In 1890 a companion machine went to the De Vinne Press for this purpose:

> Sixty-four plates of THE CENTURY, truly bent to the proper curve, are firmly fastened on one cylinder sixty inches long and about thirty inches in diameter; sixteen inking rollers, supplied with ink from two fountains, successively ink these sixty-four plates with a delicacy and yet with a fullness of color never before attained. The shafts of the impression cylinder and the plate cylinders, 4½ inches in diameter, do not spring or give under the strongest impression. Although rigid in every part, in the hands of an expert pressman it can be made responsive to the slightest overlay. . . .[31]

Joseph Adams a half century before could not have done better. The Hoe "Rotary Art" press had superseded the Adams platen for speed as well as quality.

As Hassard warmed to his writing chore he thought that, with the substitution of rotary for reciprocating motion, "upon one surprising revolution press-making went back to antique simplicity and compactness."[32] His expectations would not hold. Although the early post-Lightning machines were indeed smaller than the ten-cylinder—the

1878 *Tribune* perfector was only 24 feet long (including folder) and 5 feet high—miniaturization was not a necessary condition. Succeeding presses inevitably grew complex and increased in size, but they would do so much more even Hassard could not object. Newspaper proprietors everywhere began to clamor, as had Lloyd, for whatever special operations they wanted—and the company delivered them.

The *St. Louis Republican* in 1880 ordered a web that folded and pasted (along the centerfold) an eight-page paper at 22,000 per hour. The following year the *Chicago Daily News* took two machines with a folding apparatus (after a British patent of Archibald Sandeman, Edinburgh) which for the first time made the initial fold parallel to the web; it was a very popular press. The St. Louis *Globe-Democrat* (an 1876 merger) in 1882 wanted its Hoes to fold from 2 to 12 pages at 24,000 hourly, a feat (with Crowell's 1876 and 1879 patents) accomplished with a supplemental web introduced between two plies of the main web:

> [T]his three ply web passed down over a folding triangle of a rounded form and received a longitudinal fold. It continued on down to a pair of combined cutting and folding cylinders which cut it at page length into twelve page sheets and folded these transversely to half-page size and delivered them to a fly-cylinder which laid them on a slow moving apron. . . .[33]

From now on these presses performed more bindery work than printing.

A web machine called the "Presto" in 1883 was designed for moderate circulation papers up to eight pages at 10,000 hourly. The "Three-page-wide Press" was introduced 1888 to print, perfect, and deliver two-page papers at 60,000, or up to 12 pages at 12,000. Two of these in 1888 went to the Baltimore *Sun*. These types were largely byproducts of Crowell's intensive folding experimentation which, as Hoe lamented, had been very costly to the company. But they soon paid for themselves in ample sales—if only to Pulitzer's *World*.

As a Hungarian youth Joseph Pulitzer had become interested in America and boldly accepted a recruiting agent's offer of passage to the United States to serve in the Union army. Upon reaching Boston, however, he jumped ship and in the fall of 1864 enlisted in New York. He found himself in a German battalion organized by Carl Schurz.

Himself a German, Schurz had similarly found it necessary to fight for his adopted country before settling down to private pursuits. His enthusiasm for this obligation perhaps exceeded that of many native sons:

> The government had to create an army of several hundred thousand men to be put into the field without delay. The youth of the country responded with enthusiastic alacrity to the President's call. In an

> incredibly short time the ranks were filled with young men from every conceivable station in life and every grade of intelligence. . . . Nor was the foreign-born citizen less zealous than the native. . . . There are interesting stories told of the tricks resorted to by some patriotic youths to smuggle themselves into the ranks. . . .[34]

After a term of office in Congress Schurz in 1881 joined the New York *Evening Post* (staying only two years); he wrote books and became editor of *Harper's Weekly,* 1892-1898. His fellow trooper fared much differently.

In the Civil War Pulitzer served only nine months, saw little combat, and in fact soldiered poorly. But he had discharged the noble duty others could not seem to manage.

After the war, his German accent still a hindrance, Pulitzer went West to find work. In St. Louis he accidentally fell under Schurz's influence once more, becoming a reporter on a German-language newspaper founded by his old army commander. He was soon deep in politics and journalism. In 1878 he formed the St. Louis *Post-Dispatch* from two earlier papers, and with its gradual success began to dream of conducting a newspaper in New York—pinnacle of the American press. While in New York on a trip to Europe he learned of a rare opportunity to realize his desire.

The New York *World* had started 1860 as a religious daily, and a year later absorbed the *Courier & Enquirer.* Despite the investment of $200,000 for building and plant, including a ten-cylinder Lightning, and copious telegraph, syndicated, and shipping news, the paper did not prosper: its moralizing forbade the tantalizing gossip and rumor which made journalism popular. As the Harpers were doing well with a monthly and a weekly, a daily seemed obvious, but they declined to buy both the *Sun* and the *World* (Fletcher instead started *Harper's Bazar*).

The *World* limped along. Under Manton Marble as editor it struggled through 1868-1876 as a small, respectable paper. When it fell into financiers' hands and continued to decline, Pulitzer bought it (1883) as an extraordinary bargain. Within four months he doubled its circulation to 40,000, and a year later it was at 100,000—at which point he grandly fired 100 guns in City Hall Park and gave every employee a silk top hat. Pulitzer had not altered the paper's format very much, but he put pictures on the front page and demonstrated a passion for accuracy and irresistible headlines, leads, and editorials. The whole newspaper fraternity would benefit from his innovations and refinements. But most of all he strove for circulation, and explained why:

> If a newspaper is to be of real service to the public, it must have a big circulation, first because its news and its comments must reach

> the largest possible number of people, second, because circulation means advertising, and advertising means money, and money means independence.[35]

Pulitzer consequently took a critical look at his mechanical department.

Having bought two Hoes for his new *Post-Dispatch* in 1881, Pulitzer began to appreciate the latest machinery. When he acquired the *World* he found the Park Row basement similarly filled with Hoes two years old: these were six single eight-page machines, and although another was added, they were proving inadequate. Like other papers with groaning batteries—the *Sun's* Bullocks and the *Times'* Walters, the *World* was scarcely equipped for the surge that would characterize the frantic era to come. When a fire in 1882 destroyed the *World* building, Pulitzer had to ponder his equipment anew. He immediately bought two new machines and ordered two more for later delivery.

In March 1884 the first pair was installed, necessitating some adjustment to the paper's format. A *World* editorial announced:

> These annoyances of growth [format alterations] will soon be obviated. Two of our new Hoe presses are now in our building, and in short time the other two will be delivered. With these four new presses we will be able to handle our various editions with facility—at least for a season. . . .[36]

But there was no "facility" to speak of—in fact, very little production. Strangely, the two machines "required the constant presence of Hoe mechanics to keep the going." The *Journalist* explained:

> They will go along all right for a little while, carrying off the edition as fast as possible. Then something will go wrong, tapes will break, and the whole machinery will get into such a hopeless tangle that there is doubt for some time whether the work can be finished early enough to get the paper out at its proper hour. Consequently there is a constant strain upon the managers in the *World* office, who hardly dare to leave the place until the whole edition is printed.

A month later the situation was the same:

> [One machine] would run right along at a clipping gait until six or seven thousand copies were struck off, and then it would suddenly stop. From that time out no human ingenuity could get it to work again. . . . Although there have been half a dozen of Hoe's most expert mechanics fairly living in the press-room of the *World* since this press was put in, they have found it totally impossible to reduce the machine to anything like reliability. . . .[37]

The *Journalist* (a trade paper absorbed in 1907 by *Editor & Publisher*), enjoying this dilemma, urged that R. Hoe & Company be sued, as the paper's inability to turn out sufficient copies to meet circulation was substantial. "The end of this naturally is that they [newsdealers] soon sell out, and that not two-thirds of the inquiries for *Worlds* are satisfied," *Journalist* claimed.

By October 1884, however, the mysterious bugs were driven out, and the new Hoes together (now six in all) produced 1200 copies per minute—72,000 eight-page papers or 150,000 four-page papers per hour, printing, folding, and delivering in counted bundles. To the public the *World* again proclaimed:

> These machines are the finest, best and fastest that the ingenuity of man in this inventive age can produce. . . . No newspaper establishment in the United States is better equipped than the *World* to-day, and yet we shall further add a double Hoe Perfecting Press to our present array in a few months, making, in effect, eight in all. These presses cost a fortune, and they are not indulged in as luxuries. They mean business of the most active kind. . . .[38]

Five months later the *World* indeed added the double-width perfector —equivalent to two conventional machines. This press Robert Hoe III considered a Darwinesque "connecting link" between the ordinary "single" or two-page-wide machine and the "Double Supplement" that was to follow. Pulitzer was now ordering so many—and so fast, the Colonel reportedly granted the *World* a $25,000 discount on each machine "because it bought in bulk."[39]

This experience had nevertheless been a horror to Pulitzer. Not long before he had been the first (at the *Post-Dispatch*) to install one of Crowell's new folding attachments; it performed poorly. "Its vagaries nearly drove him mad," Seitz wrote, "and bred in him a strong prejudice against new devices. Quoting this experience, he would ever after add: 'Let the other fellow pay for the experimenting. . . .' "[40] Pulitzer may have been all the more exasperated because the rival *Globe-Democrat*'s new $80,000 web press with Crowell's triangle folder "operated so successfully," said Tucker, "that the proprietors soon ordered a duplicate machine."[41] Curiously enough, perhaps, the *World* remained editorially most enthusiastic about technological progress.

Pulitzer nevertheless soon had an opportunity to test the sincerity of his resolve. The Colonel offered him the first Double Supplement in 1885—and Pulitzer declined emphatically. The press (No. 140) went instead into Bennett's *Herald,* where it worked perfectly. Pulitzer promptly bought the second of its kind.

Crowell's 1884 Double Supplement or improved double-width machine delivered up to 12 pages at 24,000 per hour. It was a very simple press and "several hundred" (Hoe's words) were employed in the United States and Great Britain to produce the popular two- and four-page supplements inserted automatically into the usual daily paper. The first one at the *Herald* (1885) created what Hoe called "for a third time, a complete revolution of the methods of fast newspaper printing."[42] Characteristic of this revolution was overwhelming attention to the folder rather than the printing mechanism. George Isaacs wrote:

> In the history of newspaper press development the folder was another revolution in more ways than one, and led the way to the far wider field of development that now began. . . . Turning bars marked another very great step forward in press development [1882] to be credited to the pioneering work and engineering efficiency of R. Hoe and Company. Whilst the folder made it possible for faster running presses to be put into commission, the turning bars led to the production of larger papers.

The highly polished steel bar, fixed at a 45 degree angle to the web's direction of travel, enabled the integration of a second web, from a second printing section, into the folding mechanism along with the main web. This required extraordinary coordination and necessitated a new system of roll tension control:

> It must be remembered that there were numerous factors, some old, some new, affecting the "pull" of the sheet. There had to be sufficient grip on the reel itself to prevent it running too freely and causing the sheet to "wander" from side to side. . . . Then the relative size of the impression and printing cylinders needed careful regulation. Sometimes the first impression or inner forme would feed more paper than the second impression or outer forme could take away, and thus lead to spoilt papers. . . .
>
> Then also there required to be perfect harmony between the pull of the sheet, the impression and the propellers at the top of the [triangular paper] "former," and the propellers and the drawing rollers at the bottom of the "former." Then again there was the need for equality of tension on each of the sheets when they began to come together over the "sup," for should one of the three travel tighter or slacker than the others faulty copies were the result. . . .

The roll stock itself added problems—badly wound reels, split or damp edges, mill splices. These difficulties were regulated, for the first time on any press, by mechanically adjustable tension control, further improved in 1887 with Hoe endless belt automatic paper feed control.

In 1887 too a pair of "Double Sups" was installed at Lloyd's, where Isaacs as a youth met a new fraternity of specialized pressmen:

> I first saw a double sup in the year 1895, when at the age of 12 years I was taken by my father to see the production of "Lloyd's Weekly Newspaper." He was then a "sup" hand on "No. 6." . . . It was then I saw the wonderful new type of press, the Hoe three roll straight line, in course of erection. . . .[43]

The added tension problems of the Double Supplement required a separation of crew duties, the "tension hand" controlling the paper, the "machine minder" watching printing quality. It heralded the start of the modern newspaper press.

Stephen Tucker had happily participated in all these developments. In 1877 alone over a dozen patents were issued to him, all relating to web printing and delivery mechanisms. Over the following years, however, with Luther Crowell playing an increasing role in the company's experimentation, Tucker began to enjoy a supervisor's post rather than a workman's. He now had time to correct Whitworth's faulty wheel cutter, tinker with Thorne's typesetter, and pursue other projects in addition to continued folding trials. He remained so busy in and out of his Experiment Room he hardly noticed the passage of time.

By 1888, after interminable years with the firm, Stephen Tucker was long beyond commemorations and anniversaries. When his foreman Luther Crowell came upstairs one day to tell him a delegation of workmen was waiting to address him, the busy partner visualized the start of new labor problems. The Grand Street workforce had been removing machinery from a large section of the first floor and by 12:30 about 700 men had found room to stand—facing a little table at which sat Tucker's fellow officers, Peter Hoe, Robert Hoe, and Theodore Mead. As Tucker approached he was given a rousing cheer "which almost raised him from his feet," and Crowell—so long and so often a quixotic and irksome employee—stepped forward:

> Mr. Tucker, you have been here longer than any of us. You started as a 'prentice boy and went through all the grades to the partnership. It is to your inventive genius, as master mechanic, that much of the perfection in the art of printing is due.
>
> We have tried to show in some simple way how greatly we respect and esteem you, at the same time wishing you long life, health and prosperity. . . .

A draped cloth was then removed from the table, revealing a sterling silver dining table centerpiece, two feet high, its four baskets filled with bouquets. On it was the inscription:

Presented to
STEPHEN D. TUCKER
By the Employees of R. Hoe & Co.
January 28, 1888.

It was Tucker's 70th birthday and his 55th year with Hoe. A "tall, portly man, as straight as an arrow [with] thick hair and full beard . . . iron gray," noted a reporter, he was completely surprised by the ceremony. He managed later to respond to reporters:

> Yes, I started at the foot and worked my way up. I've been there a long time. When Col. Hoe was alive he and I always worked together, as we were both of a mechanical turn of mind. In 1848 I went to France and England and spent two years introducing our presses. I expect to go over again during the coming summer. The presentation this afternoon was a complete surprise to me. I did not even suspect that they knew it was my birthday.[44]

"He is at work every day," said one of the papers covering the event, "and has charge of the entire mechanical department of this, the largest printing press foundry in the world." Tucker got no splendid printer's dinner—those Waygoose times were already fading away—but he had been fittingly remembered.

Stephen Tucker retired 31 August 1893, transferring his share of the business to Robert Hoe. He had been 33 years a partner, 59 years with the company. Yet to the Hoes he was never an employee, and for him Hoe was not a firm, but a family. His own kin had almost deliberately become estranged: among three of his four older brothers one had died a Confederate soldier in a New York state prison camp at Elmira; Tucker searched frantically to find traces of him. Another brother was a forelorn prospector stumbling about the Southwestern United States, making only one or two visits to New York in a lifetime. The third was an explorer in Brazil. Tucker welcomed the large Hoe clan as if it were his own, and remained close to it all his life.

Among Hoe ventures outside of press building Tucker was most interested in a nearby bank with which the brothers had become involved. The Oriental Savings Bank, East Broadway and Grand Street, dated to the 1850's with some of Hoe's printing friends among directors. Apparently rechartered in 1871, the Oriental attracted Robert Hoe II (who became a trustee) and Eugene Smith (attorney). Peter Hoe, William Colby, and Tucker invested in it.

> [*Tucker, Munich, to New York, 8 May 1873*] . . . *I am very much gratified at the result of last year's business. On account of the Strike and less hours of work per day I had not anticipated so large a sum. No doubt piece work contributed in no small degree to this very favorable result. I of course agree to whatever sum it is thought advisable to divide. Although you do not state the reasons for retaining such a greater proportion than usual (one half!) I have no doubt they are very good ones. . . .*
>
> *It is gratifying to know that the O. S. Bank continues to gain. With time and good management will come confidence and with confidence plenty of business. . . .*

The following year (18 September 1874) he regretted that "Our little Savings bank has not got up to the point reached before the panic of last year," but the thrifty little enterprise apparently did not survive. Receivership proceedings followed in 1878. Aside from a home in

New York which Tucker bought (1876) with funds advanced from the partnership, he engaged in no other speculations.

A severe illness in 1894 prompted a visit to Europe for 18 months, and in 1902 Tucker was again traveling abroad with his wife and two daughters, enjoying the summer in Switzerland. He died suddenly in London, age 84, an event sadly reduced (like Conquest's notice not long before) to a whisper in Old Thunderer:

> Tucker—on the 9th Oct. 1902, at Sackville Hotel, London, Stephen Davis Tucker of New York City, U.S.A., born January 28th, 1818. Mr. Tucker was prominently connected with the firm of R. Hoe and Co. for nearly half a century.[45]

He was buried in Brooklyn's Greenwood Cemetery.

Tucker's collection of about 62 sundials must have been a sublime joy to him. He always traveled with one or two small ones, carefully setting them up wherever the sun could be coaxed to make them work. In 1903 his widow gave 25 prize specimens to the Metropolitan Museum of Art (of which Tucker and daughter Aimée were fellows in perpetuity). Intricately made of ivory, gold, silver, ebony, the ornate masterpieces went unnoticed until 1967, when they were put on display. One, an ivory diptych made 1602 in Nuremberg, tells time in London or Constantinople and 31 other places in between, and can give it in Babylonian hours (beginning at sunrise), Italian hours (beginning at sunset), and Jewish hours (dividing the day in twelfths); it can also show the direction of prevailing winds and calculate the date of Easter. A skilled mechanic could certainly appreciate the handcrafted precision of such exquisite timepieces.

"Father used to speak of himself as a 'machinist' or 'mechanic,'" wrote his daughter Aimée, "but I should call him an 'inventor and manufacturer of printing presses' as that was his life work."[46]

LATHE CLASS, HOE APPRENTICE SCHOOL

C. 1900

Courtesy Bronx County Historical Society

10.

Human Relations Around the Shop

PATERNALISTIC FACTOTUM / PIECEWORK AND LABOR STRIKES / THE HOE APPRENTICESHIP PROGRAM / A YOUNG MACHINIST'S CAREER / FACTORY FORTRESS / LOWER EAST SIDE TURMOIL

Throughout its existence R. Hoe & Company had little difficulty finding good workers, despite the early shortage of labor in the United States and the novelty of the machinist's trade. Skilled mechanicians arriving steadily from Europe found employment with Hoe in New York to be almost all that could be claimed for prosperity in America. The composition of the workforce was variously reported as containing Swiss, German, and French, and soon, among immigrants from "all" nations, according to *Graham's Magazine,* "several Armenians, said to be persons of great intelligence and excellent deportment." [1] (It may be no coincidence that the industrious Armenians a generation later would dominate the American photoengraving industry.)

The company consciously made every concession it could to its employees, perhaps as today's corporations cater to engineering and computer personnel, so laboriously assembled and expensively outfitted. Tucker reported:

> The various shops were well heated and ventilated and otherways made comfortable for the workmen, and it was the policy of the firm to employ the better class of men and the most skillful mechanics and pay them accordingly, being convinced that they were the most satisfactory and in reality the cheapest workmen. As a consequence, good mechanics always sought employment in the factory and remained there, and R. Hoe & Co.'s works became noted as one of the best equipped in mechanics and tools in the country.[2]

From the beginning of its cylinder press manufacture the firm was obliged, in many instances, to furnish a knowledgeable pressman along with a machine, especially for remote printshops. These men, sometimes literally detached from Hoe's own staff of mechanics and erectors or purloined from mechanical departments of nearby newspapers, commanded very decent wages for such assignments, although their disappointment with living or working conditions in the hinterlands usually insured their return. Like itinerant compositors, whose skills were similarly sophisticated enough to command employment wherever they went in small midwestern towns, some pressmen were given to drinking or wanderlust to escape the drudgery. Most employers had little recourse but to write for a replacement:

> [*To Nelson, Wright, Ferris & Company, Cincinnati, 3 July 1850*] . . . *Good feeders here obtain from 8 to $9 a week—we can procure a good pressman for 12 to 14$*† *per week. It is doubtful whether you would gain by having the pressman feed as he would neglect the press and could not see his work while so employed.* . . .

> [*From Savannah Daily Courier, July 1853*] . . . *Mr. Weilten sent me a Pressman yesterday* [*who might*] . . . *answer my purposes until I could get another through your kind assistance. He has just arrived in a state of beastly intoxication & wholly unfit for duty.* . . . *For a competent man—one who fully understands his duties & will perform them, I will give $20 per week & pay his passage & provided he remain in the situation six months.* . . . *I would expect him to learn*† *a colored boy to sheet the press.* . . .

Occasionally, when requests mounted, Hoe placed advertisements for pressmen. One response came from the former employee who had coupled a cylinder machine to Boston water power three years before:

> [*From H. Ditmars, Buffalo, 13 June 1853*] . . . *I not*[*e*] *even your Advertisement in the Tribune & I am Happy to inform you that I am at your cirvice*† *should you Desire it, as far a*[*s*] *preses*† *is concirned.*† *I Have Had some considerable Experience cince*† *I left you, & should you Employ me as a Pressman I trust that you will not Have cause to regret.* . . . [*P*]*lease let me no*† *to what part of the south you would wish to send me.* . . .

> [*From John Hudson, Richmond, 14 June 1853*] . . . *Will Messrs. R. Hoe & Co. please to (if not already suited) name the place & state the compensation to be received by the Pressman taking the Situation mentioned.* . . [?] *References from this place* [*Daily Dispatch office*] *& from Ledger Office, Philadelphia (if required) can be given.* . . .

Such craftsmen in obvious demand were understandably choosy at times, and seldom afraid to ask questions. "I am desirous of knowing the regular rates in book & jobbing offices paid in New York to compositors, also the scale of prices demanded by the N.Y. Typographical

Society," an Albany inquirer demanded in July 1853. "If not taxing your civility too much, . . . forward me the above information by to-morrow evening's mail. . . ." Most, however, were not pretentious:

> [*From J. Musick, Philadelphia, 7 October 1857*] . . . *I am out of work and no prospect of any in this city. I send*† *word to my Brother in Baltimore and he says I had beter*† *let you know; he says you often know of a situation when nobody else does.* . . .

> [*To M. Potter & Company, Cincinnati, 15 April 1862*] . . . *We have inquired among our own men and elsewhere* [*for pressmen*] *but all object to the labor of the pressroom such as carrying up paper, washing rollers, cleaning up, &c which in large offices* [*is*] *done by the feeders and an assistant. We will advertise if you wish but we do not think it will be of any use. The person whose letter we return herewith we know but cannot recommend.* . . .

Hoe might have declined to recruit compositors, a branch of the business it had consistently avoided, but press customers could not be offended:

> [*From George Knapp & Company, St. Louis, 5 January 1865*] . . . *We ask the favor that you will get some competent man to procure for us six or eight good and steady compositors accustomed to work on morning papers. We have already several from New York, and we learn from one of them that Hunter Bradford and David Breckenridge in the "Tribune" newsroom are willing to come out.* . . . *You can furnish them tickets to St Louis at our expense. We will give permanent situations at fifty five cents per thousand ems to competent men.*
>
> *When they arrive at St Louis, let them tell no one their business, because a committee from the union are endeavoring either to bribe or frighten men from our office.* . . .

(A telegram two days later clarified no need for pressmen: "We are full of printers dont send any. Geo. Knapp & Co.") A few years later (28 November 1872) Knapp's *Missouri Republican* was shopping again: "Will you be kind enough to see T. Stefanazzi, assistant stereotyper at [New York] Herald & get him to come out at once. Wages $35. a week. If necessary get him a ticket out. Our Col. Knapp saw him when in New York. . . ."

At one point the New York shop could call upon the Boston branch for men:

> [*From Theodore Mead, Boston, 8 February 1866*] . . . *We have not an old Adams press hand that you would like to spare but can send you a young man of twenty one, a German-Yankee, son of one of our old hands, who I think would work into your ways better than an old man. He is a stout, industrious handy fellow, not afraid of hard or dirty work, and he has had considerable experience—without being as yet by any means a first class hand.* . . .

As late as the 1880's the company paid close attention to such letters of request for labor, recognizing its machinery to be only as successful as the men running it.

[Chicago] Dec. 31st 1877

Gentlemen

Yours dated the 22nd to hand, to day. We have made inquiries respecting F. Primrose and find that he has had 1 year experience, and is incompetent in every respect for the Situation, and you must not on any consideration send for him unless for 2nd Man. We called at the "Tribune," "Inter Ocean," "Times," "Mail" and it seems as if they [pressmen] have been canvassed before. They all refuse to take the job unless Transportation is paid for, and if you decide to send for anyone you must not omit this, as it will cause delay & inconvenience to you. And you cannot get a man to leave a Situation, unless this is all right. We had a long talk with Lucas at the "Tribune"; he is now getting $22 p. week, and he is a first class man, and will work hard and no doubt make a success of it. We recommended in our last letter a Mr Sampson; he used to work for the "Tribune" and he got his hands (both) in the Shaving Machine which was very careless, and they were glad to get rid of him. There doest not seem to be any out of work, and that is why at present you cannot get anyone at less than $30 p. week from Chicago and Transportation. Although if you want a good man you will have to pay this. We have expected the "Mail" and the "Inter Ocean" in to day, as they promised to come down if they could find a good man, but they have not turned up, & we suppose they cannot find anyone.

We think you ought to get a good man at the first, and then work some of your own men into it.

If we hear of anything further will write you by first Post.

Yours Very Truly
R. Hoe & Co.
Crowther

Mess. The "Times Mirror"
Los Angeles, Cal.

There were apparently few repercussions from plants touched by such recruiting tactics, which culminated, perhaps, with Hearst's outrageous theft of an entire Sunday edition staff. It was evidently an acceptable part of the labor situation; money talked loudly for good pressmen as for crack journalists. But the other side of the matter, massive layoffs when times were lean, balanced joy with sorrow.

Visiting the Hoe shop in July 1857, a *National Intelligencer* reporter was surprised to find it a formal model of efficiency:

> It is worthy of remark that, in each room of every building occupied by the concern, one sees, in large letters staring him in the face, two mottoes: "Order is Heaven's first law," and "A place for every thing, and every thing in its place." [3]

Indeed, in the twenty years since the 1837 depression—undoubtedly the company's lowest point, the brothers' industriousness and attention to detail had proved themselves. The manufacture of precision cylinder presses became sufficiently routine as to be uneventful; the rotaries up to ten cylinders were enthusiastically received in both American and British markets; saw work was lively (despite increased competition); occasional contract work still filled empty periods in the work week. Tucker had installed his tool crib and the shop was as modern as any in the country.

Financially, too, the firm was sound. Debts to the Sandersons had not quite been repaid, but there were prospects soon of doing so. The company had no other notes outstanding. But other businesses in a recklessly expanding America were not faring as well: the national economy was once more shaken by panic and reaction. In 1857 another financial collapse was, said the Colonel (15 September 1857), "sweeping over the country, breaking up every unsound concern & some that have thought themselves rich but two months since. . . ." He took drastic steps to protect the family investment.

In September the Colonel began by discharging 50 men; although he admitted to Robert the "financial revulsion" had not yet reached their own customers he was preparing for the worst. His ruthlessness was born of fear of the unknown: he noted excellent grain and cotton crops for the year and could not understand why out-of-state banks were suspending specie payments. He nevertheless reduced wages and by month's end had cut the workforce by over one half, employing the rest only 7 hours a day, then even less. "We are in a chaotic state," he wrote, his earlier confidence having completely vanished:

> [*Richard Hoe to Robert, "on the continent," 26 October 1857*] . . . *We are running the engine from 8 to 1 o'clock making five hours or half time for the men, & having the boys work on from 2 to 5 o'clock, not running the engine for them. Our wages this week will not be over $1300 & next week I shall bring them down under $1000. We are selling literally nothing, & I really dread the months of December & January for the great amount of suffering the poor must endure. It will be a season of unparalleled distress & I fear starvation in many cases. . . .*

The Colonel's analysis would sound much like that of the dismal 1930's:

> . . . *The whole course of this revulsion shows the extreme folly of parties with a capital of 4 to 500,000 dollars doing a business on long credits of three to five million per annum, or in that proportion.*
>
> *The Harpers show a surplus of over a million, owe about 700,000—$200,000 confidential to friends & relatives, $100,000 in Bonds & Mortgages, $400,000 in notes outstanding; on these they ask an extension of six, twelve, & eighteen months which I think they will*

get, & I further think they will find it very difficult to pay up so fast. Cyrus W. Field & Co. are paying in paper or giving notes at six, twelve, eighteen, & twenty-four months. So they go. . . .

In November cylinder press orders dropped almost to nothing, and the printing industry seemed at a complete standstill. Yet by March 1858 the Colonel returned to a full working day, the saw business in particular leading the way to a resumption of lively business. (This the Colonel attributed to a new sales catalog.) But the 1859 winter looked as bleak to Hoe as the one before. Early that year his spirit nevertheless returned with his ambition, and he soon turned the shops upside down to improve them.

[*Richard Hoe to Robert, Paris, 4 January 1859*] *. . . We have been making great changes in Gold St. Have moved all the* [*wood*] *finishers from the back building with the tools into the front and put Curtis & his men in their room & Curtis' floor we have filled with wood tools with Deprey in charge. We shall have them all in full operation in about a fortnight & then* [*with special tools built a few months before*] *shall make* [*type*] *cases with them, as well as pretty much—if not all—the printers furniture, &c. I feel quite safe in saying we shall reduce the cost of making cases from 50 cents per pair to 25 or 30. . . .*

This creative momentum extended well beyond the factory, too.

Richard's enthusiasm carried him to Boston in the spring of 1859 where he bought the Adams concern; the yankee brothers had perhaps also been victims, if only psychologically, of the recent depression. The Colonel himself was alternately discouraged by the dragging settlement with Whitworth, and obviously cheered by happy prospects at the Adams shop. He remained, nevertheless, a confirmed New Yorker:

[*Richard Hoe to Robert, London, 21 June 1859*] *. . . The question of continuing the manufactory at Boston or selling out there or removing the manufacturing* [*of*] *the presses here may well wait till you come home even if you do not come before next year. . . . I sometimes think how pleasant it would be if we could find a suitable piece of property somewhere on Center street, below Canal, where we could build a factory & sell out both up town & down; but then again the great expense stares me in the face. . . .*

In November 1859 the Colonel had had his leg broken in a carriage accident, and even long recuperation at his country home did not slow the company's impetus:

[*Richard Hoe to Robert, London, 13 December 1859*] *. . . It is a fortunate season for me to be laid up—we are having comparitively*[†] *few customers calling at this season of the year, & at the Factory I believe the foremen all look more strictly after their workmen as I have had them all up here to see me, & urged the necessity of looking closely*

> *to their men &c &c. It keeps brother Peter at it pretty briskly but he seems to thrive on it. . . .*

The brothers' usual reluctance to be optimistic about the future this time would be rewarded by the worst possible developments in American history.

The puzzling crash of 1857 had further, deeper repercussions. It drove Northern manufacturers to cry for Federal aid, although President Buchanan's Cabinet was dominated by Southerners. The agricultural South viewed the depression as proof of industrial instability, and began to project for itself a safer economic existence without the North. The conflict became part of the growing feud between two major sections of the United States which could not be reconciled. Civil War tensions at home and bloody combat of shocking proportions left little room for America's routine business interests. It was a long time before R. Hoe & Company could return to its work uninterruptedly. Providently, European affairs blossomed.

In 1871 Richard journeyed to London for perhaps the third time, to look over what was becoming the company's most lucrative market. When he mentioned his enjoyment of the daily attentions to business his brother responded:

> *[To Richard Hoe, 29 April 1871] . . . My minister the other day preached a sermon on being "diligent in business" & my experience quite agreed with his teachings; that it is a very safe & comfortable thing to do, many other good things following in its train. There is so much to do here by our increased business & competition outside that we have to be diligent. As we have the satisfaction that it pays in so many ways, & all of us are of one mind, we have only to take the comfort of it all. . . .*

The Colonel frequently had thoughts of forming a stock company; the additions of Tucker, Colby, and Robert Hoe III to the business had posed complex partnership problems. But the three senior members, truly "of one mind," had little desire to make drastic alterations in the successful venture—even when conditions were again lean:

> *[Robert Hoe to Richard, London, 12 July 1871] . . . Like all other kinds of business we must expect to be pinched as to our profits when economy is the order of the day, as it is here & more especially if competition demands it. The question of a reduction of wages has been considered by us. It seems difficult to do it generally throughout the works without creating a good deal of dissatisfaction, & to limit it to a portion only would hardly pay. All things taken into account, therefore, we determine not to try any doubtful experiment of that sort at present. . . .*

Robert wisely knew that in 1871, unlike 1834 or even 1857, one could not easily discharge large numbers of men, or arbitrarily reduce wages, for that matter. The shop now performed machinework of a considerably sophisticated description, and the men capable of doing it deserved more than that. Too, the labor force in America was increasingly vocal, increasingly demanding. Had it not won the war for the North?

But Robert's sense of justice, of industriousness (and diligence) would not let him accept the slightest compromises. He would ever reward the deserving handsomely but would not tolerate "impositions." He now had an idea how to insure excellence:

> *. . . All agree that the best & most costly hands would like to work by the piece, or job, & we have determined to adapt that plan gradually in all the shops, until we can bring it to a system with most of the work. It seems to me that it is the surest remedy against too high wages & the best for all interests concerned. Mr. Colby showed me a letter in which you had urged it for consideration in his department. The fact is both William* [Colby] *& Mr. Tucker begin to feel the need of some let up from incessant labor. If Mr. Tucker goes Colby will of course be quite sure to follow his example. . . .*

In this letter Robert described "very close, sultry, dry, dog days" and a very sticky atmosphere—weather that usually makes men irritable and pugnacious. The police were in fact ready for a disturbance between Catholic Irish and Protestant "Orangemen" whose parade that very day evoked some objection. Robert returned to his philosophy:

> *. . . With careful judicious management in the works & in every department we may be sure of a good pleasant & lucrative business in the present & future. Perhaps we may make on the average nominally less percent on sales, but with an enlarged business, more in the aggregate than ever before. We must be careful in selling, especially when going back to hard money times. This requires constant attention & very careful, sound judgment. . . .*

All departments in 1871-1872 were busy (more so because of the Chicago Fire) but inflation was cutting into profits, and printers everywhere wanted more credit. Yet the growing workload impelled the company to solve the problem of increasing output without hiring more hands.

But piecework was still a dirty word. Although Theodore Mead had completed a study favorable to it, Colby objected strenuously even to piecework trials, as did the shop foremen:

> [*Robert Hoe to Richard, Paris, 2 April 1872*] *. . . The foremen, naturally, are not pleased to find that the hands are thriving so much more by the system than themselves, whose duties are rather increased by it, with no advantage directly. We told them that if it worked well they should receive a present at the end of the year. This pleased them for a while but not entirely. They want a percentage. . . .*

Exposure of the notorious Tweed Ring had just made all New Yorkers leery of bonuses and percentages. The resulting city administration was now a "skeleton," said Robert, who also noted that railroads were watering their stocks and bonds (again), and general labor strikes—for an 8-hour day—around town were all threatening to "break up the fabric of business." These outside influences—reminding Robert of 1837 convulsions—made it difficult for R. Hoe & Company to mind its own business, and restlessness spread:

> [*Robert Hoe to Richard, June 1872 (day unclear)*] . . . *Our men continue to work in the new shop, the only one running since the breakdown of the large engine. Thursday of this week has been set down by the trades for the machinists to strike. But for the strike pressure there would be no excitement of strike in our shop. I still keep a slight expectation that the gale may blow over yet without injury to us. Our finances are snug, considering the vast extent of our business. . . .*

In June paradoxically the Hoe plant was so glutted with work, finished and unfinished, there was no place to store completed machinery. But the men struck.

> [*Robert Hoe, 4 June*] . . . *The carpenters and pattern makers at the works uptown, not having their demands complied with, took up their hats and put on their coats and left in a body one day last week. The outside excitement fanned the flame within the shops.*
>
> *Much talk and excitement increased in the machine shops as well, and matters were looking quite serious and we determined to fix on nine hours for a days work and—we gave notice on Saturday A.M.—to be without change of wages. This has seemed to stop the strike and cause a return of good feeling. . . . There are some dozen or two noisy new men who are doing all in their power to commit the hands to 8 hours. . . .*

But there is usually little management can do when mob psychology goes to work.

> *New York, June 11th, 1872*
>
> *My Dear Brother;*
>
> *Since posting my letter for this Steamer I went up town to the works and met a committee of some six of the hands, one from each shop who desired to present their demands on the part of the hands, for 8 hours and 10 hours pay. Tucker, Theodore* [*Mead*], *Stephy* [*Hoe*], *and I were present in our private office. We went over the whole question in a friendly way but* <u>*positively*</u> *declined to accede to the 8 hour rule and 10 hour pay.*
>
> *We offered to add 10% additional pay at 10 hours work or to work nine hours at present pay. They went right off expressing much pleasure at the interview, and quite enlightened on some points. It was evident that they believed we would yield and seemed to be*

somewhat alarmed at the prospect to themselves and families if we stand firm. They have posted a notice up to-day on the bulletin board for a meeting this evening of all hands to "consider very important business." This is in anticipation of the meeting of all the trades for to-morrow evening.

If you were here it would be a great relief, but I am sure that all we have said and done would meet your views. We find that the large shops in all trades are determined to resist the 8 hour demand. The [union] procession of yesterday is pronounced by the Newspapers and all a failure, some 3000 only turned out in it.

We shall now have to let the men learn by bitter experience if they will. There is no scarcity of machinist hands here. I dont believe they will resist us long, especially as the piece workers are so contented. I thought it best to give you all the latest news and you may be quite sure we will act with all prudence and consideration. I think the men all know we have acted kindly and for the best. The outward pressure is all they dread.

Your affectionate brother

Richard M. Hoe [Europe] *Robert Hoe*

But then the worst—

New York, June 17th, 1872

My Dear Brother:

When I wrote to you last week there was reason to fear that the machinists at the factory would strike for 8 hours, notwithstanding the concessions we offered to make of nine hours and present pay.

On Friday all the Iron workers struck in the City, and ours among the rest. On Saturday the repairs on the Engine were completed and we gave notice that all shops would run to-day. There were a few hands in each department at work to-day, and I am inclined to think that there will be a very strong demonstration by many more of the best hands, to return.

There has been a pressure from outside organizations through a few of our men, to deter any from accepting any but their own terms. This afternoon when we came to pay them off for the last weeks work there was no excitement, only the presence of several policemen caused some remarks to be made. The Gold St. Shop goes on as usual.

I am persuaded that there has been too little attention paid during the last 10 or 15 years to recruiting the shops with foremen of character and ability and filling the shops with workmen upon whom we can depend. . . . The strikers are giving the police so much extra work there is very little mercy given to any who attempt any breach of the peace or show a mob spirit. If nine hours had not been yielded in England we should have stood firm for 10. . . .

June 18th. The men still stand out. . . . The Machinists affirm that they have plenty of money to continue the strike. . . . Certainly the effect will be to make Machine labor more plenty† *than ever, as business must be very much curtailed in consequence. . . . All well, fine weather and all comfortable.*

Your affectionate brother

R. M. Hoe [Europe] *Robert Hoe*

[*To Richard, 25 June 1872*] . . . *As we telegraphed you to-day the men have returned to work in all the shops. Until yesterday only about 200 men were out, and to-day nearly all have come back. We work nine hours per day. . . .*

The strike and the breaking of the Engine have given all hands about all the holiday and play spell they wanted. Certainly we start to-day stronger with our men than ever, and I believe the character and dignity of the firm have been thoroughly maintained. . . . The piece work must also be advanced as far as possible. All those hands however left with the rest. While we depend upon foreign labor so much in our shops, and have so few [*native*] *hands brought up and advanced to places of trust, it will be difficult to prevent combinations among our workers. There is a society of Communists here who advocate the wildest theories, such as division of property &c. . . .*

The torrid season seemed fit only for long letter writing.

New York, July 8th, 1872

My Dear Brother:

With such mid summer heat it is difficult to cogitate upon abstruse questions, such as that of labor as now presented by the strikes. . . . For myself, after the exhibition made by our workmen and by others, I do not feel disposed to waste much sympathy on [*men*] *who are so able to take care of themselves and to live so well. Whatever education they and their children have has been given to them gratis by the State, chargeable mostly to Capital. Then modern science has opened vast fields which the nation gives away to all who will till them, thus relieving the Cities on the sea board of great numbers of skilled mechanics or such as would soon become so. With the demand for skilled labor so far in the aggregate exceeding the supply in our vast and prosperous country it is very fair to let the natural law of supply and demand alone. . . .*

Good wages promptly paid, it seems to me, is the best thing for the working man. I do not think it to his interests to have any part of his wages of labor dependent on the profit which may never be realized. My sympathies would rather be extended to the sick and the widow and the orphan. . . .

So far as any threats go, we made none except to one or two men like Preston, who petitioned for 8 hours. I said to him, "if you strike on that ground, such as you will not come back." We now have several of the old workmen sick and unable to do any thing, such as Blair, Curtis, Bailey, &c, and who receive their wages notwithstanding.

When we met the Shop Committee no threats were made and no pressure put on any one, except a plain quiet expression of our opinions and statement of facts. No one feels personally aggrieved and as I have written to you before there is no kind of demoralization or dissatisfaction. . . .

This nomination of [Horace] Greely† *by unprincipled men, just to get hold of the purse and power of the Nation, . . . is a rotten spectacle of degeneracy and political wickedness. . . .*

The fearfully hot weather of last week was allayed by a succession of grand thunder storms. . . . We have an order for a four cylinder for California, new. Give my love to all the family.

Your affectionate brother,

Richard H. Hoe, London *Robert Hoe*

Robert without Richard nearby tended to a softness in dealing with employees which the Colonel, coming home, would soon rectify. But again national economic calamity had a strong bearing on press fabrication: a crash and depression in 1873 was punctuated by long and deep corruption and scandal—both in New York and in national politics. If the Hoes could ever be criticized for their apparent civic inactivity it must be noted that their complex and fragile enterprise needed every managerial attention through these frightening nationwide cycles of prosperity and paucity.

[Richard Hoe to Robert, London, 7 November 1873] . . . The time to make up stock for future sales is when everything is dull, just such a time as the present. Now we have had no difficulty in returning to 10 hours & besides reducing wages 10%. I suppose if the men find things very brisk next spring we shall have to add on the wages, but we shall by that time have them all working by the p[s] *[piece] or so nearly all that they will not talk of 9 hours for a day except to get more wages.*

I suppose Mr. Tucker was anxious to come home & so is coming by the St[r] *[steamer] to leave to-morrow. I shall be glad to see him. We are running 10 hours a day full time with a little over half our force. In fact we have given orders that no man is to be kept at work except on ordered work. If it were not for the Web machines I do not know what we would do for work. This dull time will I think be of use to us in breaking up many abuses, cleaning out useless men, &c &c.*

Each department has been in the possession of a set of laborers, now we have cleared them out & require the laborers to be at the call of that part wanting them at the time, & old useless things are being put into the cupola instead of buying scrap. . . .

Having by chance—for the moment—beaten the labor situation, the Colonel turned to other money-saving activities. Reorganization of the shops was a nagging requirement, as the plant had never ceased growing haphazardly with new equipment. After the Civil War the brothers periodically bought properties adjacent to the Broome and Sheriff

Street "up town" factory, sometimes adapting existing structures to their use, sometimes building anew. More ambitious plans—a grand, unified multiple-storied design topped by a massive mansard roof and clock tower—had been postponed by the recent depression and by the Colonel's increasingly anachronistic attitude about corporate debt.

> [*Richard Hoe to Robert, London, 12 December 1873*] . . . *I think we shall by the time have them all working by the* p^s [*piece*] *or so months from now, & I should very much like to see every mortgage we owe, either individually or the firm, paid off before we undertake any important or serious change or improvement. I think if we stop ordering or buying new tools or incurring any expense not absolutely required we can clear off all these pests in from three to four years, & then if we have a surplus we can build a new foundry.* . . .

Under new state-chartered corporate structures becoming popular in America after the Civil War, owners of businesses were liable only to the extent of their holdings. A certain financial daring ensued—a willingness to incur unconscionable debt on even the most chimerical and long-range schemes. Such practices of course were partly responsible for the periodic economic crashes, although limited liability was frequently an ameliorating factor even for the worst gambles.

But the Hoes took no such risks; they borrowed money, to be sure, but always with caution. They lent personal money to the firm; they invested soundly in real estate and government bonds; they borrowed from little neighborhood banks (one of which they themselves had formed). These penny-wise policies were the single most pervasive lesson they passed on to Robert Hoe III, and ultimately spelled the end of the family firm. Its officers had no experience thinking "big" about corporate financial affairs although the establishment was indeed large enough to warrant it.

The Colonel's rearrangement was thorough if only temporary:

> [*To Robert, London, 5 January 1874*] . . . *We have now grouped together 13 large planers; they occupy the ground floor of new part under office &c. We have all the large lathes brought down to ground floor of new shop on Sherriff*[†] *Street. This arrangement & having all the small cylinder lathes together will enable us to do a great deal more work with less men.* . . . *I can assure you our shops have a pleasant & business like look. Much obliged for the handsome substantial Umbrella.* . . .

Toward mid 1874 the country was again on its way to recovery and the Colonel's insistence on piecework and the construction of an imposing new headquarters seemed to pay off. The company was competing nervously with Bullock and other entrepreneurs on both sides of the Atlantic—a long and feverish technological fight for which, to win, the firm would need the eccentric brilliance of Luther Crowell. The early shop reorganization at least made it possible for the brothers to concentrate on their external problems.

> [*Richard Hoe to Robert, 9 June 1874*] *. . . I have it planned to use the room which was occupied by the Adams Presses for a room to box them, and finish* [other] *cylinders. By doing that I feel quite sure we can do the work for one half the cost heretofore & very much more work can be done by the same tools. We should have . . . every thing arranged so that no one can beat us in price & quality. It is a very pleasant thing to see how well the piece work system is working. . . .*
>
> *I suppose we could after all the changes have been made inside find room for upward of 1000 men, & with the great supply of tools we should turn out from 2½ to 3½ millions* [of dollars] *of work.*

The brothers, however, were never greedy for profits, and their degree of concern for their workers—indeed demonstrated repeatedly by the admirable tenure of individuals—was more sincere and effective than the 1872 strike suggested. R. Hoe & Company could easily have been considered a victim of the times for both the Civil War unrest in Boston and that in depressed New York shortly after.

Having been impecunious workmen themselves, the three Hoes knew what laboring to keep families fed was about. Richard had deplored the French aristocratic disdain of workingmen in 1849 and, like Conquest at Whitworth's, was not beyond removing his coat in the shop when his skills were needed. These democratic attitudes did not escape the men, as son followed father into the factory and promotions were made from within the ranks. The Hoes sensed that this collection of draftsmen, patternmakers, machinists, and erectors—among the most professional for their time—were not to be treated with the common contempt shown laborers and itinerants. Beyond a skill in one of the metallurgical trades the company required of its employees a yankee's ingenuity, an artist's flair, or a craftsman's patience to make these printing machines run with rare synchronization. The men did not disappoint their Colonel, and he rewarded them periodically with the little tokens which represented far more than their worth suggested:

> *Broome Street Works*
> *Jany 3rd 1865*
>
> *Messrs. R. Hoe & Co.*
> *Gentlemen,*
>
> *Nearly a Years experience has again proven to me that I have kind, thoughtful employers. I thank you for Fifty dollars just received, and trust that by my entire devotion to your interests I may be enabled more fully to express my feelings.*

My work for you has been a pleasure to me. I Feel that for the future it will not be less so.

I am, gentlemen,

Yours Gratefully
John Cochrane

Tucker remembered Cochrane as building, in 1843, a ship steering mechanism for Messrs. Woodhull and Minturn, in which case his service with Hoe was lengthy; he was no doubt related to Thomas Cochrane, the firm's hearty traveling agent. Such Christmas bonuses were frequent.

Other fringe benefits, perhaps unusual for the time, required some contrived enforcement:

> [*To Dry Dock, East Broadway & Battery Rail Road Company, 28 May 1874*] . . . *We are paying out weekly to our men from $10 to $12 in car fare for travel on your line, and are confident they, in many cases, put the money in their pockets and walk. To obviate this and get an equivalent for time for money expended, we suggest your supplying us with tickets to be used exclusively by our employee's.*† *We would buy them in lots of 1000.* . . .

Such paternalism, once begun, became overwhelming in just a few years. Although described to the trade as "broad comprehensiveness," there was something oppressive in nineteenth-century attempts to create a pleasant job environment:

> The Mutual Relief Society of the [Hoe] Works, which aims to relieve members disabled by sickness or otherwise and in various ways to promote the health and comfort of their families, also receives the generous moral and pecuniary support of the employers. The funds which the society keeps at its disposal are raised by membership fees of twenty-five to seventy-five cents, and three to eight cents weekly dues, according to the grade of fellowship; and a disabled fellow receives as many dollars a week as he pays cents a week. The fellows of these three grades are all employees, but there is another grade composed of members of the firm who pay membership fees of five dollars and weekly dues of one dollar each, and receive no benefits.
>
> A Board of Representatives, consisting of the fellows who are members of the firm and ten others who are employees and are elected annually, manages the society's affairs; the president and treasurer being always members of the firm; the vice-president and secretary, employees. There are standing committees on House Accommodations, Food and Fuel, and the Prevention of Sickness. These have done excellent work in securing reasonable rents in desirable neighborhoods, buying supplies at wholesale, and in guarding against defective drainage and ventilation and standing ready to enforce legal measures against offending landlords.[4]

In 1882 the society bought 400 tons of coal, 300 barrels of flour, and

8000 pounds of coffee, among other staples. The company felt well repaid for its efforts. Comprehensive employee welfare programs, however, were not simply a hedge against turnover and union agitation. They were in fact common around the industrialized world among the largest manufacturers whose employees, usually obliged to settle in one-industry "company" towns, required them out of geographic necessity.

With a plentiful labor market and no shortages of food and shelter, however, New York imposed no such requirement on the Hoes, who traditionally paid above-average wages as well. Although good business in the long run, Hoe's paternalistic practices were generous and genuine reflections of the Colonel's good will. Happily, his workers accepted them in this spirit, and almost effortlessly found themselves life-long company employees. Examples abounded: George C. Gill, general superintendent, an employee for 56 years; Charles Carpenter (who became a partner), 40 years; William Spalckhaver, chief engineer, 56 years; Otto Raabe, 49 years. There were many others.

R. Hoe & Company sometimes seemed indifferent and powerless to satisfy all its many workers—individually or collectively—whose strength in numbers was dimly perceived (by a few) to be right made by might. Conditions dictated that the union, whether weak or strong, had to demonstrate its ability to talk back to management. The strike was its most powerful language.

Machinists at the factory in 1898 sought determinedly to eliminate the firm's piecework system of calculating salaries, and at the same time announced their intention of obtaining a minimum scale of wages of $2.50 per day. Most workers, the company was reluctant to admit at the end of the century, were not privileged relatives like Mead and Seymour, thankful and devoted employees like Cochrane and Bowen, oblivious tinkerers like Tucker and Spalckhaver, nor even a new breed of self-oriented achievers like Charles Stelzle. In a time of unrest punctuated by muckraking journalism and Spanish-American War fever, the labor unions across the country had found an opportunity to protest long abused working conditions. At Hoe 600 machinists threatened to strike 19 July 1898, hoping the pattern-makers and the molders would join them.

Although spokesmen for the company's 1700 workers had frequently expressed their intentions, Charles Carpenter, speaking for the firm, doubted the disturbance would have much impact. Hoe had always paid slightly above prevailing rates, and there were few malcontents. There were few union members, too, although two powerful groups, the Amalgamated Society of Engineers and the International Association of Machinists, would have been happy to represent the men. Borrowing his senior partner's mantle of quiet arrogance, however,

Carpenter announced that Hoe had "decided to ignore entirely the existence of the labor organizations. . . ."[5]

With fatalistic determination—and some weary workers possibly visualizing a short holiday with their families—all but 50 nonunion machinists failed to report for work. Pattern-makers and molders did not strike; the factory remained open. Three weeks later the men were reminded of the solid and unassailable front of the impassive company; they had poor prospects and sinking morale. Samuel Gompers, himself a product of the lower East Side, appeared personally at Grand Street on behalf of the men, but to no avail. An officer told him directly that the firm would deal only with its own employees. Furious, Gompers and his committee stormed out the iron gate. All was lost. On 30 August—after six costly weeks—the strikers returned without significant advances to their cause.

Again in 1901, along with workers in many other industries, Hoe employees struck, this time for a nine-hour day. About 700 machinists went out, but now the company conceded. It agreed to put helpers, blacksmiths, and a few other groups on nine hours too, as their production was related to that of the machinists. But the firm reserved the right to continue hiring nonunion men, and still refused to bargain with non-company union officials. Exactly 10 years later the machinists would strike again, this time for an eight-hour day; now less munificient, the company (under new management) threatened to move the plant out of the city, thereby arousing many civic-minded people who appreciated the impact of Hoe operations on New York's economy. But the firm suddenly capitulated, instituting a two-year program to reach 8 hours by June 1913. A Hoe statement betrayed some bitterness:

> Although our strike is settled, it is clear that we must eventually quit New York. We may leave in two years or it may be ten or more. There is no immediate hurry. We are leisurely looking about for a new site. We would rather stay in the East, than go West, but that preference may not determine our final decision. . . .[6]

This discomforting news came ironically almost exactly 100 years after Robert Hoe had opened his Manhattan shop. The 1911 strike and the company's rash reaction were just the first episode of new policies following Robert Hoe's death in 1909. Without echoes of the old Colonel's cheerful whistling in the aisles, the firm's employee relations were quickly deteriorating.

By 1923 the company was paying a wage scale 12 percent higher than "peak" 1920 and 1921 rates, which had been inflated by wartime living expenses. Yet 1100 Hoe machinists (Eureka Lodge 434 of the now very powerful International Association of Machinists) in July had passed a resolution calling for a minimum increase to $1 per hour.

They had been receiving 85 cents hourly for a 44-hour week in a verbal agreement valid until May 1924. A three-week strike, through August 1923, was necessarily sanctioned by the Chicago-based parent union because of the impressive unanimity, although some of its officers were somewhat embarrassed by the greedy New Yorkers. The strike had strange repercussions.[7]

Throughout the United States 16 newspapers were affected; erection of Hoe presses in their offices was interrupted by sympathizing union erectors. The Boston *Sunday Advertiser* was effectively stymied from completing the installation of a giant multicolor machine it had bought six months before; a Tampa *Times* sextuple and the Philadelphia *Public Ledger* were similarly stopped. The company felt it had no choice but to fight—hard. Its resolution was uncompromising:

> *WHEREAS* the machinists and tool makers employed by R. Hoe and Co. quit work in a body on August 2nd, 1923, in violation of their agreement with this Company, and whereas it has been found upon investigation that the Company's machinists and tool makers, to whom two advances in wages have been given during the present year, are paid higher wages for similar labor than in any other of the leading machine shops in this section of the country and that the working conditions generally are fully as good as in any other similar shop, it is hereby
>
> *RESOLVED* that the Officers of this Company be and they are hereby instructed and directed as follows:
>
> *First:* To offer the employes of the Company who have suspended work the opportunity to return to work at the wage rates and conditions prevailing. . . .
>
> *Second:* That if the old employes do not return to work on or before that date, then the Officers of the Company are
>
> (a) To take whatever steps may be expedient to replace them with new men and thereafter operate the works under open-shop conditions. . . .
>
> (b) To pay such new employes a fair wage, similar to that paid by other manufacturers in this district. . . .
>
> (c) To institute a system for handling all questions of labor and working conditions directly with its employes, without the intervention of any non-workers.

To this the firm added a special notice to old employees:

> While the management would be sorry to lose the services of its old employes, there will be, if they refuse to return to work by the date set, no alternative. . . . No compromise is possible. . . .
>
> The Company has endeavored to treat all of its employes with the utmost fairness, paying the highest wages of any large machine shop. . . . Management hopes that the old employes, appreciating this and realizing that they have been poorly advised by outsiders who have no real interest in them individually or in the business which

has given them steady employment for so long, will return to their places before it is too late.[7]

Newspaperdom, a trade paper, was remarkably supportive:

> Let's all stand by the Hoe Company in its battle with unfair union labor. . . . It was well known by the professional union mouthpieces that the Hoe Company was far behind in filling its orders when the unwarranted strike was called, and their mustard-seed brains regarded this appeal as a weakening instead of conciliatory effort. . . .
>
> As a consequence of the pernicious influence exercised over the men by their well-fed union officers, the Hoe Company has gone about the work of converting its factory into an "open shop" without discrimination between union and non-union labor, and under this program the places of the machinists who broke their contract are being rapidly filled, while the rebels who have so ignorantly remained stubborn will be locked out for all time.
>
> Honest union labor does not violate sacred understandings except at the risk of losing all they have previously gained, and there can be little sympathy anywhere for the bone-heads who have allowed themselves to be led around by the nose by red agitators who have nothing to hazard, not even the comfort of their wives and families. . . .[8]

By December 1923 the firm had indeed replaced the strikers and become an open shop; the International Association of Machinists was completely enjoined from picketing or harassing employees. R. Hoe & Company, however, would have little time to enjoy its manpower situation.

In 1926 labor difficulties shifted to England, where a management consortium, Engineering Employers' Federation, voted to lock 900 striking employees—acting independently to win a £1 per week increase—out of the Hoe works. This retaliatory action involved seven unions, but upwards of 500,000 men in the country's entire engineering and allied industries were potentially affected—work in all of England could have stopped. The Federation promptly postponed its drastic action one week to give local union leaders a chance to think; various British engineering union officials thereupon urged the Hoe men to drop their "trifling grievances" (a handful of nonunion workers had been hired). The Hoe dissidents finally returned, peaceably, after nine weeks. The following month, when a foreman was dismissed, they walked out again, but settlement was quick.

Strikes and squabbles were always destructive of schedules and profits—they were meant to be. Perhaps no American firm—even one with hundreds of truly loyal old-timers—could have avoided the inflammatory activities of a growing labor movement. The company, from 1909 without the quiet but imperious rule of a Hoe, and from 1886 without the cheerfulness of the Colonel, saw no alternative but to continue high pay scales—a questionable policy some believed contributory to its eventual demise in just a few decades. Yet it was a small price to pay had other company activities remained fairly constant or improved. But from 1924, upon the public incorporation of the firm, stockholder and leadership disputes and litigation began to occupy management personnel to the almost complete exclusion of all else. In some ways the Colonel's attention to detail and his capacity to organize the factory into a superbly functioning entity, almost self-governing, made it seem to require no attention. Time would tell.

The technical nature of Hoe's machine-shop business had been complex almost from the beginning, and untrained men, however strong and willing, would never be adequate. At first an apprenticeship program brought the company a small supply of young lads anxious to work and bright enough to succeed. The traditional apprentice was a minor bound by law to a master who taught him an art or skill in exchange for labor—which was often trivial, frequently servile, and usually quite irrelevant to the promised technical instruction. But that was truer of nonindustrial crafts and skills; in factories where manpower was scarce and production was closely scheduled it was less likely to occur.

An early apprentice arrangement demonstrates Robert Hoe's attention to one such lad:

> [*To A. Freyre, Puerto Principe, Cuba, 25 November 1844*] . . . *Mr. Owen . . . introduced the young Mr. Isadore Amaiz to our acquaintances and we forthwith gave him such employment as was suitable to his years and experience. He is much younger than we had expected and for several years will not be able to be of much service to us. We shall hope however to make a first rate machinist & engineer of him more so as he appears to be industrious & disposed to learn.*
>
> *We have engaged board for him at a very respectable private family where he will be likely to form good habits. We shall keep an exact account of his expenses and at proper intervals render you a statement of them as well as an a/c of his earnings. It will afford us pleasure to receive your further orders and commands, and it would much gratify us to accept of your kind invitation to visit your island this winter and it is possible that one of us may do so. . . .*

The customer aspect of the situation was probably not a mitigating indication of what the firm would do for any novice. Just a year later Peter Hoe discussed a local boy with the same general concern:

> [*To E. Pustch, Duffield? 11 March 1845*] . . . *The boy was found in the streets selling papers. The* [*job*] *offer was made to him and he accepted without asking any questions* [*although he did not take the work, with a Hoe customer*]. *The lad has never been in our employ and at present* [*we*] *could not employ him having more boys than we know what to do with. . . .*

Similarly binding, perhaps, were requests from family:

East New York July 15th 1878

To Robert Hoe Esq.
My Dear Cousin,

My son John has now arrived at the age of 14 years and I am very anxious to place him with you to learn some mechanical trade. When my son John was about 2 years of age you promised me that you would do anything you could for him, when you was† visiting at "Cross River" [*Mead estate*].

I would consider myselfe† very much under obligations to you, if you would take your promise under consideration and set him to work at a trade in your establishment. I remain

Yours Respectfully
Mrs. E. H. Jennings
formerly Emaline Reynold[*s*]

Many relatives, whether close or distant, opted to serve some time with the company; not all showed great promise, nor wished to remain beyond exploratory positions. The Colonel knew all too well how to keep an incredible number of boys (nepotism notwithstanding) fully occupied.

About midcentury the Colonel's interest in the welfare and progress of the youth of the community had involved him increasingly in philanthropic work with boys at the New York House of Refuge, located on Randall's Island, in the East River. For the last three decades of his life he was a manager at that institution, of which his friend Benjamin H. Field, was President. When in 1882 the New York Senate was considering legislation to curtail the hiring of children under 16 years of age from orphanages and reformatories, the Colonel made a rare statement:

> I favor the retention of the present system of labor at the House of Refuge until a better plan can be devised. This has not been done. But it is said that the children should be taught trades instead of working for contractors. Where is the money to come from? It would cost thousands of dollars to get the machinery and materials for conducting a trade or trades, and for paying the skilled workmen that would be necessarily employed as keepers. This would be a large additional expense to the State, and it is hard enough to get money for the institution now, although I believe no institution is carried on more economically.
>
> But aside from the question of expense, the shortness of the time during which most of the children are at the House of Refuge would prevent their learning more than the rudiments of a trade. They come for two or three months, six months, a year perhaps, or more, and then their parents want them back, or a place is procured for them. And I may say that I have known of no instances where boys from the Refuge have failed to get work, notwithstanding the charge

that boys who have been doing one particular subdivision of work under a contractor here would be unable to get anything to do outside.

I consider it not of so much importance to try to teach the boys a trade in the short time they are confined in the Refuge as to impart to them habits of industry, punctuality and order, to train their hands to work and to inure them to just discipline. For this purpose useful labor is necessary, and it has been found, after a long experience, that the best results are obtained by letting the labor of the inmates to proper persons. The boys learn how to work, and it is their advantage, not the pecuniary returns to the Refuge, that we have constantly in view.

If we had the requisite money and the boys the requisite time for learning trades, it might be preferable, although I have always thought it unwise to allow the labor of reformatory institutions to enter into direct competition with that of the outside world. To a certain extent this happens now, it is true, although this letting-out of labor is different from turning the Refuge into a manufactory. It might be possible to find some light trades which the boys could carry on without a large expense, although this question of competition would arise. It has been proposed to have a hothouse at the Refuge, in which the boys could learn the care of plants and gardening, and the plan was also broached of getting some small farms on Long Island on which the boys could work. It is much better than teaching trades here to have the boys leave the institution as soon as possible and go on farms, where they can learn to become farmers.

Again, the advocates of this bill speak of the bad character and cruelty of the contractor's men in charge of the children. This is ridiculously untrue. First, none of the contractor's men can work in the institution without being examined and admitted by the superintendent, who has power summarily to discharge any of them. And, secondly, the boys are not under the charge of these overseers except for inspection of their work, and the regular attendants of the Refuge are constantly present. The superintendent, of course, refuses to admit any overseers who he thinks may exert a bad influence on the boys.

As regards charges of cruelty, I may say that any case of misbehavior is reported to the superintendent, who alone has power to order punishment. Among 900 children of this class of course there are some who occasionally deserve and are benefited by a whipping; but in all my experience I have known of only one instance in which excessive punishment was inflicted. That was through the carelessness of an assistant superintendent, who was promptly discharged.

The hours and the nature of the business are specified in the contract. The children themselves are often allowed to judge of the tasks set them. They are asked if the work will be too hard, and if they think so it is lightened. When they accept a task after trial this is their daily work, and by doing the work of seven or eight hours in five or six they gain so much more time for recreation. But the hours for

> rest and meals are never interfered with. None of the contractor's employes have any power to discipline or overwork the children. None of the very young children are allowed to work, unless at some slight task for a short time simply to occupy them.
>
> If the hiring of the labor as we conduct it now be prohibited by law, we have no means to purchase machinery and help to do work by the piece, or to buy materials to manufacture in the institution, nor have we the ability to dispose of the products of this labor profitably, should we take the risk of entering upon manufacturing. Therefore the inmates would be forced to remain idle and lose the instruction now obtained in their work, and the revenue of the Refuge would be reduced about $10,000 yearly. This amount it would be necessary to ask the Legislature to make up. This House of Refuge, I may say by the way, is the oldest chartered reformatory institution in the world, and it has always been conducted on philanthropic as well as economical methods. I wish Mr. Gerry and Mr. Letchworth, instead of constantly pulling such institutions to pieces, would come and join us and aid us in finding a better way of conducting the management.[9]

Elbridge T. Gerry was president of the American Society for the Prevention of Cruelty to Children; businessman Josiah Letchworth was also active in social affairs. By 1882, however, the Colonel himself had amassed considerable experience with youthful workers. His unique training program had long won the admiration of all who visited the Hoe establishment.

The usual apprentice system—training with board and keep in exchange for labor—at first had served Hoe well enough, Stephen Tucker himself a notable product of that arrangement. When it became impossible to house great numbers of boys with any pretense of family discipline, financial substitutions for food and lodging probably followed. Hoe apprenticeship training—clearly recognized as something of value with its promise of full employment—insured a continual supply of steady hands. But the intricacies of the shop and foundry had multiplied almost beyond what conventional employers could teach, and the easy way out was to hire those already skilled. A trade journal as late as 1906 still lamented the practice:

> It is true that many American factories have installed mechanics of foreign birth and training in responsible positions, but it is humiliating to the American national spirit that native industries should not be manned by native workmen. . . .[10]

A few Hoe employees brought their diversified training to Hoe and were indeed capable of responsible positions.

Louis A. Schmidt was such a European trained machinist. Born 1855 in Germany, he came to America 15 years later and worked six years in various machine shops. He returned to Germany 1876, attended a technical school in Chemnitz (now Karl-Marx-Stadt) and traveled throughout Germany, Russia, Holland, Denmark, and France, studying machine methods. In 1886 he returned to the United States and came to Hoe, where he remained. Schmidt proudly described his work from then on:

> I have, of course, grown with the Hoe Company since that time, and have at present full charge of the construction of the newspaper presses. Naturally I have invented a number of improvements and patented them, and these are in use to-day on newspaper presses. I have constructed roll [fed] presses with a capacity of ten thousand eight-page papers up to octuple presses with a capacity of two hundred and eighty-eight thousand eight-page papers. . . .[11]

Inland Printer in 1917 soon had him writing a monthly column on newspaper press design and construction.

Like Schmidt, John M. W. Jordan came to the United States in 1878, when he was 17. He joined Hoe five years later and remained until his death in 1921—nearly forty years with the company. Jordan specialized in the electrotyping and stereotyping branch of the business, where he was recognized as an expert in the field.

But these men, largely self-trained, were few in number. Some proper technical education in America had to be devised if large-scale manufacturing operations were to continue. Yet the public schools, even in metropolitan New York, aspired to no more than rudimentary coverage of reading, writing and arithmetic, at which point children were often thrust into menial jobs, their education complete. Since such boys were already working full time to support their families, someone in industry had to show the way to make them technically trained lads. Although the factory itself was considered the best teacher, additional technical night-school was an obvious necessity. The Hoes had faced all this decades before.

On his first trip to England in 1838 the Colonel had already developed some progressive ideas about education. Only 26 and with his own adolescence fresh in his mind, he clearly strove to correct deficiencies he had felt in his own education. His journey diary expounded:

> *August 18th* [1838] *Evening. Have had a conversation with Mr. Cha*[s] *Sanderson respecting matters & things in general & spoke to him of a plan of a school which I have had in my mind for several years which would educate persons according to the branches they intend to follow.*
>
> *Now my plan has been but vague & I have endeavored to obtain as much information on the subject as possible.*
>
> *I intend to have applied the principle to Apprentices & to have made a night school for them where they might be able not only to learn to write & the common rules of arithmetic but all that is in any*

way connected with the particular branch of business they are engaged in; but, I see no reason why this system should not be introduced more fully & extensively into our Public Schools. Why can not these institutions be made such that it will be an honor to be educated in a public school that the rich will send their children[?]

If a system of this kind were adopted instead of the flimsy plan which is now in operation, taking care that a boy should not be forced into any thing that was repulsive but that the best of his intellect, that his "natural talent" was consulted, what a difference there would very soon be among our mechanics who are now in a better condition than the same class in any other country. In fact every branch of scientific knowledge would increase & it would promote the virtue & Morality of all concerned, for where the mind is employed in good, evil vanishes. Since I have been in England I have understood that there is a school of this kind in Manchester & I intend to visit it and get a history of its size & progress & the plan on which it is conducted.

Mr. S told me he proposed to the Austrian Government when he was in Vienna the Establishment of a school on the following plan—

That boys intended for mechanics or for scientific men who had received a good foundation in the first branches of Education & had arrived at the age of 10 years should be taught pure mathematics until they arrived at the age of 12 years, then they should be taught the mixed sciences till they arrived to the age of 17 or 18, & should then be brought to the study of that particular branch which they intended to pursue—say for instance they were designed for manufacturing iron. There should be shown them specimens of iron ore that they should be taught how to determine the existence of iron in what soil it was to be found—that there should be models of the different kinds of furnaces for fusing & casting melting &c &c & that they should be made competent to give plans & superintend the building of these furnaces & their working after they were built & so on in every other branch of business.[12]

Somehow he must have known he would get a chance to try.

About 1849 Richard Hoe determined that more must be done by the company to attract and keep a reliable force of bright, capable workers. He had already effected policies to make the shops modern, relatively safe, and (not incidentally) comfortably warm, and he paid well for all types of skilled labor. Hoe now instituted a free evening technical school for apprentices, where instruction likely to be useful both to the individual and to the company would be given. The Colonel attended to the classes personally at first, but his other cares soon kept him away.

Within three years Hoe had engaged two instructors for teaching mathematics, science, drafting, French, and English, and a visiting professor who discoursed on Experimental Philosophy. *Graham's Magazine* was impressed:

> Every one of their many apprentices is required to give a punctual attendance at the school, which is also open to such adult members of the establishment as choose to attend. . . . [The Philosophy lectures] were attended by nearly all the workmen as well as the apprentices.[13]

Another surprising instructor turned out to be the Colonel's own brother-in-law, Moses W. Dodd, who, upon retirement from publishing, still felt lively enough to handle the squirming boys. The program soon became a model of good industrial relations as well as a community benefit.

Tucker, recollecting the scant education with which his own career began, could not help but find it methodical, stimulating, and generous:

> Evening school for the firm's apprentices, six months in the year, has been established for many years, and the good results have been so encouraging that great interest is taken in it. The school room is well supplied by the firm with suitable books and appliances and competent teachers are employed. Each boy attends two evenings a week and quits those evenings one hour earlier. They are instructed in the various branches of a common English education according to their needs and capabilities, and those who are sufficiently advanced are also taught mathematics and mechanical drawing. Many of the boys make rapid progress, and some of them in after years express the great advantage they have since derived from their apprenticeship studies.[14]

Countless lads from far and near hoped for an opportunity:

> *Apr 6th 1858*
>
> *Mr. Hoe Dear Sir*
> *I hardly know how to address you as I have never done so before but have heard of you frequently and have seen your name connected with printing presses and other Machinery. I would like to enquire if you would like an apprentice. I want to learn the Machinist trade and Father told me to write to you and see what you said. I would like to have you let me know what your terms is[†] and whether you would like to have me or not. I am 16 yrs of age, will be 17 the 1st of Sept. Direct yours to*
>
> *Havana, Schuyler Co., N.Y.*
> *Henry E. Rowley*

The school's reputation had spread even to quarters where honest work was not generally admired:

> *New York March 20th 1855*
>
> *Dear Sir,*
> *Mr. Russel Glasier of the 7th ward, a Brother of Mr. G. G. Glasier, late Commissioner of Streets and Lamps, has a son a smart intelligent boy about 17 years of age whom he would like to have learn the profession of Machinist & for that purpose would be pleased to have him enter your establishment under the rules which govern therein. I*

should not trouble you with this matter but that Mr. Glasier is for the most part of his time confined to his house by indisposition and is therefore unable to give the matter his attention personally. I would take the success of his application as a personal favor to

Yours very truly
W. M. Tweed

Infrequently too there is notice of a termination from the program—and the job. An Albany mother wrote (22 October 1877) asking if her son's dismissal "is on account of any misconduct on his part. . . . He has never had a father to govern him, & I condemn myself for a foolish indulgence; and now he has to be taught by strangers that he must submit. . . ." Some boys the Colonel felt to be totally intractable:

Feby 1st 1876

Dear Sir,
In front of our premises No. 3 Columbia St., much to the annoyance of our tenants, a crowd of boys and loafers congregate every evening from about seven to ten o'clock. If you will kindly give this attention and see that the crowd is dispersed you will much oblige

Yours very truly
R. Hoe & Co.

To Capt. 13th Precinct
178 Delancy St.

"It is impossible," said *Harper's Weekly* in 1876, "to estimate the influence which such a school may have had in developing that skill without which the production of such marvelously perfect specimens of mechanical ingenuity . . . would be impossible." The magazine agreed with the Colonel's fond dictum, "The diffusion of knowledge among the working classes—it makes the man a better mechanic, and the mechanic a better man."[15]

Toward the end of the century the training program, acknowledged the first of its kind in the country, had put about 500 apprentices into the Hoe shop—a considerable workforce at reduced wages (1500 skilled workmen were also employed). The discipline became a little more rigorous:

> For the benefit of the latter [apprentices] the proprietors furnish an evening school during the winter months, in which mechanical drawing is taught two evenings in a week and mathematics one evening a week to each boy. Every apprentice is compelled to attend this school. No charge is made for their tuition; on the contrary, each apprentice receives a good lunch, consisting of sandwiches and coffee, after leaving work and before going into the class. The head teacher in this evening school devotes his whole time to this work and in visiting the sick and poor among the workmen.[16]

(About 1875 it had become awkward for the boys to return to the plant after supper, and the hours were changed to late afternoon.)

Yet so rushed was the company for complex printing machines, a few years later, in 1902, the firm had 2000 skilled workers and 300 apprentices. The five-year program was now formally structured:

> Almost the only manufactory about New-York which follows out old apprenticeship lines is the big one in Grand-st. which makes printing presses. The system has been in vogue there for thirty-five or forty years, and with the growth of the factory the number of apprentices has increased until there are now three hundred of them. Even more striking is the apprentice school, in which the boys are taught the theoretical side of their work. Presided over by a headmaster and four assistants, this school does in two hours what other schools would need a full day to accomplish. . . .
>
> So great is the interest that more than half [of the novices] . . . are following regular courses. Boys who desire to become apprentices come to the factory with their parents, who are required to sign an "apprentice engagement," binding the boy to serve for five years. . . . The boy [between 16 and 18, with at least four years' schooling] agrees to perform his duties with fidelity . . . and to attend the night school. The company agrees to pay $2.50 a week for the first six months, and to increase the wages yearly until the apprentice receives $7 a week for his fifth year work . . . [and] to give him every opportunity to learn his trade by moving him from shop to shop. . . . An effort is made to keep the boys for the full . . . term, and then to give them employment in the factory at full wages. . . . The rules which govern the apprentices are rather severe. . . .
>
> At 5 o'clock a large bell rings out the closing hour. The workmen hurry into street clothes and leave the works. The apprentices rush off to the washroom, and in five minutes are in line before the storage building, where each evening the company provides a luncheon of milk and of meat sandwiches. There is no doubt about their hunger. At 5:30 they go to their desks in the several schoolrooms and for an hour and a half work over their books.
>
> There are seven classes, and the average boy is expected to complete the work in each of them in four months. . . . A mark of at least 65 per cent is required in all examinations. . . . [Subjects taught are] English, mathematics, mechanics, free hand and mechanical drawing, . . . and few . . . fail to pass. The practical work which they do each day in the shops is of great help to them. The company keeps close watch of the apprentices, and has a man, an old mechanic, who visits their homes when they are sick. . . .
>
> A couple of years ago the boys made trouble for some of their masters. They threatened to "chuck 'em out the window," and for a night or two had things their own way. Then a new headmaster appeared —a Columbia graduate. He is a big man and an athlete of gridiron fame. It was necessary for him to stoop to get through the schoolroom door, and he looked strong enough to knock apprentices' heads together at will. There has been no trouble since.[17]

On more than one occasion a tour of duty as Hoe instructor was ample qualification for a public school teaching career.

The *New York Tribune* reporter stayed for the 1902 commencement, held at a nearby church. A score of apprentices, with their proud families looking on, received diplomas. At another time Charles Carpenter addressed the graduates and Robert Hoe (IV) presented certificates, and a gold watch for particular excellence. Gifted boys were encouraged to continue their technical training at advanced schools like Cooper Union.

By 1920 the remarkable Hoe institution was furnishing the company's "manufacturing administration force" with 90 percent of its people, and about half the "high-grade" outside Hoe men—erectors and repairmen—were also apprentice school graduates. A ratio of one apprentice to every five journeymen was fixed by the union.

But the colossal Hoe training program, turning out perhaps 2000 in all, was scarcely considered a philanthropic gesture. The boys themselves never thought it so, although some actually acknowledged their frank appreciation. Hoe had had no other recourse to obtain skilled workers in such numbers; the city of New York was unable to furnish such technical education until the 1920's, when a public highschool for printing apprentices was founded. And as Gompers and other union advocates described the dignity of labor, Hoe employees were not more the "Company men" for their in-house education. As one century turned to another labor threw off its humble mien and considered work a right, not a privilege. A few intrepid Hoe employees—new to America and unacquainted with the labor concept—experienced this transformation with rare acuity.

The lower East Side of New York in the last quarter of the 1800's was a miniature urban enclave so intense and self-centered it seemed to shut out the rest of the city. Yankees, Irish, Germans, Bohemians, Russians, Italians, Greeks—all put their "foreign" mark unmistakably on the area only to find the fickle community did not claim to belong to any of them. During its proud occupancy of the East Side R. Hoe & Company for decades had remained prosperous, haughty, guarded, unaffected, indifferent to the changes in tenement complexion around it, much as the medieval castle loomed above the manorial village to which it deigned to owe nothing. It was a bastion of free enterprise in America, to be sure, but for a new wave of immigrants nestled in its shadow it did not always symbolize—as it did for Robert Hoe 70 years before—what a determined man could do with only his hands and a dream.

Otto L. Raabe (1861-1922) in 1873 graduated with honors from the old Fifth Street grammar school not far from the Grand Street shop. As was its custom with promising neighborhood boys, Hoe offered Raabe employment, starting him in the shipping department and then the wareroom. He was soon moved through several offices and departmental positions (and probably received the apprenticeship training along the way). He demonstrated a great interest in mechanical matters, and, thanks to long personal study and application, he developed an "almost unequalled" experience in printing and platemaking machinery. He had a number of patents to his credit.

For many years in charge of sales of electrotyping and stereotyping equipment, Raabe shortly before his death was elected assistant secretary of the company. He was also chairman of the engineering committee, particularly concerned with development of intaglio (engraving) printing processes. Raabe (despite a period with Goss) had been longer with R. Hoe & Company than any other officer. He typified the uncomplaining, loyal worker the company had always sought when it accepted untrained, uneducated—but bright—youths from the community. Another local lad was expected to do as well.

Charles Stelzle's German parents, proud and industrious, had refused charity and fought desperately to exist on a tiny sewing and washing income. Shabby tenement living, with the noise of street brawls, domestic quarrels, and boisterous drunks to accompany a supper of stale rolls sprinkled with salt, shaped many such young lives surprisingly well for later careers and families. Stelzle started work early, his first job (afternoons, 50 cents a week) stripping tobacco leaves in a cigar factory. He then hawked newspapers: "I stood on the corner of Fifth Street and Avenue B, and yelled at the top of my voice; 'Pepper—pepper Daily Noos!'" He had little trouble selling the four-page evening sheet, widely read by East Side laborers.

During school vacations Stelzle worked in a restaurant, and during the Christmas holidays he peddled oranges and candles. About the age of 12 he was empowered to join the public library; reading for him reached passionate proportions. But Stelzle had already left school by then to work full time. Boyhood was not a long play period for the poor.

For about three years Stelzle was employed in an artificial flower factory, where two of his sisters were also engaged. Although happily occupied, he soon began to entertain visions of more meaningful employment:

> Very early in my life I had an ambition to run a newspaper. I used to sit on the end of the docks of the East River and build castles in the air; how I would become a newspaper editor in a small city and shape public opinion, and finally, of course, run the whole town.

One of Stelzle's cousins, a pressman for the New York *Daily News,* had served an apprenticeship at R. Hoe & Company, and offered to get Charles enrolled. Thinking press manufacture close enough to a newspaper career, Stelzle heartily agreed, only to find he was still nine months too young to begin. The company nevertheless put him in the warerooms until he reached the age of 16. With some trepidation he had given up the flower factory paying $6 per week for Hoe's weekly $2 for four years and $7 thereafter, in order to learn a machinist's trade. His mother, foresighted enough to see the greater value behind the tiny salary, encouraged him. The coveted moment arrived in June 1885:

> It was my good fortune to spend my apprenticeship days in one of the greatest shops in America, R. Hoe & Company, printing-press manufacturers. It had a splendid organization and the highest possible standards of workmanship. . . . No man who ever wore a uniform did so with greater pride than when I put on my first pair of blue overalls as a machinist's apprentice. To me they were a badge of honor. I had achieved a distinction to which I had been looking forward for years.

Stelzle welcomed the Hoe apprenticeship training program as well as the good meals that went with it. To his surprise, his fellow students one day started a "strike" to protest the food prepared by the company kitchen. "In a very few minutes," Stelzle reported, "it was proved the meat we were eating was the same cut and quality as that served to our millionaire bosses in their private dining-room at noon." The boys were not satisfied, however, and the company agreed to give them each a check which would permit them to buy whatever food they preferred.

The 16-year-old lad had much to learn of business methods and employee relations, but he wisely sought the opportunity to do so:

> I was soon made an assistant to the secretary of the mutual benefit society, and in this capacity I visited every part of the shop, becoming acquainted with the men in the various departments. It was a great surprise to me that so few American workmen, unlike the English and German machinists in the shop, knew how to read a drawing or work to a scale. . . .

Stelzle quickly realized it would take very little extra effort to rise above an ordinary machinist's job. In addition to the company's modest apprenticeship requirements, he bought a $30 set—then two months' wages—of technical engineering books, which he studied faithfully, as well as other tomes on construction work and machine shop practice. He took pains to keep his foreman appraised of this outside reading, and at the same time boldly tried to set some records on the job for speed and quality. The test of his skill and its awful, inevitable, aftermath followed:

> One day word came that the big impression cylinder of a rotary press that had been sent to Australia had proved to be defective. No time was to be lost in making a new one. It was my job to cut two one-

inch key-ways on the four-inch shaft, besides two long one-inch slots on the cylinder itself. It was a piece-work job: six dollars for what was ordinarily a twelve-hour job. I finished it in five hours, using some of the clever "kinks" I had evolved in the course of my studies, principally on tool shapes and cutting qualities of various kinds of steel.

It was true that I actually risked my life standing over the top of the job, which was done on a big planing machine. The cylinder itself weighed about three tons. If the belt had broken, I might have been cut in two. The work was done on Saturday night, too, from seven o'clock until midnight, when I was extremely fatigued.

How that steel did curl off as I dug the tool into it, feeding it by hand! Somehow, when a machine acts that way you feel like petting it, as you would a horse. It becomes almost human, part of the man who is running it.

When I came into the shop on Monday morning, I was highly elated. Everybody in the department was watching the progress of that cylinder. There was a big bonus to the firm if it got through in record time, and I felt that I had given it a big boost. The superintendent was greatly pleased, and told me so. He was particularly gratified because I, though still an apprentice, had made the mechanics in the shop look rather sheepish.

But I was compelled to live and work with those mechanics, and they soon showed me that I wasn't going to get away with this record-breaking business so easily, especially on a piece-work job.

"You think you're damned smart, don't you?" my nearest fellow-worker fired at me, after the superintendent got out of the way.

"Sure I do," I replied, with a grin.

"Well, you won't think so long," he said to me. "Wait until those damned efficiency experts in the office hear about it, and they'll cut your piece-work price so that you'll have to kill yourself to make a decent week's wages."

This hadn't occurred to me, but it probably wouldn't have made any difference if it had, and I comforted myself with the thought that these experts would understand that the job had been turned out under very great pressure, which could not be sustained very long at a time. But the men were sore at me, and all during the day they showed it by throwing hardwood driving blocks in my direction and bunches of oily waste at my head when I wasn't looking. I saw that this was no time for arguing, so I said nothing, but tried as good-naturedly as I could to stand their insults and injuries. Wasn't I a free man? I consoled myself. Who had a right to dictate to me how much work I should turn out?

But in a few weeks a new piece-work schedule came down from the office. The price of the cylinder had been cut thirty per cent! Then the men gave me the horse laugh. How humiliated I felt! What was the use? These fellows were right. It didn't pay to hustle. You get just as much money by not rushing.

That's the way I argued for a while; but I soon got over it, and went back to breaking records, or trying to. And it paid in the end. For before I had finished my apprenticeship I was promised that some day I would be given the foremanship of that department. I often wondered, however, what would become of the ordinary workman if high-pressure methods should prevail in all industries. What about the man who couldn't stand the pace? Ordinarily, that cylinder was a twelve-hour job. I had made it a five-hour job through extraordinary methods which the average workman knew nothing about.

Stelzle learned quickly about many other aspects of contemporary work ethics and human nature:

I had heard a great deal about the snobbishness of the bosses and of the rich in general. And I saw some evidences of it. There were some men in the office who undoubtedly looked upon the workers in the shop as an inferior order of human beings. Even the clerks regarded the shop men with contempt. This attitude resulted in a bitter hatred of the men toward whatever came from the office, men or messages. A notice posted upon the bulletin board was regarded with the greatest suspicion. "I wonder what those fellows have got up their sleeves now," was a common comment. The workmen felt that nothing good could come from behind that glass door that led into the office. And if perchance a worker in the shop should graduate into the office he was considered a renegade, a traitor to his class. And, as he knew the "tricks" of the shop, it was felt that he would soon turn out to be a common spy.

But strangely, snobbishness was more prevalent among the men themselves than it was between the office men and the shop workers; and with less excuse on the part of the men in the shop. For example, at lunch time the skilled mechanics would not think of permitting the laborers to eat their sandwiches and drink their beer in the same corner in which they ate. The draughtsmen considered themselves much superior to the pattern-makers, the pattern-makers thought they were better than the machinists, the machinists looked down upon the tinsmiths, and so it went on. There were at least half a dozen different grades of "society" among the men in the shop. . . .

Impressionable personalities were also a source of wonderment:

There was a big Yankee in my department who was probably the most unpopular man in the place. He not only always stood with the bosses whenever controversy arose between the men and the office, but he was always looking out for the bosses' interest in the routine of the daily work. And this, of course, was unpardonable. For instance, he was always going about the shop turning out the gas-jets which thoughtless workmen had left burning.

But there was another habit for which they hated him most cordially. He always came into the shop at 6:30—we began at seven in those days and worked until six—and he filled his oil can and trimmed his oil lamp, grinding his tools, and making ready all that he could before seven o'clock. Usually the engine started at about 6:45, so as to get a good start before the strain of hundreds of machines was placed upon it. And the Yankee mechanic invariably threw on the belt of his machine as soon as the engine was fairly under way, amid howls of derision from all over the shop. He was perfectly oblivious of it all. He just wanted to be industrious and economical. But the men thought he was an ordinary "sucker," although that was the mildest term which they applied to him.

Needless to say, he was not a member of the union. Indeed, he was the only man who worked when the men went out on strike; and I remember that he was badly beaten up one day as he went out at noon hour to buy his can of beer.

It is really a question whether a man of this general type is a useful man in the average shop. Some of his practices were undoubtedly commendable, but, on the whole, his conduct bred discord and hatred because it was unreasonable to expect that his fellows in the shop could or would follow his thrifty example. . . .

Stelzle gradually appreciated the workers' uneasiness and mistrust:

The average workingman is more afraid of being out of a job than he is of going to hell. The possibility of losing my job in the Hoe press works constantly hung over me, although there was no particular cause for me to have feared that catastrophe. Nevertheless the feeling that for any one of a number of reasons the boss could fire me if he felt so disposed made me almost bitter toward him. Furthermore, any one of a number of minor bosses could have fired me if they had really wanted to.[18]

One day the dreadful thing happened. Times had become very dull. The men were being laid off in relays, and some of them never came back to the machine shop. It was a good time to get rid of undesirable hands.

I hated that designation. Why not "men"? I wondered whether my millionaire boss ever thought of them in that way. Apparently nobody ever knew. . . .

Inevitably the end came for Stelzle, whose personal approach to work and his feeling for shop justice and compassion indicated no common laborer.

A short time before I was laid off, on a cold winter morning the janitor in the office failed to appear, and I was sent into the big boss's private office to build a grate fire. I had never seen a grate before. It was so big that I could almost get inside of it. Finally, just as I had got the fire started, I heard a step behind me, and there was the big boss, with a smile on his face, watching me.

"It's rather cold this morning, isn't it?" he said in a high-pitched, though not unpleasant voice.

He hadn't a very good reputation in the shop. He rarely was seen there, and he never stopped to speak to anybody when he did rush through any of the departments.

I'm afraid that I was so surprised that he spoke to me at all that I didn't have sense enough to agree with him about the weather, but the chill in my own heart toward him disappeared.

"You fellows are all wrong about the boss," I declared with vigor, when I returned to the shop. And I began praising him because of his friendliness toward me, perhaps beyond what was his due, considering the slight evidence he had given of what I vehemently said was his natural disposition. But, what was most important, my own mind was changed, because, for once, I had had a real human contact with the man who had heretofore stood in my mind for everything that was autocratic and despotic.

And so when I was laid off I took it as one of the things which just naturally comes into a workingman's life—particularly as I lost only two weeks' time.

During those two weeks I spent my time wandering about the city interviewing the bosses of other machine shops, asking for a job, but hoping that I would not find one, because I liked my own job best of all. I wanted to see what it was like really to hunt a job, because I had long been familiar with the bitterness which filled men's hearts when they were out of work.

Scores of other men haunted the gates of the shops that I visited, sometimes literally hundreds of them. I am afraid that some of the old resentment that the boss could hold my job in the hollow of his hand came back, aggravated by the treatment given me by many of the foremen or employment managers, who not only spoke gruffly but frequently swore at men who, because of their hunger or need of money for their families, were too aggressive in trying to find something to do. . . .

There were frequent occasions when Stelzle fell prey to other propagandizing influences in the shop:

To strike is to many employers of labor the unpardonable sin. I recall in the main office of the shop in which I worked a photograph. Beneath the picture, in the narrow margin of the card-mount, was this legend:

HARRY JONES
Oldest employee in the Works.
Born in Wales Sept. 4, 1827
Entered our employ Oct. 1, 1843
HE NEVER WENT OUT ON STRIKE

That was undoubtedly a fine record. To have been with the same concern so many years meant a great deal both to the firm and to the workman. But I confess I never could see the supreme virtue of remaining at work when a strike was justifiable—and there were undoubtedly occasions when it became necessary to protest in this fashion in order to better the conditions of workingmen, unless, of course, some fair means had been agreed upon whereby industrial differences could be arbitrated.

Reminiscing about apprentice days, Stelzle revealed what a powerful —if undisciplined—force were those boys when they were spoiling for trouble:

> Anyway, a psychological situation is developed during an industrial conflict which inevitably gets practically every workingman concerned. One day the conductors and drivers of the old Grand Street line went out on strike. The utmost confusion reigned. According to the law, the company was compelled to run at least one street car a day over the line of its franchise. The strikers were determined that this should not be done. They piled trucks upon the tracks and in every way possible obstructed the course of the street car. In spite of police protection furnished by the city, every street car that attempted to make the trip had its windows smashed. In several cases the street cars themselves were turned upside down and laid upon the tracks.
>
> This, of course, was highly exciting to us apprentices. One day all the apprentices in the shop resolved to take the afternoon off to see the fun. They followed the street car across the city, trailing behind the mob of strikers. I recall, once when the strikers were chased by a large force of policemen, that I was thrown down and my thumb was mashed by the heel of a husky driver. But I continued with the rest of the apprentices until we saw that street car through to its finish.
>
> We had been warned by the bosses that if we left the shop that afternoon we would all be discharged, but the fever had got us; we cared nothing about being discharged, under the circumstances. Actually, none of us lost his job. When the superintendent was asked why he did not make good his threat, he simply smiled. He was a good psychologist. He understood that it was simply impossible for the boys to have remained at their machines overlooking Grand Street and see the mob pass by, with all the thrill of a baseball game, a fight, and a struggle to get more wages combined calling them to come on and have a part in it all.

Stelzle was, after about eight happy years, nevertheless reluctant to remain a machinist; his restlessness expressed itself in a most unorthodox way:

> I have rarely met a workingman who envied the rich man his wealth; but I have known many of them who coveted his leisure, the chance to see things and to enjoy them. During the last few years that I worked in the machine shop I ran a planing-machine down in the basement underneath the sidewalk. The only daylight that I saw came through a tiny window, except when occasionally the back door leading into the yard was opened. I worked ten hours a day, with no half-holiday on Saturday.
>
> Living in East New York, which was then the extreme end of Brooklyn, across the river, I was compelled to get up at 5:30 in the morning, and I never reached home until about 7:30. Although I thoroughly enjoyed my job as a machinist and truly was eager to get

back at it in the morning, because there was a constant thrill about the new jobs that came along and because of my eagerness to excel, I greatly missed the opportunity to visit places in and about New York which I wanted so much to see.

So one day I broached the subject to the boss and told him that I wanted two afternoons a week off. I realized that this was an unusual request to make, but I saw no other way of getting the time unless I left the job. Finally he agreed. Furthermore, he astonished me by saying that he would raise my wages sufficiently so that there would be no actual money loss on account of the time taken off those two days.

When the old-fashioned bell rang at twelve o'clock on those days and I had pulled off my overalls, put them into my tool chest, and quickly eaten my lunch of bologna sausage, rye bread, and three or four apples, I followed my prearranged schedule, visiting art galleries or museums, going to lectures, or just roaming about the city. It was a grand and glorious feeling to realize that I actually had some time which was my own, so that I could go where I pleased.

Stelzle's final recollections of his association with R. Hoe & Company were of long personal satisfaction:

The only diploma which I have ever received from any institution is a finely engraved apprenticeship certificate, signed by the head of the firm, indicating the departments of work in which I was employed during my apprenticeship and giving my personal rating as "superior workman." It holds a prominent place in my office.

It states that I spent five years in my term of apprenticeship, from June 15, 1885, to June 15, 1890. Those years, together with my three years as a journeyman, constituted a background of education in the day-by-day problems of workingmen which has since been most valuable to me. That big machine shop actually became my training school, my university, my seminary. I am rather proud of the title "superior workman." I think I earned it fairly. Whatever else I may have done since the day I received my diploma as a graduate machinist has been based upon the broad education received among two thousand highly skilled workmen, among whom there was just as great a variety of opinion and human experience as one can find almost anywhere else in the world. . . .

Charles Stelzle (1869-1941) became a Presbyterian minister through Chicago's Moody Institute—even though he was promised the impressive position of assistant foreman had he stayed at R. Hoe & Company. He returned to New York to found a "Labor Temple" on 14th Street, and gave special counsel and assistance to countless workers and their unions. He arbitrated labor disputes, especially in the graphic arts industry, wrote copiously, and lectured frequently on the role of the church in the labor community. His particular camaraderie with the common workingman had begun at Hoe, although not, as might be expected, as an anti-establishment agitator:

> I first became interested in the labor question in a practical way when I went out on strike with about five hundred men. I learned some strange things during that strike; that is, they were strange to me then. I had always supposed that my employer was a Christian gentleman. I discovered that he was a fiend incarnate. At any rate, the walking delegate said that he was. . . .

Stelzle's simple experience lighting a winter fire for the big boss—Robert Hoe III—had changed his opinion. Similarly, his status as apprentice was not sufficient to make him think like one. He observed:

> We apprentices were ungrateful fellows. Our employers made provision in an evening school for the technical education of every boy during the five years of his apprenticeship, providing a wholesome supper on every evening that he attended school, without charge to the apprentice. How we fretted and fumed all through the course! How we complained about the meals! The coffee was made of chicory, we declared, and the corned beef had been driven before a wagon in palmier days. We did not know at that time that it was the same kind of corned beef and coffee which was being served to our employers at their midday meal in their private dining-room. Perhaps if we had known it we would have been disgusted because of the "miserliness" of the millionaires who ate the same kind of meat that we ate. I fear that some of us to this day misinterpret the motives of the men who were anxious to turn out the best mechanics in the country.[19]

Stelzle's natural sympathy with the workingman's problems—despite his genuine gratitude to the company for its training program (and liberal treatment of his personal idiosyncracies)—made him a perceptive employee and a dedicated social worker. Others soon came along who, with the same sympathies channeled their activities not into religion but into the dramatic new labor movement in America.

One employee—not a participant in the apprentice program—remembered his days at the Hoe works quite differently.

> Through a newspaper "ad" I succeeded in landing a job in the typecase department of the Robert† Hoe Company (on Cliff or Gold Street—I don't remember which). My task was to nail the backs of typecases to the frames. The work was not difficult and the pay was four dollars and fifty cents per week to start with—not a royal wage for a fifty-eight hour week, but bananas were then very cheap and so were bread and cheese. I felt quite happy at my work the first day. On the second day, however, I was compelled to consider the social atmosphere of the shop. There were forty or fifty young workers of both sexes on my floor. Their manners and speech were what is called "free and easy." To me they seemed too free and easy. As my knowledge of the English language was rather scant, I had to consult my pocket dictionary very frequently. Many of the words used by the

> girls I could not find in my lexicon at all, and some which I did find were too shocking for my modesty to repeat. . . .

Gregory Weinstein, later a civic worker and publisher, had, like so many Jews in the 1880's and 1890's, fled from the Russian pogroms to enjoy America's promises of opportunity and equality. He did not bargain for prejudice:

> My failure to appreciate the "humorous" remarks thrown at me by my side-partners did not improve my status in the shop. I was soon made to feel that a "greenhorn" was a pariah, with no human rights anyone was bound to respect. Every kind of annoyance and mental torture was heaped upon me:—my frugal lunch package would disappear; a shower of fine nails would come down on my head from nowhere; my work would be spoiled. The climax was reached when I discovered, one afternoon, just before passing the time-keeper, that my coat pockets were filled with case nails. Exasperated by this mean attempt to expose me as a nail stealer, I took out the nails by the handfuls, and showered them on the jeering youngsters, and I quit the shop in none too leisurely a manner. . . .[20]

This microcosm of antisemitic feeling inside the Hoe shop was a taste of what transpired daily in the boisterous East Side streets as well. Yet few cared—or even saw the problem.

Samuel Gompers, an old friend of Weinstein's, along with his concern for the poorer workers of New York, had become interested in the plight of the Russian immigrants since their earliest arrival. "Those wretched, hunted people," he wrote, "seemed for a while just as likely to be exploited in the New World as in the Old."[21] Gompers and Weinstein were instrumental in bringing many Russian Jews into the U.S. labor movement, but they were almost powerless to control the community's social reaction to their efforts.

R. Hoe & Company, without which the country's newspaper industry could not meet its obligations to millions of readers, without which the nation as a practicing, living democracy could not discharge its decision-making tasks for millions of voters, did not see the muckraking pages of its clients nor the changing faces of its East Side neighbors. When so much more was necessary in a large twentieth-century organization it had become an isolated fortress.

Hoe's new shop structures in 1869 had innocently elicited the admiration of all who applauded efficiency, safety, and industry:

> The 80 large windows, without blinds, in the outer wall of the new factory, give promise of abundant light within, and the 31 six-pointed iron stars which stud its surface, hint at the strength of the structure which they help to secure. . . .

The interconnected complex of departments and shops seemed strangely in need of massive defenses:

> An iron picket fence runs along the smooth-sloping Sheriff-st. sidewalk, a few feet from the basement windows it is designed to protect, and vertical iron rods shield the lower windows of the office building. On the Grand-st. side a high brown iron grating, topped with bristling spearpoints, guards the glass front; the number, 504, is painted in red and black figures on the horizontal pane above the double recessed doors; the name, R. Hoe & Co., appears in raised letters on the iron door-step; and various signs are neatly inscribed on the slate-colored window shades. . . .

Abutting the new sections was a series of ponderous bulwarks:

> From the same [older] building five pairs of slate-colored doors (one pair in each story) open into the factory. Each pair consists of a light wooden door used during the day, and a heavy iron door, hollow and filled with plaster. These are all shut at night, and form, with the brick walls, a perfectly fire-proof barrier between the buildings. . . .

Indeed, the new Sheriff Street factory embodied all the fire-fighting (and insurance minimizing) expertise of three brothers long in the business:

> The beams, girders and columns being all of wood, a means of rendering them fire-proof has been devised, so ingenious and effectual that the owners do not consider it necessary to obtain any insurance whatever upon the building. This plan, which is patented by Hoe & Co., consists simply in incasing the columns, girders, and beams in sheet iron and plaster of Paris. . . . As further protection against fire, there is a room for hose, immediately under the water tank, which is filled by a powerful steam pump; each floor is supplied with a [water] delivery pipe and hose reel. . . .[22]

Somehow unobstructed windows, spear-pointed fences, iron doors and steps, and steam-powered water pumps and hoses, although clearly associated with the threat of conflagration the Hoes had rightly learned to fear, instead became weapons against quite another threat. They were defenses aganst the East Side itself; a manifestation of total disengagement from the community of neighbors, friends, and business colleagues who, once, had all labored together in their separate pursuits. Those early associations no longer existed; in their place were the frequent cries—in strange tongues—of bearded street vendors, and the occasional drone of strange street processions. The lower East Side by the turn of the century had crystallized as a ghetto the Hoes could scarcely appreciate. But the sides were clearly established; conflict was almost inevitable.

Wednesday, 31 July 1902, was a hot, sultry day in R. Hoe & Company's Manhattan domain. A monumental Jewish funeral was in progress. Almost 1000 Jewish schoolboys led the procession, chanting the *Thiltm*—Promise of David—and more than a thousand mourners, men and women, followed. The marchers crowded each other so densely they looked (to some) as though they were fighting to avoid trampling. Behind them the somber black hearse picked its way slowly among hundreds who sought to touch the unpainted pine coffin. Chanting and lamentations continued among thousands more of mourners trailing behind the hearse. Still further to the rear were 200 funeral carriages carrying distinguished participants, including 62 rabbis from the entire metropolitan community. Some newspapers reported 50,000 persons in all. This was the solemn funeral of New York's Chief Rabbi, Jacob Joseph.[23]

The lower East Side was everywhere alive to the magnitude of the ceremony. Thousands of Russian Jews had settled here only ten years before, their recollectons of the Czar's brutality and oppression fresh in their memories. To demonstrate their profound respect and veneration Jewish merchants had closed their shops, and from daybreak the bearded men, their wives and children, had lined the stoops and sidewalks, windows and doorways, fire escapes and rooves, block after block to see the procession. As the impressive cortege turned eastward into Grand Street there was for many spectators a growing apprehension of what might follow.

Jacob Magidoff, a reporter for the *Jewish Morning Journal,* remembered many outrages over the months—and how futile it was for Jews to complain to the police. He knew it was common for Hoe employees to insult and pummel Jews with impunity; he feared the funeral would turn into massacre.

Sigmund Schwartz, an undertaker, dreaded passing the Hoe factory; at a funeral just three weeks before one of his drivers had been struck in the face by an apple thrown from a window.

Dr. G. Wonson Vandergrift, visiting his father's office opposite the Hoe plant, heard the 1 o'clock whistle signifying the end of the lunch hour. Most Hoe employees and clerks had already passed through the spiked front gates, but some were still outside.

Edward A. Collins, a general foreman at the Hoe works, was disturbed by the milling crowds outside; he had a feeling some deliberate mischief against Hoe property would materialize.

Dr. Joseph Barsky, a resident of the east side for 26 years—and witness to many local indignities upon Jews—felt that in any altercation aggrieved Jews would get little justice from Irish policemen and judges.

Edward L. Shepard, another general foreman, agreed with Collins. Young neighborhood hoodlums had frequently made trouble; broken windows were costing the company $100 monthly. Hoe apprentices had occasionally been involved in scuffles, but they were seldom at fault.

Herman Serle, himself a Jew and a Hoe employee, was running a screw machine on the fourth floor when his fellow workers began jeering at the window. A man Serle did not recognize soaked an old rag in oil and tossed it out; soon others were throwing water and missiles. Although a big, husky fellow, Serle knew it would be unwise to interfere.

Walter Nevers, another Hoe worker, was watching the funeral from a window when a bag of water came down from the floor above.

Benjamin Asht, a storekeeper who considered himself an American Jew without prejudices, was just opposite the Hoe entrance. As the hearse passed he saw melon rinds descend from second and third floors of the press factory.

Abraham Postman watched bread scraps being thrown from a Hoe window onto the solemn procession. In rage he ran to a sidewalk cluster of garbage cans and threw them, one by one, through the Hoe ground floor windows. He later paid a $12 for misappropriation of the six cans.

When the marching Jews turned to shout their protests against the rain of refuse, Asht saw two streams of water pelt them from Hoe's second and third floors. Old bolts and iron scraps followed. Soon seven hoses were clearly visible at the windows, as were five men heaving pails of water from the roof.

George A. Duncan, a Hoe clerk, was still in the street when the crowd threatened the building; while he rememberd occasional snowball fights with neighboring Jews, he had never expected so heated an exchange as this. He ran inside and turned on a water hose, but after a few minutes Robert Hoe made him turn it off.

Marcus Singer stood waiting in East Broadway to see the cortege. Suddenly he saw buckets of water drenching the heads of people below—even before the main body of vehicles arrived.

Lena Weiner was standing near the Hoe works, where she was scalded on the arm by hot water thrown from a window.

Mary Greenfeld, watching from a stoop opposite the Hoe building, noticed pieces of old paper and oily waste drifting down from the factory onto the mourners; those on the street were soon throwing missiles back at the windows. Then she witnessed water pouring from almost every window—she was nearly blinded by the dirty spray.

When his men were driven from the windows by a storm of incoming missiles, William May, a Hoe foreman, made an attempt to throw a stream of water out the window, but his hose was too short.

Joseph Kaplan, a livery stable keeper, was driving a carriage in the procession when Emil Adams, a Hoe employee, turned a hose on the vehicle, drenching all its occupants.

Richard Fichte, another Hoe foreman, looked out the window to see neighborhood women, their aprons full of stones, supplying men with missiles with which to bombard the factory.

Edward Maloney, inspector for the street car company, had been posted on the front platform of a Grand Street horse car when the trouble started. The funeral procession had passed but spectators still milled in the streets, immobilizing service. He watched a committee of protesting men rush into the Hoe office; when they stormed out angrily the crowd began stoning the building.

Rosie Goldberg was doused with water as she passed the factory. She rushed to the office entrance to complain. Two men met her on the steps and struck her so severely she had to be carried away in an ambulance.

Rudolph Gartenburg, a pallbearer, was struck in the side by a brick. Somebody pointed out a laughing man at a Hoe window as the culprit. Gartenburg stormed into the Hoe office but was kicked back into the street.

An aged Jew managed to enter the Hoe office to beg a company official to stop the water disrupting the sacred Hebrew ceremony. "Oh, I don't care anything about the Jews," he was told, and ordered out of the building.

Albert Levine, a city marshal, saw Hoe employee George Church on the ground floor playing water on people in the street. With others, he rushed into the Hoe office protesting, only to be stopped and grossly insulted. A company officer drew a pistol and ordered them all to leave the premises.

Wilfred Hector, a clerk, saw people wildly disrupting the Hoe office —he agreed that water was the only recourse to protect lives and property.

W. E. Carver, a Hoe official, was at his desk when a noisy party of Jews burst in. They were obviously excited but they spoke a language Carver did not understand. He ordered them to leave immediately.

Vandergrift, still watching from a window, saw one crowd of Jews apparently following employees into the Hoe building. He sensed trouble. As he expected, clerks soon pushed the intruders out. One of the mob broke a front window with his umbrella. Hoe men were then trying to close the outer iron gates.

Just as he got the irate Jews out of the office, Carver discovered the entire Hoe building to be under bombardment—within the next few minutes most of the first and second story windows were shattered.

Ten-year-old Nathan Sherman, singing in front of the coffin, watched rags and iron strike the funeral hearse; mourners were furiously returning the debris as police arrived to disperse the crowd.

"Is this Russia?" shouted an old man with a foreign accent. Increased mayhem was his unwelcome answer.

Raphael Greenfeld, 8-year-old son of Mary, was standing in the street when he was struck by a missile; his hair was covered with blood.

Inspector Adam Cross had detailed 60 patrolmen to the funeral that morning, and dispatched additional reserves when the tremendous size of the populace became apparent. He raced to the scene with 200 men and six patrol wagons and at 1:30 p.m. went directly to Robert Hoe's office. They conferred for about 10 minutes.

As the hearse passed the factory Singer witnessed a shower of rubble rain down from the windows. A machine screw struck little Raphael nearby and knocked him down. Singer ran with him to a drug store for first aid.

Rushing to the scene with Inspector Cross and expecting a full riot, Sergeant Cohen found that order was already being restored. He accompanied the inspector into the R. Hoe & Company office where he posted himself while Cross talked with Hoe and his son.

Mrs. Greenfeld suddenly watched the police enter the street, swinging right and left with clubs. As fighting erupted all around her, she ran for her life—at that moment a man beside her was knocked senseless by a policeman.

W. M. Weinberg, of Jewish extraction but not of that faith, saw groups of fleeing Jews followed by police beating their clubs on the backs and heads of men, women, and children. Many were dropped to the ground. Acting with the police was a man in civilian clothing —straw hat, shirt sleeves—who shoulder to shoulder with a squad was also wielding a heavy stick against those in flight.

Singer returned to the scene in time to find the crowd throwing missiles back at the factory; suddenly he saw the police emerge from patrol wagons, swinging and striking with their clubs. He started running with another man who, after just a few steps, was struck by a policeman and fell on his face.

Fichte thought he saw a woman hit a policeman four or five times; he feared the police were being overpowered. He readily condoned the use of water to protect company property and restore order.

Alfred Brafan watched several people ask the police to stop the deluge of water; they replied by resuming their clubbing.

Detectives Stransky, Manning, and Rosenberg arrived with the relief. Manning was instantly struck unconscious by a stone, and his associates, who were also pelted, got him into the Hoe office where an ambulance surgeon was at work. Detective McCafferty, assigned earlier to catch pickpockets, was also at the aid station with a cut on his head.

Samuel Schneppes, Hyman Silverman, and Samuel Shuslik were marching together in the procession when scrap metal and other missiles rained down on them. As they sought to escape, they were battered by police; Shuslik was struck, sustaining an abrasion of one ear, but he managed to dodge and get away.

William Spalckhaver, chief Hoe draftsman and engineer, was

startled to see an officer being attacked; as more policemen rushed in there was a general melee. Recollecting that (of a 2100 work-force) more than 60 Jews were employed by the company, Spalckhaver could hardly bring himself to believe that race feeling was to blame.

After a nasty drenching with water, Kaplan saw fragments of iron come down as well. He approached a man at a nearby window, shouting to stop and telling him he ought to be ashamed of himself. "Get out, you Sheeny, or I'll douse you with water," was the response. Kaplan was furious—"This is free America!" he screamed.

Zachariah Feingold was accused by a Hoe clerk of throwing a brick into the factory. He was promptly arrested.

Maloney was stunned to see a man standing atop his wagon, inundated by a stream of water, shouting something about "free America."

Brafan saw Roundsman Jackson run toward the Hoe building—clubbing as he went. He saw the officer knock a man down, grab him by the collar, club him again, and drag him to the patrol wagon.

"The man is dead," people shouted in Singer's ear, but the officer did not stop the beating. The man—Rosenblum by name—was not dead; he rose to run again, when another officer dropped him for good and dumped him into a wagon. Singer had many times seen Jews harassed near the Hoe factory, but he knew this day's terror was the worst of all.

Roundsman Jackson caught Henry Rosenblum, a tailor, throwing stones; he subdued the man with his stick and at the same time sustained injuries to his own head. Rosenblum was thrown into a patrol wagon.

Maloney saw several boys take scrap metal off a junk wagon immobilized near the Hoe plant, and throw the iron at the Hoe windows. They also threw pieces at Roundsman Jackson.

George Stillgenbauer and Henry Stockhusen were arrested by Patrolman Maher for playing water on the crowd.

Officer Keenan collared Harry Grinsky, 17 years old, throwing a brick at the Hoe works; as he and another policeman were bringing Grinsky to a patrol wagon the other was struck by a stone.

Mrs. Bolde, Mrs. Perry, and Mrs. Sullivan watched from parlor windows on East Broadway. They saw Roundsman Jackson bleeding from wounds.

Jacob Rubel, a City College student, saw a man chased by the police for three blocks to be knocked down and clubbed.

Asht saw a child's head cut open by a piece of a gear wheel. Incredulous, he picked up the missile and put it in his pocket. He remembered that Grand Street opposite Hoe's was usually avoided by Jews during the lunch hour, and he knew why.

Rosenblum, beaten and arrested on Grand Street, was struck again in the patrol wagon by Roundsman Jackson; Samuel Ellenstein, similarly apprehended, watched the assault.

Jane Hitchcock, connected with the Mercy Settlement House, in particular found the police unnecessarily rough on women fleeing in the street.

Kaplan looked around in amazement. The owner of a stalled junk wagon—an Irishman, someone said—was himself hurling scrap at the Hoe plant. "Even we have the Irishman with us!" a street fighter exclaimed.

J. R. Gastenberg suffered bruises and internal injuries and E. Lipman a badly sprained leg; an old woman was knocked down by a policeman's club and a dozen other women had fainted. The ambulance crew was feverishly busy.

Frederick Levy, trying to get a balky team of horses moving, began swinging his crop recklessly; he struck Patrolman McKean with the whip and was immediately arrested. Three boys threw stones at McKean to impede the arrest, but he apprehended them too.

Weinberg had studied the face and general appearance of the strange civilian assisting the police, hoping to identify him later. At the end of the confusion he saw the man fraternizing and laughing familiarly with the patrolmen. His club was broken.

Robert Hoe III admitted to Inspector Cross that there had been frequent minor scraps between apprentices and the Jewish community, but he had issued orders to put a stop to such incidents. Hoe told the inspector that men upstairs, seeing their fellow workers being mobbed in the street, couldn't be blamed for responding as best they could.

Cross did not get back out to the street until mid afternoon, when the violence had diminished. He heard no complaints of clubbing, but he thought the squads had been badly utilized. He had the distinct impression the attack on Hoe was premeditated.

Somehow the weary procession succeeded in reaching the Grand Street ferry to Williamsburg, Brooklyn. Inspector Cross remained to return the east side community to some semblance of order. Enclaves of Jews at many gathering spots threatened retaliation, especially against their obvious tormentor, R. Hoe & Company. Street-car service was restored about 3 p.m. and police on hand everywhere kept traffic —foot as well as vehicular—moving briskly. Cross next approached the problem of getting Hoe employees safely out of the factory. At 5 o'clock, with policemen forming a double line in front of the entrance, the 1800 men and 300 boys filed out, protected for several blocks along Grand Street toward the Bowery, where they were finally able to disperse without police protection.

Late that afternoon Robert Hoe gave a formal statement to the press:

> At about 1 P.M., just as the workmen and clerks of R. Hoe & Co.'s establishment were returning to their work, a number of men in the funeral procession of the rabbi, which was passing in the street, were seen fighting each other with their umbrellas, and a portion of the crowd, to get away from the fighters, rushed up the steps to the entrance of R. Hoe & Co.'s building, 504-520 Grand Street. Somebody threw a piece of cotton waste out of one of the windows. Thereupon the crowd rushed to get into the door of R. Hoe & Co.'s office, making a great noise and disturbance. They were forced out by the hands inside, and simultaneously the mob began throwing stones and other missiles at the windows, and crowded against the entrance with the view of again forcing admission.
>
> In order to disperse them and save employes, who were outside, from being injured by the mob, some of the hands in the works turned on the fire hose, many of the employes not having succeeded in getting in and some of them having been badly injured. The mob continued throwing stones, bricks, and other missiles and breaking windows. The men upstairs, seeing their fellow workmen being mobbed in the street, threw missiles at the crowd. Stones and bricks and other missiles were thrown at the plate glass windows, breaking twenty-five in the old office and twenty-four in the new office, and of the small windows fifty in one department and twenty-five in another.

Hoe estimated the total damage to the plant at $1200 (his later claim against the city for $870 in damages was officially rejected). Reporters asked if his employees had not in fact frequently harassed Jews in the neighborhood. He conceded:

> We have 1,800 employes and about 300 apprentices, boys, in the building. You know what boys are. Some of them have a dislike for the Jews, many of whom live around here. We have done everything in our power to prevent these boys from annoying anybody, although we have no jurisdiction over them when they are in the streets. Many of them have been discharged from time to time on account of complaints received, and we are ready always to listen to any complaints from responsible parties. . . .[24]

Americans had long grappled with a steady intrusion of European immigrants who, like the first Robert Hoe long before, had confidently made the sober but hopeful decision to reconstruct their lives totally anew. Guided by the newspapers, aloof New Yorkers were often laudatory about the individual achievements of gifted foreigners, but cruelly hostile to the common masses of them pressing—in the tenement districts—for decent jobs and living accommodations. Russian Jews, the most recent wave and the most alien, ethnically, experienced the most determined prejudice and discrimination of all. But although in minor affairs complaints were indeed useless, a disturbance of this magnitude could not officially be overlooked. The mayor appointed a committee to investigate. It responded to the call:

New York, Sept. 12, 1902

To the Honorable Seth Low,
Mayor of the City of New York
Sir:

. . . On July 28th, 1902, Jacob Joseph, the Chief Rabbi of the Orthodox Jewish Congregation of the City of New York died at his residence, No. 263 Henry Street. He was a man universally respected and beloved among his people. His death was sincerely mourned not only by those to whom he had ministered during his long and useful life, but also by the entire Jewish community, as was shown by the touching manifestations of grief and sorrow which were exhibited throughout the East Side. . . .

The committee . . . finds on undisputed testimony that the attitude of some of the Hoe employes toward the Jewish residents of that vicinity had been for a long time hostile, and that this was due to the fault of those employes or their employers, or of both. The Hoe factory had been known for years to be a place from which emanated various forms of annoyance upon passers-by. . . .

Many complaints have also been laid before us that the police have for a long time past been insulting and brutal in their treatment of the Jews of the lower East Side. . . . We find that instances of uncivil and even rough treatment toward the people of this district by individual policemen are inexcusably common. . . .

It remains to apportion the responsibility for these occurrences, so far as it is possible for us to do so:

. . . The primary responsibility clearly rests upon persons in the establishment of R. Hoe & Company employed in the upper stories at the corner of Grand and Sheriff Streets. Their insults to the procession and bystanders led to all that followed.

. . . While there is no reason to suppose that the members of the firm of R. Hoe & Company were aware at the time of the funeral that their employes were insulting and assailing the mourners, nobody seems to have been discharged, or even censured, by them after the occurrence. The only step taken by them seems to have been for each foreman to canvass the men under him for signatures to a statement that they had seen nothing thrown from the building. To us there seems to be every indication of a concerted effort to hush up the affair and to protect all the inmates of the factory from discharge or prosecution. The members of the firm have not availed themselves of the opportunity to appear before us. They have, however, seen fit to prevent the success of the attempt made by us to secure the identification of some of the offenders. . . .

Respectfully submitted.

Edward B. Whitney   Louis Marshall   W. H. Baldwin, Jr.[25]
Thomas M. Mulry   Nathan Bijur

Whitney was a lawyer, formerly with the U.S. State Department; Mulry was from the St. Vincent de Paul Society; Baldwin had been president of the Long Island Railroad and was a city government reformer; Marshall and Bijur were Jewish community leaders.

When pressed by a *New York Times* reporter Charles Carpenter stuck to the company's guns:

> We were placed in a very peculiar condition on the day of that funeral. This shop was stormed by a mob of men intent on killing us. We turned the hose on the mob, of course. It was the first thing at hand. It was the only thing to do. . . .
>
> As to whether any missiles were thrown from our upper floors, I do not know. I was sitting here in my office on the first floor. . . . There may have been some jeering, but I do not believe it went beyond that. . . .
>
> This [committee] report is full of inaccuracies and statements made for political effect. . . .[26]

A genteel uptown public was embarrassed by the whole affair—perhaps little more, although some society people (Hoes included) were indeed at work urging greater professionalism for the New York City Police Department. More irate and appalled witnesses to the extraordinary East Side display of intolerance were distressed by its sad commentary on man's spiritual progress. One of them was particularly lucid:

> To the Editor of the *New York Times:*
>
> . . . I desire to say that if Agassiz or Darwin were alive to-day, they would have found the missing link between monkey and man, for the ruffians, looking through the windows and throwing pieces of iron, paper wads, wood, &c., on the people below, reminded me of a jungle in South America or Africa. . . .[27]

ROBERT HOE III

1839–1909

Courtesy Bronx County Historical Society

11.

Multicylinders and Perfecting Machines

HOE AND GUTENBERG IN PARTNERSHIP / MONSTROSITIES AND COLOR MADNESS / MORE PRESS COMPETITION / WOOD AND THE AUTOPLATE / MUNSEY AND THE GLORIFIED PEDDLER

When Robert Hoe III assumed command of the company in 1886 he dreamed of creating a new chapter in the history of printing technology—one that would elevate him (earlier Hoes notwithstanding) to the level of Gutenberg himself. He wished, in fact, to identify himself so intimately with the engineering progress of the firm he professed to construe his career as a mechanical component of the company's output. Of this he convinced James Bridge in 1902:

> To describe this machine [the Double Sextuple] is to delineate the personality of Robert Hoe the Third, for in a way it is himself. It is to him the embodiment of ancestral thought, effort and ingenuity. The fifty thousand pieces of metal, of which it is built up, are the perfected conceptions of his father and grandfather, and the ideas of those among whom they lived. His prenatal influences can be traced in rachet and wheel, crank-pin and roller. . . .[1]

Something about his qualifications for this high post prevented him from telling Bridge he was a person, not a machine.

Recuperating from his broken leg in 1860, the Colonel had had ample time for contemplation. "Robert I suppose desires to follow his illustrious predecessors," he queried his brother (24 January 1860), and so began the lad's preparation in that direction. Robert Hoe, Jr., then 21, was indeed disposed to serve the company, after a good education and broadening travels. He had made the grand tour of Europe (more than once), dutifully paying his respects to MacDonald of the London *Times* with frequent calls. Whisked to the other side of the printing world he had similarly discharged his assignment exploring the market in Chicago. He was soon party to all the company plans and gradually realized he would become its sole heir.

Toward the end of the 1870's the two senior brothers ruminated increasingly about advancing age and its infirmities. Young Robert grew impatient and restless, venting his annoyance over the Colonel's clock tower, which had now undergone delays and alterations for years. A large white marble dial was changed to black with gilt figures, housed under a square exaggerated mansard roof.

> [*Robert Hoe III to Robert, Paris, 24 May 1874*] . . . *Fortunately he* [*Richard*] *has become disgusted himself with the awful clock tower* [*we are*] *putting up on the new office & just before I left* [*for Chicago*] *consented to stop it, & put up a smaller dial in the Upper window. What was done to the tower must have cost $1500-- & I think it would have blown over some day if erected.* . . .

But the junior partner had a good point: "I don't believe in making hoistways, or clocks, or calendering machines or pumps &c in the Shop," he added, "but we stick to *making presses* & buying these articles."

At the same time the Colonel wrote his brother that the new building was fully occupied, but spacious—to him "a splendid establishment" in every respect. Its crowning feature, however, still irritated his nephew:

> [*To Robert Hoe, London, 23 December 1874*] . . . *By last mail I sent you a photograph of the new building with the unfinished tower. I wish (please do not allude to this in your letters) it had never been commenced, for it spoils the building to my eye, not to speak of the expense. The cost will be at least $10,000 & the clock arrangement before it is done will be $5000 or more. Uncle R. has stuck to this part of the "improvement" in opposition to all our entreaties.* . . .

Two years later the mechanism was still at the clockmaker's but it eventually got installed.

When the Colonel a year later admitted he couldn't "rough it" as he had twenty years before, and tired easily, Robert in Paris—like Sereno Newton many years before—began to reminisce with a view toward passing on his observations and experience in company management:

> [*Robert Hoe, Paris, to Robert III, 26 February 1875*] . . . *Looking back as far as the crisis of 1837, which your Uncle Richard can tell you all about, and in its long felt evils, arising from overtrading on extended credits and an inflated currency so like the present,* [*the difference is that*] *the paper in circulation then to a great extent was worthless.* . . . *Our business through all those years succeeding that crisis was flourishing and fair, for we had not been over producers so far as the supply and demand was concerned.* . . .

Knowing that periodic depressions had continued to be fateful for American business, Robert must have wondered at recollections of "flourishing and fair" conditions. Although his father too was getting old, Robert had decided he would buy his uncle's share of the business.

Robert brought up the matter several times, and in 1881 the Colonel expressed no objection, as long as the other partners concurred. Altogether the members of the firm were pretty content with themselves, their fears of economic slumps and devastating competition proving little more than chimeras:

> [*Richard Hoe, Paris, to Robert, 24 May 1883*] . . . *I wrote to the boys* [*Robert, Theodore Mead*] *that I hoped they would not declare any dividends for last year till my return, so you see we are agreed. Before we do so I hope we shall consider the interests of the firm and not any individual interests; for you and me perhaps it would be more interesting to declare all that could possibly be fairly done. . . . I feel my aims and trust for the future of our business are being realized. I told Theodore I considered the Concern fully worth what we had set down as its nominal Capital and felt quite sure of the result then but I did not anticipate such a collapse of our opponents. . . .*

Robert understandably grew anxious about the actual transfer of ownership. After discussions his father counseled him (6 May 1884) that "time and patience" were needed with the Colonel, as "his mind does naturally work slowly in matters of account." At last in September their transactions commenced, but were not completed when the Colonel died in 1886.

When his own father died in 1884 Robert had inherited much of his personal property, giving him resources with which to bargain. Two years later, negotiating with J. Henry Harper (acting for the Colonel's executrix, his wife Mary), Robert offered $160,000, then $192,500 for his uncle's complete interest in the firm. This figure was probably accepted, and Robert's possession of a multi-million dollar enterprise was undisputed. He looked at it with smug satisfaction.

But the big boss' inner office was now a very lonely place. There were no brothers with whom to discuss problems, to share responsibilities, to warn and comfort. Robert had made few friends, took little outside counsel, and was too remote from colleagues, suppliers, customers, or bankers to discuss everyday conditions. Instead, he had more remote comforts—a compelling avocation in books, to which he repaired more and more. He nevertheless kept up with his engineers.

The conception of multicylinder web-fed presses with folders, at the end of the century, was no longer the brainchild of a gifted wizard or tinkerer. It was committee work; the joint effort of printing engineers, paper specialists, electrical technicians, cost analysts. Robert, none of these, well appreciated this, but could not permit himself to get lost in the incidental paperwork of his vast enterprise. He—much more than his father or uncle—found it necessary in the annals of printing history to paint himself into the record in whatever guise suited him. First, of course, he had to seem vital to the firm's creative and developmental work:

> It appears that when Mr. Hoe gets an idea that something should be done by machinery which has hitherto been done by hand, he selects some one of his sixty draftsmen to outline on paper a small first part of his conception. This he turns over to some specialist, in the factory, to develop. . . .

When several such little steps are completed, their assembly is ready.

> If at any stage difficulties arise, a general conclave of experts is held, and the problem discussed by all until solved. Then the idea is patented, and soon after becomes a part of the Hoe printing-press. In this way Mr. Hoe finds justification for the disclaimer which he makes to the charge of being an inventor. "It is not I," he says: "it is the corporation." Which goes to show that, contrary to the legists, corporations have souls, and Robert Hoe is the soul of this corporation.[2]

Thus Robert could be an innovator of ideas without (correctly) pretending to be their inventor. It was said he "developed" color printing for newspapers but "refused" to take out any patents in his own name.

Robert replaced the company's quaint shop signs on neatness with two vague maxims he was fond of quoting. One, "Concentration is the first condition of success," seems only the sophomoric observation of those who (lacking strict seclusion) concentrate poorly. The other, "Get behind a thing, and push it: don't put yourself in front, and pull," tended to explain (if the first maxim did not) why the proprietor was seldom seen in the plant by its 2000 employees.

Bridge found Robert Hoe graceful and fluent as a talker, surrounded by mechanical experts who described his methodology with the enthusiasm of disciples. He spoke publicly only once, delivering a Grolier Club lecture on bookbinding, although a talk may similarly have become the manuscript for his privately printed history of presses. Nevertheless, with these few simple artifices he established himself as the infallible head of a mammoth factory which was, truly enough, still turning out incredible machines. More than that, under his management the business increased to five times its employee strength of 1866. Yet, fearing anonymity, Hoe sent his image with every job:

> Under Robert Hoe's regime, every year in the history of the firm has been marked by the perfecting of some new style of newspaper press that involved a new principle, or combination of principles, increasing efficiency and speed to meet the pressing requirements of newspaper proprietors. For forty years the career of Robert Hoe, the Third, has been marked by a singleness of purpose that has resulted in making his name familiar in every country. His autobiography is in almost

every newspaper office in the world. It is written in massive shapes of iron and steel. . . .[3]

This surprising record of achievement continued the work the busy company had started earlier. The concept of the supplement press had opened a whole new world of mechanical ingenuity.

In 1887 Luther Crowell was at last ready—reliably—to plunge the company into "mass" manufacture with his Quadruple machine:

> This was a step in advance of anything heretofore attempted. The first one was constructed in 1887 and placed in the office of the New York "World." The same principles were embraced in this as in the "Double Supplement," but developed to a greater extent. The supplement portion of the press was increased in width. By means of ingenious arrangements and manipulation of the webs of paper this press was made to produce eight-page papers at a running speed of 48,000 per hour; also 24,000 per hour of either ten, twelve, fourteen or sixteen page papers; all delivered with great exactness and perfection; cut at the top, pasted and folded ready for the carrier or the mails.[4]

The Quadruple was actually a double supplement where the supplement-printing portion was the same size as the main press, with a second folder added. It was the last machine to have one component working at a right angle to the other, but it became the exciting beginning of a large new family:

> Henceforward for some twenty years, newspaper presses took on a new nomenclature, ranging from quadruple, through sextuple, octuple, decuple, double sextuple, and double octuple, describing the various steps upward of 16-page units in press capacity.[5]

Pulitzer's advisors acknowledged the prototype Quadruple, "the base of all the monsters of today," to be perfect. "Mr. Pulitzer," wrote Seitz, "was compelled to forego his desire to let the other fellow pay for the experiment and installed the first to come from the factory." In Hoe tradition he promptly bought another. "Soon," Seitz concluded, "there was always a press on the stocks in the Hoe factory designed for the *World* establishment."[6] Pulitzer next prepared for the grandest operation R. Hoe & Company had ever experienced.

The New York *World's* need for new and larger facilities erupted in 1890 with the announcement of a colossal headquarters. In little more than six years the immigrant German had discharged all his debts and still would manage to pay cash for a new brass-domed building on Park Row, 16 stories high—the tallest edifice in New York city. The southeast edge of City Hall Park, Park Row had become Manhattan's "newspaper street" because reporters—before telephone and rapid transit—had to be near politicians (City Hall), the courts (federal and municipal), and financiers (on nearby Wall Street). At the same time they found themselves surrounded by the worst poverty in New York, a source of human interest material they had learned to be of perpetual value.

Noting that nearly all of the city's crime occurred in the adjacent Bowery, Allen Churchill described this geographic microcosm:

> Behind the Bowery, and feeding its ruthless demands for new humanity, was the Ghetto district of the lower East Side. Already outrageously overpopulated, with a vivid agglomerization of Irish, Italians, Poles, Russians, Bohemians and other definite national groups, the New York Ghetto was incubating some of the worst slums the world has ever known . . .—so unclean, unsanitary and generally appalling [it was] . . . "not fit for men and women, much less children. . . ." [7]

The *World* would nevertheless capitalize handsomely on its prestigious new address.

In the fall of 1889 Pulitzer had dictated much of the architectural and mechanical details of his project. He wrote his staff from Paris:

> I am anxious that [architect George B.] Post should keep to the original agreement, and put in a temporary roof over the pressroom, with a view to its earliest possible completion and the actual removal of the press room long before the building is finished. . . .
>
> The big point to be made with Hoe, and the one that is truly of the greatest importance to him, is that, in getting the new order for new-sized presses, he not only receives the most extraordinary order ever received by his firm but that it will lead to a quasi revolution in press machinery, forcing the other leading papers to imitate this change, as they have imitated every change introduced by the *World*. It will mean a great many presses to Hoe, besides those ordered by the *World*. This point ought to be very strongly used. . . .[8]

The result of Pulitzer's genius in running a large metropolitan newspaper, goaded by Hearst, was an inevitable recourse to the devices of sensationalism—the contrary rationale that a paper had to manufacture news in order to maintain or increase circulation. Indeed, another Pulitzer biographer held Hoe's big machines partly responsible for the success of yellow journalism:

> The seeds of sensationalism are found as much in the new mechanical wonders of the late nineteenth century as in social movements: in such inventions as the telephone and typewriter; in press folders that could cut, paste, fold, count and deliver twenty-four page papers at the speed they were printed; in the Hoe Company's web feed quadruple press, capable of printing 24,000 sixteen-page papers per hour. . . .[9]

It may be possible to postulate that, as social protest today often occurs because agitators know they will get television coverage, so also did

America's labor movements, women's suffrage, and other great reforms succeed because reporters were sure to be at the scene. In 1894, at any rate, the public had a fine chance to see the Quadruple: at the Columbian Exposition it earned a first prize as "most comprehensive illustration of the highest state of newspaper perfecting machines."

In a smaller way the company again served its old friend, the Baltimore *Sun.* On 7 February 1904 a great fire began with an explosion in the heart of Baltimore's business district; it would burn for a week. Despite its Hoe-approved fireproof structure, the *Sun's* Iron Building, fully half a mile away, was bombarded with sparks, briands, and smoke, and firemen were using dynamite nearby to contain the flames.

Twenty minutes short of going to press the historic building was abandoned, with only a hope the wind might change:

> But by midnight the heat of the burning *American* Building [a paper across the street] so greatly expanded the confined air in the Iron Building that some of the upstairs windows blew out, and in twenty minutes the heavy linotype machines on the fifth floor were falling through the blazing editorial and counting-rooms to the pressrooms in the cellar. . . .[10]

Fortunately the *Sun* had a second (unaffected) mechanical shop which continued to publish. In 1906 a new home was ready with four giant Quadruple machines in the cellar. With five seven-deck Hoes in the 1930's the Baltimore *Sun* proudly passed its centennial anniversary, and it continues to this day.

For Pulitzer, who willy nilly paid for much of Crowell's experimentation in making the Quadruple a smashing success, there were only diminishing personal pleasures. "Tall, blind Joseph Pulitzer was a tyrannical genius," wrote Swanberg, "so nervous that the mere sound of rattling paper could react on him like an explosion, causing him to weep or curse."[11] It was said he so feared libel suits (engaging Roscoe Conkling to handle them) he read every paragraph of his paper, costing him his eyesight. From Paris he embarked on a trip around the world, and died in 1911.

With its tremendous fund of experience, its modern machine shop, and a world-wide reputation for solving press problems, Hoe was now on top of the newspaper printing situation. Although the Double Supplement and Quadruple would supersede almost all other machines in the United States and Great Britain, still another giant was attempted. After about 18 months of construction (with 16,000 component pieces) the firm in 1889 completed a new machine for the New York *Herald.* It was the "Sextuple."

Chiefly responsible for the Sextuple was William Spalckhaver (1837-1919), born in Denmark and educated in Germany, France, and England. Arriving in the United States in 1860, he was swept almost immediately into the Civil War and commissioned a lieutenant after two enlistments in New York units. He resigned from military service in 1863. Spalckhaver had married a fellow immigrant, Anna Brussen, the year before, and they were to have six children. At this time the young man started at R. Hoe & Company, quietly performing drafting and supervisory work; he soon because chief engineer.

During 56 years with the firm Spalckhaver accumulated over 60 patents for printing mechanisms, of which the first significant one—a masterpiece of mechanics which old Robert Hoe would scarcely have comprehended—was the Sextuple. Highly esteemed by colleagues, well regarded by Hoe management, courteous and considerate to (nearly) all, Spalckhaver quietly developed a major breakthrough:

> The new Hoe press which is being set up in the "Herald" building is nothing less than a miracle of mechanism. To say that it is the only one of the kind ever built and that it throws all previous inventions into the background are facts which the following figures abundantly prove.
>
> Its consumption of white paper is so astounding that even the imagination grows tired and sits down to catch its breath. It is fed from three rolls, each being more than 5 ft. wide. When it settles down to show its best work it will use up in one hour nearly 26 miles of this paper, or, to make the matter more significant, it will use up about 52 miles of paper the ordinary width of the "Herald" every 60 minutes. . . .
>
> Before this press was built the fastest presses in the world were Hoe's "quadruple" presses, of which the "Herald" has two. These presses turn out 48,000 four, six or eight-page papers an hour, 24,000 ten, twelve, fourteen or sixteen-page papers an hour, and 12,000 twenty or twenty-four-page papers an hour, all cut, pasted and folded.
>
> Our readers will be startled to learn that [the Sextuple] can print and fold, count and deliver complete, 48,000 ten or twelve-page "Heralds" in one hour, which is equivalent to 800 a minute and a fraction over 13 a second.
>
> This is lightning work with a vengeance, and yet it is possible that there may be some who read this who will live to call it slow. That will probably be when they have found out all about how to put a harness on electricity. No one can predict when inventive genius will reach its limit in the printing press. . . .

This was a splendid occasion to applaud the work of a half century:

> The printing press is synonymous with progress, with the diffusion of knowledge and the spread of ideas. Without the great improvements that have been made in it within the memory of many men now living the modern newspaper, the best friend of liberty, and the greatest foe of tyranny, would be an impossibility. . . . In 1829 the Washington Hand Press was introduced and regarded as quite a mechanical triumph. . . . Now, a little over sixty years later, a ma-

chine is brought out which, when the number of papers alone is compared, . . . does over 2,000 times as much work.[12]

In 1900 the *Los Angeles Times* took a Sextuple. Founded 1881 by General Harrison G. Otis, a veteran of the Civil War and Spanish-American War, it would long reflect his character and integrity. Declining to describe the workings of so monstrous a machine, *Fourth Estate* concentrated upon some promotional copy not unlike Hoe's preoccupation with a plaque for the *Times* machines in London:

> The *Los Angeles Times* is about to install a new $45,000 Hoe Sextuple combination color and half-tone newspaper press, and in connection therewith is preparing a most unique and artistic name-plate.
>
> It will be inclosed within an ornamental iron frame, which will be located appropriately on or near the great machine. This frame will contain a fac-simile of Will H. Low's magnificent Columbian Exposition diploma, a steel engraving which was awarded to R. Hoe & Co. for the superior excellence of their newspaper perfecting presses. . . .
>
> Extending down the sides [a series of designs traces] the development of the printer's art from the old Franklin press, . . . through the era of the Washington hand press to the magnificent product of mechanical skill, the new Hoe press about to be installed, whose name, "Uncle Sam," is blazoned across the bottom of the mat. . . .[13]

The *Times,* despite labor troubles in the early 1900's, prospered.

The New York press-makers were now alive with new ideas. In 1895 the London *Times* abandoned its Walter presses for Hoe three-roll "Double Supplements," but Hoe that year was busy with another novelty, the "Octuple" (or "Double Quadruple"), using four folders and doing 96,000 eight-page papers hourly. Thankfully, said Isaacs, this was a "roomy" machine:

> As newspapers became larger and circulations increased, the demand for larger presses increased also. Press manufacturers were faced with applications for quart-sized machines to fit into pint-sized rooms. . . . Newspapers were unable to get the additional space required to meet their rapid growth, so press development was hampered, "monstrosities" were constructed, men were compelled to work under ridiculous conditions, and the list of injured and maimed operatives took on alarming proportions.[14]

"In the decade 1900 to 1910," Arthur Robb commented, "pressmen had to crawl, climb, stretch and heave with no little danger to life and limb in plating and threading operations,"[15] a condition which declined about 1912 in America. But as late as the 1930's in England, Isaacs added, there were still presses that "ought to be dynamited as the best way of clearing them off the earth to make room for something more modern, sane, and safe."

Hoe proudly advertised the Octuple as "largest newspaper machine in the world," and reported its own progress with some detachment:

> They manufacture over one hundred kinds of Printing Machines, large and small, including Multi-Color and Half-Tone Printing Presses, producing papers in eleven different colours in one operation at Newspaper Press speed. Their machines are to-day printing over one thousand of the leading Newspapers, as well as all the most important Magazines. . . .[16]

There was no stopping the company now.

The Hoe Sextuple was improved in 1901 to become the "Double Sextuple," and the Octuple was enlarged a year later to "Double Octuple" size. Capacity was now up to 144,000 16-page papers hourly.

The company's old friend *Lloyd's Weekly* in 1901 had abandoned three-roll machines in favor of Sextuples, but the following year even these were replaced. In 1902 Lloyd ordered the first of seven Double Octuples, called (in Isaac's words) "the largest and most powerful presses in the world":

> The double octuple dimensions were: Length 50 ft., height 27 ft. from sub-basement to top of folders, but only 14 ft. 10 in. from pressroom floor to top of press proper. Thus the process of compression of quart size into pint pot space began. Width 12 ft. over all, weight 120 tons, with approximately 100,000 separate parts. . . .[17]

The Octuple's production was truly staggering, fittingly described by an old-timer who, like a succession of reporters, marveled at the sight:

> I had recently the privilege of seeing the latest triumph of the famed builders of printing machinery, Messrs. Hoe and Co. of London and New York, the double octuple press, at work at the office of *Lloyd's Weekly News,* nearly a million and a half copies of which newspaper it prints every week. It is an elaborate piece of intricate mechanism, with apparently hundreds of wheels of varying size and numerous cylinders . . .—a veritable giant, a mighty engine indeed compared with the Napier two-feeder with which the writer became familiar when he first entered a newspaper office some fifty years ago.
>
> It was about that time that Mr. Richard M. Hoe, of New York, invented the rotary press with horizontal cylinders, but the speed which might have been attained was restricted by reason of the sheets of paper having to be separately fed into the machine by manual labor. The first Hoe printing press erected in England was put up in the office of *Lloyd's Weekly News* in 1857 and the proprietors of the weekly newspaper with the largest circulation have ever since been faithful to the Hoe principle, adopting from time to time its many developments and improvements, culminating in the production of the double octuple, which prints a twenty-eight page paper at the rate of 48,000 copies an hour. . . .
>
> The double octuple, as the name indicates, performs the work of sixteen ordinary rotary presses, massed together for economy of space and rapidity of production, and is one of the most remarkable inventions in connection with the art of printing of the present day.[18]

Isaacs, who had referred to tin cans thwarting the truancy of Lightning feeders, now reported the greater hazards faced by later pressmen:

> Each of the eight decks of these presses was a complete press in itself in every respect. Any deck could be taken out of gear without any drawback or hindrance to the rest of the press. With eight decks in commission a 32-page paper was produced on each folder. Should a 28-page, 24-page, or less number of pages be required, any deck could be left out at will. The objection to the deck type of press from the operatives' point of view was the inadequate space allowed to perform the essential work of blanketing, plating, washing-up, and entering sheets . . .—"washing-up" was certainly an awkward task, [and] plating was a fine art if the plate-hand wished to avoid the risk of rupture or damage to the head. . . .[19]

The versatility of these mammoth machines gave them an incidental capability to handle up to 11 colors—6 on one side, 5 on the other—on the perfected sheet. The Sunday editions and supplements used color sporadically and people liked it; mass magazines also needed two and four colors on covers, text, and advertising pages to attract sufficient attention. Robert Hoe himself took the next step:

> After satisfactory experiments it was ascertained by the writer that, with the inks properly prepared, and suitable plates to print from, colors could be printed almost simultaneously upon the paper, without mingling; in short that the supposed necessity, in much of the work done, of drying the sheets after the impression of each color on the paper, was not necessary for the production of a good quality of printing. Further experiments also proved the mechanical possibility of obtaining most accurate register in printing from a roll and that the number of impressions, or colors, could be increased to advantage. These various experiments resulted in the construction by Hoe & Co. of color presses which were almost simultaneously installed by the proprietors of the New York "Herald" and the New York "World." . . .[20]

Before long another newspaper giant called at Grand Street.

William Randolph Hearst had acquired the San Francisco *Examiner* in 1887 as a 23,900 morning newspaper with a "creaky single-web press." He promptly ordered a new Hoe and with handsome inducements collected a good staff. A year later his father, then a U.S. senator, gave him another press as an anniversary gift, part of a $300,000 investment necessitated, said W. A. Swanberg, by "the tremendous cost of new presses and equipment, higher salaries, a larger staff, and above all huge promotional expenses. . . ."[21] Hearst had already observed Pulitzer's winning techniques and their stunning results, and he determined to match or better the old German. One of Hearst's earliest personal acquisitions was a brilliant fellow who engineered much of the mechanical wizardry the Californian would need.

George E. Pancoast (1862-1939) was born in New Hampshire and began as a printer's devil at Welsh, Bigelow's University Press. After some six years as typesetter he left Cambridge to barnstorm around the country as a comedian in a traveling theatrical group. Stranded in San Francisco in 1889, Pancoast presented his union card for work at the *Examiner*—and remained with the Hearst organization thereafter.

On one occasion filling in at the telegraph desk and finding one of his headlines altered—to his immense displeasure, Pancoast tacked a sarcastic note on the bulletin board. Hearst noticed it, met its rash author, and promptly engaged him as private secretary. Although never having operated a typewriter, Pancoast was not about to lose the rare promotion: he rearranged the keys more like a compositor's type case and performed for seven years as Hearst's confidential aid.

Traveling around the world with Hearst, Pancoast introduced his chief to the marvels of amateur photography, acquainted himself with newspaper mechanical problems, and was soon in charge of the Hearst plant. When the time came to build new facilities, his boss said:

"George, if I have the stereotyping superintendent lay out our new plant it will be the way he wants it. If I ask the pressroom superintendent to do it, it will be best suited for him. I don't want that. So, I think you're the boy to do the job. Here's a pencil and paper. We'll do it together." [22]

After that, Pancoast was in charge of all of Hearst's mechanical operations, and they soon accumulated. One came from a strange source: Albert Pulitzer. Joseph's younger brother had started a New York penny newspaper in 1882 which quickly needed Hoe's attention (its circulation in 1887 would reach 225,000):

OFFICE OF R. HOE & CO.
504 Grand Street

New York, Nov 13th 1884

Albert Pulitzer Esq.
"Morning Journal"
Dear Sir,

We have to thank you for your order received to day for the new [Double Supplement?] press, which we will put in hands at once and complete with the utmost dispatch.

This machine will produce perfect Morning Journals at a speed of 24,000 copies per hour, and with the other three presses you have of ours, your office will possess a capacity of turning out about 96,000 Morning Journals per hour.

If you run these machines for two hours and a half they ought to give you from 200,000 to 240,000 papers in that time.

Congratulating you upon your success and wishing you a continuance of it, we remain

Very truly yours
R. Hoe & Co.

In 1894 Albert doubled the paper's price (to 2 cents) so as to compete with his brother's *World*—and quickly began losing money. Hearst snapped it up a year later.

As Hearst and Joseph Pulitzer competed for attention over the Spanish-American War they exercised control over public opinion more compelling and stirring than the government's own propaganda. "The presidency of the United States," said Swanberg, "was being preempted by batteries of cylinder presses."[23] Hearst (with unlimited money) and Pulitzer (with the industry's best talent) fought ruthlessly for supremacy in the Sunday editions as Bennett and his foes had scrapped 60 years before. When the *World* offered free "chromos" suitable for framing or a new color supplement, the *Journal* promised "Eight pages of iridescent polychromous effulgence that makes the rainbow look like a lead pipe."[24] Incredibly, something wonderful emerged from these nasty spats.

In 1893 the *World* had bought a four-color newspaper press only to find that, with untried inks and unreceptive web papers, its four-color artwork always looked dismal. Growing interest in political cartoons, however, suggested that colored caricatures for entertainment might print much better—and so was launched a great institution, the Sunday comic section. But the inks still gave trouble; Charles W. Saalburgh, a *World* foreman, explained to his artist:

> I want you to color the kid in this drawing yellow because I've got to find a way to prevent that yellow ink from running and sliding around. . . . I want a good test area which will be printed in yellow, so that I can easily check on the results. . . .[25]

So on 16 February 1896 Richard F. Outcault's "Yellow Kid" first saw light in yellow in a sequence showing a squalid "M'Googan's Avenue" —which might very well have traversed Hoe's lower East Side.

At one point Hearst succeeded in coaxing Outcault to the New York *Journal,* and Yellow Kids appeared in both papers, giving rise to what —for the new variety of newspapers' ungentlemanly conduct to follow —was condemned as "yellow journalism." Hearst nevertheless was determined to out-perform all his rivals. Hardly the rational nonpartisan press that sensible people hoped would emerge from the political turmoil decades before, American journalism suffered all over again. The maverick Californian seemed to succeed without any diligence whatever in collecting news, writing perceptive analyses, or guiding serious thought. Hearst had in fact found another formula:

> In the strict sense, the Hearst papers were not newspapers at all. They were printed entertainment and excitement—the equivalent in newsprint of bombs exploding, bands blaring, firecrackers popping, victims screaming, flags waving, cannons roaring, houris dancing, and smoke rising from the singed flesh of executed criminals. . . .[26]

For these theatrics Pancoast began to perform a series of tricks with his presses which sorely taxed his ingenuity as well as that of his ink-makers, papermakers, and machinists.

Turning his attention to the printing of color in newspapers, Pancoast, with guidance from Hearst and R. Hoe & Company, developed equipment to rival Pulitzer's. It was costly:

> Pancoast smashed three machines in an effort to perfect a[n electric] motor-driven press. Horrified at the loss, Hoe & Company advised Hearst to halt the expensive experiments. Hearst merely grinned and told his mechanical superintendent: "Smash as many presses as you have to, George. Only give us what we need." . . .[27]

A Hearst color press was finally ready in November 1895 to confront the *World,* and a year later Pancoast was first to electrify all his presses, at the *Journal.* The amateur engineer next designed and patented the "Pancoast Multi-Color Universal Unit Hoe high-speed web press," building the first one in 1897. Although not entirely successful, he continued tirelessly to experiment in this area; his last color press had 24 cylinders, printing one color almost instantaneously on top of another. The Yellow Kid ran and slid no more.

When Hearst in England noticed that rotary printed papers had hand-set typematter ready for last-minute news, clamped into a "stop-press" device and printed without stereotyping, he urged Pancoast to do the same on his machines. Working for some time, Pancoast finally developed the "fudge," a type area first employed by the New York *Journal* 17 March 1897 to report the Corbett-Fitzsimmons prize-fight (alas, no more crucial than the *Tribune's* Saratoga races). The outcome of the fight was nevertheless on the streets within minutes of the wire dispatch. About the same time Pancoast printed the first half-tone pictures in New York, bending conventional etched (screened) plates to be cast as curved stereotypes.

In 1900 Hearst decided to start a paper in Chicago. He rented an old building and waited for his presses. Because of a boxcar shortage, however, the equipment remained stranded in New York. Hearst instructed Hoe to ship the machinery "first class" in Pullman parlor cars, and the Chicago *American* began. When his *Examiner* plant was lost in the 1906 San Francisco earthquake Hearst again demanded unorthodox speed. He learned that a Hoe had just been shipped to a Salt Lake City newspaper; he got it for the *Examiner* simply by paying double.

In 1914 as plans for the Panama-Pacific Exposition in San Francisco materialized, the Chief was annoyed to find his *Examiner* staff contemplating a modest display quite like the other participating newspapers. He suggested to Pancoast that they build a new color press for the Exposition and print the paper's comics and magazine section right before the public. This called for a special 12-cylinder machine. The fair's opening date—only four and a half months away—made the request almost an impossibility, but the plucky inventor would try.

Pancoast rushed with his plans to New York and persuaded R. Hoe & Company to give him every priority; the company worked night and day on the new equipment. Two weeks before the exposition commenced Pancoast arrived at the site with eight railroad carloads of printing machinery. Laboring around the clock mechanics erected the press in time for the opening. Pancoast collapsed with fatigue.

In his later years the inventor's eyesight began to fail, and he was almost blind for two years. Through Hearst's generous concern operations were performed to restore his sight, and in 1938 the veteran press designer was honored by a New York dinner at which 750 newspaper executives gathered to celebrate his 50th anniversary with the Hearst Corporation. He died the following year.

Working from New York, Hearst indeed became the great factotum he visualized himself to be:

> Hearst was running a huge enterprise. He ran it himself, with an intimate knowledge of all its phases—circulation, advertising, typography, features, comics, pressroom routines. Minutes were precious to him. . . .[28]

At the height of his career Hearst owned 26 newspapers, 13 mass magazines, 8 radio stations, a newsreel company, a feature syndicate, and a wire service—influencing 30 million American readers and viewers daily.

With principal customers like Hearst, Pulitzer, Bennett, and Munsey, it is no wonder Robert Hoe III developed strange business practices and reclusive personal habits. But with the newspaper industry such as it was, the resolve of the original partners to specialize in this area seemed nothing short of prescience. When the company solved the problems of on-press folding it had eliminated all remaining barriers to speed and productivity. The firm's progress was charted in stunning annual contributions to newspaper technology:

1871—R. Hoe & Co. brought out first roll-fed continuous-printing rotary press with gathering and delivering cylinder.

1874—First four-page-wide newspaper press made by R. Hoe & Co.

1875—Rotating blade folding cylinder patented by Stephen D. Tucker . . . ; Newspaper press with paper rolls above the printing units first designed by R. Hoe & Co.

1876—First newspaper rotary web perfecting press with rapid folding and gathering mechanism exhibited by R. Hoe & Co. at the Centennial . . . ; R. Hoe & Co. acquired patents for devices by which webs of paper could be turned over after being printed. . . .

1880—First rotary magazine web perfecting press printing from curved stereotype plates made by R. Hoe & Co.; Tapeless delivery introduced. . . .

1881—R. Hoe & Co. introduced the triangular former folder thus making possible the modern high-speed newspaper press.

1882—First supplement-inserting press and first machine with two printing sections at right angles to each other made by R. Hoe & Co.

1885—Angle bars for turning webs of paper introduced. . . .

1887—First quadruple (or 32-page) press, with two folders . . .; Endless belt automatic paper feed control introduced . . .; Combination collecting and cutting cylinders invented and patented by R. Hoe & Co.

1888—First three-page-wide press made by R. Hoe & Co.; Automatic spring tension device for equalizing strain on paper web. . . .

1889—First sextuple (or 48-page) press made by R. Hoe & Co.; Staggered cylinders introduced . . .; Transverse collecting introduced. . . .

1891—Rotary wire-stapling mechanism for fast-running presses. . . .

1892—First newspaper press with four-roller ink distribution, for fine half-tone printing, made by R. Hoe & Co.

1893—First multi-color rotary press made by R. Hoe & Co.; Hoe quadruple newspaper press given highest award at the World's Columbian Exposition.

1895—First octuple (or 64-page) newspaper press made by R. Hoe & Co.; First combination newspaper and color rotary press made. . . .

1897—First rotary newspaper press five-plates-wide made by R. Hoe & Co.; Adjustable spring-seated propellers introduced. . . .

1899—Late news devices introduced by R. Hoe & Co.; Reversible cylinder for color printing invented . . .; V paper brakes invented.

1900—Central folder design of rotary press invented . . .; Adjustable segments for folding cylinders invented and patented by R. Hoe & Co.

1901—First double sextuple (or 96-page) newspaper press brought out . . .; Gold Medal, Award of Highest Merit, granted to R. Hoe & Co. at Pan-American Exposition. . . .

1902—First double octuple (or 128-page) press brought out . . .; X-design newspaper press invented. . . .

1906—Tubular cylinders introduced and patented by R. Hoe & Co.

1907—First combination double octuple newspaper and color press, with reversible cylinders and offset rolls, made by R. Hoe & Co.

1908—High-speed rotary camless folder perfected . . .; Safety locking knife box invented . . .; Automatic self-oiling boxes for high-speed presses. . . .

1912—R. Hoe & Co. perfected and introduced "Simplex" type newspaper press designed especially for smaller city dailies; First combination of an intaglio printing unit with an ordinary newspaper printing unit . . .; R. Hoe & Co. first used independent horizontal nipping roller drive, independent horizontal drive for folding cylinders and . . . for delivery fans; First intaglio rotary web press in the United States. . . .

1913—Cross association of webs or products from one row of presses to another introduced by R. Hoe & Co.; First roll-fed rotary offset press with folder and flat delivery. . . .

1914—First combination color and black newspaper press of the patented universal unit-design built by R. Hoe & Co.

1915—Hoe 12-cylinder universal-unit combination color and black newspaper press awarded the Grand Prize and Gold Medal of Honor at the Panama-Pacific Exposition; R. Hoe & Co. developed and patented automatic pump system of ink distribution for high-speed newspaper presses . . . ; First "Superspeed" rotary newspaper presses equipped with ink pumps, solid forged steel cylinders, running in roller bearings . . . ; Heavy-duty "Superspeed" rotary camless folder introduced by R. Hoe & Co.

1916—Newspaper press with folders underneath and paper rolls above the printing units . . . ; Alternating forged steel and fabroil gears for newspaper press drives introduced. . . .

1919—R. Hoe & Co. perfected and introduced improved low-type unit newspaper press with "Superspeed" folding mechanism; Multiple line arrangement of printing press units in parallel rows introduced. . . .

1920—Tubular folding cylinders introduced and patented by R. Hoe & Co.

1921—Page valve cut-off on ink pumps introduced . . . ; R. Hoe & Co. improved ink motion on universal unit color presses by designing ink cylinders and form rollers so that no roller would be of the same diameter or an even multiple of the same diameter as any other; Improved new-style hydraulically-operated paper roll feeding mechanism designed. . . .

1922—R. Hoe & Co. introduced and patented improved method for removing stereotype plates from printing cylinders; Patented spring-seated bearers in folding cylinders introduced . . . ; R. Hoe & Co. introduced and patented new-style late-news device for printing in center margins of Japanese newspapers; R. Hoe & Co. developed extensively balcony type of press with the printing units supported on an integral cast-iron substructure and the paper rolls and folders on the floor level below the printing units.

1923—First 24-cylinder combination multi-color and black newspaper press of the patented universal-unit design, the largest newspaper printing press in the world, built by R. Hoe & Co.; Improved folding-roller drive for heavy products introduced by R. Hoe & Co. . . .

1924—New-style late-news device for printing late news on cylinder-collected products introduced . . . ; First underneath paper-roll racks with automatic blanket tension introduced by R. Hoe & Co. . . .[29]

To these accomplishments may be added a few significant developments which brought the company to the eve of World War II:

1928—Hoe Linear-type press with printing units end to end behind the folders; Hoe vertical-type newspaper press designed for installations where space is limited.
1929—"Super-Production" press introduced by R. Hoe & Co.
1933—First Hoe unit-type press.[30]

Appreciable developments up to 1924 illustrate that, lack of popularity notwithstanding, Robert Hoe maintained an unwavering and uncompromising domination over the firm which gave it unmistakable purpose, direction, and vitality. True, the reputation of his father and uncles long hovered over the firm. That these were the blossoming years of the newspaper and its (letterpress) technology is also true, but at the same time R. Hoe & Company had almost singlehandedly made them so, and was justified in reaping a good share of the market. By 1910 it claimed "over one hundred different models of our rotary presses" to be in operation around the world; as Hoe customers were invariably the biggest daily metropolitan newspapers, this was a tremendous dollar-volume segment of the industry.

Although the 95-year-old company at century's end had only three new competitors—Scott (dating to 1874), Duplex (1884), Goss (1885)—in the newspaper area, competition would be keen. Technologically, perhaps, the race was a matter of sheer growth rather than mechanical revolution, simply to serve the wildly expanding press market. Hoe had solved (for the entire industry) the dilemmas of machine shop production, employee training and customer receptivity, patenting precedents, and marketing strategies. Bravely these four major American newspaper press-makers faced a nation of self-styled publishing moguls whose eccentricities became classic:

> Newspaper publishers are a special breed. Out of this world. They are arrogant. They are used to advising the world how to run its business—kings, governors, mayors. They are difficult to negotiate with. They're not practical people. Some of them operate at a loss, for reasons of prestige, power, or personal ambition. Their personal opinions enter in, rather than a concern for what their business requires. Newspaper publishers know very little about their own business. I knew one who had never been in the mechanical department of his paper. . . .[31]

Countless press builders would nevertheless try to attract such capricious purveyors of the printed word. New platen jobbing presses, cylinder flatbeds, and fast rotaries—now all equipped with automatic feeding as well as a variety of delivery options—were available. Two important new names had widespread appeal.

Robert Miehle had begun in 1887 in Chicago to manufacture a

two-revolution flat-bed machine invented by S. K. White, although the principle (cylinder making two revolutions for each printing cycle) was not new. Himself a pressman, Miehle had made it efficient: strong, accurate for register work, adjustments easily accessible. It dominated the flat-bed field for half a century, excellent for process color and other fine work. "You never heard of a Miehle being scrapped,"[32] was the company's homely slogan. In 1920 Edward Cheshire of Milwaukee patented a vertical flat-bed machine which he sold to Miehle the following year. The firm made 25,000 of them over the next forty years, the last in 1962.

Similarly successful was William M. Kelly with his 1914 flat-bed cylinder. As early as 1876 Hoe had invited Kelly to send a diagram of his press improvements, but he was apparently not yet convincing. Showing "how wide a knowledge an inventor may need,"[33] said Collins, Kelly was a compositor, pressman, salesman, and repairman in India, Australia, and South Africa; he needed a press to accompany his 1912 composing machine, and simply set to work. Built by American Type Founders, Elizabeth, New Jersey, 5000 Kellys were running by 1927, doing a considerable amount of book and job work.

To fill the gap between slow flat-bed machine for small newspapers, and large, expensive fast stereotype rotary press, two brothers in 1884 developed a reel-fed flat-bed machine printing from type at 3500-4000 impressions per hour. A year later Joseph (inventing from the age of 20) and Paul Cox brought out their "Duplex," featuring two flat beds, each for four pages. Duplexes were popular in rural areas where steam or gasoline engines remained even after the invention of electricity; *Editor & Publisher* considered the Duplex a major step in newspaper progress. The Duplex Printing Press Company, Battle Creek, Michigan, organized in 1885 and headed by Irving L. Stone, himself once a pressman, put the successful machine on the market, and by the end of the century (by adding a third bed which increased output to 5000-5500 for 12 pages) made it the most popular printing machine in its class.

Henry Bechman was born 1858 in Iowa and at 12 apprenticed in a machine shop. Soon working in Chicago, he rebuilt presses for George E. Lloyd Company, and began patenting his ideas. In 1891 he joined the Goss organization but went to Duplex the following year, where his flat-bed perfecting and rotary stereotype machines were very successful. He invented the 1909 Duplex tubular plate ("Stove Pipe") rotary press.

Upon the occasion of a suit in 1922 by Duplex, alleging infringements, R. Hoe & Company took the occasion to pontificate:

> In 1910 we made the statement in several trade papers that the bringing of this suit by the Duplex company was a "false alarm" and we

> reiterate that statement now. At the same time we wish to emphasize, as we pointed out then, that for over one hundred years R. Hoe & Co. have been the pioneers in the designing and improving of printing machinery and that stripped of Hoe inventions and Hoe mechanisms, including the triangular former folder, angle bars for associating webs of paper, transverse collecting apparatus, tapeless folder and delivery mechanism, combination cutting and folding cylinders, rotary blade folding cylinders, and other exclusively Hoe features too numerous to mention, the modern newspaper press would not exist. . . .

The suggestion that a pioneer manufacturer could not possibly infringe upon a small firm was bolstered by more recent history:

> We were first to introduce (in our celebrated Double Supplement and Quadruple Presses) the Low-Design-Unit Press, as well as the other principal designs of press construction now in use, including Double-Width Presses, Supplement-Inserting Presses, Superimposed Unit Presses, Central Folder Presses, Multi-Color Presses, and combination color and black presses, many of which have been copied by other manufacturers. . . . Our latest development, the Superspeed Press with automatic-pump system of ink distribution, is an epoch-making one and conclusive evidence that we do not have to copy others. . . .[34]

Curiously, Duplex's 1910 suit had been abandoned in 1913, when Robert Hoe IV was on his way to Battle Creek to assume the company's presidency; in 1922 the new court action was, according to R. Hoe & Company, based on the same old claims. The Duplex nevertheless enjoyed considerable success in the United States and England. Goss took over the Duplex firm in 1947.

Walter Scott (1844-) was born in Scotland where he learned the machinist's trade. He came to the United States in 1869, settling as a mechanic in Chicago and becoming pressroom foreman in a printing shop. He soon began to demonstrate his ingenuity. While chief engineer of the Chicago *Inter Ocean* he invented a newspaper folder, and two years later developed a complete press. He formed Walter Scott & Company in 1874 and erected a factory in 1884 at Plainfield, New Jersey.

Scott was, said Charles Francis, "a man of most striking appearance, weighing 250 pounds and having a handsome beard that came to his waistband."[35] An inventor of considerable reputation and popularity, he retained his Scottish accent and (tucking his long white beard into his coat for extra warmth) enjoyed exchanging jokes with his many Park Row customers; one of his more subtle (on himself) was to be photographed before one of his large machines—his own 6½ foot height made even a two-story press seem smaller, unimposing. In 1891 Scott produced a reel-fed flat-bed perfector to compete with the Duplex, and his machine for the Philadelphia *Inquirer* in 1894 was considered the largest press in the world.

By 1890 the Plainfield plant was making 117 different models of presses when the Scotsman began concentrating on newspaper color: using five benday tints for each of four colors, yellow, red, blue, black, over 30 combinations of shades were possible. It was he who had equipped the New York *World* in 1893 to print four pages in such colors, leading the way for the bright Sunday supplements—the comics—that followed everywhere. Scott's sons David and Walter continued the business, and in the 1930's their well received equipment was capable of up to 60,000 newspapers hourly. The firm later merged with Wood Industries.

C. B. Cottrell & Sons in 1890 developed a high-speed perfecting periodical press for the popular Boston *Youth's Companion.* It used a "shifting" tympan, a slowly unrolling sheet of white muslin to absorb smudging between impressions on each side of the web—a scheme Welsh, Bigelow had years before tested for Hoe. The publisher was so skeptical about Cottrell's claims for speed and quality, however, he made Cottrell's pressmen run the machine for a full week and then withheld payment for another week until his own men could master it.

In 1891 Cottrell hired Howard M. Barker, who remained chief engineer for 55 years, with his staff amassing 400 machinery patents. Barker in 1898 improved a web perfector for *Munsey's Magazine,* then two- and three-color presses with automatic folders for *Munsey's, Youth's Companion, Everybody's, McClure's, Cosmopolitan, Ainslee's, Saturday Evening Post, Ladies Home Journal, McCall's, Metropolitan, Delineator, Good Housekeeping, Modern Women, Street & Smith's,* and other mass magazines.

Also at work at Cottrell's was Milton A. McKee, a pressman and tinkerer of great skill. McKee developed a system for making four-color process printing plates, putting the firm foremost in this area. His system was first used 1907 on *Ladies Home Journal.* Cottrell intensified its work in color printing technology, bringing out a four-color web perfecting press in 1926 which became famous for its mass production: Curtis Publishing Company, Philadelphia, alone took 72 of them, Crowell Collier, New York, 19, and many others were placed around the country. Cottrell was soon able to print full color on its fastest machines. In 1953 the firm merged with Harris-Seybold.

Three British brothers, Samuel G., Fred L., and William T. Goss started in Chicago to manufacture their "Clipper" and "Monitor" printing presses in 1885 for small Midwest newspapers. All former newspaper printers, they brought experience and genius to the rapidly expanding inland market, giving many small towns their first four-

and eight-page dailies. Their first such order in 1886 came from Grand Rapids, Michigan. The Goss Printing Press Company was soon delivering its largest newspaper machinery throughout the United States and Europe.

Working at Goss was New Yorker Joseph L. Firm (1837-), an experienced designer. Firm had apprenticed at Harpers; at Frank Leslie's in 1859 as press foreman he introduced much new equipment, including a rotary letterpress using an offsetting cylinder. In 1889 he invented a "straight-line" press for Goss which printed 72,000 eight-page papers hourly from five rolls of paper fed simultaneously. (Hoe later bought the Firm patents.) In 1897 the New York *Herald* ordered several six-roll or Double Sextuple Gosses, each with six folders, and a still larger machine for the Sunday edition. The 1910 "six-deck duodecuple Goss Straightline press" was claimed the largest in the world; its 12 folders did 48-page papers at 25,000 hourly.

With steadily increasing European orders, the Goss brothers, like the Hoes, were soon persuaded to erect a plant in England. Their new 1889 shop at Hayes, Middlesex, 10 miles from London, employed G. A. Menendez as works superintendent (he had been foreman of the U.S. shop), and as chief engineer, Otto Raabe (after 36 years with Hoe). Like the Hoes too the Gosses had their British moment of triumph: in 1909 Lord Northcliffe reformed the London *Times* plant with Monotypes and Gosses, the latter capable of 32 pages at 25,000 per hour.

The London *Daily Mirror* four years before had been the first large British paper to take a Goss, which Isaacs compared most unfavorably to the Hoe Double Octuple. As a spokesman for England's pressmen, he was understandably concerned for their welfare, but he repeatedly substantiated his claim that the big machines were monstrosities rather than marvels:

> This Goss ["four roll straight line single"] was of the deck type, then rapidly coming into favour, mainly on account of the small floor space it occupied in comparison with the ordinary or right-angle "quad." . . . The press looks neat and compact, but appearances are oft-times deceptive. First note how close to the floor the bottom deck cylinders are placed, and it will be apparent that plating the bottom deck was a job that was possible only in a posture of devout reverence—on the knees. On the third deck this difficulty is more apparent, and whereas the first and third decks required a similar posture, so plating the second deck required stretching. Oilers and platers on these presses, therefore, had a good deal of physical jerks. . . . Note also should be made of the meagre space between the top of each set of cylinders and the tape pulleys at the bottom of the deck above, saving of space effected at the cost of many cut heads. . . .[36]

Another London paper, the *Daily Mail,* in 1927 apparently indecisive, ordered its new equipment from both Hoe (London) and Goss (Chicago) for its Northcliffe House pressroom. Goss today is a component of MGD Industries, Inc.

About 1904, according to Isaacs, Messrs. R. W. Crabtree & Sons, Leeds, England, began by manufacturing "fudge box" devices for late news. Their plans for a complete rotary newspaper press were interrupted by the world war and not resumed until 1919.

The first Crabtree was a three-roll deck-type machine for the Dublin *Irish Independent,* installed 1921. It featured worm and worm wheel drive (instead of gears) for smooth, quiet operation, a "waving cam" ink distribution system, improved roller setting (without tools), roller bearings throughout, completely automatic oiling, and hydraulic brakes. Some of these innovations were subsequently abandoned because of their cost, but the company continued to introduce many modern features. Similar progress of course continued in the United States at the same time, especially in the hands of yet another gifted New Yorker.

Henry A. Wise Wood (1866-1939) seemed ordained to create a great era for the printing industry. At the age of 6 he was presented a small case of type and a toy press—and became so adept at setting type by hand (and reading it backwards in the stick) he had to be retaught to read correctly when his schooling began. Sent to an academy in Media, Pennsylvania, where he again obtained a small press and started issuing a school paper, the boy bought additional fonts of type with modest profits and soon ran a thriving printery.

When Henry learned that the best shortstop on the school's baseball team was a pressman for the town's weekly paper, he obligingly filled in for the boy on Saturdays so the star could practice with the team. In this way he became familiar with an old Campbell and other larger presses. When his father died the family had no money to send Henry to college, and the lad went to work with what he knew best: printing machines. In 1883 he joined the Campbell Printing Press Company in New York, but he had a myriad of interests.

Like most creative people of his time, Wood was fascinated by much of the machinery that characterized the social progress of the nineteenth century. In the early days of flying he accompanied many pioneer aviators as engineer; he founded *Flying Magazine* (1911), edited it for eight years, and was first president of the American Society of Aeronautic Engineers and vice president of the Aero Club of America. "In those days," he once recollected, "I had many offers from airplane firms to join them as engineer, but I was too much wrapped in Newspaper presses." He also studied marine engineering and was an enthusiastic yachtsman, in 1915 a member of the U.S. Naval Consulting Board. He belonged to the Society of Naval Architects and Marine Engineers, New York Yacht Club, and Cruising Club of America. But newspaper work again kept him, like Crowell, from a life at sea.

In later years Wood expressed himself vociferously through speeches, pamphlets, and letters to newspapers on subjects of great social impact: farm problems, immigration, prohibition, technocracy, currency reform—even advocating use by married women of their maiden names. Having written a book of poetry (*Fancies,* 1904), and two volumes of prose (*Symbols,* 1904; *Money Hunger,* 1908), Wood was inevitably romantically associated with the printed word.

> He viewed the steel of the presses and the whirling white web of paper with a poet's eye, centered his creative and analytical powers upon them, always with the idea of refinement, greater speed, and a more perfect product. . . .[37]

But Wood, like William Morris, desperately needed that romantic escapism, finding himself in a most difficult personal situation.

Son of Fernando Wood, three times mayor of New York but remembered more for his scurrility, Henry was profoundly disturbed by his father's unenviable reputation. He set for himself an irreproachable standard of behavior which permitted him, ironically, to scold his own generation unmercifully. His *Money Hunger* reflected a disdain for the greed and moral decay he found (still) rampant at the end of the century, especially in politics and business; but his stern chiding in exquisitely polished language often lost its point. He criticized the press:

> That function of the newspaper having the most profound influence upon the popular character is exerted in its role of intimate friend, wherein it comes into close and daily touch with the lives of the people, and sympathetically deals with their troubles and joys, their hopes and fears [all too often] . . . in order the more easily to be able to exploit its public for private ends. . . .[38]

Long before his writing and philosophizing, however, Wood had made for himself a genuinely impressive record to place alongside his father's ignominy.

Henry Wood amassed over 600 patents, most of them for printing equipment. He had become president of the Campbell Printing Press Company, which he bought (from his father-in-law) and in 1909 renamed the Henry A. Wise Wood Company; then in 1914 he merged it with the Autoplate Company of America (established 1910) to become the Wood Newspaper Machinery Corporation. Wood proved in effect the genius to inherit Richard M. Hoe's mantle as the country's foremost press designer. He continued as president of his establishment until 1935 still—at 69—actively engaged in printing research. Upon his death in 1939 *Editor & Publisher* remembered the early state of affairs:

> When Wood began newspaper work more than half a century ago, the printing art had advanced little in principle from the days of Gutenberg, Caslon, Bodoni, and Caxton. Type was still set largely by hand. Stereotyping was new and wholly a manual process. Steam power for presses was still not a universal and electricity had been applied only experimentally. Woodpulp paper was in its infancy. News and advertising seldom forced newspapers beyond eight pages. A circulation of 100,000 was about tops.
>
> The same era brought Pulitzer with a new formula for mass circulation; Mergenthaler with his revolutionary invention that gave one typesetter the power of five; Hoe with a press that used curved stereotype plates; a few years later Hearst with his intensification of the Pulitzer formula. The public was ready to take all the newspapers that the presses could produce. Presses multiplied and pushed the ceilings skyward. But the bottleneck was stereotyping. . . .[39]

Robb described the labor of manually pouring molten lead into a boxed matrix:

> It was hard, hot work, involving the lifting of more than 50 pounds of hot metal in a heavy ladle from melting pot to casting box for every page plate cast, and the shop that could cast 30 plates in half an hour was exceptional. After casting, the plates had to be cooled, trimmed and made-ready by hand. . . .[40]

Incidental to the older hand techniques were almost medieval practices: "[M]atrices were prepared from simple materials by a process held in deepest secrecy by members of the stereotypers' union," Robb added.

About 1894 Henry Wood, still working at Campbell, determined he could produce machinery to make stereotyping more automatic. There had indeed been no advance in the equipment for twenty years. With some newspaper offices by the end of the century requiring production of 1 million papers hourly, it became unbearable to allow up to an hour for the making of stereotypes by hand. It was equally ridiculous for many New York papers, finding stereotyping costly, slow, and cumbersome as late as the 1890's, to continue printing from type locked up in turtles. Wood studied the situation at the New York *Herald* and in 1900 saw his brainchild adopted by that paper. His solution was a machine which *Scientific American* considered "almost human in its ingenuity."

The "Standard Autoplate" was a casting mechanism at one end and a plate finishing machine at the other. In four minutes it cast, shaved, trimmed, and finished a curved plate, but as it did these four things simultaneously, it could deliver one plate per minute. The plates were far superior to any processed by hand, and although the first machines were not entirely trouble free, they were quickly improved. "A single machine does work that could not be so well performed by thirty-five men,"[41] said *Scientific American.*

The cauldron of the Autoplate held 12,000 pounds of type metal (a mixture of lead, tin, and antimony) at 550 degrees Fahrenheit. A paper maché matrix was positioned on the machine. For each plate 40 to 50 pounds of molten metal was pumped into the shell mold. When cast, the water-cooled plate passed between knives and shavers which trimmed it to exact dimensions, shaved it precisely type high, and added slots for clamping it to the press. It had safety features to preclude molten metal accidents, and automatic stopping devices if foreign objects fell into the mechanism.

Wood's first machine, built for the New York *Herald,* was so successful it prompted an order for three more. The New York *World, New York Times, Brooklyn Daily Eagle,* Philadelphia *Bulletin,* Philadelphia *Telegram,* Boston *Post, Boston Globe,* Chicago *Tribune* also took early machines. They reported savings of $500–700 weekly, faster press start-up, and postponed press deadlines so as to get late news. Under tests the printed product from the Autoplate, whether type or halftone, was better. In 1905 a simpler, faster "Autoplate Junior" was introduced, also very successful. Wood's energy was boundless:

> He never stopped dreaming, planning, inventing. A few years ago, he predicted hundred-page newspapers in full color, whirled off at speeds not attained by any machine of today. And while all modern presses now perform at speeds with precision that was only a vision 20 years ago there seems little likelihood that Wood's prediction of 100-page newspapers will be realized soon, if ever.
>
> No one man contributed more to the contemporary newspaper miracle, and many of his 450 patents will keep his name green, even among his competitors. Advancing age and his own temperament made him impatient with the enforced standstill of the past decade, and his association with the press company he formed three decades ago had not been active in recent years. . . .[42]

Scientific American too treated him well: the three greatest printing inventions of the nineteenth century, it opined, were the 1860 Hoe rapid rotary press, Mergenthaler's Linotype, and Wood's Autoplate.[43]

In 1914 one reporter followed the story of a giant newspaper going to press with erie visions of the Autoplate stereotype room:

> As one looks through the door [to the foundry] one sees ponderous and groaning machines, lit by weird flames, and surrounded by a black troop of acolytes, who seem to produce continually from nowhere bright sheets of newly cast metal, bent into the form of half-cylinders. They hurry with these plates, which they hold with hand-guards of felt, to a [cooling] tank, into which they plunge them for a few seconds before sending them down the lift to the basement.

Gradually the first hurry subsides; we may now go in to be introduced to the marvels of that almost human monster the auto-plate machine, which is the last word of speed and efficiency in the founding of stereotyped plates for printing. Each of those mighty machines on the floor of this department has cost £6,000, but it pays itself over and over again in time-saving, for, incredible as it may seem, it can cast, and turn out finished, no less than six of those ponderous plates a minute.

It is very hot in this room, for each machine carries a tank containing nine tons of molten type metal. It is not easy at first to believe that the seething liquid is not some kind of soup. Its surface colour seems in the imperfect light to be almost purplish, and it is covered with bubbles and scum. There is something horrible about this fiery mass. The whole effect of these tanks, as well as the weird shape of the machinery, with its flaring gas furnaces and heating appliances outside the furnace, make the scene like some Dantesque vision rather than a scene of everyday wage-earning. . . .

The autoplate is like the barrel of a "twelve-inch gun," in which the plate is cast, douched with water (at a rate of 80 gallons per minute), its edges trimmed, and it emerges a finished plate. A single edition [of a large metropolitan daily] needs 1,200 plates, on average.

The intrepid reporter then ventured into a similarly unreal pressroom:

Long before we reach the basement the brool and clash of machinery surges up to us, for already some of the machines have got to work and the edition is flying to the four quarters of London.

As we finally reach the basement floor, we are stunned by noise and bewildered by the vast and towering structure of complicated machines, which, although so colossal, are yet as delicate in their mechanism as any watch, and require the most perfect tuning in order that they may do their work aright. Yet amid this apparent chaos of whirling rollers and gliding shafts there is the same order and purpose that has guided all the other departments through which we have passed, and if the necessity of the case entails that there shall be more apparent bustle and hustle at the last moment of the paper's production, it is nevertheless very far removed from anything like confusion.

The last machine to start towers above us for nearly twenty feet or more, and its entire range from end to end is nearly sixty feet. On the platforms at the top of the apparently wild chaos of machinery the men are already fixing the plates that we have seen cast not five minutes ago. The work does not take very long, but until the moment comes when yet another clashing contribution shall be made to the sum total of the awful din that is already stunning the ears in this Babel, and making human speech or hearing impossible, there are many interesting things to be seen which it may be well to notice now, lest haply our unaccustomed brain begin to reel and lose all power of receiving definite impressions. Such, at any rate, is the first threat of the machinery.

The huge webs of paper are already in their place on the end of the machine, and every one of the machines already whirling around us is being fed in like manner, and is reeling off miles and miles of paper every hour. Every reel contains a continuous roll of paper five miles in length and in one London newspaper office more than 300 of these reels are used in every twenty-four hours, or nearly 10,000 miles of paper a week. . . .

There are signs, however, that the great Gorst press [of Sir John Gorst, *Morning Standard*], the quickest in the room, is ready to move. Engineers and machine-men go from platform to platform, assuring themselves that all is well. . . . Suddenly above the din a whistle shrieks. It is the signal either to stand to attention or to stand clear, according to what your work may be, and then, with a gentle, purring movement at first, the tremendous combination of mechanism glides off and falls into its stride. The reel revolves, the tapes guide the paper on its way, the cylinders clank around, and a multitude of subsidiary mechanisms click and turn and slide and check and countercheck, while over all rises at length the individual song of the machine; for every machine, like every express train, has its own note and plays its own tune. . . .

Standing where the web, with its fresh impression, was rushing down to be cut, one could for the first second or two distinguish individual copies of the paper. Gradually this became blurred, until at last it seemed only like a great waterfall of parallel columns; yet still the dark mass of an illustration could be distinguished, appearing and appearing again many times per second. But the note of the machine rose still higher, and as it rose the swift succession of images of that illustration drew nearer and nearer as it seemed. At last the eye could no longer single out individual impressions of the picture. . . .[44]

The press was now producing 96,000 papers per hour. The creaking printery music of a century before was now orchestrated cacophony.

Henry Wood had of course familiarized himself thoroughly with the manufacture of printing presses in the pursuit of his stereotyping machinery. His incentive, he recalled, was truly the stuff of romantic dreams:

I was standing one day looking into the press room of the New York *Herald* through its plateglass windows and watching the manipulation of the several colossal presses when the thought occurred to me, were I privileged to live thirty years longer, I would not be confronted with the same presses, running at the same speed or in the same manner—that someone would have developed machinery that would produce papers three or four times as fast! Why shouldn't I be the man instead of waiting for the other one?

This gave me the inspiration to begin the construction of printing presses, analytically, to accomplish this result. . . .[45]

Devoting considerable time and resources of his own as well as those, again, of the *Herald* (and other papers), Wood soon developed some novel gadgetry from his experiments.

Wood discovered, for example, that printing quality actually improved with speed; that breakdowns usually occurred because all the parts of a large press were not in "harmonious relations"; that newsprint webs tore from erratic and unequal strain, not from high tension itself. He patented many of his developments singly before a modern machine—30 years ahead of its time—could be put together.

Wood produced a mechanism to lead the web through the press with "minus tension," pushing it instead of pulling; his impression cylinders had special high-temperature bearings, eliminating any off-axial aberration in cylinder rotation. This prevented "jumps" (jarring starts) and "friars" (light or broken patches; "monks" were black, overinked areas). Improved inking equipment followed. One of the first Wood presses was installed 1915 at the New York *Herald;* another, a 1917 machine for the Philadelphia *Evening Bulletin,* printed sixteen pages at 120,000 per hour.

The first press incorporating all of Wood's modifications was erected at the *New York Times* plant in 1928. It was a 16-unit machine with five folders—containing 372 tons of steel in its total weight of 541 tons. Wood himself commented dryly, "it need not be added that its printing and folding units are unbreakable."[46] "It performed satisfactorily in most respects," Robb added, "delivering a high-quality product at speeds not hitherto known in newspaper plants. . . ."[47]

Wood also produced the Autoclamp, a system whereby clamps holding the plates to the cylinders could be opened and closed automatically. This positive control avoided the error of starting a machine with some or all clamps open or poorly closed or plates tearing loose from the centrifugal force of the really high speeds becoming possible. These new devices and the greater attainable speeds ushered in a truly modern concept in printing production and management everywhere. The magazine industry in particular began to talk of a "new breed" in its printed product.

Men like De Vinne, Munsey, and other publishers of mass magazines at the turn of the century almost instinctively recognized the paradoxical economy of the most expensive new equipment. Street and Smith, foremost New York publishers of pulp fiction and other paperbacks, demonstrated this prescient feeling:

> [N]either of the partners ever skimped when it came to buying the most up-to-date presses, the newest engraving equipment, or folding machines to bind the two million books and magazines which spewed from the building each month. They were especially proud of the thirty presses, ranging from the huge cylinder affairs to smaller flat presses which turned out covers and advertising sheets. . . .[48]

One peripatetic printer finally revealed the logic of this attitude.

Charles Francis was working at a New Zealand printery when in 1865 the firm bought a Wharfedale. As no one knew how to set it up or work it, Francis volunteered to put it together, and got a slight raise. He soon traveled to London, New York, Chicago, finally buying an old New York shop in 1894. When he looked at his newly acquired battery of machines an old foreman warned him, "Don't pull that lever! If you start that press I won't be responsible for what will happen."

"If it's as bad as that," Francis replied, "I want to know it now, and throw it out. Let her go!"

Obviously in the market for new equipment, Francis had strong feelings about its replacement:

> I see to it that my old machinery is broken up. I don't want it set up in the next block. . . . I regret to say that a press manufacturer tells me that I am the only printer he knows who does break up his old machines; others agree that it is a good thing—and then sell them for what they will bring. . . .

Francis urged machinery manufacturers to require larger down payments and stiffer credit terms, so that exchanged equipment "went to the junkman, instead of being 'rebuilt' and put to work again to depreciate prices."

More importantly, perhaps, Francis thought two factors alone insured success in the printing business. The first was a combination of "brains and ability," which, if present, would seem sufficient to obviate further prerequisites. But he made a special case for the other:

> The second essential is an up-to-date equipment. A good printer may maintain an existence with a lot of old machinery and type, but the money is made with the newest and best presses, composing machines, feeders, folders, stitchers and supplies. When you see a machine that will produce ten per cent. more than the one you have, it pays to make the change and drop the old servant. Printing is a ten per cent. business, and with machinery that is ten per cent. behind the times you are just at 0.[49]

In 1916 Charles Francis Press won several important magazine contracts, *Field & Stream* and *Printer's Ink* among them, for which its proprietor bought a $20,000 Hoe web press. Francis had found that what was true for newspapers was just as vital for magazines; provided there were few format changes, and with all equipment designed for these formats, the newest—the fastest and the best—spelled the best production. But a greater genius had known this for years.

Frank Munsey, as a boy an expert telegrapher, in 1882 came to New York from Maine to seek a fortune in publishing. He adroitly manipulated several popular magazines (*Argosy,* 500,000 circulation, *Scrap Book,* 500,000, *All-Story,* 300,000, *Munsey's,* 800,000—by

1905) into so fluid an enterprise he was able to do the same with newspapers. His colossal New York printing plant, containing all the latest and fastest equipment, at one point became a pawn of almost ludicrous proportions.

To escape difficult labor conditions in Manhattan, Munsey in 1895 had decided to move his entire shop to New London, Connecticut. He built an eight-story headquarters there and began shipping his presses and bindery equipment to their new home. But as the ponderous machines arrived two obstacles arose: postal regulations would not permit him to keep a New York address on publications mailed from New London; and he refused unequivocally to pay overtime as Connecticut unions understood it. Munsey would not be defeated.

> By six o'clock the same evening the first of the big new presses was lifted off its foundation and set onto a boat; before midnight the greater part of the equipment was afloat on Long Island Sound. Before the next monthly publication time, everything was back in New York, set up and ready to go.[50]

(The New London building became the home of a vast Munsey supermarket chain.) By 1905 Munsey owned no less than 18 newspapers, ruling them from his suite in New York's prestigious Flat-Iron Building, America's most famous business address. He reigned supreme:

> Throughout his career he never took a partner, nor created else than a dummy board of directors, nor yielded any of his authority to a powerful legal advisor. Nor did he marry. He never held staff conferences, seldom took suggestions, and supervised all details himself. . . .[51]

In 1916 Munsey bought the venerable New York *Sun,* and caring little for its long traditionally independent life, combined it with the New York *Press,* which he bought four days later. He purchased the New York *Herald* in 1920 and merged it too with the *Sun.* The New York *Commercial Advertiser* (whose name had changed to the *Globe* in 1904) next fell to Munsey; he bought it in 1923 and also merged it with the *Sun.*

True, the price of newsprint, labor, and machinery had risen so much these famous papers could make no profit on circulation alone; it now became necessary to appease the advertisers who had long been treated almost contemptuously. But Munsey, a "ruthless, glacial man," said Churchill, saw newspapers as business properties and acted without regard to their "soul."[52] To journalists everywhere this was a crime. He became detested for his callous treatment of papers, many over a century old—the sheets that had made the American press greatest in the world. In New York alone he reduced 14 famous banners to 9.

Munsey had few friends and none of the private joys of art and book collections such as would delight generations of Hoes. His only apparent love—a mania—was for commercially sound mechanical gadgetry; he craved modern equipment, big and expensive, and his magazine printing and binding facilities were the pride of his operations. Munsey, perhaps more gullible than any other Hoe customer, had fallen completely for the company's punning 1899 catalog:

> The publisher who wins success—who is always prepared to take advantage of his opportunities—is
>
> The Man with the Hoe
>
> It is the most important factor in any publisher's prosperity, and represents the latest and greatest achievements in printing machine construction. . . .[53]

Munsey had already bought eight Hoe 96-page magazine presses. He completed a new manufacturing facility in 1911 with a particularly radical machine; taking 18 months to build, it was unrivaled.

The new Hoe multi-color magazine press, largest in the world, was named "The Frank A. Munsey" by R. Hoe & Company, "—and it bears this name proudly on a brass plate, just as a crack cruiser floats its admiral's ensign from its topmost peak," said the firm. Really two separate presses in one and workable together or singly, the machine turned out with each revolution 384 pages (equivalent to three ordinary novels, according to the company) from electrotypes which alone weighed half a ton. It printed, folded, and delivered any one of the following:

144,000 eight-page signatures per hour	in 2 colors
72,000 eight-page signatures	in 4 colors
72,000 sixteen-page signatures	in 2 colors
72,000 sixteen-page signatures, half in 3, half	in 1 color
36,000 sixteen-page signatures	in 4 colors

"*Wunderbar,*" said an amazed Johann Gutenberg, resurrected by a Hoe copywriter to marvel at the new behemoth. Elsewhere in the Munsey shop the other Hoes produced 1,680,000 sixteen-page signatures daily, or a total output of about 50,000 magazines hourly. "*Alles ist Hoe; . . . Ich habe nicht um sonst gelebet* [All is Hoe; I have not lived in vain]," [54] murmured the bewildered master of printing. The novel reemployment of Gutenberg was not accidental.

Robert Hoe III seemed to enjoy his conceited kinship with the Mainz inventor. His view of himself—not his great uncle or lucky grandfather; surely not Koenig or Napier—in the history of printing technology as the old German's direct successor as "humble builder of presses" was well timed. In 1900 the world had celebrated the 450th anniversary of Gutenberg's invention of printing. Hoe commissioned (for $1000) sculptor Ralph Goddard's "At the Dark Tower," a "heroic" gothic full-length statue to be mounted above the

Grand Street entrance, not unlike patron saints on medieval church façades. According to Goddard, the name signified Gutenberg's "challenge to Death." Hoe also directed Scharf, a Viennese sculptor, to strike medals for the occasion, which he sent unannounced to the company's vast clientele.

Although Robert had spent little time at work in the shop, had associated with few employees, and had complacently reduced company advertising to barest essentials, he felt the image of Hoe grandeur he had manufactured to be satisfactorily established—perhaps even beyond his death. As if to commemorate Hoe's recent passing with a fitting epitaph, the inauguration of the Frank A. Munsey called for a courtesy advertisement in *Munsey's Magazine.* Robert H. Davis of the publisher's editorial staff "on the spur of the moment" dictated a hymn to his stenographer which duly occupied one page:

I AM THE PRINTING PRESS

I am the printing press, born of the mother earth. My heart is of steel, my limbs are of iron, and my fingers are of brass.

I sing the songs of the world, the oratories of history, the symphonies of all time.

I am the voice of today, the herald of tomorrow. I weave into the warp of the past the woof of the future. I tell the stories of peace and war alike.

I make the human heart beat with passion or tenderness. I stir the pulse of nations, and make brave men do braver deeds, and soldiers die.

I inspire the midnight toiler, weary at his loom, to lift his head again and gaze, with fearlessness, into the vast beyond, seeking the consolation of hope eternal.

When I speak a myriad people listen to my voice. The Anglo-Saxon, the Celt, the Hun, the Slav, the Hindu, all comprehend me.

I am the tireless clarion of the news. I cry your joys and sorrows every hour. I fill the dullard's mind with thoughts uplifting. I am light, knowledge, and power. I epitomize the conquests of mind over matter.

I am the record of all things mankind has achieved. My offspring comes to you in the candle's glow, amid the dim lamps of poverty, the splendor of riches; at sunrise, at high noon, and in the waning evening.

I am the laughter and tears of the world, and I shall never die until all things return to the immutable dust.

I am the printing press.[55]

Within a month, according to the Munsey company, over 2000 newspapers reprinted the "prose poem" with hearty approbation, and other advertisers, notably the National Cash Register Company (from whom Davis earned a $500 "royalty"), paraphrased it for their own mechanical products. The Davis work and its imitations appeared in newspaper editorials and advertising all over the world, although R. Hoe & Company, alas, was never credited. But strangely enough, there was much more in Munsey's career that would become poignant for Hoe, the New York *Herald* purchase fateful indeed.

James Gordon Bennett II had assumed control of the paper upon his father's death in 1872. Unimpressed by the skill and loyalty of his editorial staff, Bennett on the other hand had treated his mechanical department personnel with utmost deference. Shop foremen were well paid and respected. "Their several domains of authority were never invaded," said Don Seitz. "They were given the best of tools to work with, and rigid rules as to closing forms were enforced upon the editors to serve their convenience." [56]

In 1894 Bennett had architects McKim, Mead & White design a new plant at Herald Square. The building, imitating the Palazzo del Consiglio of Verona, had a great clock and thermometer, a giant bell and two mechanical workmen (now in Herald Park) who sounded the time. The pressroom, visible from the street through great plate glass windows, impressed many more than Henry A. Wise Wood with its giant machines. But inside, the shop was cramped and unhandy. Two years later a new Quadruple was nevertheless installed to meet the frenzy of the Spanish American War. But Bennett in France little followed the paper's daily operations, and after the turn of the century the whole enterprise fell to pieces.

In 1918, when the younger Bennett died, trustees of the estate resolved to sell the declining New York *Herald* to raise funds for projects (for example, a memorial home for newspapermen) specified in the expatriate's will. Edwin A. Potter, Jr., vice president of the Guaranty Trust Company, New York, was executor of the Bennett estate. Potter (1878-1940) and his fellow trustees, Charles H. Sabin and Francis H. Sisson, were called by George Britt "glorified peddlers" for the mercenary estate work they would perform.

Hawking the formidable newspaper property, Potter approached William Randolph Hearst, Cornelius Vanderbilt, Herbert Hoover, and then Frank Munsey, meeting little success with each. But then the banker shrewdly introduced an element of rivalry millionaires often cannot resist:

> Potter, a competent trader and the most active agent, matched one prospect against another. He told Munsey that Hearst was interested and Hearst what Munsey was offering, and confided his out-of-town bidders to each of them, and as eagerness mounted the $2,250,000 valuation of the appraisers came to appear very meager indeed. Potter made a last call on Munsey.
>
> "I'll give you $4,000,000 for your papers and that's final," Munsey said.
>
> "Very well," said Potter. "We'll accept, on one condition—that all

current assets [$67,000 in cash, $643,566 in other forms] must be reserved to the estate."

"Sold," said Munsey.

In January 1920, for the auspicious consummation of the *Herald* purchase, Munsey had already set the stage dramatically in the paper's own offices: "He sat up straight in James Gordon Bennett's chair," wrote Britt, "at Bennett's small old-fashioned French desk," as Potter, carrying the *Herald* stock certificates, was ushered into the room. Potter too sensed the drama of the sale and ceremoniously passed the documents to Munsey. Witnesses were much impressed.

"You're a tough bargainer," Munsey remarked to Potter when the transaction was completed. "I had no idea a trust company would fight so hard for somebody else's money. I think I'll make the Guaranty my executor to handle my own estate. . . ." [57]

Ironically, upon his sudden death in 1925 Munsey left an unsigned testament; his earlier (valid) will of 1921 strangely gave New York's Metropolitan Museum of Art—in which all his life he had expressed no interest whatsoever—the sum of $40 million (the balance of his estate—then the largest money gift to any museum). It was an unforeseen development in which the Guaranty Trust Company had no opportunity to participate. For Edwin Potter, nevertheless, the extraordinary compliment from a powerful if unfathomable man was priceless. He would perhaps confide the remark to other businessmen contemplating similar arrangements. As events proved, Potter would buy as shrewdly as he sold—and despoil with the rarest finesse.

When Gutenberg had returned from his fantastical visit to Munsey's plant he called the great printers of history around him to describe the state of twentieth-century printing. And he modestly made a flattering observation:

> Brothers, I have yet to tell of the greatest revelation. That great successor of ours—Robert Hoe by name, blessed be his memory—who built the far-famed Munsey presses, and the hundreds of other throbbing presses that proclaim the splendor of our discovery throughout the world, was a collector of books. Not long ago he was gathered to his fathers, and all of his treasures were sold. On the very day that I beheld his masterpiece in steel they sold the pride of his collection. And what do you think it was? It was my own forty-two line Bible that I made with my hands in Mainz in those beloved days gone by. It was handled like a sacred thing, and it brought—you will never guess—the sum of 125,000 gulden. Think of it—a prince's ransom for my poor work! [58]

THE CLUB BINDERY

c. 1903

Courtesy The Huntington Library,
San Marino, California

12.

The Hoes in Private Life

NEPHEWS DODD AND MEAD / ROBERT, RICHARD, AND PETER AT HOME / A BIBLIOPHILE MARRIES OFF HIS DAUGHTERS / FABULOUS BOOK AUCTION / BACHELORS AND FEMININE WILES

"About seven years ago, I stepped from the cars in a country town," recalled the peripatetic Grant Thorburn, neglecting to mention that prior arrangements had been made to receive him at the Bloomfield, New Jersey, railroad terminal. "Among them who were looking on, stood a man of genteel appearance; said I, 'Sir, I wish to stop here for a week, I don't like to put up in a hotel; can you direct me where I may lodge in a private family?'

"He said he could. We entered the next street, he stopped in front of a respectable tenement; on the front stoop sat a comely matron. She might have seen twenty-eight summers; on her lap sat a babe. Said my friend to the matron, 'Gude wife, this is Mr. Thorburn, from New York; he wishes private board for a week, can you accommodate him?'

" 'Yes,' says she, 'for a year or a lifetime, if it is his wish. Oft has my father told me, when he was sick, and a stranger, that Mr. T. took him in, and administered to his wants.'

"What was your father's name?" Thorburn inquired, with mock surprise not unlike his bedside strategy many years before.

"Robert Hoe," she replied, sweetly.

Ever thankful for Thorburn's merciful hospitality, Robert Hoe had delivered standing instructions throughout his prolific clan that the eccentric seedsman be welcomed wherever he traveled. Once, about 1845, Thorburn had occasion to journey to the country town where resided Hoe's daughter Rachel. He now met her with joy.

Rachel had done well: about 1840 she married Moses W. Dodd (1814-1900), an 1837 graduate of Princeton University who dropped out after two years at Princeton Theological Seminary. With stern Presbyterian training, however, Dodd found he could proselytize just as well publishing and selling religious books and tracts. He bought a share of John S. Taylor's bookstore business in New York but struck out for himself a year later with a tiny shop sequestered in a corner of Brick Church Chapel in City Hall Park. Here began the throes of an exciting new publishing career for himself and for his family as well.

The young couple lived in Doddtown, New Jersey, a family community eventually lost in greater Bloomfield. Rachel bore Moses "six sons in reasonably rapid succession," wrote Edward H. Dodd, Jr., her grandson, "and she imparted to them a charm and sprightly intelligence which went a long way toward leavening the sturdy moral character they inherited from Moses." To restore decorum Dodd was accustomed to fetch a harness strap and, with little warning, put his noisy sons—and their visiting friends—across his knee with unsparing thoroughness. The old seedsman and raconteur would find the Dodd household memorable.

"And is this your child?" Thorburn continued his coy questioning.

"It is."

"I held the babe in my arms, it smiled on my face," Thorburn concluded, his voice rising dramatically to the magnitude of this solemn occasion.

" 'Now,' says I, 'Madam, this day my prophecy is fulfilled in your eyes; it's just forty years, at a critical moment in your father's life, when I told him that I hoped to hold his grandchild in my arms.' " [1]

Thorburn proudly acknowledged that in his many travels he had received countless tokens of kindness from Hoe's "worthy family of sons and daughters." In a way money couldn't buy, the successful press manufacturer had munificently repaid the first debt he incurred in the New World. So Thorburn's curious contributions to the Hoe family were completed. With temperate habits and modest home comforts (including two or three wives), he enjoyed 90 years and died in 1863.

Frank H. Dodd (1844-1916) was unquestionably the babe who smiled on Thorburn's face. He not only survived his father's leathern justice, but, after a preparatory education at uncle Ebenezer Seymour's Bloomfield Academy, he also entered papa's establishment in 1859 with a ten-year apprenticeship to discover he was, like his father, "a natural born" book publisher. Another influential uncle helped:

July 23rd 1863

Messrs Welsh Bigelow & Co.
Boston
Gentlemen,
A young friend of ours in the publishing business desires to know whether you will print a work for him and use the type head pieces and initial letters in the Poems of [*Oliver Wendell*] *Holmes in Messrs Ticknor & Field edition of this year printed by you. The book would be quite different in character from that series & intended not to conflict with it in any way. We would like to know what your prices are in case you could do the work; an early answer is requested. A series would probably be published if you should undertake it. The party is entirely responsible, as we can satisfy you, but desires at present as we do that this communication should be treated strictly confidential by you.*

Yours truly
R. Hoe & Co.

The impulsive young friend could scarcely wait for a reply. He quickly got uncle Richard to write a personal introduction and rushed to Boston.

[*To Welsh, Bigelow, 27 July*] . . . *This will be handed to you by Mr. Frank Dodd, a nephew of ours, and the young man who we wrote to you as about to publish a book.*

We hope you will be able to carry out his wishes, and as it is his first venture in a business way on his own account, we are particularly desirous that it shall be very creditable every way as well as successful. He will explain his wishes & we commend him to you. We have told him to say to you that we will be his security for bills of paper, printing, & binding of the Edition of Paradise Lost. . . .

How auspicious to commence publishing with an edition of Dante, and how fortuitous to have unquestioned credit on a first venture!

In 1870 Frank Dodd acquired his father's firm and invited an old chum to join him as partner—Edward S. Mead (1843-1894), third son of Elizabeth, and younger brother of R. Hoe & Company's Theodore Mead. Frank was 26 and Edward 27 when the cousins merged forces. They needed all the help they could get.

Dodd and Mead began their operations in a nation-wide economic depression with a tired backlist and few lucrative author connections. Their youth and enthusiasm were a major stock in trade, however, and their talents complimented each other magnificently: Frank in the acquisition of important literary works, Edward (when not writing dozens of books, especially juveniles, on company time) apparently handling the manufacturing, taking dubious credit for the introduction in America of hideous multicolor stamped ink bindings. The company's first sensational bestseller (E. P. Roe, *Barriers*

Burned Away), by a minister who went to Chicago after the great fire, capitalized dramatically on that 1871 disaster; other "radical" fiction, including almost 20 other Roes, soon made the company quite prosperous. But early profits were elusive; in desperation a letter was posted to Richard:

April 8, 1874

Dear Uncle

I send herewith a statement of our affairs on Feb. 1, 1874 at which time we took our last account of stock.

In these statements the steretype† plates are reckoned at cost. Those bought of father ought to be worth cost for they were put at a very low figure. Many of them at their value for old metal.

The new plates made by us are not worth cost or at least could not be sold at such a rate. It is hard to say how much real depreciation has taken place on an average but probably 30 per cent would not be an unreasonable deduction. This would reduce our assets by $7617.00. I cannot account for the unfavorable result of last Year's business as here shown. We cannot as yet find any error in our books on stock account. We seem to be $5568.00 worse off than on Jan. 1, 1873 which, as we have drawn out in the thirteen months $6141, leaves the profit on a business of nearly $134,000 but about $500.00. Every calculation, our own past experience and the experience of others would make the gross profit at least 30 per cent on sales or $40,200. Our expenses including interest and all other items were $32,288 (equals 24 per cent on sales). We thus should have made at least $8000 which would have left us $2000 above our living expenses to add to assets. In ordinary times we can do our business at 20 per cent expense, & should be able to make 10 per cent at least on sales.

We are analyzing our sales for the present quarter & will take account of such gain on May 1st and thus hope to discover where the leak is.

I know no other way to get at it as it would be almost impossible to work over the mass of detail involved in the whole of last year's sales. If we are selling at too small profit in any department we shall now find it out & if stock is being stolen we shall also discover it.

We can in a few days give you a full statement to April 1st which will show a very considerable reduction in indebtedness.

Affe. your Nephew

R. M. Hoe, Esq.[2] *F. H. Dodd*

Although the cousins had at least one early instance of embezzlement by a young clerk (whose father made partial restitution), this "leak" proved to be a mere bookkeeping error, and required no further Hoe advice. But they got some anyway: on one occasion their uncles criticized the ordering of books by the half dozen when only one copy was wanted. But in addition to substantial loans to get the boys started the Colonel also let them charge manufacturing at Lovell's Lake Shore Press and other shops to the Hoe account. These obligations they would repay handsomely—if confidentially—when another addition to the firm was made.

In 1873 Frank's brother Robert H. Dodd (-1934) had entered Dodd, Mead. He started a rare book department in 1889 which became the largest and most prestigious in America. He managed his retail and antiquarian book business with rare acuity, training many assistants who ultimately embarked upon similar careers. Robert Dodd's activities indeed proved most fortuitous to his cousin Robert Hoe III. Lehmann-Haupt wrote of Dodd:

> His excellent personal and business connections placed him in a strategic position at a time when some of the great [book] collections were being made in this country. He had a substantial share in the building up of the famous Hoe collection. . . .[3]

The Dodd, Mead rare book department was a providential creature of the times. As machinery everywhere threatened to spew out countless copies of all manner of artifacts, the rich (having become wealthy by just such industrialization) began to demonstrate an interest in the precious handmade things of old. Such a treasure were books, especially incunabula—those printed before 1500—the most expensive intellectual symbols of all.

The Hoes for two generations in America had soiled their hands in the grease, graphite, and ink of printing machinery, the Colonel himself awakening only in the 1870's to the bookman's mania. His nephew had started collecting long before, but—despite the wonders of his own machines making things as deftly as any human fingers—he was beyond championing the industrial age. In his 1886 lecture on bookbinding Robert Hoe opined:

> At the risk of being set down as an ignoramus by some practical binder, I do not hesitate to say that it is as impossible to produce a fine bookbinding through the agency of machinery as to produce a fine painting by such means. . . .[4]

Somehow Dodd, Mead and other proud entrepreneurs, struggling to show that book publishing was a new profession with lofty aims including both literary and esthetic excellence, would be determined to try. Frank Dodd himself spent 55 years at his desk, seeing the old replaced by the new, retaining a capacity, in poet William Dean Howell's words, for "delicate and generous things" [5] for which all in American publishing knew him. He died in 1916, age 72, leaving a modest estate of $150,000. Dodd would have been a fine son in the old Hoe tradition.

Robert Hoe (II) was, like his father, perhaps, not inclined to be graphically oriented or literary. With a "mercantile" rather than a "mechanical" education, he was from early life destined to walk a separate path from his brother Richard, who at 15 had already demonstrated his cleverness for the firm, and from brother Peter, content to supervise the company's internal affairs. Robert was, in effect, a superb salesman, whether writing as a youth day after day to printers about steam presses never before seen in most American towns, cajoling clerks and agents to travel and sell, sell, sell, or in later years making transatlantic trips to vend colossal machines to cranky publishers. Robert's one weakness—if it can be called that—was real estate.

With ambitious expansion of the United States to the north, south, and west, and constant reports of virgin timber, fertile acres, and subsurface treasures beyond imagination, the promise of appreciable real estate profits seemed foolish to ignore. Robert was early drawn into land ventures, many of which he accepted on behalf of the company in lieu of legal tender:

New York May 27, 1837

Messr. Baggs Barnes & Co.
Detroit
Gents.

We have this day shipped the press as per your instructions, a bill of the same we annex together with our dft at 4 mos for the amount vis. $242.32 which you will please accept and return us by first mail. . . .

We have now a favor to ask of you if it would not be putting you to too much inconvenience. Your neighbors Messrs. Snow & Fisk have at length sent us a deed of the land we purchased of them but not accompanied with any description as we requested. If you could find time to go upon it and make yourself perfectly acquainted with its locality, and give us a description of its its† advantages & disadvantages, or in case it would be inconvenient to go yourselves to send some trusty person it would confer an especial favor. We also would like to have you examine the records & courts and learn whether there are any incumbrances or judgments upon it which could in any way effect† this property & all expenses attending it we will liquidate upon hearing from you. We annex a copy of the deed. The land is about 13 miles we believe from your city.

Very Respectfully
Yr Obt Svt
R. Hoe & Co.

Copy of Deed. All that certain parcel or tract of land consisting of 90 75/100 ft being the east half of north west fractional quarter of section N° one, in township N° 1, north of range N° twelve east being the same purchased by said Snow & Fisk of US government at land office Detroit Michigan.

Such purchases by proxy and default, understandable considering the vast distances involved, were nevertheless risky:

New York Nov. 4, 1837

Rev. Esra Styles[†] [*Ely*] *Esq.*
Dr Sir
A year ago last spring I purchased of government through yourself and Professor Agnew a half section of land. The premises which Mr. Agnew held onto as to the pains which should be taken in entering for me a tract of the very best quality were of the most decided character, and I have his written receipt to that effect. He was an entire stranger to me but coming among us as he did endorsed by you I felt that I might put every confidence in his statements. . . . It was not my desire to trouble you until I had ascertained beyond question the truth of these statements. Accordingly I have employed Saml Agnew Esq. at an expense of $18.-- to survey it for me. He did so in September last with a friend and in his letter of 25th Sept. says "they were of the opinion that the land was none of it better than third rate & much of it no quality," that is, valueless. I presume you are acquainted with that gentleman and will not wonder that I am surprised at this statement. . . . It was my intention one day to settle upon it myself and you may imagine what my disappointment would have been had I gone to the far west and found it in the condition above described. My feelings are the best nature toward yourself and the institution over which you preside. In justice to myself however I must request you to send me the am^{t} I have paid vis $500. & with 10 pct interest as per agreement in case of dissatisfaction. I would be pleased to receive it in fund in this city by return mail. I hope you will not delay to reply to this as I am very desirous of hearing from you immediately.

Yours very Respectfully
Robert Hoe

To Rev. E. S. Ely, D.D.
Marion College
Palmyra

P.S. My land lies on the west line of Township No. 55, Section No. 19 South half, North of base line Range No. 7 west of the 5th principle[†] *meridian. I understand that it was in Rolls County but Mr. Agnew informs me it is in* [*undecipherable*] *County. R.H.*

Reverend Ely and another, Abner S. Ely, were apparently relatives of Giles S. Ely, who married Emeline, the brothers' second oldest sister. The Elys lived in Brooklyn.

Robert's real estate operations, aided by his brothers, nevertheless grew to extensive holdings, many of them on behalf of the company. When Robert Hoe III died the accumulated real property would be so involved a separate estate-run corporation had to manage it.

Robert in 1838 married Thyrza Mead, evidently a relation of Enoch, husband of his sister Elizabeth. The Hoes, Smiths, and Meads were well established in South Salem, where the ceremony took place:

[*To Richard Hoe, London, 14 May 1838*] *. . . On the 23^{d}, brother*

Richard, I would be very glad to meet you among my other brothers & sisters, cosens† & friends in So. Salem. The occasion you will doubtless imagine. So dear Dick you see my youthful days are drawing to a rapid close. . . .

Robert was 23. "[W]e all intend to go up to the wedding and return the next day," Matthew Smith added (15 May). "Things are looking better at the B&D [bank], but business is extremely dull. . . ."

The Robert Hoes had three children: Robert junior (born 1839), Arthur (1841), and Laura (1845-1942), who did not marry. While Robert really harbored no intentions of settling in the "far west," he succumbed early to the infatuation of a country estate (in addition to a townhouse), although his role as salesman would conflict. His farm at Stoney Brook, Long Island, was a pleasure fast turning into another problematical investment:

[*To Benjamin Brown, 11 July 1850*] . . . *I have not made any progress on my farm yet. The season is getting so far advanced that I shall have to let some one else gather the crops, I fear, tho I have not lost relish for a place. Business this week has been hurrying us up uncommonly. . . .*

If Gen[l] Taylor knew too little about politics perhaps Mr. Fillmore may not be liable to any such charge. I am afraid that the Democrats instead of the Whigs are Tylerised this time. . . .

Robert soon regretted the inconvenience of traveling to Long Island and again confided in Brown:

[*31 October 1851*] . . . *My Stoney Brook farm has cost $7200, including it all. How much ought I to ask for it? I think it should bring not less than $7500 as it is in perfect order. . . .*

Robert and family occupied his 11 East 36th Street home for many years, welcoming various members of the growing clan for visits, festivities, and weddings. The Hoes were accustomed to address distant cousins and in-laws—and their offspring—as "brothers," "sisters," and "cousins," confusing the exact relationship of several individuals.

Always near to cholera (a common non-epidemic form) and other seasonal and fatal diseases which had taken Newton, the Smiths, and several infants, the Hoes embraced life through hosts of active children; large families were a refuge and a comfort. The brothers encouraged the young to visit and play, stay overnight or for the summer, join them while traveling, or exchange news by mail.

Robert had expectations his second son Arthur would help run the shop. The lad was anxious to participate, but his uncertain health did not improve. A warmer climate was indicated, but Arthur did not survive. He was suddenly seized with a paroxism:

[*Robert Hoe, Barbados, to Robert III, 25 December 1865*] . . . *I was present in a moment & he immediately said, "Father, I am dying."* . . .

I feel sure that his prayers and ours are answered and that he was prepared to go when God his Heavenly Father called him. . . .

Considering the firm's growing obligation to entertain, Robert's Manhattan townhouse was undoubtedly meeting place for pleasure and business—until the Lightning in England opened a vast new territory for the company. This development sent Robert abroad so frequently he seemed expatriated: "one of the partners, being a resident in that city [London], attending to the business there," [6] a reporter confirmed.

In addition to real estate transactions, Robert functioned in several insurance companies and savings banks (some of which he may have been instrumental in forming). He was a supporter of the Metropolitan Museum of Art as well as member of the National Academy of Design. *Harper's Weekly* said of him:

> [H]is friends knew Robert Hoe as a man who liked to hide his life, who preferred to perfect himself rather than to build a reputation, and who was endowed with a singularly happy moral balance, and a genuine gift for the good and the true. Like all such natures, he had a real love for living a life of his own, apart from the unsympathetic lives of others. . . . The pleasures of the club and of fashionable society did not entice him, partly because of a certain frailty of health, partly because of his fondness for the country, partly because of his extremely happy domestic relations, and partly, also, because of his taste for what is administrative rather than for what is imperatorial. . . .

In 1871 Robert and his family moved into their new country estate at Tarrytown, New York, of which he was very proud. He named it "Lothar," possibly after the hero Lothair of Disraeli's popular 1870 novel. It was in no way a "farm," but Robert hoped its country charm would sustain him. But one fine moment of public glory was still ahead. *Harper's* continued:

> When, therefore, the wholesale thievery and incipient communism of the Tweed *régime* were to be met and smitten, who of all that now historic Citizens' Committee of Seventy was more fit for the work than he? It was the sole instance in his long and masterful career when he consented to assume a conspicuous civic position, and the assertion may safely be ventured that none of his comrades did less talking. A man who makes presses that print, cut, and fold twenty thousand large eight-page daily newspapers in an hour is, perhaps, professionally indisposed to loquacity. . . .[7]

Robert's committee work was nevertheless demanding. Announcing to his brother (13 September 1871), "There has got to be some hard work done before such scoundrels, as rule us now in the city, can be brought to justice . . . ," he went to Albany three times in 1872 to help present a new city charter featuring proportionate representation, which he considered appropriate for New York's "promiscuous population." But with the summer strike and other business pressures, he was anxious to resign this civic chore. (In 1873 he was reappointed a school commissioner for three more years, his duties undisclosed.)

In 1878 Robert had his 36th Street townhouse completely refurbished to signal his intention of quitting Europe and retiring in some comfort near the plant. In his characteristically precise way he instructed his contractor:

> [*To Mr. Van West, Katonah, 10 May 1878*] . . . *Paint the whole of the kitchen walls and Kalsomine ceilings and stain the floors.*
>
> *Room next* [*to*] *kitchen, Kalsomine walls and ceilings and stain floor.*
>
> *Waitress' pantry, paint the walls and Kalsomine Ceilings.*
>
> *1st story Northeast room, paper walls, Kalsomine ceiling and stain floor.*
>
> *Hall, paint walls, Kalsomine ceiling and stain floor.*
>
> *1st story South room, paper walls, Kalsomine ceiling and stain floor.* [*2nd story, four rooms, the same treatment.*]
>
> . . . *All walls to be painted three coats.*
>
> *You to furnish all materials except the paper, which I will supply. The rolls of paper I have are 21 ins. wide (twenty-one) and 8 yards long. Let me know the exact quantity each room will require, as all will be different. Make an accurate estimate, that there may be no waste.*
>
> *I will send you the red paint, already mixed, for painting stairs in old and new parts of house. You are to put it only on the steps and risers and string pieces, letting it stop where string piece joins the base. All the bases in halls, including bases on stairs next* [*to*] *wall, are to be painted white in new house and a special tint in old house, & shade darker than walls. . . .*
>
> *I do not feel willing to pay over $200 and think this should be an outside price. . . .*

In 1881 Robert nevertheless made another trip to England, but psychologically he was indeed ready for a rest. After a minor illness a year later, back in New York, he gave his brother a chance to travel. The Colonel (in France) reminded him of his lifelong preference for homeopathy:

> [*To Robert, 16 December 1882*] . . . *I hope you have done with Doctors, but I would never have either of those two you had in New York—their pretentions are infinite. A poor miserable Homeopathic Doctor who attempts to ameliorate your condition with the millionth part of a drop of some poison is much the safer man to trust in. You see I could not forego the pleasure of a fling at the Medical Fraternity. . . .*

"What do you think of my sending you over a Try† Cycle[?]" Richard added. "I read an article [saying] one can propel himself on one of these machines 4 miles with as little fatigue as walking! . . . just first rate to go around your place and mine on. . . ."

Robert soon retired to his home at Tarrytown, where he died, 13 September 1884, aged 70. "Malarial fever" was blamed. He was buried in a new family vault at Sleepy Hollow Cemetery. He left his brothers Richard and Peter, and his son Robert to manage the company. His death was called the first serious test of the family structure since his father's in 1833, but that was mostly rhetorical. The Colonel was still very much in command.

Richard March Hoe had long worked so hard and steadily he scarcely knew how to relax. Office pressures were staggering, and there seemed little time for unrelated social or civic affairs:

> As a business man Colonel Hoe was fully as successful as in inventing. Up to his death he retained the dominating influence in his firm, and was vigilant, if not as active, as ever in the management of its business. He was a man of public spirit and of a warmly charitable disposition.[8]

The Colonel's public spirit and charity (and hence the company's image) was one of sober—limited—generosity. Although the machine works prospered with every passing year its officers did not seem to notice the ghetto and slum which the lower East Side had become, nor did it work extensively for better government, whether local or national. Hoe public efforts took many little forms:

> *New York, January 28th 1854*
>
> *Messrs. R. Hoe & Co.*
> *Gentlemen*
>
> *At a meeting of the Board of Directors of the "Printers' Free Library," held on Saturday evening, 21 inst., it was unanimously resolved, that the Chairman and Secretary acknowledge, and return the thanks of the Board for, the donations received at the Banquet on the 17th inst. in aid of the Library.*
>
> *Agreeably to said resolution we would hereby acknowledge, and tender you the thanks of the Board, for your liberal donation of $100, in furtherance of the objects of the "Printers' Free Library."*
>
> *In behalf of the Board*
> *James O. Taylor, Secretary* *Peter C. Baker Ch[airman]*

Baker in 1850 had started a flourishing printing business with Daniel Godwin and John Thomas. Hoe occasionally donated a cylinder press to colleges for student use, one of which was the Black Hampton Institute, Virginia.

Perhaps the Colonel had many more requests for charities than he could possibly honor. Widows and children of old friends and employees were particularly successful in reaching him:

New York November 12, 1876

Mr. Co† Hoe

I feel it my duty to call and see you & now it is not very Plesant† for you to have me to call at your office and as you ask me where I could pay you at that presant† time I could not say. I have let the Lower part of my House with 2 rooms on the third floor for Fifty Dollars per Month. Times is so Dull I could not get any More; the same apartments has† let for eighty Dollars. [I]f you will accept of the fifty Dollars every month I will send it to you or I will bring it myself. Mr. Bailey has Put it [the house] in Several Agents hands for Sale; he is now trying to hire Money to pay the interest and taxes and insurance. [H]e leaves here to morrow morning for North Port to see wat† he can do to git† some money.

He thinks if he can hire that a Mount† for a few Months so that it would not be Sold at Auction we Would Git our price. We had 2 Calls to look at it—they see† it in Good order. I have spaired† no time in trying to git the Money myself but it seems it† impossible to git anny† Monny† and if he Dont Suckcede† in giting it we have to let the Law take it† corse.† I do not want you to loose† one cent of your Monny if you are willing I will Make the House over to you rather [than] have it Sold for less then† it† worth of.† I could keep it till the first of May. We git our Price and it will let at lease† for twelve Hundred per year. You have done more than I ever expected for us and I hope you will be rewarred† for all you† kindness.

Your† in return

Please excuse this poor writing. *Mrs. Sarah Bailey*

This Bailey was probably the pensioner of Robert's letter to Richard, 8 July 1872, during the company strike.

The Colonel was usually receptive to such pleas, quietly making a payment here or canceling a note there. He was also very generous with letters of recommendation which were quite valuable for former domestic employees seeking new positions, and for travelers uncertain as to distant accommodations. All things considered, Richard's life style became fairly rigid and consistent: not much time for entertainment, not much patience with mediocrity, not disposed to much generosity—but a "soft touch" to old friends.

Hoe reportedly commanded the Washington Greys, a prestigious New York National Guard unit (still in existence) during the 1830's, although after about 1842 he participated in so few military functions reporters unaware of his service usually considered the commission merely gratuitous. He must also have commanded, if only from his busy desk, the "Hoe Guards," a company of employees whose appearance in the 1850's the *New York Times* considered "as respectable a muster as the City numbers among its impromptu corps."[9]

The Colonel was a member of the Union League Club and later in life a warden of New York's 24th Ward, as well as a governing member of West Farms' Home for Incurables. He was an officer of the

Central Gas Light Company of New York city—among a few honorary positions he could not successfully escape. But his most important community work centered around the House of Refuge, which, like his apprenticeship program, gave him immense satisfaction in raising and training young lads to be responsible workers and citizens.

Richard Hoe was twice married. Lucy Gilbert, of South Salem, New York, had given him two daughters, Emily (born 1834) and Adeline (1836). After Lucy died in 1841 the Colonel was married again, to Mary Say Corbin, of Philadelphia. Friends were delighted for him:

> [*From William Swain, 23 November 1842*] . . . *I have been to Philad and heard of your prospective pleasure* [*engagement?*], *and am glad. Should like to be in her place; that is, stand in her shoes, sleep in her bed, &c. You understand me, I should like to be the woman herself, under such favorable circumstances. . . .*

> [*From Edward Sanderson, 17 August 1843*] . . . *I take it for granted that nothing has occurred to prevent . . . the auspicious & happy event you announced was to take place on the first of this month. Permit me therefore to offer to you & Mrs. H. my sincere and hearty congratulations. . . .*

With Mary he had four more girls: Anne (1852), Mary (1854), Fannie (1855), and Helen (1858). Six daughters (and no sons) begged for comment: "Philosophers, lawyers, editors, poets, literary men and brain workers generally have a large excess of daughters,"[10] the New York *World* noticed in 1883. The Colonel might indeed have been flattered by this inclusion in intellectual society. He was never disappointed by his girls.

Rarely indulging in formal or prolonged entertainment (except for an occasional concert or opera), Richard and Mary somehow found time to be very cordial to old friends:

> [*To Charles Van Benthuysen, 25 April 1850*] . . . *We have received a line from our friend & correspondent in Charleston. He says "In reply to your enquiry I should say that the weather would not be too warm for an invalid up to the 1st June. We consider April & May our pleasant months, and think your friend would be pleased. . . ." The Southerner leaves pier N° 4 North River at 4 o'clock P.M. Thursday, May 3rd. Certainly it would be a pleasure to have you & madame with your son when you come down direct to 309 East Bway & she* [*Mary*] *will strive to make your good lady alter her opinion of New York. I shall expect you. . . .*

Occupying the East Broadway home from 1835, the Colonel himself soon hungered for the bucolic life praised by so many nineteenth-century gentlemen. About 1851 he made his move.

For his large, fair-haired family Hoe purchased a handsome estate in Morrisania, Westchester County (later—1874—absorbed into the Bronx), three miles north of Manhattan's village of Harlem. A homestead dating to the 1760's, he named it "Brightside." Nearby was the "Mount Fordham" farm of Lewis G. Morris, at this time the most influential short-horn cattle rancher in the United States, as well as other gentlemen farmers whose animal stocks were local pride and joy. Here the Colonel applied his customary efficiency in a twenty-year program he enjoyed immensely—except for the "enormous" cost.

Hoe drained and graded the rolling hills which swept down to the East River; he lined his lanes and paths with fruit trees and hedges, planted grapes and strawberries, and cultivated varieties of flowering bushes and shrubs. Occasionally there was time to relax:

> [*Theodore Mead to Robert Hoe III, Paris? 31 July 1856*] . . . [S]*eated around the table they* [*Richard, Mary, Peter, Hannah*] *all partake of ice-water tempered with a little "Old Queens," Uncle Richard cautioning me between the sips against giving way to smoking or imbibing. . . .*

The next day "Old Queens" talked recklessly:

> [*1 August*] . . . *The first few letters of Uncle Robert's written on thin paper were so difficult to read that Uncle Richard declared that he should write him to use thicker paper if it cost $24 instead of $.24. He probably forgot it. . . .*

As trees, plants, and vegetables bloomed and blossomed, Brightside was ready for an animal population. The Sandersons sent over some cattle, as did another country gentleman:

> [*Edward Sanderson to Richard Hoe, September 1856 (no date)*] . . . *I am in receipt of yours* [*of*] *13th ult. asking me to look out for a "famous Cow" for your place in Westchester, so that you can "brag" about her with your neighbors &* [*have*] *the right sort of stuff for milk punch. . . .*

Having just become acquainted with Edward Lloyd, the Colonel was delighted to meet a fellow enthusiast eager to share experiences:

> [*Richard Hoe to Robert, London, 29 June 1857*] . . . *I am inclined to think Mr. Lloyd would be a better hand to aid you in finding one* [*a young heifer*] *than any one else you know. I should like a pure red Devon if I have any. The one received from Mr. Lloyd is considered a very fine animal, said to be equal to any in the neighborhood. . . .*

The Colonel now had plenty of fresh vegetables and fruit, and was busy inviting friends and family to visit. Within just a few years Lloyd had sent him a veritable menagerie:

New York Febry 7th 1859

Edwd Lloyd Esq.
12 Salisbury Square
London

My Dear Sir,

Your esteemed favor of the 14th ulto came to hand last week & the fowls a few days afterwards; they are certainly fine birds & much admired by all who have seen them and especially so by a brother in law of my wifes who is one of the best judges of fowls in the country. He obtained for me a few rare birds from a friend of his in Philad^{a}; they are there called the Billy Beard—games-fine plumage & have yellow legs & bills. It would give me great pleasure to send you a cock & two hens if you would accept them.

The cock & hens you sent me last summer are doing well. I raised one brood from them & have three pullets that are the exact counter part of the largest of the hens & five (two cocks & three pullets) from the smaller hens.

The Durham heifer went to bull 4 times last season but did not take. I have had her examined & she is to be operated on, . . . which the vet. says will obviate all difficulty. My neighbor Mr. Faile who has the best herd of Devon in the U.S. has been to see the[†] *Mayflower* [another heifer]. *He pronounces her perfect in every respect except that he thinks her rather small. We are expecting her to calve next week.*

My place is very small only a little over 16 acres but the land is strong & we are getting it in thorough order. I have laid down about three miles of tile drain pipe from 3 to 5 feet deep & what was very wet is now in fine condition. I am planting and transplanting trees & one of these days mean to build a good new house. The old cottage I put in good repair & it is very comfortable but the rooms are not large & the ceilings are low. It was built some ninety years since.

I should be most happy to see you out here & think we could make the trip a pleasant one at least after your arrival here. I have made various efforts to obtain some of those wild plants you wrote about last year but have not as yet been successful. . . . Please accept my warm thanks for your attention to my commissions & believe me

Yours very truly
Richd M. Hoe

[To Lloyd, 10 May 1859] . . . *I have within a day or two had Mr. Edwd G. Faile, who is one of the first breeders of Devon cattle in our country, up to see the Devon Heifer* [Mayflower] *& her calf; he said I was very fortunate in getting so very fine an animal & pronounced the calf which is a heifer to be perfect in every respect—a beauty.*

My Durham cow although sent many times to bull last season did not take. . . . She has now gone to a place about nine miles from here to one of the best Durham Bulls about.

I had bad luck with the last game cock you sent me as he died from some cause. . . . If without much trouble to yourself you can procure for me another cock of the same breed & size as the last I shall feel very much obliged. He was an uncommonly fine one; I suppose he would come better with a couple of hens.

The Dorkings [fowl] you selected for my brother are very much admired & are considered by Judges to be the best we see in the country. Should you find the game chickens I should like them sent as soon as possible so as to get a brood from them this summer.

If you want a pair of trotters of moderate size, say about or not over 15 hands . . . I have heard of a pair that can be bought for 800$ & another pair . . . for 1000$. They are good and sound. . . .

While he loved horses and racing, the animals often let him down.

[To Robert Hoe, London, 8 November 1859] . . . I shall write to Mr. Lloyd by next steamer. I have had bad luck in getting horses for him, the team I expected to send him were taken sick with a distemper the day they were to have been tried on the track. I then went up to the State fair (agricultural) but could find nothing. . . .

[To Robert, 30 March 1860] . . . I wish you would tell Mr. Lloyd that my friend Mr. Faile . . . says I need a more particular pedigree with my heifer, one with the herd book numbers for sire, grand sire, &c, dam, grand dam, &c. . . . Mr. Faile says should I wish to sell any of the stock it will make a difference of one half in the price by having the pedigree perfect. . . .

Over the years the Colonel labored tirelessly at his country tasks. Shrubs and hedges, fruit and flowering trees, vegetables and strawberries (a Hoe specialty) and flowers followed the seasons with almost commercial success:

[To Robert, 26 January 1875] . . . While I was home sick [with a cold] I had a look over the Pig business with [caretaker] Mr. Harris & found he had orders for more young pigs than we could produce, say about 100 ordered at 15 dollars each, 10 to 12 weeks old, so I wrote for 20 young sows in pig. . . .

My Camelia house looks splendid, and Mr. H. is now selling the blooms for $10 per hundred. . . . I had just in time for the cold weather completed the new chicken house. . . .

He had little time to reflect, but when he did so he was pleased:

[To Robert, 23 April 1876] . . . Brightside has run away with so much money that sometimes I feel as if I would part with it & thus end the constant drain, but it would seem as if it could be run on a moderate sum. . . . It is a delightful home. . . .

As he had hoped, Brightside was as perfect for raising a family as it had been a country farm.

The girls grew to be perfect credits to their parents. Mary Hoe married Joseph Henry Harper in 1873 and would share a publishing life. Her husband, son of Fletcher Harper, at the age of 14 in 1864 had been sent to European schools to learn French and German so as to handle foreign affairs for *Harper's Bazar*. He received a delicate compliment from Harper editor George W. Curtis upon his marriage:

> [Y]ou may be sure that I shall think of you and your bride, and pray that your life may be as full of blessings as June is of roses. I do not know her, but from what my friend Mrs. Cranch tells me, I am sure of your happiness: and I am more of hers—for I do know you! . . . And let me hope that your wife will not reject the good wishes of a friend whom she does not know. . . .

Joseph became the historian of the Harper family and remembered his father-in-law with some affection:

> Personally Richard M. Hoe was a man of exceptionally cheerful temperament and gentle ways. He was devoted to his life-work, but at the same time was essentially domestic. His name is ranked with those of Fulton, Whitney, and Morse; he may indeed be styled the modern Faust; such an impetus has the art of printing received from his marvelous inventions that he has almost created it anew. Their branch house in London is an extensive establishment, and the Colonel spent much of his time abroad. He was almost as much at home in London as he was in New York, where he was often mistaken for [English statesman] John Bright, whom he resembled. . . .[11]

The Hoe-Harper marriage seemed so blessed (producing six children) it was enthusiastically duplicated four years later: John Harper, son of Joseph A. Harper (Joseph's cousin), married Fannie Hoe, Mary's younger sister. They had four children.

The two girls' older sister Anne had married Charles Platt but had no children; semi-invalid all her life, Anne was accommodated by an ingenious folding wheelchair developed by her father for their frequent European museum visits. She died 1887. Helen, the youngest of the Colonel's offspring, died in infancy: he had sadly lamented her loss as he fumed over Whitworth's intransigency. Brightside was nevertheless a haven for the women as it was for its creator.

Constantly troubled with colds and other nagging family maladies, the Hoes at first feared the climate of the Bronx, but they finally judged it perfect. The Colonel participated in the work incorporating the southern tip of Westchester County into New York city (1874), and in several plans to get fast railroad service to Morrisania and other communities. He may have persuaded his friends the Schieffelins, who (in 1894) moved their chemistry shop from Pearl Street to Southern Boulevard and St. Ann's Avenue, very near Brightside. Despite a taxing work schedule the Colonel enjoyed good health and was seldom incapacitated by his rheumatism.

In November 1859 Richard had hitched his wagon to post a letter, and as he started off one of his two colts panicked and sent the vehicle into a tree. Thrown under the wreckage, he crawled out, his left leg broken above the ankle. He was a surprisingly good patient:

> [*To Robert Hoe, 28 November 1859*] . . . *Here I am on my bed, a borrowed (patient) one, which can be elevated so as to make a back to it, in fact almost into a chair. My leg is slung from the ceiling which is an arrangement of my own or rather my idea executed by my country Doctor Naudane & my son in law Cyrus. It relieves me from an extra amt of pain which was fast becoming too great for my fortitude.* . . .

The Colonel could expect to remain in bed about a month, at which time he might try crutches. He sent for his foremen periodically for first-hand reports and was soon immersed in shop problems as before. In January he was back at Broome Street, with an armchair and pulley system to hoist him between floors. To some extent, nevertheless, more sedentary activities for a while were prescribed.

Why not collect modern oil paintings—an excellent investment! his son-in-law must have suggested. Cyrus J. Lawrence (1832-1908) had married the Colonel's daughter Emily in 1856, taking her resolutely out of the drab factory world and into the aura of banking. A year later DeWitt Clinton, Cyrus' younger brother, married Adeline, Emily's younger sister. The clan immediately recognized the Lawrences as a different breed:

> [*Theodore Mead to Robert Hoe III, 18 October 1860*] . . . *De Witt & Cyrus were there* [*a family gathering*], *too, looking as gaunt & hungry as possible. Making haste to be rich perhaps doesn't agree with them.* . . .

"Able, upright, and sagacious," Cyrus (and his brother) joined the financial community to become one of the "oldest, most conservative, and most respected members" of the stock exchange. He was a director of the mammoth Bush Terminal Company, Brooklyn, and of several railroad companies, a member of the Metropolitan Museum of Art, American Museum of Natural History, American Archaeological Society, Bibliographical Society of Boston, Municipal Art Society (New York), Grolier and Union League Clubs. He began to collect paintings as well as Chinese porcelains and Phoenician glass. Although poorly educated in art, said the *New York Times*, "In middle life he developed the taste and gift for aesthetic appreciation, which became more marked with experience until it attained an unusual degree of certainty and refinement." [12]

Richard—who for a time rented office space to his sons-in-law and relied frequently on their financial advice and services—could not ignore Cyrus' enthusiasm for art. He resolved to form a picture collection, although he admitted to little expertise. His association with Samuel F. B. Morse and his portrait in Christian Schussele's massive

1862 "Men of Progress" (hanging in Cooper Union) incorporated with those of Colt, McCormick, Cooper, Goodyear, Ericsson, Morse, Howe, and other giants of American industry, had not equipped him to evaluate professional artists. Yet, more than all the other Hoes, Richard had so long busied his eyes with blueprints and patents he may have longed most for the painter's world of color and fantasy.

The Colonel's early selections, however, did not stand the test of time. In 1882 he decided to discard some of his first purchases. George A. Leavitt & Company, Broadway, auctioned 42 pieces (along with 115 works of his friend Levi P. Morton, later vice president of the United States and governor of New York) which the public also found disappointing. Represented were the artists Church, Kensett, Toulmouche, Chavet, Le Jonghe, Col, Fichel—not, alas, all lasting names. The *New York Tribune* art critic diplomatically complimented the sellers rather than condemn the art:

> [O]n the contrary, it is to their credit, as men of taste and discernment, that they should wish to disembarrass themselves of works acquired in the course of that painful and expensive process known to all collectors as "education," and which they have now outgrown.[13]

Although Lawrence reportedly guided him in the purchase of Cassatts which would indeed have withstood any scrutiny, the Colonel is not on record as ever having owned one. Nevertheless, toward the end of his years he had put his collection in fine order: "His art gallery," said a later critic, "is filled with many valuable paintings and marbles, and his private library is one of the largest and richest in New York." [14]

Indeed, the Hoe library was more curious than any critic could know. Some time in the late 1870's the Colonel commissioned a London rare book dealer to collect for him a library of books on printing and related graphic arts. Strangely, neither the noisy commencement, decades before, of his nephew's collection, nor his own many business trips to Europe had encouraged the delights of deliberate and selective book hunting. His earlier private library, although "large and rich," could only have included what must have fallen his way over the years. Now, however, he would rectify this discrepancy, as he had his taste for art.

The printing library was duly assembled and fully described in an 149-page catalog exquisitely printed in England by the Chiswick Press: *The Literature of Printing: A Catalogue of the Library Illustrative of the History and Art of Typography, Chalcography and Lithography, of Richard M. Hoe* (1877). He busily made plans to house this new acquisition, but because of delays the boxed library was instead put into storage. The Colonel would never see these books; they were uncrated and auctioned posthumously.

Many artists and writers recognized an inspirational story in the Colonel's long life, his mechanical wizardry, and his impressive family business. This growing fame had not escaped him, but he needed some encouragement:

> *[Robert Hoe to Richard, Rome? 2 March 1872] . . . You inquire whether we would like to have a bust of you in marble. Of course . . . yet, if I had my choice a fine painting of you by Charles Landelle of Paris would be more to my liking. . . .*

The Colonel had probably already met a sculptor named Summers in Rome and promised himself to carrara marble. The resulting bust was given by Peter Hoe in 1890 to the New York Typothetae, where it "graced" their rooms for many years.

Although 74 in 1885, Richard still attended daily to business affairs, commuting from his Brightside estate. The hour's trip did not dampen his spirits, said a reporter, who found he was

> . . . accustomed to walk through the long aisles of his workshops whistling the latest popular operatic airs and chatting pleasantly to his employes, many of whom have spent the best years of their lives in his service. . . .[15]

Upon a recent meeting with Arunah Abell, his intimate friend of some fifty years, at Guilford, the latter's "country seat," the Baltimore *Sun* commented, "They have ever taken the tide at the flood, and gone on to fortune. They both are ornaments of American life. May their hairs whiten peacefully. . . ."[16] Abell indeed retired graciously with his era: when Mergenthaler (then a Baltimore jeweler) succeeded with his typesetting machine Abell refused to buy it—he would not throw old hand compositors out of work. But the Colonel considered himself no ornament; there was still much to do.

In the late 1870's and early 1880's he inspected European printing plants, studied foreign patents and machinery (especially lithographic developments), and took his family on ambitious side trips to Tunisia, Greece, and Turkey (where he saw three of his Adams presses, purchased 20 years before, still printing government forms). Again in October 1885 he escorted the group through England and the continent. He was in good health and fine spirits for a long tour, his party not reaching Italy until the following spring. In Florence Charles Platt, husband of his daughter Anne, became ill; while thus detained the Colonel himself was suddenly struck with heart disease and died, Monday, 7 June 1886.

The Western world was profoundly shocked, and acknowledgments of Richard Hoe's greatness accumulated rapidly. The New York Typothetae's tribute to the Colonel, "whose trademark on his presses have[†] been everywhere accepted as the synonym of honest, thorough workmanship," gave unreserved honor "to the public-spirited citizen and genial friend, who never wearied in kind word and good deeds. . . ."[17] "Those who worked with him held him in affectionate respect," said Henry Bullen, "—a quiet, assiduous, friendly employer, appreciative of talent and good workmen. . . ."[18]

St. Ann's Episcopal Church, Morrisania, was so crowded for the August funeral hundreds had to stand outside on the grassy slope. Many Hoe employees were present, as were members of the General Society of Mechanics and Tradesmen: for most of these lower East Siders it was a unique visit to this newly incorporated section of their sprawling city. Pallbearers were his distinguished friends and associates: Stephen Tucker, Joseph A. Harper, Lewis Morris, Aaron H. Burlingham (Baptist minister), Samuel C. Caldwell (*New York Tribune*), William R. Beal (businessman, philanthropist), Benjamin H. Field, Theodore De Vinne, and Charles T. Galloway. He was unobtrusively interred in a vault under the church.

The Colonel's will was revealed about the same time. As executed by his wife Mary, Joseph A. Harper, and Stephen Tucker, the stipulations were surprisingly simple (if ambiguous years later). To Tucker he left a three-volume edition of Crocker, *Dictionary of Arts and Sciences;* to his nephew Richard, Peter's son, he left a microscope. The remainder of the estate went to Mary, and on her death was to be divided into five parts: Emily A. Lawrence (wife of Cyrus), Anne C. Platt, Mary S. Harper, Fannie B. Harper, Adeline's issue (DeWitt Lawrence) were each to receive a fifth. Because Annie died only a year after her father, the will would not go unchallenged.

Upon the death in 1901 of Mary Hoe, the Colonel's widow, the remaining daughters stood ready to divide the estate. The Platt family, however, did not see itself disqualified so quickly: $150,000 was at stake. Hoe's will specified "that if any of the children died before the widow that child's share should be bestowed as though Mr. Hoe died single,"[19] explained the *New York Times.* Arguing over whether "single" meant "bachelor" or "widower," parties to this litigation in 1903 became final witnesses to the dissipation of the Colonel's fame and fortune.

The fine Brightside library, including cases of rare books on printing, was first to suffer a strange fate—summarily auctioned under incredibly deplorable conditions. In May 1887 the tomes were released at minuscule prices: in 1433 lots, they sold for 10 and 25 cents per lot, one going for 1 cent. The important parcels of printing and graphic arts fared no better: a copy of Bodoni's 1818 *Manuale Tipografico* brought $11. Except that most of the buyers were representing various college libraries—who may indeed have conspired in the higher interests of their academic patrons—the Hoe typographic collection was squandered mercilessly. (Remarkably, despite this first-hand example of auctioneering injustice, Robert Hoe III had probably already decided to dispose of his own library the same way.)

In June the Colonel's prize cattle were auctioned. Again a disaster! Some of the animals had been boarded at his nephew's New Brunswick farm, where an epidemic of "pleuropneumonia" erupted. Shipped back to Brightside in December, the returning Hoe animals jeopardized the entire herd, necessitating a six-month quarantine. Although in June the time period had not quite expired, the suspect cattle were nevertheless put on the block; there were few buyers. Theodore Havemeyer and Levi Morton took a few animals. The carefully registered stock—Hoe's life-long pride—went for about $64 a head, the price of common cattle. In all, the auction brought only $2430.

Despite the loss of its greatest motivating force, R. Hoe & Company seemed unaffected. Robert Hoe III quietly assumed complete control, but as the business ran smoothly he was seldom needed for daily operations. The company's great prosperity—and workload—apparently put further homage to the Colonel out of mind. In 1901, upon the death of his widow, it appeared she had wished to correct the omission of history's notice of her husband. Mary had left instructions that he be honored in some fitting way.

Again fate stepped in. In November 1927 four memorial stained-glass windows at St. James' Episcopal Church, Madison Avenue, New York, were installed in memory of the Colonel. At the dedication Bishop William T. Manning spoke, taking the opportunity to condemn, like old Bishop Wilberforce flailing at Darwin, new anti-God forces threatening the world: "Powerful influences are at work in our land," he declared, "seeking quite openly to discredit Jesus Christ and challenging his right to be our guide. . . ."[20] But even here something miscarried: nothing was said about the Colonel's influence on world literacy and suffrage, or his foundling work. Fortunately, the great contributions of Richard March Hoe had already been recognized during his lifetime, and millions of newspapers continued to acknowledge him every day.

Peter Smith Hoe, the third and last son of Robert Hoe, born in 1821, married Hannah M. Smith (possibly also from the early partners' family) in 1846. They had five sons: Stephen Smith (1847-1887), Richard March (1849-1925), Charles Robert (1852-1926), Edward Sanderson (1856-1921), and Frederic Howard (1858-1897). Like Peter, perhaps, the family was blessed with only conventional ambitions and modest tastes to match them. Peter and his boys at first lived near the Colonel's Brightside residence, but when his son Edward began farming in New Brunswick, the father enjoyed New Jersey more and more.

Peter Hoe's particular fascination was ornithology, although there is little to describe. His letters frequently asked for specimens:

> [*From George Van Bokkelen, San Francisco, 4 June 1858*] . . . *I have not forgotten yr quail. What we here call the Mountain Quail are very hard to get, they are much prettier and larger than the Valley Quail. I shall strive to get them for you*. . . .

Interest in wild birds may have involved hunting:

> [*To a Maryland inquirer, 10 October 1859*] . . . *My pointer dog is second to none in appearance, is very staunch and I have no objections to have him tried by any of your friends here. I have not shot over him this season, and as there is no probability of my being able to use him at all, I have concluded to part with him*. . . . *I advertise the dog in James Yetman's name to avoid the visits of our New York dog pedlars*. . . .

After almost twenty years of animal farming he was still dabbling:

> [*To an Illinois inquirer, 21 February 1876*] . . . *I have nothing to sell at present, in fact do not nearly sell any thing. I would refer you to my nephew Mr. Herbert Mead, Goldens Bridge P.O.* [*Post Office*], *Westchester Co., N.Y. who has stock for sale and who can probably give you good bargains. Any statement he makes you can rely on*. . . .

Toward the end of his life Peter became concerned and confused over the just distribution of his wealth. His remaining four sons (after Stephen's death) had given little apparent cause for alarm. Richard started as a grain merchant and became a banker and real estate developer; upon his death he left an estate of $1 million, of which his widow in 1926 gave $100,000 to Teachers College of Columbia University to endow a memorial professorship. Frederic, like Stephen, entered the company and stayed until his death—the last Hoe to remain on the job. Charles, a credit checker, embarrassed himself by borrowing money against a munificent inheritance from his uncle, the Colonel, which did not materialize. But time does strange things.

In 1900 the father went to court to revoke trust deeds of $110,000 in property to each of the three boys with wives and offspring; Richard, married but without children, as trustee was to get only $55,000. Believing "there is a prospect of the alienation of his estate," the old partner of R. Hoe & Company seemed determined to fight his own family. The outcome was not revealed. Peter died two years later.

Long before, however, the ominous mantle of Hoe management had already changed hands with trepidation:

> Robert Hoe [III] is now ill at his home, No. 11 East Thirty-sixth-st. He has not been informed of his uncle's death lest it should have an unfortunate effect upon him. . . .[21]

Robert himself was not the kind of company man his predecessors had been. He knew that. As a boy he did not loiter around the shop or apply his ingenuity to mechanical toys. On the contrary, his father encouraged him in social and intellectual pursuits, preserving much of his earliest scholarship:

> The Loss of the Arctic
>
> She was built for Mr. E. K. Collins of this city a few years since and was considered one of the largest and handsomest steamers in the world. . . . She had often crossed the ocean before, and braved unflinchingly all the dangers of the sea, and those that took passage in her on her last fatal trip felt perfectly secure. . . . [They] little thought that before the expiration of two weeks many of their number would have found a watery grave. . . .[22]

Robert had been 15 when the vessel sank in 1854. Such attentions to and veneration of the written word by the Hoe clan—even to saving the tritest ephemera—evoked and reinforced in the boy a strong regard for literary accomplishment. But instead of appreciating the sacred mission of producing machinery by which the printed word—and democracy, its proudest child—would flourish, the lad's interest went awry. Yet no one seemed to notice.

Robert's summers of 1854 and 1855 had been spent with the Meads at "Cream Ridge," their South Salem farm. In 1856 his father fortified himself for the historic Lloyd installation and the *Times* confrontation by bringing his wife Thyrza and their family along. Robert, brother Arthur, and sister Laura divided their time between sightseeing and social amenities. Theodore Mead was not really successful recalling earlier pleasures of farm life:

> [*Mead to Robert Hoe III, Paris, 30 June 1856*] . . . *Now turn your mind's eye homeward though your face is still set the other way; take in each well remembered feature of the little* [Cream Ridge] *landscape, the orchards, the row of cherry trees, the brown old house and best of all the three old elms now shaking in the wind, and tell me don't you feel a quicker, warmer thrill than ruined abbeys and moldering sepulchres could ever cause—beautiful though they are. . . .*

Far more sociable than his cousin, Theodore gadded about New York and Bloomfield as a young bachelor ought; despite weather "like a burning fiery furnace" he still managed (at cousin Emily's wedding) "two or three quadrilles and a schottische." But, working at his uncle's shop, he also asked about business:

> [*Mead to Robert, 8 July 1856*] . . . [Y]*our mother refers me to you for an account of the charming Misses Sanderson, concerning whom you are strangely silent. . . . We are anxious to hear further from the* [Lloyd] *press—of its final success. The readiness of the* Times *people* [to negotiate] *is quite surprising & perhaps they are not so John Bullish after all. . . .*
>
> [*Mead to Robert, Interlaken, 29 August*] . . . *I am almost ready to believe with Uncle Peter that your appetite for novelty will not be*

put on short allowance—that a half year more of travel will be necessary to satisfy it, leaving even then a relish for a dessert. . . .

Theodore in 1857 joined the Hoes on the continent and cousin Henry Seymour took his place in the office. When Theodosia Leavitt, the brothers' youngest sister, made the pilgrimage to Europe it was a family treat:

> [*Seymour to Robert, 1857 (no date)*] . . . *Aunt Theo seems to have enjoyed her touch* [*of seasickness*] *very much, at least she describes it in such a comical manner that one cannot help laughing to hear it. Tell her that her letter is going the rounds—Brooklyn* [*the Elys*], *Bloomfield* [*Seymours and Dodds*], *Bright-Side*† [*Richard and Peter*], *Port Chester* [*the Meads?*]. . . .

Robert took his cousin Theodore to the British Museum library to ogle rare books and manuscripts, but Mead's interest was elsewhere:

> [*Robert Hoe II, Paris, to New York, 28 October 1858*] . . . *We should like a little more society at evening. The Misses Johnson have reached Paris & we have exchanged calls. The boys go to see them occasionally. . . .*

A high point for the young men in February 1860 was their presentation at the court of Napoleon III. Although the yankees could not subscribe to the Emperor's policies, the senior Robert must have appreciated the absurdity of the event—the cousins had to be in uniform. "So," said Robert, Jr. (27 February 1860), "we applied to our tailor, who fitted us out in gold embroidered military dress, cocked hat, &c, &c. . . ."

The happy tourists returned to New York shortly thereafter where they found—of all people—the Johnson girls:

> [*Mead to Robert, 27 October 1860*] . . . *Don't you think it might benefit your health to take a drive in the Central Park this afternoon with certain three idle damsels*[?] . . .

Theodore was soon assigned to rescue the Boston shop and (in December 1863) finally wedded one of the irresistible Johnsons.

Robert returned to Europe and in England apparently met Olivia James, another pretty lass, whose wealthy American parents lived in London. They were married there on 12 August 1863, with Uncle Richard proudly in attendance. The bride's parents gave the couple a New York townhouse as a wedding gift.

Educated in New York city public schools and Harvard College, member of Union League, Century, Players, Fencers, and Engineers Clubs, summers at Lake Waccabue, New York, socially prominent wife, addicted to art, print, and book collecting—Robert was not a driving factory boss to spend 16 hours a day on the job. He had acquired too much culture, too much affluence, to take routine corporate affairs compulsively. In their efforts to raise him beyond their own modest level, his father and uncle had in effect made him too good for what R. Hoe & Company would need to run its future well.

Yet, for the sake of the company Robert took great pains to perpetuate the Hoe image; he gave his imprimatur to an analysis which almost overstated his false position:

> In the third Robert Hoe the inventive genius of his race was especially pronounced. He identified himself with the creative work of the family as soon as he had mastered the simple curriculum of the schools of the day; and since then his life has been lived in the factory. It is true that in his moments of leisure he has established a family, collected a great library of rare and valuable books, developed a model farm, served as one of the original founders of the Metropolitan Museum of Art, and done much for the Industrial Art School connected with it. But these have been merely the incidental diversions of a career devoted to the advancement of the art of printing. . . .[23]

The contradictions of Hoe's performance and his philosophy would not go unnoticed.

Worse for the company, however, Robert's own family of seven daughters—flattering his "brain power," to be sure—and two sons left slim hope of family control of the business. Despite *Scientific American's* opinion many years before that dynasties were best for business continuity, an industrial firm at century's end was far from a family affair. Even Robert knew that incorporation (a relatively new practice in America) was the answer, but he could not bring himself to inaugurate it.

Unlike the fabulously wealthy who, despite commercial beginnings, commanded society's attention with their ostentation, the Hoes and their friends had toiled all their lives (and occupied very modest townhouses)—almost without social recognition in mind. For the plucky original Robert Hoe, said Alexander Flick, "innate ability, training, contacts and luck, rather than origin, counted most," a comment true of many immigrants content only with hard work.

Richard, Robert, and Peter Hoe had found the world so full of physical wonders they were unimpressed by the subtleties of parlor and ballroom. They were indeed welcome at many social functions, but they always seemed to prefer less ostentatious companionship. After some little cause for dissatisfaction in 1869 both the Colonel and Peter resigned from the American Jockey Club, without explanation. They lent their name to many worthy causes and organizations, but seldom participated in social activities.

The dilettantish Robert Hoe III called attention to family affluence and wealth as had no other family individual. His impeccable education and club affiliations were at last useful; his creation of the

Grolier Club a masterful social, snobbish balm to remove any final traces of grease and grime. Flick continued:

> Gradually, however, the bars were let down, and the Havemeyers [sugar], the Stuarts [refining], the Colgates [soap], the Allaires [iron], and the Hoes mingled in the drawing-rooms with the Livingstons, the Schuylers, the Fishes, the Van Cortlandts and the Stuyvesants. . . .[24]

Theodore A. Havemeyer in particular was a friend of the Hoes: he had a large cattle ranch at Mahwan, New Jersey, and a Newport, Rhode Island, villa once belonging to William L. Andrews. Thus it was a third generation Robert and his family who first began to enjoy the social civilities of New York.

Robert and Olivia produced seven daughters and two sons: Elizabeth (born 1865), Caroline (1866), Olivia (1868), Laura (1870), Ellen (1872), Thyrza (1874), Robert IV (1876), Ruth (1877), and Arthur (1880). When his girls reached marriageable age Hoe was at the height of wealth and social position, and the weddings promised to be splendid in every sense. The first bride, Elizabeth, was 21 in 1886 and unquestionably received the most sumptuous attention. The invitation was simple:

> Mr. & Mrs. Robert Hoe request your presence at the Marriage of their daughter, Elizabeth Woodbridge, to Mr. Frank Sherman Benson, on Thursday, November the eleventh, at half past three o'clock. Church of the Covenant, Park Avenue & Thirty-fifth Street.[25]

Social reporters covered the affair with appropriate detail:

> Every seat in the church was filled. The back of the church, including the organ, was almost completely hidden behind a wall of plants and graceful festoons of similax. Palms, ferns and autumn flowers in boxes were arranged around the pulpit, giving to the whole church a beautiful effect. The bride entered the church on the arm of her father, by whom she was given away. She wore a costly gown of white satin with veil. . . .
>
> The Rev. Dr. Marvin R. Vincent, pastor of the church, performed the ceremony, after which a large reception was held from 4 until 5 o'clock at the house of the bride's parents, No. 11 East Thirty-sixth Street. The spacious parlors were filled with a constantly changing throng of guests, calling to congratulate the newly-married pair. The house was elaborately decorated with plants and flowers. Between three and four hundred guests called. . . .[26]

But this auspicious marriage was not to last.

Elizabeth died in 1889, just three years after the wedding. Her husband Frank (1854-1907) as a result of her loss may well have resented her money. Although he graduated a lawyer, he never practiced. He gave $40,000 to a Brooklyn church to eradicate its debts, and began to lose himself among his clubs: Numismatic & Archaeological Society, English Society for Hellenic Studies, University, Hamilton, and Grolier Clubs, Sons of the American Revolution, Colonial Wars Society, Historical Society of Long Island, Brooklyn Institute of Arts and Sciences, Brooklyn Hospital, Society for the Prevention of Cruelty to Children—he somehow made a career out of memberships for which the Hoes had never found time. He died at 52, leaving a daughter, Thyrza, and a privately printed treatise on ancient Greek coins.

Caroline Hoe, also at 21, was next to marry in 1887:

> Mr. & Mrs. Robert Hoe request your presence at the marriage of their daughter, Caroline Phelps, to Mr. Léon Marié, on Thursday, January the nineteenth, at half past three o'clock. Church of the Covenant, Park Avenue & Thirty-fifth Street.[27]

From the similarity in invitations it appears Caroline's wedding was but a copy of Elizabeth's. Caroline also died early, in Germany, 1904. It was perhaps the reluctance of the next girls in line to repeat the formula that prompted them to do something unusual.

> It will be a genuine surprise to friends of the family to learn this morning that Miss Olivia Hoe and Miss Laura Hoe, daughters of Robert Hoe, were married yesterday morning at St. Mark's Church without notice, it is said, to their family. . . .

Olivia and Laura were 23 and 21 respectively.

> Miss Olivia Hoe was married to Mr. Henry Lewis Slade of the Calumet Club and the Lacrosse Club, the Rev. Dr. J. H. Rylance, rector of the Church, performing the ceremony. Her sister was married by the Rev. Richard Cobden to Ernest Trow Carter. Mr. Carter is a young graduate of Princeton, class of 1888, and is a lawyer at 49 Broadway. . . .

Olivia and her fiance had called at the church a few days before to arrange the wedding. Slade admitted there was some Hoe opposition, "but," said the rector, "since it is not in the province of a clergyman to interest himself in family troubles," the plans were not questioned. On their wedding day they brought another engaged couple to act as witnesses.

> The work of recording the marriage and receiving the signatures of the witnesses Dr. Rylance had left to his assistant. He was very much astonished when he turned around presently and found that the "witnesses" were making arrangements to get married themselves by the Rev. Mr. Cobden. Then for the first time Dr. Rylance noticed that the young woman who had been a witness was Miss Laura Hoe, a sister of Olivia. . . .[28]

The newly married Slades then returned the favor by witnessing the second ceremony.

The two sisters had accompanied their mother from their summer cottage at Katonah, New York, to open their 36th Street home for the winter. They told her they were going shopping, "and," the *Tribune* added wryly, "their friends say they succeeded in obtaining bargains whose genuineness and merit cannot be gainsaid." One of the sisters' friends added:

> Miss Olivia and Miss Laura I have known for some time. They are two very pretty girls, and in no way eccentric, as some people might think from the incidents connected with their wedding yesterday morning. They are not only beautiful women, but they are both accomplished and sensible girls. . . .[29]

"The whole thing is most extraordinary from beginning to end," the rector remarked, "and could only happen in America." The simple weddings actually solved some awkward social problems:

> At an uneventful season the romantic wedding of Robert Hoe's two daughters furnished society with something of a sensation, but not the sort that the average parent is fond of, even when it is taken into consideration what a deal of fuss and worry was saved in the way of trousseaux, wedding cards and breakfasts. The Misses Olivia and Laura Hoe had been seen seldom in society, owing to the death of their sister, Mrs. Benson, which threw the family into mourning just as the brides of Tuesday were about to be formally introduced. The death of Mrs. Hoe's brother, Frank James, in Africa, again throwing the family into mourning, prevented on the part of Mrs. Hoe anything but the simplest kind of entertainments last winter. The elder of the two brides passed some years at school in Germany, returning to New-York shortly before the death of her sister.[30]

The young Slades settled snugly into the Hoe townhouse at 45 West 11th Street. Slade, from an old family tracing itself to the *Mayflower,* did well in New York real estate, but his activities were overshadowed by his wife's new wealth, reportedly about $1,195,000. Such money should not have been a problem; she had always lived comfortably. But possession made a different person of her; she grew detached from Slade, who was obliged in 1920 to obtain a separation on grounds that "after she had inherited a fortune from her father she drove her husband out of the home she purchased at Mamaroneck." Slade, 59, died in 1924.

When Olivia died a few years later, her will directed that half the residue—$250,000—be used to establish a home for disabled veterans, to be called the "Robert Hoe and Olivia Hoe Memorial Home for Crippled Soldiers and Sailors." Her family were staggered and long contested the testament.

After their marriage in 1891 the Ernest Trow Carters had much to do. Having studied law, Ernest practiced for a year and then abruptly turned to his real love: music. He taught music in California but in 1894 studied organ and piano in Berlin, Germany; four years later the Carters returned and Ernest resumed lecturing and composing. Laura Carter, active in the Colonial Dames of America and music organizations, died in 1951 at their Redlands, California home; Ernest died two years later, leaving $10,000 to the Princeton University Music Department which had made him a Doctor of Music in 1932. The Carters had two daughters and a son.

Robert Hoe's fifth daughter, Ellen James Hoe, in 1899 married Dr. Samuel Morley Evans, a physician. She too died early, but Evans remained close to the parent family to insure that his children continued to benefit from Hoe business income.

The sixth Hoe girl, Thyrza, had died at the age of 9.

Ruth Lancaster Hoe, the last, married Robert D. Sterling in 1905. Her interests included soil and wildlife conservation and gardening, and she and her husband were among the donors of the Hall of North American Forests in the American Museum of Natural History, New York. As last surviving child (she died 1966) Ruth had retained much of the company stock her father left her; together with her husband, a real estate broker and banker, she would appear frequently at stockholders' meetings to protest the disintegration of the old firm. The Sterlings left two sons, 6 grandchildren and 4 great grandchildren.

Raising and giving away his daughters had not been a consuming chore for Robert Hoe. He found time to act as trustee of the New York Childrens' Aid Society and was a member of the American Society of Mechanical Engineers. His model stock farm at Lake Waccabue, New York, his summer residence, featured a choice dairy herd improved by the importation of thoroughbreds from England and the Channel Islands. His animals and dairy products were often exhibited at county fairs and elicited "the praise of connoisseurs, as well as the wonder of the country people."

Through Mrs. Hoe, perhaps, Robert found himself participating in some social causes in which he—like others before him—had only passing interest. A close friend was John Jay, a lawyer who managed his family estate, a founder of the Union League Club, member of the Metropolitan Museum of Art, National Academy of Design, and other organizations. The Jays and the Hoes were curiously interested in one branch of city affairs:

11 East thirty-sixth Street

My Dear Mr. Jay,

Will you allow me to suggest as a new member of the "First Circle" and an active member Mr. Albert Gallup. Mr. Hoe has seen a good deal of Mr. Gallup in connection with his "Grolier Club" & feels that he will give us very efficient help. Mr. Gallup has expressed his willingness to join and to do some work. He has attended the lectures & feels their values, and realizes the importance of the training for policemen as the life of an agent relative was lost lately through the ignorance of all the policemen in one of the precincts. . . .

Hoping I am not troubling you unnecessarily, I am Sincerely yours,

Olivia P. Hoe

Nov. 23rd./[18]*84*

Gallup, a lawyer, became New York City Park Commissioner. The city's police force, along with another unscrupulous municipal administration, was plagued with inefficiency and dishonesty, conditions the "First Circle" hoped to correct through professionalization.

For his part, John Jay was connected with a local farmers' club as well as the Royal Agricultural Society, which Robert hoped to join. He wrote Jay (24 December 1888), "As my business may call me often to England I think I might gain both instruction & entertainment by belonging to this society. . . ." Hoe's activities probably gave him little time for farming lectures.

Robert Hoe's taste for art befitted his social position. He supported the New York Art Students League periodic sales of paintings, adding many canvases to his collection. He was a founder-trustee of New York's Metropolitan Museum of Art, serving 1870-1892. Only 31 in 1870, he was of necessity thrust into the close professional company—perhaps for the first time—of fellow millionaires with a uniquely (almost democratic) common purpose, but even among close business associates he was not likely to have been gregarious or friendly. Nevertheless he made substantial contributions to start the museum collections:

METROPOLITAN MUSEUM OF ART
128 West 14th Street

New York, December 1 [*1876*]

To the Hon. Hamilton Fish
Secretary of State
Washington, D.C.

Dear Sir,

In the absence from town to-day of the President of the Museum, Mr. [*John Taylor*] *Johnston, I take the liberty of addressing you in reference to an application the Museum has made to the customs authorities for permission to remove from Memorial Hall, Philadelphia, the Castellani collection of works of art, and exhibit it in the New York Museum.*

Mr. [*Charles*] *Castellani is prepared to leave the collection with our Museum, for exhibition during one year, with the privilege of purchasing it at the expiration of that time, if the Museum so desires.*

The retaining here for purposes of study and instruction [*of*] *such a collection, even for one year, is a matter of importance to the country. . . .*

Yours very respectfully
Robert Hoe Jr.
Chairman [*Gift Committee*]

Hoe also donated some of his own artwork to the museum, and Olivia, like other directors' wives, could periodically be found dusting the rambling accumulations.

In 1891 the trustees, at Hoe's home for a special meeting, voted apprehensively to open the museum on Sunday afternoons "as an experiment only," although—as with Sunday newspapers 50 years before—there was much opposition on moral grounds. Ironically, according to Lillian Wald, famous nurse and social worker and founder of the Henry Street Settlement, the museum had strong immigrant friends: "It is interesting to recall," she wrote, "that of the 80,000 signatures to the petition for this privilege, 50,000 were of residents of the lower East Side."[31]

In 1878 Hoe had welcomed into the distinguished group of founders his close friend, William Loring Andrews, with whom he shared a love of books—an activity in which his museum associates seemed less interested. That love had become a consuming mania.

Robert, who often said that the love of books was born in him, was ideally situated to collect on a grand scale. Although Wesley Towner considered him "no moldy bookworm and perhaps in no proper sense a bookworm at all," Carl Cannon disagreed:

> He possessed almost every qualification necessary in a first-class collector. He had a deep love for books which he manifested at an early age; he came of a well-established family of printers† and press manufacturers; wealth came to him as he grew older, and at all times he was shrewd and aggressive. . . .

As early as 1850—at 11—Hoe was buying rare tomes from his lunch money—and describing with awe to his father those he could not quite afford. He particularly liked William Pickering editions, then Aldines, Elzevirs, and gradually still older works.

In 1860 when he came of age to be an officer of the firm—and began receiving profits to confirm that prestigious status—he immediately realized that his literary buying habits could at last be amply supported. His father, at the same time seeking to commemorate the boy's entry into the partnership, proudly stopped at a bookstore to buy some coveted volumes as a gift.

"I'll take those books my son saw here and wanted," he said to the bookseller.

"Sorry, Mr. Hoe," the dealer replied, "but your son was in about twenty minutes ago and took them with him."

William Andrews, an equally enthusiastic collector about Hoe's age, pooled his resources with Hoe to enter bids by mail at London auctions. When they opened parcels of books they had bought they took turns picking. Once when Hoe had first choice he selected a Shakespeare first folio for which they had paid about $2400; it later brought $13,000 at auction.

By 1866 Robert, 27, was quite knowledgeable about book rarity, condition, binding style, and other criteria. He made recommendations to publishers on reissuing out-of-print or scarce items, and edited Maberly, *The Print Collector,* a classic treatise on the subject, adding notes, bibliography, and comments on the state of contemporary etching and its practitioners. It was published 1880 by his cousins, Dodd and Mead.

Hoe soon began to develop a "continental" taste in books, a fondness for European illuminated manuscripts, incunabula, and important early printed works. These he occasionally "beautified" by engaging leading French hand bookbinders to work in sumptuous compatible styles, so that his library presented the unique aspect of both literary value and aesthetic appearance. Said Cannon:

> With rare judgment and taste in estimating a volume or manuscript, Hoe combined the most aggressive tactics in pursuit of new game, and often carried off the prize of being first on the ground. He also paid liberal prices. . . .[32]

This happy combination Hoe brought together in his "illustrated" or grangerized books (named after Granger, *A Biographical History of England,* first given this treatment), a sad affectation of the time.

Precious editions of arty works were carefully disassembled, blank leaves inserted, and rare illustrations—prints, woodcuts, lithographs—often ransacked from their own fine origins—even if only vaguely associated with the books' subject, surmounted thereon. Hoe's edition of Bray, *Life of Thomas Stothard,* the "Raphael of England," contained numerous extra Stothard designs, among them artist's proofs, india paper impressions, and pencil and colored originals.

Stepping beyond mere embellishment Hoe was called the "Prince of Waltonians" for his grangerized edition of Walton's *Angler* which "cannot be equalled by anything of its kind in Europe or America." It was a two-volume Pickering (1832-1836) on which the publisher had already spared no expense to make it a "typographic monument." Then followed Hoe's metamorphosis:

> And when such a book is still further beautified at the hands of a bibliomaniac who has wealth, taste, and time to lavish upon it, who shall attempt to describe it? Enlarged from two to ten volumes, it fairly teems with beautiful engravings of piscatorial subjects, rare old views, and portraits of the prominent characters mentioned in the text. . . .[33]

Despite this artificial brush with nature, Robert Hoe and Charles Darwin were not kindred spirits. A biography of the naturalist elicited a *New York Times* review:

> [Darwin] was no respecter of books, as far as paper and print went. He would cut a man's book, not with his pen, but with a knife, declaring the volume too big to use, and to have a single tome made into two by the Darwin process was even a compliment, for it showed he used it. Pamphlets he chopped all to pieces, only keeping the useful fragments. . . .[34]

"He had none of the reverences of the bibliophile," the *Times* added.

At Andrews' home one evening (5 February 1884) a group of nine bibliophiles and book-makers met to consider forming an organization.* The idea was probably Hoe's, fresh from France, where collecting fine bindings was already a passion and where similar groups had begun a decade before. Robert selected "Grolier" as its name and became the club's first president—a post which pleased him like nothing at the stuffy museum or the crass press factory.

Although only one of the founding members, Edward S. Mead, was at all closely related to Hoe, within the first ten years the club became a haven for various Hoe relations: Peter S. Hoe, Richard M. Hoe (Peter's son), J. Henry Harper, Theodore S. Mead, Robert H. Dodd, Richard Hoe Lawrence, Cyrus J. Lawrence, Léon Marié, Peter Marié, Frank Sherman Benson, Edward G. Kennedy. (The Philadelphia Kennedys were in-laws of the Colonel: [Samuel] Baker & [W. Henry] Kennedy were Chestnut Street printers; W. Henry's father David was a sometime inventor.) It would not be surprising if these were all "gift" memberships to swell the expensive club's ranks. Close friends included, among founders, Theodore De Vinne, William Andrews, and Albert Gallup. Within a few years the Grolier Club enlarged its headquarters and finally settled into an elegant home on New York's east 67th Street.

But Hoe's consuming interest in books far outdistanced all the others', who may have hoped occasionally for some social—or business—rewards of membership. The club's originator became intimately concerned with hand binding and repair, not only for the faddish grangerizing, but for proper restoration of incunabula. Hoe editions were often so meticulously restored that, despite their ages they were crisply clean and unblemished. This was the result of his observations in Europe.

Beverly Chew, another book-collecting friend, recalled a conversation with Hoe on the state of the craft abroad. At a library in England Hoe had asked to see some ancient tomes and manuscripts; they were brought out with leaves stained and abused, bindings broken, covered with dust. He was shocked. He told Chew:

* See *Books for the Millions.*

> This confirms me in the conviction that those who love books should have them in their custody and will take the best care of them. If the great collections of the past had not been sold where would I have found my books? [35]

Robert's own massive wood bookcases, with glass doors and ebony trim, were built double in depth; Hoe nevertheless knew the location of every hidden item, and sometimes, when showing an obscure specimen to a visitor, would glance at a page and—with photographic memory—recite it verbatim.

His reaction to British book storage had much to do with a grandiose plan forming in Robert's mind. Rather than continue to send books to Europe for definitive binding, why not bring the best European craftsmen and supplies to America, to work close by? There was no question about where they would come from; the English were not to be trusted. The organization of the Club Bindery, a subsidiary of the Grolier Club, with French master binders and the most exquisite French tools and supplies, quickly followed in 1895. It began to execute some of the most superb bookbinding work in America.

Within a few years, however, the bindery was doing more and more work for Hoe, who, by 1902, was commissioning lavish, time-consuming designs with tools made in France for his exclusive use. Many of his binding designs were painstakingly drawn by an artist quietly sequestered in the R. Hoe & Company drafting department, and Léon Maillard, Paris' best finisher, worked for months at a time at Hoe's home. As Hoe dominated the little shop other clients diminished; expenses increased prohibitively. It could not continue.

The Club Bindery was finally dissolved in 1908 when Robert Hoe excused its failure by telling fellow members, "We have been interested in obtaining fine bindings, not profits!" [36] Each embellished with Robert's monogram, Hoe's impeccably bound volumes accumulated steadily. (A copy of *Bookbinding as a Fine Art*—originally a Grolier Club lecture of Robert's—at Bangs & Company, New York auctioneers, went in 1894 for only $30; it had apparently escaped lavish Club treatment by scarcely a year. Hoe's own copy, executed by Maillard, would be worth hundreds.) Auspiciously, at the end of the project Hoe made the shy French binders pose for photographs, the camera throwing more light on their secretive workshop methods than paying members had ever been permitted to see. The workers remained in the United States for a while, but the magnificent era was over.

Robert employed a staff of librarians to catalog and care for his great collection. Beginning in 1903 he authorized the publication of its contents in about 16 catalogs, compiled by James O. Wright and Carolyn Shipman. These library guides were privately printed, usually 100 copies of each. On a special bibliography Wright completed his task and had some unimportant works left over.

"I have put aside thirty or forty books, Mr. Hoe," he said, "since they are not worth cataloguing."

Hoe looked at the titles. "Keep them and put them in," he replied. "They were the first books I bought and I went without lunch money to buy them."[37]

There were indeed an insignificant number of mundane books among a collection of gems which would, by the time of his death, reach 20,962 major works making 141,030 volumes. Thousands of his acquisitions were printed before 1700, and most, despite their age, were in "mint" condition, washed leaf by leaf, blemishes removed, and damage repaired before rebinding. Where most collectors aspired to own one or two Grolier bindings, Hoe had eleven, six of them Aldines. Of many works he possessed the only known copy.

In 1895 Astor librarian Oscar A. Bierstadt had written a popular illustrated description of some major titles in the Hoe collection. He was especially impressed by the number of works once belonging to European nobility, a feature which had not escaped their owner:

> Entering Mr. Hoe's house one is confronted in the hall with a large old full-length portrait of Marguerite de Valois [Queen of Navarre, 1544-1549], and in his library it is pleasant to meet with some of her books, for she was bookish as well as beautiful. . . .

Bierstadt tried hard to understand the collector as well as the library. He was partly right when he speculated, "With such a magnificent private library the bibliophile might well say: 'I will bury myself in my books, and the Devil may pipe to his own,' " but he could not really have believed this captain of industry actually so detached. His estimation of Hoe was no more than flattering:

> Assuredly it is possible to know a man by his books, and his library may justly be regarded as the *speculum mentis suae* [mirror of his mind]. . . . [The collection's] chief characteristic is its many-sidedness, and it is as cosmopolitan as the metropolis of America. . . .—a library so nearly perfect as to make it a heaven upon earth to the bibliophile.[38]

But Hoe reflected by his books was hardly many-sided, he was merely bibliomaniacal; like them he was not cosmopolitan, but protected, isolated; like them he was perfect—cold, antiseptic, inhuman, with one surprising exception. The man who compared himself to Gutenberg—to a printing machine itself, in the privacy of his London home (far from his books) had a curious, tender, secret love affair. The woman was a modern-day matriarch also named Margaret.

At the turn of the century Robert Hoe made frequent trips to England, ostensibly supervising conditions at the big shop: "He was anxious to bring his London works up to the same standard of perfection as his factory on Grand Street,"[39] said the *New York Times,* helpfully. With 800 hands in London there was much the same expansion problem the New York plant had experienced years before. But strangely Hoe was seldom at the London works at all, his staff there often unaware of his British activities, and quite uninformed as to his serious illness.

In September 1909 the septuagenarian was incapacitated with a kidney infection. His son Robert sailed immediately to be with him in England, but too late. In London his housekeeper quietly described Hoe's last hours to the dutiful son, and when he reembarked for New York (sending the body on another vessel) she found herself suddenly very much alone and apprehensive. The years with Robert had been hectic indeed.

Claude W. Johnson, for many years mayor of Lexington, Kentucky, and former director of the government's Bureau of Printing and Engraving at Washington, D.C., had joined the London branch of R. Hoe & Company about 1900. His daughter Margaret was married to Lewis F. Brown, an agent of the Chesapeake & Ohio Railroad. One day she chanced to meet Robert Hoe while on business in New York for her father. They experienced an immediate attraction for each other, something compelling they knew shouldn't be encouraged. Perhaps hoping nevertheless to see more of the girl, Hoe gallantly offered to be of service to her.

Married in 1895 and just arrived in New York from Lexington shortly after the bride's parents returned to London, the Browns (including a small child, Claudina) were lost in the big town. Margaret saw her opportunity: she asked Robert to give her husband a job in England so they could be together with her parents. He obliged, and in 1903 the Browns were on their way to new careers. Lewis worked for his father-in-law for about a year, but suddenly, without apparent reason, was fired.

"It's a bad sort of business bringing us over here," Margaret fumed to her husband, "and then discharging you without a thing to do in London."[40]

She was irate at Hoe's perfidy, but she contrived to do something about it. She sent Lewis and Claudina out of the house one evening and invited Robert Hoe to dinner. When Lewis returned Margaret assured him he would be working once again. The Browns were to manage Hoe's West End residence at 38 Brunswick Square, W.C. 1, about a mile north of the company's office; Margaret would be housekeeper, and Hoe would get Lewis employment with a printing ink firm. The perplexed husband was distinctly unhappy about the unusual arrangement, and, unknown to him, the dangerous alliance between Hoe and Margaret grew.

Robert often met Margaret surreptitiously at the National Art Gallery in Trafalgar Square, and when he traveled he secretly wrote her.

Lewis was quite unaware of these trysts and the correspondence until a servant showed him one of the letters. When confronted by her husband Margaret confessed to receiving many such intimate notes. She obligingly suggested that she and Lewis move to a secluded apartment adjacent to the principal quarters, where they might avoid Hoe's domineering presence. The situation seemed salvageable, but, alas, there was a secret passage between the two households.

When Kentucky relatives of the Johnsons became ill and in need, Margaret's mother felt compelled to return for a visit, and Lewis obtained two months' paid leave to accompany her home. To his utter surprise, in New York city he met his wife conducting herself quite independently—apparently newly clothed and supported by Robert Hoe. Margaret advised him not to return to London; they would separate. Brown was furious.

Lewis decided regretfully his only recourse was to sue Robert Hoe for alienation of affection. He took some of the revealing letters, "without reading them," to his lawyer for study. Shortly thereafter he met Margaret at a Broadway hotel to discuss the distressing developments. She told him that Hoe himself wished to terminate the affair; Robert would give Lewis $20,000 to set up a business in Kentucky. They would live again as if they had never met Hoe. Margaret cried, and Lewis almost hated to tell her about the damaging correspondence in his lawyer's hands.

Lewis retrieved the letters and gave them to Margaret. But the distraught husband and wife could not so easily restore their former bliss. In 1907 Lewis sued for divorce in Kentucky, charging Margaret with abandonment; at the same time he engaged a local lawyer to institute the alienation of affection suit against Hoe he still deemed necessary. The parties involved in the pending court action journeyed to Washington, D.C., to discuss settlement, but Margaret cried again and said Hoe would pay $20,000 out of court. Lewis wanted $60,000. Thoroughly disillusioned and broken, Lewis took a $20,000 check Margaret had ready for him and returned alone to Kentucky. He subsequently married again, but died soon after.

Margaret returned dutifully to her uneasy position in London, but the house on Brunswick Square became somber and quiet. Herbert L. Jones, foreman of Hoe's London plant, noticed his employer's increased secretiveness about the address. Hoe kept his name out of the London City Directory; on one occasion when the name appeared inadvertently Hoe confiscated the entire edition and had the name "Brown" substituted. He forbade London employees to communicate with him there, and he particularly did not want family to call. Jones received many messages from Hoe asking for money, some of them delivered by a "mysterious" messenger; when Jones at last tried to see Hoe servants turned him away because he was indisposed.

Jones first learned of Hoe's illness via a cablegram from the United States. He rushed to the Brunswick Square residence and was most surprised to meet Claude Johnson, Margaret's father, who told him Hoe might not live another 24 hours. Jones knew that for some time Hoe had been unhappy with Johnson's work—they had had frequent stormy scenes. (Jones finally got a note from Hoe telling him to remove Johnson's name from company stationery; he was discharged.)

The foggy London idyll for Margaret and her parents was fast disappearing. She hoped that in his last days Robert would remember her fidelity and attention. She began to believe that he would provide adequately for her—perhaps by letting her retain the London house for which she had already made such painful sacrifices.

Robert reportedly suffered a short illness; the London *Times* understood him to have been at his office only nine days before, and still attending to business within three or four days of his death. The end came on 22 September 1909.

Margaret continued to maintain the coveted residence, although she shortly met a most chivalrous gentleman, Arthur Johns, a New York lawyer, with whom she fell in love. They were quickly married.

As attorneys began to inventory Hoe's estate, Mrs. Johns made sure she would not be overlooked. Her lawyers appeared in New York Supreme Court, May 1910, to present some startling testimony. Margaret revealed that Hoe had become deeply attached to her over their many years together, and her father added some curious details:

> After my daughter had nursed him through several illnesses his own daughter died, and he wanted to make my daughter his ward. He did this with the entire consent of myself and my wife. . . .[41]

Actually, three Hoe daughters died between 1904 and 1907, indeed a heavy loss for a devoted father. Over these years Hoe had given Margaret $250,000 worth of various stocks and bonds, gifts he later considered imprudent. He asked that the securities be returned, and in July 1908 had a document prepared to acknowledge their receipt and to promise she would instead receive £30,000 ($150,000) upon his death. The New York court thereupon duly overruled a demurer brought by Hoe's executors Frederick Crane and Phineas Chew, to frustrate Mrs. John's claim against the estate.

The London property looked more and more as though it would soon be hers. Tidying the household in December 1910, Margaret conducted an auction of furniture, curios, artwork, and antiques which New York executors protested she had no authority to sell. Margaret had to make another decisive move, and she had to be swift.

Early in 1911 Margaret Johnson Brown Johns brought suit against the Hoe estate to force a transfer of some bank stock and a $30,000 mortgage of the Brunswick Square property which she believed rightfully hers. The executors discovered that she had already "obtained access to Hoe's strong box and procured the valuable papers which Hoe had assigned, but not delivered. . . ." With considerable reputation at stake—family integrity and company name, and with strong resources both in New York and London, the executors were not to be teased or threatened. Margaret's father, meanwhile, ready for the press, in May made another curious announcement:

> In view of the publicity which is being given to the suit brought by my daughter, Mrs. Arthur Johns, against the executors of the estate of the late Robert Hoe to recover securities which he voluntarily presented to her through certain documents drawn up by his own solicitors in London, I think it proper in his and my daughter's defense to make the following statement:
>
> Mr. Hoe was my personal friend of nearly fifteen years' standing, and my daughter, with the full consent of her former husband, Lewis F. Brown, and her parents, assumed the management of the house adjoining and opening into the one occupied by her mother and myself for Mr. Hoe. These facts I feel should protect his memory from the aspersions which are the result of an effort to save a paltry sum of money when compared with the millions which by his industry, intelligence, and ability he amassed.
>
> The securities referred to were left by Mr. Hoe with his own London solicitors with the exception of a mortgage, which he assigned to my daughter in 1907. As a matter of fact this assignment was acknowledged before one of his executors as a notary public and prepared by Edward G. Pringle, employed at that time by Mr. Hoe as foreman over the apprentice boys in his printing press establishment.[42]

The Hoe executors had nevertheless prepared for a full seige upon Margaret's impertinent position. But the proceedings went poorly.

In February Margaret had been obliged in court to confess an intimacy with Robert which had been so genuine and so uncompromising she was compelled to leave her husband. In June two servants from the London household were subpoenaed to testify. One, a chambermaid, said that Hoe and Margaret appeared to have close relations: they exchanged caresses and endearing remarks—when they thought they were alone. The other, a cook, gave similar testimony. It seemed that additional evidence only substantiated Margaret's contention that Hoe cared dearly for her; the plucky girl, no more reprehensible than her millionaire employer, had a persuasive case.

Suddenly Margaret's new husband, Arthur, with an unexplainable impulsiveness lawyers deplore among clients, believed it necessary to defend his wife's honor with a gesture not seen since the newspaper days of Webb and Bennett. On the evening of 6 June 1911 he waited outside the 54th Street home of Robert Hoe IV until the latter returned from work. Johns confronted him:

> He struck him in the face and chased him around his car for a lap or two before the defendant in the action escaped into the house. . . .

Arthur was formally charged with disorderly conduct, and at night court he too had a statement for reporters:

> Robert Hoe has been abusing the name of my wife and has caused the executors of the will to make scandalous accusations against her. When I saw him to-night I decided I would thrash him. . . .[43]

Hoe appeared at court with his lawyer, Edward Pringle, but bore no marks of a scuffle; he had wanted to charge assault but settled for the lesser offense. Pringle reminded reporters that this Robert Hoe had absolutely nothing to do with the suit over his father's estate. Johns posted a $500 bond to keep the peace for 3 months.

In England the Hoe solicitors had discovered another method of attack when a foundling home, actual owner of the Brunswick Square property, innocently sent the Hoe estate a bill for six months' rent. Hoe lawyers deliberately refused to honor the item, at which point the hospital appealed to the British court for satisfaction. The court decided in the landlord's favor and forced payment—thus conveniently identifying the Hoe estate, not Margaret, as the legally responsible custodian of the property. With this happy precedent, the New York court could hardly rule in Margaret's favor.

The executors of the estate would now fight on the grounds that Hoe's gifts and assignment were for a consideration—love and affection—contrary to public policy and hence unenforceable; they were an attempted testamentary disposition before Hoe's decease, which they would show to be illegal. This would take time.

By January 1912 Margaret had added a $30,000 slander suit against Hoe executors Chew and Crane to earlier claims now totaling $175,000, not including 66 shares of Metropolitan Trust Company stock she believed hers. She identified herself to the American press as the "former nurse" of the great book collector, and contended that the basis of her complaint were the securities amounting to $230,000 left to her by her employer but never delivered. She engaged a law firm which quickly saw the futility of battering the Hoe position.

On 2 March 1912 Margaret's lawyer announced quietly that the suits were settled, the ingenious defenses of the executors massive indeed but unavailing:

> The claims of Mrs. Margaret Johnson Johns against the estate of Robert Hoe have been adjusted out of court, and Mrs. Johns has received a satisfactory sum. The defenses set up by the executors involved some interesting questions as to the law of England, upon which eminent English barristers had delivered contrary opinions, and involved also some rather complicated questions as to whether the law of England or the law of this State should, under the circum-

> stances, govern. However, the executors of the Hoe Estate finally made an offer of settlement, which was acceptable to Mrs. Johns, and these intricate questions will not be passed upon by the courts. . . .[44]

The executors may have realized that, although justice was with them, prolonged exposure before the public of their patron's private life was not in the best interests of the family. At exactly the same time, the auction of Robert Hoe's library had also come to public view—never had the family experienced such notoriety.

The grand Hoe collection of art and incunabula had to be appraised before the will could be settled. The artwork was disposed of at public sale 15 February-3 March 1911, Thomas E. Kirby auctioneer. In three parts—(1) paintings; (2) oriental and European porcelains, enamels, silver, watches, arms and armor; (3) watercolors, drawings, etchings, engravings—the event did not attract much attention. Book collectors waited nervously.

"At auctions many goods are disposed of, which would not in other ways find purchasers," Robert had once written in an elementary school composition. "These sales give the auctioneer a fine chance to display his wit and humor." [45] But the mature book collector had learned to respect the auction as a sound literary institution, uniquely obliging a rabid collector to demonstrate his enthusiasm anew over each and every title he wanted. The challenges of this arena Hoe loved dearly, and its use now was his one great gift to fellow bibliophiles.

Book people everywhere were delighted to learn that the entire library would be auctioned. The colossal sale took two years to organize and, upon chauvinistic considerations would be conducted in the United States; the Anderson Auction Company, New York, would handle it. John Anderson, Jr., had been a rare book dealer since 1900, but the Hoe transaction would now make his firm world famous. The library, second only to that of John Pierpont Morgan, constituted the greatest book sale in America until the 1960's. It was divided into 14,588 lots and scheduled for four separate sessions. The gigantic sales catalog was personally compiled and written by Hoe's old friend, Beverly Chew.

On Monday, 24 April 1911, 400 seats were reserved for distinguished clients at Anderson's little theater-like gallery. The great illuminated manuscripts and the best early printed works were scheduled first for the gavel, so as to call attention to the sale and give prestige to the entire venture. Invitations had gone out around the world and were answered auspiciously. From Europe came foremost dealers, led by Alfred Quaritch and Ernest Maggs. From around the United States came Shakespeare specialist Henry C. Folger, bibliographer Henry F. De Puy, collector Peter A. B. Widener (pioneer in metropolitan transportation, heading three generations of bibliophiles), collectors Beverly Chew and Brayton Ives, magazine publisher Robert J. Collier, rare book dealers Dr. A.S.W. Rosenbach, Belle da Costa Green (buying for Morgan), and George D. Smith, who started at Dodd, Mead (buying for Californian Henry E. Huntington)—and to everyone's surprise, railroad millionaire Henry E. Huntington himself.

The auction room, draped with great dark velvet curtains, was filled to capacity; when seating and introductions were accomplished, voices diminished to whispers. A newspaper reporter covered the event:

> When No. 229, the Gutenberg Bible, finally was announced, the crowd of men and women, many of them in evening dress and decollete gowns, sat silent as if in a church, nervously fingering their catalogs or the arms of their chairs. Gasps of hysterical laughter came from several of the women collectors as they caught their first glimpse of the most wonderful book in the world.
>
> "How much am I offered for this, the first Bible printed?" came from Auctioneer [Daniel R.] Kennedy.
>
> Back in the rear of the gallery a voice murmured, "One hundred dollars." The auctioneer frowned and repeated his question. Then came ten thousand dollars, the second bid, from George D. Smith, most active of purchasers. A ripple of applause greeted his crisp utterance, which promptly gave way to a tense hush when the representative of Quaritch, of London, exclaimed, "Fifteen thousand dollars."
>
> "Twenty thousand dollars!" shouted an agent for Dodd & Livingston, acting for [businessman] P. A. Valentine.
>
> An instant afterward, before the auctioneer had repeated the bid, Harry Widener, half rising to his feet, fairly shouted, "Twenty-one thousand dollars!"
>
> His excitement seemed to spread thru the room with lightning rapidity. From every side the bids, one thousand at a time, came tumbling so fast, Kennedy, practiced as he is, had difficulty in keeping track of them. At twenty-nine thousand dollars there was a brief lull. Then Smith, sitting just below the auctioneer's rostrum, put in quietly: "Thirty thousand dollars!"
>
> Again there was whispered applause, and again Widener, Quaritch and the other bidders lifted their offers higher and higher. The excitement finally became so unrestrained Kennedy was obliged to ask for less noise. It was Smith, however, who raised the price to forty thousand dollars. He and young Widener were the only bidders left. They kept at it desperately, leaping up one thousand dollars or five hundred dollars at a time until the grandson of the traction king almost screamed:
>
> "Forty-nine thousand dollars!"
>
> "Fifty thousand dollars!" was Smith's instant rejoinder.
>
> There was a pause as young Widener, pale-faced and apparently unnerved by the struggle, dropped down into his chair. Then Kennedy exclaimed: "This is the last chance you will have to purchase this book. Do I hear any advance?"

> Again he waited for a second or two.
> "Sold for fifty thousand dollars!" he finally said.[46]

Edwin Wolf, Rosenbach's biographer, reviewed the frenetic activity:

> From the outset it was apparent that the Hoe sale was a battle of Titans; mere demigods and heroes were outclassed. Thousand-dollar bids were piled up like Ossa on Pelion. Huntington took the Gutenberg Bible on vellum away from P.A.B. Widener at the record price of $50,000. At $42,800 Morgan beat out Huntington for the excessively rare first printing by Caxton of Malory's Morte d'Arthur. At $21,000 Mrs. McCormick beat out Huntington for the unique Wynkyn de Worde Edition of Helyas Knyght of the Swanne printed on vellum. At $13,000 Quaritch outdistanced the field for the [Shakespeare] First Folio. But Huntington won the Book of St. Albans [1486, actual title *Book of Hawking, Hunting and Blasing of Arms*] for $12,000, and so much else all the way through the sale that, when the proceeds were totaled, it was found that George D. Smith had spent for the Californian over half the nearly one million dollars the sale had realized.[47]

By his physical presence at the auction Huntington had pushed prices higher than expected, although Hoe's exquisite bindings may also have been partly responsible. At the close of the first session $809,140 had been collected. The newspapers, of course, were delighted to describe the millionaires' book-buying contest. " 'Absurd' Prices at the Hoe Sale," said a *New York Times* headline, prompting Carolyn Shipman (now Mrs. Whipple), Hoe's former librarian for eight years, to emphasize the uniqueness of each item:

> . . . Only one who was in constant association with Mr. Hoe for many years could realize his infinite grasp of details connected with his library. With one exception, he did not buy en bloc, like some recent collectors. . . .[48]

Hoe might be criticized as a "gatherer" regarding paintings, she insisted, but not in books. Between the first and second sales countless newspaper and magazine articles kept the public tantalized about the colossal event.

In subsequent sessions Huntington stayed away, however, so that an "at-any-price" fever did not again result. The second occurred in January 1912. The third, 15 April, coincided with the announcement of the *Titanic* disaster; the tragedy brought listless bidding, as many friends had been on board (George and Harry Widener, Peter's son and grandson, both perished; Harry had planned to attend in April). Part four took place in November. When all receipts were counted, the 14,588 lots had brought in $1,932,056.60. Hoe had paid perhaps $750,000 for his treasures, adding to them about $250,000 in bindings—thus his investment had actually doubled in value. But money was not the point; rare contentment was. A book dealer who professed to know him well admitted:

> Mr. Hoe had a rough and unpleasing exterior and an overbearing manner, but back of that and back of the restless, energetic business-man there was the true collector, the true bibliophile, and we may be certain that the happiest hours of his long and busy life were those spent among his books.[49]

For U.S. rare book collectors Hoe's auction produced another benefit—it began a trend wherein Americans achieved a domination of the market never before contemplated. European buyers took home for their patrons only what yankee collectors had not cared to keep. Hoe's reluctance to see his library, in *Inland Printer's* blunt words, "consigned to the cold tomb of a museum and subjected to the stupid gaze of the careless passer-by,"[50] proved a reward for American scholars of all time.

Robert Hoe III's will was finally settled in June 1916, with Phineas Chew (Beverly's brother) and Frederick Crane as executors. Initial (1910) estimates of his personal wealth were somewhat inflated, reckoned at more than $4 million and later (1916) at $3.5 million. His real estate holdings, including many valuable New York city properties on Broadway, Lexington Avenue, and 28th Street amounted to $1,887,386; stocks and bonds were $804,091; Chinese and Japanese ceramics, $167,830; paintings $152,540. His European art collection, exclusive of paintings and statuary, was worth $395,712. R. Hoe & Company holdings were $4,258,546—$7.5 million in all.

Five Hoe children, Robert Jr., Arthur, Ruth, Olivia, and Laura, each received $1,195,445, as did a granddaughter, Thyrza Benson Flagg. Three other grandchildren, Ellen James Evans, Robert Trogwy Evans, and Samuel Hoe Evans each received $398,481. Arthur, the younger son, hopelessly spoiled by his mother, was probably bored by so much money talk. Indulging in a popular New York pastime, he could scarcely keep his mind on such inheritance figures, astronomical though they were. A girl-watching episode had become a delicious nightmare.

"Didn't we meet at Atlantic City?" a young gentleman reportedly asked, as he introduced himself one September evening in 1911 at a New York intersection. "I was walking on Second Avenue, and at Twenty-third Street I saw two young girls who were without an umbrella," Arthur Hoe later explained, without the slightest apology for the trite introduction. "It had begun to rain, and I offered to see them home or to let them have my umbrella. We all walked to the Subway station, where I left them at the entrance."[51]

Hoe had been fascinated by their youth and fresh beauty. Mae was

16, petite, with a bright, round pretty face; Dorothy, somewhat older, was frail, quiet, with dark eyes. Both were obviously working girls—they wore dresses reaching only to their shoe tops, as fashion dictated for very young ladies. Mae, with long hair in braids down her back, was especially appealing, thought the 31-year-old businessman.

Arthur was supervising the movement of his townhouse from 53rd to 86th Street that month. He had married Evelyn Perry in April 1910: a quiet affair, his brother Robert as best man, Laura Carter and Ellen Evans, his nieces, as flower girls. Evelyn, once an actress, and their eleven-month-old daughter were now staying at their Westchester County home in Bedford Hills during the move. Consequently alone in New York, Arthur welcomed an occasional stroll. It was Sunday, however, and foregoing friends and associations (the Harvard Club was nearby), he had difficulty passing the time. The impulse to engage two pretty girls for a few moments' conversation was too sweet to pass.

No, said Mae, she had never been to Atlantic City. But the introduction was nicely accomplished nevertheless. As the trio approached the subway entrance Arthur managed to make another date for the following evening.

On Monday night Arthur took Mae and Dorothy to the Cafe Boulevard, surely an impressive—if not ostentatious—place to dine; would not a millionaire know just how to beguile two modest working girls? Although all three were seated on the same side of their table, Arthur found no difficulty chatting with his guests, who seemed similarly entertained with him. Mae certainly was attractive, Arthur thought. He told the girls he was married and had a baby; perhaps Mae did not hear. At 10 o'clock after dinner he escorted them to the subway entrance once more, gave them his name, address (504 Grand Street), and telephone number. They parted pleasantly.

Well aware that her gentleman friend was not the usual flirting bachelor, Mae had nevertheless perceived Arthur's extra glances and sensed the empirical truth of two as company, three a crowd. To her this meeting—with Dorothy present—had not counted. Indeed, she considered Dorothy only a friend, and not very close at that.

Mae and Dorothy met in 1910 when they worked at the same Fifth Avenue office. Both were orphans; Mae, educated in a Catholic school, was a ward of the court since 1908 and under the guardianship of a lawyer. She had become a telephone operator. Dorothy lived with an older sister on 19th Street and Mae, who had just quit work and given up a flat, was staying with an aunt on Shelter Island. The two girls had met Sunday evening for a window-shopping walk when they attracted Arthur Hoe. Despite a telephone operator's usual attention to detail, Mae had ignored the most routine facts of the relationship.

Was not their street-corner meeting already beyond the wildest expectations of a poor orphan?

A few days later Hoe met the girls again for dinner. They had decided to take a flat together, and asked Arthur to help furnish it. He rose to the task magnificently, giving them $200 for furniture. Gleefully they went apartment hunting and quickly secured a little place for $35 monthly on Seventh Avenue. They dashed to a department store and bought $700 worth of furniture, using the $200 as a down payment. Proudly they invited their new benefactor to see their homemaking handicraft.

Hoe was most lauditory and encouraging. He paid for the furniture in full, gave the girls $40 or $50 for incidentals, and on another occasion brought some old curtains (from his mother's residence) for the windows. When Mae expressed interest in a short vacation Hoe financed the trip and sent Dorothy along for companionship. Over many visits through the month of October Hoe managed to spend no less than $4000 on the girls and their flat. But despite these luxuries, the roommates' relationship grew difficult.

Dorothy constantly urged Mae to find work. Mae had only a $5 weekly job offer with a milliner which—under Hoe munificence—seemed quite unrewarding. Unfortunately Dorothy also lost her job, and both girls suddenly seemed in each other's way—both were always home when Arthur came to call. Although he usually stayed only a few minutes and often talked about his family and business, Dorothy's presence proved most irritating to Mae. Dorothy, for her part, was not particularly conscious of being in the way, but she certainly noticed something else: Mae's beautiful new wardrobe.

"I asked her where she got all her new dresses," Dorothy said, "and she told me that she had an account in a department store run by a man who had married one of her cousins." Dorothy checked this story and found it to be untrue. Furious, she ordered Mae to leave the apartment, threatening to report her misbehavior to her aunt. About the end of the month Dorothy also revealed to Hoe that Mae's name was not "Smith," as she had purported, but Sullivan. Surprised at such unnecessary subterfuge, Hoe stopped calling—for a while. Mae left to take another apartment nearby, glad to get away. The girls saw each other very seldom after that.

Mae A. Sullivan had somewhat cruelly been exposed to more affluence than an orphan girl could ever manage. She sought desperately to fill the void in her life that only orphans must feel. But Hoe's wealth alone could not heal the emptiness; she accepted it, to be sure—but she welcomed Arthur almost in spite of his money. Mae sought too the respectability and status of marriage and family; she yearned to fill the days and nights with laughing voices and squealing children. Her subconscious had shut out many aspects that did not contribute to her fantasy, and created many that abetted it.

To establish her capability as married woman and mother, Mae in November 1911 thought of visiting an acquaintance, Nellie Glassberg, on whom she would test herself. She introduced herself as Mrs. Jack Burr (a cousin's name), and later, unclearly, to Nellie's husband as Mrs. Jack Brannigan (Dorothy's surname). She called on Mrs. Glassberg one afternoon to show off a fine bouncing baby boy as her son. Both women fussed over the infant, as women usually do. The baby, however, was actually the child of her cousin, Mrs. Dailey, to whom she paid $100 for the privilege of renting the infant.

These encounters with Nellie were reinforced over the next few weeks by letters from Mae:

> *. . . The name mother means a whole lot, and I wish I had a dozen like the baby, because he is a little lamb. Baby joins me in sending love to you. . . .*

And again,

> *. . . The boy is just as sweet as he can be. He refuses to go to bed. I don't know what to do about it. I want to have a conversation on baby talk. He has three teeth. Billie is going to start in as a business man. By the way, Billie has been away since Monday. I miss him a great deal. . . .*[52]

It seemed unimportant that "Billie," her purported husband (perhaps a disguise for "Arthur"?), had a first name strange to the Glassbergs, or that in one note she signed herself "Mary B." Somehow Mae thought that posing as a married woman would at the same time make it easier for Arthur to visit; somehow she believed it was Arthur's idea.

In December 1911, after three months of agonizing trysts, Arthur Hoe thought of protecting himself from whatever repercussions might materialize. He had endured the embarrassing newspaper exposé of his father's strange affair with his housekeeper—a relationship that, like his own, had begun innocently enough but which grew quite out of bounds. He would not let this happen to him; his child, his wife, their home, the company—and most of all, his aged mother—would never understand.

Hoe presented Mae with a paper drawn by his attorney, relieving him of any and all responsibility, financial and otherwise, concerning Mae. It stipulated that Arthur's generosity was only an act of compassion for her as an orphan, and clarified, legally, his particular desire to deny his name any association with hers. Other matters were occupying Arthur's thoughts: if not the pressure of business, then his father's important library, in which he had some interest.

Auction sales were resuming again in January, and Arthur wanted a few things, perhaps memorabilia, perhaps for his own collection, perhaps as gifts. The Hoes surprisingly enough had not found the commercial event too painful to endure. Towner wrote:

> Hoe's family had come to see the fun: his children and his grandchildren, the bewildered widow with her hands primly folded, as for an evensong devotion. Even the law-happy Mrs. [Margaret] Brown lurked somewhere in the nether regions of the theater.[53]

Arthur won the bid for the Pembroke *Hours* (breviary of the Earl of Pembroke, beheaded in the War of Roses) for $33,000, although he was buying it on behalf of a friend. He also paid $850 for his father's *Lecture on Bookbinding as a Fine Art*—obviously the pampered copy. Thyrza Benson, a Hoe granddaughter, took the *Hours of Anne de Beaujeu* (daughter of Louis XI) for $24,000. Dodd and Livingston also bought at the auction, but that was purely business.

When Arthur returned to thoughts of Mae he found her reluctant to sign his document, having shown it to her lawyer uncle. He dispelled her fears of legal entanglement—and, besides, gave her a gift of $3000 for her pretty signature. Somewhere too about this time Mae returned to Arthur (without compensation) all the letters he had written her; she really loved him, she once said, and trusted him.

Mae next moved in with what she considered a "proper" family, a Mr. and Mrs. Stade, Cathedral Parkway, so as to regain a semblance of respectability. While at Stade's, purportedly at Hoe's instigation, Mae mysteriously ordered wedding announcements printed with the name "Mrs. Edward T. Moriarty," and spent a great amount of time addressing envelopes. She slyly suggested that Arthur pretend to be her uncle during his visits, but he could not play the part. He innocently called on Stade at his place of business, one day, and described the unusual relationship. Stade was so exercised by the irregularities of the situation Hoe was obliged to give him $3000 as a calmative. Mae too began to warn that people were talking about them; he gave her $4000 to ameliorate any embarrassment.

Thanks to Hoe's confession, Stade realized he could capitalize handsomely on the delicate matter. To a demand for $23,000 Hoe responded with $10,000, and later, $8000. These gifts, however, did not sour his feelings for Mae at all; he gave her another $10,000 gift—which Stade deposited in a bank for her but which registered "no funds" when she wrote her first check. Hoe also gave Mae $8000 for an automobile; Stade bought her a $3000 second-hand car and kept the change, claiming she owed it to him. The automobile sped Mae headlong into new complications.

Mae loved her car. She made wild excursions into the country and out to seaside resorts; she drove so much (Dorothy thought) she began putting on weight. To one motoring companion Mae showed a complete wedding trousseau, explaining she was just about to marry William Burr, a cousin. Many times her travel escorts were gentlemen friends only incidentally known to Hoe; Carl Pye, once her fiance, was one, as was Burr, with whose lawyer father she had begun to discuss Arthur's reluctance to marry.

Mae claimed that Arthur had proposed marriage many times, once in the quiet intimacy of the Hotel Astor tea room. She had actually gotten him, at one point, to fix a day in February 1912 for the ceremony. But Mae remembered so much of this bewildering affair very differently from anyone else.

Mae realized sadly that Arthur was indeed married to another. Although she had understood him to be getting a divorce for her sake, she threatened to seek legal advice from her uncle if he was not quick about it. This was a new danger to Hoe, who, despite his careful document, now reckoned he could not chance going to court even though justice was on his side. Another dramatic family trial would be a shock from which his elderly mother would not recover. Arthur responded as usual: with money. He took Mae out of the Stade household and put her in a dignified Harlem hotel, where she remained for a month. At this point she became hysterical about their relationship and Arthur rashly promised her $52,000 yearly for life. When Mae was ready for the details of this extraordinary munificence Arthur duly began to furnish her $1000 weekly and set up for her a handsome Riverside Drive apartment with four servants.

The Schuyler Arms (part of Hoe's father's estate) was a sumptuous upper West Side apartment hotel. Although Arthur had been disguising himself as "Mr. Smith" or "Mr. Hanley" to visit Mae, he had to avoid the Schuyler—then on sale and frequented by Hoe agents—altogether. But in his absence he provided her an impressive entourage to keep her entertained. Mae and her maid soon indulged in one childish prank after another to keep amused. To share such social activities she invited another cousin, William Dailey, his wife (née Burr) and baby, to move in with her. To the Daileys she described Hoe as "a very charitable man [who] gave her money because he was sorry for her," possibly unaware of the simple truth of the statement. Dailey, unemployed for years, resigned himself agreeably to running Mae's errands, chief of which were trips to Grand Street for money.

Arthur made Dailey's acquaintance and soon employed him (and Burr as well) for his own purposes. For several months they traveled together wherever Dailey could be useful. All together Hoe gave Dailey about $7000 over and above Mae's expenses, constituting payment for services rendered toward providing evidence (should it be useful) that Mae was insane. There were ample signs: rented baby, frequent confusion of names and facts, unneeded wedding invitations and trousseaux. The hired cousin would shortly sue for $18,000, the balance he claimed Hoe had promised him. Dailey's position—spying on his cousin—was mitigated by financial rewards, of course, but must have appeared justified. He later sent Hoe a note:

> *. . . Please phone me at 9 o'clock in the morning at 6810 Audubon. I did everything I could to help you, and you always told me you would never forget anybody who helped you. Please tell Mrs. Hoe all about it, and then she will not object to your seeing me. Please tell her that my wife turned against her brother* [Burr] *because you were falsely accused. . . .*[54]

Although Evelyn Hoe apparently disliked Arthur's strange new cronies, he could not bring himself to reveal the magnitude of the involvement. No matter what he did, complications multiplied.

Arthur Hoe next resolved to defend himself against the breach of promise suit he suspected was brewing. He called on Mrs. Glassberg, whom Dailey apparently described as being aware of Mae's strange maternal behavior. He explained the unusual liaison, admitted to having a wife and (now) two children, and an aged mother; Mrs. Glassberg could certainly appreciate how distressed Arthur's family would be if they learned of the affair. Did she still have those curious letters Mae had written? Mr. Glassberg had them.

Manny Glassberg, like Stade, recognized a good thing. He offered the letters to Hoe for $2000, but Arthur, maintaining an innocent mien, was oddly reluctant to pay. Glassberg then offered them to Mae's attorney for $1000; they served Mae no purpose, however, and Glassberg was again repulsed. He ended up giving them to Hoe, apparently hoping that such a kind gesture would be well rewarded at some future opportunity.

In May 1913 Hoe met Mae for what he hoped was the last time. He told her he wanted nothing more to do with her and wished never to see her again. His chivalrous offer of an umbrella when they met almost two years before had started an incredibly expensive escapade: in only one year he had spent $15,000 to $20,000 on the girl, surrendered some $14,000 to Stade in various extortions, paid $7200 to Dailey, and about $7000 to Burr. The entire affair would cost Hoe $100,000—even if he escaped court fees, penalties, damages, and settlements. Arthur left on a business trip and considered the bittersweet affair at an end.

The following month, however, when he returned, Mae was waiting for him. Hoe was annoyed. He told her he had no more money for her and could no longer help. In utter frustration he sarcastically offered to put Mae where she could make all the money she wanted—in a "disorderly resort." Mae spurned the suggestion and put her ambitious legal plans into motion.

The New York Supreme Court trial of Mae A. Sullivan versus Arthur Ingersoll Hoe commenced 25 May 1914 in New York City. Mae was suing for $150,000 for breach of promise, and was ready with another claim for $75,000 against Hoe's failure to provide $1000 weekly for life. Mae was an extremely beautiful girl and Arthur was an extremely rich man; the trial was very well attended by the public, many spectators obliged to listen from outer corridors.

Arthur Hoe (1880-1957) reportedly inherited $8 million from the estate of his father. He had graduated 1905 from Harvard University and dutifully joined the family concern. His marriage in 1910, just a year before meeting Mae, ultimately brought him five children: Caroline (born 1911), Arthur (1913), Evelyn (1918), and twins Laura and Thyrza (1921). Arthur's wife Evelyn had received an anonymous telephone call in November 1911, informing her that her husband was furnishing a flat for two young girls, but perhaps suspected nothing more; she nevertheless supported him faithfully throughout the trial.

The *New York Times,* enduring the mountains of evidence and testimony, observed loftily that Arthur's indiscretions assailed the very democratic ideals Hoe presses had for a century worked to support:

> An annuity of $52,000 in weekly payments of $1,000, five $100 bills as a casual gift, frequent gifts of from $100 to $300, an advance of $8,000 to buy a motor car and the purchase of a second-hand one for $2,000† and no change returned. A bank deposit of $10,000, just pocket money, made by a kind friend, and no money to meet the first check drawn thereafter. Payments of $5,000 here and $10,000 there, representing the squandering of wealth that is beyond the dreams of nine-tenths of the hard-working and intelligent population of this city. We have noted, without prejudice, some of the statements of pecuniary transactions made in the course of a trial for breach of promise. It seems that there is more to inflame the anarchistic mind in such testimony than in all the revolutions, true or false, concerning high finance. . . .[55]

The proceedings were long and tense. Mae frequently shot angry glances at Arthur, who seldom regarded her, but Evelyn Hoe often returned them meaningfully. When called to testify, Mae spoke simply and quietly at first, but soon had long periods of uncontrolled weeping. On one occasion she pointed a finger at Arthur and spoke directly to the jury:

"I was a good girl till he met me, and I always wanted to be good." She collapsed in the witness chair and cried violently. Composing herself, she went on with what might have been the climax of a popular play:

"I don't want any of his money. All I want is my good name. Let him give that back to me. Let him build a monument to the woman who is sitting beside him, and I will go to my Maker with a clean heart."

The judge was not impressed by tears. (There had been no allegations or intimations that Arthur had been familiar with Mae.) "Put on another witness," he said.

Hoe's attorney described him as wealthy and philanthropic; he had felt sorry for a girl without home or family, and tried to make her comfortable. "Demands upon him then began to be severe," the lawyer added. "It was a case of helping a young lady in the giant city of New York and that is the most dangerous thing anyone can get caught at. . . ."

Arthur Hoe under examination was nervous. When asked why he had paid so much in blackmail because of his relationship with the girl, he replied:

"I wanted to keep the matter from my family. My mother is 77 years old, and I was anxious for her not to think things that might worry her. About the time of my father's death her memory began to fail, and she was not strong. There was no other reason why I should try to keep my acquaintanceship with Miss Sullivan quiet." [56]

The jury was asked to notice criminal ramifications of the case, as Arthur's lawyer saw them:

"Don't be carried away by a girl with locks of hair hanging down her back and wearing short dresses. Miss Sullivan was a telephone girl and she had been a cash girl, and so she knew something of life. A girl of 17 knows as much as a man of 25. . . . Burr was to marry Miss Sullivan, but Burr was bankrupt. His business was gone, he had no money, and so he couldn't marry. What was he to do? Well, they all combined to get money from Mr. Hoe. . . ." [57]

To describe Arthur, Mae's lawyer in his summation returned to the language of the old penny newspapers: he was a "vulture," a "boa constrictor" who with the "mouth of a beast of prey" led Mae along the "gilded and illuminated path of sin known as Broadway." Hoe's "forked tongue [was] heavy with pestilence" brought about by "the vulgar wealth of this wretch."

The judge was not entertained by bad rhetoric either. "Come down to simple English and analyze the evidence," he warned, coolly. In his own remarks to the jury Justice Edward J. Gavegan revealed a tender understanding of the pathos that was Mae's emptiness:

"You may consider what was in Miss Sullivan's mind. Consider the crew by which she was beset, the financial difficulties, the pressure under which she was living, the loss of her parents, and her estrangement from her living relatives. . . ." [58]

The jury deliberated for nearly 14 hours, emerging at last on the morning of 4 June 1914. Its members utterly failed to reach an agreement and were discharged.

A year later Mae sought to reinstate her suits, but the court would not entertain them, as she had not paid any of her prior legal expenses. "You know now I can go wherever I please, as I am my own boss," Mae had once remarked with hollow defiance, after the deaths of her parents. In the great cosmopolitan city of New York she had her day in court—to the delectation of millions of newspaper readers—then quietly disappeared. But Arthur, for his wealth and position, would long remain in view.

Shortly after the trial the Arthur Hoes moved to Ontario, California, and soon thereafter also established residence (aboard a palacial motor launch) in Paris, France. In October 1927 Arthur engaged Anderson Galleries to auction his private collection of 155 items, many of them large, rare pre-1400 A.D. Chinese cloisonné temple koro vases and incense burners. He told reporters he was "spending too much time in Europe to take proper care of it." [59] But he had not lost his personal and compelling interest in R. Hoe & Company, and would not yet retire in happy luxury.

Unlike his brother Arthur, Robert Hoe IV (1876-1960) was a cautious lad. He knew the boldness of his great grandfather's entry into the manufacture of cylinder presses, and his grandfather's and granduncles' resoluteness and genius in building the sturdy firm. He also knew his own father's passion for books, paintings, and erudition that had almost completely shut all else out of his later life. Robert would have to make his own peace with the conflicting forces of industry and wealth.

Robert's initiation into the firm had been accomplished with his own ink-pump distribution system; such original invention at the turn of the century no easy task for an individual surrounded by flurries of work by engineering committees. With a pleasant personality and an endearing sincerity, to everyone's satisfaction, he represented the firm with simple dignity. The company seemed in stable hands.

But when Robert commuted to Barker's Point, Port Washington, Long Island, he was on shakier ground. (In October 1910—as he and a small party watched from his offshore yacht—his home was completely consumed by fire.) He had married Ethel Louise Dodd in 1900 when he graduated from Harvard. A minister's daughter, Ethel developed rash and forward attitudes about women's roles in marriage and society, and yearned to exercise them. In the summer of 1916 the bewildered husband joined the "divorce colony" at Reno to establish residency. After 17 years of marriage Robert obtained a divorce (January 1917), claiming cruelty and desertion. He alleged that "his wife was a student of modern philosophy, and that she told him she didn't want to be bothered with a married life." [60] She did not oppose the suit and asked for no settlement.

Robert was not one to enjoy bachelor life. After about two months he was once more in love; the girl was Grace A. Sedgwick, daughter of Arthur G. Sedgwick, a prominent legal authority (who had unfortunately committed suicide just two years before). Grace was emotionally a delicate girl—"brain fever," her father had once described her condition. The happy couple were married in April 1917 and quickly off to a honeymoon.

The wedded couple's first encounters were disastrous. Accustomed to an independent and resourceful wife, Robert found himself with a fragile personality with which he had had little experience. Grace became "mentally unbalanced" and broke down on the honeymoon and Robert terminated the marriage unceremoniously.

In his second divorce that year Robert Hoe alleged that his wife "has not been restored to sound mind since said marriage, and is still a lunatic, and [he was acting] on information and belief that the lunacy of the defendant is of a permanent nature." [61] The Court appointed a guardian for her, but there was no contest; Grace and Robert were duly separated.

In January 1918 Robert's wounded ego had healed sufficiently to try marriage again. Mary E. Thompson, daughter of Edward L. Thompson of Baltimore and New York, was the girl with whom he formed a union that lasted.

But Robert's professional career had become equally uncertain, for a time. He resigned as president of R. Hoe & Company in 1913, the following year taking—of all things—the presidency of the Duplex Printing Press Company, Battle Creek, Michigan, for two years. Apparently his usefulness to the small press-maker was one of tremendous psychological impact; while there he issued a 16-page pamphlet, "The Modern Press," perhaps to supplement his father's classic history. But he was obviously unhappy. In 1919 having returned to New York city he established the Hoe Corporation, a wood-working firm which his own son Robert soon acquired.

Robert Hoe IV settled in Poughkeepsie, New York, where he became president of the First National Bank (1930-1952) and was long a director. In 1931 he was appointed member of the New York State Bridge Authority, holding the post of chairman 1954-1957. He was also a trustee of Vassar Brothers Hospital, Poughkeepsie, and a member of many professional and social organizations. He died in 1960, age 63, at Miami Beach, Florida, survived by his wife, sons Robert and Edward, and six grandchildren.

Mrs. Robert Hoe III, Olivia Phelps James Hoe, had been 72 in 1909 when her husband died. She assumed control of the family as its senior member and discharged her duties with dignity and propriety. Her problems were indeed extraordinary those first years: the tumultuous auction of her husband's literary properties, the disappointing marriages of her son Robert, the absurd trysts of her son

Arthur—and, of course, the incredible embroglio of Mrs. Johns in England. And the company itself would soon need her attention.

Olivia was born 1837 in England, where her American father, Daniel James, a partner in the New York mercantile and banking firm of Phelps, Dodge & Company, was the Liverpool representative. Her brother Frank had long been close to her, although physically quite remote: as adventurer and traveler through Africa he was among that early wave of men for whom the "dark" continent was unspoiled sports land. He described Africa in two books and died there suddenly in 1890—killed by a rogue elephant.

In 1880 and 1884 Olivia received bequests from her own father's estate and set up a $450,000 trust fund for her children. She was a founding member of Woman's Infirmary of New York and the First Aid Society (forerunner of the Red Cross). Always busy in social work, she nevertheless found time, reportedly, to advise and encourage her husband in the acquisition of books for his great collection.

In her later years Olivia made her home on lower Park Avenue, then a superb residential thoroughfare bordering the family's several townhouses nearby. But New York was changing once more; Olivia's happy world was rich with powerful recollections. In 1927 a lawsuit, brought by the J. Pierpont Morgan family, to preserve this fashionable Murray Hill neighborhood, sought to block the Lutheran Church in America from converting a 35th Street dwelling into its headquarters. The New York *Evening Post* reminisced:

> Here in this section, came home of an evening from their offices, for upwards of eighty years, captains of industry and finance who were among the most prominent men of their time in New York. There was Robert Hoe [II], who founded† the great printing press business of that name; William E. Dodge, the Phelps family, and any number of families whose lives are identified with the social and economic growth of the city. . . .
>
> A little further distant was the home of Theodore Havemeyer, very definitely to be taken into account in the life of the city. . . . The life in that house expressed the topnotch of everything that was recherche, . . . a port cochere, . . . an organ. . . . On thirty-sixth Street, between Madison and Fifth Avenue, was also the house of Robert Hoe [III]. . . .[62]

All of this soon faded away.

As the second decade of the century ended, Olivia alone remained to retrace these fond scenes. Her son Robert moved to upstate New York; Arthur, to California, then Europe; the girls—well, they were no longer Hoes. Although blind for the last seven years of her life, Olivia saw Murray Hill clearly, remembering the old mansions as only girls would—the high staircase of Dodge's Madison Avenue house giving the ladies a glimpse of the "comparative trimness of each other's feet as one would precede the other in making the ascent. . . ." How serene were those old days.

Despite her frailty and loss of vision, Olivia had for years been keen and active as ever. In May 1935, however, returning from her seventy-seventh trip to Europe, her health declined and she died in August at Lake Placid, New York, 98 years old. Her two sons, and daughters Laura Carter and Ruth Sterling, survived her. She left an estate, after taxes, of $187,917, according to newspapers, and a $4000 property at Sea Cliff, New York, which she gave to the New York Infirmary for Women and Children.

With consummate poignancy in 1935, harking to more than a century of Hoe industry, quiet affluence, and social responsibility, the *New York Times* closed its tender obituary of Olivia Hoe with abject depression:

> Twenty shares of R. Hoe & Co., Inc. class A stock [in her estate] were appraised as of no value. . . .[63]

It was true; nothing was left but an empty name.

The Hoe factory, Bronx, New York, about 1933 /
Last standing walls and final notices, 1977

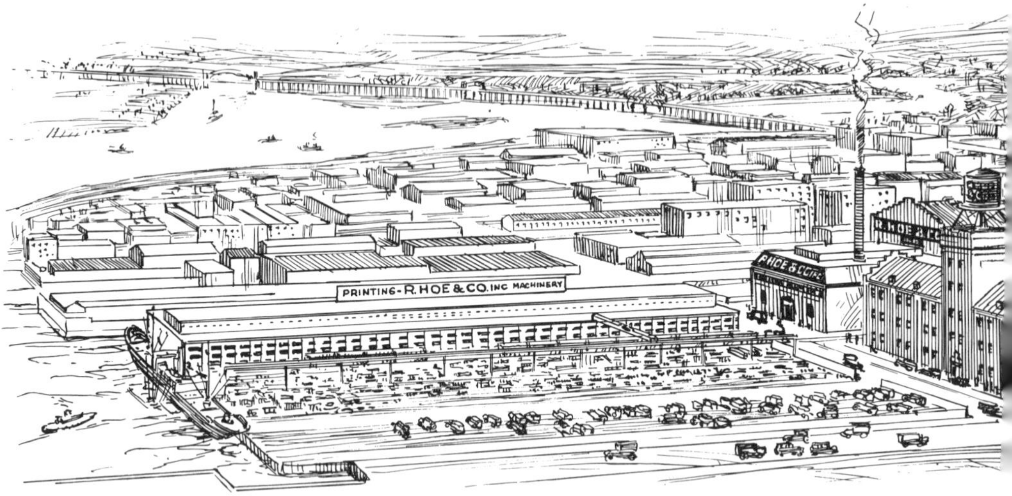

Part IV

NEWSPAPERS IN TROUBLE:

THE UNHAPPY INSTITUTION

The invention of lithography in 1796 by Aloys Senefelder (1771-1834), a Bavarian actor, might have caused a more dramatic revolution in printing had not his countryman Koenig at almost the same time so thoroughly revolutionized the technology of letterpress. Koenig's feat was (for 150 years) more pervasive, perhaps, only because it involved the use of existing types, inks, and other conventional printshop paraphernalia and simpler—mechanical—principles. Senefelder's process was more an artist's medium—a chemical novelty, and as a new and strange technique was much more difficult to disseminate.

By the inventor's own admission, the discovery was a lucky happenstance. Lacking a sheet of paper, young Senefelder scribbled his

mother's laundry list on a smooth stone slab, and later (dabbling to find a way to print music) he "etched" the stone and pulled a proof. The etch, a mild acid reacting with the chalky block, did not so much reduce the height of non-printing areas, it made them (kept moist with water) non-receptive to ink when a brayer was passed over the stone's image. To Senefelder's satisfaction the greasy writing was perfectly transferred to paper. As an art form lithography ("stone printing") could be quick and spontaneous, completely free from the confines of type columns, lock-up chases, and other letterpress restrictions.

George Englemann brought the novelty to Paris and in the 1820's applied colored inks—he invented the word "chromolithography."

French graphic artists found it useful in popular periodicals, cartoons, fashion plates, and music. But in their haste to get such works onto the street printers paid less and less attention to the fussy chemistry of the process, and by the 1850's the technique was acknowledged cheap and tawdry. In Germany the fad continued as well, and throughout Europe artists both great and small experimented with it. They particularly liked the opportunity to sketch—with special lithographic crayons—in a range of tones and textures on paper and transfer the drawing to stone; they also enjoyed drawing directly on the stone itself. In England fine lithographic albums and portfolios were issued from about 1845 to 1865, in addition to scattered use by penny papers and magazines.

William S. and John B. Pendleton, Boston, claimed to have introduced lithography to the United States when they set up shop in 1824. Four years later they moved to Philadelphia, where they continued to practice the new art. In Boston too Louis Prang, a victim of Europe's 1848 revolution, began as a lithographer in 1856 and soon enjoyed considerable popularity with a portfolio of Civil War sketches.

Nathaniel Currier had been apprenticed in 1828 to the Pendleton brothers at the age of 15, and after several starts opened his own shop: he succeeded famously with a print, "Ruins of the Merchants' Exchange N. Y. after the Destructive Conflagration of Decbr. 16 & 17, 1835. Sketched and drawn on Stone by J. H. Bufford."[1] Bufford, a Boston printer, executed many such scenes; Benjamin Day, Jr., was similarly a Currier artist. As it had strangely helped James Bennett and American journalism, the 1835 fire had also catapulted lithography into public attention.

Currier's younger brother Charles commenced manufacturing lithographic crayons and one day in 1852 sent his 28-year-old brother-in-law, James M. Ives, to the family concern as bookkeeper and draftsman. Ives became a partner in 1875. Currier & Ives soon had a five-story factory on Spruce Street turning out thousands of hand-colored prints. Mechanical chromolithography did not occur until 1888, by which time a wave of better lithographic machinery had been developed. The firm quit in 1907.

In France a steam-driven lithographic press appeared in 1850. Hughes & Kimber, London, manufactured it for domestic and U.S. markets. About 1867 the Hatch Lithographic Company, New York, introduced the British press to America, finding it capable of 4000-5000 impressions daily. George W. Hatch (long a Hoe tenant) also imported the Paris prototype, but was not happy with either of these early machines.

Many years before, thanks to the endeavors of his friend, Charles Van Benthuysen, Colonel Hoe in 1842 had (apparently in some token way) been apprenticed to the "Lythographic Business." But the greater demands of letterpress, of course, for which an eager clientele already existed, kept him from indulging his curiosity about this arty technique. Another old friend, Hippolyte Marinoni, furthermore anticipated him commercially. Tucker explained:

> Attempts had been made for several years, both in this country and in Europe, to build a suitable machine press for lithographic printing with more or less success. The older attempts generally retained the scraper for taking the impression, but in some of the later ones an impression cylinder was tried with promising result. Mr. R. M. Hoe, who was in Europe at that time, saw a machine of the latter description made by Mr. H. A.† Marinoni, a press builder of Paris, and he was so pleased with its performance that he proposed to Mr. Marinoni that he, as the inventor, should make an application for a United States patent, and for a certain consideration, assign the patent to him. . . . One of Marinoni's machines, and a set of drawings, were sent here for our guidance. . . .[2]

Within a few months a model was ready for the American trade:

> [*To E. Mendel, Chicago, 17 February 1869*] . . . *We have the first Steam Lithographic Press nearly ready, & are now taking orders for these machines, having already received several.*
>
> *The size we are making takes on a stone 19 x 25 ins. The price is not yet fixed, but will be somewhere in the neighborhood of $3000.--. This machine is strongly made & adapted to the requirements of the trade in this country. . . .*

The Hatch firm probably put the press on trial, and soon found it ideal:

> The iron hands of the [Hoe] machine took hold of the paper with the delicacy of touch of a woman's fingers, and it was adjusted to the picture on the stone with accuracy surpassing the skill of the best workman, while the whole machine moved with the perfection of a watch. . . .[3]

By 1883 Hatch would have 12 such presses in constant operation. In general, however, the printing fraternity conceded the technological development of lithography to be at least fifty years behind letterpress. It moved along fitfully.

As with this early lithographic experimentation, R. Hoe & Company was also obliged to maintain an interest in yet another printing process—gravure. Printing from the recesses (which filled with ink) of an inscribed or etched plate, the intaglio (Italian, "cut in") press needed to wipe away completely the surface ink and to pull with tremendous pressure the ink from the grooves. The Hoes had made hand-operated gravure presses from the 1830's, but, serving a tiny banknote industry, they were not an important item. Richard had duly looked them over:

> [*To Thomas Jones, 24 August 1837*] . . . *I have also made an improvement in the copper plate press which is simply this—the applying* [*of*] *a friction roller under the journals of the under Roll. I have never heard of this being done and have a model & drawings made which I send to your address. . . .*

Ironically, the gravure process would succeed commercially with giant high-speed rotary machines before it became feasible for smaller job applications.

An Ohioan, Robert Neale, in 1858 commissioned Hoe to build a steam-operated copperplate (gravure) printing press. Tucker:

> This was a large and expensive machine. The printing plate bed was attached to a pair of endless chains which carried it along horizontally with the printing plate face upwards under the impression cylinder and an inking apparatus. . . .[4]

The wiping mechanism did not clean the plate adequately and the press was not successful.

About 1862, when the U.S. government was authorized to issue engraved paper money, three New York banknote companies each ordered 50 Hoe (hand) copperplate presses to enlarge their facilities, and subsequently bought dozens more. The intricate banknote scrollwork, the meticulous cutting of plates, and the careful pulling of sheets had kept the process a hand-operated skill, but with this new impetus its mechanization was again examined.

In 1876 Neale returned with a radically different design for gravure printing, this time so promising he had already interested a large publisher. "The arrangement for inking, wiping, and polishing the plate," Tucker wrote, "was said to be very effective, and the machine was set to work in the office of Messrs. D. Appleton & Co., Brooklyn." When the apparatus came to the attention of the Bureau of Engraving and Printing, Washington, D.C., Neale and William S. Appleton looked forward to heavy government endorsement.

Large expenses incurred with R. Hoe & Company, however, had (again) to await the convenience of bureaucratic printing committees:

> [*From Neale & Appleton, Washington, 10 January 1878*] . . . *Our relations with the Government are such, that we cannot get money until the press has worked a few weeks to prove its capabilities for the purpose. This, we are informed, is customary in trying other new presses. . . .*

The Colonel, however, was not disposed to wait for the blessings of officials who had not long before been so disappointing with regard to his patent extension request. He proceeded to sue Neale and Appleton (whose family certainly had considerable resources), but his Washington lawyers found this objective frustrated:

[From Worthington & Heald, Washington, 16 July 1878] . . . After suit was filed against W. S. Appleton and Robt Neale your Agent Mr. Smart learned certain facts relative to the ownership of the plate printing machine which rendered further prosecution of the suit useless. . . .

Apparently the two entrepreneurs had safeguarded themselves against such an eventuality. They nevertheless realized that if they succeeded they would have to return to Hoe for more machines. They kept the Colonel informed:

TREASURY DEPARTMENT.

November 19th 1878

Messrs Hoe & Co.
Gentlemen,

You will be pleased to know that the Steam Plate Printing Press which you made for me and which has been on trial in this department, works in a satisfactory manner, and produces impressions of excellent quality.

Some of my friends, Engravers, and Experts, in the Treasury Department, insist that the printing done by the steam press is superior to that usually done by hand; in some respects, I certainly think so too.

The Committee of five, appointed to examine and report upon the press, and the printing done by it, complete their testing to day, and from some expressions of satisfaction, we anticipate from them a favorable report.

Under these circumstances, I trust we shall soon receive orders for more presses, which, in some degree at least, will make amends for the various unforeseen delays attending the introduction of the press.

I remain very Respectfully
Your Obt. Serv't
Robert Neale

A few days later formal—printed—invitations were issued:

For some weeks past, a committee appointed by the Secretary of the Treasury, has been testing a recently invented steam press, intended for printing steel and copper ingravings† of the highest order in the engraving art.

The press has produced a revolution in that art, kindred to that achieved by the steam press in typographic printing, and has successfully printed the finest steel engravings published by Messrs. D. Appleton & Co. in "The Art Journal," 'Picturesque America' and 'Picturesque Europe.'

The last named gentlemen and the inventor of the press, Mr. Robert Neale, maintaining that their machine can print the revenue stamps, greenbacks, and other securities of the Government, in an equal if not higher degree of perfection than has been attained by the comparatively slow and much more expensive mode of hand roller printing now in vogue, the Secretary of the Treasury, by his committee of experts, is making the test of the adaptability and fitness of this new invention for printing the Government issues.

Messrs. Neale and Appleton respectfully invite you to call and see their press at work on Saturday next, November 23d [1878], at 1 o'clock, P.M., in the Machine Shop of the Treasury Department on Fifteenth Street, south of the Treasury Building.

(Please present this invitation at the door.)

Messrs. R. Hoe & Co.
Gentlemen,
I shall be happy to see any friends of yours in Washington who would like to see the press in operation on Saturday next.

Very Respectfully
Your Obt Serv't
Robert Neale [5]

These successful operations and happy prospects were smashed in an undisclosed turn of events:

> In 1878, the Neale-Appleton press and another plate printing press, the Milligan press, were placed in operation on trial in the Bureau of Engraving and Printing. After finding that 'the presses were suitable for printing certain classes of work in a satisfactory manner with a considerable savings to the Government (the Milligan type more so than the other),' the Bureau offered to construct five additional machines of each type at $500.00 per press and royalties of $1.00 per 1,000 impressions. 'Owing to disharmony between the owners of the Neale-Appleton press, the proposition was not accepted by them and, consequently, no further use was made by the Bureau of that machine.' . . .[6]

The Colonel probably shook his head in dismay, and returned to other pressing demands. In the area of lithography one of these was a novel system in which the heavy stone itself could be eliminated.

In 1819 John McLoughlin, a Scots coachmaker, had come to the United States to seek his fortune. Meeting Robert Hoe in New York he became fascinated with printing and soon opened a shop to produce children's books and other literature. His son John, Jr., developed a technique (his own invention) of printing from zinc plates, handling the grainy metal surface much like that of lithographic stone. Although the plates had a fairly short life, the idea of wrapping a sheet of metal around the cylinder of a large press soon led to the complete elimination of the cumbersome blocks.

Hoe's two-color "Rotary Zincographic" and "Aluminum" presses were in service by the end of the century. The speed of these machines, without the dead weight of the stone, increased considerably, but the number of impressions obtainable was still relatively low. The great comic weeklies *Puck* and *Judge* (and others) were soon printed by "litho rotary machines" which gave them a spontaneous life and riotous color enjoyed by all; they sometimes recaptured the mirth and satire of the earlier French caricaturists.

A New Jersey paper manufacturer and printer of office forms, Ira W. Rubel (-1908) at this time was producing bank deposit slips on a stone lithographic press employing a rubber-covered impression cylinder:

> As would happen with any hand-fed flatbed cylinder press, the feeder occasionally missed a sheet; the [Rubel] press had no impression throw-off so the print would be made on the rubber blanket, then the next sheet through would have a direct print from the stone on one side and an indirect—offset—print on the other. . . .[7]

Rubel thought the offsetted reproduction actually superior to the direct, and in 1904 designed a lithographic press that deliberately transferred the image from an aluminum plate to rubber blanket and then to paper. The metal-to-rubber impression reduced plate wear remarkably.

The inventor's first machine, nicknamed "Sherbel" (his partner's name was Sherwood), was built in 1905 by Potter, who began making offset presses for other printers. A. F. Harris (who also discovered the offset phenomenon) began a year later, acquiring the Potter interests in 1927. R. Hoe & Company introduced its own offset machines in 1908, but this new kind of printing, unlike the firm's near monopoly in letterpress, was not to be Hoe's alone.

Manufacturers almost everywhere in America spotted the offset technique as a new and viable way to approach the printing industry without Hoe domination. Its one-hundred-year tradition in letterpress had given the firm a formidable psychological advantage, if not a real one, but no such commercial heritage existed for offset lithography. Twentieth-century machine shops now had ready design and engineering facilities to produce precision machinery quickly, and they were pressed into action. The expansion of the business was phenominal: in 1912 eight firms were offering offset presses in 560 sizes and styles, and four years later printers had over 800 models to choose from. The Harris Automatic Press Company, Cleveland, in 45 years would alone build 75 sheet-fed models in 29 sizes—over 2000 different machines.

Undaunted, more and more contenders would hang out their signs:

> In addition to Harris, Hoe, Potter and Scott, we recognize John F. Webendorfer as one of the pioneers in offset press design, specifically as the inventor of the first web offset in 1933. Later, American Type Founders, Miehle, Miller and others were to build offset presses, and Goss and Fairchild were to take the lead in manufacture of offset presses for newspaper printing. For the record, we insert the fact that Hoe, as early as 1939, built a web offset newspaper press of four per-

> fecting units, with cylinders carrying four page-size plates across and two around, with automatic paster and folder. . . .[8]

Lithography had also made possible, as early as 1875, the printing of sheet metal (a branch termed "specialty" work) later shaped into trays, canisters, and boxes. Hoe, Harris, and Rutherford Machinery Company, East Rutherford, New Jersey, were (and are) principal manufacturers of this equipment.

Oblivious to these incursions, the practitioners of letterpress printing continued to nurture their metal monsters almost without examining the principles and objectives involved. With the sophistication of American engineering observers thought the trend to be without diminishing returns. Said Ringwalt:

> Each new mechanical facility supplied to newspapers enlarged their sphere, cheapened their cost, and increased their attractions, until now they well-nigh absorb all other forms of printing, and embody every description of intellectual effort. . . .[9]

By 1934 such engineering was indeed a major accomplishment:

> The past half-century . . . has seen machine design, which started with a contraption whose every revolution shook the building on its foundations, turn out presses weighing more than 500 tons, with bearings like those of a watch and upon which a coin can be stood on end without a tremor for hours of full-speed operation. . . .[10]

Significantly, R. Hoe & Company was in the engineering business, and proudly so, but like many firms unperceptive of pragmatic twentieth-century—McLuhanesque—questioning of means and goals it was trapped by its rank technology. The big presses were not an end; they were tools for mass communications. The firm might have done better thinking itself in the communication industry than in machinery.

The complexities of letterpress mounted one upon the other. The setting of lead typematter and the casting of lead stereotypes (when one only wanted a weightless image on almost weightless paper), the onerous and time-consuming chore of makeready (which—despite experimentation with rosins and chalks—couldn't be mechanized), and the ruthlessly pressing demands of overskilled workers (who could never be placated) made it very vulnerable indeed. Only the immense nation-wide plant investment alone seemed to keep relief printing alive.

The major addict to letterpress was, of course, the newspaper, quietly groaning for a long time and unable to find a cure—or a diagnosis. Advertising revenues were falling, newsprint doubled in price, and union demands became insufferable. Oswald Villard in 1944—despite the war—found newspapers changing from purveyors of information and news into entertainment media, caused largely by the increased use and popular acceptance of syndicated materials. He charted the shrinkage of the industry as a trend: of 2042 English-language dailies in 1920, and 1933 in 1930, there were only 1754 in the early 1940's (104 disappeared between 1941-1943 as he wrote). Without attributing much of this decline to depression and wartime tensions, alarmists blamed publishing moguls for their manipulations of papers as if they were artificial and perverse acts. When even the great giants of American journalism vanished, the whole newspaper industry seemed doomed.

The final issue of the New York *World* had gone to press on 27 February 1931—the last paper remaining on Park Row. The *Sun* (its own days numbered) recollected with a fine sense of history:

> Night after night and year after year, until last night hale old ghosts foregathered in Park Row to marvel at these new times and sometimes curse them roundly. They were grand marvelers and grand swearers, too, those great old boys who came flitting out of the hereafter—Dana of The Sun, the elder Bennett of the old Herald, Greeley of the Tribune, Henry Raymond of The Times, and Pulitzer of The World. . . . With the passing of the New York World went the last newspaper of Park Row.[11]

Visiting the United States in 1900, Lord Northcliffe had used the *World* to introduce his concept of a tabloid newspaper (his own word), hastening to add, "Let me explain at once that this is merely a twentieth century suggestion on my part . . . closely resembling the newspaper of the future." New York in the 1920's was soon bombarded with several of these popular, picture-oriented sheets which seemed to bring a new vitality to the business.

But the more literary press institutions did not benefit. Celebrating its centennial in 1933, the *Sun* tried to look ahead cheerfully:

> To all such contemporaries this newspaper, observing its hundredth birthday, offers congratulations on the completion of a century of free and independent journalism, and with its gratulatory spirit is mixed curiosity as to what the newspapers of the world will be like in 2033 A.D. . . .[12]

But in 1950 the *Sun,* a victim of mounting costs, was combined with the *World-Telegram,* which itself would soon pass away.

The New York *Herald-Tribune* (a 1924 merger) had bravely tried to cope too. Its ability to respond quickly to news events was in 1936 deemed the epitome of all Bennett-Greeley and Hoe genius put together:

> In this lofty apartment [the pressroom] are fourteen presses—three decuple, nine octuple and two color presses—which can turn out thirty-two-page *Herald Tribune's*[†] at the rate of 240,000 an hour or 4,000 a minute. Every sixty seconds more reading matter comes

through the machines than was in the whole first edition of the *Tribune* ninety-three years ago. In fifteen minutes from the time that an edition reaches the stereotypers, the first copies will be printed and folded, ready for distribution.[13]

Yet the grand old paper was long ailing, with readers ever slipping away to the suburbs, inflation growing ominously, and debilitating contract negotiations with up to ten independent unions. It expired in 1966. What followed was almost comical if it were not so expensive:

The [New York] strikes continued with a 25-day tieup in 1965. Thereafter, Hearst's *Journal-American* and Scripps-Howard's *World-Telegram & Sun* combined in a single evening paper and joined John Hay Whitney's New York *Herald Tribune,* a morning paper, in what they hoped would be a 24-hour-a-day operation in a single plant. However, the unions never gave the merger a chance. On the opening day of publication, April 25, 1966, they struck before any papers of the new combine appeared. When the struggle dragged on through the summer, Whitney folded his *Herald Tribune* and agreed with his partners, Hearst and Scripps-Howard, to put out a single evening paper, the *World-Journal-Tribune,* if the unions would permit it. . . .[14]

This too collapsed the following spring. Efforts to keep the proud old names alive (including a New York *Trib* in 1978) did not take into account the fact that not only was the traditional paper an anomaly, so also was its means of production.

Many modern—often electronic—processes had made their appearance in the twentieth century, but, surprisingly, they were not to be found in newspaper offices. An editor asked:

If such technical capabilities [photocomposition, satellite printing, electronic picture transmission] exist for modernizing newspapers, why aren't they being used? On the whole, the technology which produces a great majority of newsprint in this country is over a hundred years old, and organized labor has been blindly obstructive thus far in preventing a move toward any of the most elementary techniques to improve even the printing and setting procedures. . . .[15]

A Britisher also commented:

The newspaper industry has been remarkable in the past for its lack of any radical technological innovations over a long period. Much of this has been due to the cautious attitude of management, with vast investments in heavy conventional equipment, backed by the conservatism of the trade unions who have been suspicious of the effects of new methods on security of employment. However, technological development is upon all of us, whether we welcome it, as I do, or no. . . .[16]

In 1919 Oscar Roesen, a vice president of R. Hoe & Company, had traitorously admitted that "monster presses, complicated linotype machines, matrix making and sterotype† plates, [and] casting boxes" were becoming an impediment to modern dissemination of the news. Inspired by Henry A. Wise Wood's stimulating gaze through the old *Herald* windows, Roesen similarly visualized a futuristic newspaper office:

> No press room, but a large room filled with telegraph instruments and operators, each instrument connected with thousands of wires forming a cable running underground to the houses of all subscribers, a bulletin board in the home of every subscriber with a stock ticker printing arrangement connected by a wire with the main instrument in the publication office.
>
> When any news was ready and the copy given to the telegraph operator it is flashed over the wires, possibly one instrument printing through 20,000 wires to 20,000 subscribers, on the order of the telephone system, all the news as fast as it comes in. Before the subscriber goes to bed he puts in a fresh sheet of paper and while he sleeps the news is printed on the sheet and read at his breakfast. . . .[17]

In 1970 this concept was developed by a Japanese journal; in the subscriber's home an electrostatic stylus reproduced a newspaper page in five minutes. The *Los Angeles Times* also experimented with printed news sent via television. But these attempts were more a confusion of technological capabilities and misunderstood public preferences than a real step forward.

Letterpress printing methods were nevertheless seen to be so archaic they were beyond salvaging. Historically, newspapers had indeed found that big presses were expensive to install (although there was no other viable system) and suitable only for a narrow range of circulations. The addition of more machines only confounded folding and collating operations. When a paper prospered and its circulation multiplied, the progressive modifications to stereotype, to rotary, to web, to perfecting—in units of 8, 16, 32, and larger numbers of pages—spelled gigantic, constant plant alteration and expense. Hoe's unending work for Pulitzer's *World* was only the best manifestation of this dilemma. Now it seemed the end had arrived. Letterpress printing died, said James Moran, "approximately 1940, when it may be said to have reached its zenith. . . . After five hundred years, the relief printing press has nearly had its day."[18]

Frustrated newspaper publishers looked around for signs of life elsewhere. The New York *Daily Graphic* had, toward the end of the nineteenth century, demonstrated that photo-journalism could be a dynamic, urgent news tool—more so than the older illustrated newspapers—but itself in the grip of strong letterpress technology it furnished no model for others. Not until 1939, with Hoe's four-unit offset press, was the first carefully designed lithographic newspaper plant a reality, but it too, thanks to the intervening war, attracted little attention. As late as the 1960's therefore, large installations of offset machinery were heralded as major breakthroughs, although the public—even printers—thought the phenomenon had occurred long before.

For the weekly paper *Grit,* Williamsport, Pennsylvania, R. Hoe & Company in 1962 built a 350-ton offset newspaper press (partly designed by Aller Press Ltd., Copenhagen). Publishers of this nationally distributed paper wanted rich colors and detailed black and white formerly available only in slick magazines. The resulting new equipment enabled *Grit* to print 72 pages in four colors, at 50,000 per hour (most offset presses are somewhat slow by letterpress standards—inks are tackier and slower drying). Two years later the first British-built web-fed offset machine was erected for a Colchester newspaper, a five-unit balcony-type Hoe-Crabtree. Although Hoe in 1946 had bought a Pittsburgh lithographic press plant to strengthen its position, these installations, although significant, did not auger a firm hold in the offset area. Too, Hoe's mania for ROP color was seen as another chimera.

Whether offset or letterpress, the forced introduction of four-color process printing into newspapers—ostensibly to compete with television—was reckoned a colossal blunder. Editor Carl Lindstrom in 1960 criticized science (not journalism) for maintaining the credibility gap: "Scientists are determined not to be understood and describe their research in a more opaque argot than that of the bureaucratic directives. . . ." But more critically he saw the quest for life-like color on newsprint a waste of time and resources:

> Run-of-the-paper color printing is probably the greatest technical delusion to which American daily journalism has subjected itself in recent history. It can be likened only to the diversionary adventure stumbled upon at the turn of the century, *The Yellow Kid,* a comic strip, and after him the deluge. . . .

Lindstrom judged newspaper colorwork in poor taste, poorly registered, and inevitably a failure on cheap paper. "The greatest threat journalism ever faced was the invention of the printing press," he remarked enigmatically, but he was not disturbed: "The printing press is on the way to obsolescence; yet journalism marches on."[19] He was transferring his allegiance and enthusiasm to electronic media.

At the start of the 1970's nevertheless, the lithographed paper was a growing phenomenon. Almost 500 American dailies were offset-printed, many of them offering occasional full color. More than 350 composing rooms, furthermore, were using computerized (photo) composition. By 1978 1260 of 1786 U.S. dailies (70 percent) were offset, but they were small papers, accounting for only 31 percent of national readership. But they would grow in circulation as well as number over the next decades. There was yet a future for journalism via the offset newspaper if only the industry could support—and survive—the necessary technological revolution. One firm had approached the problem methodically.

MGD Graphic Systems, Chicago (acquired by Rockwell International in 1969), today the world's largest manufacturers of printing presses, commissioned the Battelle Memorial Institute of Columbus, Ohio, a research firm, to study the graphic arts market. In 1971 its report stated that the newspaper as a mass-produced news and advertising medium will remain essentially unchanged through 1990—but that every aspect of its preparation, printing, and distribution will change.[20] The Institute similarly believed that other well entrenched products—books, for example, will also endure, but it insists that by 1990 letterpress will indeed be gone.

MGD has responded by designing families of new equipment for this exciting graphic arts future. Its line of sheet-fed offset machines (Miehle "Super 60") far outdistances its sale of letterpress machines. Its newer web offset newspaper presses (Goss "Metro" and "Cosmo-Offset") and considerable ancillary equipment have already become immensely popular among publishers. The prestigious *Wall Street Journal* as early as 1968 switched to Goss Metros and prints 60,000 32-page papers hourly on each. So many owners—over 75—have converted to the Metro alone they have formed an association: the Metro Users Group. Roughly two out of three newspapers in the United States are now printed on Gosses.

Surprisingly enough, the letterpress industry has not been sleeping, and various alternatives to stereotyping—with some fifteen different plate-making processes now under study—are beginning to convince printers and publishers that letterpress is not dead after all. There is much room for improvement and for new concepts as well. Erwin Jaffe, a newspaper research director, has commented:

> The design of our 1977 newspaper presses is almost as antiquated as newsprint mill technology. It is hard to believe that a 400-ton monster is still needed to put one-fifth of an ounce of ink on 64 pages of newsprint. A 16-page press unit costs over $500,000. We also require hoards of manpower to operate our century-old designed presses, be they letterpress or offset. . . . Newspapers cannot enter the world of fiber optics, facsimile, and satellite receivers and still continue to use these outmoded 19th century press designs. . . .

With such ultra-modern news collection agencies and ultra-rapid typesetting facilities a reality, pessimists fear there is little more to look forward to in improved speed and economy. Jaffe had one further point to make:

> We are confident that in approximately five to seven years, there will

> be compact, low-cost, double-width press units to print from quality letterpress plastic plates. These press units will consume considerably less electrical energy, and need much less manpower than the age-old press designs we currently use.[21]

Far from doomed, letterpress is still faster, more reliable, and better in quality of product wherever faithful color and dense blacks are needed. The process may have a promising future after all.

Wood Industries, Inc., Englewood Cliffs and Plainfield, New Jersey, has inherited the Hoe mantle as a champion of letterpress methods. Although it has sought to expand only moderately and diversify with caution, it has some firm tenets and sound properties on which to build. For the offset market it was, as early as 1968, inviting printers to "investigate our 100 years of technical know how . . . Go with Wood . . . The Total offset installation. . . ." In 1969 the firm introduced its "Colorflex 250," a single-width web-offset press designed for intermediate-size newspapers. It can manufacture 22,500 papers hourly up to 96 tabloid pages; with a color unit added it can print three colors, one of them by direct lithography.

But at the same time Wood seems reluctant to abandon its Hoe-inspired letterpress potential. After its acquisition of the Hoe properties it boasted "12 uninterrupted years" of research on the (letterpress) Hoe "Colormatic" and on the "Hoe Double Five Color Press," offering customers "Hoe versatility, Hoe dependability, Hoe productivity."[22]

In 1974 Wood-Hoe announced its "Book-O-Matic"[23] in-line production system which issues complete books or magazines in one continuous operation. Developed originally by Grafisk Maskin, a Swedish engineering firm, for a large book printer at Falun, Sweden—B. Wahlstroms Bokforlag, A.G.—two years before, the system was dramatically improved upon by Wood-Hoe engineers, separating the printing process from the machine's other operations. Thus the Book-O-Matic is available with letterpress, offset, gravure, or flexographic (aniline dye) options, in one, two, or four colors. With a variety of impression cylinders—116 to 180 inches in circumference in half-inch increments, governing trim size of book, and by employing from one to four of the impression cylinders, books from 48 to 768 pages can be handled. The extraordinary versatility of the machine is reminiscent of Colonel Hoe's family of Lightning presses.

After printing is accomplished the Book-O-Matic web is slit into ribbons, folded, cut, and collated into signatures ready for binding by a separate adhesive binder (for paperbacks) coupled to the press. The mammoth installation is capable of 24,000 paperbacks hourly, slightly less for books of larger dimension. The machine delivers conventional signatures (with centerfold intact) so that, if desired, Smyth-sewn books can also be manufactured. The first American installation was at the Watchtower Bible and Tract Society, Brooklyn, New York.

As print shops across the country ponder the merits of the three major methods of printing, one giant newspaper has quietly demonstrated its ability to accept such dramatic changes in its internal procedures.

The proud old *New York Times,* had, with "All the News that's Fit to Print," long banked heavily on its reportorial integrity to carry it across occasional periods of stress and strain, such as moves to new headquarters:

> January 1, 1905, was a Sunday and, as the [Times] presses at 41 Park Row stopped rolling off the big Sunday edition, an army of men from the Mergenthaler Linotype and Hoe Printing[†] companies took over, together with the entire Times staff and hundreds of special workers hired for the event. Everything usable from the old office was piled into horse-drawn drays and hauled the three and one-half miles uptown to the New [Times Square] building. . . .[24]

The mammoth presses were sunk on Manhattan's bedrock right by the subway, so bundles of papers could easily be hustled throughout the city.

From its 42nd Street vantage the *Times* chronicled 70 decades of man's activities both domestic and foreign. The Linotype composing room had changed little in all that time:

> The machines were great hulking metal monsters the size, weight and vintage of Model T Fords. Long lead ingots, or "pigs," each weighing almost 19 pounds, hung from chains, and turned molten deep in the machine's recesses. Lights flashed. The machines thrashed and rattled and clanked and clattered and spat out little slivers of lead at the feet of the men who operated them. The room was dark and dirty and it reinforced a mind's-eye image implanted by a dozen grainy black-and-white movies of the 1940's.

So it was until 1975, when the paper began converting its procedures to computerized composition. One by one, various ponderous sections of the newspaper were modified, until 3 July 1978, when the main news and editorial columns were given the final treatment. "Five whirring machines, each no bigger than an executive desk," said the paper, "now do the work once done by 129 Linotype machines." The computer-set composition is "photographed" by a scanning laser beam and transmitted electronically either to the *Times'* basement printing plant or to its satellite shop at Carlstadt, New Jersey, six miles away. But this was only part of the *Times'* revolutionary conversion.

In a feature article harkening to the London *Times'* announcement of steam presswork, the New York paper now had its long awaited moment. Despite its modernisms, the *Times'* language is perhaps as mystifying (for its physics and chemical allusions) as John Walter's mechanical description of a cylinder press:

> The invention of a sticky, clear plastic about the consistency of honey was responsible for the next innovation in Times printing. When exposed for about one minute to intense ultraviolet light, this liquid plastic coalesces into a substance so hard it can be used to print hundreds of thousands of newspaper pages.
>
> In the basement of the Times building, the gooey stuff is smeared in an even layer over thin steel sheets and exposed to ultraviolet light. The light is first passed through the negative mask prepared by the laser system. That way, only the plastic that corresponds with the printed images on the page gets hardened.
>
> The rest, still liquid, is blown away with an air jet, and recycled.[25]

At the Carlstadt plant the negative mask can be used to make offset plates, if desired. The light-weight plastic plates are installed on the *Times'* Goss letterpress printing machines, where "The New York Times will still cascade from the presses in a stream of about 120,000 copies an hour."

Had R. Hoe & Company survived a little longer it might have participated in these exciting twentieth-century revolutions in printing processes. But the company perished along with hasty predictions of the end of letterpress. The myopic giant would—for its final half century—dissipate its tremendous energies and resources in senseless, primitive—animal-like—infighting among management, employees, and stockholders, a beast almost incapable of reacting, adapting, to external technological change of any proportions. Had Charles Darwin observed the struggles of corporations as if they were living creations, he might have found R. Hoe & Company old, tired, feeble—unfit for survival. But its example remains nonetheless as provocative and rewarding a study as nature itself:

> [W]hen we contemplate every complex structure . . . as the summing up of many contrivances, each useful to the possessor, in the same way as any great mechanical invention is the summing up of the labour, the experience, the reason, and even the blunders of numerous workmen; when we thus view each organic being, how far more interesting—I speak from experience—does the study of natural history become![26]

PROGRESS
IN
PRINTING
THE TECHNOLOGY OF NEWSPAPER PRINTING, AS DEVELOPED OVER FIVE CENTURIES—FROM GUTENBERG TO THE COLORMATIC.
—A PORTFOLIO OF TWELVE PAINTINGS COMMISSIONED FOR THE HOE COLLECTION.
1456 JOHANNES GUTENBERG in Mainz, Germany, produced the Latin Bible, which is generally accepted as the first book to be printed in the Western World from movable type. Through his methods, Gutenberg established printing as an industry as well as a craft, and is thereby recognized as the father of commercial printing.
SEYMOR CHWAST, New York artist, newspaper designer and publisher of "Monthly Graphic", selected the woodcut and hand illumination techniques of Gutenberg's time to symbolize Gutenberg—the man and the method.

1958 THE HOE COLORMATIC*... year of a giant stride... The planned simplicity of Colormatic* design permits unparalleled ease in the use of ROP Color coupled with maximum speeds up to 70,000 pph.
*Trademark
DICK SHELTON, gifted fine artist and illustrator, depicts the massiveness of the Colormatic* while successfully evoking the quiet efficiency of this great press.
PROGRESS
IN
PRINTING A CONTINUING RESPONSIBILITY OF R. HOE & CO., INC. HOE
910 E. 138th Street, New York 54, N.Y.

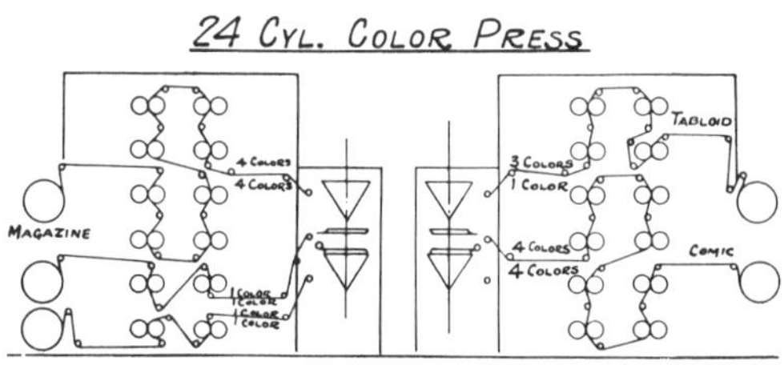

HOE 24-CYLINDER MULTI-COLOR PRESS

c. 1923

Largest Printing Press in the World

13.

Family Reorganization of the Firm

THE CONTENTIOUS BULLEN / R. HOE & CO., INCORPORATED / SWARTZ AND THE BRONX / THE GREAT DEPRESSION / BANKERS AND BROKERS CONSPIRE / ZIRN FIGHTS A PHONEY BANKRUPTCY

At the end of the century American Type Founders, Inc., a multi-million dollar reorganization of some twenty foundries across the country, hoped (in the face of several mechanical typecasting systems already at work) to salvage a declining foundry industry.

ATF as a corporation started in 1892 and soon consolidated its sprawling activities on Communipaw Avenue, Jersey City, New Jersey. That same year Henry Lewis Bullen (1857-1938) was appointed New York manager for the new enterprise. The extraordinary amalgamation could indeed use this unusual man. Bullen was born in Victoria, Australia, where he apprenticed in typography, lithography, and bookbinding. He came to the United States in 1875, working as compositor in—among many places—Davenport, Iowa, St. Louis, Boston, and New York. In 1882 he entered Golding & Company, Boston, a printers' supply house, becoming sales manager a year later and head of the Boston office after five years. Through his witty advertising the company was happily selling, among 36 platen presses on the market, its "Little Pearl." But Bullen was restless and homesick; he returned to Australia for three years.

Back in America, Bullen, about 35, labored hard and long among the many subsidiaries at ATF. This sales work, like that at Boston, undoubtedly put him continually in competition with Hoe agents, whose advantageous position must have impressed him considerably. He also learned much of Hoe's reputation as he collected, for his own amusement, books, advertising pieces, memorabilia, and other printing keepsakes. At ATF he began to store a library of graphic arts materials which gradually became a refuge (perhaps like Robert Hoe's library) from the performance-oriented salesman's world. In 1899, however, Bullen accepted a still more challenging position.

As manager for F. Wesel Manufacturing Company, New York, Bullen began to market printers' numbering machines and other accessories in a very competitive environment. But he moved on. About 1904 the United Printing Machinery Company, a merger of three small supply houses, offered Bullen the position of general manager. He accepted the post, brought some old friends along with him, and entered his new office with his usual enthusiasm and drive.

The pressure of his new duties, however, quickly told on him, his puzzled superiors expressing their growing concern:

> [T]he eccentricities of Mr. Bullen, his frequent and erratic journies, his devotion to inconsequent details at a time when the company required all of his experience and attention in organizing and placing it on a solid footing, became a source of anxiety. . . .

Remonstrances from his friends and reprimands from his employers had little effect. Bullen's thoughts were elsewhere—perhaps in his quiet library, or perhaps in Victoria, where life had been simpler. In September 1905 United Printing Machinery abolished his job and reassigned him to lesser duties. In the course of this action some irregularities in Bullen's accounts were called to his attention, but he would not discuss them. Suddenly he disappeared completely.

Missing too were $2750 in funds from the firm's account as well as some important company papers. A private detective found him:

> Secured by the Pinkerton detectives after a long chase, captured with passage to Honolulu in his pocket, he positively refused to return to talk with any one connected with the company or accept any terms in the nature of a fair arrangement that would have prevented the matter from becoming public, and thus save himself.

The company had no choice but to press charges; bail was set at $5000, but neither Bullen nor his friends posted it, and he remained in custody until his trial. It was succinctly covered in the trade press:

> On Monday, October 9, 1905, Mr. Henry L. Bullen, formerly general manager of the United Printing Machinery Company, pleaded guilty in Part IV General Sessions Court, New York, to the charge of grand larceny, first degree, in the sum of twenty-seven hundred and fifty dollars ($2,750), preferred by the United Printing Machinery Company, 12 Spruce Street, New York.[1]

The judge, finding the case "a very sad one," reduced the usual ten-year sentence to two. Noting his great energy, high ability, and nervous, forceful temperament, *Inland Printer* clearly saw that Bullen's "mental faculties were temporarily clouded." During his imprisonment he had time to reflect, and comforting new schemes entered his thoughts.

Bullen's love for books and printing transcended his interest in making money, much as had Robert Hoe's, but unlike Hoe he had no personal means to pursue such activities. In 1906 (from jail) he proposed publicly the establishment of a typographic historical society, an organization which would grow around a collection of graphic arts materials such as ATF's and his own. His premise was sound: with typefoundries and printing institutions everywhere threatened with extinction, with machinery evolving so fast there were few inducements to save old models, and with the increasing costs of preserving old records the industry needed active, alarmed protection for its valuables. Bullen knew that many archives, patent records, and obsolete type-casting equipment were already in jeopardy.

In 1907 Bullen began to solicit graphic arts publications of all kinds, and at the same time advertised for a sponsor to support the colossal bibliographic task. He got eighteen replies, chief of which was from Robert W. Nelson, his old employer and associate at American Type Founders. Nelson, president of the company, encouraged Bullen to form his library at ATF, starting him with the old Bruce patents and records already on hand. To these Bullen added a few hundred of his own books, and Carl Schraubstadter of the Inland Type Foundry, St. Louis, donated a like number. These were the nucleus of the world famous ATF typographic library.

Bullen soon amassed 70,000 items, many of them advertisements and ephemera, type specimens and house organs, printing and paper samples, periodicals, complimentary copies—hardly a costly library to assemble, but with the passage of time irreplaceable at any price. In 1920 he bought the complete office library of Theodore De Vinne; five years later he toured Europe and brought back many graphic arts materials, putting them on public display the following year. Along with printed works came artifacts as well—medals, dies, matrixes, molds, alphabet designs, renderings, clay tablets, marble inscriptions, and stained-glass windows (possibly inspiring the Colonel's belated church memorial).

Bullen lectured occasionally ("Printing and Civilization," Stationers Hall, London, November 1924), and also collected the biographic details of scores of American printers who became famous—not for their craft, but as editors, legislators, soldiers, and merchants: Lehmann-Haupt considered the civic success of these men "one of the great sagas of nineteenth-century American life."[2] Bullen soon became the authority on all typographic matters, conducting an *Inland Printer* column to expose new members of the profession to their proud heritage. Like the Colonel Bullen too was personally interested in the technical training of the nation's youth.

Despite rumors that tons of old Hoe materials and records were lying unwanted in dank company vaults, Bullen did not (at first) get very far with the Hoes:

Dogwood, Port Washington, Long Island
Telephone 91 W. Port Wash.
April 17, 1918

Mr. Henry L. Bullen
Jersey City, N.J.
Dear Sir

I have received your letter of the 8th. I regret that I have no extra copies of the [Scharf?] medal you refer to but if you will write to Mr. Crane the President of R. Hoe & Co. I think they have or did have some extra copies and I have no doubt he will be glad to present them to the museum.

Some time when I am in Jersey City I am going to give myself the pleasure of visiting your museum and looking over your interesting collection.

Yours very truly
Robert Hoe[3]

But within a few years, no doubt coming to light during the 1920's when Hoe began to move from Grand Street, the old company press books and other handwritten documents were shown to him.

It was miraculous they survived at all—a corpus of some 100 volumes dating from 1833 to 1905. They were moldy, faded, water stained, some irretrievably stuck together page after page. Bullen studied them and bought only a few portions he deemed of particular interest: Lightning press development, Civil War sequences, Conquest's letters from England. Along with a copy of Tucker's manuscript history of the company (from the Typothetae, whose library —and Hoe bust—he acquired), Bullen had plenty of source material.

Beatrice Warde, of Lanston Monotype Corporation, London, had been a life-long Bullen disciple, calling him, with affection, "that great and still largely unrecognised character." As a student of Barnard College, New York, Beatrice had once traveled to Jersey City to see an old manuscript; she began to study calligraphy and yearned to associate with those who appreciated the graphic arts as she did. Bruce Rogers helped: with a letter of introduction from the designer (whom Bullen admired greatly) she presented herself in 1922 at Communipaw Avenue—where Bullen was already processing 90 applications for an assistant. Bullen took Rogers' recommendation and immediately introduced the girl to the collection's 14,000 tomes.

"Take this [a feather duster], and dust all these books," he told her. "I shan't mind if you stop and read anything that interests you. . . ."[4]

Beatrice stayed three years, and with few interruptions managed to catalog and read thousands of books. In that short time too she acquired an extraordinary appreciation for types and graphic design that enabled her to qualify just as quickly for publicity and advertising work at Lanston. Upon visiting the United States a decade later she found Bullen ill with three strokes and dying. He delivered a trust to her:

"You know, I always tried to tell the lads [his readers], 'If you're going to be just a pica-thumper, you might as well have been a plumber, they use lead too.' "

"Yes I know," she answered to his request. "I'll tell them. . . ."

Beatrice Warde (1900-1969) similarly dedicated herself to the task of acquainting young people with the importance of the printed word. When she wrote her famous tablet, "This is a Printing Office," she thought, "Henry Lewis Bullen, you would have loved this thing! It's right up your street. And if it hadn't been for you and your vision, I could never have written it. . . ."

Mrs. Warde had correctly appraised her master's advice and counsel. "Bullen had the vision to realise that [his library] was merely part of an enormous whole," she wrote. "His most catholic mind looked over fences and saw how, for example, the art of journalism affected typewriters and typefounding, calligraphy and writing—all the things which have to do with the passing of ideas from one mind to another."[5]

Some time in the 1930's Hellmut Lehmann-Haupt also journeyed to Jersey City for a visit. He found his host most amenable: "and there, amidst the books and broadsides, the old presses and composing sticks, the shrewd, wise, kind, human face of Henry Lewis Bullen, welcoming and at once conducting you to his treasures. . . ."[6] Bullen's librarianship had been thorough: in 1936 he issued a catalog of duplicates amounting to hundreds of titles, reassuring impecunious readers that they were indeed for sale, and cheap. That year the 80,-000 item library was donated to Columbia University, New York, where it lay in boxes for the many years of World War II.

Unlike Robert Hoe, whose exquisite book collection was designed largely for his own delectation, Bullen courted the public through his bibliographic activities with aggressive proselytizing always in mind. He thrust his salvaging mission and his tastes on contemporaries, sometimes unmercifully. He lambasted Edmund Gress (editor of *American Printer*) for errors in his book, *American Handbook of Printing,* and he also criticized type-designer Frederick Goudy. But he especially disliked Robert Hoe III, taking particular pleasure in contrasting him with his uncle, the Colonel.

Although Bullen ultimately characterized Richard Hoe a "plain American citizen [with a] useful but uneventful career,"[7] he could not avoid superlatives in his description of the inventor:

> Richard March Hoe (1812-1886) [was] the man who made R. Hoe & Co. famous. He invented the first successful fast newspaper press, the type revolving press. . . . Hoe, assisted by Stephen D. Tucker, made R. Hoe & Co. the greatest printers' supply house in history. It built the most advanced machinery for letterpress, lithographic and copperplate printing, and for platemaking and bookbinding. It made everything, small and large, needed to equip a printing house, except types. Its output excelled not only in ideas but in quality. Richard March Hoe loved his business. . . .

Here was a man Bullen could respect—perhaps not so much for conducting his business prudently, but for something else. He recognized unstinting devotion to duty and excellence, to be sure. But he admired, most of all, the typographic perfection made possible by Hoe presses, and (strangely) he considered the Colonel's typographic library to be evidence that both thought alike:

> [Hoe] showed this [love] by collecting a library relating to printing, the most extensive ever made by one man. He was public spirited and popular, and advanced those who helped him achieve his ideas as an inventor and manufacturer. . . .

It is possible that Bullen somehow acquired many of the Hoe typographic volumes. In 1922 and 1926 *Inland Printer* articles Bullen retold the Hoe stories and anecdotes he had heard and read from the records. He deplored the fact that the Colonel was often confused with his nephew, for whom he had more hostile thoughts:

> This nephew was quite a different type of man. It is true he accumulated a great library, which was sold at auction after his death for more than $1,900,000, but this library had very little in it about printing. The nephew cared little about printing as an occupation, and less about printers. He believed that almost everything made in his establishment, when he succeeded to it, had arrived at perfection. Neither he nor any representative of his was ever seen at a printers' meeting. He ceased to advertise. He was inhospitable to new ideas. Fortunately, the momentum of the prestige which his uncle had earned carried the business along, but when Robert the Third died, the prestige of his house was in eclipse. He forgot the printers, and they forgot him. . . .[8]

Upon Robert Hoe's death in 1909 the *Inland Printer* (and other journals) had almost automatically eulogized him uncritically:

> With the passing of Robert Hoe the graphic arts lost one of their most eminent friends and uplifters. By his enterprise and inventive genius, he added greatly to the sum of human happiness and justified his right to live. He followed in the footsteps of his father and grandfather, and did his share toward winning immortality for the family name. So long as men absorb knowledge by reading, the simple Anglo-

> Saxon patronymic of Hoe will be linked with that of Gutenberg, Manutius, Caxton and other worthies who live on and on. . . .[9]

The *New York Times* had also editorialized magnificently: "Robert Hoe, the greatest of his line, is dead; but all men live more completely by reason of his work among them. . . ."[10] Bullen hated these lies.

The self-appointed chronicler of printing history held Robert Hoe III accountable for much that was wrong in the printing industry, its frenetic business practices (and his own sad entrapment in them) having become a yawning abyss between the two men. They were at diametrically opposite ends of life's duties and pleasures.

Bullen considered himself a practicing printer, a knowledgeable salesman, a technical advisor, (and in a small way an inventor of printing equipment). He was a designer of printing, an editor, an author, a librarian. His collecting and related bibliographic work, however, he largely performed unaided on his own time and at his own expense, running the ATF library after other technical duties were accomplished. He advertised to make his library available to students and apprentices, and he wrote countless printing articles.

Bullen saw Hoe opposite him in every respect. Hoe was imperious, aloof, socially afluent, rich. He played with books and incunabula during the normal working day, but showed them to few young friends and expressed no interest in introducing the beauty of the printed word to practitioners of the art. Despite the colossal resources of a giant printing press firm he exercised none of the teaching the poorer Bullen considered an essential duty to the craft. With Hoe's death in 1909 the competition between them ended, but Bullen kept watch.

The *Fourth Estate* (later merging with *Editor & Publisher*) substantiated Hoe's imprudent advertising policy, in 1924 claiming to have participated dramatically in remedying the situation:

> R. Hoe & Co. had, in 1894, as it has today, the largest press building plant in the world. Robert Hoe, the second† of the name, was then at its head. [He was] a man of erect and patrician bearing, who blushed furiously when an advertising campaign for his presses was suggested to him. He believed the name of Hoe was so well known throughout the world that the insertion of an advertisement would be almost a sacrilege, and that the only impression it could possibly convey would be one of admission that the business needed bolstering up.
>
> Finally, through the co-operation of Charles W. Carpenter, one of the partners in the firm, Mr. Hoe was persuaded that only a product of quality can advertise with hope of success, and that every year a new group of young men were entering the publishing field who had to learn about presses and machinery of all kinds, and who would be interested in knowing, what he already knew, that his name stood for everything of which a manufacturer might be most proud.

> Today the house of Hoe uses page after page in THE FOURTH ESTATE and other trade publications and finds it difficult to supply the demand for their presses thus created. . . .[11]

Upon the death of Robert Hoe III the big firm faced routine liquidation in order to dispose of assets and liabilities to the satisfaction of both creditors and inheritors. The partnership was considered to be in such perfect condition—thanks to Robert's administration of both U.S. and British offices, said Joseph McCabe a year after his death,

> . . . that when the hand of fate touched him and called him from his labors these great enterprises, although jarred by the personal loss, soon recovered their business stride, and, as most competent judges think, are now going on to still greater and wider spheres of usefulness. The great name of R. Hoe & Co. has suffered no eclipse since its start first began slowly and modestly to ascend the business heavens over a century ago, and the place it now holds at meridian is strong and secure. . . .[12]

Reporter McCabe, who at close range had failed to see the "eclipse" Bullen gleefully spotted a decade later, judged the name "Hoe" on a piece of equipment to be as reliable as money issued by the United States. At any rate, dissolution of the firm would be folly indeed.

The deceased Hoe was found to own 19/21sts of the company; the balance held by Charles Carpenter. To forestall any disintegration, Hoe's sons Robert and Arthur together bought the two shares for $350,000, thus restoring entire ownership to immediate family and heirs. For their part, the interested branches of the Hoe clan promptly agreed to preserve the company, its good will, and its identity (and income) as a going establishment. Few of them, however, had any idea what such a commitment would involve.

Chief spokesman for the Hoe family was its matriarch, Mrs. Olivia P. Hoe, widow of Robert Hoe III, 72 years old in 1910. She proudly identified herself as the mother of Olivia Slade, Laura Carter, Ruth Sterling, Robert and Arthur Hoe, and the grandmother of Thyrza Benson and Ellen, Robert, and Samuel Evans. In the family's plans for reorganization Mrs. Hoe—apparently not yet aware of her husband's damaging situation in England, nor of her son's flirtations in New York—expressed her constant concern for the harmonious comfort and welfare of her brood: she was "desirous of perpetuating the Robert Hoe name and business [and anxious that] all my descendants shall unitedly and upon an equitable and equal basis continue the business which has been under the control of my deceased husband and his father and grandfather before him, since the early part of the last century."[13] It seemed unfortunate that Robert himself did not restructure the company as he would want it (much of American business had already made this adjustment), availing himself of full legal counsel and deliberation on every aspect.

In his egocentric final years, of course, this complex undertaking may have been much too disagreeable for Robert Hoe. Too, from another point of view there was no need to do so. Many years before, when the Colonel looked at it, the firm's ownership was disturbing:

> [*Richard Hoe, London, to Robert, 8 May 1866*] . . . *There is one subject which often forces itself on my mind in relation to the increase in the numbers of Partners in our concern—which is that all are human & to err is human & the errors of one of our partners might be of very serious detriment to our business & all concerned & this has led me sometimes to consider that it might be at some time advisable to make if possible what is called here a limited liability company of the establishment. Such a proceeding would perhaps simplify the copartnership & avoid any difficulty on the decease of a copartner. I think it would be possible to get a special charter from our Legislature for our concern & if not perhaps it would be possible to make a stock company under the general manufacturing laws. I wish you would give this subject some thought.* . . .

Twenty years later, however, the situation had changed dramatically; the company's six or eight copartners were reduced to one—Robert himself, eliminating any fears that parties acting independently would embarrass the proprietor. Now, in 1910, the superannuated partnership was in dire need once again of many helpful hands.

There was every reason to believe that the concerted interest of the heirs, under the stern maternal gaze of a determined woman, would accomplish the needed modernization. Hoe's will specified that a corporation be formed. To effect this, a trusteeship was established which elected Robert Hoe IV president and his brother Arthur vice president, and Frederick Crane as secretary. Each taking one preferred and one common share of new stock were Olivia Slade, Laura Carter, Robert Hoe, Arthur Hoe, Ruth Sterling, Thyrza Benson, and Samuel Evans. Apart from family position, the officers of the new concern were at least adequately qualified for tasks ahead.

Robert and Arthur, only 34 and 30 years of age, respectively, had both attended Harvard and been put through "various technical courses" by their father to acquaint them with the engineering details of the business. Both were already working at the plant. Robert began in 1899 at the manufacturing department's accounting books and then functioned successively as foreman in one department after another; in 1903 his father placed him in full charge of the circular saw division, whose volume doubled during his management.

Robert was credited with the development of the Hoe automatic ink pumping system for newspaper presses, although it did not become operational until after he left the firm. Introduced to the trade in 1914 (and first used on the New York *American*), the mechanism pumped ink from a large tank and sprayed a fine stream onto the inking drum. Until that time ink had still been poured by hand from drums, picking up dust and paper lint in the process, and distribution on the press had been regulated by fountain handscrews very dangerous to adjust while machinery was running. Robert would do well as president of the complex company.

Arthur also joined the firm after graduation in 1904, and for three months dutifully took instructions from a Hoe clerk. He was next put in charge of press erection, then transferred to the purchasing department. He also had charge of preparing the company's sales catalogs. Arthur soon undertook the reorganization of the pattern department, and after six months was able to dispense with half the 75 patternmakers employed. But it was more a ruse than a credit.

Finding that finished patterns, usually complicated wood forms from which sand molds were made, could be imported into the United States duty free, Arthur had the company transfer a large part of this expensive—but routine—carpentry to its London shop, where wages were lower. About 35 New York patternmakers were discharged, the remaining 40 weakly protesting the massive importations. But the patternmakers' union succeeded in getting the Treasury Department to put a 35 percent duty on the forms, and Hoe was obliged to increase carpenters' salaries from $3 per day to $3.25-3.50. Any savings to the company vanished.

Frederick W. H. Crane (1845-1925) had joined Hoe in 1878 as accountant, and worked all his life under the company banner. As he discharged larger responsibilities he came to Hoe's personal attention; Robert soon entrusted him with confidential and private matters, and ultimately gave him very broad jurisdiction in the large company:

GENERAL ORDER NO. 666

> Hereafter no clerk or other office or shop employe is to be engaged without the approval of Mr. Crane.
>
> R. Hoe & Co.
>
> May 10th, 1907 [14]

Making his home in Old Saybrook, Connecticut, Crane was long a faithful friend of the Hoe family.

The mechanics of modifying the partnership into a corporation involved a company indebtedness of some $3 million to the Robert Hoe estate. The new firm was authorized to transfer $3,025,437 in preferred stock (of a $3,250,000 issue) to the trustees for distribution to the heirs. The company was also permitted to issue $2,625,000 in common shares, for a total capitalization of $5,875,000. No business in 1910 could have been more sound.

Justice Edward B. Amend of the New York Supreme Court (coincidentally himself a product of the East Side and a large property owner there) heartily approved the plans 11 January 1910. His consent included a petition of Dr. Samuel M. Evans, on behalf of his three children, to participate in the family agreement (his wife, Ellen Hoe, having died three years before). Each relative receiving a seventh of the estate (worth $432,200) were: Mrs. Robert Hoe III, her sons Robert and Arthur, the Evanses, Laura Carter, Olivia Slade, Ruth Sterling. The two brothers also retained Carpenter's shares, which they had bought personally.

Named as directors of the new firm were Otto T. Bannard (banker, 1909 mayoral candidate, Yale fund-raiser), president of the board of trustees, John S. Hoyt (a family friend, of Stamford, Connecticut), Robert and Arthur Hoe, Arthur C. James (financier, nephew of Olivia P. Hoe), Charles W. Parsons, William W. Carman, trustees.

In May 1910 Robert Hoe inaugurated an annual banquet—a company waygoose!—at which the senior officers of the firm and their chief lieutenants could enjoy themselves and discuss matters of importance. In a private room of the Waldorf-Astoria twenty places were set at an immense circular table. Robert gave a hearty welcome and an informal talk, and for a silent toast to his late father all rose to honor his memory. Those present were Crane, Spalckhaver, F. Crowther, Otto Raabe, Oscar Roesen, Phineas Chew, Edward Pringle, Richard Kelly, Addison Gallien, Ralph C. Seymour, Howard Reynolds, Rozier Louis, William Nixon, J. Carruthers, E. C. Melledge, E. P. Sheldon, H. S. Mount, and Harry V. Ball. Arthur Hoe, the twentieth, could not attend. Telegraph greetings were sent to agents too far away to warrant invitation.

To all appearances, with this ceremonial gathering the firm had bridged a tenuous gap in its long life and had pledged itself to continue its century of service to America. Joseph McCabe boldly stated what many still wondered about:

> There are very few business enterprises in the United States that have an unbroken history of over a century's continuous business success. They nearly all begin to decline in the second generation, and go entirely to pieces in the third. But the Hoe family has proved a unique exception. The business has already passed through three generations, each succeeding one adding greatly to the prestige and prosperity of the firm. . . .

The assumption of a new Robert Hoe to the presidency of the corporation was another record McCabe could not overlook:

> While there was only one "Glorious Ben [Franklin]," there have been three Robert Hoes, each of which has contributed largely to the science of machinery building, the combined efforts of the three resulting in the wonderful business organization that today bears the name of R. Hoe & Co., a name that represents the very best products

> of mechanical skill and is known throughout the whole civilized world. . . .[15]

The public would soon notice, however, that the new Robert Hoe was not to remain, decade after decade, in the company's annals.

Despite his contribution to press inking, and despite strong family obligations, Robert Hoe IV was not comfortable in the uncertain position he held. Although the Colonel had once warned that nepotism was risky, Robert now found himself a corporation president strangely working for all of his family relatives who—admittedly—knew almost nothing about running a press or a machine shop. Quiet, unassuming, modest, he was not suited temperamentally to ride roughshod over a directorship apt to be querulous and stubborn. At the end of 1913 he resigned, consoled, perhaps, that his brother Arthur was still on hand, as were knowledgeable members of Peter Hoe's family.

But Arthur and the other Hoes had demonstrated no evidence of dynamic leadership; none were disposed to take aggressive action. Robert nevertheless continued as a director until 1924, and remained the firm's largest stockholder. Arthur also continued as a director for many years, gradually occupying himself more and more with his art collections and social activities. Henry Bullen, in a Bryant-like anti-Darwinian mood, had a curious explanation for such disappointing progeny, using the once great type houses as his example:

> The earlier letter-founders were all . . . of strong physique, robust health, and intellectual-looking countenance; and all were long-lived. In noting the fact that their descendants exhibited a decadence both physically and mentally, with few exceptions, I have formed a theory that the excessive labors of their sires and grandsires was the cause. . . .[16]

In October 1914 R. Hoe & Company announced completion of incorporation proceedings in the state of Illinois, apparently only to reflect its increasing interest in the Midwest market. Frederick Blackall became president and F. A. Cole became secretary and sales manager. Frederick Crane, now vice president, assured stockholders that the Chicago address, 544 Clark Street, was still only a branch office and would not become the corporate headquarters.

Frederick Blackall (1865-1928), like so many loyal Hoe men, was a rare asset. Born and raised in New York, he completed his education in Berlin and returned to establish a firm, Blackall & Blackall, machine contractors, of which he was president, 1881-1903. He then joined the Taft-Pierce Company, Woonsocket, Rhode Island, and came to Hoe in 1911. In 1916 nevertheless the presidency reverted to Crane. These officers, apparently selected by the family board of directors, although capable, similarly lacked that extra drive and spark to elevate the company above routine management. But these were long years of trial and testing for all.

The April 1922 annual meeting, commemorating a decade of family control, was nevertheless a pleasant, dignified affair, much like a clan reunion. Reelected members of the board were Robert and Arthur Hoe, Dr. Samuel Evans, Robert Sterling, George E. Hite, Jr., Olivia Slade, Laura Carter, Richard Kelly, and Addison J. Gallien. A widely respected printing press expert, Gallien had been elected a director and vice president the year before, upon the death of Oscar Roesen. Gallien and Kelly, although not related to the Hoes, were old family indeed; both had been with the firm for over 35 years.

At a subsequent meeting Richard J. Kelly was reelected president and general manager for a third term; Gallien as first vice president, Charles MacInnes as treasurer, and Harold M. Tillinghast as secretary. Otto Raabe (apparently after a short stay at Goss) was back as assistant secretary. They too were all old-timers: MacInnes, 10 years; Tillinghast, 12; Raabe, 47. Despite the post-war business depression Hoe had installed more newspaper and magazine presses in 1921 than during any previous year in the company's history, and had more unexecuted press orders (the industry's gauge of future prosperity) than ever before—over 300 sixteen-page units. Expansion of manufacturing facilities and improved machinery and equipment were contemplated. Despite the small adjustments of personnel, all seemed to go just as old Robert Hoe would have wanted it.

In April 1924 R. Hoe & Company bought the Hall Printing Press Company, Dunellen, New Jersey, with intentions of operating it on a larger scale with 800-1000 employees. Phillip W. Hall, Sr., remained as manager. This manufacturing facility was only lightly integrated into Hoe methodology, and some of its irksome working conditions might have been promptly rectified had the Colonel or young Theodore Mead gone to look at it. Correspondence to a 1928 *Daily Worker* was embarrassing:

> DUNELLEN, N.J. (By Mail)—My address and name you will, of course, omit, but my co-workers await the publication of this letter with interest and, if you advise me of the date of publication, my friends, thru me, will remit for 100 copies. Then the lukewarm few will see that, even tho the International Association of Machinists are dead in this locality, the thinking few in this shop are going further in an attempt to remedy their condition.
>
> I represent the so-called radicals in the Hoe and Co. shop at Dunellen. We, a scab shop, assemble the printing presses that are handled by the elite union men of the Printers' Union. Little do they know or perhaps care how our downtrodden 470 workers sweat and toil to build the machines they draw their princely wages to operate.
>
> The chief slave driver here is Graf, the works manager. Known as the clean-up man of Hoe's, his job is to cut wages and make two men do the work that three men staggered under before.

> Fifty wage slaves of the white collar class, production clerks, timekeepers, etc., all married men, exist on $25 a week. Threats of moving the plant keep the machinists cowed, and stool-pigeons from the Metal Trades' Association weed out the so-called malcontents.
>
> The Hoe Co. has a $2,000,000 order for presses for the Chicago Daily News, yet even union men in Chicago without fuss or worry assemble scab presses for union men to operate. Let us wake these chaps up. It's their sweat and toil that makes the millions for the Hoe Co.
>
> We slaves who patronize the restaurant remember army days, slum, slum, every day. But it is eat here or starve, for we are fifteen minutes from the nearest lunch wagon. Eat slum or go hungry is the Hoe motto. —RUDY BENITO.[17]

The Dunellen plant was long underdeveloped and only partially used; on several occasions it was to be phased out of Hoe operations but was never discarded. It became more important after World War II.

The operation of R. Hoe & Company by a board of directors consisting of members of the family could not last. It is strange, perhaps, that Robert Hoe III and his wife, themselves so reserved, isolated, egocentric, could visualize a democratic or communal family directorship among in-laws and children accustomed to and raised in affluent, protected, circumstances. The family, far from homogeneous or especially cooperative—most no longer with Hoe surnames—grew less and less interested in corporate affairs rattled by war, strikes, depression, unionism. Their recollections of family pride were diminishing, their emotional connection evaporating. Only the tremendous earning capacity of the firm could not be dismissed. It had paid for so many nice things. But family affiliation, however congenial and mutually dependent, was no qualification to run such a mammoth enterprise.

For the company's own good, too, reliance on the personal dispositions of relatives was no way to manage a 3000-employee firm. Expansion had been deemed advisable as early as 1922, but financial support was not available. The family, perhaps sensing the earlier partners' aversion to debt, also seemed reluctant to borrow boldly or mortgage the property. Yet new capital was needed—capital that the family directors, without powerful banking or securities connections, simply could not raise. And even as family members discussed the situation, the net value of their investment, if unable to grow, was necessarily declining. The directors decided they had no choice but to sell the magnificent property while it was still in prime condition.

In September 1924 a special stockholders' meeting was announced to pass on a proposal to sell R. Hoe & Company, its assets and good will. To some members of the family this was perhaps a surprise and a relief. Those who had long been detached from family affairs and removed from business pressures had become vaguely aware they no longer controlled their company's destinies. What cannot be controlled often becomes the object of fear or contempt.

Ruth Hoe Sterling received her copy. Upon her father's death fifteen years before she had inherited $1,195,445, apparently in securities of the great old company. These in 1924 amounted to 2716 preferred shares and 3464 common, of which Ruth owned 782 preferred and all the common, her husband owned 734 preferred, and as trustees they jointly held 1200 shares for two children, Edward and Oliver. The Sterlings very much protested the company's falling into strange hands. Robert Hoe's last daughter had perhaps inherited the sentimental feelings of the clan and its great fighting spirit.

At the meeting company officers described their plans for the sale. In exchange for $7,750,009, the entire business would transfer to the Press Securities Corporation; the firm would continue without visible interruption under a new name—R. Hoe & Company, Incorporated. With shocked interest, the Sterlings watched the tally of 15,251 preferred and 22,786 common shares—more than the required two-thirds—voted in favor of the transaction. Sterling objected: ". . . both before and after the decision of the result of the vote," said the *New York Times,* he "protested against the sale and conveyance, and demanded payment for all stock standing in the name of Ruth Hoe Sterling."[18] To the Sterlings' amazement, perhaps, they were the only members of the family officially opposed to the sale.

Robert Sterling on October 17 addressed a letter to Richard Kelly, company president, protesting the transaction and again demanding payment for their holdings rather than accede to quiet conversion of the old family enterprise to public control. Kelly responded to *Mrs.* Sterling that as stockholder she was entitled, upon surrender of the certificates, to collect $1,085,979 "as a first and final liquidating dividend in the dissolution of R. Hoe & Company." The money had been deposited for her at Guaranty Trust Company, payable upon demand. The Sterlings, however, believed the stock to be worth considerably more.

The laws of the State of New York permitted a minority stockholder to have the value of the stocks appraised—and paid for—when an owner opposes an action to dissolve a company. The Sterlings in November 1924 petitioned the New York Supreme Court to appoint appraisers to examine their stock. Meanwhile, negotiations began.

The directors moved ahead. They arranged for—of all people—Guaranty Trust Company of New York to handle the transaction. What a fateful choice: had Edwin A. Porter's encounter with Frank Munsey and others come to the Hoes' attention as evidence of extraordinary banker loyalty to a client, however impersonal? Had they no fear that the grand old company could suffer in the hands of men who, like themselves, were not engineers, nor pressmen, or printing salesmen? Would they care, as long as they got their own investment back—or might they suppose that nothing could hurt so substantial an enterprise?

President Kelly was duly instructed to describe the giant concern to the banking syndicate which would underwrite a $4.5 million bond issue.

R. HOE & CO., INC.

New York, N. Y., October 4, 1924.

Guaranty Company of New York,
Edward B. Smith & Co.,
Dominick & Dominick.

Gentlemen:

In connection with your offering of $4,500,000 First Mortgage Gold Bonds, Series A, 6½%, of R. Hoe & Co., Inc., I take pleasure in submitting the following information:

HISTORY AND BUSINESS

The business of R. Hoe and Co. was founded more than 100 years ago, about 1818, and has been developed into the largest manufacturer of newspaper printing presses in the world.

The majority of large newspapers throughout the world are printed on Hoe presses, and of the leading papers in the English language more come from Hoe presses than from all other makes combined.

For more than a century the Hoe name has been associated with most of the important developments in the printing press. About 1827 the Company produced the Washington hand press which then led competitors in turning out a few hundred single-page sheets an hour, and is still in use for taking fine proofs. Today the Hoe Superspeed Double Octuple Rotary Press will print, cut and fold 160,000 sixteen-page newspapers hourly. Such a press consumes the equivalent of a strip of newsprint six feet wide at the rate of nearly two miles a minute.

Unfilled orders (largely for newspaper presses) now on the books represent more than a year's output. As a result of the increasing demand during the past few years for Hoe Superspeed newspaper presses, the Company has largely devoted to their manufacture the facilities formerly used in the production of other important types of printing machinery, which it has done much to develop. These other lines include magazine, book, intaglio or rotary gravure and lithographic presses, electrotyping, stereotyping, photoengraving, cutting and folding machinery, all of recognized Hoe quality. The Company also has a large and stable business in repairs and in renewal parts.

During the early period of the Company's history presses were largely of wood and the Company made its own saws. As an outgrowth of this, the Company has become one of the leading manufacturers of circular wood saws, principally of the inserted-tooth type. This branch of the business contributes substantially to the net earnings of the Company.

PLANTS

The main plant of the Company is located on Grand Street, New York City, covering two entire city blocks and parts of three others. The Company also manufactures a line of Hoe products in Great Britain through its subsidiary, R. Hoe & Co., Ltd. In response to the demand for increased production, the Company has recently acquired the plant of the Hall Printing Press Company at Dunellen, N. J. This addition, together with the improvements to the British plant nearing completion, increases total manufacturing space about one-third, but has not yet been reflected in earnings.

The total floor area of the three plants is over 750,000 square feet. Employees of the Company in the United States total about 3,000, or considerably more than all other domestic press manufacturers combined. The London plant employs over 800 additional and, including the new extensions, has a capacity approximately equal to the aggregate of all British competitors.

RECAPITALIZATION

The proceeds of these Bonds and of 80,000 shares of Class A Stock and 160,000 shares of Common Stock will be used for the retirement of $800,000 of mortgages and for the purchase of the assets and business of R. Hoe and Co. by a successor corporation, R. Hoe & Co., Inc., organized under the laws of New York, which will issue these Bonds and such stocks.

The present executive management, which has developed for the Company the largest business and greatest profits in its history, will continue in charge.

PROVISIONS OF ISSUE

The First Mortgage Bonds are to be secured by a first mortgage upon the plants in New York City and Dunellen, N. J., and by pledge of all the stock (except directors' qualifying shares) of R. Hoe & Co., Ltd.

The Trust Indenture will provide, among other things, substantially as follows:

Additional authorized First Mortgage Bonds may be issued to the extent of 70% of the cost of additions and betterments to the properties, but only if earnings available for interest in two years preceding have averaged at least twice interest requirements upon all first mortgage obligations then outstanding and those proposed to be issued.

The Company will covenant not to pay cash dividends upon its Common Stock except out of surplus earnings accrued subsequent to June 30, 1924, or if after the payment of such dividends current assets would be reduced below 200% of current liabilities (all as defined in the Trust Indenture).

SINKING FUND

The Bonds will have the benefit of a cumulative Sinking Fund payable semi-annually beginning October 1, 1925, of $135,000 annually plus interest on all Bonds held in the Sinking Fund, to be used for purchase or redemption of Bonds. This Sinking Fund is calculated to retire more than $1,500,000 of Bonds by maturity in 1934.

FINANCIAL

Upon completion of the proposed financing the capitalization will be as follows:

	Authorized	Outstanding
First Mortgage Bonds (this issue)	$6,000,000	$4,500,000
Class A Stock (no par)	100,000 shs.	80,000 shs.
Common Stock (no par)	160,000 "	160,000 "

An excess of $11,300,000 of tangible assets over all liabilities except these Bonds is shown on the accompanying consolidated balance sheet of June 3, 1924, prepared and adjusted to this recapitalization by Messrs. Barrow, Wade, Guthrie & Co. This figure includes more than $6,100,000 for the estimated present depreciated values of real estate, plants, equipment and patterns as appraised by Messrs. Sanderson & Porter, and makes no allowance for valuable patents and good-will. The excess of current assets alone over all liabilities except these Bonds is more than $5,200,000. Current assets are 5.4 times current liabilities.

EARNINGS

Consolidated net earnings of the Company available for all interest, after depreciation, but before Federal and New York income taxes, have been reported as follows by Messrs. Barrow, Wade, Guthrie & Co., New York, and Messrs. Charles Comins & Co., London:

Calendar Years		Calendar Years	
1918	$1,137,276	1921	1,198,812
1919	1,425,425	1922	1,454,600
1920	815,243	1923	1,473,281

Average net earnings for the six years were $1,250,000, equivalent to 4.2 times annual interest charges on these First Mortgage Bonds, and 1923 earnings were 5 times such charges.

Net earnings for the first six months of the current year were $523,893, somewhat below the average owing to the delays in shipment and the inclusion of high-cost materials resulting from the strike in the latter part of 1923. In this strike the Company was completely successful and as a result has established its plants on an open-shop basis.

The Company should soon benefit from its enlarged capacity, which is expected to result in further increases in earning power.

GENERAL

The Company is the principal world factor in the manufacture of newspaper printing machinery. The size and circulation of newspapers has substantially expanded during the past ten years, with constant demand for larger and faster presses. There is also a noteworthy tendency on the part of the smaller papers toward consolidation into larger units with improved machinery.

The Company has throughout its history maintained its reputation and the quality of its product, so that it now has behind it a century of good-will. As the producer of the leading newspaper presses, the Company is particularly favored by the trend in the industry, and future prospects appear excellent for a continued large volume of profitable business.

Yours very truly
(Signed) RICHARD KELLY,
President.

[Consolidated Balance Sheet not reproduced] [19]

To furnish capital with which the company would be purchased, a syndicate headed by stockbrokers Edward B. Smith & Company and Dominick & Dominick, both of New York, on 3 October 1924 (date of incorporation in New York State) announced the offer of 80,000 shares of Class A participating and preferred stock, as well as 160,000 shares of common. The preferred shares were entitled to cumulative dividends of $4 per share yearly before any common stock dividends were declared; after the common stockholders were paid $1 per share, furthermore, both classes would divide (in a proportion of ⅓ Class A and ⅔ common) whatever earnings remained. Reminding the public that the company at the moment had unfulfilled orders amounting to 12 months' work, the Class A securities at $50 sold out in one day—they had in fact been oversubscribed.

The common stock, unfortunately, was "pure 'water' from the outset," said a critic, designed to net a high cash profit. No less than one-third of it was sold—for a considerable sum—to Arthur Hoe, the only member of the family, apparently, interested in the firm's affairs. As his name lent appreciable respectability to the proceedings, the bankers set him up only to discard him later at their convenience. While the A stock had partial voting power (for example, it could elect a majority—6 out of 10—of directors), the common had exclusive voting jurisdiction for the remaining directors and for other company matters. Through a voting trust engineered by Guaranty, the banking group retained the remaining two-thirds of the common stock, giving it more than enough strength to control the company. Although many employees and friends, out of pride and loyalty, perhaps, had bought the Class A preferred stock, the firm was without question in the complete possession of the three-headed banking syndicate.

Four days later the syndicate offered for sale the parcel of new bonds, $4,500,000 R. Hoe & Company, Inc., Series A 6½ percent first mortgage gold bonds due October 1, 1934. They were quickly sold at $99.50, with interest calculated to yield 6.55 percent. A day later the bond offer was closed. Application was made to list the stock on the New York Stock Exchange, and until such permission was granted, it was offered on the curb exchange. Obtaining $4 million from stocks, upon the completion of bond sales the syndicate had its $7,750,000—and the proud old company as well.

On 21 December the revitalized firm declared an initial dividend for the last quarter of 1924 of $1 per share of Class A, payable 15 January 1925. In addition to this pleasant Christmas news the directors of the firm were at last revealed. They were impressive:

Harold Stanley, President, Guaranty Trust Company; holding key positions at several banking firms, Stanley (1885-1963) would win the court's praise in the 1940's for "absolute integrity" as principal witness for 17 large financial houses charged with monopolistic practices.

John W. Cutler, an officer of Edward B. Smith Company, stockbrokers.

Wallace E. McGraw, Vice President, Proctor & Gamble Companv.

Charles A. Sackett, President, Mutual Bank of New York.

Richard J. Kelly, President, R. Hoe & Company, Inc.

Boudinot Atterbury, Assistant Vice President, Guaranty Trust Company.

Frederick G. Coburn, Sanderson & Porter, real estate appraisers: long an engineer with the U.S. Navy, then at Bethlehem Shipbuilding, Coburn specialized in engineering consultation work.

Morris W. Kellogg, President, M. W. Kellogg Company, worldwide builders of oil refineries and later pioneers in nuclear technology (designers of Oak Ridge, Tennessee, and Hanford, Pennsylvania, atomic installations); Kellogg (1873-1952) was also a director of Guaranty Trust and other corporations.

Phillip G. Gossler, Chairman of the Board, Columbia Gas & Electric Company: also a director of Guaranty Trust and other firms, Gossler (1871-1945) in 1930 would be named one of the 59 men who "rule" America.

Arthur I. Hoe, R. Hoe & Company, Inc.[20]

The Guaranty Trust Company had extraordinary representation on the Hoe board of Directors, to be sure. Starting in 1864 as New York Guaranty and Indemnity Company, the early firm made a business, according to its elegant corporate history, "mainly lending against warehouse receipts, a field of financing previously not much explored by banks." The company was reorganized in 1891 and earned a fine reputation in investment as well as commercial banking; it also built a large volume "in the growing fields of managing other people's money as trustee and giving professional advice on investments."[21]

The new Hoe management, strong in finance where the Hoe family had been weak, was not overly impressed with the long human and personal history of the firm as a monument to individual genius and hard work. Where the Colonel had kept batteries of lawyers busy with patent cases, the corporation officers would find themselves in far different law chambers. Before long the first sign of difficulty appeared in the front office.

The first full year of incorporation of the new company came to an end with the announcement, by Wallace McGraw, that Richard Kelly, president and general manager, and chairman of the board since 1920, had expressed the desire to resign. Officers discovered that Kelly had served 40 years with the company, and—perhaps to their quiet amazement—they learned something more of employee fidelity:

> A large part of the credit for the splendid position that the company has maintained for a period of so many years is attributable to the unusually long service of its employes, of whom no less than six have been associated with it for fifty years or more, thirty-one for forty years, eighty-seven for thirty years and one-hundred sixty-nine for twenty-five years. . . . The board accepted his resignation with sincere regret.[22]

Richard Kelly seemed the last of the Hoe management to have considerable printing press and machine-shop experience, an observation the bankers must also have made. They searched well for a successor, and found the one man who seemed absolutely right for the job: Harry Swartz.

"There must be some special destiny that guards over the fortunes of printers' 'devils,'" began a *New York Times* feature on Harry R. Swartz. "One of them has just become head of the largest printing machinery manufacturing company, making the great presses that turn off newspapers by thousands wherever the printed word is known."

Swartz (1876-1945) reminisced fondly of the early days that shaped his dynamic career. As a lad of 12 he began "carrying a route" for a newspaper in a little Pennsylvania town:

> I couldn't very well forget it, for I began delivering papers in the year of the big blizzard—1888. The Hazelton Plain Speaker thought well enough of me to intrust 110 of its customers to my care and I resolved right then that they never should go a day without the paper. But I didn't quite know what I had resolved upon. The blizzard descended and cautious folk were afraid to go out of doors. I told my mother that I had promised myself to deliver 110 papers every day, blizzard or no blizzard. She just looked at me and said, "All right," and I went out to tackle the job.
>
> Cold? Whew! I can feel the chill yet. Our office was almost frozen up, along with the rest of Hazelton, but the men had gone to work

> by one way or another and the papers were waiting. No vocation in modern life seems to call forth quite so much devotion as that given by a newspaper worker to his paper. On the first day of the blizzard I don't believe any of the other boys delivered all of their papers, but I got through with mine. . . .

Swartz soon moved inside to become the *Plain Speaker's* devil, a post he felt equivalent to "general factotum." "A printer's 'devil' must be bathed in ink at all times and he properly should wear a tall paper cap," the *Times* reporter explained, "although the cap usually is missing in practice. It is his special task to keep the handle of the office broom well polished by dint of much sweeping. He must be prepared to run errands and find lost galleys of type. Mr. Swartz did all of these things. . . ."

One day a circus came to town and Swartz yearned to see it, but he lacked the 50 cents admission. To earn a ticket he agreed to a grueling day's work keeping the calliope boiler stoked and filled with water; he enjoyed the show, to be sure, but the next day was summarily fired—a fellow reporter's coverage of the circus had included his clandestine toiling at the organ.

Swartz gravitated to Philadelphia, where he learned of a YMCA course in discussion and debate. He applied for permission to attend the class provided he could listen without participating.

"I will tell you a secret," said the secretary; "those chaps don't know any more than you, and perhaps not as much. Come on to the classes and start talking for yourself."

Swartz did, apparently without difficulty. By the time he was 24 he was treasurer of the Sprague Electric Company, a budding New York firm in serious financial crisis. Swartz applied his knack for corporate management and organization, and within four years the company revived and was absorbed into the General Electric Company. Swartz went along with the merger, but he soon feared he would be lost in the bureaucracy of the large corporation.

"I had always cultivated my own way of doing things and my own way of thinking. I suppose you might call me an individualist. I feared to surrender my freedom of action. So I embarked in new channels and never have regretted the step."

In 1917 Swartz was elected president of the Intertype Corporation, a firm originally founded 1912 as the International Typesetting Machine Company, New York, for the express purpose of producing a competing machine when Mergenthaler's Linotype patents expired. The resulting "Intertype" was very similar in appearance and operation, and the two manufacturers long had bitter battles for a market almost too small to justify mass production—a factor conspicuous in Hoe operations as well. The Intertype Corporation had also faced troubled times, and Swartz' abilities soon turned it too into a thriving organization. Swartz was unquestionably a forceful influence in difficult company situations.

In its second year as a corporation, however, R. Hoe & Company, Inc. was far from periled. But because of its tremendous size and world-wide operations, because of its uniqueness of product and foremost reputation in the industry, and because of loose state and federal laws regarding incorporation and financing, there was no point waiting for difficulties to arise. Men who worked with Swartz found him a "humanist in industry," giving a "fair deal" to everyone in return for cooperation and loyalty. These were concepts not always found in bossism and unionism at the turn of the century. Said Swartz:

> The first duty of a boss is to obtain the confidence of his men and establish himself as their leader. The boss necessarily is placed on a plane above his men and should be an example for them to emulate. Sometimes it will be required of him to fight for his men and sometimes for his company. He should be fair and impartial and try to practice justice toward others before he thinks about himself. . . .[23]

Swartz came to Grand Street with rare energy and enthusiasm. His familiarity with the graphic arts industry was a serendipitous asset of great value. His analyses were practical:

> Europe has bought virtually no printing equipment since 1914, and her presses are wearing out. They must be replaced before long. I believe that financial improvement warrants the expectation of a large European business. The Locarno treaties have given a new meaning to peace and the nations of the Old World are looking to the future with firmer confidence.
>
> We are equally certain of an expanding business at home. How could it be otherwise? Everybody reads today and the newspaper is the textbook of the masses. I would like to affirm my confidence in the newspaper as an institution. We hear from time to time that it has deteriorated or fails of its mission, but I believe the American newspaper is an accurate reflection of life and a strong, wholesome influence of social improvement. Where else should we get so many ideas about government, international affairs, and intellectual development?

Swartz had indeed seen his assignment at Hoe in terms considerably broader than that of shop boss or even company officer of the old school. He talked in scope never attempted by any of the old Hoe patriarchs themselves. Their concerns had been timid and self-conscious: assuring quality, promising delivery, dispatching services. The company, to be sure, had become great on these foundations, but they were no longer the issues. Swartz would not offer the world efficient printing; he offered viable democracy in an arena of free expression and progress for mankind:

> The newspaper sold on the corner for a few pennies is one of the greatest influences of progress and perhaps the greatest safeguard of liberty. Any nation is likely to be fairly safe and assured of its progress so long as its citizens read an untrammeled press. It is the privilege of our business to serve this institution of thought and I deem that privilege an obligation.[24]

The company seemed in safe and capable hands, but was soon paying dearly for a man of Swartz's foresight. He had been hired at three times the salary of Richard Kelly, his predecessor, and he bargained successfully for yearly bonuses based not on net sales but on gross. Years would pass before the effect of the directors' mysterious generosity to their printer's-devil president came to light.

Adjustments a year later were routine. In April 1927 Sackett and Gossler were removed as directors, Fred G. Kent, Harry Swartz and Tillinghast added. Swartz was reelected president, and vice presidents were Tillinghast, Kent, Harry Ball, Charles Gaskell, Addison Gallien. Allen W. Lishawa (auditor, accountant) was elected treasurer, John M. Masterson, secretary, and Louis Roehm, assistant secretary. The following month R. Hoe & Company received the Chicago *Daily News* order for 68 sixteen-page superspeed units with 17 double folders—said to be the largest sale ever made. It would take 18 months to fill.

Soon business was as good as ever. Undelivered orders in October 1928 represented 6 months' work; new jobs would include an installation of twelve 16-page units for the Philadelphia *Record,* five additional 16-page units for the Louisville *Courier-Journal,* six units for the Boston *Globe,* and three for the Montreal *Star*. The British subsidiary too was busy with a full year's backlog. Much of President Swartz's happy predictions for the industry were materializing. Literally there was singing throughout the shop: R. Hoe & Company at this time was one of the first factories to test the theory that "canned" music would improve efficiency.

In July 1928 the company announced acquisition of property in the Bronx, to which it would transfer all its manufacturing operations. After fully a century on lower Manhattan—with some structures still in use since 1830, R. Hoe & Company had become a landmark: its Grand Street clock tower, visible from the tall newspaper offices around City Hall Park, seemed to establish an enduring link to the important industry it served. The lower East Side had seen farms turn into thriving urban communities; it had watched fancy town houses yield to immigrant ghettos; it witnessed rowdy gangs and sweat shops tamed by settlement workers and fairer working conditions.

To all these neighborhood changes the Hoe enterprise had been a constant landmark—welcomed at first as a new settler, long unnoticed as it quietly minded its business, then hostile like an old homesteader resenting the intrusion of squatters. Now abandoning its crusty territory, the company perforce accepted its new corporate dynamism. President Swartz confirmed the gossip:

> There were good grounds for these rumors, since the officials of the company for some time had been contemplating such a move. Indeed, they had gone so far as to have plans prepared for buildings where all manufacturing operations could be consolidated in one location. Recent developments, however, necessitated immediate expansion and the purchase of a great deal of new machinery.[25]

The company in fact had a most ambitious project on the drawing boards and wanted real space to develop it.

Speed and economy were what newspaper publishers had always clamored for: faster production on more compact machines in smaller press rooms. As publishers' headquarters in the largest cities became more costly real estate properties, they deplored the floor space allotted to giant presses despite their essential services. R. Hoe & Company had made a twentieth-century breakthrough.

Hoe engineers had all long agreed that most machines throughout the printing industry were never run at full speed. Said Harry Ball, a Hoe vice president:

> Press men, we know, are always anxious to run the presses as fast as possible. The only reason they don't is because they figure that delays resulting from web breaks will lower production due to the difficulty in rethreading and cleaning up. . . .

Concentrating on the path of the paper through the press, designers were able to place cylinders in pairs one above the other, allowing easy access to webs and rollers. This enabled simplified externally threaded webs which, even if breakage occurred, could be repaired so quickly, "there would not be sufficient loss of time to tempt a pressman to reduce speed," said Ball. This single concept had effected the speed and economy they sought.

Named the "Super-Production" press, trial units were capable of production up to 50,000 64-page papers an hour, equivalent to a speed of 25,000 revolutions per hour—25 percent faster than anything Hoe had done before. The units with heavier and stronger frames were smaller and cheaper to install, with savings in foundation work, electrical conduits, ink system, and maintenance.

Two trial units were shipped to the Chicago *Daily News,* where they were tested for two months. Walter A. Strong, the paper's publisher, even before the equipment demonstrated full capacity, was so satisfied he ordered 60 Super-Production units for a new plant to be constructed in 1929. Strong visualized his new pressroom to be only one-third the size of that for conventional straight-line machines.

(The unhappy Dunellen worker had picked a poignant example; the *Daily News* ended in 1978.) Orders totalling 40 units also came in from the Toronto *Telegram,* Boston *Globe,* Louisville *Courier-Journal,* and the Montreal *Star*—papers whose press superintendents had seen the trials.

Although production of other Hoe machines would continue as long as there were requests for them, the company believed the Super-Production model would eventually replace them all. With orders on hand and promising prospects for more, the company decided to move fast. Swartz continued:

> We had planned to construct new buildings of our own, but it was soon apparent that no time could be lost in waiting for such construction, and we decided if possible we must find an existing plant —one suited for the manufacture of machinery along strictly modern lines.[26]

In 1906 the company had bought a large lumber yard on the East River at Broome Street for use as a foundry. This was its last expansion in Manhattan. By 1929 all the Hoe structures were comparatively old and costly to convert to the modern assembly line techniques inspired by the automobile industry. The company had begun to visualize an ideal new plant, but time ran out.

The new blueprints were thrown out. The search for an existing structure, where a modern machine shop could—like Munsey's presses —be emplaced at a moment's notice, was probably a well kept realtor's secret. Said Swartz:

> A plant was finally located which met the exacting requirements of the company. It covers three complete city blocks, running from the New York, New Haven & Hartford Railroad tracks to the East River and from 137th to 138th Street. A railroad siding runs directly through the entire length of the plant. Thus the delivery of raw material and the shipment of finished product is provided for by both rail and water.[27]

This was the property of the old De La Vergne Machine Company, a refrigeration plant and machine shop, located almost at the southeastern tip of the Bronx in "Port Morris," a large industrial sector. Hoe would commence movement to its new site as soon as the present occupant left. Although it was expected that Dunellen operations would also be consolidated under "one roof" so as to discard the New Jersey plant, this was not later feasible. As real estate brokers tackled the ponderous transaction, Hoe directors faced the problem of cost.

In connection with the gigantic program to move and equip a plant housing over 3000 workers, R. Hoe & Company in August 1928, through the original three-banker syndicate, issued $800,000 worth of new 7 percent notes and 16,000 new shares of Class A Stock. Both were offered to holders of Class A stock on the basis of $100 worth of notes and 2 shares for every 10 shares of A stock held; thus the "A" stockholders subscribed to 85 percent of the issue—a "privilege" they later had good cause to regret.

The new notes were dated 10 October 1928 and were callable on 30 days' notice at $105 plus accrued interest up to 1 October 1929, and thereafter at 1 percent less each year. The new stock (like old Class A stock) would carry $11 in accumulated dividends, accrued and unpaid since 15 October 1925. By October the stock, increased in quantity to 20,000 shares, was selling briskly on the New York market. It was not clear at the time that the receivables against which these notes had been drawn were already pledged to Guaranty Trust. They too were "pure water." But the arrangements for a grand move were under way.

The fate of Hoe's Manhattan property was not without importance to the entire East Side community. It was one of the largest pieces of private real estate to appear on the market for a long time; although the buildings were of various ages and sizes, they constituted 489,109 square feet of floor space on 111,000 square feet of land. The location was rendered all the more advantageous by current city proposals to extend the subway system (a Houston Street east-west connection from 6th Avenue to Essex and Rutgers Streets) nearby. Sale of the property was expected to be lively.

Appointed by the company to dispose of the parcel by private sale was Joseph P. Day, a real estate broker with an impressive reputation. Day had sold the Third Avenue Elevated in 1910 for $26 million, the largest New York auction that dealers could remember. On 26 February 1929 Day received for the plot a single bid of $601,000 from "Hoe interests" who were never identified, but he would not honor it; the City of New York had assessed the property at $1,312,000 and Day was aiming even higher. In the halcyon days of fortunes for everyone, there was no need to sell at distress prices. A month went by.

Henry Mandel, a builder and operator of housing developments—he engineered New York's fashionable "London Terrace" in the Chelsea area, had immediately recognized the Hoe property as an ideal site for housing. At the end of March he began negotiations to buy, but at the same time the same idea was being discussed in other quarters. Lillian Wald, of the nearby Henry Street Settlement, long hoped for modern housing on a grand scale in the lower East Side. She had seen the overcrowded ghettos where outsiders could scarcely penetrate its quaint—but painful—Old World customs and conditions, as well as the substandard tenements where friction between peoples was tearing away those customs and replacing them with bitterness and disillusion. The East Side was New York's most poignant example of toughness and pathos.

Miss Wald had frequently appealed along humanitarian lines to her friend, Lieutenant Governor Herbert H. Lehman; finally, when Governor Roosevelt brought Lehman, a banker, together with Aaron Rabinowitz, another real estate developer, something could be done. Rabinowitz as a youth had participated in East Side settlement activities, and Lehman had gotten valuable social work training there. They were quick to respond to the governor's personal encouragement.

Together Lehman and Rabinowitz bought the property, paying only $500,000—"an exceedingly low price," said the *New York Times*, recollecting the city's assessment and the spurned bid from "Hoe interests." The figure, in fact, was one third of prevailing real estate prices—Mayor Walker himself had to abandon a similar housing scheme the year before because property owners had held out for "as much as the traffic could bear." Having refused a $601,000 bid only a month before, Hoe officials could only be called exceptionally charitable and civic-minded in accepting the Lehman-Rabinowitz offer. It was possibly the last charity the great press-maker could have afforded.

Lehman and Rabinowitz were working close to charity themselves. The model houses would be erected under a "limited dividend" plan of the state housing law, prohibiting the two from receiving any profit at all. Other investors would be limited to a 6 percent return on their money; the apartments would be rated a maximum of $12.50 monthly per room. These conditions merited a little public information, said Rabinowitz:

> There has been a great deal of talk about the development of model housing on the east side, but obviously there has been little action. The sale of the Hoe property offers an irresistible opportunity to acquire a large parcel of land in the very heart of the crowded east side available for immediate development, without the necessity of displacing any families resident on the property. As soon as the present occupants have removed, the buildings will be torn down and not long afterward construction will be started by the limited dividend company.[28]

The first East Side project to be built under provisions of the new state housing law was expected to cost $2,850,000 for land and construction. A cluster of five and six-story brick and steel houses, with large garden courts, would accommodate 400 families. Elevators, tenants' meeting and recreation rooms, and—of all things—a sparkling swimming pool were visualized (Hoe had fortuitously left behind 10 artesian wells for a limitless supply of fresh water). But these were mostly press-agent dreams. Springsteen and Goldhammer, the Lehman-Rabinowitz architects, instead produced a handsome seven-story building with a great central court to contain (only) trees, grass, flowers, and reflect these brightly into cheerful rooms. Occupancy was expected 1 May 1930.

Lehman and Rabinowitz received congratulations from many civic groups and individuals. Governor Roosevelt telegraphed: "I am greatly gratified at this initial undertaking in the rehabilitation of the lower east side in the city of New York. . . ."[29] For a little ceremony at the closing of the real estate transaction Hoe's Robert Swartz was invited to participate. Hoe, after all, had been an important community neighbor for a century, although (with the 1902 riot almost forgotten) its recent civic activities, as always, had been unobtrusive where possible. Rabinowitz was careful to remind all that "Neither Colonel Lehman nor myself expects to make a penny of profit out of this model housing project. . . ."[30] In those carefree investment years an intentionally nonprofit operation must have seemed the epitome of civic endeavor.

The public received the news of the housing development with great enthusiasm. For Russian Jews in the neighborhood the sublime irony of turning the bigoted factory into homes for the oppressed was too rich to overlook. By 1 June 1929 the East Side Chamber of Commerce received 500 applications for the new apartments, mostly from families then living in the old cold-water flats nearby. When registration began, credit was extended to those without cash; no applicant would default, and all pledged to keep the beautiful grounds spotless. Wreckers moved into the courtyards on 5 October 1929, and began smashing the Colonel's red brickwork and iron girders. But by the end of that fateful month all America came tumbling down.

The stockmarket crash of 29 October was bad news for R. Hoe & Company. It need not have been more. Still negotiating its property sale—now viewed almost as a calamitous pre-depression "giveaway," the firm was still moving from the Grand Street premises which were not finally vacated until 30 April 1930. The company had reported a small loss in 1928—only an adjustment to modern conditions, but enjoyed a profit for 1929 despite its disastrous last three months. As the impact of the economic depression on America became fully apparent, however, the directors, almost all of whom were financially oriented businessmen, seemed to doubt their ward could be in excellent health, incredulous that the old firm was not suffering nearly as much as the business community as a whole. Perhaps they were embarrassed, uneasy, with their solvency. Considering the panicky times, however, they may hastily have regarded a robust condition as chimerical, illusionary, unreal. Pessimists seemed to know better what to do for bankruptcy than for prosperity—professionals always manage to make more money for themselves when the client is ailing. The Hoe directors in their discomfort soon made unfortunate confidential decisions which can only be called reprehensible. (Harry Swartz, with principles much as he had described them in his interview not long before, would resign as president in 1930; the directorate proceeded without him.) A scheme capitalizing on the depression would make some tidy money for Guaranty Trust officers—and probably not hurt the company very much at all.

A year passed with Hoe, along with most of the business world, in stunned silence. There was no indication of company profits or loss, no word of orders obtained or backlog of work. The silence gave birth to fears of the worst, and it came. In March 1932 the directors announced soberly that they would be unable to meet interest and sinking fund requirements, due 1 April, on the company's 7 percent notes and first mortgage 6.5 percent gold bonds. This bald declaration of insolvency meant that not all creditors could expect to collect—an incredibly inflammatory statement at a time when hundreds of businesses were similarly failing. With the collapse of so many money empires few inside observers seemed surprised that the fine old name of Hoe was in serious trouble; this too could happen, pessimists said. The public record was indeed confusing:

1924 Annual Report (issued April 1925) showed a profit of $729,880 after interest, depreciation, and taxes; total assets $14,773,025.

1925 Annual Report (March 1926) showed a profit of only $3734, assets only $3,753,772. In December 1925 the company had suspended class "A" dividends.

1926 Annual Report (March 1927) showed a net profit of $475,962, and assets of $7,383,027; $146,000 first mortgage bonds were retired through a sinking fund.

1927 Annual report not released to public.

1928 Report (August 1928) for six months ending 27 June showed $141,215 income but a net loss of $187,308; at year's end the loss was only $79,281.

1929 Report (August 1929) for six months ending 30 June showed a net profit of $202,649; Annual Report (March 1930) showed a net profit of $609,478. No dividends having been paid, the accumulated Class "A" earnings amounted to $17 per share.

1930 Report (August 1930) for six months ending 30 June showed a net profit of $196,980, assets of $4,213,906; Annual Report (March 1931) showed a net profit of $115,326.

1931 Report (August 1931) for six months ending 30 June showed a net loss of $117,867.[31]

In what were American businesses' best (pre-crash) years in general, the Hoe record seemed to jump up and down without explanation.

Upon hearing of the difficulties, however, security holders galvanized to action. Three separate committees formed with amazing speed for protection against possible loss through corporate mismanagement or favoritism, and to take an active part in reorganizing the indebtedness along as remunerative terms as possible. These were a bondholder's committee, Charleton B. Hibbard (an old associate of the

Wright brothers), chairman; a note committee, Ranald H. MacDonald, Jr. (real estate operator and investment banker), chairman; and a stock-holder's committee, Harold G. Hathaway, chairman. Strangely, almost all these committee "officers" were members of the very brokerages and banks comprising the original syndicate; curiously too they all chose to employ Guaranty Trust as repository for collected securities. The great Wall Street firm was still fully in charge, with no one to answer to—and the depression, so costly to many, hardly a bother for clever bankers.

In its own internal affairs the Guaranty Trust Company had made no hasty, careless moves. World conditions had surprised the banking firm as it had others, but the directors were not distressed. Although the collapse of international trade in 1921 had been somewhat costly to Guaranty, it pulled through "with all commitments honored and its credit untarnished." In 1933 a federal banking act would require the separation of investment and commercial banking; throughout the depression nevertheless Guaranty Trust (which in 1959 would merge with the banking firm of J. P. Morgan & Company) had not a moment's difficulty: "the bank's soundness," said the company historian, "was never in question."[32]

Many of the investing public had innocently thought too that the soundness of R. Hoe & Company was beyond question. News of its financial embarrassment was staggering to stockholders, most of whom—employees and their friends, and other middle-class investors—had considered the firm good as gold. Few of them knew they had any legal recourse, and fewer still knew what could be done.

Upon learning of R. Hoe & Company's financial plight, George Polak, a silk merchant holding only five bonds, thought by the end of April that the laws of bankruptcy would work best for the company and its creditors, designed as it was to protect all parties from deliberate or circumstantial pillaging. He filed a suit asking the New York Supreme Court to appoint a receiver to guide the firm through potentially troublesome waters. Justice Ernest L. Hammer, looking into the company's legal records, found to his surprise that R. Hoe & Company was already in voluntary federal receivership, declared (but not published) just a week before, 21 April.

Perhaps because the company at its confidential receivership hearing had vigorously asserted its solvency, Swartz (on hand as a director) and the Irving Trust Company, New York, had been duly named equity receivers by the court—giving them new and extraordinary control of the great company. Judge Hammer was annoyed; he accused the firm and its counsel of "dilatory acts" and "seeming affront"[33] to the court. The company's officers had in fact raced Polak to court with a sad but "rosy" presentation just to get custody of the receivership.

Although Polak suggested vacating the earlier receivership and beginning anew, the court dismissed his suit without prejudice so that proper legal investigation might proceed unencumbered. Swartz, his colorful presidency already ended, resigned as receiver in July.

By June other interested parties were also investigating. Samuel Zirn, a New York attorney and a stockholder, announced his intentions of forming an independent security holder's committee. He regarded the three "protective" committees as dangerous complicities—formed, he claimed, a month before the receivership itself. Zirn represented a group of friends and relatives possessing $85,000 (par value) 6-year 7 percent notes and approximately 2700 shares of Class A stock. He hoped to "protect the rank and file . . . against the so-called 'Protective' committees," and, furthermore believing the company "abundantly solvent" in the first place, raised some curious points:

> The actual accounts payable are comparatively small, and the chief current liability consists of loans owing exclusively to the Guaranty Trust, which on January 31 last, stood secured by these gilt-edge notes and accounts receivable from the leading publishers, in an amount equal to about 130% of the bank loans,—*not* a dollar unsecured. The following day, this Bank compelled a $250,000 cash reduction of its loan (*twice* the sum required to pay the April 1st interest on the Bonds and Notes), without releasing *any* of this collateral, thus retaining about 140% of security. These loans have further been reduced by about $165,000 during the first month of the Receivership, again without releasing any security.
>
> This same bank is the *Depositary* for all three Committees, whose various secretaries are all connected with the original Bankers' Syndicate, as are *eight* of the ten gentlemen making up the three Committees. There is a great deal more to be told. . . .[34]

Zirn felt no need whatever for receivership. As he collected more facts he was emboldened to present a nasty picture: that the original syndicate had, upon incorporation of the firm, sold securities to the public at prices 12 to 50 times their present market value purely "for the purpose of forcing a judicial sale and reorganization in which they would be the underwriters." (The $50 class A stock, for example, as Zirn spoke, had dropped to 3/8 bid, $1 asked, and the notes were at 4 percent of par value.) He accused the Guaranty Trust, Dominick & Dominick, and Edward B. Smith & Company group of seeking temporary federal receivership "for its own selfish interest and advantage."[35] By the end of July Zirn had determined that the original receivership was permitting Guaranty to retain $85,000 from its seized Hoe account and was directing the receivers to pay Guaranty $143,226 collected from notes and accounts receivable, as well as paying Guaranty all sums thereafter collected on illegally pledged collateral. The intrepid lawyer continued his probing.

In January 1933 Zirn made additional observations. He pointed out that the federal receivership decreed by Judge Frank J. Coleman gave creditors no opportunity to present their position; that the court could not have been fully aware of the realities of the case. In the deplorable world economic situation, Zirn maintained, every means should have been exercised to keep a company—even when in honest difficulties—out of receivership, with its destructive psychological repercussions. At last, on 21 February 1933 the U.S. Circuit Court of Appeals—for the first time in 20 years—reversed a court order. It declared the 1932 receivership open to reexamination.

Justices Augustus N. Hand and Thomas W. Swan concurred that the earlier court erred in not letting creditors be heard. They found that the company had purported (but not made public) losses of $594,000 for 1931, although under scrutiny the sum was more like $95,000. Too, the company had not waited until the end of a 90-day grace period before declaring itself unable to pay interest for 1932. Suspiciously, it had rushed into receivership all too quickly. While it did not require all of the settlement arrangements between Irving Trust and Guaranty Trust to be put aside, the ruling did stop some wholesale despoilment of the bewildered old company.

Like the old Harper horse plodding its circular path long after retirement, R. Hoe & Company seemed to attend to its press manufacturing duties more by habit than by active leadership in the shop or board room. In 1933 it paused a moment to help celebrate the New York *Sun's* centennial, an event toward which the company had long played an instrumental part. The paper now had no less than 15 Hoe Octuples and a new 12-unit line of Hoe "unit-type" presses capable of 135,000 32-page papers per hour. Each unit, with two plate cylinders, printed 16 pages; the total 192 pages of stereotype plates, weighing 6 tons, hourly impressed 188 miles of a 68½-inch paper web. For a courtesy notice in the *Sun's* commemorative issue, however, an exhausted Hoe copywriter simply dug out "I Am the Printing Press," adding, with some oversimplification, "Written by Robert Harding Davis for R. Hoe and Company, founded in 1805 and builders of all the *Sun's* presses during its lifetime." With offices now in Boston, Chicago, San Francisco, New Orleans, and London, the company was heralding its "Patented Universal-Unit Multi-Color and Black Rotary Web Perfecting Presses" (essentially a Pancoast design), the 24-unit model of which, turning out 720,000 pages an hour, was still labeled "The Largest Printing Press in the World." [36] Hearst alone had 12 of them. But the company (again) was busy elsewhere.

Justice Coleman at Federal District Court consented to reopen the Hoe receivership hearings. First, however, he ordered Irving Trust to report on its year-long stewardship of the company. The bank published its response on 19 June, indicating a fairly sound constitution for the old organization. Although the receivers in 1932 had had to borrow $25,000 immediately to meet the payroll, and another $250,000 for operating expenses, the firm remained continually in business since their appointment. The company's outstanding bank loans in April 1932 of $1,863,868 in the year's time had been reduced to $362,472. Hoe was practicing some economies (monthly overhead, for example, down $47,000) and it had $518,224 in unfulfilled orders on its books. With the Irving report duly distributed to stockholders and creditors, Judge Coleman, forewarned to look at corporate matters very closely, began to ask pointed questions.

Coleman called to the witness stand Kenneth A. Christian, East Orange, New Jersey, a prospective son-in-law of Hoe treasurer Allen W. Lishawa. Under oath, Christian reconstructed some amazing company history. When interest on a Hoe note had been defaulted, two years before, the resulting bankruptcy would have been disasterous to the company; a friendly receivership, however, brought by someone outside the state, was deemed much better. Lishawa had telephoned the young Christian and asked him to petition for receivership. He consented to do so.

Former president Harry Swartz testified that the Hoe board of directors had indeed decided equity receivership to be the proper course of action. On the other hand, Leighton H. Coleman (vice president, Guaranty Trust), attorney for the Hoe bond-holders' committee, told the court he had advised against it. The scheme unfolded.

Jones & Laughlin, Pittsburgh, was the original holder of the defaulted Hoe note, but the big steel concern, perhaps considering almost a century of association with Hoe, had declined to take action. Incredibly, Christian bought the tiny $3,226.96 note—with Hoe (London) funds supplied by Lishawa. When called to the stand, however, the Hoe treasurer, on advice of counsel, declined to testify. Nevertheless, the ugly scheme had at least been revealed. Assignment of a claim against the company, purchased by a petitioner for receivership with Hoe money and at the request of Hoe officials, was enough for Judge Coleman: "I am convinced from what has been testified that this suit is fraudulent and that the receivership in it is at least voidable. . . ."

Coleman ordered a transcript of the testimony to be sent to the United States Attorney for possible grand jury investigation. At the same time, however, he extended the Irving Trust receivership another four months, rather than return the company immediately to its suspect corporation officers. Yet the judge was optimistic enough to feel that no real harm had been done. "I am convinced," he concluded, "that the receivership has, from the point of view of economics, been most beneficial to creditors and stockholders."[37]

Unhappy that the case had been returned to Justice Coleman (whom he considered "banker-minded"), Zirn sought unsuccessfully to disrupt the continuing receivership by charging the court with prejudice. He had a very formidable dossier of facts which sponsors of the receivership (including an old peddler, quietly on hand) did not deny:

1. The charge that Guaranty Trust participated in arranging for this ex parte receivership and in procuring the assignment of the claim to Christian, for the purpose of instituting this suit.

2. That 9 of the 10 Hoe directors were officers of or partners in, or in close business association with, Guaranty Trust and this Bankers' Group, and the only independent directors † was in California when these crucial meetings of March 17th and April 18th, 1932, were called on two days' notice by mail.

3. That Hibbard, Chairman of Bondholders' Committee is a Vice-President of Guaranty Co.; MacDonald, Chairman of Noteholders' Committee is a partner in Dominick & Dominick; and Hathaway, Chairman of Stockholders' Committee, is a partner in Edw. B. Smith & Co., composing this Bankers' Group. And all original members of these three Committees, except one, are officers or employees of these banking houses and another, which participated in marketing these securities.

4. That said Hibbard and MacDonald, and E. H.† Potter, brother of Guaranty's President, were all Directors of Hoe, who had voted in favor of the resolutions to default, to ratify the illegal modification of the Loan Agreement with Guaranty, and to consent to a federal equity receivership.

5. That Guaranty is the Depositary for all three "Protective" Committees, and the Secretaries of all Committees are officers or employees of Guaranty or these underwriting firms.

6. That Guaranty at all times since organization and at receivership, held voting control of Hoe through a voting trust holding ⅔ of the unlisted Common stock, with exclusive voting power,—in recent years through the inaction and trust reposed by the Class A stockholders.

7. That Guaranty Company's President selected Swartz as President in January, 1926 at a salary and bonus over twice that of his predecessor (who had brought Hoe its greatest prosperity); and that Mr. Todd's firm [Baldwin, Hutchins & Todd] could not have been general counsel for Hoe [against George Polak] unless Guaranty approved their selection and continued their employment.

8. That since March, 1926 (2 months after Swartz took office) Guaranty Trust took and has retained a monopoly of Hoe's bank loan and deposit business, thus putting it into the hands of the one Bank which absolutely dominated it. That Guaranty loaned up to $1,750,000 without security up to January, 1928 and then suddenly took 120% of collateral on existing bank loans, and Swartz participated in concealing that fact, in exhibiting the Balance Sheet of June 30, 1928, attached to his letter as President, dated Aug. 29, 1928, inviting the "A" Stockholders to subscribe at par for the 7% Notes (unsecured).

9. That at the Annual Meeting of Hoe on April 12th, 1932, Swartz presiding, he denied any knowledge of a contemplated receivership, while the next day, the Minutes show, he stated that he had several times recently advised seeking a federal equity receivership and again urges it; and then became the beneficiary as receiver, 8 days later, of what he had advised. He then obtained approval of the Receiver's petition on June 14th (with the aid of Mr. Todd's firm, as Receiver's counsel) to allow Guaranty to retain the $86,000 Hoe bank balance seized on day of receivership, and to turn over to Guaranty $143,000 collected by the Receivers on pledged accounts and notes, and similar large sums thereafter, while having effectively tied the hands of all other creditors.

10. That Mr. Todd's firm drafted the resolutions adopted on March 17th—to default and to ratify the illegal modification of the Loan Agreement with Guaranty, on January 30th, 1932; and their letter of April 13th strongly urged at once seeking a Federal equity consent receivership which "we have repeatedly advised," and enclosing a form of resolution for that purpose, adopted at meeting of April 18th and that both these crucial meetings were held in Guaranty Trust's office.

11. That not the slightest efforts were made by Hoe's Board of Directors or officers, or these "Protective" Committees—all undeniably controlled by Guaranty, to obtain any waivers of sinking fund or postponement of interest from the Bond or Note holders, and such advice urged by me at the Annual Meeting and at the conference in Guaranty's office—postponed by it until the morning after Receiver's appointment, were ignored and frustrated.

12. All efforts to obtain the Bondholder's List, so that the scattered Bondholders, without actual notice of the origin of the receivership or the situation existing, might be communicated with to co-operate in protecting our common interest in the situation, was blocked and frustrated by the agents of Guaranty Trust, which utilized its exclusive information acquired in a fiduciary capacity, to serve its selfish interest, by giving it only to its own Bondholders' Committee.[38]

The bankers' group, Zirn concluded, unquestionably had but one aim: "a permanent Receivership with a judicial sale and reorganization, to provide huge allowances to everybody helpful to their purposes. . . ." The healthy old company was to be a fat carcass for the delectation of greedy vultures.

Zirn had not alone reached the damning conclusion that bankers generally were plundering much of what they should have been preserving. He quoted a contemporaneous New York *World-Telegram*:

> From facts coming to the surface in various cities it is evident that the receivership racket is one of America's major scandals in high places. . . . Great sums are finding their way into pockets of favored lawyer cliques . . . in equity proceedings; estates are being looted,

corporations that might be saved are being carved up in fees before the eyes of the courts; shareholders and creditors are left impoverished. . . .[39]

Richard M. Hoe, who had long eschewed indebtedness, had after a century been sold out to the slickest of operators—the company's own management-bankers who so effortlessly inspired trust and stability. Henry A. Wise Wood, whom the Colonel might scarcely (at first) have considered an ally, had just a few decades before contemplated such events. He advocated better legislation against corporate evils:

Conceive for example a statute, relentlessly enforced, under which a corporation, its officers being liable in heavy criminal penalties for nonfulfilment,† is bound semi-annually to issue to its share-holders and the public authorities the transactions in detail of its various governing bodies, and of its major and minor officers. . . . Could directors have manipulated for their own gain the funds or securities of their share- or policy-holders? [40]

Maybe not, if relentless enforcement were possible. But as economists and jurists frequently noted of America's glorious two-century history, the business of the United States was business; the corporation was seldom seen to be at fault when attacked by private citizens. And even if corporate management were as flagrantly dishonest as, say, the Tweed Ring three-quarters of a century before, would not the enormity of the crime (again) simply boggle the public (and the courts) into despair? But there were still fighters like Zirn.

In another missile Zirn tirelessly recapitulated the grisly details of the fraud to illustrate its audacity:

. . . The undisputed evidence showed that the $3,200 claim on which the receivership suit was brought by a dummy plaintiff [to whom it had been assigned a few days before by a great steel concern which had done business with Hoe for thirty years or more], had been actually paid in full with Hoe's funds before assignment, when that concern refused to sue; but the payment was made by a concealed method through checks manipulated right in Guaranty Trust's office—by a Guaranty cashier's check made to Hoe and endorsed over by some of its officers to this dummy plaintiff, from funds which Hoe's London subsidiary had cabled over, through Guaranty's London branch, to Guaranty in New York; and that Guaranty Trust's Vice-President-Treasurer caused this cashier's check to be certified, and even the signatures of Hoe's officers to be guaranteed when they endorsed the check.

He continued:

That very day Hoe's directors held the special meeting consenting to the Federal receivership right in Guaranty Trust's office—one of which directors, E. A. Potter, Jr., was also a vice-president of Guar-

> anty and a brother of its president, and voted in favor of this receivership resolution. Another director so voting, was Mr. Hibbard, a Vice-President of Guaranty Company and Chairman of the Bondholders' protective Committee, and another director so voting was Mr. MacDonald, a partner in Dominick & Dominick, part of this Bankers' Group, and Chairman of the Noteholder's Committee.[41]

In November, when the receivership was to expire, the Judge nevertheless ordered it continued; by this time more than 300 creditors had registered claims aggregating $988,555. Zirn retired to reorganize.

In March 1934 Samuel Zirn publicly asked the holders of Class A stock to submit their proxies to him for voting. Needing only 10 percent to achieve a small voice in corporate matters, he received, by month's end, proxies on 40,000 shares from 400 holders—almost 42 percent of the shares outstanding. As dividends on this highly prestigious preferred stock had not been paid for seven years, Zirn was getting considerable sympathy and support. At the same time Irving Trust issued another report (to 31 December 1933) showing the original $1,863,868 debt to Guaranty Trust fully paid, continued reduction of overhead expenses, and a small profit. Having received only $12,000 in fees, however, Irving Trust also requested an additional $100,000, half for itself and half for its attorneys. These various disclosures, of course, were footwork for the April stockholders' meeting. It promised to be interesting.

The annual meeting of R. Hoe & Company took place 10 April 1934; it was a short and neat defeat for management. Richard Kelly, Harry Ball, E. M. McIntosh, Edwin L. Weise (an attorney), and Samuel J. Kane (of Tide Water Oil Company, New York) were elected directors. Atterbury, G. L. Austin, and Lishawa resigned, as did Hibbard (who also resigned at Guaranty Trust). Frank Shepard was elected to the board to succeed E. A. Potter; Tillinghast and Ball were reelected vice presidents. It was a triumph for Zirn; his group had elected five of the six directors.

Creditors and stockholders, however, had still gotten no real satisfaction. They pressed for financial reorganization that might bring them some security, if not money. In July the company sought to benefit under new bankruptcy legislation: in a petition brought by Tillinghast seeking reorganization under Section 77B of the Federal Bankruptcy Law, the firm showed assets of $9,347,676, liabilities of $9,271,054, but current obligations of $441,278 versus liquid assets of $391,803. Although bondholders were unhappy with it, Tillinghast's plan was approved 30 July by Federal Court Judge John M. Woolsey, who appointed Irving Trust anew as equity receiver and temporary trustee during the reorganization.

Zirn continued to bombard security holders with informative missiles, furnishing a historical perspective as well:

Around Sept. 1924, the Bankers' Group bought the old Hoe concern—then the world-wide leader in newspaper and magazine presses and continuously in business with an unblemished record for 106 years, from the Hoe family. They immediately organized the present Company and issued and sold to the public the present Bonds at 99½ and the "A" stock at $50 per share—prices netting a huge cash profit. Some of this "A" stock was also then sold to Company employees who still retain it. They additionally imposed this issue of Common stock, which was pure "water" from the outset, retaining two-thirds of it without cost, after having sold one-third to Mr. Arthur Hoe for a very large sum. To this Common stock they gave the exclusive voting power for the election of all directors and on nearly all other matters; while the "A" stock was merely given the contingent and temporary right to elect a majority of the directors (6 out of 10) after an accumulation of dividend arrears, and on certain special propositions. Through a voting trust on this ⅔ of the Common stock, controlled by Guaranty officials, it retained control until last April.

The prospectus they issued to the investing public showed that the Company's net consolidated profits during the preceding 6 years, under the presidency of Mr. Richard Kelly, averaged nearly $1,000,000 per year,—enough to pay this Bond interest nearly four times. About 80% of these earnings were by the American Company (Hoe), and the rest by the British subsidiary. Mr. Kelly is now one of the "A" Directors elected to the new Board, and voted in favor of this Plan.

Read this record of comparative earnings of the American Company before the Bankers took control and after that, and then note how the British subsidiary in London, operated independently and 3,000 miles away from that Banker-control, has prospered amazingly. These figures tell a remarkable story.

	BEFORE BANKERS' CONTROL (Mr. Kelly, Pres.)	
Year	*Net Profit American Co.*	*Net Profit British Co.*
1919	$ 743,000	$210,480
1920	585,000	62,628
1921	836,000	77,348
1922	1,050,000	234,662
1923	976,000	304,284
1924	444,000	285,233
	AFTER BANKERS' CONTROL (Mr. Swartz, Pres. since 1926)	
1925	446,000 Loss	450,020
1926	140,000	335,020
1927	363,000 Loss	225,042
1928	579,000 Loss	499,471
1929	76,000	533,278
1930	458,000 Loss	573,673

In 1931 Hoe again lost heavily, while the British subsidiary made substantial profit. The first two years of the "equity" receivership, started in April 1932, resulted in a loss for Hoe of about $1,000,000, including interest and estimated receivership allowances; while the British subsidiary (about one-fourth the size of Hoe) in 1933 earned $336,000 net, after paying over 20% British income tax. . . .

In Nov. 1925, they suddenly discontinued dividends on this "A" stock sold at $50 a year before, on the plea of economy and conservation, but immediately engaged Mr. Swartz as the new President at three times Mr. Kelly's salary when the Bankers took control, plus a bonus based on gross instead of net profits. And that contract was negotiated by Guaranty Co.'s then President as a Hoe director. Note the operating results that followed.

In August, 1928 the Bankers' Group "underwrote" the Note issue of $800,000 and had it offered to the Class "A" stockholders by a circular in the Company's name, as a "privilege," annexing a balance sheet showing a large amount of accounts and notes receivable, which receivables were the chief protection the Notes could look to. The "A" stockholders subscribed for over 85% of these notes. But the bulk and choicest of these receivables had been already pledged to Guaranty Trust on continuing loans, but not disclosed, and was first discovered after the receivership started. . . .

Zirn then recounted the details of the fraudulent 1932 receivership, and of the subsequent resignation of Swartz as receiver and other Guaranty persons as protective committee officers. He concluded:

Please remember, that during this Receivership, voted for by Hoe's old Board right in Guaranty Trust's office, Guaranty's bank loans of about $2,000,000 at the moment of receivership, have been paid off in full, principal and interest, while holders of the First Mortgage Bonds (for whom it was then acting as Trustee), which its syndicate marketed at 99, who saw them go down to 8 right after receivership, have not received one cent during that period, nor have the Noteholders or other creditors.

While the Bankers' Group pulled out their investment long ago, and Guaranty bank loans have been paid in full during the receivership, they still want to resume control of the property in which your money and ours is still invested. I still have a cash investment of about $25,000 in the Hoe concern, including $14,000 of its Notes as an original subscriber. Having abandoned the ship and left it sinking, they seemingly resent the efforts of the new Directors and myself, genuinely representing those whose money is still in the enterprise, to rebuild the business and put it back to its former highly profitable and commanding position, in co-operation with its old-time competent and loyal officers and employees.

Generally speaking, the writer has never had any prejudice against bankers who are really bankers in the respected sense of that word. But what has all this to do with real banking? . . .[42]

Still somewhat unhappy (he had wanted Richard Kelly named permanent trustee), Zirn in September described another reorganization plan which was more favorable to bondholders. Upon its presentation to the court, however, Judge Alfred C. Coxe gave it short shrift; he construed the plan as a holding tactic against court instructions to reorganize speedily, and threatened to order a sale to liquidate the company if there were further delays. Irving Trust, meanwhile, was ready with a comprehensive final report of its work up to the new petition for bankruptcy.

Irving Trust Company had bravely approached several technical matters in which it admittedly possessed little expertise. In pursuance of an existing Hoe policy to continue press development and patent work (the firm was, after all, in the highly competitive printing press business), it spent $160,193 to obtain several "important devices" and realized some significant orders as a result. The bank was greatly relieved to find it had in its financial administration made some decisions almost worthy of a Hoe patriarch:

> Trade reports would indicate that in the sales of its large printing machinery more than the ordinary average proportion of the corporation's business has been obtained, and according to the published reports, the operations of the receiver will compare very favorably with the operations of competitors and also the operations of companies in allied lines in the printing machinery industry during the period of receivership.[43]

These fortuitous management decisions had been complicated by the diversified interests of the company, which, in the depression environment of the NRA program, required cooperative dealings with no less than three price-sensitive agencies: the Newspaper Printing Press Builders Association, the association of the printing equipment industry, and that of the saw and steel products manufacturing industry.

Unfulfilled orders had totaled $1,197,905 as of April 1932 and were now $1,566,835 as of 26 September 1934; orders procured during the last year were $4,949,968. Orders totaling $2,381,723 were completed and shipped (cancelled orders amounted to $199,318). Current assets of $4,356,534 at the start of receivership were now in July $2,537,519; current liabilities were $2,436,512 and $795,137 for the same dates, respectively. With the company's own reorganization plan to be announced momentarily, the bank's work was done. In July 1935 Irving Trust submitted bills for its services—amounting to $250,000. Bondholders claimed the exorbitant charges were further inflated by agent and attorney fees which were being paid elsewhere in security transactions; they also objected to fees which committee members (and Zirn) had awarded themselves.

The Hoe reorganization plan was not actually approved by the court until February 1935. It called for replacing $3,171,000 worth of 6½ percent bonds payable 1 October 1935 with new bonds maturing October 1944. Holders of Class A stock would receive 6 percent dividends on the basis of 1 to 3 until dividends on preferred stock were paid in full. The individual creditors' committees soon became reconciled to the arrangement and similarly the busy Securities and Exchange Commission, examining the original voluntary receivership, quietly took no action. It seemed that all hostilities had been played out.

In August 1935 the stockholders' meeting was called to elect officers to carry out the new plan. In an eight-hour session Zirn and his independent group joined amicably with management's forces to place eleven men on the board of directors. They were Henry Holmes, Robert J. Lewis, Samuel J. Kane, Edwin L. Weise, Tillinghast, Claude V. Callister, Richard Kelly, John T. Harrison, E. D. Timberlake, Corcoran Thom, and Maxwell M. Upson.

The New York Stock Exchange approved a listing of the new 6½ percent bonds due October 1944. But suddenly there was still turmoil: a consolidated report (10 months ending 31 July) showed net loss and another report (year ending 30 September) showed a net loss for the British subsidiary as well. The bonds were removed from the stock exchange listing; Weise and Tillinghast resigned as directors. Fred L. McCarty, vice president and general manager since June, was elected president in December. It was hard to find harmony anywhere.

As a new stockholders' meeting approached in 1936, the board of directors applied to the New York Stock Exchange to list 100,000 new shares of Class A ($10 par). They had on hand from recent bookings (October 1935 to April 1936) large machinery orders amounting to $2,188,000; unfulfilled orders totalled $4,173,000. These figures brought the company back to where it had been fully five years before: from 1 October 1929 to 1 April 1930 its bookings had been $2,163,000 and at the end of the same period total orders had been $4,653,000. The depression had surely been damaging to the company, but its own mismanagement had been incalculably costly too. Recent loss statements were disappointing—no one was really happy.

On 14 April 1936 in the main offices of R. Hoe & Company, 138th Street and the East River, the annual stockholders' meeting commenced. Samuel Zirn had beforehand accumulated proxies of one-third the eligible vote, but in the last 10 days before the meeting the management had mobilized fully one-half of the total vote. Because of extraordinary qualification requirements, out of a total of 104,568 shares only 80,561 were allowed to be counted. The *New York Times* explained:

> The situation was complicated by the fact that the voting stockholders are divided into two classes, prior preferred and preferred, one class to elect six directors and the other to elect five; while a

voter in either class might, under a cumulative voting system, cast all his votes for one director or spread them over the vacancies to be filled.

Further, no prior preferred stock could be voted unless the holder actually owned three bonds for each share of preferred. And finally, the outstanding prior preferred entitled to six directors totaled about 8,574 shares, less those disfranchised by selling their bonds; while the preferred shares entitled to five directors totaled 95,994.[44]

Both Zirn and the encumbents came to the contest equipped with "tables of combinations and permutations" to determine what voting strength each could muster based on the tellers' counts of the proxies. Both forces raised objections at each step of the attorneys assigned as tellers, obliging them, said the *Times* reporter, "to look repeatedly at every piece of paper attached to every vote." It was tense and grueling; 14 hours of the first day were expended in this meticulous counting, only to be continued the next.

The second day of the meeting plodded well into the evening with continued vote examination. By 9 p.m. the original voting strength of 86,286 had by qualification technicalities been reduced to 80,561, although neither side had lost its relative position. The deliberations of the two harassed attorneys, inspecting the proxies and hearing conflicting arguments, nevertheless went on. A third day of dreaded negotiations seemed inevitable when Fred McCarty, president of the board, spoke:

"Let us get to the balloting for directors," he commanded.

"We are not sure we can take our instructions entirely from the chair," said the tellers nervously. "Since we have started to allow each side to examine the proxies which we have accepted from their opponents, we feel the examination should be completed."

"Proceed with the balloting," boomed McCarty more forcefully.

Zirn jumped up. "You can't railroad stockholders out of their rights," he shouted. "We'll wash this election up in court."

The balloting nevertheless commenced. Three of Zirn's candidates were elected: Kelly, Holmes, and John Keating. Eight were elected by management: Upson, Lewis, Corcoran Thom, Tillinghast, McCarty, Timberlake, Harrison, and Clifford N. Bradley.

"The balloting is closed," McCarty said, simply.

"Thank God," shouted Boudinot Atterbury, expressing the relief of most of the participants.[45] He was now just a stockholder.

Zirn, however, was not placated. He insisted that the proxies and ballots be impounded, pending a Supreme Court suit to have the election declared invalid. As the meeting adjourned Zirn declared that both he and his three directors would continue the fight. Management, exhausted, quietly hoped that one or two of Zirn's men would at least be more reasonable.

As matters cleared up later that hectic week, McCarty, reelected president, announced that Tillinghast had been promoted from vice president, sales, to vice president and assistant general manager, John Masterson from secretary to vice president and general sales manager, and Lishawa reinstated as treasurer and secretary. It was almost like old times again, as if nothing had changed. The financial picture for a depressing 1936 was hopeful: the company had standing orders of 6 months' duration—slightly higher than unfulfilled orders for the entire previous year. An attempt, by a newly disgruntled stockholder group in September, to set aside April's furious election of officers, faded away like an unpleasant reverberation.

As the company made plans in December 1936 to put into motion its stock and bond reorganization scheme, it occurred to someone in the office that a word or two about printing technology might be appropriate. A company letter went out:

> . . . The volume of newsprint consumption in the United States is running at the highest rate experienced in the last six years. Newspaper and magazine advertising linage has shown a steadily rising trend during the last several years. The widespread demand for additions and replacements in newspaper and magazine printing plants and equipment reflects not only the need for larger, higher speed production, but likewise the obsolescence which has been brought about by recent improvements in this field and the current development of daily multicolored printing. The company, well prepared by recent production and research work to meet the accumulated demand by offering modern equipment of efficient design, has already begun to capitalize this opportunity. . . .[46]

Slowly profits and unfulfilled orders grew. But war, not prudent company management, was responsible. Never before had the firm's prosperity been so blatantly linked to world misfortune. Crisis, panic, dictatorship, refugees, (and soon) appeasement, blitzkrieg were the circulation vocabulary of newspapers. And never before had company officers, so preoccupied with stockholder squabbles, so callously observed—if indeed they noticed—the cause and effect relationships at work. The company issued another arrogant letter in January 1937:

> . . . The current increase in our sales and unfilled orders is to some extent the result of generally restricted placement of orders for heavy machinery during the depression years, while, in the meantime, improvement in the type of equipment offered has aggravated the obsolescence of existing plant and thereby contributed to the heavy volume of replacement orders. . . .[47]

Announced the following month was a large assignment for seven presses from Conde Nast Publications, part of a $500,000 improvement for its Greenwich, Connecticut, printing plant; included were four new high-speed color machines. R. Hoe & Company found such orders "gratifying," and talked of presses and service once more. But old echoes returned: in March it was back in court.

Justice Albert Conway in New York Supreme Court, Brooklyn, was approached by that disgruntled group of five stockholders who charged, once again, that the management had placed the firm in "unnecessary and unwarranted Federal Court receivership so that the conspirators might profit in various ways in a reorganization of the corporation."[48] The court denied a motion to dismiss the action; it would reappear.

The annual meeting of April 1937 was long but quiet. In 16 hours (ending 3 a.m.) management directors were reelected: McCarty, Harrison, Timberlake, Lishawa (now a director as well as secretary and treasurer), Upson, Lewis, Tillinghast, Frederick M. Schlater, and Neil P. Cullom (a lawyer). Zirn's group managed to reelect only Kelly. The general picture improved slightly for 1938, with one surprising exception.

After its many profitable years of service despite the world economic situation, the London branch came to attention. In 1936 the City of London renamed Mansfield Street, Southwark, "Rotary Street,"[49] to honor Hoe's contribution to newspaper presswork. But the plant was not thriving. Old, scattered, inefficient, the London works were beginning to falter because of war conditions and material scarcities. R. W. Crabtree & Sons, Ltd., established 1895 and manufacturers of printing equipment since 1905, were interested in the property and ultimately paid £1 million for it.

Had the company a bright young energetic heir to study the European market for presses, it might have decided that dilapidated buildings were hardly reason enough to jettison a viable enterprise. But the systematic depletion of financial resources at home made almost $3 million too irresistable to consider rehabilitation. The money was welcome indeed. After three-quarters of a century the London shop, a proud jewel for the introduction of American technology to all of Europe, was ended. Certain agreements between the two companies, however, would make this transaction, like recurring stockholder suits, another invitation to appear in court.

In January 1939 Fred McCarty resigned; he had been with the company 7 years, 5 of them as president. To replace him four directors were appointed as an operating committee: Tillinghast, Foster, Bradley, Lishawa. Business for 1938 had, as in England, deteriorated badly in the United States. Rumblings of world war were unsettling rewards for the years of drought and bickering. The February 1939 Hoe annual report groaned:

> This year was one of the worst in the printing equipment industry, the net consolidated loss of the Company was $669,971. There was

a decrease of 74% in press sales and 35% in saw sales. Unfilled orders as of September 30, 1938 were only $560,100.[50]

All the verbiage about "heavy volume of replacement orders" was just that. Loss of the London shop—a European sales window—spelled the end of Hoe's world expansion. With a dismal domestic picture utterly beyond its control there seemed little to do but cut up the small pie the company found in its hands.

Disposition of the proceeds from the subsidiary sale became a looming issue. In March the company proposed to use 60 percent of the $2 million to retire bonds, and 40 percent for operating costs. In April Tillinghast became president, George G. Carnegie, Jr., R. M. Tierney, Arthur Dressel (executive vice president and general sales manager), and Arthur I. Hoe were elected directors.

Many years had now passed since the original directors had sold the company into receivership in 1932—enough time, in fact, to consider the six-year statute of limitations then in effect. The disgruntled stockholders, patient and persevering, finally got court attention in March 1939. They had charged (1) excessive legal fees, (2) settlement of certain stockholders' suits at exorbitant prices, (3) excessive payment for the purchase of the Bronx property—all of which were dismissed by the court for expiration of limitations. Two other charges, (4) unnecessary receivership, and (5) excessive interest payments on loans by Guaranty Trust, were not dismissed.

By 1940 the company had moved well into government orders, very largely reflected in profit and loss statements. Elections became more routine as press work yielded to military contracts—and profits, such as they would be, became more reliable. Government war work, never before so demanding for R. Hoe & Company, ironically had a truly calming effect on the company's daily affairs.

In July 1941 the New York Supreme Court made a final pronouncement on the nagging stockholder suit that was indeed so close to the heart and life of the company. Begun in 1936, the action had been expected to recover $3 million from Guaranty Trust Company; in its last phases it had taken 7 weeks and 6000 pages of testimony, 500 exhibits, and 800 pages of briefs. Justice Charles W. Froessel in Brooklyn opined on six counts: (1) no wrong-doing was involved in the receivership action; (2) the original protective committees did not constitute a conspiracy—they represented the bank's continuing responsibility to protect the company's financing; (3) unfair domination by Guaranty Trust was not proved; (4) a 28 January 1928 loan by Guaranty Trust was not unlawful; (5) the price paid for the Bronx property was not excessive; (6) interest payments were not unusual.

With a sinking feeling of doom the records of the company finally closed on one of the most heart-breaking, inexplicable losses in American business. Disappointed stockholders were crushed. Again Henry A. Wise Wood, like an old friend, had foreseen such defeats and tried to commiserate:

> No legal creation of the people [that is, a corporation] should be permitted to grow beyond the power of the people to control it: nevertheless, through his legislative grant, unwisely unaccompanied by checks adequate to safeguard his interests, the citizen, in opposing the aggressive corporation, seems to be helplessly confronting a Frankenstein. . . .[51]

As if to put aside the questionable victory, corporate attention was again turned to a moment of nostalgia, an element seldom present on the battered Hoe court agenda.

The Hoe shop at Port Morris, the Bronx, lay coincidentally only about a mile and a half south of Brightside, Colonel Hoe's residence (a curious building still standing today, although nothing else remains except a small thoroughfare called Hoe Avenue). Also nearby was St. Ann's Episcopal Church, where Richard M. Hoe had been quietly buried beside members of the distinguished family of Revolutionary War patriot Governeur Morris, who founded the church. Parish records revealed that Hoe, interred 55 years in 1941, had performed as vestryman for two decades; his greatness to the community and his service to the church had not in all that time been properly noticed.

On 26 October 1941 Colonel Richard M. Hoe was honored at St. Ann's by the installation of a commemorative bronze tablet. Invited to the ceremony were church and municipal officials, and to represent the Hoe family and the business firm bearing that proud name were Mrs. Robert D. Sterling and President Harold M. Tillinghast. The auspicious tablet was unveiled by Ogden Reid, president and publisher of the New York *Herald Tribune,* who judged the "great inventor, great humanitarian and great American . . . not only a genius as an inventor, but . . . a good business man." The tablet reads:

> In Memory of Colonel Richard March Hoe,
> Inventor of the Rotary Printing Press,
> Twenty years Vestryman of St. Ann's Church
> Born September 12, 1812—Died June 7, 1886—Buried in the Crypt of St. Ann's[52]

Perhaps those who had followed the company's history for the last fifteen years had seen the poignancy in Reid's recognition of a "good business man." No matter. Soon after Tillinghast returned to his office the whole world exploded.

JOSEPH L. AUER
1898–1963
Courtesy Mrs. Joseph L. Auer

14.

Paradoxically Unfit for Survival

WAR PRODUCTION REDEEMED / STOCKS AND BONDS, SUITS AND COURTS / AUER TRIES TO MAKE PRESSES / FINAL COLLAPSE / NEW BANKRUPTCY PROCEEDINGS / WOOD-HOE INDUSTRIES

In December 1941 America was suddenly at war. Although newspapers everywhere had covered the excruciating details of Axis conquest for two years, the public—and American industry—was far from ready for the horrible involvement to come. Pearl Harbor had caught the nation by surprise. With skilled personnel and vast tool and material resources, however, the machine industry to some extent was ready for impressment into war duty. Like other shops, R. Hoe & Company had already learned—in an earlier conflict—how to convert its communication capacity into sinews for fighting.

Prior to 1917 the nation had similarly watched the powers of Europe deploy for an insane war, quite expecting it would not last, and equally reluctant to participate in it. Most of all, the United States in 1914 was woefully unsure just where its strength on behalf of democracy lay. For Hoe the business of manufacturing sophisticated printing machines—then in their zenith, thanks to world conflict and tension—so that newspapers could pursue their propagandizing, seemed of sufficiently high priority to continue almost uninterruptedly. Indeed, through 1914-1916 Hoe's list of technological accomplishments was as full as ever. Upon America's entry into World War I, however, nothing less than engineering miracles were needed. Fortunately Hoe and other major firms could perform them.

Although both Army and Navy bureaus of ordnance had developed methods of acquainting civilian industry with the requirements of war, such plans were limited and very late. As the technology of warfare raced beyond the dreaded fears of civilian and military observers alike, most commercial shops were poorly prepared to make quickly whatever might be needed. In some cases fanatical secrecy prohibited civilian contractors from working on government orders; in others even experienced ordnance personnel had never before seen the equipment they were expected to fabricate. Time for design, tooling up, requisitioning (later called "lead" time) and testing were known factors, but only vaguely considered in overall plans.

World War I was clearly an artillery war. If nothing else, the Allies needed great quantities of guns of all calibers—although only France had a perfected system. The French 75 millimeter field gun, thanks to intense government propaganda, had acquired such a mystical reputation it was believed that no other country could manufacture anything like it. Yet America had no choice but to try: it adopted the French materiel, especially its highly complicated hydro-pneumatic recoil system, without opportunity to "Americanize" it either for rapid production in yankee shops, or for expert handling of the piece by yankees in the field.*

In 1917 R. Hoe & Company was awarded contracts to manufacture U.S. Naval three-inch anti-aircraft gun mounts. The firm, of course, was unqualifiedly capable of doing the job. The plant had become one of the finest manufactories in the world: "Their equipment," noted James Bridge in 1902, "is of the character that, until a few years ago, could be found only in the shops of some European governments." This veiled comparison to national arsenals was further masked by the supporting claim, "Hoe factories, in Grand Street, could build a locomotive or construct a watch as readily as they turn out a printing-press."[1] The Grand Street works, with batteries of heavy tools ("so completely automatic as to call for the merest pretense of supervision") themselves fabricated on the premises to permit adjustments to a thousandth of an inch, had undoubtedly long been on the government's list of potential suppliers—but not for instruments of peace.

Many of the company's wartime output were the Mark XI Model 2, and Mark XVI Model 1 gun mounts, massive foundations for large naval cannon. Perhaps the most complicated piece of equipment in both army and naval arsenals, the mount had to withstand the constant pounding and slamming of its gun in all conditions and climates, and still operate with clocklike precision. Hoe tackled its assignment carefully, within a year almost completely converting to war work. Upon declaration of the Armistice in November 1918, however, the company still had two large unfulfilled contracts, whose cancellation (like the Civil War rifling machines) became thorny legal issues.[2] Hoe was relieved of its assignment nevertheless and happy to return to press production.

* See *Age of Great Guns.*

Despite two years of war again raging in Europe (and the Orient), America was once more slow to tool up for combat. Not until well into 1940 did the government release a number of large "educational" contracts for materiel. For this work Hoe once again was included.

In July 1940 the War Department announced the award of $2 million in military contracts, of which $732,000 went to R. Hoe & Company for the manufacture, as before, of artillery recoil systems. German forces with large quantities of anti-aircraft guns were (in the absence of air targets) using their high-velocity fires against Allied tanks and ships with devastating effect; the British and French had almost nothing comparable. In America, as before, the problem was one of fast manufacture, even before employment could be approached.

Now even more acutely than in 1914, the recoil mechanism had become the most complicated and exacting assignment a machine shop could receive; historically it was still a novelty—a classified secret only two decades before. The hydro-pneumatic recoil system was designed to absorb the shock of firing, stop the rearward movement of the heavy barrel, and return it smoothly into firing position; this had to be done rapidly, repeatedly, without fail, and under all likely field (and sea) conditions. The mechanism was a combination of delicately honed pistons, hydraulic lines and valves capable of withstanding great pressure, and machined parts possessing almost no allowable tolerances. Most components, in fact, were so critically related to one another mass production was not feasible to make them. The company quietly studied the blueprints and bravely approached the job.

The firm tried to communicate its boggling assignment to old customers: "[A] mechanical yoke . . . lies near the heart of an anti-aircraft gun and helps to give that magnificent piece of American Ordnance its swift maneuverability and unfailing accuracy," said a 1944 advertisement. Hoe equated war work to what printers knew of graphic arts: "The machining of this yoke requires the same exacting skill as the manufacture of printing machinery. . . ."[3]

From 1941 on, of course, the company experienced mounting shortages, crippling losses of personnel to the armed forces, and irritating government directives and priorities. Press manufacture fell behind and unfulfilled orders doubled from the year before. President Harold Tillinghast announced to stockholders in April 1942 that, although the firm was already on a six-day, two-shift work week, only a limited number of spare press parts could be expected, and the completion of some large machines would have to be deferred for the duration. But the scrapping stockholders and haughty management had providently put their differences aside in the national interest, and with good results.

In September 1942 the company received the coveted Army-Navy "E" Production Award. At a ceremony at the Bronx plant General

Ralph K. Robertson unfurled the efficiency pennant for Tillinghast and Lieutenant John D. Lodge (USNR) presented blue lapel pins to employees. Among distinguished speakers was Mayor Fiorello Laguardia, who congratulated the factory's 1500 workers while at the same time warning, paternalistically, "any interruption of work at this time is not only a crime against the country but against humanity itself. . . ."[4] Employees did not need to be told: they would win the Army-Navy "E" four more times, and for his masterful part in orchestrating the immense shop Joseph L. Auer, vice president and general works manager, was later awarded a U.S. Army Ordnance citation for distinguished service.

The war had brought to Hoe new solidarity, new direction, new purpose. Stockholder squabbles and legal battles disappeared in flurries of meaningful and determined production. The proud old company seemed back on course doing what it always did best: working hard. To celebrate its happiness and rededication R. Hoe & Company sponsored a "Victory Rally" only remotely connected with America's war progress—after years of retreat, retrenching, and remarshaling, the Allied cause early in the fighting had never seemed so hopeless. Insiders, however, were aware the occasion was really a Hoe testimonial to its own future. Some of the company officers revealed this sentiment in their speeches:

> [Arthur Dressel, Vice President, R. Hoe & Company] . . . Our victory rally is dedicated to the workers of R. Hoe & Co., Inc. Building ordnance equipment for our nation's armed forces is a tradition of our company. There is an entry in an old Hoe Company record that tells how Hoe built, nearly 100 years ago, in 1847, a 3-inch cannon under the direction of a now unknown army ordnance officer. That was long before the Civil War. Later, when that epochal struggle broke out, our company built the equipment that was used to make the rifles for the Union Army, and before the end of the war Hoe was producing about 50 to 60 different kinds of machine tools for war purposes. So it was again in the Spanish American War, and the Hoe record of service to our country continued into World War No. 1, when the United States assumed the burden of stamping out autocratic militarism in Europe. In that war our company's main products were mounts for the Navy. Today Hoe is again serving our country to the utmost of its ability and capacity. . . .

The company was indeed to be complimented for its concerted efforts, at last. But it was also a good opportunity for all to participate, including the United States government both as customer and ultimate user:

> [Colonel J. K. Clement, Deputy Chief, New York Ordnance District] . . . I take no little pride in having recommended that R. Hoe & Co. receive its original educational order. You got that order and went to work and did such a good job that when the Ordnance De-

partment again had more money and could place a production order for recoil mechanisms, it naturally came to this company. That order you have finished and now you have finished another large order for a different recoil mechanism. As a result of this excellent work you have done you are going to be required to expand your facilities and to expand your production. You have large additional orders. You will receive more orders. Not only are you producing ordnance of a high order, without any single rejection, but you are designing for us. We have made contracts with your company to do engineering work and you are now engaged, as you know, in the design and improvement of various items of ordnance. . . .

[Colonel Rodney H. Smith, Anti-Aircraft Artillery Command, Eastern Defense Command] . . . The work you are doing is most vital and I am especially pleased that you are going to be able to see today a completed 90 mm. gun in operation; to note the important part your product plays in this grand weapon. I think this gun is one of the finest pieces of anti-aircraft ordnance in the world today. . . .

Like nineteenth-century toasts to Colonel Hoe or testimonials to his fine old printing presses, the firm's officers and industry guests retorted to the military with enthusiasm:

[Joseph L. Auer, Vice President, R. Hoe & Company] Now, what are *we* going to do for this? I am supposed to make a promise and I will, and I know that you will support it. Colonel Smith, I promise you and all anti-aircraft commanders, that our mechanisms will not fail you in the hour of need. Colonel Clement, I promise you the same whole-hearted cooperation to you and your office, and what perhaps is more important, to your men assigned in this plant. I promise that we, as and when directed by the government, will cheerfully adhere to their wishes, even though it means heavier burdens, and in general I promise that we will maintain delivery schedules even though they are tough, and that we will do everything within our power through the medium of management and labor committees to do a real job. . . .

[Walter M. Dear, President, American Newspaper Publishers Association] . . . As to Hoe's place with reference to the manufacture of newspaper presses in the future. It requires no particular gift or foresight to affirm that Hoe readily will resume its former place in the world of commerce. It cannot be otherwise or we must confess, which I certainly do not, that there is no place for a newspaper in the future. Like every other business, the newspaper business has felt the enforcement of priorities and the deprivations of new mechanical equipment. Astute management will be looking for the earliest opportunity for replacement and new installation, and that demand will be increased in proportion to the length of the duration. At that time quality and service, which have always been the standard of this organization will reap the benefits of the splendid record of

> the past, and if this government and nation will endure, as endure it must, the firm of Hoe and you faithful workers need only concentrate on the importance of the war work you are doing today and then it will be said of each of you—A man he seems of cheerful yesterdays and confident tomorrows.
>
> [Harold M. Tillinghast, President, R. Hoe & Company] . . . This victory rally is being held to celebrate and commemorate the successful completion, many months ahead of time, of the second large contract which the War Department entrusted to your Company. . . . I want to close by thanking you men in the shops—you men who have done the actual work—you men for whom this celebration is held. Without you and your energy and enthusiasm and your fine craftsmanship, all the other things I have mentioned would have been valueless, so I say to you "Well Done—keep up the good work on the even harder jobs ahead." . . .[5]

The company was soon making parts for the unique 75 millimeter cannon mounted in the B-25 "Billy" Mitchell bomber; although some pilots claimed the firing of this gun stopped the plane in midair, it was an engineering triumph. The company also made parts for American tanks, and for anti-aircraft sound tracking devices, search lights, patrol-torpedo boats, and pursuit aircraft as well. By 1944 the list included parts for the 155 millimeter howitzer, 90 millimeter anti-aircraft gun, and 5-inch naval gun.

In April 1943 as the novelty of its exacting assignments wore away, and, in fact, as changes in the course of the war altered materiel requirements, the firm underwent major reorganization. This apparently caused some uneasiness: representing 1000 Hoe workers, the International Machinists Association appealed for a 5 percent pay raise. The government's War Labor Board studied the situation to find that, whereas journeymen machinists in metropolitan New York were averaging $1 hourly, Hoe employees were already at $1.25. As their wages were considerably higher than the prevailing rate, the Board refused to grant the union demand.

At almost the same time the War Production Board announced that Hoe manufacture of recoil mechanisms had surpassed the Army's need for them, as the Axis threat began to diminish in intensity. The original Hoe order had amounted to $12 million in these components, the firm as early as November 1942 already turning out an impressive 500 recoil mechanisms per month for the U.S. 90 millimeter gun. Production of this war materiel had, for a company of superb machinists, proved no particular problem at all.

War Production Board "trouble shooters" visited the Bronx plant to examine its resources, much impressed by the firm's equipment and personnel. They redirected its war effort; recoil mechanisms were cut back to 240 per month, and new contracts were drawn, amounting to

$10 million in July 1943, for the manufacture of urgently needed airplane parts, combustion engines, and naval gun mounts. Still at full war capacity, Hoe was turning out these new items with a minimum of adjustment and at a billable rate of $2 million monthly. The company had acquitted itself with rare commendation after a 16-year period of agonizing corporate bickering and administrative nightmares.

At the end of the war the U.S. Army acknowledged the magnitude of Hoe's participation. General G. M. Wells, Office of Chief of Ordnance, wrote:

> [I]t is only natural that our thoughts would turn to those who have done so much to help bring this war to a successful conclusion.
>
> It is hard to believe that we once doubted the ability of anyone outside the Arsenals being able to produce satisfactory recoil mechanisms. Your company not only showed the way but you also proved that it could be done better and at lower cost. Many times you were called upon to meet seemingly impossible production schedules which you and your associates always managed to accomplish. . . .[6]

Still impressed with the engineering performance of the company, the government in 1948 organized an ordnance maintenance depot at R. Hoe & Company for the repair and reconditioning of materiel. This was the first arrangement of its kind. Four years later the Army returned with another order, even more prestigious. Atomic engineers had succeeded in designing a field artillery piece capable of firing a nuclear projectile—the 11-inch "atomic cannon."[7] Because of its long experience "in extreme precision on massive equipment," the company was chosen to make the weapon's new double-action recoil system, even more involved than earlier ones.

Mounted on a long carriage propelled by two automotive tractors, one on each end, the 43-foot tube was literally housed in its elaborate recoil mechanism. The enormous shock of recoil was absorbed in a primary hydraulic system which in turn relied on an arrangement of two pistons, both of which were returned to normal firing position by special hydro-pneumatic devices. Barrel and recoil mechanism in the firing cycle traveled back and forth on stainless steel guides, kept—despite a total weight of over 22 tons—in such delicate balance the gun could be completely pointed by hand, if necessary. Parts manufactured by Hoe weighed up to 1600 pounds, some of which needed a mirror finish accurate to 4 micro-inches (engineering measurements of which Whitworth may only have dreamed); a soft copper ring one foot in diameter had to be machined to an accuracy of one-half thousandth of an inch. Hoe's enlarged honing machinery was especially useful for the project.

Although the atomic cannon in many parts of the world served as a deterrent, its nuclear capability was never needed. The heavy weapon had a short life in the nation's arsenal, thanks to rapid development of guided missiles and other nuclear-capable systems. Its prompt and successful development was nevertheless a major step in ordnance engineering, and R. Hoe & Company had shown once more that it could do the job. Someone else had thought so too.

In 1948 the Industrial Relations Section of Princeton University, Princeton, New Jersey, described the results of an analysis in which it found R. Hoe & Company "conspicuously successful" in its employee-employer relations. The University admired "the conviction of the company, based on experience, that collective bargaining is superior to an open-shop for efficient, profitable and harmonious operation in its type of business." Two factors were involved:

> The realization by both sides that each has the latent power to injure seriously, if not destroy, the other, and the consequent determination to work out settlements by mutual discussion and compromise without resorting to the use of that power. [The second:] The maintenance of relatively high standards of wages and working conditions.[8]

The industrial group noted that Hoe employed 2000 persons, of whom 1200 were members of the International Association of Machinists, with the remainder divided among four other trade unions.

Indeed, no one could deny the company's relationship with workers was ideal—for the workers. But with management affairs so largely consumed by stockholder problems—indeed, with its officers so one-sidedly experienced at financial affairs—the worker status was a matter more of default than superior government. The company had always paid high wages, principally to get good workers, but also to forestall employee agitation; because of its unique product it could afford to do so. But living on the verge of real bankruptcy since 1932—a condition workers could not overlook—the company never gave the appearance of a fat profitmaker thriving at the expense of workers. Yet the Hoe unions nevertheless greedily wanted some of the spoils they had watched management take with impunity. The war years had unquestionably colored the harmony which the academic group found thriving. Its report served largely to ameliorate a tenuous union-management relationship as the country geared up hungrily for post-war competition. A multi-million dollar corporation that collapses only 20 years later cannot truly be called "conspicuously successful"!

Throughout the war years R. Hoe & Company had taken advantage of a relatively quiet managerial situation to examine its financial structure. Frequent large loans had been expensive and threatening to corporate independence; the two classes of stock ("A" and common) were considered complicated and burdensome. At the 1943 annual meeting stockholders approved a plan to issue new 5 percent bonds (entirely subscribed by the Metropolitan Life Insurance Company, New York), the proceeds used to retire 1,405,000 old outstanding 6½ percent bonds. In September the company agreed to pay dividends in arrears on "A" stock—amounting to $79.83 per share—dividends going back to 1937. In 1944 the election of directors was reorganized into three staggered slates so that three-year terms could provide some overlapping continuity. Tillinghast occasionally whetted security holders' appetites with talk of exciting post-war prospects in the printing press industry, and at last the apprehensive moment was at hand.

Stockholders had indeed desisted in their hostility and opposition during the war. But in March 1946 Joseph Auer, as the firm's new president, was obliged to announce that dissident elements, lead by three new directors, Thornton C. McCune, Harold W. Danser, and Morton A. Jenks, had already begun sniping. Auer announced:

> We regret to advise . . . that a stockholders committee has been formed to solicit proxies against the management, the third such committee formed in the last decade. . . . None of these committees, in the opinion of the present management, has ever contributed anything of value to the company, but each of them has been generous in its abuse of the management. The present committee, we understand, is composed exclusively of stock brokers and their associates.
>
> Based upon our record of performance as outlined in the company's annual report . . . we ask your support and think we are entitled to it. In our opinion, nothing could be more detrimental to the welfare of this company during the crucial months just ahead, than to turn over the management to an untried group of stock brokers who have had no experience in the business of manufacturing printing presses. No one of them has been connected with the company before.
>
> Only as late as March 5 this group delivered an ultimatum to the management demanding control of the board of directors at this time as consideration for withdrawing the threatened contest. The management declined this demand.[9]

Joseph L. Auer (1898-1963) was a fighter and persistent worker for the company of whom the Hoes would have been proud. Born in the Bronx and educated (at night) at Columbia and New York Universities, he was apprenticed at the old De La Vergne plant whose property the company had bought. By hard work he learned the machine industry; having joined Hoe in 1937 as works manager he had been vice president and general works manager very effectively during the war. Auer skillfully carried the company through its meticulous government work, serving at the same time on two important industry production committees. He was also a director of the Bronx County Trust Company, member of the Bronx Chamber of Commerce, the American Society of Metals, the American Society of Tool Engineers, and the American Ordnance Association. A well known Roman Catholic layman, Auer was made a knight of Malta by Pope Pius XII, and an executive member of several religious, interfaith, community service (Salvation Army), and educational institutions. "Joe" Auer was not a genius, perhaps, but a very capable and likable guy.

Jenks, McCune, and Danser were elected to the board of directors nevertheless. Arthur I. Hoe, like a spectre from the past, was on the ballot but not elected. The company was able to announce in June that a final bank loan of $750,000 was fully repaid, leaving the firm no debts other than current obligations. The company seemed at last ready to tackle the promising post-war world of affluence and abundance. Some early signs were very hopeful.

Well in advance, the Hoe company had formulated elaborate plans for reorganizing press fabrication upon termination of the war. "A year ago," said Auer, "this new set-up was merely a picture on paper. Right up to V-J Day we were working up to capacity on war orders. Then we shut down for two weeks and made the changes."[10]

Scores of old machines were scrapped and more than 70 new ones —valued at a million and a half dollars—were purchased, mostly surplus from the United States government. New, large rooms were assigned for offset and metal decorating presses, and newspaper, rotogravure, and magazine presses. The Dunellen plant was expanded as a machine shop and foundry, and virtually all heavy machining would be done there.

The first post-war press equipment, according to Arthur Dressel, vice president and general sales manager, was shipped to the Washington, D.C., *News* in October 1945, and manufacturing was being resumed on several machines on which work had stopped before the war. Dressel also announced that production had begun on newer post-war models, featuring streamlined design and convertible features making color printing easy if wanted at a later date.

In June 1946 the directors had studied a grandiose recapitalization plan to eliminate remaining dividends in arrears on class "A" stock (now amounting to $82 per share) and convert all "A" stock to common stock. By voting time (September 1946), however, the directors opted to retain the cumbersome "A" and common stock arrangement (96,000 "A" shares, 160,000 common outstanding), although each old "A" was to be converted into a new "A" and four shares of a new class "B" ($20 par value). Nevertheless, the "A" arrears would indeed be paid: by January 1947 Auer was able to announce that, like the eradication of outstanding loans, the payment of all stockholder obligations, dating to 1924, had also been cleaned up. Involved at the same time with these stock matters was the merger with Pittsburgh Litho Press Corporation, which became a wholly owned subsidiary.

R. Hoe & Company in June 1946 announced the receipt of a $4 million printing press order from *American Weekly* (Hearst Consolidated Publications, Inc.), apparently the largest printing press order in the company's long history. Almost a year was needed for joint planning by the publisher's mechanical staff and Hoe engineers. Nine color presses, each with 24 printing units (each unit prints one side of a sheet in one color), 4 automatic non-stop paper roll changing and tensioning devices, and 2 folding machines—totaling 270 major elements, "embody the most modern and advanced engineering principles for high speed quality letterpress printing,"[11] said the company. The plant jumped to the task with the care and attention it had demonstrated on government work. The order paved the way to a golden future.

In December 1947 Auer in an interview hoped to offer the printing industry some cheer for the holidays. Printing press output was still hampered by material shortages, especially pig iron, but consumer conditions indicated big changes in the wind: need for increased circulation to combat rising labor and materials costs, competition from other media, especially television, and pressure from advertisers to offer color, all meant new machines for newspapers in most cases. Many newspaper shops were old and inefficient, and orders from them were already coming in.

Statistics the company had begun to assemble showed that 20 percent of the nation's dailies already had facilities for ROP (run-of-press) color, and 95 percent of Hoe's new orders were for color presses or units that could be converted to color. Quality letterpress color printing spelled a big future for the firm. Color was also making strides in the printing of magazines. Auer said:

> The influx of new publications and the demand for increases in circulation of those already established are putting a serious strain on press equipment already in publishers' printing plants. Furthermore, the relatively low volume of new press equipment installations during the war, and for several years previous to the war, has resulted in depreciation of much machinery. Finally, publishers are always alert for faster, more efficient and more versatile equipment to meet rising labor and material costs and the demand for more and more color printing.[12]

Also promising was the tremendous increase of color in metal printing—for food containers, toys, barrels and drums, kitchenware. Auer finished by announcing that the company in 1947 recorded its highest net profits on record and highest sales for any peacetime year. He was most sanguine for bright prospects for the future.

With the annual meeting of 1950 Chairman of the Board Tillinghast, after 40 years with the company, in poor health, stepped down in favor of Neil P. Cullom, Hoe's general counsel, who had been on the board of directors for 13 years. Although Auer continued as chief executive, Cullom soon visualized himself in complete control of the company. First, however, he faced another nagging stockholder action.

R. Hoe & Company's plans in July 1950 to merge its wholly owned subsidiary, Hoe Export-Import Corporation (a calculated move in connection with the elimination of the company's "B" stock), were met with understandable general surprise. Few were aware the subsidiary existed. It was, in fact, created only three days before the announcement of its demise, described as having a capitalization of 200 shares of no par stock. One stockholder saw the manipulations of the 1930's all over again.

Nathan Katz, a New Yorker holding 500 shares of Hoe stock, in August brought suit to prevent the merger. He charged the company and its officers with a "bold, unabashed and fraudulent plan"[13] to dilute the sound stock of the company by absorbing the worthless shares of a "dummy corporation"; as a stockholder he found himself faced with a fait accompli on which he was not invited to vote. The court agreed to examine the case on 25 September.

The company nevertheless went ahead with a meeting it had scheduled for 12 September, at which time the merger was approved, 73,398 shares for it versus 9226 against. The plan had to be held in abeyance, of course, pending the outcome of the court action. Appearing before Justice Dennis O'Leary in New York Supreme Court, Cullom explained that the program was "conceived in an honest and sincere effort to reconcile the conflicting interests of the various classes of securities."[14] Secretary of the company Charles E. Littleton described the awkward stock situation as beginning in 1935 when, during difficult conditions, it was generally agreed that holders of Class A stock should control the company although dividends could not be paid.

Katz and Alfred Martin, a Hoe director, also appeared, the latter insisting the scheme was "a sham and a blind." Auer testified dutifully that "each and every allegation of fraud, misconduct or wrongdoing" was unfounded; he recommended the merger strongly. After five days of hearings the case was dismissed. At a U.S. District Court, to which Katz appealed, Judge John F. X. McGohey concurred: "A less worthy appeal to the court's equity power is impossible to imagine,"[15] he said. Indeed, the courts had never believed that the great old company could possibly have been in danger, least of all at the hands of its own apparently competent officers. Management went ahead at last with its merger intentions. (A year later a New York State Appeals Court denied any further action.)

In December 1950 the company filed a certificate with the New York State Department of State expressing its intention of exchanging four shares of new Class A and two shares of new Common Stock for each share of old "B" stock, and, finally, one new Common for each old Common Stock. This apparently innocent program proceeded without difficulty.

Suddenly in May 1951 the United States government filed a civil antitrust suit against R. Hoe & Company for restraint of trade practices in connection with its sale of the London branch in 1938—thirteen years before. Among the improper agreements included in the purchase, the suit alleged, were pacts with R. W. Crabtree & Sons, Ltd., Leeds, that the two principals would not sell printing presses in each other's territories. The government sought to cancel such agreements and enjoin the defendant from any similar agreements in the future. The company was quick to deny the allegations.

The problem, according to Cullom, was that Crabtree, the buyer, "covenanted not to sell printing press equipment under the name of Hoe in North and South America, territories previously served by the vendor, and correlatively, the seller agreed not to sell such equipment under that name in the United Kingdom, territory previously served by its subsidiary." Cullom continued:

> Such a covenant was quite obviously necessary to protect the buyer's enjoyment of the good will of the Hoe name, which constituted an important element of the purchase price. It was equally important to prevent intolerable confusion on the part of the purchasing public here and abroad.[16]

Since in the original agreement there was no attempt at price fixing, or division of world markets, no exclusive patent licensing arrangements nor prohibition against competition under any other name, Cullom denied that the contract had created a cartel. He was only about half correct.

Cullom did not care to refer to a more recent agreement with Crabtree which may actually have triggered the government's inquiry: there had been a 1945 "close working agreement [to] exchange drawings and designs, as well as full information on all the machinery which they produce for the graphic arts industry," and, in addition, to grant manufacturing rights on each others' products. "It will, therefore," said an R. Hoe & Company news release at the time, "be possible for any purchaser throughout the world to secure any of the many products manufactured by these companies."[17] That certainly sounded very much like patent licensing and world market division. Four years later, in 1955, the United States Attorney General ordered Hoe to withdraw from its "cartel" arrangement, a decision President Auer claimed would not affect the company's domestic sales and manufacturing policies at all. Hoe's foreign or world potential was nevertheless ended as decisively as its British influence had ended in 1938.

By 1952, meanwhile, shareholders had begun to experience the staggering depreciation Nathan Katz had heroically tried to prevent. As the company's stock exchange plan unfolded, holders found that for about \$23 worth of old stocks they were getting only about \$7.50

of the new. One disgruntled group, still rankling over the Export-Import merger, controlled over 66,000 shares (now sold only over the counter at 6½ bid, 7⅜ asked, on 3 March 1952) stood to lose two-thirds of its investment. It now brought suit to force the company to honor the earlier "investment value" of the securities. Supreme Court Justice Carroll G. Walter simply advised the plaintiffs to recognize that the market value of stocks was their most realistic worth:

> Whether they be called market value or investment value is immaterial. They demonstrate that petitioners' expert's appraisal of $23.30 a share, despite the elaborate circumlocution with which he states it and the appearance of a scientific approach with which he has been able to color it, is nothing but an expression of his opinion that investors ought to have thought something different from what the evidence shows they actually thought.[18]

Again, Hoe stockholders left court with something less than victory. They had good reason to curse their misfortune in always bringing the company to court on flimsy charges, despite the undeniable evidence that, since 1932, they were being almost systematically extorted of millions of dollars. When fate seemed gloomiest, there was some relief.

The following year the New York Supreme Court Appellate Division did in fact decide in favor of the dissenters. The 9226 common stockholders (constituting 42 percent of the vote) who had objected on 11 September 1950 to the Export-Import consolidation were awarded a price of $10 per share instead of the $7.50 market value. It was slight consolation.

At a July 1953 board meeting a new and debilitating struggle erupted. Joseph Auer, who had been president 7 of his sixteen years with the company, was removed by a vote of 7-4, although he kept his place on the board. Appointed acting president was Arthur Dressel; he had been 28 years with the company. Considering Auer's excellent record and reputation, however, many felt the act unwarranted.

Although the Philadelphia *Bulletin* in 1953 auspiciously—appropriately—celebrated its 106th birthday with new Hoe presses for its new headquarters, Auer had at the same time been obliged to note that acute shortages of strategic materials—because of the Korean War—were plaguing production. Net sales had dropped $2 million as increased government machine tool and ordnance work caused some interruption of press sales programs. These were unfortunate but not extraordinary business conditions.

Neil Cullom, Chairman of the Board, later claimed that Auer's dismissal was not personal, yet perversely citing, among other things, "inefficiency, waste, extravagance, and an utter failure to comprehend the requirements of an executive in these circumstances." Two factions, Auer's and Cullom's, began to face each other squarely.

Auer would not take his dismissal gracefully; he was determined to resume his post. *Business Week* had great admiration for his pluck:

> He knew there was only one way: get control of the 11-man board. Five places on the board (Auer's and Cullom's not included) were due to be filled in the annual meeting this April. Auer figured he had to win all five spots to get control. . . .[19]

An "R. Hoe & Co., Inc. Stockholders Protective Committee" was formed to restore Auer. In September 1953 it filed a proxy statement with the Securities and Exchange Commission seeking to remove four codirectors and reinstate their man. The committee accused Cullom, Harry K. Barr, William L. Canaday, and Edwin L. Munzert of "conduct inimical to the best interests of the company," calling into action a New York State law permitting stockholders themselves to take dismissal action. They added that the four had improperly offered Auer $50,000 to resign the previous July. The pro-Auer committee, owning or representing more than 20,000 "A" shares, included G. Jarvis Geer, Russell Hartranft, Jr., T. Percy Rathbone, Howard E. Rein, and Richard F. V. Stanton. Littleton had resigned as secretary of the company in protest over Cullom's action, and was now secretary for the opposition stockholders' group. The committee called for a special meeting.

Quickly eight defensive board directors mailed a letter urging all stockholders to resist the committee's tactics. Although legal action had to be employed to order the special meeting to convene, the company made no particular arrangements for it; when the court ordered another date set (19 January 1954), several important officers of the company announced they would resign if Auer were reinstated. Still avoiding a special stockholders' meeting, the incumbents in February elected James D. Mooney president, but all factions were looking forward to the April annual meeting for the final confrontation. Persistent court action to force a special meeting finally got another date —13 May—well after the annual meeting instead of before it. As in years before, when crusading individuals like Samuel Zirn dared to tackle Guaranty Trust's powerful management, there was considerable maneuvering preparatory to the showdown. And this time there was more.

On 25 January 1954 1300 union members of District 15, International Association of Machinists, AFL, struck the Bronx plant, and 100 more did the same at Dunellen. They sought wage hikes and other increases. This strike lasted six weeks, and after intercession by the Federal Mediation and Conciliation Service, finally ended with a substantial 85 cents per hour raise, a third vacation week after 15 years of employment, and other benefits. Tension relaxed, but the impending proxy fight, said *Business Week,* still "had all the ingredients of a heavyweight match."

The "protective" committee announced in March that it was nominating, for the vacancies, a full slate of five candidates, all pledged to reinstate Auer. They were Rein, Stanton, John Kadel, Mitchell May, Jr., and Erastus H. Munson. The committee took no chances: it hired a public relations firm, Berens Organization, Inc., New York, to find —and bombard with literature—uncommitted stockholders whose proxies would be essential to victory. A *Business Week* reporter paid a visit:

> Berens became like a regimental command post during a big drive. . . . A climax was reached in late March when Berens sent out a one-page "newspaper," entirely devoted to the Cullom-Auer fight. As might be expected, the newspaper allowed itself considerable editorializing. The headlines on the two lead stories were: "Cullom Has Received More Than Half Million Dollars In Legal Fees From Hoe," and "J. L. Auer Dismissed As President By Split Board Despite His Great Record."

The Hoes would have found such newspaper tactics and liberties much in keeping with earlier times, but how poignant it was to find such techniques now at work within the company!

Cullom and his associates, for their part, engaged a management engineering firm, Cresap, McCormick & Paget, to make a detailed study of the great ailing corporation, now more than ever before looking like a blind old bull plagued by hungry jackalls. The consultants had indeed found many things wrong with the company, and with them, with Auer. Any observer following the firm's trials and tribulations since 1924 would also have seen many things—and long before Auer arrived. Such a study was more insulting than instructive.

At the long awaited annual meeting the "protective" committee won handsomely, and without resorting to all-night voting sessions. Five pro-Auer candidates won places on the board of directors and promptly voted as intended. Auer was elected president and acting chairman of the board, and Littleton was reinstated as secretary along with him. Far from finished, Auer's hardest work at that point would just begin. *Business Week* commiserated with him:

> His immediate task, however, is to hold the stockholders' confidence. This may be difficult right from the start. Sales this year are expected to drop around 25%, to around \$14-million or \$15-million, according to Auer. He believes that net income after taxes will still pass \$1-million; it was over \$1.1-million last year. The sales drop will be blamed on two things: a six-week strike, which closed down the entire company last winter, and the termination of defense contracts to build machine tools. . . .[20]

But first there was unfinished executive business. Cullom had refused to resign from the board of directors. He planned to challenge the proceedings of the special May meeting, and resume control. The fight was not yet over.

Some corporate matters of a constructive nature were successful. In 1955 the company bought a promising asset, the Carolina Saw and Tool Company, High Point, North Carolina. Overshadowed by its printing press problems was the comforting fact that, as the country's oldest manufacturer of saws and files, Hoe had managed to remain very successful in that area. The Carolina plant, making carbide tip saws, would be designated "Carbide Saw and Tool Division," with Clarence Goodman, its head, to remain as general manager. (By the end of the year, however, William A. Harloe, a Hoe vice president and director, slipped into the general manager's post.)

A year later the company also bought the Jones and Orth Cutter Head Company, Seattle, Washington, makers of cutting tools for making moldings and grooves. This gave R. Hoe & Company very broad coverage in the cutting tool business. But for 1955 Auer was forced to accede that income was down substantially, due largely to terminating defense contracts and decline in printing machinery orders. The attorney general's cartel decision too had been a psychological blow, if not a real one.

In 1954 Hoe had announced a new press for small and medium papers. This was "Printmaster," capable of 40,000 per hour, and easily converted to color printing. The machine had new inking apparatus, lifetime ball-bearing seals, forced feed lubrication, and a folder which could be changed while running. The following year Hoe finished its work as one of 70 subcontractors on the U.S. Navy Martin P6M Sea Master, a $2 million project. Two prototypes of the new aircraft had crashed, and a third was urgently needed to complete the flight test program. (Hoe also had a large order for Chance-Vought F8U aircraft parts.) The company added 40 aircraft tool designers to its staff who translated aircraft terminology to language familiar to Hoe machinists. The skilled Hoe men worked with 3500 aircraft tools and jigs to make the parts routinely, as they would components for a printing press. Hoe always "had to live with precision work,"[21] said a happy Martin spokesman, who recommended the subcontracting plan for other capable firms not familiar with aircraft work.

During his debilitating proxy fight Auer had observed that such diversification might be advisable for the company, well beyond the saw industry. Hoe was probably one of the few companies of its size and composition to rely so heavily on custom-made machinery of such limited flexibility. Printing presses have little application outside the printing industry; worse still, the biggest newspaper presses have no value whatever outside the diminishing realm of metropolitan journalism. Although World War II had ended fully ten years before, material shortages, personnel readjustments, international tensions (including the Korean War), and other world-wide problems were making it difficult for all firms to appraise public trends and tastes, restart their production lines, and field effective sales forces. During this time television and other mediums that would develop in competition with the newspaper were suffering as much as the world of journalism.

One of Hoe's own developments in tool-making was a device to reduce time-consuming and costly handwork: in 1957 the company announced development of the "tracer lathe," a machine capable of "bridging the gap between mass production and laborious hand machinery."[22] With a template for a particular piece of work the machine needed only the push of a start button to turn out 20-30 units in a single operation. The machine automatically followed the template and duplicated printing press parts with great efficiency.

Another innovation for Hoe, traditionally concerned only with its own products, was an agreement to function as American agent for a foreign pressmaker. In 1956 Hoe announced it would handle machines of the West German firm, Maschinenfabrik-Augsburg (MAN), makers of advanced stereotyping equipment and web-offset and gravure presses. MAN dated to 1840 as a printing press factory. Hoe's first sale in the United States was an automatic plate caster for Tenafly Printing Corporation, Tenafly, New Jersey.

As world conditions settled, newspaper orders increased, to be sure. Some were very big: a $2 million order from four Brazilian and one Venezuelan paper in 1958 involved 20 Hoe color-convertible units and related equipment. Similar were $2 million in 1960 orders from two Swedish newspapers, handled by Ampress, S.A., a wholly owned Hoe subsidiary in Europe.

But various sales promotion schemes and products were only half hearted (or half baked) attempts to call attention to a company resting on its laurels and unsure of future direction in which it had little control. R. Hoe & Company two years after the war should have been well in command of its fate; like other companies it saw rough depression years, it had war years which paid the bills handsomely even though they left the firm without salable peace-time products, and (some companies did not even have this) it enjoyed a labor force of skilled mechanics and machinists. If only it had a top management geared for peace, cognizant of future needs in the graphic arts industry—and relieved of harassing squabbles and pressures—it might be on top.

Coming out of its workshop in 1958 was a Hoe development in which the company had truly invested its future. This was the Hoe "Colormatic" press—"Year of a giant stride," said advertisements with grand style. "The planned simplicity of Colormatic design permits unparalleled ease in the use of ROP color coupled with maximum speeds up to 70,000 pph. . . . Progress in Printing—a continuing responsibility of R. Hoe & Co., Inc." To ward off any impressions that Hoe was worried by competition or pinching pennies, a touch of lavishness was added in 1965:

> See the Hoe Colormatic. In style. A chauffered limosine will pick you up anywhere in the New York Metropolitan area . . . hotel, airport, railroad station, dockside, heliport. . . . And bring you to 910 East 138th Street. . . . The car's waiting. Give us a call. . . .[23]

A potential customer was pictured in a sleek Rolls Royce and later shown inspecting the plant.

To call steady attention to the lucrative commitment for color in newspapers, the company began its "Hoe Report," a compilation by Media Research, Inc., New York, of the use of color by newspaper advertisers. It was based on space measurements at 411 papers in 132 cities. At first, as rates of color usage climbed, it seemed a useful sales tool, and figures were faithfully reported monthly from 1958 to 1961. About this time it became apparent that other factors too were influencing newspaper production. Run of press color was still—for a long time to come—a future desideratum.

Appealing to the April 1961 meeting of the American Newspaper Publishers Association, Joseph L. Auer said:

> The peoples of the world have come to look to the newspaper as the necessary medium for information on the progress of life in the universe, in the world, in their own country, in their city or town, even in their own neighborhood.
>
> But the newspaper offers more than information. It entertains, it educates, it relaxes, it interests the readers in new horizons, in art, music, books, sports, any number of new and fascinating things that the reader may not come into contact with during his day-to-day life.
>
> Above all, though, the thoughtful man agrees with George Bernard Shaw when he wrote, "Let those who may complain that it is all on paper remember that only on paper has humanity yet achieved glory, beauty, truth, knowledge, virtue and abiding love."
>
> The newspaper of today is the result of centuries of achievement in reportage, in analysis, in literature, and art—and in printing techniques.[24]

No one would have quarreled about the primacy of the newspaper in American life, but there were a few who sought new ways to put one together. Like *Grit* in 1962, some were contemplating switches from letterpress to offset, and others were studying computerized "cold" composition. The Hoe management did not seem enthusiastic about such technological novelties.

R. Hoe & Company, Inc. slumbered quietly through the 1960's, almost thankful for a viable existence. But as before it was a sleeping volcano. Between 1963 and 1969 the firm had four presidents. In 1963 R. Palmer Hollister (son of a Hoe machinist) succeeded Auer, who had died; in 1966 Thomas F. Hanley had replaced Hollister.

In January 1969 George F. Scanlon resigned from the Hoe board; Arthur Gordon was elected president and chief operating officer, and W. R. McKillip, general manager of the saw division, was made a director. These developments were independent of the annual meeting, which had been postponed until 27 February. At the unexpectedly stormy meeting, ending in "wild confusion,"[25] management quickly adjourned and disappeared, but 100 unhappy stockholders refused to leave the premises. They held a "rump" session to form a 15-member grievance committee headed by Fred Werner, a New York lawyer, and several of them independently filed suit against the company and certain officers and directors for financial misstatement.

On 23 January the firm had announced earnings, for the year ending September 1968, to be mysteriously "somewhat less" than the $2.9 million reported for the first nine months of the period. Just a day before the February meeting it revealed a further drop in income, citing, among other things, "the inability to allocate certain inventory production to specific orders due to the failure of customers to make final decisions as to whether to place certain orders which had been anticipated."[26] (Where were the old-time Hoe salesmen, homely enough in grammar but who knew how to camp on customers' thresholds and throw their rumpled—but victorious—hats on customers' desks?) Waiting for customers' decisions indeed!

Furious were officers of no less than seventeen mutual funds who had bought one-third of Hoe's outstanding 2.4 million shares. They had read Hoe financial reports correctly enough, but in their haste to make extraordinary profits for their clients they banked on legendary company deliveries that had not really materialized since the nineteenth century. There was renewed shuffling of management and hasty preparation of legal briefs.

In April Burton C. Polglase, vice president for press engineering, was elected chairman of the board, president, and chief executive, replacing Hanley as chairman and chief executive, and Gordon as president, both of whom resigned. Hanley and Gordon, both accountants, had quietly, a year before, disposed—through a privilege they couldn't ignore—several thousand shares of Hoe stock in their possession. (Two more stockholders promptly brought suit against the company for misstatements in recent financial reports and improper insider trading.)

Hoe, on 4 July 1969 for the second time in four decades was obliged to file a bankruptcy petition in Federal Court, Manhattan. It reported assets on 30 April 1969 of $50.4 million and liabilities of $29.4 million; it had about 3700 holders of common stock (1,937,412 shares) and 1400 holders of the Class "A" (448,704 shares). By filing under Chapter X, in which a court-appointed trustee takes title to the property and makes all decisions (in contradistinction to Chapter XI where the debtor himself remains in control), Hoe officers were admitting they could not solve the company's imminent problem—insufficient working capital—nor its other ills, either. Directors were understandably not available for comment; the Securities and Exchange Commission quickly suspended trading (stock closed after the announcement at 13¾; 1969 high had been 57⅜).

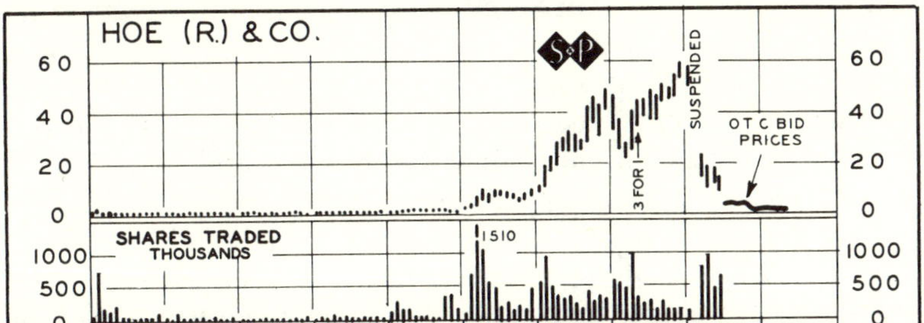

	1961	1962	1963	1964	1965	1966	1967	1968	1969	1970	
*Sales	19.47	20.81	16.86	20.63	24.83	31.93	46.12	48.27	----	----	Yrs.
*Net Inc.	0.41	0.44	d0.31	0.61	1.02	1.31	2.40	2.08	----	----	Ended
¹Per Sh.	d0.03	d0.007	d0.53	0.11	0.40	0.59	1.25	0.88	----	----	Sept. 30
¹Divs.	Nil	Nil	Nil	Nil	Nil	Nil	Nil	Nil	Nil	----	Cal.
¹Range	2⅜-¾	1¼-½	⅞-⅜	1⅛-½	2⅝-1	10⅜-2⅛	49⅜-9¼	59⅝-21⅞	57⅜-¼	3½-¼	Yrs.

*Million $. ¹On com. shs.; adj. for 3-for-1 split in May 1968. ²In arrears, see Dividend Data. d Deficit.
†O-T-C bid quotations.

Shares traded, R. Hoe & Company, 9 July 1970
Courtesy Standard & Poor's Corp

The firm received hearings beginning 8 July 1969, at which Justice Sylvester J. Ryan presided with considerable sympathy. As if to signify his awareness that the beleaguered company, so often in court, need not be treated with the law's usual haughtiness, he made an "informal town meeting" of the sessions. Coming down from the bench, Judge Ryan seated himself and the principal lawyers at the court reporters' table where he got right down to matters. While he readily approved the Chapter X petition, he was especially concerned to continue the life of the 150-year-old firm and its "skilled craftsmen." He ordered a "freeze" on all Hoe bank accounts so that creditors could not get at them. "I am concerned with saving this company and 2,000 jobs and I'm not interested in seeing that Talcott [principal creditor] gets paid,"[27] said the judge.

The Court noted the recent succession of corporate officials, but expressed no criticism of the incumbents under engineer Polglase, all of whom had volunteered to accept a 25 percent reduction in salary during the company's distress. The president's salary had been $60,000 and other officers' had been $25,000, which Judge Ryan considered modest enough for a $42 million business. Justice Ryan was shortly disappointed to receive "most unsatisfactory" reports of company operations and no records whatsoever of its sales and expenditures during the critical months of May and June.

The judge learned that Hoe had about $493,000 in bank accounts (largely assigned to James Talcott, Inc., financiers who had advanced Hoe $12 million since April at 13 percent). The company was in arrears with the United States government for about $250,000 in payroll withholding taxes, overdue for about $2 million in employee vacation credits, and behind in its pension fund account payments. The Bronx property and factory, surprisingly unencumbered from mortgages, were worth about $4 million. A union officer told the judge that a labor contract, with about two years remaining, could be considered another company asset.

Attorneys for several Hoe printing and publishing customers were most anxious about equipment they had ordered and for which their firms were seriously handicapped. Some of the presses involved were at that moment finished and sitting on Hoe shipping platforms, ready for delivery. Judge Ryan agreed these machines should be released. A spokesman for Triangle Publications, Philadelphia, offered to advance some capital immediately if the court could guarantee such delivery, and other publishers also offered similar plans. A representative from Merkle Press, Inc., Washington, D.C., who expressed his company's interest in purchasing R. Hoe & Company, suggested that mortgages on the Bronx real estate might be suitable guarantees. At this point Judge Ryan "passed the hat like a minister" and collected enough pledges of immediate cash—approximately $2 million—from customers to insure the company's staying in business. He then assigned a trustee, John J. Galgay, former chief of the United States' New York antitrust division, to reorganize the firm.

The Court also urged a thorough investigation of recent Hoe financial activities to determine if there were any criminal wrongdoings that led to the puzzling collapse of a company worth millions of dollars on the books. Justice Ryan had already alerted the Securities and Exchange Commission, whose attorney was among the 100 interested parties at the hearings. There were plenty of curious trading occurrences to consider.

From its deficit in 1963 (with stock selling as low as 41 cents) Hoe shot up to a $2.4 million profit in 1967; in the same five-year period sales went from $16.9 to $46.1 million, and there were expectations of the same performance in 1968. Hoe's announcement in 1967 of multi-million dollar press orders from several major communications firms had sent its $1 and $2 stocks spiraling upward; by December 1968 over 40 percent of the revitalized stock was held by mutual funds, most of them buying greedily at the innocent recommendation of Blair & Company, New York stock brokers. The erratic fluctuations ($60 per share after a three-way split early in 1968, dropping to $9 and rebounding to $14 just weeks before bankruptcy was filed) however, had prompted the American Stock Exchange in January 1969 to suspend trading.

The company had been delinquent with its 1968 figures, and when they appeared they described extraordinary losses: earnings were under $2.1 million—$800,000 less than the company had already announced in its third-quarter report. Investors who expected yearend sales of $60 million (they were actually only $48.3 million) were stunned. Again, when trading was stopped as the Hoe stock hit $9, it was the biggest loser on the stock market for the first half of 1969. These were not hallmarks of a stable, successful business.

Judge Ryan called all Hoe customers to a meeting to consider issuing formal orders requiring them to pay some advances to alleviate Hoe's cash situation. Those whose presses would be completed by 8 September—about a month hence—would pay immediately the amounts due, plus 10 percent additional on the contract price (the older contracts lacked provision for inflation incidental to the extended production time involved) when the press was delivered. Among the firms were Orlando *Sentinel Star,* Charlotte *News & Observer, Fresno Bee,* Newark *Star-Ledger,* and *San Antonio Light.* On a second list were 64 firms (35 of them newspapers) expecting delivery after 8 September who would be permitted to renegotiate their payment plans. The Hoe clients agreed to the scheme and all returned to work with some confidence.

Speculation grew, nevertheless, over the causes of Hoe's 1968 troubles, especially as there seemed to be no reason for them. Management blamed its independent accountants, Lybrand, Ross Brothers & Montgomery, who in the final quarter rashly charged off $1.2 million development and improvement costs the company had wanted to amortize over five years; it also set aside a $700,000 reserve to cover unanticipated increases in manufacturing costs. The company, however, also admitted sales in the last quarter had been "substantially reduced due to heavy production of inventory parts not allocated to any particular order, which were produced principally in anticipation of new orders under negotiation." (Another stockholder suit was accusing the company of treating these "anticipated" orders as if they were actually sales.) *Business Week* was not impressed:

> While business conditions may explain Hoe's troubled marketing picture, they don't explain the sharp difference between nine-month and yearend reports. Clarification of that tangle depends on the stockholder suits, the trustee's investigation to see whether any conduct may constitute a criminal act, and the Securities & Exchange Commission's inquiry into Hoe's fiscal 1968 results.[28]

Another mitigating factor in the sales picture was the increasingly hostile competitive situation. While Hoe's letterpress machines were unquestionably excellent, other manufacturers' were equally good. Hoe's lagging position into the offset lithographic market gave it no opportunity to be flexible. "Under happier business conditions, Hoe might have muddled through," said an analyst, "but . . . it was no time for a financially over-expanded company like Hoe."

For the six months ended 31 March 1969 the situation worsened. The company lost $1.2 million (against $1.8 million profit the year before); sales dragged at $16.1 million. Frantic at the drop in stock values, the involved mutual fund executors joined forces to find potential partners with which R. Hoe & Company might merge. But the company was not interested in merging. As hearings proceeded officers of Hoe, Talcott, and Lybrand, Ross Brothers & Montgomery—as well as exdirectors Gordon and Hanley—were all incomunicado. One concerned broker expressed his condolences:

> This thing is all involved. It was all very much unnecessary, and I didn't want any part of it. I don't want to say anything that will hurt anybody. I just think it's too bad for the stockholders. . . .[29]

How very much like the situation in 1932! How doubly incredible that so healthy a firm should twice enter ruinous bankruptcy procedure.

On 22 July 1969 U.S. District Judge Sylvester J. Ryan formalized his plan whereby press customers would pay in anticipation of delivery. Ten companies on the "A" or pre-September 8 list (five of them newspapers) quickly put up the money. Despite the 10 percent price escalation, the court claimed no intention of making a profit on the completed machines, satisfied only that the factory remain in operation. Judge Ryan praised Talcott for advancing payroll money without interest (to be repaid from the earliest customer pre-payments) and asked the firm to lend another $750,000 at 7½ percent (against accounts receivable of $1.2 million). He urged that "everyone trust the other" as the difficult work of rehabilitation got under way. The judge hoped that several customers would keep alive their original plans for enlarging their Hoe orders.

Desperate for prompt financial information both to determine any possible culpability and to obtain production costs for pricing purposes, the court appointed an independent accountant, S. D. Leidesdorf & Company, New York, an old firm with what Judge Ryan called wide "ferreting" experience. The last official audit of Hoe books had been in September 1968, and the judge had reckoned the books since that time "a mess." When a New York bank reported it was holding $1.5 million in Australian notes awaiting the completion of equipment for Mirror Newspapers, Sydney, the judge was surprised to learn that the transaction had been engineered by Ampress, S.A., the Hoe foreign sales subsidiary. He declared he wanted to learn more about the international aspect of Hoe business—and reminded the new accountants not to overlook Hoe's more sequestered saw division.

Justice Ryan several times reminded observers that six or seven substantial parties were interested in acquiring the Hoe assets and continuing the business. "It might even come as a pleasant surprise," he said, with more enthusiasm than seemed warranted, "that we may save the Hoe company for itself if we succeed in straightening out its financial situation. . . . We might even wind up with a new factory where the orders for Hoe equipment could be produced more reasonably than in the present factory." "Our aim is to break even for a while," the judge added later, "maybe turn a profit on the next batch of orders."[30] Judge Ryan had heard too that Hoe's very desirable tract of land in the Bronx was wanted for—of all things—a housing development.

In order to finish as many presses as possible, Hoe was obliged to restrict its sales of spare parts to what it had in inventory. Parts were always a profitable item, but customers were suddenly ordering them only to build up supplies against Hoe's failure to survive the bankruptcy. The company in July closed for two weeks' vacation, although a crew of 178 men remained to assemble machines for delivery by 8 September. Prepayments from the "A" group customers now amounted to $3 million—some impertinently considering them "ransom"—but they were compelling fast action.

Despite the Hoe vacation schedule, workers and their labor organizations were vitally concerned. A committee of fourteen trade unions had formed to preserve the New York company and its 2000 jobs; Harry van Arsdale, Jr., and Theodore W. Kheel, both veteran labor arbitrators, were named as co-chairmen. The New York chapter of the International Association of Machinists and Aerospace Workers (an AFL-CIO affiliate), long Hoe's nemesis on labor matters over the years, now raised the specter of monopoly if Hoe failed: one remaining giant, Miehle-Goss-Dexter, would get all the domestic press business. At present the workers had received two weeks' paid vacation, and with other economies the plant was working efficiently.

Also concerned with preservation of the firm was a new class A stockholders' protective committee, headed by Ralph H. Haas, representing about 28,000 shares, as well as a creditors' committee representing about $4.5 million in debts. The judge welcomed such committees, offered to pay their legitimate expenses, but could give them no special voice in the proceedings. John Galgay, meanwhile, was meeting with a group of American and foreign representatives interested in buying the company in whole or in part. Judge Ryan said there would be no "finder's fee" nor any brokerage commission if the firm were sold, but felt "we are sailing along very well," as he himself took two weeks for a vacation in Puerto Rico.

In September, returning from his Caribbean stay, Justice Ryan resumed hearings with renewed vigor. A native of the Bronx, the judge seemed chauvinistically determined to salvage the neighborhood company whose "cupboard was bare," as well as its 2100 skilled employees—most of whom actually lived on Long Island. He saw a "tremendous future" for the 150-year-old company. Although he continued to conduct sessions without his robe and in "town meeting" fashion, he quickly made his authority clear. When asked if the old board of directors was still in power, Judge Ryan replied vehemently, "No! I am the director general."[31]

Even with such undisputed authority, however, the judge recognized—symbolically, perhaps—a more desperate, forgotten one. When a coatless, tieless man wandered into the courtroom Justice Ryan barked, "Who are you and what do you want?"

"I'm a stockholder," said the intruder.

"Well," said the judge mildly, "come right in, find yourself a seat and listen to what's going on."

The visitor settled within three feet of the judge. Some 200 other interested spectators sat in benches and stood along the walls. One of them asked what a shareholder was likely to get out of the situation.

"You'll be lucky if you get anything," the magistrate replied. Shares that had sold for $140 were now $2 or $3, over the counter. Stockholders and creditors, fearing the worst, listened carefully, hoping against hope.

Judge Ryan reported that the Hoe unions were cooperating to save the company, production was actually increased, and he was restoring the 25 percent salary cuts that ten key executives had voluntarily taken a few months before. Burton C. Polglase, Hoe president, was one.

"We are lucky to have a man like Polglase heading up the operations," said the justice. "He is a very capable and imaginative executive who envisions a great future for the Hoe products that are under development." "And," he added, "neither Polglase nor the other members of the management team are being overpaid in a company that presumably has assets in excess of $50 million."[32]

Galgay in January 1970 had concluded that the Hoe divisions would best be sold singly, rather than try to market the company as a whole. He believed this fragmentation to be more beneficial to stockholders and creditors alike. At the same time, however, he discouraged the idea of internal reorganization unless purchase or lease were made by an outside party—thus squelching Justice Ryan's frequent optimism about putting the company back on its own feet. Several Hoe officers supported Galgay: Polglase, Raymond R. Dittrich, and Richard Woessner. The reason, however, was a surprise to all.

Acknowledging the press division of Hoe to be responsible in 1967-1968 for 80-85 percent of sales, Galgay explained to the court: "We cannot continue with reorganization because there are no new orders. . . ."[33] Dittrich, vice president, sales, admitted that, even though the company had about $50 million in proposals and quotations, only one new order for presses had been received in the past year. Orders had ceased fully six months before the July 1969 Chapter X petition, at which time the court officially suspended their solicitation. Judge Ryan explained he had been unable to authorize the acceptance of continuing business because of "the uncertainty of the future" for the company. But Dittrich was asked why orders had stopped so much earlier. He admitted that customers as early as January 1969 had begun to sense the company was in trouble.

Hoe salesmen had revealed that a number of former customers had refrained from placing orders with Hoe competitors while awaiting the company's rehabilitation, but they could not wait much longer. In January 1970 several newspaper publishers testified bluntly that they would not buy or recommend that anyone buy anything from Hoe in its present financial condition. They added, however, that if Hoe were proven stable, through purchase by a reliable firm, they would certainly consider doing business. The publishers pointed out that the important thing in ordering new press equipment is not merely getting a guarantee of refund in case of nondelivery (which, in the form of trustee certificates, was all the bankruptcy court could promise): "We don't want to wait around 16 months to two years," said one, "only to get our money back and no press."[34] A hundred years before Hoe had assured domestic delivery when original Napiers were a thousand ocean miles away; now, solvency had to be assured.

Woessner, Hoe's chief engineer, warned the court that the company would reach a work gap in the summer of 1970, and be forced to lay off many valuable hands. His staff, numbering 112 before bankruptcy, was now already down to 73. As bids and talks continued the press division's operations in the Bronx were slowly phasing out as orders were completed. Judge Ryan admitted the plant could not continue even as a "repair shop" because of overhead costs and lack of new business. He began to see only the profitable saw division as a nucleus around which the Hoe name could be saved.

Edwin Wesely, counsel for Galgay, explained how the court-appointed management was handling the completion of work in progress. Contracts were being renegotiated with customers who expected delivery in the fall or early 1970. Those who had been first in line (the "A" list) had, with one exception—Orlando *Sentinel Star*—gotten their machines on or before the deadline, and the crews were hard at work on presses for the American Can Company and Nor'west Publishing Company, Portland, Oregon. Dayton Newspapers, Inc., anxious for delivery by January, could expect to pay 15 to 17 percent more on its order for 12 Colormatic units, although it hoped to increase its order to 22 units. Dayton was part of Wesely's "near group 2," to be ready at year's end, as was *Long Island Press* (Newhouse

Group) and *Atlantic City Press.* These renegotiations by November 1969 brought in $2.4 million, or $1 million more than the original contract prices.

Also nearing completion were machines for Sydney's Mirror Newspapers, who obligingly released $278,000 for some remaining parts and equipment. Because of problems of international currency, however, foreign purchasers could not easily advance payments to rescue their presses; renegotiations of contracts with some of them, Brazil for example, required governmental issuance of revised import licenses.

Wesely's third, "far group 2," for delivery in 1970, would require concentrated effort on the part of Hoe employees. "We need you and you need us," said Judge Ryan, addressing the Hoe workers abstractly. "We expect a full day's work."[35] In connection with continued production, spokesmen for the many unions at Hoe had neither been sympathetic nor fully persuaded the company was in financial trouble. After long meetings in private with Justice Ryan and Wesely union representatives had yielded little; although prepared to reexamine clauses dealing with the company's right to subcontract (to outside, possibly nonunion shops), the unions would permit absolutely no other alteration to their contracts, and were also adamant about keeping the Hoe shops just where they were—Bronx and Dunellen—as though they might slip away in the confusion.

In October 1969 the accumulating bids for acquisition of R. Hoe & Company were announced by Wesely:

Hoe Company complete:	White Consolidated (King Press)
Metal printing press division:	Sun Chemical
	Miehle-Goss-Dexter
	Harris-Intertype
Printing press division:	Harris-Intertype
	Cutler-Hammer Company
	United Industrial Syndicate
	Amerofina
	Nohab (A.B. Bofors)
Saw division:	Eltra
	Black-Clawson
	Vermont American
	Strauss du Parquet[36]

Nohab was later described as a syndicate of two American firms, Joshua B. Powers, Inc., and A. G. Becker, along with the Swedish armament firm. Wesely explained there were others whose negotiations were not sufficiently advanced to announce. Two of these a month later were revealed to be Fasco-Andreotti, interested in the press division, and Wagner Lithographic Company (National Standard Company), interested in the metal press division. In January Consolidated International sent a complex reorganization plan for the press division. The proposals of all these giant corporations indicated considerable variation in what they thought they were buying; somehow all they knew for sure was that the stamp "Hoe" on equipment was valuable.

The Sun Chemical Corporation (Rutherford Machinery Company) offered to pay $750,000 in cash for the exclusive right, for ten years, to use the Hoe trademark on metal decorating presses, and would pay $1,250,000 for Hoe's inventory in this line.

E. Wharton Shofer, president of Powers, Inc., thought his firm could get Hoe operating again within a month. With its newly merged resources (ATEC Corporation, designers and erectors of newspaper plants; *Hablemos*, a widely circulated Latin American newspaper magazine supplement; Editors Press Service, a large feature syndicate; and Powers itself, advertising representatives), which Joshua B. Powers called the world's first "total service organization for publishers,"[37] such quick action almost seemed possible. The Powers-Bofors-Becker offer was partly a purchase and partly a lease with option to buy. Bofors, a minority partner, had already been manufacturing Hoe presses under a special licensing agreement. Shofer admitted that in its new Hoe Company a president would be brought in.

Harris-Intertype, successors to the original Intertype Corporation, presented an offer that included initial capitalization of $1 million upon establishment of a New Hoe Corporation which would buy certain of the old company's assets by issuing 50 percent of New Hoe's common stock to the R. Hoe & Company trustee, callable in four to six years for up to $1,650,000. In addition, there would be an issuance of $3 million in 7 percent New Hoe notes payable 1973, and $7 million in 6 percent New Hoe notes payable in 1979. Thus the Harris-Intertype bid involved a ten-year period for completion of the sale—a distinct disadvantage to the court wanting to terminate the case with more dispatch. Although Harris-Intertype also made printing presses there seemed to be no serious question of competition.

Harris-Intertype would lease the Bronx and Dunellen properties with options to renew; a Harris-Intertype representative would head the new firm. George S. Dively, Chairman of Harris-Intertype, felt Hoe's labor contracts "abnormally strong," claiming they would "penalize us before we could start." He suspected they were part of Hoe's original problem. Dively would eliminate union clauses limiting subcontracting, cost-of-living increases, and mandatory union membership; he would reduce pension contributions and holidays (from 13 to 11). All these modifications, he said, were in line with industry practices elsewhere. Harris-Intertype also wanted the metal decorating division, for which it would pay $1,250,000 in cash and warrants to purchase 30,000 shares of Harris-Intertype stock. Dively said his company would not take one division without the other.

Another prime prospect for taking over R. Hoe & Company's press division was Fasco, A.G., an Italian firm with headquarters in Liechtenstein. Owned by Michele Sindona of Milan, Fasco (a holding company) had controlling shares in two Chicago firms and interests at times in other American companies. One of Fasco's holdings was Andreotti, S.P.A., Ceprano, Italy, manufacturers of medium and small gravure presses popular throughout Europe, Soviet Russia, and South America. The Hoe line of heavy duty newspaper, magazine, publication, and packaging presses would complement the Italian company's own machines handsomely.

Fasco described its intentions to the court in November 1969. It would form a Delaware Corporation, to be called New Hoe, which would issue to Fasco 5 million shares of common stock, $1 par value, in exchange for all outstanding capital stock of Andreotti. For the transaction New Hoe would receive 530,000 shares of 5 percent convertible noncumulative preferred stock, $10 par value, each share of which would be converted in 1973 into 3.125 shares of common. New Hoe would also receive $8.1 million aggregate principal of 6 percent subordinated notes. In all, Fasco's proposal amounted to an offer for Hoe of $17,448,000—and there could be more, if necessary.

Dr. Gaddo Quilici, managing director of Andreotti, said New Hoe would retain all present Hoe personnel, it would assume all employment contracts, and it would keep the Hoe name—as well as lease the Bronx and Dunellen plants (with option to buy). The New Hoe board of directors would include Quilici as president, and Dittrich and Polglase, who were presently at Hoe. With Galgay's recommendation, the court accepted Fasco's (and Sun Chemical's) bid and scheduled a hearing in January 1970. Galgay, meanwhile, was free to accept other bids, although Fasco (and Sun Chemical) had the option to meet any new competition. Quilici was unafraid of such new bids: J. H. Whitney & Company, New York, a private investment firm, agreed to purchase a million and a half shares of Fasco's New Hoe common (and would put two partners on the board of directors); the Mellon Bank of Pittsburgh similarly agreed to extend Fasco $2 million in credit if its bid were accepted, and $2 million more if collateral were offered. Judge Ryan urged Fasco to apply for the extra $2 million.

Complicated by changes, late bids, and announcements of interrelated operations among bidders, the sale of R. Hoe & Company by the end of January 1970 seemed like E. A. Porter's sharp peddling, if not the lower East Side street haggling of a century before, and uncertain indeed. Harris-Intertype announced it was willing to raise the call price for common shares offered and, in lieu of $4 million in ten-year debentures, agreed to pay the Hoe trustee 5 percent royalty on net sales for 10 years, with minimum yearly guarantee of $500,000. Fasco was considering a new tender making $5.5 million immediately available to Hoe. Continental International changed its offer from interest in the company as a whole to the press division only. Crown Cork and Seal, Inc., entered a revised bid ($950,000 and $2 million for inventory) for the metal decorating division, and Sun Chemical withdrew.

An agreement was announced between Fasco and Nohab wherein the two firms would share equally the operation of Ampress, the Hoe international sales unit, when Fasco set up its new company. An exchange of stock and a joint five-member board (favoring Fasco) would ensue. The court also noted that both Nohab and Andreotti were presently Hoe creditors; Justice Ryan tried to get these firms to drop their claims if selected, but unsuccessfully.

In the face of such bewildering complications the judge was discouraged, but he refused to conduct a "distress sale." Yet, with Hoe's limited working capital (crippled by having to pay cash for materials) and mounting creditor and stockholder claims, Justice Ryan abandoned any hope too of internal self-reorganization—although his enthusiasm for a quick sale was also thwarted. "I am disappointed in the amount [of the bids]," he said, "and I am disappointed in the whole thing [the bidding competition]. It is a question of whether anything is acceptable."[38] He postponed hearings until March.

The court in March 1971 tentatively approved sale of the metal decorating division to Crown Cork and Seal Company. It still had two valid offers for the printing press division: Consolidated International and—of all people—Wood Industries, Inc., a late arrival at the convoluted proceedings.

Wood Industries, with plants in Plainfield and Middlesex, New Jersey, was no stranger to R. Hoe & Company. The firm founded by prolific inventor and business philosopher Henry A. Wise Wood had proved surprisingly flexible and substantial despite periodic hard times and changing technologies. It now comprised the former Wood Newspaper Machinery Company and the Scott Printing Press Company—both long-time Hoe competitors. It had been making quality pressroom and stereotype equipment as well as automatic lithographic platemakers, for years. In World War I it manufactured ammunition fabricating equipment, and in World War II it participated in the super-secret atomic Manhattan Project. Wood had diversified into converting (that is, web control) and packaging machinery (its Stanford Engineering Division, Salem, Illinois) and in business-form machinery (Ashton Press Manufacturing Company, Montreal, Canada). It had adopted its present name about 1968 to reflect its successful diversification.

Although Wood was a Hoe stockholder and long quite aware of the company's predicament, the New Jersey firm first appeared officially at the Hoe hearings in January 1970. An officer of the company later confided that Wood had "for some time" entertained thoughts of merging with Hoe, "but the demise made it easier, but not necessarily cheaper." [39] Its offer for the press division only, excluding real estate, machinery, and equipment, was for some 100,000 shares of Wood common stock and $500,000 in cash. Wood also offered to pay 7½ percent royalty on the sales price of the first $5 million of annual sales and 5 percent royalty on the next $10 million of annual sales (excluding saw and decorating metal sales). It would purchase no less than $5 million of inventory usable to Wood over a four-year period, and it would also act as subcontractor to the trustee to finish Hoe presses on order.

Daniel McColley, president and chairman of Wood Industries, had estimated the assets of the Hoe press division between $20-25 million, of which he thought $3-8 million accounted for tools, dies, and patents relating to press manufacture. He increased the offer subsequently from $5 to $8 million for inventory and reduced the royalty scale, and made other adjustments. As some bidders dropped out and others were disqualified by the court, Judge Ryan was somewhat relieved to find the Wood proposition survive; he gave it his approval.

With a welcome sense of history, President McColley announced to the public that Wood's successful bid did not mean the "death knell" for the grand old firm, but instead a continuation of the Hoe name and products. The New Jersey company would within a few weeks put Hoe presses into production at its two plants, to fulfill its subcontracting role. It had an option on buying or leasing Hoe's foundry at Dunellen, surprisingly located only about a mile away. The firm also expressed interest in Hoe's arrangement to build German MAN stereotype equipment, although it had its own brand.

While Wood continued to offer customers its Colorflex 250 offset press (and had just received an order for one) as well as its original Wood and Scott presses, it was now prepared to take orders for Hoe machines. By June 1970 the Wood Industries-Hoe merger was essentially completed. Wood payed $500,000 cash and a $1.7 million note later to be replaced by 105,000 shares of Wood common. Wood also agreed to buy $5 million of Hoe inventory and pay 5 percent royalty on the first $15 million of annual sales of Hoe products for 10 years, with a total guarantee of $8 million in inventory and royalty payments. President McColley was pleased:

> This action should be looked upon as a merger of two great firms which for more than a century each have pioneered in the development of high speed printing presses and equipment and have contributed notable advances in production to the greatest of all media of mass communication—printing. We wish to make it clear that this is another step of our planned expansion program, not the elimination of a competitor. Combined engineering, sales and support

> personnel are being housed in new facilities in Englewood Cliffs, New Jersey. This new organization is known as the Wood-Hoe Division of Wood Industries, Inc.[40]

At the Wood-Hoe closing officers also signed an agreement with Nohab, enabling the latter to sell and manufacture Hoe presses in Europe with engineering and technical assistance from Wood-Hoe. In June the merged organization already had several large orders for press equipment, and Polglase and Dittrich were moving into their Englewood Cliffs offices to get them started. There was again a cheerful whistle in the air.

In the Fall of 1970 John Galgay, acting as trustee for R. Hoe & Company, Inc., submitted requests for $1,108,446.70 as interim fees for his work ($149,536.76), that of his counsel ($722,131.26), and the auditors ($236,778.68). Stockholders immediately protested the figures as exorbitant for a year's work, Judge Ryan reading into the record the names of about 25 who had so objected. Edwin Wesely, Galgay's counsel, nevertheless explained that during the year the three sums had financed considerable company guidance in money matters, including renegotiation of existing sales contracts and consummation of the two Hoe division purchases. S. D. Leidesdorf & Company, the auditors, taking careful inventory at both Bronx and Dunellen shops —a chore Hoe had not performed for over 22 years, would work another six months at it.

Wesely also reported that only the Hoe saw works was still functioning; the priority press construction had been completed and remaining orders and repair parts would soon be fulfilled by Wood-Hoe. By mid October Hoe's press division—once the largest in the world—would have a strength of just 146 men, and ultimately only 24 as maintenance personnel. Judge Ryan added that sale of the Bronx property was not propitious at that moment, although he expected an existing urban housing plan for the area to include the Hoe parcel.

After six months' deliberation the justice reduced the three-part fee request to $571,746, a little more than half the original figure. There was no comment in the courtroom. Leidesdorf meanwhile had completed an unaudited balance sheet dated 30 September 1970, showing the company to have had liabilities of $1,184,559 in excess of assets. Outstanding financial claims against Hoe, however, amounted to an additional $10 million, much of which, the accountants believed, involved contract penalties. Considering Hoe assets at over $30 million, the discrepancy did not seem overwhelming, although the assets by this time included $1,500,000 of Wood Industries common, in part payment for acquisition of the press division.

Because of a number of inquiries about it, Judge Ryan reversed his decision not to sell the saw division. In May 1972, nevertheless, the profit-making saw shop was still seen as a possible nucleus for a reorganized R. Hoe & Company—"It will be the financial miracle of the day,"[41] the judge admitted, if it could be done. Its success was contingent upon the settlement of all suits—well over a colossal $40 million.

As required by bankruptcy law, a New York law firm, Townley, Updike, Carter & Rodgers, was engaged to investigate the causes for Hoe's failure. Attorney John P. Reiner of that firm in September 1970 announced he would take testimony from former and present officers and management as well as third parties, to determine if misdeeds or impropriety were involved. Arthur Gordon had resigned as president in April 1969, four months before the company filed for bankruptcy—an act Judge Ryan likened to leaving a sinking ship. Thomas F. Hanley, past chairman of the board, had left about the same time; both gave depositions, the latter at his home because of illness. Considerable additional testimony was also heard.

In July 1971 civil suits totalling more than $40 million were filed against Hanley, Gordon, and the Hoe accounting firm, Lybrand, Ross Brothers & Montgomery. The 39-page document, filed in the name of John Galgay as Hoe trustee, listed five counts:

> Against Lybrand, Ross Brothers & Montgomery—
> (1) Negligence in auditing the year-end Hoe financial statements of 1967 and 1968;
> (2) Recovery of $132,400 in auditing fees charged for the two years
>
> Against Hanley and Gordon—
> (3) Corporate waste and mismanagement amounting to $40 million;
> (4) Return of $160,000 from Gordon and $138,000 from Hanley for insider profits from sales of stock in 1968
>
> Against all three parties—
> (5) Indemnity from the settlement of pending stockholder class-action suits (in which Hoe was named codefendant along with the former officers), should the stockholder suits be successful. (A cross-claim indemnity suit against all former officers and directors, on the same charge, would follow in two weeks.)[42]

In September three additional persons were charged with taking insider profits from the sale of Hoe stock just prior to the company's collapse. Robert Marcus, sales manager for Plant Installation Associates, New York, his wife Mildred, and son Joel were charged with taking illegal profits amounting to $160,000. Marcus, as a Hoe consultant and contractor, had been privy to confidential information on the company's desperate situation. The Marcus family offered an out-of-court cash settlement of $50,000, which attorney Reiner considered "very favorable." It was accepted (pending public notice, as required by the Securities and Exchange Commission).

In March 1972 Hanley, Gordon, and Lybrand, Ross Brothers & Montgomery similarly agreed, by out-of-court negotiation, to pay to the company trustees $3.25 million in restitution. Their respective insurers added $1.25 million. This fund in 1974 was used to reimburse many legitimate claims. But there was still more to be done.

The principal Hoe creditor was James Talcott, Inc., a factoring firm from which the company had borrowed over $12 million. Claiming to be unaware of Hoe's financial predicament, Talcott insisted on being a secured creditor, and managed to bargain for $1,482,000 in cash and $4,435,000 in administration claims (first priority notes), giving Talcott an assured percentage of Hoe income until the debt limit was reached. Among liabilities in the 30 September 1970 balance sheet was an earlier Talcott loan of $12,530,920 (including almost $2 million in accrued interest).

In September 1971 legal objections were filed against Talcott and five other creditors (two banks, two insurance companies, and the Swedish firm, Nohab) regarding their "secured" status. Their loans, amounting to $14 million, had been made in April 1969, a time when Hoe's insolvency should have been known to them. About the same time another creditor's claim for $296,000, by a firm owned by Joel Marcus, was disallowed, the actual amount on Hoe books being only $54,000.

Several stockholders' class-action suits, and one derivative suit, had been filed against Hoe and its former management. Four mutual funds in 1969 had jointly filed against Hoe and its accounting firm, charging they bought stock based on false and misleading financial statements for fiscal 1967—their claimed loss $6 million.

It was Galgay's hope that all these many and varied complaints, including the Reiner five-count suit, could be negotiated as a "package." The stockholder actions would be settled with a partial payment of $1.25 million. Talcott and its co-creditors would be paid $5.4 million, for a total of $10.7 million. Talcott, however, now insisted upon interest (13½ percent) on the original loan, which Judge Ryan considered asking too much. Talcott agreed to restudy its demands.

Negotiations with Wood Industries did not at first go smoothly. Proxies on the 105,000 shares of common stock (soon worth close to $2 million and representing 15 percent ownership) were a problem; an agreement that the Hoe trustee sign them over to Wood management was found to be contrary to law in Virginia, the state in which Wood was incorporated. The company nevertheless consolidated its own press facilities with its newly purchased Hoe manufacturing properties at Englewood Cliffs, a location which accommodated the many former Hoe employees living on Long Island. Included in its growing line of products were Hoe's Colormatic (letterpress) and Lithomatic (offset) presses, publication presses, rotogravure, and other auxiliary equipment. The Wood-Hoe Division also contained Wood's own Colorflex web offset machines, Scott presses, and their related items. But the word "Hoe" in the name was a problem.

> When Wood Industries bought the press manufacturing division of R. Hoe & Co. in April '70, the last thing the company wanted was for Wood to be confused with the financially distressed press and saw manufacturer—but the report is that people aren't bothering to differentiate between the old Hoe and the newly formed Wood-Hoe division of Wood Industries. . . .[43]

Although Hoe's quality design and workmanship were thought to be a tremendous sales asset, Wood for a while contemplated dropping "Hoe" from its name to end the confusion.

Gradually, however, orders began to come in, restoring the company's faith in its auspicious hyphenization. R. R. Donnelley in 1971 ordered $4.8 million in Hoe Colormatic equipment; a Brazilian newspaper, *O'Estado,* Sao Paulo, also placed a large order. Soon several other newspaper orders materialized, including another in 1972 from Brazil (*Journal do Brazil,* Rio de Janeiro) for $2 million in Colormatics. By March 1971 Wood already had a $15.3 million backlog of confirmed orders, twice its figure of the year before. This business seemed to justify the merger, but the paperwork dragged on.

In its final arrangements Wood Industries had payed the Hoe trustee $500,000 in cash and 105,000 shares of stock; it also agreed to a 5 percent royalty on the first $15 million of yearly Hoe equipment sales until 1980, and to buy not less than $5 million in Hoe press inventory prior to 1977. Wood had declined (although retaining an option) to buy both the Bronx and the Dunellen plants, as well as the old company's tools and machinery. While the large Dunellen foundry would have been a bargain, Wood preferred to subcontract its intermittent foundry work. (JTC Inc., a group of used machinery dealers headed by Charles Hood, formerly a Hoe foundry foreman, in January 1971 offered to buy all scrap and materials, machinery and equipment, and—at one point—the real estate, but the court was not impressed by the proposal.)

Wood had also not originally been interested in Hoe personnel under the original union contracts, but it eagerly took many capable individuals, Burton C. Polglase (voted an officer of Wood in 1971) as vice president, engineering, among a score of top design and sales people. Polglase and his design crew were already at work on new newspaper web offset machines to compete with MGD [Miehle-Goss-Dexter] Graphic Systems, Inc., now the largest U.S. manufacturer of printing presses. The Wood management continued a program of cautious expansion, but thorny problems still kept calling company officers back to the courtroom.

In January 1972 Wood Industries in three show-cause motions was asked to explain "violations" of the purchase agreement. The Hoe trustee wanted Wood (1) to designate promptly the unwanted portions of Hoe inventory which could then be discarded; (2) to return certain Hoe engineering drawings for presses under contract in 1969 but never completed, given for examination only; (3) to produce its books so as to reveal sales of Hoe-related equipment for royalty calculations. With Wood's basic acquisition and integration hardly completed, trustee Galgay seemed in a great hurry to generate money for Hoe, but Judge Ryan—for his own reasons—was anxious to keep the mammoth transaction moving along.

A Wood attorney explained that by their agreement Wood "from time to time" could designate non-usable inventory, but that what seemed useless now might very well be valuable later. (The Hoe trustee was required to maintain the inventory for up to 10 years—a costly drain.) Conferring with President McColley, the Wood representative supposed it would take about two more years to decide; the judge thought 30 days enough time. Wood also insisted the disputed drawings were now part of the purchase, and Judge Ryan himself dismissed the showing of books as unnecessary. The whole matter was adjourned pending the submission of briefs.

In August Wood Industries and Hoe met again in court, this time on a more positive tack. For judicial approval trustee Galgay proposed an amendment to the original agreement, setting Hoe inventory, real estate, and other valuables in which Wood was interested, at $4,250,000—about half of their former assessment of over $8.4 million. By paying the $4 million Wood could also obviate the earlier royalty and inventory purchase arrangements; the real estate was the 8.4-acre Dunellen tract. Certain other minor adjustments in storage charges, interest payments, and other fees were also included. Said Galgay:

> Although the monies which the Debtor [Hoe] will receive for the assets which it would convey to Wood under the proposed agreement might appear disproportionate when compared with the value of those assets . . . I believe that the proposed agreement is of importance. . . .[44]

He recommended the new amendment be accepted. Two groups, however, protested: some Class A stockholders and the Hoe creditors, both of whom had seen so many of Hoe's assets slip away (they were also still opposing Talcott's interest claim), appealed for a better than 50 percent price.

Judge Ryan, wearying over negotiations about which he had been so patient, so cheerful at one time, got down to practicalities. He observed that outright sale of the inventory would save Hoe over $1 million in 10 years of projected maintenance costs alone. He accused the two groups of "tying up the sale" which would bring an equitable and welcome end to three years of litigation. The shareholders and the creditors, equally weary, perhaps, conceded. On 21 February 1973 the amendment was approved by both parties (with slight final modifications, for example, a royalty of 1 percent was retained). Daniel McColley, president of Wood Industries, Inc., was pleased once more. The printing press manufacturer was at last free to return to his work.

Justice Ryan, having officiated since 1969 at the Hoe Chapter X bankruptcy, to the surprise of all, expressed his intention, upon conclusion of the proceedings, of taking senior status (semi-retirement), leaving him free to take as many or as few cases as he wished. He was anxious to close the whole affair, which now awaited only the after-action report required by the bankruptcy law.

Robert Corrao and James B. Kilsheimer III succeeded John Galgay as trustee in the spring of 1973 and began immediately to weigh the causes of the company's failure. They first studied a crucial 1963 change in Hoe accounting procedure, from reporting sales (profits) as contracts were fully executed, that is, when each press was delivered to its customer, to reporting a percentage of total profits as unfinished machinery progressed through the plant. A fully acceptable system, the "percent of completion" alternative had serious implications for the firm. The trustees noted:

> However, this method requires that the manufacturer be able to estimate accurately the cost to complete such a long-term press contract at the various stages of manufacture, and is unworkable unless the necessary information to prepare the estimates is available from production and accounting data. During the investigation conducted by the attorneys for the Trustees, they concluded that Hoe did not have adequate production records and systems of reporting. . . .[45]

The result of this was not immediately apparent, but it soon emerged.

Several factors combined to drain the company of its financial vitality, enumerated by the trustees:

(1) When the company in 1966 began to subcontract parts manufacture in Italy and Sweden, its unions demanded (and got) exorbitant overtime concessions, each worker at least 10 hours a week.

(2) When the company in 1967 began to announce results based on percentage of completion, it described an unreal increase in profits which thrilled the public to large sales of securities; the firm responded by splitting its stock 3 for 1, exacerbating its indebtedness.

(3) In the conversion to percentage of completion accounting, 1968 profits from three large (Latin American) press orders were highly inflated because they included inventory parts already reported under the old system.

(4) For the period ending April 1968 a decline in incoming orders —from an anticipated $24 million to about $3 million—produced a serious cash shortage.

(5) Deceived by the purported prosperity of its own 1967 financial statements, the company in 1968 gave excessive bonuses and salary increases, paid current dividends in full as well as certain dividend arrearages, bought unneeded parts, materials, and real estate, entered hastily into some unprofitable press contracts, and borrowed money at unnecessarily high rates.[46]

These blunders served to effect a shortening of working capital the firm could not do without. Yet many of these developments had been noted by company officers themselves, their warnings unheeded by management. On 15 May 1968 the Hoe treasurer had issued a memorandum almost predicting the firm's collapse:

> The cash squeeze has been felt in many areas. Now appearing are many signs of financial strain that this company is experiencing at a time of prosperity. If this condition continues, we will not only lose our competitive position in the market, but we will experience such a tremendous setback financially that recovery would take several years. . . .
>
> [O]ur sales position has deteriorated two and one-half times that of 18 months ago based on the growth trend we are under.
>
> The swelling in inventory occurring during this period of time is one of the primary causes of cash drain.
>
> At September 30, 1966, 65% of the customers advances was invested in inventory leaving Hoe approximately $6,700,000 of customers money for operating needs. This ratio, financially, was always considered a good indication as we normally operate on half of the customer's money requiring the difference, I repeat, for our own needs since we were considered to be undercapitalized. This trend has slowly disappeared whereby at September, 1967 90% of the customers' money was covered by inventory, and at March, 1968 Hoe invested all the customers' money in inventory, plus an additional $7,000,000.
>
> In conclusion on this point, we went to the stock market to obtain money which was to improve our capitalization position. This relief was offset by improper inventory controls, since the spread we normally enjoyed between customers advances and inventory was totally eliminated, as we went from $6,700,000 unapplied advances at September 30, 1966 to $6,600,000 overapplied. This also means that we are buying inventory substantially faster than we are building presses and collecting advances from customers.
>
> The financial strength which at one time showed in our Balance Sheet is diminishing as the current ratio, as reported to insurance companies, at August 31, 1967 was 4.82. This ratio has slid to a current ratio of 2.1 as of March 31, 1968, after giving effect to $3,500,-

000 received in stock proceeds. (A current ratio below 2.0 would put us in default of the loan agreement.)

The Balance Sheet also reveals a doubling in our accounts payable and accrued liabilities, and accounts receivable when compared with statements of only 6 months earlier (September 1967 versus March, 1968).

An explanation of what possibly can transpire when a company receives excessive capital (or new capital) if generally, not closely controlled, leads to loose policies—such as, excessive inventories and poor credit and, furthermore, encourages unnecessary expenditures for unproductive assets. Some of this we witnessed to a point where now we have problems of inadequate capital for which this company will experience the following:

1. Inability to take advantage of cash discounts.
2. Chronic pressure on cash position—which will impair credit standing.
3. Inability to make necessary discretionary expenditures without cash pressure, i.e., capital expenditure and dividends.
4. Inability to take advantage of sudden changes in business opportunities.
5. Inability to weather minor reversals in business conditions.

With a current cash strain, high overheads, idle time, etc., this company must review the current position and, only after this has been properly ascertained, corrective measures may be taken to reverse all the above-mentioned points. We must be sure that we did not reach our optimum profitability last year and that we are now generating sales at a greater cost thereby losing profitability at a greater rate than the incline.

Finance takes a strong position at this date and we are desirous of being heard. The consequences to Hoe will be far too great if we were to be ignored. . . .[47]

Corrao and Kilsheimer noted sadly that the treasurer's memo was ignored. They then described the net effect of these excruciating difficulties:

Despite the increase in working capital brought about by the loan from Talcott, the Debtor's financial condition continued to deteriorate. As a consequence of the delay in reporting, the marked decline in reported earnings, the sharp drop in the price of the stock, the suspension of trading, the uncertainty as to the Company's earnings and financial condition, certain lawsuits commenced by shareholders, . . . and considerable adverse publicity, there was a widespread public loss of confidence in the Company and its management.

Potential customers for presses were unwilling to accept the risk that the Debtor might not be able to remain in business. Presses take a year or more to manufacture, cost more than a million dollars in many instances, and require parts and servicing. The slightest doubt about the manufacturer's ability to stay in business would deter the

> placing of new orders. The long-term decline in volume of new orders for Hoe presses accelerated sharply. Major new press orders were not received. Without a sufficient volume of new business, Hoe was unable to continue normal operations. . . .[48]

The two trustees turned their attention to resuscitating the saw division, which would become the future R. Hoe & Company. The rest was scattered and destroyed.

Through the 1970's, as the Hoe factory was systematically dismantled and razed, it could also be said (once again) that much more than a few old buildings were toppling. Only coincidentally, perhaps, New York's borough of the Bronx, especially its southeastern portion—Hoe's domain since 1930, was a decaying "inner city" slum. Middle-class families had fled to the suburbs leaving old apartment houses to a new wave of desperate immigrants—Blacks and Puerto Ricans. Uneducated, unemployed, and unacculturated, and intimidated by young addicts who scavenge voraciously (selling copper plumbing and even manhole covers for scrap) and the jobless (who obtain better welfare treatment only when burned out of their homes), the new residents watched theft and arson literally consume their community. Skirting an open manhole, a reporter in 1977 picked over the ruins as had Bennett in 1835:

> This is the South Bronx. Once a happy enclave, mostly of Jews and Italians who had escaped the tenements of Manhattan's lower East Side, today it is a scorched monument to social planning gone awry. . . . And the battle still rages. When darkness comes the torches go to work. . . . On all sides stand the wrecked shells of apartment buildings, many charred by fire, all eerily quiet. . . .[49]

The scene was worse than the sudden conflagrations of New York, Chicago, or Boston. Between 1975 and 1977 alone 7000 fires—averaging ten each night—destroyed 2000 square blocks containing over 1200 abandoned buildings at a loss of perhaps $50 million and homes for 250,000 people. The story of the borough's urban renewal is still not yet written.

These insidious fires did not reach the Hoe premises, instead methodically leveled by bulldozers by the end of 1977. On an outside wall a blue and white Hoe sign, "Put lunch scraps in refuse cans," ludicrously addressed itself to the silent rubble. It too fell unceremoniously, a last vestige of management's misguided concern. In April 1978 a federal court order confirmed R. Hoe & Company's reorganization under Chapter X bankruptcy and declared the old common stock cancelled, of no further value.

As the last consignments of Hoe inventory were conveyed to Wood Industries, two curious objects of art came to light in the old Bronx factory: a brass model press and a bronze statue of Johann Gutenberg. No valuation could be set for them.

REFERENCES
BIBLIOGRAPHY
INDEX

For complete description of published sources appearing only by surname in the text and below, refer to the Bibliography, which contains all heavily utilized materials, usually of interest in their entirety. The following abbreviations are used here:

DAB	Dictionary of American Biography
DNB	Dictionary of National Biography
E&P	Editor & Publisher
INP	Inland Printer
NCAB	National Cyclopedia of American Biography
NYT	New York Times
SAM	Scientific American

Abbreviations for sources of unpublished materials, where identification is feasible, are as follows:

ATF	American Type Founders Typographical Collection, Special Collections, Columbia University.
JGB	James Gordon Bennett Papers, Manuscript Division, New York Public Library.
WHB	William H. Brown Collection, Special Collections, Yale University.
GLF	Gordon L. Ford Papers *(New York Tribune)*, Manuscript Division, New York Public Library.
Hoe-B	R. Hoe & Company Collection, Bronx County Historical Society.
Hoe-C	R. Hoe & Company Collection, Special Collections, Columbia University.
Hoe-F	R. Hoe & Company Collection, in possession of Hoe family.
Hoe-L	R. Hoe & Company Collection, Manuscript Division, Library of Congress.
Hoe-Y	R. Hoe & Company Papers, Special Collections, Yale University.
Jay	John Jay Family Papers, Special Collections, Columbia University.
SDT	Stephen D. Tucker Collection, Bentley Historical Library, University of Michigan.

References

I. EVOLUTION IN PRESSWORK: THE FIRST MECHANICAL ART

1. Joseph Moxon, *Mechanic Exercises on the Whole Art of Printing,* 2nd Ed., Herbert Davis and Harry Carter, Eds. London: Oxford University Press, 1962, 252-253, 255-256.
2. Robert Hoe, 7.
3. Clair, 205, 212.
4. *Mechanics' Magazine* (New York), Feb. 1837, 73.
5. Pasko, 317-318.
6. *Times* (London), 29 Nov. 1814.
7. Robert Hoe, 16.
8. Grant, Vol. 1, 452.
9. *Times* (London), 3 Dec. 1814.
10. "Printing by Steam." *London Literary Gazette,* 26 Oct. 1822, 672 ftnt.
11. Clair, 213.
12. Morison, *History of the Times,* Vol. 1, 117-118, postscript omitted.
13. Howe, 14.
14. Morrill Wyman, "Memoir of Daniel Treadwell." *Memoirs of the American Academy of Arts and Sciences,* Vol. 11 (1888), 333-335, 337, 339, quoting letter of 3 May 1820 to Dr. John Ware, Boston.
15. Wilson and Reader, 17.
16. Wilson and Reader, 18-19.
17. Thomas C. Hansard, *Typographia: An Historical Sketch of the Origin and Progress of the Art of Printing. . . .* London: Baldwin, Cradock & Joy, 1825, 661.
18. Hansard, 710-712.
19. Robert Hoe, 17.
20. "The Early American Improvements in Printing Presses [Dow]." *SAM,* 28 Aug. 1869, 130.
21. Silver, *American Printer,* 54.
22. Wyman, 344.
23. "Phineas Dow." *Printers' Circular,* May 1869, 91-92.
24. Ellen B. Ballou, *The Building of the House: Houghton Mifflin's Formative Years.* Boston: Houghton Mifflin, 1970, 22, quoting an 1893 Carter letter to Houghton.
25. Exman, *Brothers Harper,* 17.
26. McNamara, *INP,* Oct. 1884, 5.
27. Wyman, 473, 491.
28. McNamara, *INP,* Oct. 1884, 5.

29. Robert Hoe, 11-12.
30. Lehmann-Haupt, 165-166.
31. Ballou, 24-25, from an 1849 Boston *Traveller.*
32. J. Henry Harper, *I Remember.* New York: Harper, 1934, 8.
33. Ringwalt, 20-21, from an 1864 New York Typographic Society publication.
34. Theodore L. De Vinne, *Printing in the Nineteenth Century.* New York: Lead Mould Electrotype Foundry, 1924, 7, reprinted from New York *Evening Post,* 12 Jan. 1901.
35. *Mechanics' Magazine,* June 1833, 302.

ADDITIONAL SOURCES FOR PART I

Samuel Smiles, *Men of Invention and Industry.* New York: Harper, 1885; Nicholas Barker, "Richard Taylor: A Preliminary Note." *Journal of the Printing Historical Society,* 1966, 45-48; *DNB,* Vol IV, 1310-1311 (Cowper); "John F. Trow." *American Bookmaker,* Sept. 1886, 89; June 1912, 477.

1. ROBERT HOE & COMPANY, NEW YORK

1. Thorburn, 56-57.
2. Thorburn, 55.
3. Thorburn, 55.
4. Thorburn, 57.
5. Du Bin, 33.
6. George Hickman, Hoe sesquicentennial (Hoe-F).
7. Thorburn, 57.
8. Cornelius S. Van Winkle, *The Printers' Guide: Or an Introduction to the Art of Printing. . . .* New York: Van Winkle, 1818, 229.
9. Munsell, 62.
10. Lewis M. Allen, *Printing with the Handpress.* New York: Van Nostrand Reinhold, 1969, 21.
11. McNulty, 47.
12. "Hand Presses." *SAM,* 22 Jan. 1853, 147.
13. Silver, *American Printer,* 51, from Johnson's *Typographia* (1824).
14. Robert Hoe, 8; Tucker, 359.
15. Tucker, 359 ftnt, quoting Ralph Green, *The Iron Hand Press in America.*
16. Emery, Ed., 40, from New York *Evening Post,* 5 Nov. 1825.
17. Robert Hoe, 8-9.
18. Tucker, 359 ftnt, quoting Jones (1829).
19. Robert Hoe, 9.
20. Karolevitz, 19, from a Hoe catalog.
21. Moran, 85.
22. Rollo G. Silver, "An Early Time-Sharing Project: The Introduction of the Napier Press in America." *Journal of the Printing Historical Society,* 1972, 30-31.
23. Silver, "Time-Sharing," 32, 34.
24. Mott, 179.
25. Goldberg, 226, from a letter, 2 Oct. 1828.

26. Goldberg, 73, 229.
27. McNamara, *INP,* Aug. 1885, 478.
28. Pasko, 267, 446.
29. McNamara, *INP,* March 1885, 250; Sept. 1885, 523.
30. New York *Herald,* 9 June 1886.
31. "Hoe's Single 'Large' Cylinder Press." *Printer,* Jan. 1859, 214.
32. "Robert Hoe & Co." *Mechanics' Magazine* (New York), June 1833, 335, 336.
33. Tucker, 353-354.
34. Tucker, 358-359.
35. Alexander C. Flick, Ed., *History of the State of New York,* New York: Columbia University Press, 1934, Vol. 6, 238.
36. Tucker, 359 ftnt.
37. Elizabeth F. Baker, *Printers and Technology: A History of the International Printing Pressmen and Assistants' Union.* New York: Columbia University Press, 1957, 18, from a 1901 convention souvenir.
38. New York *Sun,* centennial issue, 1933.
39. Hudson, 582.
40. Tucker, 361.
41. New York *Herald,* 19, 21 Dec. 1835.
42. James Parton, "The New York Herald." *North American Review,* April 1866, 379.
43. "Hoe & Company's Printing-Press and Saw Manufactory." *SAM,* 25 Dec. 1858, 126.
44. Burlingame, 394.

ADDITIONAL SOURCES FOR CHAPTER I

"Smith's Printing Press." *Mechanics' Magazine,* April 1833, 199; "Rutt's Printing Machine, Made by Napier (Hoe's Improvement)." *Mechanics' Magazine,* May 1833, 238; Grant Thorburn, *Forty Years' Residence in America: Or the Doctrine of a Particular Providence Exemplified in the Life of Grant Thorburn. . . .* Boston: Russell, Odiorne & Metcalf, 1834; Charles N. Haswell, *Reminiscences of an Octogenarian of the City of New York (1816-1860).* New York: Harper, 1897; A. Frederick Collins, *A Bird's Eye View of Invention.* New York: Crowell, 1926; John J. Scanlon, *The Passing of the Springfield Republican.* Amherst: Amherst College, 1950; James Moran, "The Columbian Press." *Journal of the Printing Historical Society,* 1969. *DAB,* Vol. XIII, 534-535 (Noah); Vol. IX, 105 (Hoe); Vol. IX, 490 (Thorburn); *Appleton's Cyclopedia of American Biography,* Vol. III, 225-226 (Hoe).

2. BUILDING THE FAST PRINTING MACHINES

1. New York *Herald,* 9 June 1886.
2. New York *Herald,* 9 June 1886.
3. *New York Tribune,* 6 July 1869.
4. J. C. Derby, *Fifty Years among Authors, Books and Publishers.* New York: Carleton, 1884, 106-107.
5. James Collins, 252.
6. Tucker, 353-355.
7. Tucker, 355, 356-357, 357-358, 361, 362, 369.

8. Philip Hone, *The Diary of Philip Hone, 1828-1851,* 2 Vols., Bayard Tuckerman, Ed. New York: Dodd, Mead, 1910, Vol. 1, 185, 257, 258, 261.
9. R. Hoe & Company stationery (Hoe-L).
10. Tucker, 376, 388.
11. "City Courts." Unidentified clipping, Hoe scrapbook (Hoe-L).
12. *Graham's Magazine,* 8.
13. *DAB,* Vol. V, 203 (Jones), quoting Francis Fowler, "Memoir of Dr. Thomas P. Jones." *Journal of the Franklin Institute,* July 1890, 6.
14. Tucker, 377.
15. Unidentified clipping, Hoe scrapbook (Hoe-L).
16. New York *Sun,* centennial issue, 1933.
17. Mott, 222.
18. Mott, 226.
19. Oswald, 213, quoting Day.
20. New York *Herald,* 9 June 1886.
21. *DAB,* Vol. II, 84 (M. Y. Beach).
22. Tucker, 371.
23. Silver, "Efficiency Improved," 344.
24. Hudson, 505.
25. Philadelphia *Ledger,* 25 March 1836 (first issue); Mott, 241.
26. Mott, 240.
27. Mott, 238.
28. Harvey W. Root, *The Unknown Barnum.* New York: Harper, 1927, 206.
29. George S. Bryan, Ed., *Struggles and Triumphs: Or, the Life of P. T. Barnum, Written by Himself,* 2 Vols. New York: Knopf, 1927, Vol. 2, 844.
30. Root, 51-52.
31. Boston *Times,* 30 Dec. 1853, clipping (Hoe-Y).
32. Brown, 5; Exman, *Brothers Harper,* 143.
33. Baehr, 13.
34. Baehr, 23-24.
35. Letter to R. Hoe & Company, 18 Aug. 1853 (Hoe-C).
36. Mott, 230.
37. John Bigelow, *William Cullen Bryant.* Boston: Houghton Mifflin, 1890, 72.
38. Henry Watterson, *"Marse Henry": An Autobiography,* 2 Vols. New York: Doran, 1919, Vol. 2, 231.
39. Davis, 8.
40. *Printer,* Jan. 1859, 214.
41. Mott, 231, quoting New York *Herald,* 31 Aug. 1835.
42. Carlson, 130-131.
43. Hone, Vol. 1, 193, 20 Jan. 1836.
44. Hudson, 460.
45. Hudson, 344.
46. Seitz, *James Gordon Bennetts,* 32, quoting the *Globe,* 29 Oct. 1832.
47. Mott, 401.
48. Tucker, 368-369, 369.
49. New York *Sun,* centennial issue, 1933.
50. Hudson, 647.
51. *Printer,* Jan. 1859, 214.

52. Moore, 241.
55. Tucker, 409-410.

ADDITIONAL SOURCES FOR CHAPTER 2

DAB, Vol. II, 82-84 (M. Y. Beach); *NCAB*, Vol. XIII, 308 (Day); 329 (M. S. Beach); *Appleton's Cyclopedia of American Biography*, Vol. VI, 403 (Webb); *SAM*, 16 Nov. 1850, 68 (Wilkinson); 27 Jan. 1855, 156 (M. S. Beach); 14 Mar. 1857, 210; 26 June 1858, 333; Edward R. Ellis, *The Epic of New York City*. New York: Coward-McCann, 1966.

3. EARLY MACHINE SHOP AND CONTRACT WORK

1. *NYT*, 19 Oct. 1855.
2. "Hoe & Company's Printing-Press and Saw Manufactory." *SAM*, 25 Dec. 1858, 126.
3. *NYT*, 19 Oct. 1855.
4. Richard M. Hoe Diary (Hoe-L).
5. "R. Hoe & Co." *New York Tribune*, 6 July 1869.
6. Tucker, 437.
7. R. Hoe & Company, 1875 (Hoe-C).
8. "Hoe & Company's Printing-Press and Saw Manufactory." *SAM*, 25 Dec. 1858, 126.
9. Tucker, 407.
10. Unidentified clipping c. July 1853 (Hoe-C).
11. Bridge, 87.
12. Tucker, 362.
13. Tucker, 364.
14. *Memoir of Hayward Augustus Harvey*. New York: 1900, 19.
15. Tucker, 362.
16. Harvey Memoir, 19-20.
17. Tucker, 377.
18. Harvey Memoir, 18.
19. Tucker, 366, 368.
20. Tucker, 385, 386.
21. Tucker, 384-385.
22. Tucker, 374.
23. Tucker, 412.
24. Tucker, 364, 367.
25. Tucker, 413.
26. Tucker, 414, 415, 417.
27. New York *Sun*, centennial issue, 1933.
28. Reid, 131.
29. *SAM*, 16 Jan. 1847, 133, quoting Morse in the Philadelphia *Ledger*.
30. Reid, 596.
31. *New York Tribune*, 6 July 1869.
32. Reid, 351.
33. Horace Greeley, *Recollections of a Busy Life*. New York: Ford, 1868, 142.
34. Gramling, 25.
35. Hudson, 612.
36. Reid, 1-2, 362, 696.

37. Tucker, 376.
38. W. T. Jeans, *The Creators of the Age of Steel.* New York: Scribner, 1884, 224.
39. *DNB,* Vol. 21 166-170 (Whitworth).
40. Tucker, 376-377, 389-390, 390.
41. Tucker, 382.
42. Tucker, 432.
43. Tucker, 436, 441-442, 442-443, 443.
44. Tucker, 444.
45. *New York Tribune,* 6 July 1869.
46. Tucker, 435.
47. Lossing, 480-481.
48. Stelzle, *Outlook,* 3 March 1926, 328.

ADDITIONAL SOURCES FOR CHAPTER 3

Jones, "Metalagraphic" letter. *Mechanics' Magazine,* May 1834, 308; Cable celebration. *Frank Leslie's Illustrated Newspaper,* 25 Sept. 1858, 259; Jacob Abbott, "Novelty Iron Works." *Harper's,* May 1851, 721-734; John A. Stevens, Ed., *Proceedings at the Mass Meeting of Loyal Citizens, on Union Square, New York, 15th Day of July, 1862. . . .* New York: Nesbitt, 1862; *NYT,* 21 May 1874 (Dyer); Joseph Illick, "The Story of the American Lumbering Industry." Waldemar Kaempffert, Ed., *A Popular History of American Invention,* 2 Vols., New York: Scribner, 1924, Vol. 1, 150-198; Edward C. Mack, *Peter Cooper: Citizen of New York.* New York: Duell, Sloan & Pierce, 1949; Maxwell F. Marcuse, *This was New York: A Nostalgic Picture of Gotham in the Gaslight Era.* New York: Carlton, 1965; C. Bradford Mitchell, "Pride of the Seas [Collins]." *American Heritage,* Dec. 1967, 64-88; Richard E. Huss, *"The Development of Printers' Mechanical Typesetting Methods.* Charlottesville: Bibliographical Society of the University of Virginia, 1973; *Diccionario Porrua,* 3rd Ed. (1971), Vol. 2, 1715 (Raphael); *DAB,* Vol. 17, 362 (W. Smith); *NCAB,* Vol. XIII, 63 (Harvey); Vol. 19, 201-202 (Saltus); Vol. 25, 242 (Ames); *Appleton's Cyclopedia of American Biography,* Vol. V, 690 (Stillman).

4. A PERSONAL BRAND OF SALESMANSHIP

1. Cornelius S. Van Winkle, *The Printers' Guide: Or an Introduction to the Art of Printing. . . .* New York: Van Winkle, 1836, Hoe advt; White & Hagar advt, 29.
2. "The Printing Press [Ruggles]." *Printer,* July 1859, 46.
3. *Printer,* July 1859, 46.
4. Bullen, "Evolution," 72.
5. Tucker, 412, 415-416.
6. Moran, 148.
7. Hoe Memorial, 41.
8. *Printers' Circular,* March 1875, 5.
9. McNamara, *INP,* March 1886, 314.
10. *New York Tribune,* 5 Feb. 1850.
11. *New York Tribune,* 9 Feb. 1850.
12. *New York Tribune,* 9 Feb. 1850.
13. Gambee, 49.

14. Fayette Copeland, *Kendall of the Picayune: Being His Adventures in New Orleans, on the Texas Santa Fe Expedition, in the Mexican War, and in the Colonialization of the Texas Frontier.* Norman: University of Oklahoma Press, 1943, 25.
15. Mott, 283.
16. Eckman, 95.
17. Tebbel, quoting Chicago *Tribune,* 10 Sept. 1871.
18. Tebbel, 49.
19. Tebbel, 49, quoting Charles Leavelle.
20. Kinsley, Vol. 2, 134, quoting Chicago *Tribune,* 12 Oct. 1871.
21. Kinsley, Vol. 2, 33, quoting Chicago *Tribune,* 12 Oct. 1871.
22. Hudson, 207.
23. Kinsley, Vol. 2, 159, quoting Chicago *Tribune,* 9 Oct. 1872.
24. Kinsley, Vol. 2, 186, quoting Chicago *Tribune,* 15 Nov. 1873.
25. Kinsley, Vol. 2, 146.
26. Telegram (Hoe-Y).
27. *NYT,* 31 July 1887 (S. Hoe).
28. *INP,* June 1886, 672.
29. Mott, 459.
30. "Valley of the Mississippi." *Printer,* March 1860, 226.
31. Ringwalt.
32. Oswald, 497 ftnt.
33. Don C. Seitz, *Artemus Ward (Charles Farrar Browne): A Biography and Bibliography.* New York: Harper, 1919, 18.
34. *SAM,* 8 Sept. 1849, 408, quoting *Athenaeum* (London).
35. Kemble, quoting publisher Colton.
36. Kesterton, 34.
37. Karolevitz, 167.
38. Karolevitz, 43.
39. "U.S. Book Trade." *Publishers' Weekly,* 25 March 1876, 409.
40. *American Bookmaker,* Jan. 1891, 12.
41. *SAM,* 6 April 1867, 217.
42. *New York Tribune,* 6 July 1869.

ADDITIONAL SOURCES FOR CHAPTER 4

Taylor explosion. *Journal of the Franklin Institute,* Vol. 49 (1850), 204-208; *INP,* Aug. 1887, 777 (S. Hoe); Feb. 1889, 421 (Taylor); *Typographic Advertiser,* Spring 1883, 785 (Painter); *American Newspaper Reporter,* 11 Dec. 1876, 910 (Lovell); Thomas E. Dabney, *One Hundred Great Years: The Story of the Times-Picayune from its Founding to 1940.* Baton Rouge: Louisiana State University, 1944; Madeleine B. Stern, *Imprints on History: Book Publishers and American Frontiers* [Lovell]. Bloomington: Indiana University Press, 1956; *DAB,* Vol. IV, 540-542 (Green); *NCAB,* Vol. 26, 190 (Philipp).

II. JOURNALISM AND DEMOCRACY: THE TECHNOLOGY OF NEWS

1. Morison, *The English Newspaper,* 4.
2. Richard D. Altick, *The English Common Reader: A Social History of the Mass Reading Public, 1800-1900.* Chicago: University of Chicago Press, 1957, 46, quoting Addison in *The Spectator.*

3. George Levine, Ed., *The Emergence of Victorian Consciousness: The Spirit of the Age.* New York: Free Press, 1967, 2, quoting from G. Kitson Clark, *The Making of Victorian England* (1962).

4. McLuhan.

5. Lewis A. Coser, *Men of Ideas: A Sociologist's View.* New York: Free Press, 1965, 8.

6. Coser, 5.

7. Mills, 41.

8. Altick, 324.

9. Vizetelly, Vol. 1, 93.

10. Altick, 354, quoting *Times* (London) in Hansard's *Parliamentary Debates* (1855).

11. McLuhan, 28.

12. Jacques Barzun, *Berlioz and His Century: An Introduction to the Age of Romanticism.* New York: Meridian, 1956, 239, 376.

13. Henry James, *Daumier: Caricaturist.* Emmaus: Rodale, 1954, 14.

14. Randall Davies, "Caricature." Warren E. Cox, Ed., *Graphic Arts.* Garden City: Garden City Publishing, 1936, 88.

15. James, 1-2, 14.

16. Roger L. Williams, *The World of Napoleon III.* New York: Free Press, 1957, 43.

17. McLuhan, 191.

18. F. Espinasse, "The House of Murray." *Harper's,* Sept. 1885, 504, 505.

19. Levine, 191, quoting from James Mill, "Periodical Literature." *Edinburgh Review* (Jan. 1824).

20. Altick, 131, quoting Henry Brougham, *Edinburgh Review* (1835).

21. Levine, 23, quoting Thomas Carlyle, "Signs of the Times." *Edinburgh Review* (June 1829).

22. Levine, 168, 170, quoting William Hazlitt, *The Life of Napoleon Bonaparte,* 4 Vols. (1828-1830).

23. Levine, 382-383, quoting Leigh Hunt, "Fiction and Matter-of-Fact." *Examiner* (Jan. 1825).

24. Levine, 74, 85, quoting John Mill, "The Spirit of the Age." *Examiner* (Jan. 1831).

25. Levine, 93, quoting John Mill, "Civilisation." *Westminster Review* (April 1836).

26. Levine, 34, quoting Thomas Carlyle, "Signs of the Times." *Edinburgh Review* (June 1829).

27. Levine, 197, quoting Charles Knight, *Penny Magazine* (1832).

28. J.A.C. Brown, *Techniques of Persuasion: From Propaganda to Brainwashing.* Harmondsworth: Penguin, 1963, 18-19.

29. "The London Newspaper." *Printer,* March 1860, 226.

30. Levine, 193, quoting W. B. Adams, *Junius Redivivus* (1833).

31. Altick, 346, 355.

32. James Bryce, *Modern Democracies.* New York: Macmillan, 1921, Vol. 1, 93.

33. McLuhan, 185.

34. Steinberg, 223.

35. James, 3, 4.

36. Asa Briggs, Ed., *William Morris: Selected Writings and Designs.* Harmondsworth: Penguin, 1962, 81, 82, 147-148.

37. Briggs, 289, quoting *News from Nowhere.*

38. William Irvine, *Apes, Angels, & Victorians: The Story of Darwin, Huxley, and Evolution.* New York: Time Magazine, 1955, xviii.

39. Andrew D. White, *A History of the Warfare of Science with Theology in Christendom.* New York: Free Press, 1965, 92, quoting Wilberforce in *Quarterly Review* (July 1860).

40. Irvine, 5, 154, letter to Charles Lyell.

41. Stephen J. Gould, "Darwin's Dilemma." *Natural History,* June-July 1974, 22.

42. Stephen J. Gould, "Darwin's Delay." *Natural History,* Dec. 1974, 70.

43. Crane Brinton, *The Shaping of the Modern Mind.* New York: New American Library, 1953, 178.

44. Harry E. Barnes, *An Intellectual and Cultural History of the Western World,* 3rd Ed. New York: Dover, 1965, Vol. 3, 979.

45. Barnes, 1109, quoting Lynn Thorndike, *A Short History of Civilization.*

ADDITIONAL SOURCES FOR PART II

Charles Knight, *The Old Printer and the Modern Press.* London: Murray, 1854; Fitz Roy Carrington, "Napoleon and the Invasion of England." *Print Collector's Quarterly,* April 1942; Edith Melcher, *The Life and Times of Henry Monnier, 1799-1877.* Cambridge: Harvard University Press, 1950; Helene Lecar, *Karl Marx.* New York: American RDM, 1966.

5. THE INCREDIBLE LIGHTNING PRESSES

1. George A. Tracy, *History of the Typographical Union.* New York: International Typographical Union, 1913, 94 ftnt.

2. Pasko, 387.

3. James Collins, 252.

4. *New York Herald,* 9 June 1886.

5. *National Intelligencer,* 22 July 1857.

6. Tucker, 380.

7. Johnson, Kent, Mencken, Owens, 87.

8. Hoe Memorial, 25, quoting John D. Flagg, letter 30 June 1868.

9. Exmam, *House of Harper,* 32.

10. "Hoe's Fast Press." *SAM,* 1 May 1847, 252, quoting *Ledger* (March 1847).

11. *SAM,* 4 Dec. 1847, 83.

12. *SAM,* 1 May 1847, 252.

13. *SAM,* 23 July 1853, 357.

14. Wayne Andrews, *Architecture, Ambition and Americans: A Social History of American Architecture.* New York: Free Press, 1964, 150.

15. Tucker, 371-372, 395

16. Tucker, 423.

17. *The Public Ledger Building, Philadelphia: With an Account of the Proceedings connected with its Opening, June 20, 1867.* Philadelphia: Childs, 1868, 164-165.

18. *SAM,* 5 Dec. 1846, 86.
19. "Color Printing Machine." *SAM,* 10 Oct. 1846, 20.
20. Jean Lipman, *Rufus Porter: Yankee Pioneer.* New York: Potter, 1968, 49-50, quoting *SAM,* 9 April 1849.
21. *SAM,* 20 Nov. 1847, 66.
22. New York *Sun,* centennial issue, 1933.
23. *SAM,* 24 May 1851, 283.
24. Isaacs, 37.
25. Oswald, 213, quoting Day.
26. J. C. Derby, *Fifty Years among Authors, Books and Publishers.* New York: Carleton, 1884, 37.
27. "Grand Supper—The New York Sun Printing Press." *SAM,* 15 Feb. 1851, 173.
28. *Graham's Magazine,* 8.
29. McNamara, *INP,* Jan. 1886, 198.
30. Hassard, 6-7, 10-13.
31. Hoe Memorial, 60.
32. Davis, 7.
33. Brown, 100, 101, quoting Maverick.
34. *NYT,* 18 Sept. 1851.
35. Augustus Maverick, *Henry J. Raymond and the New York Press, for Thirty Years: Progress of American Journalism from 1840 to 1870. . . .* Hartford: Hale, 1870, 156-157.
36. *SAM,* 13 June 1857, 317.
37. Hoe Memorial, 57, letter to Commissioner of Patents.
38. Emery, Ed., 65, quoting *NYT,* 28 March 1860.
39. *NYT,* 19 Oct. 1855.
40. Hoe Memorial, 10.
41. New York *Herald,* 9 June 1886.
42. Tucker, 439-440.
43. *Sunday School Gazette* (Boston), 19 March 1858, 53.
44. "Hoe's Mammoth Rotary Printing Press." *SAM,* 24 May 1851, 283.
45. McNamara, *INP,* Jan. 1886, 198.
46. Bullen, "Richard March Hoe," 852.
47. McNamara, *INP,* Dec. 1885, 137.
48. Tracy, 176.
49. McNamara, *INP,* Jan. 1886, 198.
50. Hoe Memorial, 56.
51. Hoe Memorial, Rev. Ed., Appendix.
52. "The Hoe Patent—Should it be Extended?" *Printers' Circular,* Feb. 1870, 450.
53. *Printers' Circular,* March 1870, 15, 16, letter to editor.
54. Moran, 178.
55. "The Printing Press." *Printer,* June 1859, 30.
56. Pasko, 395, 561.
57. Will Irwin, *The American Newspaper.* Ames: Iowa State University Press, 1969, 16, reprinted from *Colliers* (Jan.-July 1911).

ADDITIONAL SOURCES FOR CHAPTER 5

"First Type Revolving Press [Williams]." *American Bookmaker,* Dec. 1890, 184-185; George W. Childs, *Recollections.* Philadelphia: Lippincott, 1890;

DAB, Vol. II, 80-81 (A. Beach); Vol II, 407 (Bogardus); Vol. VIII, 101-102 (Porter); *NCAB,* Vol. VII, 83 (Munn); Vol. VII, 184 (Porter).

6. STEREOTYPES AND THE FIRST WEB MACHINES

1. John Johnson, *Typographia or the Printer's Instructor.* London: 1824.
2. Tucker, 397.
3. William Filmer, "Electro-Metallurgy." *Printer,* Dec. 1858, 189.
4. Tucker, 423-424.
5. Ringwalt.
6. Thomas Catling, *My Life's Pilgrimage.* London: Murray, 1911, 77-78.
7. Tucker, 411.
8. Bullen, "Richard March Hoe," 853.
9. Neipp, 231.
10. Tucker, 411.
11. Bullen, "Richard March Hoe," 853.
12. Catling, 79.
13. Ringwalt, 446.
14. Tucker, 416.
15. Tucker, 416.
16. US GPO, 42.
17. US GPO, 44, quoting Dr. A. P. Bernard.
18. *DAB,* Vol. III, 260 (Bullock).
19. *Printers' Circular,* Aug. 1870, 261, Bullock advt.
20. *Printers' Circular,* Aug. 1870, 261, quoting New York *Sun,* 14 July 1868.
21. Rowell, 200-201.
22. *New York Tribune,* 7 Aug. 1884; *INP,* 3 Dec. 1887, 129.
23. Tucker, 388, 389, 392-393, 394-395, 395-396, 403-404, 405.
24. Tucker, 429.
25. Tucker, 431, 432-433.
26. Tucker, 388-431.
27. Baehr, 67.
28. Royal Cortissoz, *The Life of Whitelaw Reid,* 2 Vols. New York: Scribner, 1921, Vol. 1, 370.
29. Baehr, 126.
30. *New York Tribune,* 8 Jan. 1875, letter to editor.
31. *New York Tribune,* 25 Jan. 1875.
32. *New York Tribune,* 15 July 1875.
33. J.M.S. Careless, *Brown of the Globe,* 2 Vols. Toronto: Macmillan, 1959-1963, Vol. 2, 135.
34. *The San Francisco Chronicle and its History.* San Francisco: 1879, 41, 44.
35. Hoe Memorial, 14, letter, 17 June 1868.
36. McNulty, 40, quoting Jonathan Swift.
37. Tucker, 446.
38. *SAM,* 30 July 1853, 362.
39. *SAM,* 27 Aug. 1853, 394; 26 Nov. 1853, 86.
40. Hassard, 14.

41. Lehmann-Haupt, 162, quoting *Paper World* (Feb. 1888).
42. Tucker, 438-439.
43. "Colonel Richard M. Hoe." *Harper's,* 9 Dec. 1876, 997.
44. Moran, 218-219.

ADDITIONAL SOURCES FOR CHAPTER 6

SAM, 13 June 1863, 374 (Bullock); "Inventor of Stereotyping—William Ged." *SAM,* 20 Feb. 1869, 114-115; *Publishers' Weekly,* 30 Dec. 1905, 1881 (Craske).

7. THE BOSTON FACTORY AND LUTHER CROWELL

1. James Collins, 255.
2. Mott, 45, quoting an early Boston *Transcript.*
3. Tucker, 369.
4. Tucker, 377.
5. "Purchase of the Adams Press Concern by R. Hoe & Co." *Printer,* May 1859, 4.
6. *Printer,* May 1859, 4.
7. R. P. Yorkston, "John Henry—Printer." *INP,* June 1889, 777.
8. *Printer,* May 1859, 4.
9. Lehmann-Haupt, 78.
10. Hoe Memorial, 58.
11. Hudson, 386.
12. Hoe Memorial, 20, letter, 29 May 1868.
13. Hoe Memorial, 16, telegram, 13 June 1868.
14. Tucker, 427.
15. Letter to R. Hoe & Company, Feb. 1858 (Hoe-C).
16. *Prospectus of the Mayall Elevated Railway Companies: Cable, Electric, and Pneumatic . . . : The Systems of the Future.* Reading: Thomas J. Mayall, 1887.
17. "Seventy Years of Invention." *SAM,* 5 June 1915, 517.
18. Tucker, 447.
19. Tucker, 444.
20. Robert Hoe, 52.
21. *NCAB,* Vol. XIII, 605 (Crowell).
22. *DAB,* Vol. IV, 575 (Crowell).
23. *NCAB,* Vol. XIII, 605.

ADDITIONAL SOURCES FOR CHAPTER 7

Crowell patents assigned to Hoe. *New York Tribune,* 2 July 1879; Horace E. Scudder, *Henry Oscar Houghton: A Biographical Outline.* Cambridge: Riverside Press, 1897; Joseph E. Chamberlin, *The Boston Transcript: A History of its First Hundred Years.* Boston: Houghton Mifflin, 1930; *Appleton's Cyclopedia of American Biography,* Vol. II, 218-219 (Dow); *NCAB,* Vol. II, 411 (T. Mead); Vol. 25, 386 (T. Mead).

8. ENGLAND AND WORLD-WIDE SUCCESS

1. Robert Hoe, 32.
2. Munsell, 177.

3. *SAM,* 17 Nov. 1849, 68.
4. Phebe A. Hanaford, *The Life of George Peabody.* Boston: Russell, 1870, 78-79, quoting Edward Everett.
5. Hudson, 540-542.
6. Robert Hoe, 37.
7. Thomas Catling, *My Life's Pilgrimage.* London: Murray, 1911, 53.
8. Catling, 53, 138.
9. Hatton (1882), 191.
10. Tucker, 398.
11. *SAM,* 24 May 1856, 289.
12. Howe, 16.
13. Hassard, 6.
14. Howe, 16.
15. P. M. Handover, *Printing in London, from 1476 to Modern Times.* Cambridge: Harvard University Press, 1960, 164.
16. *Times* (London), 29 Dec. 1848.
17. Moran, 186, quoting letter, 8 March 1851.
18. Howe, 20.
19. Ringwalt, 42.
20. Morison, *History of the Times,* Vol. 2, 348.
21. The Times, *Printing the Times,* 44.
22. Tucker, 402
23. Nathan Rosenberg, Ed., *The American System of Manufactures: The Report of the Committee on the Machinery of the United States, 1855, and the Special Reports of George Wallis and Joseph Whitworth, 1854.* Edinburgh: Edinburgh University Press, 1969, 389.
24. New York *Herald,* 15 Dec. 1856.
25. Ayerst, 130.
26. McNamara, *INP,* Jan. 1886, 198.
27. The Times, *Printing the Times,* 44.
28. Howe, 21, quoting *Printers' Register* (Dec. 1867).
29. Hassard, 18.
30. Catling, 78, 78-79, 79.
31. Tucker, 418.
32. Hatton (1882), 189-191.
33. Joseph Hatton, "Journalistic London." *Harper's,* Dec. 1881, 43, 44.
34. Hatton (1882), 191.
35. Wilson and Grey, 204.
36. *Times* (London), 8 March 1887.
37. Jackson, 307.
38. Peter Jackson, Ed., *John Tallis's London Street Views, 1838-1840.* London: Nattali & Maurice, 1969, 25.
39. Vizetelly, Vol. 1, 404.
40. Jackson, 2.
41. *Mechanics' Magazine* (London), 26 Aug. 1848, 194-196; 2 Sept. 1848, 237-238.
42. *SAM,* 6 April 1850, 228.
43. "Hippolyte Marinoni." *Bulletin Officiel,* Jan. 1904, 11-12.
44. *Bulletin Officiel,* 12.
45. Neipp, 150.
46. *SAM,* 21 Feb. 1863, 120, 121.

47. Howe, 24.
48. Moran, 211.
49. The Times, *Printing the Times,* 49.
50. Howe, 25 ftnt, quoting *Printers' Register* (1868).
51. Grant, Vol. 2, 30-32, quoting *Printers' Register* (1870).
52. Jackson, 325-326.
53. *Engineer* (London), 5 Oct. 1877, 239.
54. Hatton (1882), 68.
55. Samuel Smiles, *Men of Invention and Industry.* New York: Harper, 1885, 203.
56. Howe, 28, quoting *Printers' Register* (1873).
57. The Times, *Newspaper History,* 48.
58. Pottinger, 336.

ADDITIONAL SOURCES FOR CHAPTER 8

Geoffrey T. Hellman, "The First Great Cheerful Giver [Peabody]." American Heritage, *Great Stories of American Businessmen.* New York: 1972, 119-125.

III. DEVELOPMENT OF MASS MEDIA: THE GIANT NEWSPAPER PRESSES

1. Mott, 168, quoting Thomas Hamilton, *Men and Manners in America* (1833).
2. Bartow, 84, quoting the first issue.
3. Silver, *American Printer,* preface.
4. Mott, 169, 200.
5. Pickett, 399-400.
6. William Cullen Bryant, "Darwin and Monkeys. . . ." *Prose Writings of William Cullen Bryant,* Parke Godwin, Ed. New York: Appleton, 1884, Vol. 2, 293.
7. Alexis de Tocqueville, *Democracy in America,* 2 Vols. New York: Vintage, 1960, Vol. 1, 193-194.
8. De Tocqueville, Vol. 1, 190, quoting *Vicennes Gazette.*
9. New York *Sun,* centennial issue, 1933.
10. Walter Lippmann, *Public Opinion.* New York: Macmillan, 1922, 331.
11. Hudson, 427.
12. New York *Sun,* centennial issue, 1933.
13. Philip Hone, *The Diary of Philip Hone, 1828-1851,* 2 Vols., Bayard Tuckerman, Ed. New York: Dodd, Mead, 1910, Vol. 2, 669.
14. *SAM,* 9 May 1852, 270, quoting Boston *Museum.*
15. Rowell, viii.
16. McLuhan, 183.
17. Payne, 368-369, quoting *Review of Reviews* (Feb. 1912).
18. Payne, 60.
19. Bartow, 270, quoting issue of 4 Jan. 1851.
20. Mott, 383.
21. Hone, Vol. 2, 393 (16 Oct. 1850).
22. Alexander B. Callow, Jr., *The Tweed Ring.* New York: Oxford University Press, 1966, 18.

23. *NYT,* 16 July 1869.
24. Callow, 254.
25. *INP,* Nov. 1906, 195.
26. Rowell, viii.
27. Paul W. Gates, *The Farmer's Age: Agriculture, 1815-1860.* New York: Holt, Rinehart & Winston, 1960, 313.
28. Boston *Transcript,* 11 Aug. 1856.
29. *SAM,* 6 April 1850, 229.
30. Edward Lurie, *Louis Agassiz: A Life in Science.* Chicago: University of Chicago Press, 1960, 253, 255.
31. *Fun under the Old White Hat* [Greeley]. New York: Fay & Cox, 1872, 47.
32. William Stanton, *The Leopard's Spots: Scientific Attitudes toward Race in America, 1815-1859.* Chicago: University of Chicago Press, 1960, 183, quoting letter to George Squier.
33. Lurie, 295.
34. *Los Angeles Times,* 31 Dec. 1972.
35. Clarence Darrow, *The Story of My Life.* New York: Grosset & Dunlap, 1932, 249.
36. Mott, 412, quoting Reid.
37. Mott, 419.
38. Mott, 434, quoting Pulitzer.
39. Mott, 501.
40. Mott, 524.
41. Mott, 527.
42. Mott, 253.
43. Ingersoll, 293.
44. de Tocqueville, Vol. 1, 195.
45. Ringwalt, 317.
46. de Tocqueville, Vol. 1, 195.
47. George F. Mott, Ed., *New Survey of Journalism.* New York: Barnes & Noble, 1950, 53.
48. Bagdikian, 273.
49. James Bryce, *Modern Democracies.* New York: Macmillan, 1921, Vol. 1, 97.
50. Lippmann, 326.
51. Fred S. Siebert, Theodore Peterson, and Wilbur Schramm, *Four Theories of the Press.* Urbana: University of Illinois Press, 1963, 88.
52. Robert M. Hutchins, *A Free and Responsible Press.* Chicago: University of Chicago Press, 1947, 22.
53. Hutchins, 43, 60, quoting William Allen White.
54. Hutchins, 131, quoting Charles Beard.
55. Bryce, 71, 109.
56. Bagdikian, 46.
57. Bryce, 466.
58. McLuhan, 226.
59. *NYT,* Book Review, 11 April 1971, letter to editor.
60. Lippmann, 363.
61. Hutchins, 88-89.
62. George Orwell, *Nineteen Eighty-Four.* Harmondsworth: Penguin, 1954, 38.

63. Siebert, Peterson, and Schramm, 103.
64. Hohenberg, 490, 497-498.

ADDITIONAL SOURCES FOR PART III

John Bigelow, *William Cullen Bryant.* Boston: Houghton Mifflin, 1890; Edward R. Ellis, *The Epic of New York City.* New York: Coward McCann, 1966; *NCAB,* Vol. III, 388 (F. Wood).

9. GENIUS AT WORK IN THE EXPERIMENT ROOM

1. Tucker, 353, 355, 372-373.
2. Tucker, 379, 390.
3. *DAB,* Vol. 19, 40 (Tucker).
4. Tucker, 390-391, 391-392.
5. Robert Hoe, 22-27.
6. Tucker, 402.
7. Aimee Tucker to Florence, 23 May 1933 (SDT).
8. Robert Hoe, 42.
9. Tucker, 433.
10. *SAM,* 5 June 1915, 517.
11. Tucker, 433.
12. Tucker, 434.
13. Robert Hoe, 47.
14. Wilson and Grey, 209.
55. Theodore L. De Vinne, "Perfecting the Press." *SAM,* 14 June 1902, 22122.
16. Tucker, 424.
17. Robert Hoe, 48-51.
18. Robert Hoe, 51.
19. Tucker, 434, 434-435.
20. Telegram, 20 Jan. 1877 (Hoe-C).
21. *Engineer* (London), 30 March 1877, 215.
22. Tucker, 438.
23. Digby, 115.
24. Baehr, 27.
25. *NYT,* 19 April 1888.
26. Oswald, 222.
27. Lewis M. Allen, *Printing with the Handpress.* New York: Van Nostrand Reinhold, 1969, 11, quoting Bruce Rogers.
28. Grant, Vol. 1, 165.
29. John R. Howard, *Remembrance of Things Past: A Familiar Chronicle of Kinsfolk and Friends Worth While.* New York: Crowell, 1925, 246.
30. Robert Hoe, 51.
31. Theodore L. De Vinne, "The Printing of 'The Century.'" *Century,* Nov. 1890, 93, 94.
32. Hassard.
33. Tucker, 449.
34. Carl Schurz, *The Reminiscences of Carl Schurz,* 3 Vols. New York: McClure, 1907-1908, Vol. 2, 232, 330.
35. Juergens, 16, quoting Pulitzer.

36. Juergens, 348.
37. Juergens, 50 ftnt, 348 ftnt, quoting *Journalist* (April, May 1884).
38. Juergens, 349, quoting an issue of New York *World* (Oct. 1884).
39. Juergens, 349.
40. Seitz, *Joseph Pulitzer,* 153.
41. Tucker, 449.
42. Robert Hoe, 52.
43. Isaacs, 53, 54-55, 57, 58.
44. Unidentified clippings (SDT).
45. *Times* (London), 13 Oct. 1902.
46. Aimee Tucker to Florence, 23 May 1933 (SDT).

ADDITIONAL SOURCES FOR CHAPTER 9

NCAB, Vol. 27, 14-15 (Chambers); *NYT,* 19 April 1888 (Hassard); "Museum Dials Tell Sunny Hours." *NYT,* 23 Dec. 1967; Bruce Chandler and Clare Vincent, "A Sure Reckoning: Sundials of the 17th and 18th Centuries." *Metropolitan Museum of Art Bulletin,* Dec. 1967, 154-169.

10. HUMAN RELATIONS AROUND THE SHOP

1. *Graham's Magazine,* 570.
2. Tucker, 432.
3. *National Intelligencer,* 22 July 1857.
4. "The Old and World-Renowned Establishment of R. Hoe & Co." *Paper World,* 1880.
5. *New York Tribune,* 19 July 1898.
6. *American Printer,* Sept. 1911.
7. "Hoe May Fight the Machinist Union to Establish Open Shop." *E&P,* 25 Aug. 1923, 8.
8. *Newspaperdom,* 13 Sept. 1923, 17.
9. *New York Tribune,* 12 April 1882.
10. "A Successful Factory School." *American Monthly Review of Reviews,* Sept. 1906, 322.
11. "The Making of a Newspaper Press." *INP,* Dec. 1916.
12. Richard M. Hoe, Diary (Hoe-L).
13. *Graham's Magazine,* 12.
14. Tucker, 432.
15. *Harper's,* 9 Dec. 1876, 997.
16. Lossing, 480-481.
17. "Apprentice School." *New York Tribune,* 15 June 1902.
18. Stelzle, *Outlook,* 24 March 1926, 451-453.
19. Stelzle, *Outlook,* 31 March 1926, 489-492.
20. Gregory Weinstein, *The Ardent Eighties: Reminiscences of an Interesting Decade.* New York: International Press, 1928, 23-25, 143.
21. Samuel Gompers, *Seventy Years of Life and Labor: An Autobiography.* New York: Dutton, 1925, Vol. 2, 45.
22. *New York Tribune,* 6 July 1869.
23. *NYT,* 29 July 1902; 2-3, 5-7, 9, 12-14, 19-22, 26 Aug. 1902; 1-3, 18 Sept. 1902; 21 Oct. 1902.
24. *NYT,* 31 July 1902.
25. *NYT,* 16 Sept. 1902.

26. *NYT,* 16 Sept. 1902.
27. *NYT,* 1 Aug. 1902.

ADDITIONAL SOURCES FOR CHAPTER 10

Fourth Estate, 30 Dec. 1922 (Raabe); *E&P,* 30 Dec. 1922 (Raabe); "British Lockout Postponed a Week." *NYT,* 11 March 1926.

11. MULTICYLINDERS AND PERFECTING MACHINES

1. Bridge, 85.
2. Bridge, 87.
3. McCabe, paraphrasing Bridge.
4. Robert Hoe, 59.
5. Arthur Robb, "Modern Presses Began in Eighties." *E&P,* 21 July 1934, 198.
6. Seitz, *Joseph Pulitzer,* 154.
7. Churchill, 19, quoting Jacob Riis.
8. Seitz, *Joseph Pulitzer,* 176.
9. Juergens, 49.
10. Johnson, Kent, Mencken, Owens, 244.
11. W. A. Swanberg, *Citizen Hearst: A Biography of William Randolph Hearst.* New York: Bantam, 1963, 96-97.
12. Isaacs, 66-67, quoting New York *Herald,* 10 May 1891; Robert Hoe, 65-66.
13. "An Artistic Nameplate." *Fourth Estate,* 28 July 1900.
14. Isaacs, 72, 73.
15. Robb, 299-300.
16. *Progress of British Newspapers in the Nineteenth Century, Illustrated.* London: Simpkin, Marshall, Hamilton, Kent, 2-5, Hoe advt.
17. Isaacs, 79.
18. "Hoe Double Octuple." *E&P,* 15 Sept. 1906, quoting *British Newspaper Owner.*
19. Isaacs, 82.
20. Robert Hoe, 86-87.
21. Swanberg, 62.
22. "Pancoast, Mechanical Genius, Dies at 77." *E&P,* 18 March 1939, 12.
23. Swanberg, 168.
24. Bartow, 237, quoting New York *Journal* (17 Oct. 1896).
25. Coulton Waugh, *The Comics.* New York: Macmillan, 1947, 1.
26. Swanberg, 192.
27. John K. Winkler, *William Randolph Hearst: A New Appraisal.* New York: Hastings House, 1955, 70.
28. Swanberg, 262.
29. R. Hoe & Company, "What Publishers Say." New York: 1925.
30. Roblin, 47, 48, 53.
31. Geoffrey T. Hellman, "Printer [Bertram A. Powers]." *New Yorker,* 7 March 1970, 43, quoting Powers.
32. Roblin, 24.
33. James Collins, 261.

34. *American Pressman,* Dec. 1922, letter to editor; *E&P,* 4 Nov. 1922.
35. Charles Francis, *Printing for Profit.* New York: Francis, 1917, 44.
36. Isaacs, 85.
37. "Henry A. Wise Wood Dies; Was Mechanical Genius." *E&P,* 15 April 1939, 12.
38. Henry A. Wise Wood, *Money Hunger: A Brief Study of Commercial Immorality in the United States.* New York: Putnam, 1908, 38, 45.
39. "Henry A. Wise Wood [Editorial]." *E&P,* 15 April 1939, 24.
40. Robb, 208.
41. *SAM,* 5 June 1915, 550.
42. *E&P,* 15 April 1939, 24.
43. "The Autoplate." *SAM,* 26 Oct. 1901, 21592.
44. J. D. Symon, *The Press and its Story: An Account of the Birth and Development of Journalism up to the Present Day. . . .* London: Seeley, Service, 1914, 28-29.
45. *Fourth Estate,* 22 Nov. 1919, 12.
46. Robb, 300, quoting Henry A. W. Wood.
47. Robb, 300.
48. Quentin Reynolds, *The Fiction Factory: Or from Pulp Row to Quality Street* [Street & Smith]. New York: Random House, 1955.
49. Francis, 60, 72, 105.
50. George Britt, *Forty Years—Forty Millions: The Career of Frank A. Munsey.* New York: Farrar & Rinehart, 1935, 117-118.
51. *DAB,* Vol. XIII, 334 (Munsey), quoting O. G. Villard.
52. Churchill, 291.
53. R. Hoe & Company, Catalog, 1899, 40.
54. Houghton.
55. *Munsey's Magazine,* July 1911, Hoe advt.
56. Seitz, *James Gordon Bennetts,* 220.
57. Britt, 255, 256, 257.
58. Houghton.

ADDITIONAL SOURCES FOR CHAPTER 11

"The New Printing Machines." *INP,* Dec. 1888, 173-178; "Frank A. Munsey's New Enterprise." *INP,* Nov. 1906, 268; "Death Suddenly Summons Frank Munsey." *E&P,* 26 Dec. 1925, 3; Erman J. Ridgway, *This for Remembrance* [Munsey]. Chula Vista: 1926; *NYT,* 22 Aug. 1940 (Potter); *NCAB,* Vol. 17, 330 (Spalckhaver); Vol. 20, 47 (Munsey); Vol. 41, 26 (Wood).

12. THE HOES IN PRIVATE LIFE

1. Thorburn, 58.
2. Edward H. Dodd, Jr., *The First Hundred Years: A History of the House of Dodd, Mead, 1839-1939.* New York: Dodd, Mead, 1939, 16-17.
3. Lehmann-Haupt, 391.
4. Robert Hoe, *A Lecture on Bookbinding as a Fine Art, Delivered before the Grolier Club, February 26, 1885.* New York: Grolier Club, 1886, 9.
5. "Frank Howard Dodd—A Memory." *Publishers' Weekly,* 1 April 1916, 1108, from *Bookman* (April 1916).

6. Mackellar, 215 ftnt.
7. "Robert Hoe." *Harper's,* 27 Sept. 1884, 633.
8. "Richard March Hoe." *Typographic Advertiser,* Fall 1886 (GLF).
9. *NYT,* 19 Oct. 1855.
10. Juergens, quoting New York *World.*
11. Harper, *House of Harper,* 436, 559.
12. *NYT,* 13 Jan. 1908.
13. *New York Tribune,* 23 Feb. 1882.
14. *Typographic Advertiser,* Fall 1886 (GLF).
15. *INP,* Sept. 1885, 565.
16. *INP,* Dec. 1885, quoting Baltimore *Sun.*
17. *NYT,* 20 June 1886.
18. Bullen, "Richard March Hoe," 854.
19. *NYT,* 14 April 1903.
20. *NYT,* 28 Nov. 1927.
21. Unidentified clipping (SDT).
22. Robert Hoe composition (Hoe-L).
23. Bridge, 86.
24. Alexander C. Flick, Ed., *History of the State of New York: The Age of Reform.* New York: Columbia University Press, 1934, Vol. 6, 242.
25. Wedding invitation (GLF).
26. "Weddings." Unidentified clipping (GLF).
27. Wedding invitation (GLF).
28. New York *Sun,* 30 Sept. 1891.
29. *New York Tribune,* 30 Sept. 1891.
30. "The Week in Society." Unidentified clipping (GLF).
31. Lillian D. Wald, *The House on Henry Street.* New York: Holt, 1915, 80.
32. Carl L. Cannon, *American Book Collectors and Collecting: From Colonial Times to the Present.* New York: Wilson, 1941, 157, 158, 169.
33. "Bibliomania." Unidentified clipping, quoting New York *Evening Mail* (GLF).
34. *NYT,* 24 May 1891, reviewing Charles Holder, *Charles Darwin.*
35. Cannon, 160.
36. Elbert A. Thompson and Lawrence S. Thompson, *The Club Bindery.* Rochester: University of Rochester Press, 1954, 34.
37. Cannon, 158.
38. O. A. Bierstadt, *The Library of Robert Hoe: A Contribution to the History of Bibliophilism in America.* New York: Duprat, 1895, 4-5, 7, 216.
39. "Robert Hoe III." *NYT,* 23 Sept. 1909.
40. *NYT,* 12 May 1911.
41. *NYT,* 12 May 1910.
42. *NYT,* 26 Feb. 1911; 12 May 1911.
43. *NYT,* 7, 8 June 1911.
44. *NYT,* 2 March 1912.
45. Robert Hoe composition (Hoe-L).
46. "Old Books that Brought a Million Dollars." *American Printer,* June 1911, 444-445.
47. Edwin Wolf II with John Fleming, *Rosenbach: A Biography.* Cleveland: World, 1960, 77.
48. *NYT,* 2, 3 May 1911, quoting letter of 29 April.

49. Cannon, 169.
50. *INP,* Dec. 1909, 408.
51. *NYT,* 2 June 1914.
52. *NYT,* 27 May 1914.
53. Wesley Towner, *The Elegant Auctioneers.* New York: Hill & Wang, 1970.
54. *NYT,* 29 May 1914.
55. "Conversational Currency [Editorial]." *NYT,* 27 May 1914.
56. *NYT,* 2 June 1914.
57. *NYT,* 3 June 1914.
58. *NYT,* 4 June 1914.
59. *NYT,* 23 Oct. 1927.
60. *NYT,* 27 May 1917.
61. *NYT,* 27 May 1917.
62. New York *Evening Post,* 2 June 1927.
63. *NYT,* 20 Jan. 1937.

ADDITIONAL SOURCES FOR CHAPTER 12

John T. Cunningham, "Historian on the Double [Lossing]." *American Heritage,* June 1968, 54-64, 78-81; John T. Winterich, *The Grolier Club, 1884-1950: An Informal History.* New York: Grolier Club, 1950; *DAB,* Vol. V, 340 (Dodd); *Times* (London), 4 March 1910 (Hoe will); *NYT,* 23 Sept. 1909 (Hoe); 8, 9 Oct. (funeral); 14 Oct. 1909 (will).

IV. NEWSPAPERS IN TROUBLE: THE UNHAPPY INSTITUTION

1. Harry T. Peters, *Currier & Ives: Printmakers to the American People.* Garden City: Doubleday, Doran, 1942, 5.
2. Tucker, 426.
3. Lossing, 762.
4. Tucker, 408.
5. Invitation (Hoe-L).
6. Tucker, 439 ftnt.
7. Roblin, 71.
8. Roblin, 82.
9. Ringwalt, 317.
10. *E&P,* 21 July 1934, 300.
11. Churchill, 316, 328.
12. New York *Sun,* centennial issue, 1933.
13. Baehr, 394.
14. Hohenberg, 446.
15. Clay Felker, "Will Newspapers ever Enter the Twentieth Century?" *Dot Zero,* Spring 1967, 12.
16. Lord Thomson, "Newspapers." *Lithopinion* 12 (1968), 55.
17. *Fourth Estate,* 22 Nov. 1919, 12.
18. Moran, 13, 221.
19. Lindstrom, 7, 18, 108, 109, 245.
20. *INP,* June 1971, 75
21. Michael Chazin, Ed., "Newspaper Operations: The Future Arrives Sooner than Expected." *INP,* June 1978, 43, quoting Erwin Jaffee, director of ANPA/RI Research Center.

22. *INP,* Nov. 1968, Wood advt; April 1972, Wood-Hoe advt.
23. "What is Offset, Letterpress, Gravure, and Prints 24,000 Books an Hour?" *Book Production Industry,* July 1974, 27-29.
24. Churchill, 231.
25. *NYT,* 3 July 1978.
26. Charles Darwin, *The Origin of Species by Means of Natural Selection. . . .* New York: Modern Library, 1936, 371.

ADDITIONAL SOURCES FOR PART IV

Fred C. Shapiro, "The Life and Death of a Great Newspaper [Herald-Tribune]." *American Heritage,* Oct. 1967, 97-112.

13. FAMILY REORGANIZATION OF THE FIRM

1. "Regarding Henry L. Bullen." *INP,* Nov. 1905, 284.
2. Lehmann-Haupt, 184.
3. Hoe letter (ATF).
4. Beatrice Warde, *The Crystal Goblet: Sixteen Essays on Typography.* Cleveland: World, 1956, 187.
5. Warde, 7.
6. Lehmann-Haupt, 157.
7. Bullen, "Evolution," 854.
8. Bullen, "Collectanea," 621.
9. *INP,* Nov. 1909, 217.
10. *NYT,* 23 Sept. 1909.
11. *Fourth Estate,* 22 March 1924, 8.
12. McCabe.
13. "R. Hoe & Co. Reorganize." *National Lithographer,* Feb. 1910, 22; *Fourth Estate,* 15 Jan. 1910.
14. R. Hoe & Company book of procedures (Hoe-B).
15. McCabe.
16. *INP,* Nov. 1906, 195.
17. *Daily Worker,* 11 April 1928.
18. *NYT,* 18 Nov. 1924.
19. R. Hoe & Company, letter, 4 Oct. 1924 (WHB).
20. *NYT,* 21 Dec. 1924; 19 May 1945 (Gossler); 19 March 1950 (Cutler); 23 Feb. 1952 (Kellogg); 15 May 1963 (Stanley).
21. Morgan Guaranty Trust Company, 23 *Wall Street.* New York: 1964.
22. *NYT,* 25 Dec. 1925.
23. "Printer's Devil Goes to Top of the Ladder [Swartz]." *NYT,* 3 Jan. 1926.
24. *NYT,* 3 Jan. 1926.
25. *NYT,* 25 July 1928.
26. "R. Hoe & Company Announces New Super-Production Press." *E&P,* 28 July 1928, 9.
27. *NYT,* 25 July 1928.
28. *NYT,* 3 April 1929.
29. *NYT,* 4 April 1929.
30. *NYT,* 2 May 1929.
31. *NYT,* 4 April 1925; 24 Sept. 1925; 25 Dec. 1925; 17 March 1926;

30 Aug. 1928; 13 Aug. 1929; 1 March 1930; 5 Aug. 1930; 18 Aug. 1931.
32. Morgan Guaranty Trust Company.
33. *NYT*, 28 April 1932.
34. Samuel Zirn Letter, 9 June 1932 (WHB); *NYT*, 12 June 1932.
35. *NYT*, 24 July 1932.
36. R. Hoe & Company, *Hoe . . . Perfecting Presses.* New York: 1932.
37. *NYT*, 30 June 1933.
38. Samuel Zirn letter, 26 June 1933 (WHB).
39. Samuel Zirn letter, 26 June 1933, quoting New York *World-Telegram*, 24 May 1932.
40. Henry A. Wise Wood, *Money Hunger: A Brief Study of Commercial Immorality in the United States.* New York: Putnam, 1908, 124-125.
41. Samuel Zirn letter, 19 Sept. 1933, brackets in original (WHB).
42. Samuel Zirn letter, 14 Sept. 1934 (WHB).
43. *NYT*, 30 Oct. 1934.
44. *NYT*, 16 April 1936.
45. *NYT*, 16 April 1936.
46. *NYT*, 10 Dec. 1936.
47. *NYT*, 26 Jan. 1937.
48. *NYT*, 31 March 1937.
49. *Times* (London), 23 July 1938; 23 Sept. 1938 (Hoe-Crabtree).
50. R. Hoe & Company, Annual Report, Feb. 1939 (Hoe-B).
51. Wood, 120-121.
52. *NYT*, 27 Oct. 1941.

ADDITIONAL SOURCES FOR CHAPTER 13

Stevens L. Watts, *Henry Lewis Bullen, Printer.* Front Royal, Va.: Privateer Press, 1960; " 'I am a Communicator': A Selection of Writings and Talks by Beatrice Warde/Paul Beaujon." *Monotype Recorder,* Autumn 1970; Bullen prospectus. *INP,* March 1907, 901; "Typographical Library—A Mecca for Printers." *Pacific Printer and Publisher,* Aug. 1929, xlii-2.

14. PARADOXICALLY UNFIT FOR SURVIVAL

1. Bridge, 87.
2. *United States Daily* (Washington), 16 Sept. 1927.
3. *E&P,* 8 April 1944, Hoe advt.
4. *NYT*, 11 Sept. 1942.
5. R. Hoe & Company, "Victory Rally [souvenir]." 1942? (Hoe-B).
6. *E&P,* 13 Oct. 1945.
7. *NYT*, 26 Nov. 1952.
8. *NYT*, 10 July 1948.
9. *NYT*, 16 March 1946.
10. *E&P,* 13 Oct. 1945.
11. *NYT*, 14 June 1946.
12. "Big Strides seen in Color Printing." *NYT*, 21 Dec. 1947.
13. *NYT*, 19 Aug. 1950.
14. *NYT*, 10, 11, 12 Oct. 1950.
15. *NYT*, 9 Nov. 1950.
16. *NYT*, 15 May 1951.
17. *E&P,* 9 June 1945, 68.

18. *NYT,* 4 March 1952.
19. "He Fought his Way Back In [Auer]." *Business Week,* 1 May 1954, 70.
20. *Business Week,* 1 May 1954, 73, 74.
21. "Sea Master Tooling Nearly Complete." *Aviation Week,* 28 Jan. 1957, 78-80.
22. *NYT,* 8 Feb. 1957.
23. *American Pressman,* Nov. 1965, Hoe advt.
24. *E&P,* 22 April 1961, Hoe advt.
25. *NYT,* 4 July 1969.
26. *NYT,* 4 July 1969.
27. *E&P,* 12 July 1969, 60.
28. "The Sad Story Printed by R. Hoe." *Business Week,* 19 July 1969, 36.
29. *Business Week,* 19 July 1969, 36.
30. *E&P,* 26 July 1969, 9; 13 Sept. 1969, 9.
31. *E&P,* 13 Sept. 1969, 9.
32. *E&P,* 13 Sept. 1969, 9, 10.
33. *E&P,* 10 Jan. 1970, 11.
34. *E&P,* 10 Jan. 1970, 11.
35. *E&P,* 11 Oct. 1969, 15.
36. *E&P,* 11 Oct. 1969, 15.
37. *E&P,* 6 Sept. 1969, 10.
38. *E&P,* 31 Jan. 1970, 14.
39. *E&P,* 6 March 1971, 43.
40. *E&P,* 27 June 1970, 39.
41. *E&P,* 6 May 1972, 34.
42. *E&P,* 10 July 1971, 10; 25 March 1972, 79.
43. *E&P,* 6 March 1971, 42.
44. *E&P,* 5 Aug. 1972, 11.
45. James B. Kilsheimer III and Robert M. Corrao, *Trustees' Report and Notice Pursuant to Section 167(5) and (6) of Chapter X of the Bankruptcy Act, 1 November 1974.* New York: Winthrop, Stimson, Putnam & Roberts and Townley, Updike, Carter & Rodgers, 6.
46. Kilsheimer and Corrao, 9-11.
47. Kilsheimer and Corrao, 8-9 quoting Hoe treasurer D. Kingsley Thorpe.
48. Kilsheimer and Corrao, 10.
49. David Langford, "Fires of Urban Ruin Rage in South Bronx." *Los Angeles Times,* 13 Nov. 1977.

ADDITIONAL SOURCES FOR CHAPTER 14

NYT, 26 May 1963 (Auer); "Hoe (R.) & Co." Standard A.S.E. Stock Reports, Standard & Poor's Corp., 9 July 1970; News releases. Securities and Exchange Commission, 2, 10, 17 July 1969.

Bibliography

Ayerst, David, *The Manchester Guardian: Biography of a Newspaper.* Ithaca: Cornell University Press, 1971, 702 pp.
Baehr, Harry W., Jr., *The New York Tribune Since the Civil War.* New York: Dodd, Mead, 1936, 420 pp.
Bagdikian, Ben H., *The Information Machines: Their Impact on Men and the Media.* New York: Harper & Row, 1971, 359 pp.
Bartow, Edith M., *News and These United States.* New York: Funk & Wagnalls, 1952, 292 pp.
Berry, W. Turner, and E. Edmund Poole, *Annals of Printing: A Chronological Encyclopedia from the Earliest Times to 1950.* London: Blandford, 1966, 316 pp.
Bridge, James H., "Captains of Industry, Part VII: Robert Hoe [III]." *Cosmopolitan,* Nov. 1902, pp. 85–88.
Brown, Francis, *Raymond of the Times.* New York: Norton, 1951, 345 pp.
Bullen, Henry L., "Collectanea Typographica." *Inland Printer,* Feb. 1921, pp. 621–622.
—, "The Evolution of American Printing Presses." *American Printer,* July 1926, pp. 68–73.
—, "Richard March Hoe and the Evolution of Fast Printing Presses." *Inland Printer,* Sept. 1922, pp. 851–854.
Burke, Jackson, *Prelum to Albion: A History of the Development of the Hand Press from Gutenberg to Morris.* San Francisco: Burke, 1940, 25 pp.
Burlingame, Roger, *March of the Iron Men: A Social History of Union through Invention.* New York: Grosset & Dunlap, 1938, 500 pp.
Carlson, Oliver, *The Man who Made News: James Gordon Bennett.* New York: Duell, Sloan & Pierce, 1942, 440 pp.
Churchill, Allen, *Park Row.* New York: Rinehart, 1958, 344 pp.
Clair, Colin, *A History of Printing in Britain.* London: Cassell, 1965, 314 pp.
Collins, Irene, *The Government and the Newspaper Press in France, 1814–1881.* London: Oxford University Press, 1959, 201 pp.
Collins, James H., "The Story of the Printed Word." Waldemar Kaempffert, Ed., *A Popular History of American Invention,* 2 Vols. New York: Scribner, 1924, Vol. 1, pp. 221–261.
Cortissoz, Royal, *The New York Tribune: Incidents and Personalities in its History.* New York: New York Tribune, 1923, 78 pp.
Cottrell, Donald C., *The Cottrell Company (1855–1955): Color Press Pioneers.* New York: Newcomen Society, 1955, 32 pp.

Crouthamel, James L., *James Watson Webb: A Biography.* Middletown: Wesleyan University Press, 1969, 262 pp.
Davis, Elmer, *History of the New York Times, 1851–1921.* New York: New York Times, 1921, 434 pp.
Digby, B. C., "Printing Machinery." *Eighth Series of Annual Craft Lectures at Stationers' Hall.* London: London School of Printing and Kindred Trades, 1930, pp. 103–124.
Du Bin, Alexander, *Harper Family* [and] *Hoe Family.* Philadelphia: Historical Publishing Society, 1948, pp. 33–43.
Eckman, James, *The Heritage of the Printer.* Philadelphia: North American, 1965, 209 pp.
Edelstein, David S., *Joel Munsell: Printer and Antiquarian.* New York: Columbia University Press, 1950, 420 pp.
Emery, Edwin, *The Press and America: An Interpretative History of the Mass Media,* 3rd Ed. Englewood Cliffs: Prentice Hall, 1972, 788 pp.
—, Ed., *The Story of America as Reported by its Newspapers, 1690–1965.* New York: Simon & Schuster, 1965, 311 pp.
Exman, Eugene, *The Brothers Harper: A Unique Publishing Partnership and its Impact upon the Cultural Life of America from 1817 to 1853.* New York: Harper & Row, 1965, 416 pp.
—, *The House of Harper: One Hundred and Fifty Years of Publishing.* New York: Harper & Row, 1967, 326 pp.
Forsyth, David P., *The Business Press in America, 1750–1865.* New York: Chilton, 1964, 394 pp.
Gabriel-Robinet, L., *Historie de la Presse.* Paris: Hachette, 1961, 64 pp.
Gambee, Budd L., Jr., *Frank Leslie and His Illustrated Newspaper, 1855–1860.* Ann Arbor: University of Michigan Press, 1964, 86 pp.
Goldberg, Isaac, *Major Noah: American-Jewish Pioneer.* Philadelphia: Jewish Publication Society of America, 1936, 316 pp.
Graham's Magazine, "New York Printing Machinery, Press, and Saw Works. R. Hoe & Company." June 1852, 564–576; July 1852, 7–12.
Gramling, Oliver, *AP: The Story of News.* New York: Farrar & Rinehart, 1940, 506 pp.
Grant, James, *The Newspaper Press: Its Origin—Progress—and Present Position,* 2 Vols. London: Tinsley, 1871, 918 pp.
Green, Ralph, *"Early American Power Printing Presses."* Fredson Bowers, Ed., Studies in Bibliography, Vol. 4 (1951–1952), pp. 143–153.
—, *A History of the Platen Jobber.* Chicago: Reed, 1953, 39 pp.
Hale, William H., *Horace Greeley: Voice of the People.* New York: Harper, 1950, 376 pp.
Hamilton, Milton W., *Adam Ramage and His Presses.* Portland: Southworth-Anthoensen, 1942, 36 pp.
—, *The Country Printer: New York State, 1785–1830,* 2nd Ed. Port Washington: Friedman, 1964, 360 pp.
Harlow, Alvin F., *Old Wires and New Waves: The History of the Telegraph, Telephone, and Wireless.* New York: Appleton-Century, 1936, 548 pp.
Harper, J. Henry, *The House of Harper: A Century of Publishing in Franklin Square.* New York, 1912, 690 pp.

Hart, Jim A., *A History of the St. Louis Globe-Democrat.* Columbia: University of Missouri Press, 1961, 298 pp.

Hassard, John R. G., *The Fast Printing Machine: An Account of Recent Improvements in Newspaper Presses.* New York: Tribune Association, 1878, 32 pp.

Hatin, Eugene, *Histoire du Journal en France, 1631–1853,* 2nd Ed. Paris: Jannet, 1853, 318 pp.

Hatton, Joseph, *Journalistic London: Being a Series of Sketches of Famous Pens and Papers of the Day.* London: Sampson Low, Marston, Searle & Rivington, 1882, 249 pp.

Hoe, Richard M., *Richard M. Hoe's Memorial to Congress for an Extension of Letters Patent.* Washington: 1868, 79 pp.

Hoe, Robert, *A Short History of the Printing Press and of the Improvements in Printing Machinery from the Time of Gutenberg up to the Present Day.* New York: Hoe, 1902, 92 pp.

Hohenberg, John, *Free Press/Free People: The Best Cause.* New York: Columbia University Press, 1971, 514 pp.

Houghton, Harrison, *The Master Printer's Return: What Gutenberg saw amid the Roar of a Modern Printing Wonderland.* New York: Hoe, 1911, 12 pp.

Howe, Ellic, "Newspaper Printing in the Nineteenth Century." *Alphabet & Image,* No. 4 (April 1947), pp. 5–31.

Hudson, Frederic, *Journalism in the United States, from 1690 to 1872.* New York: Harper, 1873, 789 pp.

Ingersoll, L. D., *The Life of Horace Greeley, with Extended Notices of Many of his Contemporary Statesmen and Journalists.* Chicago: Union, 1873, 688 pp.

Isaacs, George A., *The Story of the Newspaper Printing Press.* London: Co-operative Printing Society, 1931, 287 pp.

Jackson, Mason, *The Pictorial Press: Its Origin and Progress.* London: Hurst & Blackett, 1885, 364 pp.

Johnson, Gerald W., Frank R. Kent, H. L. Mencken, and Hamilton Owens, *The Sunpapers of Baltimore, 1837–1937.* New York: Knopf, 1937, 430 pp.

Juergens, George, *Joseph Pulitzer and the New York World.* Princeton: Princeton University Press, 1966, 392 pp.

Kainen, Jacob, *George Clymer and the Columbian Press.* San Francisco: Book Club of California, 1950, 60 pp.

Karolevitz, Robert F., *Newspapering in the Old West: A Pictorial History of Journalism and Printing on the Frontier.* Seattle: Superior, 1965, 192 pp.

Kemble, Edward C., *A History of California Newspapers, 1846–1858.* Los Gatos: Talisman, 1962, 398 pp.

Kerr, R. W., *History of the Government Printing Office (at Washington, D. C.) with a Brief Record of the Public Printing for a Century, 1789–1881.* Lancaster: Inquirer, 1881, 196 pp.

Kesterton, W. H., *A History of Journalism in Canada.* Toronto: McClelland & Stewart, 1967, 307 pp.

Kinsley, Philip, *The Chicago Tribune: Its First Hundred Years,* 3 Vols.

New York and Chicago: Knopf and Chicago Tribune, 1943–1946, 1049 pp.
Kubler, George A., *A Short History of Stereotyping.* New York: Kubler, 1927, 93 pp.
Lehmann-Haupt, Hellmut, *The Book in America: A History of the Making and Selling of Books in the United States,* 2nd Ed. New York: Bowker, 1951, 494 pp.
Lindstrom, Carl E., *The Fading American Newspaper.* Garden City: Doubleday, 1960, 283 pp.
Lossing, Benson J., *History of New York City. . . .,* 2 Vols. New York: Perine, 1884, 881 pp.
Lyons, Louis M., *Newspaper Story: One Hundred Years of the Boston Globe.* Cambridge: Belknap, 1971, 482 pp.
McCabe, John F., "A Visit to R. Hoe & Company's Great New York Plant." *Printing Trade News,* Feb. 1910, 4 pp.
Mackellar, Thomas, *The American Printer: A Manual of Typography containing Complete Instructions for Beginners. . . .* Philadelphia: Johnson, 1886, 336 pp.
McLuhan, Marshall, *Understanding Media: The Extensions of Man.* New York: Signet, 1964.
McNamara, Stephen, "The Printing Press." *Inland Printer,* June 1884–June 1887.
McNulty, John B., *Older than the Nation: The Story of the Hartford Courant.* Stonington: Pequot, 1964, 231 pp.
Marzio, Peter C., *The Men and Machines of American Journalism: A Pictorial Essay. . . .* Washington: Smithsonian Institution, 1973, 144 pp.
Matthews, Thomas S., *The Sugar Pill: An Essay on Newspapers.* New York: Simon & Schuster, 1959, 221 pp.
Mills, William H., *The Manchester Guardian: A Century of History.* New York: Holt, 1922, 147 pp.
Moore, John W., Ed., *Moore's History, Biography, and Miscellaneous Gatherings in the Form of Disconnected Notes relative to Printers, Printing, Publishing, and Editing of Books, Newspapers, Magazines. . . .* Concord: Republican Press Association, 1886, 604 pp.
Moran, James, *Printing Presses: History and Development from the Fifteenth Century to Modern Times.* Berkeley: University of California Press, 1973, 263 pp.
Morison, Stanley, *The English Newspaper: Some Account of the Physical Development of Journals printed in London between 1622 & the Present Day.* Cambridge: Cambridge University Press, 1932, 335 pp.
—, *The History of the Times,* 3 Vols. New York and London: Macmillan and The Times, 1935–1947, 1999 pp.
Mott, Frank L., *American Journalism: A History, 1690–1960,* 3rd Ed. New York: Macmillan, 1962, 901 pp.
Munsell, Joel, *The Typographical Miscellany.* Albany: Munsell, 1850, 268 pp.
Neipp, Lucien, *Les Machines a Imprimer depuis Gutenberg.* Paris: Club Bibliophile de France, 1951, 444 pp.

North, Simeon N. D., *History and Present Condition of the Newspaper and Periodical Press of the United States.* Washington: US GPO, 1884.
O'Brien, Frank M., *The Story of the Sun.* New York: Doran, 1918, 455 pp.
Oswald, John C., *Printing in the Americas.* New York: Gregg, 1937, 566 pp.
Pasko, Wesley W., Ed., *American Dictionary of Printing and Bookmaking. . . .* New York: Lockwood, 1894, 592 pp.
Payne, George H., *History of Journalism in the United States.* New York: Appleton, 1924, 453 pp.
Peddie, R. A., Ed., *Printing: A Short History of the Art.* London: Grafton, 1927, 390 pp.
Pickett, Calder M., "Technology and the New York Press in the 19th Century." *Journalism Quarterly,* Vol. 37 (Summer 1960), pp. 398–407.
Plomer, Henry R., *A Short History of English Printing, 1476–1900,* 2nd Ed. London: Kegan Paul, Trench, Trubner, 1915, 276 pp.
Pottinger, David T., "The History of the Printing Press." *Dolphin,* No. 3 (1938), pp. 323–343.
Pray, Isaac C., Ed., *Memoirs of James Gordon Bennett and his Times.* New York: Stringer & Townsend, 1855, 488 pp.
Reid, James D., *The Telegraph in America: Its Founders, Promoters, and Noted Men.* New York: Derby, 1879, 846 pp.
Reitzel, William, Ed., *The Autobiography of William Cobbett: The Progress of a Plough-Boy to a Seat in Parliament.* London: Faber & Faber, 1947, 272 pp.
Ringwalt, J. Luther, Ed., *American Encyclopedia of Printing.* Philadelphia: Menamin & Ringwalt, 1871, 512 pp.
Roblin, Fred, Ed., "Development of the Printing Press." *American Pressman,* Nov. 1965.
Rowell, George P., *Centennial Newspaper Exhibition, 1876.* New York: Rowell, 1876, 301 pp.
Rutland, Robert A., *The Newsmongers: Journalism in the Life of the Nation, 1690–1972.* New York: Dial, 1973, 430 pp.
Seitz, Don C., *The James Gordon Bennetts, Father and Son: Proprietors of the New York Herald.* Indianapolis: Bobbs-Merrill, 1928, 405 pp.
—, *Joseph Pulitzer: His Life & Letters.* New York: Simon & Schuster, 1924, 478 pp.
Siebert, Fred S., Theodore Peterson, and Wilbur Schramm, *Four Theories of the Press.* Urbana: University of Illinois Press, 1963, 156 pp.
Silver, Rollo G., *The American Printer, 1787–1825.* Charlottesville: Bibliographic Society of the University of Virginia, 1967, 190 pp.
—, "Efficiency Improved: The Genesis of the Web Press in America." *American Antiquarian Society, Proceedings,* Oct. 1970, pp. 325–350.
Steinberg, S. H., *500 Years of Printing.* Harmondsworth: Penguin, 1955, 278 pp.
Stelzle, Charles, "An East Side American: The Autobiography of a Son of the City." *Outlook,* 3 March–14 July 1926.
Tebbel, John, *An American Dynasty: The Story of the McCormicks, Medills and Pattersons.* Garden City: Doubleday, 1947, 363 pp.

Thorburn, Grant, *Life and Writings of Grant Thorburn: Prepared by Himself*. . . . New York: Walker, 1852, 276 pp.

The Times, *A Newspaper History: 1785–1935*. London: The Times, 1935, 214 pp.

—, *Printing the Times since 1785*. . . . London: The Times, 1953, 195 pp.

Tucker, Stephen D., "History of R. Hoe & Company, 1834–1885," Rollo G. Silver, Ed. *American Antiquarian Society, Proceedings,* Oct. 1972, pp. 351–453.

United States Government Printing Office, *100 GPO Years: 1861–1961—A History of United States Public Printing*. Washington: US GPO, 1961, 164 pp.

Villard, Oswald G., *The Disappearing Daily: Chapters in American Newspaper Evolution*. New York: Knopf, 1944, 296 pp.

Vizetelly, Henry, *Glances Back through Seventy Years: Autobiographical and Other Reminiscences,* 2 Vols. London: Kegan Paul, Trench, Trubner, 1893, 863 pp.

Wilson, Charles, and William Reader, *Men and Machines: A History of D. Napier & Son, Engineers, Ltd., 1808–1958*. London: Weidenfeld & Nicholson, 1958, 187 pp.

Wilson, Frederick J. F., and Douglas Grey, *A Practical Treatise upon Modern Printing Machinery and Letterpress Printing*. London: Cassell, 1888, 455 pp.

Index

INDEX OF EARLY FAMILY MEMBERS

- Matthew Smith (1752-1816) / Theodosia Mead (1756-1816)
  - Matthew II (1780-1820) / Rhoda
    - Mathew III (1812-1841) /
      - Sereno N., Eugene, Peter? Smith
    - Rhoda
    - Rachel (1782-1834) / Robert Hoe
  - Peter (1795-1823)

- Robert Hoe (1784-1833) / Rachel Mead Smith
  - Mary J. (1805-1880) / Ebinezer Seymour
    - Henry, George, Edward Seymour
  - Elizabeth (1807-1884) / Henry Mudge (d. 1831); Enoch M. Mead
    - Theodore H. Mead (1837-1909) / Anna R. Johnson
    - *also* Frederick, Edward S., Herbert Mead
  - Emeline (1809-1862) / Giles S. Ely
  - Richard March (1812-1886) / Lucy Gilbert (d. 1841)
    - Emily A. (1834-1909) / Cyrus J. Lawrence
    - Adeline E. (1836-1882) / DeWitt Clinton Lawrence
  - [RMH] / Mary S. Corbin (1816-1901)
    - Anne C. (1852-1887) / Charles Platt
    - Mary S. (1854-1925) / Joseph Henry Harper
    - Fannie B. (1855-1898) / John Harper
    - Helen S. (1858)
  - Robert II (1815-1884) / Thyrza Mead
    - Robert III (1839-1909) / Olivia P. James (1837-1935)
      - Elizabeth W. (1865-1907) / Frank Benson
      - Caroline P. (1866-1904) / Leon Marie
      - Olivia P. (1868-1927) / Henry Slade
      - Laura (1870-   ) / Ernest T. Carter
      - Ellen J. (1872-1907) / Samuel M. Evans
      - Thyrza (1874-1883)
      - Robert IV (1876-1960) / Mary E. Thompson
      - Ruth L. (1877-1966) / Robert D. Sterling
      - Arthur I. (1880-1957) / Evelyn Perry
    - Arthur M. (1841-1865)
    - Laura (1845-1942)
  - Rachel (1817-1897) / Moses W. Dodd
    - Frank H. Dodd (1844-1916)
  - *also* Ira S., William M., Robert H., Charles T., Edward H. Dodd

Harriet (1819-1820)
Peter S. (1821-1902) / Hannah M. Smith
Stephen S. (1847-1887)
also Richard M., Charles R., Edward S., Frederic H. Hoe
Theodosia (1823-) / William S. Leavitt
